Models, Modules and Abelian Groups

A. L. S. Corner

Models, Modules and Abelian Groups

In Memory of A. L. S. Corner

Editors
Rüdiger Göbel
Brendan Goldsmith

Walter de Gruyter · Berlin · New York

Editors

Rüdiger Göbel
Fachbereich 6, Mathematik
Universität Duisburg-Essen
45117 Essen
Germany
E-mail: R.Goebel@uni-due.de

Brendan Goldsmith
School of Mathematical Sciences
Dublin Institute of Technology
Kevin Street, Dublin 8
Ireland
E-mail: brendan.goldsmith@dit.ie

Mathematics Subject Classification 2000: 13-06, 13-XX, 16-06, 16-XX, 20-06, 20-XX

Keywords: Abelian groups, Baer−Specker group, Butler groups, completely decomposable groups, crq-groups, endomorphisms of groups, h-local domains, locally free abelian groups, nilpotent groups, p-groups, endomorphism rings, noetherian rings, non-singular rings, $\mathbb{Q}$-algebras modules, serial modules

∞ Printed on acid-free paper which falls within the guidelines of the
ANSI to ensure permanence and durability.

Library of Congress Cataloging-in-Publication Data

A CIP catalogue record for this book is available from the Library of Congress.

ISBN 978-3-11-019437-1

Bibliographic information published by the Deutsche Nationalbibliothek

The Deutsche Nationalbibliothek lists this publication in the Deutsche Nationalbibliografie; detailed bibliographic data are available in the Internet at http://dnb.d-nb.de.

© Copyright 2008 by Walter de Gruyter GmbH & Co. KG, 10785 Berlin, Germany.
All rights reserved, including those of translation into foreign languages. No part of this book may be reproduced or transmitted in any form or by any means, electronic or mechanical, including photocopy, recording or any information storage and retrieval system, without permission in writing from the publisher.
Printed in Germany.
Cover design: Thomas Bonnie, Hamburg.
Typeset using the authors' LaTeX files: Jonathan Rohleder, Berlin.
Printing and binding: Hubert & Co. GmbH & Co. KG, Göttingen.

Introduction

On hearing of the sudden death of A. L. S. Corner on September 3rd 2006, many of his friends and colleagues felt it appropriate that Tony's contributions to algebra should be acknowledged in a volume which would reflect his wide mathematical influence. Accordingly, this volume was conceived and its title "*Models, Modules and Abelian Groups*" reflects Tony's range of research interests. For many Tony's name is inseparable from "*Abelian Groups*" but his contributions were wider: his techniques from set and model theory, his interests in representation theory and non-commutative group theory and his enormous influence on the study of endomorphism rings are all part of his legacy and this is well reflected in the contributions herein.

Some short time prior to his death, the editors persuaded Tony that his unpublished work on the classification of the finite groups which can occur as automorphism groups of torsion-free Abelian groups, should be written up for publication. Tony agreed and his lecture notes on the topic, delivered at the University of Padua in 1988, were being processed when he died. He did not see the final version which is produced in this Volume, but we believe it is a faithful record of Tony's work on the subject. We are thankful to Federico Menegazzo for his help in this and particularly for his intriguing Appendix to Tony's paper.

The editors would like to record their thanks to the authors who responded so enthusiastically to their requests for contributions and to the referees who have ensured such high mathematical standards in the accepted papers. It is also appropriate to thank Robert Plato and Simon Albroscheit of Walter de Gruyter for their support and help in this undertaking.

June 2008   Rüdiger Göbel and Brendan Goldsmith,
                                                                                    Editors

Contents

Introduction v

B. GOLDSMITH
Anthony Leonard Southern Corner 1934–2006 1

A. L. S. CORNER
Groups of units of orders in $\mathbb{Q}$-algebras 9

A. BLASS, S. SHELAH
Basic subgroups and freeness, a counterexample 63

R. GÖBEL, J. MATZ
An extension of Butler's theorem on endomorphism rings 75

D. M. ARNOLD
Locally free abelian groups of finite rank 83

A. MADER, L. G. NONGXA, M. A. OULD-BEDDI
Decompositions of global crq-groups 99

G. CĂLUGĂREANU, P. SCHULTZ
Endomorphisms and automorphisms of squares of abelian groups 121

E. BLAGOVESHCHENSKAYA
Classification of a class of finite rank Butler groups 135

A. A. FOMIN
Quotient divisible and almost completely decomposable groups 147

D. HERDEN, L. STRÜNGMANN
Pure subgroups of completely decomposable groups – an algorithmic approach . 169

O. KOLMAN
Strong subgroup chains and the Baer–Specker group 187

C. DE VIVO, C. METELLI
On direct decompositions of Butler $B(2)$-groups 199

A. MADER, O. MUTZBAUER
Diagonal equivalence of matrices 219

E. F. CORNELIUS JR., P. SCHULTZ
Root bases of polynomials over integral domains 235

G. CĂLUGĂREANU, K. M. RANGASWAMY
A solution to a problem on lattice isomorphic Abelian groups 249

P. HILL, W. ULLERY
A note on Axiom 3 and its dual for abelian groups 257

W. MAY
Units of modular p-mixed abelian group algebras 267

L. FUCHS, T. OKUYAMA
Pure covers in abelian p-groups 277

P. W. KEEF
Partially decomposable primary Abelian groups and the generalized
core class property 289

K. R. MCLEAN
The additive group of a finite local ring in which each ideal can be n-generated . 301

M. FLAGG
A Jacobson radical isomorphism theorem for torsion-free modules 309

A. L. S. CORNER, B. GOLDSMITH, S. L. WALLUTIS
Anti-isomorphisms and the failure of duality 315

V. V. BLUDOV, M. GIRAUDET, A. M. W. GLASS, G. SABBAGH
Automorphism groups of models of first order theories 325

S. THOMAS
The classification problem for finite rank Butler groups 329

T. G. FATICONI
$\mathcal{L}$-groups 339

J. BUCKNER, M. DUGAS
Co-local subgroups of nilpotent groups of class 2 351

L. SALCE
Divisible envelopes and cotorsion pairs over integral domains 359

C. M. RINGEL
The first Brauer–Thrall conjecture 369

C. GREITHER, D. HERBERA, J. TRLIFAJ
A version of the Baer splitting problem for noetherian rings 375

B. OLBERDING
Characterizations and constructions of h-local domains 385

C. U. JENSEN
Variations on Whitehead's problem and the structure of Ext 407

L. BICAN
The lattice of torsionfree precover classes 415

U. ALBRECHT
Non-singular rings of injective dimension 1 421

F. MANTESE, A. TONOLO
On classes defining a homological dimension 431

A. DUGAS, B. HUISGEN-ZIMMERMANN, J. LEARNED
Truncated path algebras are homologically transparent 445

A. FACCHINI, P. PŘÍHODA
Representations of the category of serial modules of finite Goldie dimension . . . 463

L. SALCE, P. ZANARDO
Commutativity modulo small endomorphisms and endomorphisms of
zero algebraic entropy 487

Anthony Leonard Southern Corner 1934–2006

Brendan Goldsmith

Anthony Leonard Southern (Tony) Corner, who died on September 3rd 2006, was one of the leading algebraists of his generation. His contributions to the theory of Abelian groups truly revolutionized that subject.

Tony was the only child of Leonard and Lucy (neé Southern) and was born in Shanghai on June 12th 1934; his parents had met and married in Paris before heading to China where Leonard hoped to seek his fortune as an accountant. Tony often spoke of his vivid happy memories of being brought up in the French Quarter and ascribed his love of languages (and facility in learning them) to his early cosmopolitan upbringing; indeed even in middle age he was adamant that he did arithmetic calculations in French before translating into English! This happy childhood came to an abrupt end with the Japanese invasion of China during World War II; Tony and his parents spent three years interned in a school building and his parents starved themselves to ensure that he had adequate food. The family lost everything and returned to England at the end of the war.

Tony attended Eastbourne College and then went up to Cambridge, initially to study medicine, but he soon switched to mathematics, obtaining a distinction in Part III of the Mathematical Tripos in 1957. His Ph.D. supervisor was Sir Christopher Zeeman FRS and although Tony initially studied topology, he soon changed to Abelian groups, a topic which occupied the major part of his research career. He held a Research Fellowship at Corpus Christi College, Cambridge before moving in 1962 to Worcester College, Oxford, where he remained until retirement in 2001; in retirement Tony held the position of Honorary Fellow at Exeter University. (Interestingly, despite his long sojourn in Oxford, Tony always regarded himself as a "Cambridge man"!) In 1981 Tony married Elizabeth Rapp, a well-known poetess. Tony joined the London Mathematical Society in 1960, was a Council member from 1964–67 and was Assistant Editor of the *Proceedings* during his period on Council.

Tony's first two papers [1, 2] were concerned with, what at the time was regarded as the rather bizarre behaviour of, torsion-free Abelian groups: in [1], he showed that if $N > k$ are natural numbers, then there exists a countable torsion-free group G of rank N such that for any partition $N = r_1 + r_2 + \cdots + r_k$, there are subgroups $A_i (1 \leq i \leq k)$ of G such that each A_i is directly indecomposable of rank r_i and $G = \bigoplus_{i=1}^{k} A_i$. This answered, in a very strong form, Problem 22 of Fuchs [F] and showed that there was no possibility of a Krull–Schmidt type uniqueness for torsion-free groups of countable rank. Some years later in [10], Tony exploited similar ideas to establish an even stronger decomposition property: there exists a countable torsion-free Abelian group G such that $G = A \oplus B$, where A, B are indecomposable of countably infinite rank, but

G is also equal to $\bigoplus_{n \in \mathbb{Z}} C_n$, where each C_n is indecomposable of rank 2. In [2] Tony showed that 'large' (complete) direct products of copies of the integers may contain 'small' subgroups not lying in any proper direct summand.

However it was Tony's next publication, [3], which drew him to the attention of the international group theory community. In this work Tony proved the most surprising result: every countable reduced torsion-free ring is the endomorphism ring of a countable reduced torsion-free Abelian group. From this seemingly innocent result, Tony went on wreak havoc with the theory of direct decompositions of torsion-free groups. Perhaps the simplest and most striking consequence – see [4] – is the existence of a countable torsion-free Abelian group G such that $G \cong G \oplus G \oplus G$ but $G \not\cong G \oplus G$! The importance of [3] was summed up succinctly by Dick Pierce in his review (Math. Review MR0153743): *"This paper represents some of the most remarkable results which have been obtained in the theory of torsion-free Abelian groups in the past few years."* Interestingly he also observed that *"The whole paper is a model of elegance and clarity."* Tony produced an algebra version of this result, [U2], sometime in the early seventies but never published it.

In the period 1968–76 Tony turned his attention to Abelian p-groups, producing a series of four papers, [7, 11, 12, 13], which were greatly influenced by his contact with Peter Crawley during a sabbatical in 1968 at Vanderbilt University. Since an Abelian p-group has many cyclic summands there is no sensible way to realize its full endomorphism ring as one can do for torsion-free groups. Tony however based his approach on one of his earlier unpublished works, [U1], which was to have a fundamental impact on subsequent developments in the field: endomorphisms onto summands such as the cyclic summands of a p-group are, for purposes of realizing endomorphism algebras, in some sense 'unimportant or inessential'. The situation is, in many cases, somewhat reminiscent of the situation in an infinite dimensional vector space, in that these projections actually determine the endomorphism ring! Fortunately they may be collected into a convenient 2-sided ideal. Tony's idea was to exhibit endomorphism rings as the split extension of his desired ring by this ideal of *'inessential'* endomorphisms. (The original concept of *inessential* in [U1] was very general and was subsequently simplified in my doctoral thesis under Tony's supervision and was used extensively by Dugas, Göbel, Shelah and Corner himself in many subsequent publications.) Once again the general classification of p-groups was shown to be intractable and the existence of pathologies such as a p-group G with $G \cong G \oplus G \oplus G$ but $G \not\cong G \oplus G$ was established. In [13] Tony unraveled the presumed connections between Kaplansky's notions of *transitivity* and *full transitivity*; recall that Kaplansky had stated in his "little red book" that it seemed plausible to conjecture that every p-group was both transitive and fully transitive. This plausible conjecture was completely demolished when Tony showed the existence of groups possessing exactly one of the properties and not the other. This paper contains a question which remains unanswered to this day: Corner's example of a non-transitive fully transitive p-group had a first Ulm subgroup which is elementary of countably infinite rank and Tony wondered if such a group could exist with finite first Ulm subgroup. The paper also contained a remark *en passant* which was to lead to the discovery of the strange connection between the two notions of transitivity by Files and myself [FG].

This was an extremely productive period for Tony and he also published an important work on representations of algebras, [8], expanding significantly on a simple idea of Sheila Brenner and Michael Butler. This work also contains details of how to construct indecomposable groups of power less than the first strongly inaccessible cardinal. This observation was to have significant consequences for Abelian group theory since it became one of the motivations for Shelah's work on indecomposable groups. He also produced a categorical approach, [9], to the work of Leavitt; the so-called Leavitt algebras were of interest to him right up to the time of his death.

The decade from 1976–85 saw only one publication by Tony, [14], but then in 1985, he and Rüdiger Göbel produced an extremely fine "unified approach" to the revolutionary work of Saharon Shelah. This joint work, [15], made accessible to the whole community of Abelian group theorists, the deep, new combinatorial set-theoretic techniques which are now a part of our standard "tool-kit". In 1989 he finally wrote up an idea which he had discovered many years previously, the notion of a so-called fully rigid system of modules, [16]. This notion permitted his good friend Adalberto Orsatti to transfer realization ideas of Corner's in algebra, to a topological setting – see [O].

From the early 1990's until his death, Tony worked (in various combinations) with Rüdiger Göbel and myself – [17, 18, 19, 20, 21, 22, 23]. This work was concentrated on torsion-free groups and modules; subgroups of the so-called Baer–Specker group, $P = \prod_{i=1}^{\infty} \mathbb{Z}e_i$, featured in several of these works.

Tony also had at least 20 other unpublished papers – I say at least because 20 is the number that I'm aware of, but until I complete a review of his manuscripts, I cannot say with certainty that there are not more. I would like to comment briefly on two of these, because they reveal something of Tony's nature and how this was reflected in his approach to mathematical research. In the period 1965–77, the eminent algebraist Kurt Hirsch and his pupil J. Terry Hallett produced a series of 5 papers totalling 54 pages, purporting to classify the finite groups which can appear as automorphism groups of torsion-free Abelian groups. Their proof is incomplete and Tony, using a new approach, corrected this in a deep and difficult paper, [24], sometime in the mid-seventies, but never published his work. (Plans to publish the paper were actually under way prior to Tony's death and the work is included in this Volume.) Why? Let me quote from a letter he sent me in April 1979: *"It's pretty heavy going, ..., but I've got it all fairly conceptual with the exception of one computation which I seem unable to eliminate: it's a simple enough computation, ..., but I don't like it!"* And then *"I feel sure this should generalise to infinite groups satisfying some suitable finiteness condition – perhaps compactness."* It's pretty clear from this that Tony was a perfectionist and liked to publish only results that, from his perspective, were conceptually clear and complete. A further example of this perfectionist side of him came to my attention in December 1998. I was working with Simone Wallutis on a problem when I suddenly thought "Tony must have looked at this some time". A quick telephone call confirmed my suspicions and to my utter amazement (and even more to Tony's) he was able to find a manuscript, [U4], dating from the early sixties which answered our question and a lot more! Again it was never published, why? Well, on careful reading I realized that there was one case in the classification that wasn't as neat as one might want, and I suspect that this made it just too inelegant for Tony to publish. Incidentally, I did complete the original project with Simone and Tony but only after his death. This final work with him, [25], appears in this volume.

Despite his pre-eminence as a researcher, Tony was probably at his happiest in his professional life when he was teaching. He was a talented and remarkably dedicated teacher. Much of his success in this area was due to hard work. Tony was, by nature, quite reserved, shy even, and it took considerable effort on his part to carry out the role of teacher in front of large groups. Nonetheless his lectures were greatly appreciated by several generations of undergraduates in Oxford. Indeed I know that copies of his Oxford lecture notes are carefully filed away in many mathematical departments across the globe, such was their quality.

At doctoral level Tony had just three students: Paul Peterken (*Some Problems in Abelian Group Theory*, 1971); Brendan Goldsmith (*An Investigation in Abelian Group Theory*, 1978); Loyiso Nongxa (*A Problem in Abelian Group Theory*, 1982). Paul went into industry while Loyiso and I continued as academics.

Tony's reputation for being disorganized in relation to correspondence was legendary but this also stretched to important research papers. I recall on my first visit back to Oxford after I had gone to work in Dublin, that my enquiry as to what had happened to his solution of the important problem of Kaplansky, [13], and which he had shown me quite some time previously, was met with a wry smile and the comment "I just found it again last week at the bottom of that pile and it's now gone to the Quarterly Journal!" Tony was also a great supporter of the move to try to encourage more women to seek careers in mathematics; no drum beating or posturing but a quiet commitment summed up in an email to me in December 1998: *"We've just been interviewing for undergraduate admissions ... and at last my dream of admitting more women than men has been realized."*

One of the paradoxes about Tony was that despite his natural reticence, he was an extremely witty person and no recollection of him would be complete without a reference to that famous Corner eyebrow: his ability to raise that eyebrow whilst maintaining a totally straight face, reduced many of us to tears of laughter on what should have been formal and serious occasions. His pithy asides were also legendary and often, whether deliberate or not, I shall never know, they were overheard.

My last face-to-face meeting with Tony was in June 2006. He and I, accompanied by Elizabeth and my wife, Ann, spent a marvellous few days in his beloved Padua, attending a conference and staying on for a few days relaxation afterwards. Interestingly, the organisers asked me if I could persuade Tony to give a talk, but he was adamant that he no longer had anything interesting to speak about. This, despite the fact that I mentioned earlier, that he had quite a number of unpublished results, which would have been of great interest.

He was in many respects a truly renaissance man: knowledgeable in the classics, passionate about music, widely read in many aspects of literature, a fine linguist and a lover of art. A few years ago when our mutual friend and colleague, John Lewis, died, Tony wrote: *"John was a lovely man and a perfect fit to my ideal of an academic: a man of great learning, lightly worn, and no conceit of himself."* The words apply just as aptly to Tony himself. He brought a great deal of happiness into the lives of many and will be sorely missed.

Ar dheis Dé go raibh a anam dílis.

Acknowledgements. The author wishes to acknowledge the help and input of Elizabeth Corner, Rüdiger Göbel, Dan Lunn and Robin McLean.

References

[F] L. Fuchs, *Abelian Groups*, Publishing House of the Hungarian Academy of Science, Budapest (1958); reprinted by Pergamon Press, Oxford (1960).

[FG] S. Files and B. Goldsmith, *Transitive and Fully Transitive Groups*, Proc. Amer. Math. Soc. **126** (1998) 1605–1610.

[O] A. Orsatti and N. Rodinò, *Homeomorphisms between finite powers of topological spaces*, Topology Appl. **23** (1986) 271–277.

Publications of A. L. S. Corner

[1] A note on rank and direct decompositions of torsion-free Abelian groups, Proc. Cambridge Philos. Soc. **57** (1961) 230–233.

[2] Wildly embedded subgroups of complete direct sums of cyclic groups, Proc. Cambridge Philos. Soc. **59** (1963) 249–251.

[3] Every countable reduced torsion-free ring is an endomorphism ring, Proc. London Math. Soc. **13** (1963) 687–710.

[4] On a conjecture of Pierce concerning direct decompositions of Abelian groups, Proc. Colloq. Abelian Groups (Tihany 1963) pp. 43–48, Akadémiai Kiadó, Budapest 1964.

[5] Three examples of hopficity in torsion-free Abelian groups, Math. Acad. Sci. Hungar. **16** (1965) 303–310.

[6] Endomorphism rings of torsion-free Abelian groups, Proc. Internat. Conference Theory of Groups (Canberra 1965), Gordon and Breach 1967.

[7] An Abelian p-group without the isomorphic refinement property (with P. Crawley), Bull. Amer. Math. Soc. **74** (1968) 743–745.

[8] Endomorphism algebras of large modules with distinguished submodules, J. Algebra **11** (1969) 155–185.

[9] Additive categories and a theorem of W. G. Leavitt, Bull. Amer. Math. Soc. **75** (1969) 78–82.

[10] A note on rank and direct decompositions of torsion-free Abelian groups. II, Proc. Cambridge Philos. Soc. **66** (1969) 239–240.

[11] On endomorphism rings of primary Abelian groups, Quart. J. Math. Oxford **20** (1969) 277–296.

[12] On endomorphism rings of primary Abelian groups. II, Quart. J. Math. Oxford **27** (1976) 5–13.

[13] The independence of Kaplansky's notions of transitivity and full transitivity, Quart. J. Math. Oxford **27** (1976) 15–20.

[14] On the existence of very decomposable Abelian groups, Conference Abelian Group Theory (1982 Honolulu), Lecture Notes in Math., **1006** pp. 354–357, Springer 1983.

[15] Prescribing endomorphism algebras – A unified treatment (with R. Göbel), Proc. London Math. Soc. **50** (1985) 447–479.

[16] Fully rigid systems of modules, Rend. Sem. Mat. Univ. Padova **82** (1989) 55–66.

[17] On endomorphisms and automorphisms of pure subgroups of the Baer-Specker group (with B. Goldsmith), Conference Abelian Group Theory (1993 Oberwolfach), Contemp. Math., **171** pp. 69–78, Amer. Math. Soc. 1994.

[18] Essentially rigid floppy subgroups of the Baer-Specker group (with R. Göbel), Manuscripta Math. **94** (1997) 319–326.

[19] Radicals commuting with Cartesian products (with R. Göbel), Arch. Math. (Basel) **71** (1998) 341–348.

[20] Subgroups of the Baer-Specker group with prescribed endomorphism ring and large dual (with R. Göbel), Conference Abelian Groups, Module Theory, and Topology, (1997 Padova) Lecture Notes in Pure and Appl. Math. **201** pp. 113–123, Dekker 1998.

[21] Isomorphic automorphism groups of torsion-free p-adic modules (with B. Goldsmith), Conference Abelian Groups, Module Theory, and Topology (1997 Padova), Lecture Notes in Pure and Appl. Math. **201** pp. 125–130, Dekker 1998.

[22] Small almost free modules with prescribed topological endomorphism rings (with R. Göbel), Rend. Sem. Mat. Univ. Padova **109** (2003) 217–234.

[23] On torsion-free Crawley groups (with R. Göbel and B. Goldsmith), Quart. J. Math. Oxford **57** (2006) 183–192.

[24] Groups of units of orders in $\mathbb{Q}$-algebras, this volume

[25] Anti-isomorphisms and the failure of duality (with B. Goldsmith and S. L. Wallutis), this volume.

Unpublished manuscripts of A. L. S. Corner

[U1] A Class of Pure Subgroups of the Baer-Specker Group.

[U2] Every Countable Reduced Torsion-free Algebra is an Endomorphism Algebra.

[U3] The Existence of $\aleph_1$-free Indecomposable Abelian Groups.

[U4] Anti-isomorphic Endomorphism Rings.

[U5] On the Exchange Property in Additive Categories.

[U6] The Baer Sum.

[U7] Buttresses.

[U8] Pairs and Triples.

[U9] Construction of Large Rigid Pairs.

[U10] Construction of Large Indecomposables.

[U11] On the Existence of an $\aleph_1$-free Abelian Group of Cardinality $\aleph_1$ with no Free Summand (with R. Göbel).

[U12] Structure of $\text{Ext}(M, \mathbb{Z})$, ($M$ is a monotone subgroup of $\prod_{\aleph_0} \mathbb{Z}$).

[U13] Quotients of Mixed Groups by Isomorphic Maximal Pure Subgroups.

[U14] Extensions of Torsion Groups by Countable Torsion-free Groups.

[U15] Hyperhomogeneous Groups.

[U16] On I-groups and a Conjecture of Beaumont and Pierce.

[U17] PM Cohn's Conjecture on Free Associative Algebras.

[U18] Characterization of Trivial Tensor Products.

[U19] Coset Lattices of Abelian groups.

[U20] A Simple Approach to Leavitt.

[U21] Some investigations in the theory of torsion-free Abelian groups – a dissertation submitted to University of Cambridge in April 1959.

[U22] Homogeneous Torsion-Free Abelian Groups, Ph.D. thesis, University of Cambridge, September 1961.

Author information

Brendan Goldsmith, School of Mathematical Sciences, Dublin Institute of Technology, Aungier Street, Dublin 2, Ireland.
E-mail: `brendan.goldsmith@dit.ie`

Groups of units of orders in $\mathbb{Q}$-algebras

Lectures given at the Università di Padova, April, 1988

A. L. S. Corner

Abstract. A classification of the finite (not necessarily commutative) groups which can occur as the automorphism groups of torsion-free abelian groups is given.

Key words. Torsion-free abelian groups, automorphism groups, orders of algebras.

AMS classification. 20K30.

Editors' Note: Tony Corner investigated the problem of determining which finite (not necessarily commutative) groups occur as the automorphism groups of torsion-free abelian groups in the late 1970s, but never published his results. In April 1988 he gave a series of lectures at the Università di Padova and fortunately kept a copy of these notes. This paper is essentially a copy of those notes with a few corrections of minor typographical errors. There is an intriguing *twist* at the end – see the **Remark** in the final Section O – in that Tony suggested that he had discovered a mistake in his original characterization and had replaced the result with a slightly weaker one. The Editors are greatly indebted to Federico Menegazzo, who actually attended Tony's 1988 lecture course, for re-reading the complete work and providing a proof of the "removed piece" of the characterization. This proof is included as an Appendix at the end of the main text. [RG + BG]

Introduction

Given an infinite abelian group A, consider the automorphism group $\mathrm{Aut}(A)$. In general this will be infinite. Indeed, if A is a torsion group, then $\mathrm{Aut}(A)$ must be infinite, but when A is torsion-free $\mathrm{Aut}(A)$ may well be finite: e.g. $\mathrm{Aut}(\mathbb{Z}) \cong C_2$. The main problem with which I shall be concerned is the following: to characterize those finite groups which arise as automorphism groups of torsion-free abelian groups. If we can do this, then we may ask, more generally, what can be said about a torsion group Γ which is isomorphic to the automorphism group $\mathrm{Aut}(A)$ of a torsion-free abelian group?

The finite case was considered in a series of 5 papers by K. A. Hirsch and his pupil T. J. Hallett – see [5, 6, 7, 8, 9] and a further paper by Hirsch and Zassenhaus [10] reworks [5] from a more advanced but perhaps more attractive standpoint.
Of these [5] and [10] establish necessary conditions and parts of the remaining papers show that the conditions are sufficient in certain cases. But the later papers in the series contain errors and the characterization published in [9] as the culmination of the whole is false. There are little errors in the exposition [3, pp. 268–275] of [5].

Tony Corner died on September 3rd 2006, prior to the completion of this work.

A First results

Let us start by simplifying the problem: we get rid of the abelian group A. Given a torsion-free abelian group A with automorphism group $\Gamma = \mathrm{Aut}(A)$, then Γ is the group of units of the endomorphism ring of A, in symbols $\Gamma = (\mathrm{End}\,A)^*$. The additive group of $\mathrm{End}\,A$ is torsion-free, so

$$\mathrm{End}\,A = \mathbb{Z} \otimes \mathrm{End}\,A \subseteq \mathbb{Q} \otimes \mathrm{End}\,A.$$

Then Γ is the group of units of a subring $\mathrm{End}\,A$ of a $\mathbb{Q}$-algebra.

Lemma A.1. *If $\Gamma = \mathrm{Aut}(A)$ for some torsion-free abelian group A, then $\Gamma = R^*$ for some subring R of a $\mathbb{Q}$-algebra $\mathfrak{A}$. Conversely, if $\Gamma = R^*$ is finite and R is a subring of a $\mathbb{Q}$-algebra, then also $\Gamma = \mathrm{Aut}(A)$ for some torsion-free abelian group A.*

Proof. We have already proved the first part. For the second, if $\Gamma = R^*$ is finite, then Γ is also the group of units of the subring $\mathbb{Z}\Gamma$ of R generated by Γ, and since the additive group of $\mathbb{Z}\Gamma$ is free of finite rank, there exists a torsion-free abelian group A of finite rank such that $\mathrm{End}\,A = \mathbb{Z}\Gamma$ – see [1] or [2]. And then $\mathrm{Aut}\,A = (\mathbb{Z}\Gamma)^* = \Gamma$. □

We shall see later that the second implication of Lemma A.1 is true whenever Γ is a torsion group.

We are therefore led to consider the following situation: R is a subring of a $\mathbb{Q}$-algebra $\mathfrak{A}$ such that the group of units $R^* = \Gamma$ is a torsion group. As we are interested in characterizing Γ there will be no loss of generality if we assume that R and $\mathfrak{A}$ are generated by Γ (as ring and $\mathbb{Q}$-algebra respectively). Then (assuming of course $R \neq 0$)

$$\boxed{\begin{aligned}&R = \mathbb{Z}\Gamma \text{ (homomorphic image of the group ring } \mathbb{Z}[\Gamma]),\\ &\mathfrak{A} = \mathbb{Q}\Gamma = \mathbb{Q}R \text{ (homomorphic image of the group algebra } \mathbb{Q}[\Gamma]),\\ &R^* = \Gamma, \text{ a torsion group.}\end{aligned}} \quad (\dagger)$$

Note that when Γ is finite, R is then an *order* of the $\mathbb{Q}$-algebra $\mathfrak{A}$: it is a subring of $\mathfrak{A}$ whose additive group is free abelian of finite rank, and a $\mathbb{Z}$-basis of R is a $\mathbb{Q}$-basis of $\mathfrak{A}$. (Which explains the title.) The proof of the next lemma is obvious.

Lemma A.2. *In the context* $(\dagger)$, *any subring S of R satisfies $S^* = \Gamma \cap S$.*

Proof. If $u \in R^* \cap S$, then $u^n = 1$ for some n (because Γ is torsion). Thus $u^n = u(u^{n-1}) = 1$ and $u \in S^*$, so $S^* = \Gamma \cap S$ follows. □

Lemma A.3. *In the context* $(\dagger)$,

  (i) *the involution $-1 \in \Gamma$; (so the order $|\Gamma|$ is even when Γ is finite)*

  (ii) *the only nilpotent element in $\mathbb{Q}R$ is 0;*

  (iii) *all idempotents and involutions are central;*

  (iv) $\Gamma[2] := \{e \in \Gamma : e^2 = 1\}$ *is a central subgroup of exponent 2 in Γ.*

Proof. (i) is clear.

(ii) Suppose for a contradiction that x is a non-zero nilpotent element in $\mathbb{Q}R$. Replacing x by a suitable multiple of a suitable power, we may assume $x^2 = 0 \neq x \in R$. But then $n \mapsto 1 + nx$ is a monomorphism from the infinite cyclic $\mathbb{Z}$ to the torsion group Γ – contradiction.

(iii) Let $f^2 = f \in \mathbb{Q}R$, $x \in \mathbb{Q}R$. Then $fx(1-f)$, $(1-f)xf$ are nilpotent, so vanish, i.e. $fx = fxf = xf$; so $f \in \mathfrak{z}\mathbb{Q}R$. And if $e^2 = 1$, then $\frac{1}{2}(1-e)$ is idempotent, so central.

(iv) follows at once from (iii). □

Lemma A.4 (Hirsch–Zassenhaus [10]). *Let S and T be orders in a finite-dimensional $\mathbb{Q}$-algebra. Then*

(i) S^* *is finite* $\Leftrightarrow T^*$ *is finite;*

(ii) S^* *is torsion* $\Leftrightarrow T^*$ *is torsion.*

Proof. If S and T are orders of $\mathfrak{A}$, then any $x \in T$ can be expressed as a $\mathbb{Q}$-linear combination $x = \sum q_i s_i$ with basis elements s_i from S. Thus $T + S/S \simeq T/(T \cap S)$ is torsion and tensoring the short exact sequence

$$0 \longrightarrow S \cap T \longrightarrow T \longrightarrow T/S \cap T \longrightarrow 0$$

by $\mathbb{Q}$ gives $(S \cap T) \otimes \mathbb{Q} = T \otimes \mathbb{Q} = \mathfrak{A}$. So the intersection $S \cap T$ is an order in the same $\mathbb{Q}$-algebra and we may assume that $T \subseteq S$. Then $T^* \subseteq S^*$, and it will be enough to prove that the index $|S^*{:}T^*|$ is finite. Now $S_\mathbb{Z}$ is free of finite rank and $T_\mathbb{Z}$ is a subgroup of the same finite rank, so there is a positive integer m such that $mS \subseteq T \subseteq S$. Here mS is a 2-sided ideal of S and is of finite index $m^{\mathrm{rk}_\mathbb{Z} S}$ in S. It will be enough to prove that if $u, v \in S^*$ lie in the same additive coset mod mS, then they lie in the same multiplicative coset mod T^*: for then $|S^*{:}T^*| \leq |S{:}mS| < \infty$. But $u - v \in mS$, hence $u \equiv v \bmod mS$ and multiplying by v^{-1} we get $uv^{-1} - 1 \in (mS)v^{-1} \subseteq mS \subseteq T$ which implies $uv^{-1} \in T$ (because $1 \in T$); similarly $vu^{-1} \in T$ and therefore also $uv^{-1} \in T^*$. Conditions (i) and (ii) follow. □

B The six primordial groups

These are the atoms whose chemistry we shall be studying, and we shall need to understand them well. Their genesis is the following.

Consider the context (†) of section A, and suppose that the algebra $\mathfrak{A}$ in question is a division algebra:

$$R = \mathbb{Z}\Gamma \leq \mathbb{Q}\Gamma = \mathbb{Q}R = F \text{ (division algebra } \mathfrak{A} = F). \tag{1}$$

$$R^* = \Gamma \text{ (torsion group).}$$

The following lemma is a particular case of a deep result by C. L. Siegel (see [10, Theorem 2, pages 546, 547]). The proof of this special case is fairly easy.

Lemma B.1. *In this context* (1),

(i) *the unique involution in Γ is -1.*

(ii) *if $g \in \Gamma$ is of prime power order, then $\mathrm{o}(g)|3$ or $\mathrm{o}(g)|4$.*

(iii) *Γ is of exponent* 12.

(iv) *Γ contains no element of order* 12.

Proof. (i) If $e \in \Gamma$ and $\mathrm{o}(e) = 2$, then $e - 1 \neq 0 = (e-1)(e+1)$, therefore $e + 1 = 0$.
(ii) Consider first an element $g \in \Gamma$ of odd prime-power order $\mathrm{o}(g) = q = p^k$ ($k \geq 1$). Let $n = q - 2$, which is odd and prime to q, therefore also to $2q$. Hence $nm \equiv 1$ (mod $2q$) for some positive integer m, necessarily odd.
Now $\mathrm{o}(g) = q \nmid 2n$, so $g^2 \neq 1$, $g^{2n} \neq 1$, therefore $g + 1, g^n + 1 \neq 0$ in F, and in F we may consider the elements

$$u = \frac{g^n + 1}{g + 1} = 1 - g + g^2 - \cdots + g^{n-1},$$

$$u' = \frac{g+1}{g^n+1} = \frac{g^{nm}+1}{g^n+1} = 1 - g^n + g^{2n} - \cdots + g^{(m-1)n}.$$

These are inverses, and they lie (with g) in R: so they are elements of $R^* = \Gamma$ and must be of finite order. Put

$$\zeta = \exp\frac{\pi i}{q} \; (\in \mathbb{C}).$$

Then ζ^2 is of order q in $(\mathbb{C})^*$, and ζ^2, g have the same minimum polynomial, namely

$$\frac{X^{p^k} - 1}{X^{p^{k-1}} - 1} = \Phi_q(X), \; q^{\text{th}} \text{ cyclotomic polynomial, } \textit{irreducible over } \mathbb{Q}.$$

Hence $g \mapsto \zeta^2$ under an algebra isomorphism $\mathbb{Q}[g] \to \mathbb{Q}[\zeta^2]$, and this maps

$$u \mapsto \frac{\zeta^{2n}+1}{\zeta^2+1} = \zeta^{n-1}\frac{\zeta^n + \zeta^{-n}}{\zeta + \zeta^{-1}} = \zeta^{n-1}\frac{\cos n\pi/q}{\cos \pi/q}.$$

Since u is of finite order, so is the image, therefore the image is of absolute value 1. But $|\zeta| = 1$. Therefore $\cos n\pi/q = \pm \cos \pi/q$. But $n = q - 2$. So

$$-\cos 2\pi/q = \pm \cos \pi/q.$$

This is not possible if $q \geq 5$, for then $0 < \pi/q < 2\pi/q < \pi/2$. This proves that $\mathrm{o}(g) \mid 3$ if $\mathrm{o}(g)$ is an odd prime power.
To prove the second assertion it is enough to show that Γ contains no element of order 8. Suppose then that $g \in \Gamma$, $\mathrm{o}(g) = 8$. Then $g^8 = 1 \neq g^4$ in F, so $g^4 + 1 = 0$. This means that $(g + g^{-1})^2 = g^2 + g^{-2} + 2 = 2$. Consider

$$u = 3 + 2(g - g^3) = 3 + 2(g + g^{-1}),$$

$$u' = 3 - 2(g - g^3) = 3 - 2(g + g^{-1}) \; (\in R).$$

Then $uu' = 9 - 8 = 1$, so u ($\in R^* = \Gamma$) is of finite order. As before, mapping g to a primitive 8th root of 1 in $\mathbb{C}$ shows that the image of u is of absolute value 1. But this image is $3 + 2\sqrt{2}$ (> 1) or its inverse – contradiction.

(iii) This follows at once from (ii).

(iv) Suppose for a contradiction that $g \in \Gamma$, $\operatorname{o}(g) = 12$. Since F is a division ring the minimum polynomial of g is the cyclotomic polynomial

$$\Phi_{12}(X) = X^4 - X^2 + 1.$$

Therefore $g^4 - g^2 + 1 = 0$ and $g^6 + 1 = 0$, so

$$(g - g^5)^2 = (g + g^{-1})^2 = (g^2 + g^{-2}) + 2 = 3.$$

This means that

$$u = 2 + (g - g^5), \, u' = 2 - (g - g^5)$$

are inverse elements of $R^* = \Gamma$ mapping under a suitable algebra homomorphism $\mathbb{Q}[g] \to \mathbb{C}$ to $2 + \sqrt{3}$ (> 1) and its inverse, neither of which is of absolute value 1: again a contradiction. □

We now appeal to our first *deus ex machina*.

Lemma B.2. *Every group of exponent 6 is locally finite.*

Proof. See [4, Theorem 18.4.8]. □

Corollary B.3. *Let Γ be a group of exponent 12 in which every involution is central. Then Γ is locally finite.*

Proof. The involutions (together with 1) form a central subgroup $\Gamma[2]$ of exponent 2, certainly locally finite (and abelian). By B.2 $\Gamma/\Gamma[2]$ is locally finite. The class of locally finite groups is closed under extensions. □

In particular, B.1 and B.3 imply that the group Γ of context (1) is locally finite. In fact it is finite. To see this it is enough to prove that the orders of the finite subgroups of Γ are bounded. Consider any finite subgroup G; w.l.o.g. $-1 \in G$ (because $\langle -1 \rangle G$ is finite). Let H be a Sylow 2-subgroup, X a Sylow 3-subgroup of G. Then H is a finite group of exponent 4 with a unique involution, and it is well known – see [4, Theorem 12.5.2] or [11, Satz 8.2, page 310] – that the only possibilities are

$$H \cong C_2, \, C_4, \, Q \text{ (quaternion group of order 8)}.$$

Again, X is a finite group of exponent 3. If it is non-trivial, choose $x \in {}_3X \setminus \{1\}$, and consider *any* $y \in X \setminus \{1, x^{-1}\}$. Then x, y, xy are commuting elements of order 3, and in the division ring F they satisfy

$$1 + x + x^2 = 0 = 1 + y + y^2 = 1 + xy + (xy)^2.$$

Therefore

$$(x-y)^2 = x^2 + y^2 - 2xy = -(1+x) - (1+y) - 2xy$$
$$= -(1+xy) - (1+x)(1+y) = (xy)^2 - x^2y^2 = 0.$$

So $y = x$, which means that $X = \{1, x, x^{-1}\}$ is of order 3. It follows that $G = HX$ is of order dividing $8 \cdot 3 = 24$. Therefore Γ itself is finite, of order dividing 24 and $R = \mathbb{Z}\Gamma$ is an order in the finite-dimensional division-algebra $F = \mathbb{Q}\Gamma$.

The primordial groups Γ in (1) and their associated algebras and orders can now be determined. Since $|\Gamma/\langle -1 \rangle|$ divides 12, we will distinguish cases accordingly:

$$m = |\Gamma/\langle -1 \rangle| \in \{1, 2, 3, 4, 6, 12\}.$$

The order associated with a given primordial group Γ turns out to be unique: we shall denote it by $R\Gamma$.

(i) If $m = 1$, the units are ± 1, $\Gamma = C_2$, $RC_2 = \mathbb{Z}$ with $\mathbb{Z}$-basis 1 and units $\{1, -1\}$.

(ii) If $m = 2$, then $\Gamma = C_4$, cyclic generated by i with $i^2 + 1 = 0$, $RC_4 = \mathbb{Z}[i]$, the Gaussian integers, with $\mathbb{Z}$-basis $1, i$ and units $\pm 1, \pm i$.

(iii) If $m = 3$ then $\Gamma = C_6 = \langle -1 \rangle \times \langle w \rangle$ where $w^2 + w + 1 = 0$, $RC_6 = \mathbb{Z}[w]$, the Eisenstein ring, with $\mathbb{Z}$-basis $1, w$; units $\pm 1, \pm w, \pm(w+1)$.

(iv) If $m = 4$, then $\Gamma = Q = \langle a, b \mid a^2 = b^2 = (ab)^2 \rangle$. This gives $a = bab = b^2ab^2 = a^2aa^2$, so $a^4 = 1$; the presentation is now rather obvious. Thus a, b, ab are all of order 4, and satisfy

$$a^2 + 1 = b^2 + 1 = (ab)^2 + 1 = 0.$$

We return to this in a moment and will determine RQ.

In the remaining cases $\Gamma/\langle -1 \rangle$ is of order 6 or 12. We will distinguish the two cases (only) virtually different and consider the existence ($m = 6$) or non-existence ($m = 12$) of a normal Sylow 3-subgroup of $\Gamma/\langle -1 \rangle$.

(v) Suppose $\Gamma/\langle -1 \rangle$ has a normal Sylow 3-subgroup. Then Γ has a normal subgroup of order 6, not S_3 so C_6. Therefore Γ has a normal Sylow 3-subgroup X (say). A Sylow 2-subgroup H of Γ acts on X by conjugation $H \to \text{Aut } X = C_2$; the kernel contains no element of order 4 and must be of index 2. Hence $H = \langle b \rangle \cong C_4$, and if we let $X = \langle a \rangle$ then $b^{-1}ab = a^{-1}$, $b^2 + 1 = 0 = a^2 + a + 1$.

The first of these is $-bab = a^{-1}$, i.e. $(ab)^2 + 1 = 0$. Here Γ is of order 12 with the presentation

$$\Gamma = D = \langle a, b \mid a^3 = 1, b^4 = 1, b^2 = (ab)^2 \rangle, \text{ the dicyclic group of order 12,}$$

and the generators a, b of RD satisfy

$$a^2 + a + 1 = b^2 + 1 = (ab)^2 + 1 = 0.$$

(vi) Finally suppose $\Gamma/\langle -1 \rangle$ does not have a normal Sylow 3-subgroup. Then $\Gamma/\langle -1 \rangle$ cannot be of order 6, must be of order 12, with four Sylow 3-subgroups, eight elements of order 3, and $4 = 12 - 8$ elements of all other orders: enough for just one Sylow 2-subgroup. Therefore Γ is of order 24 with a normal Sylow 2-subgroup H ($\cong Q$). A Sylow 3-subgroup $X = \langle x \rangle$ cannot normalize any $C_4 = \langle c \rangle$ in H (or it would centralise it; the product xc would be an element of order 12 which does not exist by Lemma B.1(iv)); so X cyclically permutes the three C_4 in H.
Then $H = \langle c, c^x \rangle$ and $\Gamma = B = \langle c, x \rangle$, *the binary tetrahedral group of order* 24.

Now, as $x \in \Gamma \setminus H$, also $cx \in Hx \subseteq \Gamma \setminus H$, so cx must be of order 3 or 6. If cx is of order 6, then $-cx = c^{-1}x$ is of order 3. So replacing c by c^{-1} we may assume that $o(cx) = 3 = o(x)$, $o(c) = 4$. Write $a = cx$, $b = x^{-1}$. Then $o(a) = o(b) = 3$, $o(ab) = 4$,
$$\Gamma = B = \langle a, b \rangle$$
and as generators of RB a and b satisfy
$$a^2 + a + 1 = b^2 + b + 1 = (ab)^2 + 1 = 0.$$

Finally we can now treat all three non-commutative cases ((iv),(v),(vi)) simultaneously and determine their orders S. Let $\varepsilon, \delta \in \{0, 1\}$, and consider
$$S = S_{\varepsilon,\delta} = \mathbb{Z}[a, b \,|\, a^2 + \varepsilon a + 1 = b^2 + \delta b + 1 = (ab)^2 + 1 = 0]. \quad (2)$$
The choices $(\varepsilon, \delta) = (0, 0), (1, 0), (1, 1)$ give rings of which RQ, RD, RB (resp.) are homomorphic images; similarly for the algebras, working over $\mathbb{Q}$. Certainly $a, b \in S^*$, so using the three relations we have $ab = -b^{-1}a^{-1} = -(b + \delta)(a + \varepsilon)$, hence
$$ab + ba + \delta a + \varepsilon b + \delta \varepsilon = 0. \quad (3)$$

Hence the additive group $S_{\mathbb{Z}}$ is generated by $1, a, b, ab$. (And analogously the $\mathbb{Q}$-algebra is generated by these same elements as a $\mathbb{Q}$-module.)

To prove that $1, a, b, ab$ *is a free basis* (over $\mathbb{Z}$ or $\mathbb{Q}$, resp.), we use a concrete realisation.

Let $\omega, \bar{\omega}$ be the roots (in $\mathbb{C}$) of the quadratic $X^2 + \varepsilon X + 1$; then
$$\omega + \bar{\omega} = -\varepsilon, \ \omega\bar{\omega} = 1.$$

Consider the matrices
$$\alpha = \begin{bmatrix} \omega & 0 \\ \delta\bar{\omega} & \bar{\omega} \end{bmatrix}, \ \beta = \begin{bmatrix} 0 & 1 \\ -1 & -\delta \end{bmatrix}, \ \alpha\beta = \begin{bmatrix} 0 & \omega \\ -\bar{\omega} & 0 \end{bmatrix}.$$

These have characteristic polynomials
$$X^2 + \varepsilon X + 1, \ X^2 + \delta X + 1, \ X^2 + 1,$$
so by Cayley–Hamilton α, β satisfy the defining relations of S. Since $1, \omega$ are linearly independent over $\mathbb{Z}, \mathbb{Q}$, it is clear that $1, \alpha, \beta, \alpha\beta$ are linearly independent. We may now identify $a = \alpha, b = \beta$.

Then any element of S takes the form

$$r = x1 + ya + zb + tab = \begin{bmatrix} x + y\omega & z + t\omega \\ y\delta\bar{\omega} - z - t\bar{\omega} & x + y\bar{\omega} - \delta z \end{bmatrix}$$

$$\det r = (x + y\omega)(x + y\bar{\omega}) + (z + t\omega)(z + t\bar{\omega}) - \delta(x + y\omega)z - \delta y\bar{\omega}(z + t\omega)$$

$$= (x^2 - \varepsilon xy + y^2) + (z^2 - \varepsilon zt + t^2) - \delta(xz + yt) + \delta\varepsilon yz$$

$$4\det r = (2x - \varepsilon y - \delta z)^2 + (2t - \delta y - \varepsilon z)^2 + (4 - \delta^2 - \varepsilon^2)(y^2 + z^2). \tag{4}$$

This form is positive definite. So the algebra has no zero-divisors and (being finite-dimensional) is a division algebra, with no proper homomorphic images. Therefore the homomorphisms $\mathbb{Q}S \to \mathbb{Q}R\Gamma$ ($\Gamma = Q, D, B$) already noted are isomorphisms; and we have concrete realisations of our algebras and orders.

Note that for the element r to be a unit of S we must have $\det r = 1$ (not -1 because the quadratic form is positive definite). Since $4 - \delta^2 - \varepsilon^2 \geq 2$, this requires by (4) that $y^2 + z^2 \leq 2$. Therefore $y, z \in \{0, 1, -1\}$; similarly $x, t \in \{0, 1, -1\}$. In each case S is therefore an order in a division algebra over $\mathbb{Q}$ with a finite group of units, necessarily one of the primordial groups; and we can tell which by calculating the order.

(iv) In the case of Q, $\delta = \varepsilon = 0$ and the line above (4) reduces for a unit to

$$1 = x^2 + y^2 + z^2 + t^2;$$

hence the eight units $\pm 1, \pm a, \pm b, \pm ab$ exhaust S^*. So here $S^* = Q$, as required.

(v), (vi) In the remaining two cases ($\delta = 1$ or $\varepsilon = 1$), easy calculations show that S is either the maximal order of Hurwitz quaternions

$$S_{1,1} = H = \{a + bi + cj + dk \mid a, b, c, d \in \mathbb{Z} \cup (\mathbb{Z} + \frac{1}{2})\}$$

with $H^* = B$, or an order $L = S_{1,0}$ with $L^* = D$, respectively; one uses (4). See also Hirsch–Zassenhaus [10, p. 549].

Taking these checks for granted we have

Lemma B.4. (i) *For each of the primordial groups* $\Gamma = C_2, C_4, C_6, Q, D, B$ *there exists an order* $R\Gamma$ *in a division algebra* $\mathbb{Q}R\Gamma$ *over* $\mathbb{Q}$ *such that* $(R\Gamma)^* = \Gamma$; *up to isomorphism the order and algebra are unique (given always that they are generated by* Γ*).*

(ii) *In each case there exists a* $\mathbb{Z}$*-basis of* $R\Gamma$ *in terms of which each unit in* Γ *is expressible with coefficients* $0, \pm 1$.

(iii) *Finally,*

$$\mathrm{rk}_{\mathbb{Z}} R\Gamma = \dim_{\mathbb{Q}} \mathbb{Q}R\Gamma = \begin{cases} 1, & \Gamma = C_2, \\ 2, & \Gamma = C_4, C_6, \\ 4, & \Gamma = Q, D, B. \end{cases}$$

Definition. We call $RC_2, \ldots, RB$ the *primordial rings (or orders)*, and $\mathbb{Q}RC_2, \ldots,$ $\mathbb{Q}RB$ the *primordial algebras*.

Lemma B.5. *For each primordial group Γ, the automorphism group* Aut $R\Gamma$ *of the ring $R\Gamma$ acts transitively on the set of all elements of each possible order q in Γ.*

Proof. Recall (B.1) that the only possibilities are $q = 1, 2, 3, 4, 6$. For $q = 1, 2$ there is only one element of order q, and the result is trivial. Moreover, since there is only one involution the case $q = 6$ reduces to $q = 3$. So we have only to consider $q = 3, 4$. Now any inner automorphism of Γ extends to an automorphism of $R\Gamma$, so it is enough to consider two distinct elements in a convenient Sylow subgroup. If this Sylow subgroup is not cyclic it must be the Sylow 2-subgroup Q of Q or B. But $RQ < RB$ and any element of order 3 in B cyclically permutes the three C_4 in Q. So it is enough to prove that some automorphism of $R\Gamma$ inverts some element of order 3 or 4 in Γ. When $\Gamma = C_4$ or C_6, complex conjugation is the answer. In the non-commutative cases, take the presentations (2). One obtains easily $1+\varepsilon a^{-1}+a^{-2} = 1+\delta(ab^{-1}a^{-1})+(ab^{-1}a^{-1})^2 = 1 + (b^{-1}a^{-1})^2 = 0$: there exists an automorphism of $R\Gamma$ inverting a and ab. This is enough. □

C A crucial lemma and its consequences

We now return to the context (†) of Section A:

$$R = \mathbb{Z}\Gamma \leq \mathbb{Q}\Gamma = \mathbb{Q}R = \mathfrak{A}, \ R^* = \Gamma \text{ (torsion group).} \quad (\dagger)$$

Consider any finite subgroup $G \leq \Gamma$. The subalgebra $\mathbb{Q}G$ of $\mathfrak{A}$ is finite-dimensional, semi-simple (by Maschke) (and a homomorphic image of the group algebra $\mathbb{Q}[G]$). By Wedderburn it decomposes,

$$\mathbb{Q}G = \mathfrak{A}_1 \oplus \cdots \oplus \mathfrak{A}_n,$$

where each component $\mathfrak{A}_i = M_{k_i}(F_i)$ is the full algebra of $k_i \times k_i$ matrices over a division algebra F_i. By A.3(ii) the only nilpotent element in $\mathfrak{A}$ is 0, so the same must be true for each $\mathfrak{A}_i$; and this requires each $k_i = 1$: if $k \geq 2$ the $k \times k$ matrix

$$\begin{bmatrix} 0 & 1 & \\ 0 & 0 & \\ & & 0 \end{bmatrix}$$

is nilpotent. Therefore the decomposition reduces to

$$\mathbb{Q}G = F_1 \oplus \cdots \oplus F_n, \quad (1)$$

where $F_1, \ldots, F_n$ are finite-dimensional division algebras over $\mathbb{Q}$.

Let the projections in the decomposition (1) be
$$\pi_i : \mathbb{Q}G \twoheadrightarrow F_i \ (i = 1, \ldots, n).$$

Now $\mathbb{Z}G$ is an order in $\mathbb{Q}G$. Therefore the image $R_i = (\mathbb{Z}G)\pi_i$ is an order in F_i. Clearly
$$\mathbb{Z}G \leq R_1 \oplus \cdots \oplus R_n = S \text{ (say)},$$
again an order in $\mathbb{Q}G$. We apply Lemmas A.2 and A.4. Since $(\mathbb{Z}G)^* = \Gamma \cap \mathbb{Z}G$ is torsion, therefore $S^* = R_1^* \times \cdots \times R_n^*$ is torsion. Thus each R_i^* is torsion, and by the considerations of Section B it follows that

$$R_i^* \text{ is one of the 6 primordial groups,} \qquad (2)$$

and $R_i, F_i = \mathbb{Q}R_i$ the corresponding primordial ring and division algebra. In particular each R_i^* is of exponent 12 and of finite order dividing 24. Therefore $G \leq (\mathbb{Z}G)^* \leq R_1^* \times \cdots \times R_n^*$ is of exponent 12 and of order dividing 24^n – a bound depending only on the subalgebra $\mathbb{Q}G$ of $\mathfrak{A}$.

Since G here can be any cyclic subgroup of Γ, it follows that Γ is itself of exponent 12. And since all involutions in Γ are central (A.3(iii)), B.3 implies

Lemma C.1. *In the context* (†), *Γ is locally finite of exponent* 12.

But nothing would change in the foregoing analysis if we replaced G by any finite subgroup H between G and $\widehat{G} = \Gamma \cap \mathbb{Q}G$ for we still have $\mathbb{Q}H = \mathbb{Q}G = F_1 \oplus \cdots \oplus F_n$, so that as before $|H|$ divides 24^n. Since $\widehat{G} \ (\leq \Gamma)$ is locally finite and the orders of its finite subgroups are bounded, it follows that $\widehat{G}$ itself is finite. It is natural to call $\widehat{G}$ the *closure* of G. Since $\widehat{G} \leq \Gamma \cap \mathbb{Z}\widehat{G} \leq \Gamma \cap \mathbb{Q}\widehat{G} = \Gamma \cap \mathbb{Q}G = \widehat{G}$, we have

Lemma C.2. *In the context* (†), *for every finite subgroup G of Γ, the closure $\widehat{G} = \Gamma \cap \mathbb{Q}G$ is a finite subgroup of Γ such that*
$$\widehat{G} \geq G, \ \mathbb{Q}\widehat{G} = \mathbb{Q}G, \text{ and } (\mathbb{Z}\widehat{G})^* = \widehat{G}.$$

Replacing G by its closure $\widehat{G}$ we may assume that the finite subgroup G of Γ is *closed* in $\mathbb{Z}\Gamma$, i.e. that
$$(\mathbb{Z}G)^* = G.$$

Lemma C.3. *In the context* (†), *let G be a closed finite subgroup of Γ. Then*
$$\mathbb{Q}G = F_1 \oplus \cdots \oplus F_n, \qquad (3)$$
where the F_i are primordial division algebras. Let $\pi_i : \mathbb{Q}G \twoheadrightarrow F_i$ be the projections, and write
$$R_i = (\mathbb{Z}G)\pi_i, \ G_i = G\pi_i. \qquad (4)$$
Then for each i,
$$G_i = R_i^* \text{ is primordial, } R_i = RG_i \text{ the associated primordial ring,}$$

and we have subdirect decompositions

$$\mathbb{Z}G \leq_{sd} R_1 \oplus \cdots \oplus R_n, \tag{5}$$

$$G \leq_{sd} G_1 \times \cdots \times G_n, \tag{6}$$

where the projections are induced by the π_i.

Proof. We have already noted (2) that each R_i^* is primordial and that R_i, F_i are the associated primordial ring and algebra. Since π_i induces a ring-homomorphism $\mathbb{Z}G \twoheadrightarrow R_i$, it maps $G = (\mathbb{Z}G)^*$ into R_i^*, whence $G_i \leq R_i^*$. To prove that this is an equality, note that since G_i generates the additive group of R_i and $-1_i = (-1)\pi_i \in G_i$, therefore $\frac{1}{2}|G_i| \geq \mathrm{rk}_{\mathbb{Z}} R_i$, which gives

$$2\,\mathrm{rk}\,R_i \leq |G_i| \text{ which divides } |R_i^*|. \tag{7}$$

We run through the cases in a table.

R_i^*	$2\,\mathrm{rk}\,R_i$	$|R_i^*|$
C_2	2	2
C_4	4	4
C_6	4	6
Q	8	8
D	8	12
B	8	24

From the table one sees that (7) forces $G_i = R_i^*$ except in the case $R_i^* = B$, when it is *a priori* possible that $|G_i| = 8$ or $|G_i| = 12$. In the first case $G_i = O_2(B) = Q$ and this implies that

$$R_i = \mathbb{Z}G_i = \mathbb{Z}Q = RQ,$$

giving $R_i^* = Q$ – contradiction. The other case gives $G_i = D$ which is impossible as well. The lemma follows. □

We have at once

Corollary C.4 (Hirsch–Zassenhaus [10]). *In the context* (†), *if Γ is finite then Γ is a subdirect product of (a finite number of) copies of the primordial groups.*

We want to extend this to the infinite torsion case.

Given any closed finite subgroup $G \leq \Gamma$ consider the decomposition (3) of C.4

$$\mathbb{Q}G = F_1 \oplus \cdots \oplus F_n.$$

Write

$$1 = f_1 + \cdots + f_n \ (f_i \in F_i).$$

Then $f_1, \ldots, f_n$ are orthogonal idempotents, central in $\mathfrak{A}$ (by A.3); so

$$\mathfrak{A} = \mathfrak{A}f_1 \oplus \cdots \oplus \mathfrak{A}f_n$$

and any (2-sided) ideal M of $\mathfrak{A}$ decomposes as

$$M = Mf_1 \oplus \cdots \oplus Mf_n,$$

where Mf_i is an ideal of $\mathfrak{A}f_i$, $Mf_i = M \cap \mathfrak{A}f_i$. The ideal M is therefore maximal in $\mathfrak{A}$ if and only if, for a suitable labeling

Mf_1 is a maximal ideal of $\mathfrak{A}f_1$, and $Mf_i = \mathfrak{A}f_i$ ($2 \leq i \leq n$);

and then

$$M = Mf_1 \oplus \mathfrak{A}(1 - f_1),$$
$$\mathbb{Q}G = F_1 \oplus \mathbb{Q}G(1 - f_1).$$

Since $f_1 \in F_1 \setminus M$, we have $F_1 \cap M = 0$, so

$$M \cap \mathbb{Q}G = \mathbb{Q}G(1 - f_1).$$

Therefore

$$\mathbb{Q}G = F_1 \oplus (M \cap \mathbb{Q}G) \leq F_1 \oplus M.$$

This has consequences:

(i) Given any maximal ideal M, consider any finite set $x_1, \ldots, x_k$ of elements of $\mathfrak{A}$ linearly independent modulo M. Then we may choose the closed finite subgroup G so that $x_1, \ldots, x_k \in \mathbb{Q}G \leq F_1 \oplus M$. This implies that $k \leq \dim F_1$ ($= 1, 2, 4$), a finite bound. Therefore we may choose $x_1, \ldots, x_k$ to be a $\mathbb{Q}$-basis of $\mathfrak{A} \bmod M$, and then $\mathfrak{A} = \mathbb{Q}x_1 \oplus \cdots \oplus \mathbb{Q}x_k \oplus M \leq \mathbb{Q}G + M \leq F_1 \oplus M \leq \mathfrak{A}$, whence

$$\mathfrak{A} = F_1 \oplus M,$$

$\mathfrak{A}/M \cong F_1$ (primordial division algebra).

(ii) Given any $0 \neq x \in \mathfrak{A} = \mathbb{Q}\Gamma$, choose the closed finite subgroup $G \leq \Gamma$ so that

$$0 \neq x \in \mathbb{Q}G = F_1 \oplus \cdots \oplus F_n \text{ etc. (as above).}$$

Choose the notation so that $xf_1 \neq 0$, and take M to be a maximal ideal of $\mathfrak{A}$ with $1 - f_1 \in M$. Then $x \notin M$ (otherwise $x \in M \Rightarrow 0 \neq xf_1 \in F_1 \cap M = 0$, a contradiction). This implies that the intersection of the maximal ideals of $\mathfrak{A}$ is 0, and we have an embedding

$$\mathfrak{A} \hookrightarrow \prod_{M_i \text{ max.}} \mathfrak{A}/M_i \cong \prod \mathbb{Q}RG_i \ (G_i \text{ primordial}).$$

Correspondingly we have subdirect decompositions (after suitable identifications)

$$\mathbb{Z}\Gamma \leq_{\mathrm{sd}} \prod RG_i, \tag{8}$$

$$\Gamma \leq_{\text{sd}} \prod G_i.$$

We now go back to B.4(iii): as additive groups $RG_i \cong \mathbb{Z}, \mathbb{Z}^2, \mathbb{Z}^4$, so the decomposition (8) gives an embedding (of additive groups).

$$(\mathbb{Z}\Gamma)_\mathbb{Z} \leq \prod \mathbb{Z}$$

in which each element of Γ is expressed in a 'sequence' all of whose terms are $0, \pm 1$. This makes $(\mathbb{Z}\Gamma)_\Gamma$ a group of bounded sequences in the generalized Baer–Specker group $\Pi\mathbb{Z}$: and by a theorem of Nöbeling (with an easy proof by P. Hill and G. Bergman, see [3]) this group is free abelian. So

Theorem C.5. *In the context* (†) *the algebra* $\mathfrak{A} = \mathbb{Q}\Gamma$ *may be identified with a subdirect product of primordial algebras in a way which induces subdirect-product decompositions of* $\mathbb{Z}\Gamma$ *and* Γ *with primordial factors,*

$$\mathfrak{A} \leq_{\text{sd}} \Pi F_i, \ R = \mathbb{Z}\Gamma \leq_{\text{sd}} \Pi R_i, \ \Gamma \leq_{\text{sd}} \Pi G_i; \qquad (9)$$

here the G_i are primordial groups, the $R_i = RG_i$, the $F_i = \mathbb{Q}RG_i$, and each projection $\Gamma \twoheadrightarrow G_i$ maps the involution -1 of Γ to the unique involution -1_i of G_i (i.e. -1 is represented as the 'diagonal involution'). Moreover, the additive group of $\mathbb{Z}\Gamma$ is free abelian and Γ is residually finite.

Since any ring with a free abelian additive group may be realised as the endomorphism ring of a suitable torsion-free abelian group – see e.g. [2] – the last line and A.1 imply

Corollary C.6. *Let Γ be a torsion group. Then $\Gamma = \text{Aut } A$ for some torsion-free abelian group A if and only if $\Gamma = R^*$ for some subring R of a $\mathbb{Q}$-algebra.*

If we write $\mathfrak{A}_p$ for the variety of elementary abelian p-groups, and apply Hall's notation for products $\mathfrak{A}\mathfrak{A}'$ of classes groups groups (extensions of $\mathfrak{A}$-groups by $\mathfrak{A}'$-groups) then

$$C_2 \in \mathfrak{A}_2,$$

$$C_6 \in \mathfrak{A}_2\mathfrak{A}_3 \cap \mathfrak{A}_3\mathfrak{A}_2,$$

$$C_4, Q \in \mathfrak{A}_2^3,$$

$$D = O_3(D)C_4 \in \mathfrak{A}_3\mathfrak{A}_2^2,$$

$$B = O_2(B)C_3 \in \mathfrak{A}_2^2\mathfrak{A}_3.$$

Thus each of the primordial groups lies in the variety $\mathfrak{A}_3\mathfrak{A}_2^3\mathfrak{A}_3$ and C.6 implies at once

Corollary C.7. *In the context* (†), $\Gamma \in \mathfrak{A}_3\mathfrak{A}_2^2\mathfrak{A}_3$.

D Elementary abelian direct factors, and products of squares

In view of Lemma D.3 we shall need to know a criterion guaranteeing that the diagonal involution -1 does not lie in a direct factor isomorphic with C_2. The present section is devoted to this question.

We shall have occasion to apply the well known

Lemma D.1. *Let Γ be a group with subgroups $E \leq {}_3\Gamma$, $K \triangleleft \Gamma$ such that $E \cap K = 1$. If $H \leq \Gamma$ is a subgroup such that*

$$\frac{\Gamma}{K} = \frac{EK}{K} \times \frac{H}{K},$$

then $\Gamma = E \times H$.

Proof. Certainly $K \leq H \triangleleft \Gamma$, so E and H are both normal in Γ. And $\Gamma = (EK)H = E(KH) = EH$, while $E \cap H = E \cap EK \cap H = E \cap K = 1$. □

As a piece of ad hoc notation, for any group Γ and any prime p let us write

$$A_p(\Gamma) = \langle x^p, [x,y] \mid x, y \in \Gamma \rangle,$$

$$A_p^*(\Gamma) = \langle x^p, [x,y] \mid x, y \text{ p-elements of } \Gamma \rangle.$$

These are normal subgroups of Γ, with $A_p(\Gamma) \geq A_p^*(\Gamma)$; and $A_p(\Gamma)$ is of course the verbal subgroup of Γ associated with the variety $\mathfrak{A}_p$.

The first part of our next lemma is folklore.

Lemma D.2. *Let Γ be a group, E an elementary abelian p-subgroup. Then*

(a) *E is a direct factor of Γ $\Leftrightarrow$ $E \leq {}_3\Gamma$ and $E \cap A_p(\Gamma) = 1$;*

(b) *if Γ is locally finite, the same is true with A_p^* in place of A_p.*

Proof. (a) '$\Rightarrow$' An abelian direct factor is always central, and any retraction $\Gamma \twoheadrightarrow E$ fixes $E \cap A_p(\Gamma)$ but maps $A_p(\Gamma)$ into $A_p(E) = 1$.
'$\Leftarrow$' If the conditions are satisfied, then as $EA_p(\Gamma)/A_p(\Gamma)$ is automatically a direct factor of the elementary abelian group $\Gamma/A_p(\Gamma)$, we need only appeal to D.1.
(b) Assume Γ is locally finite. Since $A_p^*(\Gamma) \leq A_p(\Gamma)$, the implication '$\Rightarrow$' is immediate from (a). To reduce also the reverse implication to case (a), it will be enough to prove that a p-element in ${}_3\Gamma$ lies in $A_p(\Gamma)$ if, and only if, it lies in $A_p^*(\Gamma)$ – in other words to prove that the abelian groups ${}_3\Gamma \cap A_p(\Gamma) \geq {}_3\Gamma \cap A_p^*(\Gamma)$ have the same p-components. Suppose then for a contradiction that there exists a

$$p\text{-element } e \in {}_3\Gamma \text{ such that } e \in A_p(\Gamma) \setminus A_p^*(\Gamma).$$

Then there exists a finitely generated (so finite) subgroup G with $e \in A_p(G)$ and clearly $e \in {}_3G$, $e \notin A_p^*(G)$. So we may assume Γ finite. Consider the quotient

$\bar{\Gamma} = \Gamma/A_p^*(\Gamma)$, $\bar{e} = eA_p^*(\Gamma)$. Certainly $1 \neq \bar{e} \in A_p(\bar{\Gamma})$ and as Γ is finite, every p-element of $\bar{\Gamma}$ lifts to a p-element of Γ, so the generators of $A_p^*(\bar{\Gamma})$ lift to generators of $A_p^*(\Gamma)$, whence $A_p^*(\bar{\Gamma}) = A_p^*(\Gamma)/A_p^*(\Gamma) = 1$. Therefore we may assume that

$$\Gamma \text{ is finite with } A_p^*(\Gamma) = 1.$$

This last equality means of course that the p-elements of Γ form an elementary abelian p-group, necessarily the Sylow p-subgroup $O_p(\Gamma)$. Then Schur–Zassenhaus (see Huppert [11, p. 126]) provides a p-complement H, say, and we have

$$\Gamma = O_p(\Gamma)H, \; O_p(\Gamma) \cap H = 1.$$

The central p-subgroup $\langle e \rangle = E$ (say) is normalized by the p'-group H (of order coprime to p). Consider the group ring F_pH and $F_pO_p(\Gamma)$ as F_pH-module, thus E is a F_pH-submodule of the module Γ with F_pH acting by conjugation. Note that the orders are coprime, hence Maschke applies. There is a direct complement F in $O_p(\Gamma)$ normalized by H.

$$O_p(\Gamma) = E \times F, \; [F, H] \leq F.$$

But then FH is a subgroup centralised by E ($\leq {}_3\Gamma$), and the easy checks

$$\Gamma = O_p(\Gamma)\, H = EFH,$$
$$E \cap FH = E \cap O_p(\Gamma) \cap FH = E \cap F(O_p(\Gamma) \cap H) = E \cap F = 1,$$

show that $\Gamma = E \times FH$. Therefore (a) gives the contradiction $e \notin A_p(\Gamma)$. □

The following easy result will be of the greatest importance.

Lemma D.3. *In the context* (†) *if either of the primordial groups C_2 or C_6 occurs among the subdirect factors G_i in* (9) *above, then $\langle -1 \rangle$ is a direct factor of Γ. In particular Γ then has a direct factor isomorphic to C_2.*

Proof. If C_2 or $C_6 = G_i$, then the central involution -1 of Γ lies outside the kernel K of the composite homomorphism $\Gamma \twoheadrightarrow G_i \twoheadrightarrow C_2$. Since K is a normal subgroup of index 2 therefore $\Gamma = K \times \langle -1 \rangle$ by Lemma D.2. □

The criterion we need is given by

Lemma D.4. *Let Γ be a subdirect product of primordial groups, and let e be an involution in Γ such that $\langle e \rangle$ is not a direct factor of Γ. Then there exists a finite 2-subgroup $H \leq \Gamma$ such that e is a product of squares of elements of order 4 in H.*

Proof. C.2 and D.2(b) imply that $e \in A_2^*(\Gamma)$, so there is a finite subgroup G of Γ such that $e \in A_2^*(G)$. But it follows from Corollary C.7 that $G \in \mathfrak{A}_3\mathfrak{A}_2^3\mathfrak{A}_3$. So $G = O_{2'22'}(G) \geq O_{2'2}(G) \geq O_{2'}(G) \geq 1$ (and here $2'$ means in effect 3). Since every 2-element of G lies in $O_{2'2}(G)$, we may assume $G = O_{2'2}(G)$. But then, if we choose a Sylow 2-subgroup H of G, we have $G = O_3(G)H$. By centrality, $e \in H \cap A_2^*(G)$. The retraction $G \twoheadrightarrow H$ fixes e and maps $A_2^*(G)$ into $A_2^*(H) = A_2(H)$; so $e \in A_2(H)$. And of course $[x, y] = x^{-2}(xy^{-1})^2y^2$. Thus e is a product of squares in H, a group of exponent 4. Squares of involutions may of course be suppressed. □

E Finite D-blocks

We have seen in Theorem C.6 that the groups Γ which concern us are subdirect products of copies of the primordial groups C_2, C_4, C_6, Q, D, B. We need a little terminology to distinguish these. For any subset $\{A_1, \ldots, A_r\}$ of $\{C_2, \ldots, B\}$ we shall use the term $A_1 \cdots A_r$-group to refer to a subdirect product of copies of $A_1, \ldots, A_r$. (For example, a C_2-group is simply an elementary abelian 2-group). We shall abbreviate $C_2C_4C_6$ to C, so all our groups are $CQDB$-groups.

A special role is played by the finite D-groups and B-groups, in particular the latter. We consider their structure in this section.

Before we start, note that the only primes dividing the order $|G|$ of a finite $CQDB$-group are $2, 3$. Therefore for any Sylow 2-subgroup H and any Sylow 3-subgroup X we have
$$G = HX = XH;$$
(i.e. for any $g \in G$ we find $h \in H$ and $x \in X$ with $g = hx$; similarly for $G = XH$) and any epimorphism $\varphi : G \twoheadrightarrow \bar{G}$ gives $\bar{G} = (H\varphi)(X\varphi)$, where $H\varphi$ is a Sylow 2-subgroup, $X\varphi$ a Sylow 3-subgroup of $\bar{G}$.

Now consider a finite D-group G with a given embedding as a subdirect power of D.
$$\boldsymbol{\delta} = (\pi_1, \ldots, \pi_n) : G \twoheadrightarrow G\boldsymbol{\delta} \leq_{\text{sd}} D^n. \tag{1}$$
Choose a Sylow 2-subgroup H of G, which will be abelian of exponent exactly 4, the unique Sylow 3-subgroup $O_3(G)$ is abelian of exponent 3. For $i = 1, \ldots, n$ the image $H\pi_i$ is a Sylow 2-subgroup of D, so is cyclic of order 4 and contains ${}_3D$ as a subgroup of index 2. The intersections
$$H \cap ({}_3D)\pi_i^{-1} \quad (i = 1, \ldots, n) \tag{2}$$
are therefore subgroups of index 2 (so maximal) in H, all containing ${}_3G$. List the distinct such intersections as $H_1, \ldots, H_t$, and partition the projections π_i into corresponding D-blocks $\boldsymbol{\delta}_1, \ldots, \boldsymbol{\delta}_t$, putting π_i into $\boldsymbol{\delta}_k$ if and only if $H \cap ({}_3D)\pi_i^{-1} = H_k$. If $\boldsymbol{\delta}_k$ contains n_k projections π_i, so that $n_1 + \cdots + n_r = n$, then $\boldsymbol{\delta}_k$ may be regarded as an ordered n_k-tuple of projections $G \twoheadrightarrow D$, or better as a homomorphism
$$\boldsymbol{\delta}_k : G \twoheadrightarrow G\boldsymbol{\delta}_k =: G_k \leq_{\text{sd}} D^{n_k}. \tag{3}$$
These $\boldsymbol{\delta}_k$ induce a subdirect embedding
$$\boldsymbol{\delta} = (\boldsymbol{\delta}_1, \ldots, \boldsymbol{\delta}_t) : G \twoheadrightarrow G\boldsymbol{\delta} \leq_{\text{sd}} G_1 \times \cdots \times G_t. \tag{4}$$
We identify $G = G\boldsymbol{\delta}$ and the factor G_k as subgroups of the direct product in the usual way.

Now $H \cap ({}_3D)\pi_i^{-1}$ is the kernel of an irreducible representation of the 2-subgroup H (over $\mathbb{F}_3$), namely the composite
$$H \hookrightarrow G \xrightarrow{\pi_i} D \xrightarrow{\text{conj.}} \operatorname{Aut} O_3(D) \ (\cong \operatorname{GL}_1(\mathbb{F}_3) \cong C_2).$$

Two such representations are equivalent if and only if they belong to the same block. (If we view $O_3(D)^n$ as $\mathbb{F}_3 H$-module *via* $\boldsymbol{\delta}$, it is clear that it is a semisimple module, whose isotypic (homogeneous) components are the submodules $O_3(D)^{n_k}$ corresponding to the D-blocks $\boldsymbol{\delta}_1, \ldots, \boldsymbol{\delta}_t$. We obtain the isotypic components of the $\mathbb{F}_3 H$-submodule $O_3(G)$ by taking intersections.) It is now immediate that $O_3(G) = \bigoplus_{k=1}^{t} O_3(G) \cap O_3(D)^{n_k}$ and it follows

$$O_3(G) = O_3(G_1) \times \cdots \times O_3(G_t). \qquad (5)$$

And from the definition of $\boldsymbol{\delta}_k$, $H_k = H \cap ({}_3 D)^{n_k} \boldsymbol{\delta}_k^{-1} = H \cap ({}_3 G_k) \boldsymbol{\delta}_k^{-1}$, whence $H_k \boldsymbol{\delta}_k = {}_3 G_k$, which is of exponent 2, and of index 2 in $H \boldsymbol{\delta}_k$, a Sylow 2-subgroup of G_k and of exponent exactly 4. Hence $H \boldsymbol{\delta}_k$ is a direct product of one copy of C_4 and a finite elementary abelian 2-group.

Consider any element $x_k \in O_3(G_k)$ and any $h_k \in H \setminus H_k$. Then for each π_i in $\boldsymbol{\delta}_k$ we have $x_k \pi_i \in O_3(D)$ while $h_k \pi_i$ is an element of order 4 in D, so $(x_k{}^{h_k}) \pi_i = (x_k \pi_i)^{h_k \pi_i} = (x_k \pi_i)^{-1}$ and $[x_k, h_k] \pi_i = (x_k \pi_i)^{-2} = (x_k \pi_i)$, while the image of x_k under any projection not in the kth block is 1. Therefore

$$x_k{}^{h_k} = x_k^{-1} \text{ and } [x_k, h_k] = x_k. \qquad (6)$$

Hence
$$[O_3(G_k), H] = O_3(G_k),$$

and it follows from (5) that

$$[O_3(G), H] = O_3(G).$$

But $G/O_3(G) \cong H$, an abelian group. Therefore $G' \leq O_3(G) = [O_3(G), H] \leq G'$, and we have proved

Lemma E.1. *For a finite D-group G with D-blocks $\boldsymbol{\delta}_k : G \twoheadrightarrow G_k$ ($1 \leq k \leq t$) and Sylow 2-subgroup H,*

$$G' = [O_3(G), H] = O_3(G) = O_3(G_1) \times \cdots \times O_3(G_t).$$

F On the structure of a finite B-group

Let G be a finite B-group with a given embedding as a subdirect power of B.

$$\boldsymbol{\beta} = (\pi_1, \ldots, \pi_n) : G \twoheadrightarrow G\boldsymbol{\beta} \leq_{\text{sd}} B^n.$$

Choose a Sylow 3-subgroup X of G. Then

$$G = O_2(G) X.$$

For $1 \leq i \leq n$ the image $X\pi_i$ is a Sylow 3-subgroup of B, so of order 3. Therefore the intersections $X \cap \text{Ker } \pi_i$ are subgroups of index 3 in X. List the distinct such intersections by $X_1, \ldots, X_r$ and partition the projections π_i into r corresponding *blocks*, putting π_i in the kth block if $X \cap \operatorname{Ker} \pi_i = X_k$. Relabel the projections in the kth block as $\pi_{ki} : G \twoheadrightarrow B$ $(1 \le i \le n_k)$, and define

$$\beta_k = (\pi_{k,1}, \ldots, \pi_{k,n_k}) : G \twoheadrightarrow G_k := G\beta_k \le_{\text{sd}} B^{n_k}.$$

Then $n_1 + \cdots + n_r = n$, and we have a subdirect embedding

$$\beta = (\beta_1, \ldots, \beta_r) : G \twoheadrightarrow G\beta \le_{\text{sd}} G_1 \times \cdots \times G_r;$$

regard β as an inclusion, and identify the factors G_k as subgroups of their direct product in the usual way.

Clearly, for $1 \le k \le r$,

$$X \cap \operatorname{Ker} \beta_k = X_k = X \cap \operatorname{Ker} \pi_{ki} \ (1 \le i \le n_k).$$

The image $X\beta_k$ is a Sylow 3-subgroup of G_k and it is of order 3, choose a generator c_k, so that

$$X\beta_k = \langle c_k \rangle, \ \mathrm{o}(c_k) = 3.$$

For $1 \le i \le n_k$ the component $c_k \pi_{ki}$ is an element of order 3 in B, and since (by Lemma B.5) Aut B acts transitively on the elements of order 3 in B, there will be no loss of generality if we suppose that these components are all equal to c, a fixed element of order 3 in B. Then

$$c_k = (c, \ldots, c) \in B^{n_k}, \ \mathrm{o}(c) = 3.$$

The conjugational action of X on $O_2(B)^{n_k}$ ($\triangleleft B^{n_k}$) passes to the quotient and makes $O_2(B)^n/(\mathfrak{z}B)^n = (O_2(B)/\mathfrak{z}B)^n$ (thought of in additive notation) a $\mathbb{Z}_2[X]$-module. The factors $O_2(B)/\mathfrak{z}B$ afford irreducible representations, two factors affording equivalent representations iff the corresponding projections lie in the same block. Now it is well known that a representation module is the direct product of its isotype components. Applying this to the image of $O_2(G)$ and lifting back, (as for (5)) we find that

$$O_2(G)(\mathfrak{z}B)^n = (O_2(G_1) \times \cdots \times O_2(G_r))(\mathfrak{z}B)^n.$$

The modular law (using $O_2(G) \le O_2(G_1) \times \cdots \times O_2(G_r)$) and a trivial commutator computation now give

Lemma F.1. *For a finite B-group G, we have with the foregoing notation*

(a) $O_2(G)(\mathfrak{z}G_1 \times \cdots \times \mathfrak{z}G_r) = O_2(G_1) \times \cdots \times O_2(G_r)$.

(b) $[O_2(G), X] = [O_2(G_1), X] \times \cdots \times [O_2(G_r), X]$.

We shall need to revert to the subject of B-groups when we come to proving sufficient conditions; but for necessity this will do. For this we shall make use of F.1(a) and the following consequence of E.1.

Lemma F.2. *Let G be a CQDB-group with a given embedding*

$$\pi = (\delta, \gamma, \beta) : G \hookrightarrow_{sd} G\delta \times G\gamma \times G\beta,$$

where $G\delta$ is a D-group, $G\gamma$ a CQ-group, and $G\beta$ a B-group. Then any Sylow 3-subgroup X of G decomposes as a direct product

$$X = X\delta \times X(\gamma, \beta),$$

where $X\delta = O_3(G\delta)$ is the Sylow 3-subgroup of $G\delta$ and $X(\gamma, \beta)$ is a Sylow 3-subgroup of $G(\gamma, \beta) \leq_{sd} G\gamma \times G\beta$.

Proof. Choose a Sylow 2-subgroup H of G. Then

$$\begin{aligned}[H, X] &\leq_{sd} [H\delta, X\delta] \times [H\gamma, X\gamma] \times [H\beta, X\beta] \\ &= O_3(G\delta) \times 1 \times \text{2-subgroup of } G\beta. \\ &\quad \text{(by E.1)} \quad (X\gamma \text{ central}) \quad (H\beta = O_2(G\beta))\end{aligned}$$

As these factors have coprime order, we have

$$G \geq [H, X] = O_3(G\delta) \times 1 \times [H\beta, X\beta].$$

We claim that $O_3(X\delta) = [X, H]$ and show that $[X, H] = [H\delta, X\delta] \times [H\beta, X\beta]$. Choose $(a, b) \in [X, H]$ (where $a \in O_3(X\delta)$ and $b \in [H\beta, X\beta]$). Then $(a, b)^4 = (a^4, b^4) = (a, 1)$, hence $O_3(X\delta) \leq [X, H]$ and equality is immediate.

But then

$$X\delta = O_3(G\delta) \leq X \leq_{sd} X\delta \times X(\gamma, \beta),$$

and it follows by the modular law that $X = X\delta \times X(\gamma, \beta)$. □

Note that if π contains no projections $G \twoheadrightarrow C_6$, then $X\gamma = 1$ and this decomposition of F.2 reduces to $X = X\delta \times X\beta = X(\delta, \gamma) \times X\beta$.

Definition. Let x be an element of order 3 in B. Under conjugation x cyclically permutes the three (cyclic) subgroups of order 4 in $O_2(B)$ ($\cong Q$). Thus if h is an element of order 4 in $O_2(B)$, then h, h^x, h^{x^2} are generators of the three subgroups of order 4, from which we get

$$h^{x^2} = (hh^x)^{\pm 1} = \pm hh^x \in (hh^x)_3 B$$

(where as usual -1 is the unique involution in B); and if $h \in {}_3B$, then $h = h^x = h^{x^2}$ and again we have $h^{x^2} \in (hh^x)_3 B$. We shall say that the 2-element h of B is *x-positive* if $h^{x^2} = hh^x$, and *x-negative* if $h^{x^2} = -hh^x$. Clearly, every 2-element in B is either *x*-positive or *x*-negative; and h is *x*-positive iff $-h$ is *x*-negative.

Lemma F.3. *For a 2-element h and 3-element x in a group G, define*

$$h \circ x = [h, x]\, [h, x^{-1}].$$

Then

(a) *for any homomorphism $\varphi : G \to \bar{G}$, $(h \circ x)\varphi = (h\varphi) \circ (x\varphi)$;*

(b) *$h \circ x = 1$ if either $h \in {}_3G$ or $x \in {}_3G$;*

(c) *$h \circ x = 1$ if $G \in \{C_2, C_3, C_4, C_6, Q, D\}$;*

(d) *if $G = B$, then $h \circ x = 1$ unless $\mathrm{o}(h) = 4, \mathrm{o}(x) = 3$, and in that case $h \circ x$ is the x-positive element of $h\,{}_3B$.*

Proof. Here (a) and (b) are obvious, so for (c) and (d) it is enough to consider the cases $G = D, B$ and to evaluate $h \circ x$ when $\mathrm{o}(h) = 4, \mathrm{o}(x) = 3$. When $G = D$ we have $1 \neq [h, x] = (x^h)^{-1}x$ which now must be $x^2 = x^{-1}$, and similarly $[h, x^{-1}] = x$, whence $h \circ x = 1$. Suppose then that $G = B$. If h is x-positive, then $h^{x^{-1}} = h^{x^2} = hh^x$, and $h \circ x = h^{-1}h^xh^{-1}h^{x^{-1}} = h^{-1}h^xh^x = -h^{-1} = h$, whence also $h^{-1} \circ x = (-h) \circ x = h \circ x = h$. □

Corollary F.4. *For $h \in O_2(B)$ and an element x of order 3 in B,*

$$h \circ x \equiv h \pmod{{}_3B}$$

and $h \circ x$ is x-positive.

We are now in a position to prove the main result of this section. First a definition.

Definition. We shall say that a group is *reduced* if it has no direct factor of prime order. It follows easily from Lemma D.2 that a group of exponent 12 can always be written as the direct product of a reduced group and two elementary abelian groups (of exponents 2 and 3), such a reduced factor will be called a *reduced part* of the group.

Lemma F.5. *Let G be a finite B-group with Sylow 3-subgroup X. Then*

(a) $G = O_2(G)X$;

(b) $O_2(G) = [O_2(G), X]\,{}_3G$;

(c) *in $[O_2(G), X]$ every involution is a product of squares;*

(d) $[O_2(G), X]$ *is a reduced part of $O_2(G)$;*

(e) $[O_2(G), X]X$ *is the unique reduced part of G;*

(f) $[O_2(G), X] = G'$.

Proof. (a) This was noted at the beginning of the section.

(b) Suppose first that G is a B-block, i.e. $r = 1$. Then X is of order 3 with generator $x = (c, \ldots, c)$, say. Given $h \in O_2(G)$ we have $(h \circ x)\pi_i = (h\pi_i) \circ c \equiv h\pi_i \pmod{{}_3 B}$ $(1 \leq i \leq n)$, by Lemma F.2(a) and Corollary F.3. Hence $h \circ x \equiv h$ modulo $G \cap ({}_3 B)^n = {}_3 G$. And since $h \circ x \in [O_2(G), X]$ we have proved that $O_2(G) \leq [O_2(G), X]{}_3 G$. The reverse inclusion is obvious.

In the general case we revert to the notation earlier in the section. By what we have just proved,
$$O_2(G_k) = [O_2(G_k), X\beta_k]{}_3 G_k = [O_2(G_k), X]{}_3 G_k.$$

Lemma F.1 now gives
$$O_2(G)({}_3 G_1 \times \cdots \times {}_3 G_r) = O_2(G_1) \times \cdots \times O_2(G_r)$$
$$= ([O_2(G_1), X] \times \cdots \times [O_2(G_r), X])({}_3 G_1 \times \cdots \times {}_3 G_r)$$
$$= [O_2(G), X]({}_3 G_1 \times \cdots \times {}_3 G_r).$$

The conclusion (b) now follows from the modular law.

(c) By virtue of Lemma F.1(b) we may assume that G is a B-block. Write
$$O_2(G)^\mathfrak{s} = \langle h^2 \mid h \in O_2(G) \rangle.$$

By part (b), since ${}_3 G$ is of exponent 2 we have $O_2(G) = [O_2(G), X]^\mathfrak{s}$. But $O_2(G)$ is a subdirect power of Q, so its derived group is central, and for $h_1, h_2 \in O_2(G)$ we have $[h_1, h_2] = h_1^{-2}(h_1 h_2^{-1})^2 h_2^2$, whence $O_2(G)' \leq O_2(G)^\mathfrak{s}$.

Still assuming that G is a B-block, the Sylow 3-subgroup X is cyclic of order 3, with generator $x = (c, \ldots, c)$, say, and as in the proof of Lemma F.1, $O_2(G)/{}_3 G$ is a $\mathbb{Z}_2[X]$-module in which X acts without fixed points. In other words $O_2(G)/{}_3 G$ may be regarded as a vector space over the Galois field $\mathbb{F}_4$. Let the dimension of this vector space be m, and let the elements $h_1, \ldots, h_m \in O_2(G)$ map to a basis. Then Lemma F.2(d) and Corollary F.3 imply that $h_i \circ x \equiv h_i \pmod{{}_3 G}$ and that the components of the $h_i \circ x$ in the factor B are all c-positive. Therefore we may assume that

$$h_i \in [O_2(G), X], \quad h_i^{x^2} = h_i h_i^x \quad (1 \leq i \leq m). \tag{1}$$

With this assumption, set
$$H = \langle h_1, h_1^x, \ldots, h_m, h_m^x \rangle. \tag{2}$$

Then $H \leq [O_2(G), X]$ and $O_2(G) = H{}_3 G$, so that $[O_2(G), X] = [H, X]$ and this clearly lies in H, because (1), (2) imply that H is normalized by X. Therefore
$$H = [O_2(G), X]. \tag{3}$$

But the defining generators of H are commuting involutions modulo $O_2(G)^\mathfrak{s}$, and they are independent modulo $O_2(G)^\mathfrak{s}$ – because they are even independent modulo the larger subgroup ${}_3 G$. Hence
$$|H : O_2(G)^\mathfrak{s}| = 4^m = |O_2(G) : {}_3 G|.$$

(Note that part (b) and (3) imply that $O_2(G) = H{_3}G$, and as the centre ${_3}G$ is of exponent 2 we have $O_2(G)^{\mathfrak{s}} = H^{\mathfrak{s}} \leq H$). Since $O_2(G)^{\mathfrak{s}} \leq H \cap {_3}G$ and we have

$$|H : H \cap {_3}G| = |H{_3}G : {_3}G| = |O_2(G) : {_3}G| = |H : O_2(G)^{\mathfrak{s}}|,$$

it follows that $H \cap {_3}G = O_2(G)^{\mathfrak{s}}$, i.e.

$$[O_2(G), X] \cap {_3}G = O_2(G)^{\mathfrak{s}}.$$

This establishes (c), because every involution in G is central.

(d) Since ${_3}G$ is elementary abelian of exponent 2, we may write

$$ {_3}G = E_0 \times ([O_2(G), X] \cap {_3}G) \tag{4}$$

for a suitable subgroup E_0 of ${_3}G$. Then (b) and modularity give

$$O_2(G) = E_0 \times [O_2(G), X]. \tag{5}$$

Here E_0 is elementary abelian of exponent 2, and (c) guarantees that $[O_2(G), X]$ is reduced.

(e) Write $G = F \times G_0$, where F is elementary abelian and G_0 is reduced. Then $O_3(F) \leq O_3(G) \leq O_3(B)^3 = 1$; so F is an elementary abelian 2-group, therefore $X \leq G_0$. Hence by normality $[O_2(G), X] \leq G_0$, therefore $[O_2(G), X]X \leq G_0$. But $[O_2(G), X]$ is normalized by X, so $[O_2(G), X]X$ is a subgroup; and it follows from (a) and (5) that

$$G = E_0 \times [O_2(G), X]X.$$

Therefore by modularity $G_0 = (E_0 \cap G_0) \times [O_2(G), X]X$; and as G_0 is reduced we must have $E_0 \cap G_0 = 1$. Then $G_0 = [O_2(G), X]$, as required.

(f) In the notation of the proof of (e) we have $G' = G_0'$; so we may assume G reduced. Then $E_0 = 1$ and $O_2(G) = [O_2(G), X] \leq G'$. But $G/O_2(G) \cong X$, an abelian group. Therefore $G' \leq O_2(G)$; and we conclude that $G' = O_2(G) = [O_2(G), X]$. □

G A further necessary condition

In Section F I defined a group Γ to be reduced iff it has no direct factor C_p, p prime. Clearly if Γ is of exponent 12, this means that neither C_2 nor C_3 is a direct factor. For our groups there is a further simplification.

Lemma G.1. *Let Γ be a $CQDB$-group containing the diagonal involution -1. Then Γ is reduced $\Leftrightarrow C_2$ is not a direct factor of Γ.*

Proof. '$\Rightarrow$' Obvious

'$\Leftarrow$' Suppose Γ has a direct factor $\langle x \rangle \cong C_3$. Choose a projection $\varphi : \Gamma \twoheadrightarrow A$ such that $x\varphi \neq 1$, where $A = C_2, C_4, C_6, Q, D, B$. Since $o(x) = 3, x \in {_3}\Gamma$ we have $o(x\varphi) = 3, x\varphi \in {_3}A$. Therefore $A = C_6$ (otherwise ${_3}A$ is a 2-group). Then Lemma D.3 implies that Γ has a direct factor C_2. □

Notation. Let $\boldsymbol{\pi} = (\pi_1, \ldots, \pi_n) : G \twoheadrightarrow G\boldsymbol{\pi} \leq_{\text{sd}} A_1 \times \cdots \times A_n$ be a homomorphism of a group G onto a subdirect product of primordial groups A_i. We shall write $R[\boldsymbol{\pi}]$ for the subring of $RA_1 \oplus \cdots \oplus RA_n$ generated by the image $G\boldsymbol{\pi}$ (so that $R[\boldsymbol{\pi}]$ = all $\mathbb{Z}$-linear combinations of elements of $G\boldsymbol{\pi}$). We will use the notations mainly when G is already a $CQDB$-group (e.g. if $\boldsymbol{\pi}$ is a subdirect embedding, so that $G\boldsymbol{\pi} = G$ and $R[\boldsymbol{\pi}] = \mathbb{Z}G$ = subring generated by G). Then

$$R[\boldsymbol{\pi}] \leq_{\text{sd}} RA_1 \oplus \cdots \oplus RA_n.$$

We write

$$\pi_i^\flat : R[\boldsymbol{\pi}] \twoheadrightarrow RA_i$$

for the corresponding projection (which when $G\boldsymbol{\pi} = G$ is induced by $\pi_i : G \twoheadrightarrow A_i$, to which $\pi_i^\flat$ reduces on $G\boldsymbol{\pi}$)

$$\begin{array}{ccc} G = G\boldsymbol{\pi} & \xrightarrow{\pi_i} & A_i \\ \downarrow & & \downarrow \\ R[\boldsymbol{\pi}] & \xrightarrow{\pi_i^\flat} & RA_i \end{array}$$

Again, if $\boldsymbol{\mu} = (\pi_{i_1}, \ldots, \pi_{i_m})$ where $1 \leq i_1 < \cdots < i_m \leq n$, we write

$$\boldsymbol{\mu}^\flat = (\pi_{i_1}^\flat, \ldots, \pi_{i_m}^\flat) : R[\boldsymbol{\pi}] \twoheadrightarrow R[\boldsymbol{\mu}] \ (\leq_{\text{sd}} RA_1 \oplus \cdots \oplus RA_m).$$

The effect of this notation is to shift the emphasis from the $\mathbb{Q}$-algebras (as in the earlier investigations, e.g. Lemma C.4) to the groups.

Now let G be a (finite) group which occurs as the group of units of an order in a $\mathbb{Q}$-algebra, so that we arrive in context (†) of Section A, and it follows from Lemma C.4 that there is a subdirect embedding

$$\boldsymbol{\pi} = (\pi_1, \ldots, \pi_n) : G \hookrightarrow G\boldsymbol{\pi} \leq_{\text{sd}} A_1 \times \cdots \times A_n$$

with primordial subdirect factors A_i, such that in our new notation

$$R[\boldsymbol{\pi}]^* = G\boldsymbol{\pi}.$$

By Lemma D.3, if any one of the A_i is isomorphic to C_2 or C_6, then $\langle -1 \rangle$ is a direct factor of G isomorphic to C_2, and G is not reduced. Assume now that G is reduced. Then no A_i is C_2 or C_6; and we shall obtain a further necessary condition on G.

As in Lemma F.2, we divide the projection $\boldsymbol{\pi}$ into a partition so that

$$\boldsymbol{\pi} = (\boldsymbol{\delta}, \boldsymbol{\gamma}, \boldsymbol{\beta}) : G \hookrightarrow_{\text{sd}} G\boldsymbol{\delta} \times G\boldsymbol{\gamma} \times G\boldsymbol{\beta},$$

where $\boldsymbol{\delta}$ contains all projections onto D, $\boldsymbol{\beta}$ all projections onto B, and $\boldsymbol{\gamma}$ the rest, which here means all projections onto C_4 and Q. Then if we choose a Sylow 2-subgroup H of G and a Sylow 3-subgroup X_* of G we know by F.2 that

$$X_* = X_*\boldsymbol{\delta} \times X_*(\boldsymbol{\gamma}, \boldsymbol{\beta}),$$

where $X_*\delta = O_3(G\delta)$ is the Sylow 3-subgroup of the D-group $G\delta$ and

$$X_*(\gamma, \beta) \text{ is a Sylow 3-subgroup of } G(\gamma, \beta) \leq_{\text{sd}} G\gamma \times G\beta.$$

But here $G\gamma$ is a 2-group, so $X_*\gamma = 1$ and $X_*(\gamma, \beta) = 1 \times X$, where $X = X_*\beta$ is a Sylow 3-subgroup of the B-group $G\beta$. Then our chosen Sylow 3-subgroup of G is of the form

$$X_* = O_3(G\delta) \times X \text{ for some Sylow 3-subgroup } X \text{ of } G\beta.$$

The proof of F.2 shows further that

$$[H, X_*] = O_3(G\delta) \times 1 \times [O_2(G\beta), X],$$

so

$$[H, X_*]X_* = O_3(G\delta) \times 1 \times [O_2(G\beta), X]X;$$

and by Lemma F.5 (e) the last factor here is the unique reduced part of $G\beta$, say $(G\beta)_{\text{red}}$. Since evidently $G \geq [H, X_*]X_*$, it follows that

G contains both the Sylow 3-subgroup $O_3(G\delta)$ of the D-part $G\delta$
and also the reduced part $(G\beta)_{\text{red}} = [O_2(G\beta), X]X$ of the B-part $G\beta$ of G.

We now change the partition of π to concentrate more closely on the B-part. Write

$$(\delta, \gamma) = \alpha, \; \beta = (\beta_1, \ldots, \beta_r)$$

so that

$$\pi = (\alpha, \beta_1, \ldots, \beta_r) : G \twoheadrightarrow_{\text{sd}} G_0 \times G_1 \times \cdots \times G_r,$$

where $G_0 = G(\delta, \gamma)$ is a C_4QD-group, and $\beta_1, \ldots, \beta_r$ are (in an obvious sense) the B-blocks (obtained by considering the intersection $X \cap \operatorname{Ker} \pi$, or $X_* \cap \operatorname{Ker} \pi_i$ for $\pi_i : G \twoheadrightarrow B$).

$$G\beta \leq_{\text{sd}} G_1 \times \cdots \times G_r.$$

By Lemma F.1(b) we see that the subgroup $(G\beta)_{\text{red}}$ of G has the unique Sylow 2-subgroup

$$[O_2(G), X] = [O_2(G_1), X] \times \cdots \times [O_2(G_r), X].$$

Now consider any involution e in G. Since G is reduced, Lemma D.3 implies that e is a product of squares of elements of order 4 in G. Therefore $e\beta_k$ is a product of squares of elements of order 4 in G_k, and as such lies in the reduced part of G_k, therefore in the Sylow 2-subgroup $[O_2(G_k), X] \leq G$. Thus the components $e\beta_k$ ($1 \leq k \leq r$) of e lie in G; hence so does the remaining component $e\alpha$. In particular, G contains the elements

$$(-1)\beta_k =: e_k \; (1 \leq k \leq r), \; (-1)\alpha.$$

By the definition of B-blocks, the intersections

$$X_k = X \cap \operatorname{Ker} \beta_k \; (1 \leq k \leq r)$$

are distinct maximal subgroups of X. So for each k in $2 \leq k \leq r$ we may choose an element $x_k \in X_k \setminus X_1$. Now, in each subdirect factor B of β_1, the image of e_1 is -1, and the image of x_k is an element of order 3, i.e. a root of $t^2 + t + 1$: thus

$$(e_1 x_k - x_k^2)\beta_1^\natural = 1 \text{ in } R[\beta_1].$$

Also e_1 and x_k lie in the kernels of α and β_k, so

$$(e_1 x_k - x_k^2)\alpha^\natural = 0 \text{ in } R[\alpha],$$

$$(e_1 x_k - x_k^2)\beta_k^\natural = 0 \text{ in } R[\beta_k].$$

Write

$$z_1 = \prod_{2 \leq k \leq r} (e_1 x_k - x_k^2) \in R[\pi].$$

Then

$$z_1 \beta_1^\natural = 1 \text{ in } R[\beta_1], \ z_1 \beta_k^\natural = 0 \text{ in } R[\beta_k] \ (2 \leq k \leq n), \ z_1 \alpha^\natural = 0 \text{ in } R[\alpha].$$

Therefore $R[\pi]$ contains $z_1 \pi^\natural$, the identity element of $R[\beta_1]$, and it follows that $R[\pi] \geq R[\beta_1]$. By symmetry $R[\pi]$ contains all the $R[\beta_k]$ ($1 \leq k \leq r$), therefore also the remaining subdirect factor $R[\alpha]$, and we have proved that

$$R[\pi] = R[\alpha] \oplus R[\beta_1] \oplus \cdots \oplus R[\beta_r].$$

Taking groups of units we obtain

$$G = G_0 \times G_1 \times \cdots \times G_r.$$

The direct factors $G_1, \ldots, G_r$ are therefore reduced B-blocks, and G_0 is a reduced C_4QD-group; and all of $G_0, G_1, \ldots, G_r$ occur as groups of units.

For $k = 1, \ldots, r$ choose any element g_k of order 4 in G_k. Then g_k is moved by conjugation by any element of order 3 in G_k; so the centraliser $\mathfrak{c}_{G_k}(g_k)$ is a 2-group. If we can find an element g_0 of order 4 in G_0 whose centraliser $\mathfrak{c}_{G_0}(g_0)$ is a 2-group, then $g := (g_0, g_1, \ldots, g_r)$ is an element of order 4 in G whose centraliser $\mathfrak{c}_G(g)$ is a 2-group.

Assume then that $G = G_0$ is a reduced C_4QD-group realised by the subdirect embedding

$$\pi = (\pi_1, \ldots, \pi_n) : G \hookrightarrow_{\text{sd}} A_1 \times \cdots \times A_n,$$

when each A_i is one of C_4, Q, D. In $R[\pi]$, we know (by D.4) that we may write

$$-1 = h_1^2 h_2^2 \cdots h_m^2,$$

where $h_1, \ldots, h_m$ are elements of order 4 in some Sylow 2-subgroup H of G. If none of the factors A_i is D, then G is itself a 2-group and for g_0 we may take any element of order 4. Assuming that at least one of the A_i is D, take

$$g_0 = h_1 h_2 \cdots h_m \ (\in H).$$

In D every element of order 4 inverts every element of order 3, and consequently has centraliser of order 4. Since in any case

$$\mathfrak{c}_G(g_0) \leq \mathfrak{c}_{A_1}(g_0\pi_1) \times \cdots \times \mathfrak{c}_{A_n}(g_0\pi_n),$$

it will be enough to prove that $\mathrm{o}(g_0\pi_1) = 4$ whenever $A_i = D$. For simplicity, suppose $A_1 = D$. Thus in A_1, we have

$$-1 = (h_1\pi_1)^2 (h_2\pi_1)^2 \cdots (h_m\pi_1)^2.$$

Here each factor is ± 1. Therefore an odd number of $h_1\pi_1, \ldots, h_m\pi_1$, are elements of order 4 in the (cyclic) Sylow 2-subgroup $H\pi_1$ of D, the remainder being ± 1. Their product $g_0\pi_1$ is then automatically an element of order 4. We summarize

Theorem G.2. *Let G be a reduced finite group which is realized as a group of units by a subdirect embedding*

$$\pi = (\alpha, \beta_1, \ldots, \beta_r) : G \hookrightarrow_{\mathrm{sd}} G_0 \times G_1 \times \cdots \times G_r,$$

where G_0 is a C_4QD-group, and $\beta_k : G \twoheadrightarrow G_k$ ($1 \leq k \leq r$) are the B-blocks. $G = R[\pi]^*$. Then

$$G = G_0 \times G_1 \times \cdots \times G_r, \ R[\pi] = R[\alpha] \oplus R[\beta_1] \oplus \cdots \oplus R[\beta_r],$$

and G contains an element g whose centraliser $\mathfrak{c}_G(g)$ is a 2-group.

Counterexample to Hallett & Hirsch.

$$Q = O_2(B) \text{ and } B/Q \cong C_3, \text{ which is } abelian.$$

Therefore

$$G := \{(g_1, g_2, g_3) \in B^3 |\ g_1 g_2 g_3 \in Q\}$$

is a subgroup of B^3, obviously a subdirect product. G contains $O_2(G) = Q^3$, and is of index 3 in B^3, so $|G| = 8^3 3^2 = 2^9 3^2$. Take an element of order 3 in B, say c, and put

$$c_1 = (1, c, c^{-1}), \ c_2 = (c^{-1}, 1, c), \ c_3 = (c, c^{-1}, 1).$$

Then $c_1 c_2 c_3 = 1$ in G, and we may take

$$X := \langle c_1, c_2 \rangle$$

as Sylow 3-subgroup of G. Let $\beta_1, \beta_2, \beta_3 : G \twoheadrightarrow B$ be the obvious projections. Then

$$\mathrm{Ker}(\beta_i | X) = \langle c_i \rangle \ (i = 1, 2, 3).$$

These are distinct, so there are 3 B-blocks. It follows that $G \leq_{\mathrm{sd}} G_1 \times G_2 \times G_3$ where $G_i = G\beta_i$; and by Lemma G.2 we have equality $G = G_1 \times G_2 \times G_3$.

In Q the only involution is the square of an element of order 4; so the same is true of every involution in G. Therefore C_2 is certainly not a direct factor of G. If G were the group of units of an order in a $\mathbb{Q}$-algebra, then G would have to be the direct product of its B-blocks, so isomorphic to B^3, of order $2^9 3^3$. Therefore G is not such a group of units.

But as B contains no element of order 12, if $x = (x_1, x_2, x_3)$ is any element of order 12 in G, then at least one of the coefficients, say x_1, is of order 3 or 6, and then $x^6 = (1, ?, ?) \neq (-1, -1, -1)$. Thus the involution $(-1, -1, -1)$ is not the 6th power of an element of order 12; so G is 'admissible' in the sense of Hallett & Hirsch.

H Enabling lemmas for realisation

Our realisations depend critically on finding sets of generators for kernels of ring homomorphisms of the type $\mu^\flat : R[\pi] \twoheadrightarrow R[\mu]$ introduced in Section G. We base these kernel computations on the following two easy lemmas.

Lemma H.1. *Let $\Theta : R \twoheadrightarrow S$ be a surjective ring homomorphism between subrings of finite-dimensional $\mathbb{Q}$-algebras. Let I be an ideal of R such that $I \leq \operatorname{Ker} \Theta$, and let $r_1, \ldots, r_n \in R$ be such that $R_\mathbb{Z}$ is generated modulo I by $r_1, \ldots, r_n$, where $n = \operatorname{rk}_\mathbb{Z} S$. Then $r_1 \Theta, \ldots, r_n \Theta$ is a free basis of $S_\mathbb{Z}$, and $I = \operatorname{Ker} \Theta$.*

Proof. An arbitrary element of R is of the form $r = i + \sum_{k=1}^n \xi_k r_k$, where $i \in I$, $\xi_k \in \mathbb{Z}$, with image $r\Theta = \sum_1^n \xi_k (r_k \Theta)$. Therefore $r_1 \Theta, \ldots, r_n \Theta$ generate $S_\mathbb{Z}$ and, as n is the (finite) rank, they must be linearly independent, so they form a free basis of $S_\mathbb{Z}$. Hence also $r \in \operatorname{Ker} \Theta \Rightarrow 0 = \sum_1^n \xi_k(r_k \Theta) \Rightarrow \xi_1 = \cdots = \xi_n = 0 \Rightarrow r = i \in I$; which gives $I = \operatorname{Ker} \Theta$. □

Lemma H.2. *Let $\pi = (\pi_1, \ldots, \pi_n)$, where the components $\pi_i : G \twoheadrightarrow A_i$ are homomorphisms affording inequivalent representations over $\mathbb{Q}$. Then $\operatorname{rk}_\mathbb{Z} R[\pi] = \sum_1^n \operatorname{rk}_\mathbb{Z} RA_i$.*

Proof. As usual the A_i here are primordial groups, so the tensor products $\mathbb{Q} \otimes RA_i$ are division algebras. Therefore the representations π_i are irreducible, and their inequivalence forces $\mathbb{Q} \otimes R[\pi] = \bigoplus_1^n \mathbb{Q} \otimes RA_i$. □

In trying to realise a group as a group of units, there is a troublesome tendency for unwanted involutions to emerge.

Example. Consider the Gaussian integers $RC_4 = \mathbb{Z}[a \mid a^2 + 1 = 0]$ where $o(a) = 4$, group of units $C_4 = \langle a \rangle$. In $(RC_4)^4$, consider the group

$$G = \langle g_1, g_2, g_3 \rangle \text{ where } \begin{cases} g_1 = (a\ 1\ 1\ a) \\ g_2 = (1\ a\ 1\ a) \\ g_3 = (1\ 1\ a\ a) \end{cases} \text{ so } \begin{array}{l} g_1^2 = (-1\ 1\ 1\ -1) \\ g_2^2 = (1\ -1\ 1\ -1) \\ g_3^2 = (1\ 1\ -1\ -1). \end{array}$$

Every element of G is uniquely of the form

$$g_1^{k_1} g_2^{k_2} g_3^{k_3} \ (k_i = 0, 1, 2, 3);$$

so $|G| = 4^3$, and obviously $G \cong C_4^3$. Therefore $G[2]$ is elementary abelian of order 8, generated by the squares g_1^2, g_2^2, g_3^2; and we note that $G[2]$ contains the diagonal involution
$$g_1^2 g_2^2 g_3^2 = (-1 \ -1 \ -1 \ -1) = -1.$$
But the subring $\mathbb{Z}G$ of $(RC_4)^4$ generated by G contains $(g_1 - 1)(g_2 - 1) = (0 \ 0 \ 0 \ -2a)$, and therewith contains $1 + g_1^{-1}(g_1 - 1)(g_2 - 1) = (1 \ 1 \ 1 \ -1)$, an involution clearly not in G.

The purpose of the next three lemmas is to suppress such unwanted involutions. The first rather bizarre result is of some independent interest.

Lemma H.3. *Let $g \ (\neq 1)$ be a unit of finite order in a ring R with free abelian additive group. Assume that $g \equiv 1 \pmod{qR}$ for some integer $q > 1$. Then $o(g) = q = 2$, and $\frac{1}{2}(1 - g)$ is an idempotent in R.*

Proof. Put $k = o(g)$, and write $g = 1 - rf$ where r is a positive integer and $\mathbb{Z}f$ is pure, therefore a direct summand of $R_\mathbb{Z}$. Expanding binomially, write
$$1 = (1 - rf)^k = 1 - krf + \binom{k}{2}r^2 f^2 + r^3 x,$$
where $x \in R$. Therefore
$$kf = \binom{k}{2}rf^2 + r^2 x,$$
and projecting into the direct summand we get $rs \mid k$, where s is the h.c.f. of $\binom{k}{2}, r$. If we suppose for the moment that k is prime, then since $1 < q \mid r \mid rs \mid k$ we must have $q = r = k$ and $s = 1$; so $\frac{1}{2}k(k-1), k(=r)$ are coprime, and the prime k can only be 2. In the general case, for any prime factor p of k, the power $g^{k/p}$ is a unit of order p and is congruent to $1 \bmod rR$, and it follows from what we have just done that $p = r = 2$. This means of course that k must be a power of 2, and $q = r = 2$. But then $g = 1 - 2f$, and $g^2 = 1 - 4(f - f^2)$. As g^2 is a unit of finite order, congruent to $1 \bmod 4R$, the fact that $4 \neq 2$ implies by what we have just proved that we must have $g^2 = 1$. Therefore $f^2 = f$, where $f = \frac{1}{2}(1 - g)$. □

Lemma H.4. *Let R be a ring with free abelian additive group. Assume that R contains no idempotent other than $0, 1$, and that R^* is torsion, and let E be any group of exponent 2. Then $R \otimes \mathbb{Z}[E]$ is a ring with free abelian additive group containing no idempotent other than $0, 1$, and*
$$(R \otimes \mathbb{Z}[E])^* = R^* \times E.$$

Proof. (On the right-hand side here we are of course making the obvious identification $(u, e) = u \otimes e$). As a tensor product of free abelian groups, the additive group of $R \otimes \mathbb{Z}[E]$ is certainly free abelian. For the remaining assertions it is enough to consider the case in which E is finite, and an obvious induction based on the identification

$\mathbb{Z}[E_1 \times E_2] = \mathbb{Z}[E_1] \otimes \mathbb{Z}[E_2]$ reduces the problem to the case in which $E \cong C_2$. Assume then that
$$E = \langle e \rangle, \text{ o}(e) = 2,$$
so that $1, e$ is a $\mathbb{Z}$-basis of $\mathbb{Z}[E]$. Given any unit u of $R \otimes \mathbb{Z}[E]$, write
$$u = a \otimes 1 + b \otimes e, \text{ where } a, b \in R.$$
The two ring homomorphisms $\mathbb{Z}[E] \twoheadrightarrow \mathbb{Z}$ mapping $e \mapsto \pm 1$ give rise to two homomorphisms $R \otimes \mathbb{Z}[E] \twoheadrightarrow R \otimes \mathbb{Z} = R$. These map u to the elements $g_1 = a + b, g_2 = a - b$, which are therefore units of R, and clearly $g_1 \equiv g_2 \pmod{2R}$. Then $g_1 g_2^{-1}$ is a unit (of finite order by assumption) in R congruent to $1 \mod 2R$, and it follows from Lemma H.3 and our assumption on idempotents that $\frac{1}{2}(1 - g_1 g_2^{-1}) = 0$ or 1, i.e. $g_1 = \pm g_2$. This means that one of a, b vanishes while the other is a unit of R, i.e. $u = v \otimes 1$ or $v \otimes e$ for some $v \in R^*$. Therefore $(R \otimes \mathbb{Z}[E])^* \leq R^* \times E$; the reverse inclusion is trivial.

Finally, suppose that $u = a \otimes 1 + b \otimes e$ is an idempotent in $R \otimes \mathbb{Z}[E]$. Arguing as before, we find that $a + b, a - b$ are idempotents in R. If $b \neq 0$, these idempotents in R are distinct, so they must be $0, 1$ in some order, and adding gives $2a = 1$, forcing $R_\mathbb{Z}$ to be 2-divisible, an impossibility for a free abelian group. Therefore $b = 0$ and $a = 0$ or 1, whence $u = 0$ or 1 in $R \otimes \mathbb{Z}[E]$. □

We continue to denote by $G^s = \langle g^2 \mid g \in G \rangle$ the normal subgroup generated by squares of elements of a group G.

Lemma H.5. *Given a primordial group A, let Γ be a group with a central subgroup $E \leq {}_3\Gamma$ of exponent 2 containing a distinguished involution e, let $\sigma : \Gamma \twoheadrightarrow A$ be a surjective homomorphism such that $e\sigma = -1$, and let $G \leq \Gamma$ be a subgroup with $e \in G$ and such that*
$$\text{if } A = B, \text{ then } G\sigma \not\cong C_4, Q. \tag{$*$}$$
Then there exists a group K of exponent 2 and a homomorphism $\eta : \Gamma \twoheadrightarrow K$ such that
$$\Gamma(\sigma, \eta) = A \times K,$$
$$G(\sigma, \eta) = G\sigma \times G\eta,$$
$$\text{Ker}(\sigma, \eta) = \text{Ker}\,\sigma \cap E\Gamma^s,$$
$$\text{Ker}\,\eta \geq E\Gamma^s \text{ with equality if } A = C_2, C_6, B.$$

Proof. Certainly $E\Gamma^s \trianglelefteq \Gamma$. Passing to the quotient $\Gamma/\text{Ker}\,\sigma \cap E\Gamma^s$ we may assume that $K \cap E\Gamma^s = 1$ where $K = \text{Ker}\,\sigma$. Then K is isomorphic to its image in $\Gamma/E\Gamma^s$, so is of exponent 2. And
$$\sigma \text{ maps } E\Gamma^s \text{ isomorphically onto the subgroup } \langle -1 \rangle A^s \text{ of } A, \tag{1}$$
so that our normalisation forces $E = \langle -1 \rangle, e = -1$ in Γ. (Recall that the primordial group A has a unique involution, -1.)

Case 1: $A \in \{C_2, C_6, B\}$. Here $\langle -1 \rangle A^\mathfrak{s} = A$, so (1) implies that $\Gamma = E\Gamma^\mathfrak{s} \times K$: let $\eta : \Gamma \twoheadrightarrow K$ be the corresponding projection, so that $\operatorname{Ker} \eta = E\Gamma^\mathfrak{s}$. Identify $E\Gamma^\mathfrak{s}$ with its image $E\Gamma^\mathfrak{s}\sigma = \langle -1 \rangle A^\mathfrak{s} = A$ under σ. Then we have the direct decomposition

$$\Gamma = A \times K \text{ with projections } \sigma : \Gamma \twoheadrightarrow A, \eta : \Gamma \twoheadrightarrow K. \tag{2}$$

But the restriction $(*)$ forces $G\sigma \cong C_2, C_6, B$. Therefore σ maps $EG^\mathfrak{s}$ ($\leq G \cap E\Gamma^\mathfrak{s} = G \cap A$) isomorphically onto $\langle -1 \rangle G^\mathfrak{s} = G\sigma$ with kernel $G \cap K$; and we have the direct decomposition $G = G\sigma \times (G \cap K) = G\sigma \times G\eta$.

Case 2: $A \in \{C_4, Q, D\}$. In this case more work is needed. Consider the images of K, G in $\Gamma/E\Gamma^\mathfrak{s}$, an abelian group of exponent 2. Using the fact that a vector space always admits a basis containing bases of two given subspaces, we obtain a direct decomposition

$$\Gamma/E\Gamma^\mathfrak{s} = A^\bullet/E\Gamma^\mathfrak{s} \times E\Gamma^\mathfrak{s} K/E\Gamma^\mathfrak{s}, \tag{3}$$

where $E\Gamma^\mathfrak{s} \leq A^\bullet \trianglelefteq \Gamma$, with the property that

$$GE\Gamma^\mathfrak{s}/E\Gamma^\mathfrak{s} = (GE\Gamma^\mathfrak{s} \cap A^\bullet)/E\Gamma^\mathfrak{s} \times (GE\Gamma^\mathfrak{s} \cap E\Gamma^\mathfrak{s} K)/E\Gamma^\mathfrak{s}. \tag{4}$$

By our assumption that $K \cap E\Gamma^\mathfrak{s} = 1$, the equation (3) implies the direct decomposition

$$\Gamma = A^\bullet \times K;$$

so $A^\bullet \cong A$. Identify $A^\bullet$ with its image A under σ; then we may once more choose a projection η to place ourselves in the situation (2). Then by (2), $\Gamma^\mathfrak{s} = A^\mathfrak{s} = \langle -1 \rangle O_3(A)$ so that $E \leq \Gamma^\mathfrak{s} \leq A$, and $\Gamma[2] = \langle -1 \rangle \times K$. In particular $E\Gamma^\mathfrak{s} = \Gamma^\mathfrak{s} \leq A$, and from (4) and modularity we obtain $G\Gamma^\mathfrak{s} = (G\Gamma^\mathfrak{s} \cap A)(G\Gamma^\mathfrak{s} \cap \Gamma^\mathfrak{s} K) = (G \cap A)(G \cap \Gamma^\mathfrak{s} K)\Gamma^\mathfrak{s}$. Intersect this with G and we get

$$G = (G \cap A)(G \cap \Gamma^\mathfrak{s} K). \tag{5}$$

But $\Gamma^\mathfrak{s} K = O_3(A)(\langle -1 \rangle K) = O_3(A) \times \Gamma[2]$, and as the factors here are fully invariant it follows that $G \cap \Gamma^\mathfrak{s} K = (G \cap O_3(A)) \times G[2]$. Again $-1 \in G[2] \leq \Gamma[2] = \langle -1 \rangle \times K$, therefore $G[2] = \langle -1 \rangle \times (G[2] \cap K)$, whence

$$G \cap \Gamma^\mathfrak{s} K = (\langle -1 \rangle (G \cap O_3(A)))(G \cap K).$$

Since the first bracket on the right-hand side here lies in $G \cap A$, we are now able to conclude from (5) that $G = (G \cap A) \times (G \cap K) = G\sigma \times G\eta$. □

I The fundamental 3-realisation

Let X ($\neq 1$) be a finite subdirect power of C_3 with a given subdirect embedding

$$\boldsymbol{\sigma} = (\sigma_1, \ldots, \sigma_n) : X \hookrightarrow_{\text{sd}} C_3^n.$$

Then $X = X\boldsymbol{\sigma}$, and the projections σ_i satisfy

$$\bigcap_{i=1}^{n} \operatorname{Ker} \sigma_i = \{1\}; \tag{1}$$

and as a group in its own right X is a power of C_3, say

$$X \cong C_3^m \text{ where } 1 \leq m \leq n. \tag{2}$$

Identify
$$C_3 = \langle w \rangle \leq \langle -1 \rangle \times \langle w \rangle = (RC_6)^*,$$
where
$$w^2 + w + 1 = 0 \text{ and } RC_6 = \mathbb{Z}[w].$$

With a slight extension of the notation of Section G, define $R[\boldsymbol{\sigma}]$ to be the subring of $(RC_6)^n = (\mathbb{Z}[w])^n$ generated by the subgroup $X = X\boldsymbol{\sigma} \leq C_3^n = \langle w \rangle^n$, and write $\sigma_i^\natural$ for the ith projection $R[\boldsymbol{\sigma}] \twoheadrightarrow RC_6$. Then

$$\boldsymbol{\sigma}^\natural = (\sigma_1^\natural, \ldots, \sigma_n^\natural) : R[\boldsymbol{\sigma}] \hookrightarrow_{\text{sd}} RC_6^n$$

is a subdirect embedding of rings, and for each i in $1 \leq i \leq n$ we have a commuting square

$$\begin{array}{ccc} X & \xrightarrow{\sigma_i} & C_3 = \langle w \rangle \\ \downarrow & & \downarrow \\ R[\boldsymbol{\sigma}] & \xrightarrow{\sigma_i^\natural} & RC_6 = \mathbb{Z}[w] \end{array}$$

Lemma I.1. $R[\boldsymbol{\sigma}]^* = X\boldsymbol{\sigma} \times \langle -1 \rangle \; (\cong X \times C_2)$.

Proof. As $R[\boldsymbol{\sigma}]$ contains both $X\boldsymbol{\sigma}$ and -1, the inclusion $\geq$ is obvious. To establish the reverse inclusion we make use of the unique ring homomorphism

$$\varepsilon = \varepsilon_3 : RC_6 = \mathbb{Z}[w] \twoheadrightarrow \mathbb{F}_3, \text{ so } w\varepsilon = 1.$$

Note then that ε maps $(RC_6)^* = \langle w \rangle \times \langle -1 \rangle$ homomorphically onto $(\mathbb{F}_3)^* = \langle -1 \rangle$ with kernel $\langle w \rangle$. Following Hallett and Hirsch, take

$$RC_6^{(n)} = \{(x_1, \ldots, x_n) \in (RC_6)^n : x_1\varepsilon = \cdots = x_n\varepsilon\};$$

this is a subdirect n^{th} power of RC_6, and it is clear from Lemma A.2 that

$$((RC_6)^*)^{(n)} = (C_6)^n \cap RC_6^{(n)} = \langle w \rangle^n \times \langle -1 \rangle \; (=: C_6^{(n)}).$$

Since $X \leq \langle w \rangle^n \subseteq RC_6^{(n)}$, it follows that $R[\boldsymbol{\sigma}] \leq RC_6^{(n)}$ and

$$X \times \langle -1 \rangle \leq R[\boldsymbol{\sigma}]^* \leq ((RC_6)^*)^{(n)} = \langle w \rangle^n \times \langle -1 \rangle. \tag{3}$$

Therefore $\langle -1 \rangle$ is the Sylow 2-subgroup of $R[\sigma]^*$; and when $m = n$ comparison of order shows that we have equality throughout (3).

In the general case, (1) means that $\sigma_1, \ldots, \sigma_n$ generate the dual group $\text{Hom}(X, C_3)$ ($\cong C_3^m$, by (2)); they must therefore contain a basis for the dual group, and relabeling as necessary we may assume that

$$\sigma_1, \ldots, \sigma_m \text{ is a basis for the dual group } \text{Hom}(X, C_3). \tag{4}$$

Let $x_1, \ldots, x_m$ be the dual basis of X, in the sense that

$$x_i \sigma_k = \begin{cases} w & (i = k) \\ 1 & (i \neq k) \end{cases} \quad (1 \leq i, k \leq m). \tag{5}$$

Write

$$\boldsymbol{\mu} = (\sigma_1, \ldots, \sigma_m) : X \twoheadrightarrow X\boldsymbol{\mu} = \langle w \rangle^m \leq ((RC_6)^*)^m,$$

$$\boldsymbol{\mu}^\natural = (\sigma_1^\natural, \ldots, \sigma_m^\natural) : R[\sigma] \twoheadrightarrow R[\boldsymbol{\mu}] \leq_{\text{sd}} RC_6^m.$$

Here $\text{Ker}\,\boldsymbol{\mu} = \bigcap_1^m \text{Ker}\,\sigma_i = \{1\}$ by (4); so $\boldsymbol{\mu}$ maps X isomorphically onto the Sylow 3-subgroup $\langle w \rangle^m$ of $((RC_6)^*)^m$, and it follows from the first part of the proof that $\boldsymbol{\mu}^\natural$ maps $X \times \langle -1 \rangle$ isomorphically onto $R[\boldsymbol{\mu}]^*$.

Note next that (4) – or indeed (5) – shows that $\sigma_1, \ldots, \sigma_m$ have pairwise distinct kernels, so they induce inequivalent $\mathbb{Q}$-representations $X \times \langle -1 \rangle \twoheadrightarrow C_6$. Hence by Lemma H.2

$$\text{rk}_{\mathbb{Z}} R[\boldsymbol{\mu}] = 2m, \tag{6}$$

because RC_6 is of rank 2. We use (6) together with Lemma H.1 to prove that

$$\text{Ker}\,\boldsymbol{\mu}^\natural = \left(\sum_1^m (x_i^2 + x_i - 2) + 3 \right) R[\sigma] + \sum_{1 \leq i < j \leq m} (x_i - 1)(x_j - 1) R[\sigma]. \tag{7}$$

To see this, consider any k in the range $1 \leq k \leq m$. By (5) the ring homomorphism $\sigma_k^\natural$ sends the kth term in $\sum_1^m (x_i^2 + x_i - 2)$ to $w^2 + w - 2 = -3$ and all other terms to $1 + 1 - 2 = 0$; and whenever $1 \leq i < j \leq m$, $\sigma_k^\natural$ kills at least one factor of $(x_i - 1)(x_j - 1)$. Consequently $\text{Ker}\,\sigma_k^\natural$ contains the ideal I (say) on the right-hand side of (7); therefore so does $\text{Ker}\,\boldsymbol{\mu}^\natural = \bigcap_1^m \text{Ker}\,\sigma_k^\natural$. To establish the reverse inclusion note that every element of $R[\sigma]$ ($= \mathbb{Z}X$) is a $\mathbb{Z}$-linear combination of 'monomials' in the commuting elements $x_1, \ldots, x_m$ of degree < 3 in each x_i (because $x_i^3 = 1$). But in view of the form of the generators of I, an obvious induction on degree shows that each element of $R[\sigma]$ is congruent modulo I to a $\mathbb{Z}$-linear combination of the $2m$ monomials

$$1, x_1, x_i, x_i^2 \quad (2 \leq i \leq m).$$

The assertion (7) therefore follows at once from Lemma H.1 and (6).

Now consider any unit $u \in R[\sigma]^*$. The projection $u\boldsymbol{\mu}^\natural$ in the first m factors is in $R[\boldsymbol{\mu}]^* = (X \times \langle -1 \rangle)\boldsymbol{\mu}$, so is of the form

$$u\boldsymbol{\mu}^\natural = h\boldsymbol{\mu}^\natural \text{ for some unique } h \in X \times \langle -1 \rangle.$$

Write
$$v = uh^{-1} \ (\in R[\boldsymbol{\sigma}]^* \leq_{\text{sd}} ((RC_6)^*)^n). \tag{8}$$

Then there exist elements $v_{m+1}, \ldots, v_n \in C_6$ such that
$$v = (v\sigma_1^\natural, \ldots, v\sigma_n^\natural) = (1, \ldots, 1, v_{m+1}, \ldots, v_n), \tag{9}$$

i.e.
$$v\boldsymbol{\mu}^\natural = 1 \text{ and } v\sigma_k^\natural = v_k \quad (m+1 \leq k \leq n).$$

Therefore $v - 1 \in \text{Ker}\,\boldsymbol{\mu}^\natural$, and
$$v_k - 1 = (v-1)\sigma_k^\natural \in (\text{Ker}\,\boldsymbol{\mu}^\natural)\sigma_k^\natural \quad (m+1 \leq k \leq n). \tag{10}$$

Now $\sigma_k^\natural$ maps each x_i to 1 or w or w^{-1} (i.e. an element of $\langle w \rangle$); so $\sigma_k^\natural$ maps each term $x_i^2 + x_i - 2$ to 0 or -3 and each $(x_i - 1)(x_j - 1)$ to an element in $\mathbb{Z}[w]$ of the form $(w^r - 1)(w^{-s} - 1)$, a multiple of $(w - 1)(w^{-1} - 1) = 3$. Therefore by (6) and (10) we have
$$v_k - 1 \in 3RC_6 \quad (m+1 \leq k \leq n).$$

But by Lemma H.3, the only unit of RC_6 congruent to 1 mod $3RC_6$ is 1 itself. Therefore $v_{m+1} = \cdots = v_n = 1$, i.e. $v = 1$ by (9); and now (8) gives us the required conclusion, that $u = h \in X \times \langle -1 \rangle$. □

J Some facts about RB

We start with the presentation of RB obtained in Section B viz.
$$RB = \mathbb{Z}[\alpha, \beta \mid \alpha^2 + \alpha + 1 = \beta^2 + \beta + 1 = (\alpha\beta)^2 + 1 = 0], \tag{1}$$

where we have written α, β for the previous a, b. We are going to introduce new generators for RB which will be more convenient, given the special rôle of $O_2(B) = Q$. To this end, write
$$c = \alpha, \ a = -\alpha\beta, \ b = a^c = -\beta\alpha. \tag{2}$$

Clearly
$$c^2 + c + 1 = 0 = a^2 + 1 = b^2 + 1. \tag{3}$$

As in Section F we know that the generators of the three cyclic subgroups of order 4 in Q are $a, a^c = b, b^c = \pm ab$; in particular $(ab)^2 + 1 = 0$, whence by (3)
$$ab + ba = 0. \tag{4}$$

Now (2) and (1) give $ac = -\alpha\beta\alpha = \beta^{-1} = -\beta - 1$, $ca = -\alpha^2\beta = \alpha\beta + \beta = -a + \beta$; hence RB is generated as a ring by a, c; and adding our last two equations gives
$$ac + ca + a + 1 = 0. \tag{5}$$

Then $c^{-1}(5)$ gives $b + a + c^{-1}(a+1) = 0$, i.e. $(b-1) + (1+c^{-1})(a+1) = 0$, i.e. by (3)
$$c(a+1) = b-1; \tag{6}$$
and $(6)a, (6)(a-1)$ yield at once, respectively
$$c(a-1) = (b-1)a, \tag{7}$$
$$(b-1)(a-1) + 2c = 0. \tag{8}$$

If we now conjugate (6) by c and appeal in turn to (5), (7), (4), we find that $b^c = 1 + (a+1)c = c(1-a) - a = -ba = ab$, i.e. $a^{c^2} = aa^c$, which is to say that a is c-positive. It follows, conjugating by c, that

$$\text{the } c\text{-positive elements of } O_2(B) \text{ are } 1, a, b, ab. \tag{9}$$

It follows from (5) that the subrings $\mathbb{Z}[c]$ ($\mathbb{Z}$-basis $1, c$) and $\mathbb{Z}[a]$ ($\mathbb{Z}$-basis $1, a$) commute, so their product is a subring, necessarily the whole of RB. So $1, c, a, ca$ is a $\mathbb{Z}$-basis for RB; and by (6), so is $1, c, a, b$. Therefore

$$1, c, a-1, b-1 \text{ is a } \mathbb{Z}\text{-basis for } RB. \tag{10}$$

Let JB be the (2-sided) ideal of RB generated by $a-1$. Then JB contains $(a-1)^c = b-1, (b-1)(a-1) = -2c$ (by (8)), therefore also $2, 2c$. Comparing (10) we deduce that $|RB/JB|$ divides 4. But from (1) there exists a homomorphism

$$\varepsilon = \varepsilon_2 : RB \to \mathbb{F}_4 \text{ mapping } \alpha, \beta \text{ to the roots of } r^2 + r + 1.$$

Certainly $\operatorname{Ker} \varepsilon$ contains $a - 1$ and therewith the whole of JB. Therefore

$$JB = \operatorname{Ker} \varepsilon, \text{ of index 4 in } RB,$$
$$RB/JB \cong \mathbb{F}_4,$$
$$2, 2c, a-1, b-1 \text{ is a } \mathbb{Z}\text{-basis for } JB. \tag{11}$$

K Realisation of a finite C_6B-block

We now consider a slight generalisation of a B-block which may be called a C_6B-block. Let
$$\boldsymbol{\beta} = (\pi_1, \ldots, \pi_n) : G \hookrightarrow_{sd} A_1 \times \cdots \times A_n$$
be a subdirect embedding in which each factor $A_i \in \{C_6, B\}$ and assume that G has a Sylow 3-subgroup $\langle d \rangle$ of order 3. Then
$$G = \langle d \rangle O_2(G), \tag{1}$$
and by Lemma B.5 we may assume that
$$d = (c, \ldots, c), \tag{2}$$

where c is the element of order 3 in RB introduced in (J2), it is convenient here to identify $RC_6 = \mathbb{Z}[c] \leq RB$.

Write
$$E = G[2] \cong C_2^m. \tag{3}$$

Now $\bigcap_1^n \operatorname{Ker}(\pi_i \restriction E) = E \cap \operatorname{Ker} \beta = \{1\}$, so the $\pi_i \restriction E$ ($1 \leq i \leq n$) span $\operatorname{Hom}(E, C_2)$ and they therefore include a basis which we may take to be $\pi_i \restriction E$ ($1 \leq i \leq m$); let $e_1, \ldots, e_m$ be the dual basis of E, so that
$$e_i \pi_j = (-1)^{\delta_{ij}} \quad (1 \leq i, j \leq m). \tag{4}$$

Write
$$\boldsymbol{\mu} = (\pi_1, \ldots, \pi_m) : G \twoheadrightarrow G\boldsymbol{\mu} \leq_{\mathrm{sd}} A_1 \times \cdots \times A_m. \tag{5}$$

Then $E \cap \operatorname{Ker} \boldsymbol{\mu} = \bigcap_1^m \operatorname{Ker}(\pi_i \restriction E) = \{1\}$, so $O_2(G) \cap \operatorname{Ker} \boldsymbol{\mu} = \{1\}$; and since π_1 maps any element of order 3 in G to an element of order 3 in A_1, it follows that $\operatorname{Ker} \boldsymbol{\mu} = \{1\}$, i.e. $\boldsymbol{\mu}$ maps G isomorphically onto the subdirect product $G\boldsymbol{\mu} \leq_{\mathrm{sd}} A_1 \times \cdots \times A_m$.

We *assume* (as we must if we are to realize G) that G contains the diagonal involution $-1 = (-1, \ldots, -1)$. Then by (4), looking at the first m components we have $(-1)\boldsymbol{\mu} = (\prod_{i=1}^m e_i)\boldsymbol{\mu}$, so that
$$-1 = \prod_{i=1}^m e_i \in G. \tag{6}$$

Now (as in Section F) $O_2(G)/E$ is naturally a vector space over $\mathbb{Z}_2[c]$ ($\cong \mathbb{F}_4$), say of dimension r. Relabeling $\pi_1, \ldots, \pi_m$ we may suppose that the first r of the projections induce an isomorphism $O_2(G)/E \to O_2(A_1)/\langle -1 \rangle \times \cdots \times O_2(A_r)/\langle -1 \rangle$ (which requires of course, $A_i = B$ for $i = 1, \ldots, r$). These r projections $\pi_1, \ldots, \pi_r$ then induce a basis of the dual space $\operatorname{Hom}(O_2(G)/E, \mathbb{F}_4)$, and we may lift the dual basis of $O_2(G)/E$ to a system $a_1, \ldots, a_r$ in $O_2(G)$. Since $a_i \circ d \equiv a_i \pmod{E}$, we may assume that the components of $a_1, \ldots, a_r$ are c-positive in the factors $A_i = B$ and are 1 in the factor $A_i = C_6$, or that (as we shall say) each a_i is d-*positive*:
$$\text{writing } a_i^d = b_i \text{ we have } b_i^d = a_i b_i \quad (i = 1, \ldots, r). \tag{7}$$

Then
$$O_2(G) = E\langle a_1, b_1, \ldots, a_r, b_r \rangle. \tag{8}$$

Finally, let F be the (central) subring of $R[\beta]$ generated by E, i.e.
$$F = \mathbb{Z}[e_1 - 1, \ldots, e_m - 1] \ (\leq R[\beta]). \tag{9}$$

The ring $R[\beta]$ may now be regarded as an F-algebra.

Lemma K.1. *Let $x, y \in O_2(G)$. Then in the F-algebra $R[\beta]$,*

(a) $d^2 + d + 1 = 0$;

(b) $xy \in F + Fd + Fx + Fy$;

(c) $dx \in F + Fd + Fx + Fx^d$.

Proof. Consider any projection $\pi_i : G \twoheadrightarrow A_i$ in β. Then $d\pi_i = c_i$ by (2), whence $(d^2 + d + 1)\pi_i^\natural = c^2 + c + 1 = 0$ by (J3). Then $d^2 + d + 1 = 0$ in $R[\beta]$; which proves (a). In the proof of (b) and (c) there will be no loss in supposing that x, y *are both d-positive*, since e.g. $x \circ d$ is d-positive and $x \circ d \equiv x \pmod{E} \Rightarrow F(x \circ d) = Fx$. With this assumption $x\pi_i, y\pi_i$ are both c-positive, so by (J9) they lie in $\{1, a, b, ab\}$ and composing π_i with conjugation by a suitable power of c we may assume that

$$\text{either } x\pi_i = 1, \ y\pi_i \in \{1, a\}$$

$$\text{or } x\pi_i = a, \ y\pi_i \in \{1, a, b, ab\}.$$

There are therefore in effect six possibilities for $\pi_i^\natural$ when $A_i = B$ which may be tabulated in the following form

	d	c	c	c	c	c	c	
	x	1	1	a	a	a	a	
	x^d	1	1	b	b	b	b	
	x^{d^2}	1	1	ab	ab	ab	ab	
	y	1	a	1	a	b	ab	
$u = (xy)^2$		1	-1	-1	1	-1	-1	
$v = (x^d y)^2$		1	-1	-1	-1	1	-1	
$w = (x^{d^2} y)^2$		1	-1	-1	-1	-1	1	
$-xy$		-1	$-a$	$-a$	1	$-ab$	b	
uvx		1	1	a	$-a$	$-a$	a	
$-vy$		-1	a	1	a	$-b$	ab	sum$=0$
v		1	-1	-1	-1	1	-1	
$(v-w)d$		0	0	0	0	$2c$	$-2c$	

Here the first column may be interpreted as covering the only relevant case when $A = C_6$. Since $ab + ba = 0$ and $(b-1)(a-1) + 2c = 0$ by (J4) and (J8), it follows that in every case $\operatorname{Ker} \pi_i^\natural$ contains the element

$$-xy + uvx - vy + v + (v-w)d.$$

This element therefore vanishes in $R[\beta]$, and since $u, v, w \in E \subseteq F$ the assertion (b) is proved.

For the final part we need only the first and third columns of the table.

	d	c	c	
	x	1	a	
$d(x-1)$		0	$c(a-1)$	sum of rows
$-(x^d - 1)x$		0	$-(b-1)a$	$= 0$

By (J7), $c(a - 1) = (b - 1)a$; and arguing as before we conclude that

$$d(x - 1) - (x^d - 1)x = 0 \text{ in } R[\beta].$$

Therefore

$$dx \in Fd + Fx + Fx^d x;$$

and (c) now drops out of (b). □

Corollary K.2. *$R[\beta]$ is generated as an F-module by*

$$1, d, a_i - 1, b_i - 1 \quad (i = 1, \ldots, r).$$

Proof. Let M be the sub-F-module of $R[\beta]$ generated by these $2 + 2r$ elements. By Lemma K.1(b), $M \cap O_2(G)$ is closed under multiplication; and as it contains E and $a_1, b_1, \ldots, a_r, b_r$, by (8) it exhausts $O_2(G)$. Thus $O_2(G) \subseteq M$, and Lemma K.1(c) and (a) now guarantee in turn that $dO_2(G) \subseteq M$ and $d^2 O_2(G) \subseteq M$. Therefore $G \subseteq M$, by (1). As $R[\beta]$ is generated by G as a $\mathbb{Z}$-module, the corollary is proved. □

Notation. Let JG (or $J[\beta]$) be the ideal of $R[\beta]$ generated by the $h - 1$ with $h \in O_2(G)$.

As $-1 \in O_2(G)$ we have at once from K.2, since $F = \mathbb{Z}[e_1 - 1, \ldots, e_m - 1] \leq \mathbb{Z} + JG$:

Corollary K.3. (a) $R[\beta] = \mathbb{Z} + \mathbb{Z}d + JG$;

(b) $2R[\beta] \leq JG$.

Note that our choice of $a_1, \ldots, a_r$ as a lifting of a dual basis means that $a_i \pi_j \equiv a^{\delta_{ij}} \pmod{E}$ ($1 \leq i, j \leq r$); our subsequent normalisations of the a_i to be d-positive, where $d\pi_j = c$ improves these congruences to equalities

$$a_i \pi_j = a^{\delta_{ij}} \quad (1 \leq i, j \leq r). \tag{10}$$

Squaring leads to $a_i^2 \pi_j = (-1)^{\delta_{ij}}$ ($1 \leq i, j \leq r$), so that $\langle a_1^2, \ldots, a_r^2 \rangle$ is a subgroup isomorphic to C_2^r in E ($\cong C_2^m$). Hence

$$1 \leq r \leq m \leq n. \tag{11}$$

Write

$$\rho = (\pi_1, \ldots, \pi_r) : G \twoheadrightarrow G\rho \leq_{\text{sd}} A_1 \times \cdots \times A_r \; (\cong B^r). \tag{12}$$

Correspondingly, $\rho^\natural = (\pi_1^\natural, \ldots, \pi_r^\natural) : R[\beta] \twoheadrightarrow R[\rho] \; (= \mathbb{Z}.G\rho \leq_{\text{sd}} (RB)^r)$.

Lemma K.4. *Ker $\rho^\natural$ is the ideal of $R[\beta]$ generated by the elements*

(i) $e_k - 1 \quad (r + 1 \leq k \leq m)$,

(ii) $(e_i - 1)(e_j - 1) \quad (i \neq j; 1 \leq i, j \leq r)$,

(iii) $(e_i - 1)(a_j - 1)$ $(i \neq j; 1 \leq i, j \leq r)$.

Proof. Let I be the ideal generated by these elements (i), (ii), (iii). As for the analogue (I 7) one proves easily that
$$\operatorname{Ker} \rho^\natural \geq I.$$
Note that by (5), we have $-1 = \prod_1^m(1+(e_i-1)) \equiv 1+\sum_1^r(e_i-1) \pmod{I}$. Therefore
$$z_0 := 2 + \sum_1^r (e_i - 1) \in I.$$

Since I contains z_0 and the elements (i), (ii), it is clear that modulo I F ($= \mathbb{Z}[e_1 - 1, \ldots, e_m - 1]$) is generated as a $\mathbb{Z}$-module by 1 and any $r - 1$ of the $e_i - 1$ ($i = 1, \ldots, r$), in symbols
$$F \leq \mathbb{Z} + \sum_{\substack{1 \leq i \leq r \\ i \neq j}} \mathbb{Z}.(e_i - 1) + I \quad (1 \leq j \leq r). \tag{13}$$

And since I contains the elements (iii), multiplying this last by $a_j - 1$ leads to
$$F(a_j - 1) \leq \mathbb{Z}(a_j - 1) + I;$$
and conjugation by d shows that the analogue with b_j in place of a_j is also valid. It now follows from Corollary K.2 that, modulo I, $R[\beta]$ is generated as a $\mathbb{Z}$-module by

1, any $r - 1$ of $e_1 - 1, \ldots, e_r - 1$,

d, any $r - 1$ of $d(e_1 - 1), \ldots, d(e_r - 1)$,

$a_1 - 1, \ldots, a_r - 1,$

$b_1 - 1, \ldots, b_r - 1;$

in all, $4r$ elements. But the components of $\rho = (\pi_1, \ldots, \pi_r)$ have pairwise distinct kernels, by (10), so they afford inequivalent $\mathbb{Q}$-representations $G \twoheadrightarrow B$, and Lemma H.2 (and Lemma B.4) imply that $\operatorname{rk}_\mathbb{Z} R[\rho] = 4r$. Lemma H.1 now guarantees that $\operatorname{Ker} \rho^\natural = I$, as claimed. □

The argument is similar, but simpler, if the block involves no subdirect factor B.

Theorem K.5. $R[\beta]^* = G\beta \ (= G)$.

Proof. Let $\varepsilon = \varepsilon_2$ denote the ring homomorphisms $RB \twoheadrightarrow \mathbb{F}_4$, $RC_6 \twoheadrightarrow \mathbb{F}_4$ with kernels JB and JC_6 (say), mapping c to a fixed element of order 3 in the Galois field $\mathbb{F}_4$ (and mapping all 2-elements in B, C_6 to 1). Then G ($\leq_{\text{sd}} A_1 \times \cdots \times A_n$) is contained in the subring
$$R_0 = \{(x_1, \ldots, x_n) \in RA_1 \oplus \cdots \oplus RA_n : x_1\varepsilon = \cdots = x_n\varepsilon\};$$

therefore so is the subring generated by G, namely $R[\beta]$. Hence

$$\begin{aligned}\langle d\rangle O_2(G) &= G \leq R[\beta]^* \leq R_0 \cap (RA_1 \oplus \cdots \oplus RA_n)^* \\ &= R_0 \cap (A_1 \times \cdots \times A_n) = \langle d\rangle(O_2(A_1) \times \cdots \times O_2(A_n)).\end{aligned}$$

To prove the theorem it will therefore be enough to prove that every 2-element in $R[\beta]^*$ lies in G. And this is obvious when $n = r$, for then $n = r = m$ and $O_2(G) = O_2(A_1) \times \cdots \times O_2(A_n)$.

In the general case, consider any 2-element $u \in R[\beta]^*$. Then $u\rho^\natural$ is a 2-element in $R[\rho]^*$ and by the special case just considered $\exists h \in O_2(G)$ such that $u\rho^\natural = h\rho$ i.e.

$$uh^{-1} - 1 \in \operatorname{Ker} \rho^\natural.$$

For each k in the range $r + 1 \leq k \leq m$, the ring homomorphism $\pi_k^\natural : R[\beta] \to RA_k$ kills all the generators of $\operatorname{Ker} \rho^\natural$ given by Lemma K.4 with the exception of $e_k - 1$, which maps to -2. Therefore

$$(uh^{-1})\pi_k^\natural - 1 \in (\operatorname{Ker} \rho^\natural)\pi_k^\natural = 2RA_k;$$

and by Lemma H.3, since $0, 1$ are the only idempotents in RA_k, the unit $(uh^{-1})\pi_k^\natural$ can only be ± 1. This means that

$$(uh^{-1})\boldsymbol{\mu}^\natural \in \langle 1\rangle^r \times \langle -1\rangle^{m-r} \leq E\boldsymbol{\mu}$$

and, absorbing an element of E into h we may assume that

$$(uh^{-1})\boldsymbol{\mu}^\natural = 1. \tag{14}$$

Now let J^*G be the ideal of $R[\beta]$ generated by the elements

$$(e-1)(g-1) \quad (e \in E,\ g \in O_2(G)).$$

Since $-1 \in E$, it is clear that

$$J^*G \geq 2JG \tag{15}$$

and it follows from Lemma K.4 that

$$\operatorname{Ker} \rho^\natural \leq \sum_{r+1}^m (e_k - 1)R[\beta] + J^*G. \tag{16}$$

But as $e_k - 1$ is central it follows from the definitions of JG, J^*G that

$$(e_k - 1)JG \leq J^*G.$$

Therefore Corollary K.3 implies that (16) simplifies to

$$\operatorname{Ker} \rho^\natural \leq \sum_{r+1}^m (e_k - 1)(\mathbb{Z} + \mathbb{Z}d) + J^*G. \tag{17}$$

Reverting to (14) we may therefore write

$$uh^{-1} - 1 = \sum_{r+1}^{m} (e_k - 1)(\xi_k + \eta_k d) + z; \qquad (18)$$

for suitable $\xi_k, \eta_k \in \mathbb{Z}$ and $z \in J^*G$. For $k = r+1, \ldots, m$ we then have

$$0 = (uh^{-1} - 1)\pi_k^\natural = -2(\xi_k + \eta_k c) + z\pi_k^\natural,$$

whence

$$2(\xi_k + \eta_k c) = z\pi_k^\natural \in (J^*G)\pi_k^\natural \le J^*A_k = 2JA_k,$$

because the only element of order 2 in A_k is -1. Therefore by (J11)

$$\xi_k + \eta_k c \in (JA_k) \cap (\mathbb{Z} \oplus \mathbb{Z}c) = 2\mathbb{Z} \oplus 2\mathbb{Z}c;$$

in other words ξ_k, η_k are *even* integers. But then the sum $\sum_{r+1}^{m}$ on the right-hand side of (18) lies in $2JG \le J^*G$, by (15); and (18) reduces to

$$uh^{-1} - 1 \in J^*G. \qquad (19)$$

For all $i = 1, \ldots, n$ we already know that

$$(uh^{-1})\pi_i^\natural = \pm 1.$$

Therefore applying $\pi_i^\natural$ to (19) we obtain

$$\pm 1 - 1 = (uh^{-1})\pi_i^\natural - 1 \in (J^*G)\pi_i^\natural \cap \mathbb{Z} \le (2JA_i) \cap \mathbb{Z} = 4\mathbb{Z}.$$

It now follows at once that $(uh^{-1})\pi_i^\natural = 1$ for all i, i.e. $uh^{-1} - 1 = 0$ in $R[\beta]$, which gives the desired conclusion that $u = h \in G$. $\square$

Remark K.6. We have just shown that a finite B-block G is realized by an arbitrary subdirect decomposition $\beta : G \twoheadrightarrow G\beta \le_{\text{sd}} B^n$ provided only that $G\beta$ contains the diagonal involution -1. In fact every finite B-block G admits such a subdirect decomposition. (Here by a B-block we mean simply a subdirect power G of B whose Sylow 3-subgroups are cyclic.) Of course this is very obvious if G is not reduced; for then if $-1 \notin G\beta$ we may simply replace an arbitrary direct factor of order 2 by $\langle -1 \rangle$. In general take an arbitrary subdirect decomposition

$$\beta = (\pi_1, \ldots, \pi_n) : G \twoheadrightarrow G\beta \le_{\text{sd}} B^n,$$

and rearrange the factors so that $\pi_1 \restriction E, \ldots, \pi_m \restriction E$ form a basis of $\text{Hom}(E, C_2)$, where $E = G[2] \cong C_2^m$. Then (as we have seen) we obtain a more economical subdirect decomposition

$$\boldsymbol{\mu} = (\pi_1, \ldots, \pi_m) : G \twoheadrightarrow G\boldsymbol{\mu} \le_{\text{sd}} B^m.$$

This $\boldsymbol{\mu}$ maps E isomorphically onto $B[2]^m \ni (-1, \ldots, -1) = -1$; so $-1 \in G\boldsymbol{\mu}$. And of course once we have found one subdirect decomposition $\boldsymbol{\mu} : G \hookrightarrow_{\text{sd}} B^m$ and an involution (call it -1) in G such that $(-1)\boldsymbol{\mu} = -1 \in B^m$, we know that we may adjoin to $\boldsymbol{\mu}$ as many extra projections $\sigma_k : G \twoheadrightarrow B$ such that $(-1)\sigma_k = -1$ as we wish.

Definition. Let G be a $CQDB$-group with a subdirect embedding $\pi : G \hookrightarrow_{\text{sd}} \prod A_i$ such that $-1 \in G$, where each A_i is primordial. A *component* of G shall be a surjective homomorphism

$$\sigma : G \twoheadrightarrow A \text{ such that } (-1)\sigma = -1,$$

where A is primordial. We may if necessary distinguish B-components $G \twoheadrightarrow B$, C_2-components $G \twoheadrightarrow C_2$, etc.

L The fundamental 2-realisation

Let G be a finite C_2C_4Q-group with a given subdirect decomposition

$$\gamma = (\sigma_1, \ldots, \sigma_n) : G \twoheadrightarrow G\gamma \leq_{\text{sd}} A_1 \times \cdots \times A_n, \tag{1}$$

where each factor $A_i \in \{C_2, C_4, Q\}$. We assume as usual that G contains the diagonal involution -1. Write

$$E = G[2] \cong C_2^m, \text{ where } 1 \leq m \leq n. \tag{2}$$

Relabeling the projections $\sigma_1, \ldots, \sigma_n$ we may assume that $\sigma_1, \ldots, \sigma_m$ restrict to a basis of the dual group $\text{Hom}(E, C_2)$. Let $e_1, \ldots, e_m$ be the dual basis of E, so that

$$e_i \sigma_j = (-1)^{\delta_{ij}} \quad (1 \leq i, j \leq m) \tag{3}$$

and

$$-1 = \prod_1^m e_i \in E. \tag{4}$$

Write

$$\boldsymbol{\mu} = (\sigma_1, \ldots, \sigma_m) : G \twoheadrightarrow G\boldsymbol{\mu} \leq_{\text{sd}} A_1 \times \cdots \times A_m. \tag{5}$$

(As before $\boldsymbol{\mu}$ is an injective homomorphism because $E \cap \text{Ker}\,\boldsymbol{\mu} = \bigcap_1^m \text{Ker}(\sigma_i \restriction E) = \{1\}$.)

Now let $1 \leq i \leq n$, and apply Lemma H.5 to the projection $\sigma_i : G \twoheadrightarrow A_i$ (with $\Gamma = G$) to obtain a homomorphism $\eta_i : G \twoheadrightarrow K_i$ onto a group K_i of exponent 2 such that

$$\widetilde{\sigma}_i := (\sigma_i, \eta_i) : G \twoheadrightarrow A_i \times K_i \text{ is surjective}, \tag{6}$$

$$\text{Ker}\,\widetilde{\sigma}_i = \text{Ker}\,\sigma_i \cap E \ (= \text{Ker}(\sigma_i \restriction E)), \tag{7}$$

$$\text{Ker}\,\eta_i \geq EG^5 = E \text{ (because } G \text{ is of exponent 4)}. \tag{8}$$

Since $E\sigma_i = \langle -1 \rangle$ is of order 2 it follows from (7) that $|\text{Ker}\,\widetilde{\sigma}_i| = \frac{1}{2}|E|$, therefore by (6) and Lemma B.4, $|K_i| = |G| / \frac{1}{2}|E| |A_i| = [G : E] / \text{rk}_{\mathbb{Z}}(RA_i)$. Defining

$$R[\widetilde{\sigma}_i] = RA_i \otimes \mathbb{Z}[K_i] \tag{9}$$

we find from what we have just proved that

$$\text{rk}_{\mathbb{Z}} R[\widetilde{\sigma}_i] = [G : E], \tag{10}$$

while Lemma H.4 guarantees that
$$R[\tilde{\sigma}_i]^* = A_i \times K_i = G\tilde{\sigma}_i. \tag{11}$$

As this group (11) is contained in the group of units of $\mathbb{Q} \otimes R[\tilde{\sigma}_i]$, we may think of $\tilde{\sigma}_i$ as determining a $\mathbb{Q}$-representation. As such $\tilde{\sigma}_i$ is a direct sum of irreducible $\mathbb{Q}$-representations. In fact these are very easy to find because K_i is a finite group of exponent 2; they are the homomorphisms

$$\sigma_i \cdot (\eta_i \circ \lambda) : G \twoheadrightarrow A_i \quad (g \mapsto (g\sigma_i)((g\eta_i)\lambda)), \tag{12}$$

where λ runs through $\mathrm{Hom}(K_i, C_2)$; and taking λ to be the principal character of K_i we see that one of these irreducible components of $\tilde{\sigma}_i$ is σ_i itself. Thus $\tilde{\sigma}_i$ *is the direct sum of* $|K_i|$ *pairwise inequivalent* $\mathbb{Q}$-*representations* $G \twoheadrightarrow A_i$ *of which* σ_i *is one*; and a glance at (12) and (8) shows that *each of these irreducible components agrees on E with σ_i* (so in particular maps $-1 \mapsto -1$). Therefore by (3), *no two of* $\tilde{\sigma}_1, \ldots, \tilde{\sigma}_m$ *have a common irreducible component.*

It is now a short step to thinking of $\tilde{\sigma}_i$ as a $|K_i|$-tuple of projections $G \twoheadrightarrow A_i$. If we do so then the apparently new definition of $R[\tilde{\sigma}_i]$ in (9) reveals itself as a special case of the notation $R[\pi]$ introduced in Section G, and we note that

$$R[\tilde{\sigma}_i] \leq_{\mathrm{sd}} (RA_i)^{|K_i|}.$$

Now let
$$\tilde{\gamma} = (\tilde{\sigma}_1, \ldots, \tilde{\sigma}_n) : G \twoheadrightarrow G\tilde{\gamma} \leq_{\mathrm{sd}} R[\tilde{\sigma}_1]^* \times \cdots \times R[\tilde{\sigma}_n]^*,$$
$$\tilde{\mu} = (\tilde{\sigma}_1, \ldots, \tilde{\sigma}_m) : G \twoheadrightarrow G\tilde{\mu} \leq_{\mathrm{sd}} R[\tilde{\sigma}_1]^* \times \cdots \times R[\tilde{\sigma}_m]^*.$$

(By (7) these are injective because γ, μ are injective.) We shall regard $\tilde{\gamma}$ as an inclusion, so that
$$G = G\tilde{\gamma}.$$

If we regard $\tilde{\gamma}$ and $\tilde{\mu}$ as concatenations then each is a finite sequence of projections of G onto copies of C_2, C_4, Q mapping -1 to -1, so the diagonal involution is still in $G\tilde{\gamma}$ and $G\tilde{\mu}$; and we may talk about $R[\tilde{\gamma}]$ and $R[\tilde{\mu}]$. And it should be clear that

$$\tilde{\gamma} = (\tilde{\sigma}_1^\natural, \ldots, \tilde{\sigma}_n^\natural) : R[\tilde{\gamma}] \hookrightarrow_{\mathrm{sd}} R[\tilde{\sigma}_1] \oplus \cdots \oplus R[\tilde{\sigma}_n],$$

where $\tilde{\sigma}_i^\natural : R[\tilde{\gamma}] \twoheadrightarrow R[\tilde{\sigma}_i]$ agrees with σ_i on G; and we have a ring homomorphism

$$\tilde{\mu}^\natural = (\tilde{\sigma}_1^\natural, \ldots, \tilde{\sigma}_m^\natural) : R[\tilde{\gamma}] \twoheadrightarrow R[\tilde{\mu}] \leq_{\mathrm{sd}} R[\tilde{\sigma}_1] \oplus \cdots \oplus R[\tilde{\sigma}_n].$$

Note that it follows at once from (13) and (10), with the help of Lemma H.2, that

$$\mathrm{rk}_{\mathbb{Z}} R[\tilde{\mu}] = m[G : E]. \tag{14}$$

Lemma L.1. (a) $\mathrm{Ker}\, \sigma_1^\natural$ *is the ideal of $R[\tilde{\gamma}]$ generated by the elements*

$$e_j - 1 \quad (2 \leq j \leq m).$$

(b) Ker $\widetilde{\mu}^\natural$ is the ideal of $R[\widetilde{\gamma}]$ generated by the elements

$$(e_i - 1)(e_j - 1) \quad (i \neq j; 1 \leq i, j \leq m).$$

Proof. (b) Let I be the ideal generated by the displayed elements, and let F be the subring of $R[\widetilde{\gamma}]$ generated by E. As in the proof of Lemma K.4, modulo I F is generated as a $\mathbb{Z}$-module by the m elements

$$1, e_2 - 1, \ldots, e_m - 1.$$

Therefore $R[\widetilde{\gamma}]_\mathbb{Z}$ is generated mod I by the products of these with the elements of a transversal of E in G, a total of $m[G : E] = \mathrm{rk}_\mathbb{Z} R[\widetilde{\mu}]$ elements by (14). Lemma H.1 completes the proof.
(a) Similar but even easier. □

Now consider a unit $u \in R[\widetilde{\gamma}]^*$. Then $u\widetilde{\sigma}_1^\natural \in R[\widetilde{\sigma}_1]^* = G\widetilde{\sigma}_1$ by (11); so there exists $h \in G$ such that $u\widetilde{\sigma}_1^\natural = h\widetilde{\sigma}_1^\natural$, i.e. $uh^{-1} - 1 \in \mathrm{Ker}\, \widetilde{\sigma}_1^\natural$. It follows from (3) and Lemma L.1 (a) that, for $1 \leq i \leq n$, $(uh^{-1})\widetilde{\sigma}_i^\natural - 1 \in (\mathrm{Ker}\, \widetilde{\sigma}_1^\natural)\widetilde{\sigma}_i^\natural = 2R[\widetilde{\sigma}_1]$. But RA_i is an order in a division algebra over $\mathbb{Q}$, therefore Lemma H.4 implies that $R[\widetilde{\sigma}_i]$ contains no idempotents other than $0, 1$, and it follows from Lemma H.3 that

$$(uh^{-1})\widetilde{\sigma}_i^\natural = \pm 1.$$

Therefore $(uh^{-1})\widetilde{\mu}^\natural = (1, \pm 1, \ldots, \pm 1) \in E\widetilde{\mu}^\natural$ by (3), and absorbing an element of E into h we may suppose that $(uh^{-1})\widetilde{\mu}^\natural = 1$, i.e.

$$uh^{-1} - 1 \in \mathrm{Ker}\, \widetilde{\mu}^\natural.$$

It now follows from Lemma L.1 (b) that for each i

$$\pm 1 - 1 = (uh^{-1})\widetilde{\sigma}_i^\natural - 1 \in (\mathrm{Ker}\, \widetilde{\mu}^\natural)\widetilde{\sigma}_i^\natural \leq 4R[\widetilde{\sigma}_i]$$

whence $(uh^{-1})\widetilde{\sigma}_i^\natural = 1$, and $u = h \in G$ as for Theorem K.5. We have proved the first part of

Theorem L.2. (a) $R[\widetilde{\gamma}]^* = G = G\widetilde{\gamma}$.

(b) *More generally, for any (finite or infinite) set π of C_2-, C_4-, Q-components of G such that $\widetilde{\mu} \subseteq \pi$, we have $R[\pi]^* = G\pi$ ($\cong G$).*

Proof. If we reinterpret $\widetilde{\mu}^\natural : R[\pi] \twoheadrightarrow R[\widetilde{\mu}]$, the same argument goes through to prove that for each component $\sigma : G \twoheadrightarrow A$ we have $(\mathrm{Ker}\, \widetilde{\mu}^\natural)\sigma^\natural \leq 4RA$. □

M Realisation of a finite D-block

From Section B with a slight change of notation we take

$$RD = \mathbb{Z}[a, c \mid a^2 + 1 = c^2 + c + 1 = (ac)^2 + 1 = 0],$$

so that $o(a) = o(ac) = 4$, $o(c) = 3$, $c^a = c^{-1}$. And we recall that
$$O_3(D) = \langle c \rangle, \quad D = O_3(D)\langle a \rangle.$$

Consider a finite D-block with a given subdirect decomposition
$$\delta = (\sigma_1, \ldots, \sigma_n) : G \rightarrowtail_{sd} D^n. \tag{1}$$

By the definition of a D-block, modulo $E = G[2]$ any Sylow 2-subgroup of G is generated by a single element a_* of order 4, and by Lemma B.5 we may assume that
$$a_* = (a, \ldots, a).$$

Automatically
$$-1 = a_*^2 \in E = {}_3G; \tag{2}$$

and we note that
$$G = O_3(G)E\langle a_* \rangle, \quad G^s = O_3(G) \times \langle -1 \rangle. \tag{3}$$

Apply Lemma H.5 (with $\Gamma = G$, and with $\langle -1 \rangle$ in place of E, so G^s in place of EG^s); for $i = 1, \ldots, n$ there exists a homomorphism $\eta_i : G \twoheadrightarrow K$ onto a group K of exponent 2 and such that
$$\tilde{\sigma}_i := (\sigma_i, \eta_i) : G \twoheadrightarrow D \times K \text{ (surjective)}, \tag{4}$$
$$\operatorname{Ker} \tilde{\sigma}_i = \operatorname{Ker}(\sigma_i \upharpoonright G^s), \tag{5}$$
$$\operatorname{Ker} \eta_i \geq G^s = O_3(G)\langle -1 \rangle. \tag{6}$$

Then
$$G^s \sigma_i = D^s = \langle c \rangle \times \langle -1 \rangle \cong C_6$$
$$\Rightarrow |\operatorname{Ker} \tilde{\sigma}_i| = \frac{1}{6}|G^s| \Rightarrow |K| = \frac{|G|}{\frac{1}{6}|G^s||D|} = \frac{1}{2}[G : G^s]$$

(in particular $|K|$ is independent of i). Therefore, writing
$$R[\tilde{\sigma}_i] = RD \otimes \mathbb{Z}[K], \tag{7}$$

what we have just proved (and the fact that $\operatorname{rk}_\mathbb{Z} RD = 4$) gives us
$$\operatorname{rk}_\mathbb{Z} R[\tilde{\sigma}_i] = 2[G : G^s]. \tag{8}$$

By Lemma H.4
$$R[\tilde{\sigma}_i]^* = D \times K = G\tilde{\sigma}_i. \tag{9}$$

Note that it follows from (9) that
$$O_3 R[\tilde{\sigma}_i]^* = (O_3 G)\tilde{\sigma}_i = O_3 D \times \{1\} \cong C_3. \tag{10}$$

Now $O_3(G)$ is elementary abelian of exponent 3, say
$$O_3(G) \cong C_3^m, \text{ where } 1 \leq m \leq n. \tag{11}$$

Therefore after a possible reordering of the σ_i we may assume that $\sigma_1, \ldots, \sigma_m$ restrict to a basis of the dual group $\mathrm{Hom}(O_3(G), C_3)$. Let $c_1, \ldots, c_m$ be the dual basis of $O_3(G)$, so that

$$c_i \sigma_j = c^{\delta_{ij}} \quad (i \leq j, j \leq m), \tag{12}$$

and let

$$\boldsymbol{\mu} = (\sigma_1, \ldots, \sigma_m) : G \to_{\mathrm{sd}} D^m.$$

(*Caveat:* it is easy to see that $\mathrm{Ker}\,\boldsymbol{\mu} \leq E$, but $\boldsymbol{\mu}$ here need not be injective.) Exactly as in Section L we interpret the group homomorphisms

$$\widetilde{\sigma}_i, \; \widetilde{\boldsymbol{\mu}} = (\widetilde{\sigma}_1, \ldots, \widetilde{\sigma}_m), \; \widetilde{\boldsymbol{\delta}} = (\widetilde{\sigma}_1, \ldots, \widetilde{\sigma}_n)$$

as families of D-components of G to obtain a ring homomorphism

$$\widetilde{\boldsymbol{\mu}}^\natural = (\widetilde{\sigma}_1^\natural, \ldots, \widetilde{\sigma}_m^\natural) : R[\widetilde{\boldsymbol{\delta}}] \twoheadrightarrow R[\widetilde{\boldsymbol{\mu}}] \leq_{\mathrm{sd}} R[\widetilde{\sigma}_1] \oplus \cdots \oplus R[\widetilde{\sigma}_m], \tag{13}$$

where here each $R[\widetilde{\sigma}_i] \leq_{\mathrm{sd}} (RD)^{|K|}$. And exactly as in Section L we obtain the crucial result

$$\mathrm{rk}_{\mathbb{Z}} R[\widetilde{\boldsymbol{\mu}}] = \sum_1^m \mathrm{rk}_{\mathbb{Z}} R[\widetilde{\sigma}_i] = 2m[G : G^\mathfrak{s}]. \tag{14}$$

Lemma M.1. (a) *The kernel* $\mathrm{Ker}\,\widetilde{\sigma}_1^\natural$ *of the ring homomorphism* $\widetilde{\sigma}_1^\natural : R[\widetilde{\boldsymbol{\delta}}] \twoheadrightarrow R[\widetilde{\sigma}_1]$ *is generated by*

$$z_1 := \sum_1^m (c_i^2 + c_i - 2) + 3, \; \text{and} \; c_i - 1 \quad (2 \leq i \leq m).$$

(b) *The kernel* $\mathrm{Ker}\,\widetilde{\boldsymbol{\mu}}^\natural$ *of the ring homomorphism* $\widetilde{\boldsymbol{\mu}}^\natural : R[\widetilde{\boldsymbol{\delta}}] \to R[\widetilde{\boldsymbol{\mu}}]$ *is generated by*

$$z_1 \; \text{and the} \; (c_i - 1)(c_j - 1) \quad (i \neq j; 1 \leq i, j \leq m).$$

Proof. Follow the strategy of the proof of (I 7) to show that, modulo the ideal of $R[\widetilde{\boldsymbol{\delta}}]$ generated by the claimed generators, the subring $\mathbb{Z}G^\mathfrak{s} = \mathbb{Z}O_3(G)$ of $R[\widetilde{\boldsymbol{\delta}}]$ is generated as a $\mathbb{Z}$-module by 2 (resp. $2m$) generators. Then the result drops out à la Section L. There are no surprises! □

Note next that it follows from (4), (6), (12) that

$$c_i \widetilde{\sigma}_j = (c, 1)^{\delta_{ij}} \quad (1 \leq i, j \leq m).$$

Hence by (10) and (11) we obtain

Lemma M.2. $\widetilde{\boldsymbol{\mu}}$ *maps* $O_3(G)$ *isomorphically onto* $O_3 R[\widetilde{\boldsymbol{\mu}}]^*$.

We are here identifying $G = G\widetilde{\pmb{\delta}} \leq R[\widetilde{\pmb{\delta}}]^*$, which is permissible by (3), (5), (12).

Now let JD be the ideal of RD generated by $c-1$. Then $JD \ni (c-1)(c^{-1}-1) = 3$, so JD contains the multiples $3, 3a, c-1, a(c-1)$ of the free basis $1, a, c-1, a(c-1)$ of RD (see Section B). But JD is obviously contained in the kernel of the surjective homomorphism $\varepsilon = \varepsilon_3 : RD \twoheadrightarrow \mathbb{F}_9$ mapping c to 1 and a to a root of t^2+1; and since $\mathrm{Ker}\,\varepsilon$ is of index 9 in RD it follows that

$$JD = \mathrm{Ker}\,\varepsilon \text{ with } \mathbb{Z}\text{-basis } 3, 3a, c-1, a(c-1).$$

But $(\mathbb{F}_9)^* \cong C_8$; so the facts that $\mathrm{o}(a\varepsilon) = 4$ and $|D| = 12$ imply

Lemma M.3. (a) $\varepsilon : RD \twoheadrightarrow \mathbb{F}_9$ *maps* D *homomorphically into* $(\mathbb{F}_9)^*$ *with kernel* $O_3(D)$.

(b) $(1 + JD) \cap D = O_3(D)$.

Now consider a unit $u \in R[\widetilde{\pmb{\delta}}]^*$. Then $u\widetilde{\sigma}_1^\natural \in R[\widetilde{\sigma}_1]^* = G\widetilde{\sigma}_1$, so $u\widetilde{\sigma}_1^\natural = g\widetilde{\sigma}_1$ for some $g \in G$, whence

$$ug^{-1} - 1 \in \mathrm{Ker}\,\widetilde{\sigma}_1^\natural.$$

Let φ be any D-component of G. Since φ maps each c_i to an element of $\langle c \rangle$, it follows from Lemma M.1 (a) that $\varphi^\natural$ maps $\mathrm{Ker}\,\widetilde{\sigma}_1^\natural$ into the ideal of RD generated by $c-1$, namely JD. Therefore

$$(ug^{-1})\varphi^\natural - 1 \in (\mathrm{Ker}\,\widetilde{\sigma}_1^\natural)\varphi^\natural \leq JD,$$

and Lemma M.3 (b) implies that $(ug^{-1})\varphi^\natural \in O_3(D)$. As this is true for each component φ in $\widetilde{\sigma}$. It follows that ug^{-1} is a 3-element. Therefore

$$(ug^{-1})\widetilde{\pmb{\mu}}^\natural \in O_3 R[\widetilde{\pmb{\mu}}]^* = O_3 G\widetilde{\pmb{\mu}};$$

and absorbing an element of $O_3(G)$ into g we may assume that $(ug^{-1})\widetilde{\pmb{\mu}}^\natural = 1$, i.e.

$$ug^{-1} - 1 \in \mathrm{Ker}\,\widetilde{\pmb{\mu}}^\natural.$$

Therefore by Lemma M.1 (b) for every D-component $\varphi : G \twoheadrightarrow D$ we now have

$$(ug^{-1})\varphi^\natural - 1 \in (\mathrm{Ker}\,\widetilde{\pmb{\mu}}^\natural)\varphi^\natural \leq 3RD;$$

and Lemma H.3 forces $(ug^{-1})\varphi^\natural = 1$. Therefore $ug^{-1} = 1$ in $R[\widetilde{\pmb{\delta}}]$, whence $u = g \in G$. And we have proved

Theorem M.4. (a) $R[\widetilde{\pmb{\delta}}]^* = G\widetilde{\pmb{\delta}} = G$.

(b) *More generally, for any family π of D-components of G with $\widetilde{\pmb{\mu}} \subseteq \pi$,*

$$R[\pi]^* = G\pi \cong G.$$

N On the structure of a finite $CQDB$-group

Lemma N.1. *The following list is closed under subgroups* (up to isomorphism of course)
$$\{1\}, C_2, C_3, C_4, C_6, Q, D, B.$$

Proof. Trivial, particularly in view of the genesis of the primordial groups (Section B). □

Corollary N.2. *Every subgroup of a $CQDB$-group is a $C_2C_3C_4QDB$-group.*

Lemma N.3. *Given a subdirect decomposition $(\gamma, \eta) : G \rightarrowtail_{sd} C \times H$ where G is reduced and C is completely reducible, then $\eta : G \twoheadrightarrow H$ is an isomorphism.*

Proof. Put $C_0 = \operatorname{Ker}\eta$. Then $\operatorname{Ker}(\gamma \restriction C_0) = \operatorname{Ker}\gamma \cap \operatorname{Ker}\eta = \{1\}$. Since C is completely reducible, the injection $\gamma \restriction C_0 : C_0 \rightarrowtail C$ splits, i.e. there exists a homomorphism $\sigma : C \twoheadrightarrow C_0$ such that $(\gamma \restriction C_0) \circ \sigma = \operatorname{id}_{C_0}$. But then $\gamma \circ \sigma : G \twoheadrightarrow C_0$ is a retraction onto the normal subgroup C_0. Therefore the completely reducible group C_0 is a direct factor of $G \;(= C_0 \times \operatorname{Ker}(\gamma \circ \sigma))$. So $C_0 = \{1\}$ because G is reduced. □

Corollary N.4. *Every reduced subgroup of a $CQDB$-group is a C_4QDB-group.*

Proof. Consider such a reduced subgroup G. By Corollary N.2, G admits a subdirect decomposition
$$\boldsymbol{\pi} = (\pi_1, \ldots, \pi_n) : G \rightarrowtail_{sd} A_1 \times \cdots \times A_n,$$
where we may assume that $A_1, \ldots, A_r \in \{C_2, C_3\}$ and $A_{r+1}, \ldots, A_n \in \{C_4, Q, D, B\}$. Put
$$\boldsymbol{\gamma} = (\pi_1, \ldots, \pi_r) : G \twoheadrightarrow C \leq_{sd} A_1 \times \cdots \times A_r,$$
$$\boldsymbol{\eta} : (\pi_{r+1}, \ldots, \pi_n) : G \twoheadrightarrow H \leq_{sd} A_{r+1} \times \cdots \times A_n.$$
Then C is completely reducible, and $(\boldsymbol{\gamma}, \boldsymbol{\eta}) : G \rightarrowtail_{sd} C \times H$: so by Lemma N.3 the projection η is an isomorphism, i.e. $(\pi_{r+1}, \ldots, \pi_n) : G \rightarrowtail_{sd} A_{r+1} \times \cdots \times A_n$. The corollary is proved. □

Theorem N.5. *Let G be a finite $CQDB$-group. Then G admits a direct decomposition*
$$G = C_2^r \times C_3^s \times G_{C_4QD} \times G_B,$$
where r, s are non-negative integers, G_{C_4QD} is a reduced C_4QD-group, and G_B is a reduced B-group. The decomposition is unique up to isomorphism.

Proof. Uniqueness follows easily from Krull–Schmidt. As for existence, since G is a finite group of exponent 12 it can certainly be expressed as a direct product $G = C_2^r \times C_3^s \times G_0$ where G_0 is reduced. And by Corollary N.4, G_0 is a C_4QDB-group. We might as well assume then that G is already a reduced C_4QDB-group. Then G certainly admits a subdirect decomposition
$$(\boldsymbol{\alpha}, \boldsymbol{\beta}) : G \rightarrowtail_{sd} G_{C_4QD} \times G_B$$

for a suitable C_4QD-group G_{C_4QD} and B-group G_B. A priori these subdirect factors need not be reduced, but a simple application of Lemma N.3 allows us to assume that they both are reduced. Choose a Sylow 2-group H and a Sylow 3-group X of G. Then by the remark at the end of Lemma F.2, G contains both $X\beta$, a Sylow 3-subgroup of G_B, and also $[H\beta, X\beta]$, which by Lemma F.5 (d) coincides with the Sylow 2-subgroup $O_2(G_B)$ of G_B. Therefore G contains $O_2(G_B)X\beta = G_B$. And a triviality on subdirect products now guarantees that $G = G_{C_4QD} \times G_B$. □

O The main theorem

Theorem O.1. *A finite group G is realisable as the group of units of an order in a $\mathbb{Q}$-algebra if, and only if, G is a CQDB-group and either*

(a) *G has a direct factor of order 2, or*

(b) *G admits a direct decomposition*

$$G = G_0 \times G_1 \times \cdots \times G_r,$$

where $G_1, \ldots, G_r$ are B-blocks and G_0 is a C_4QD-group which may be embedded as a subdirect product of copies of C_4, Q, D in such a way that it contains the diagonal involution -1.

Remark. This final condition relating to the diagonal involution is not very pretty but at least it works – unlike the corresponding condition in Hallett and Hirsch that there exists an involution in G that is *not* the 6th power of an element of order 12. For long I believed that it could be replaced by the more desirable, intrinsic requirement that there exists an element g_0 of order 4 in G such that $c_G(g_0)$ is a 2-group; but there was a silly error, and now I am not sure whether this centraliser condition is sufficient.

We have already established the necessity of the conditions in Theorem O.1, see Corollary C.4, Lemma D.3 and Theorem G.2. We base the proof of sufficiency on

Lemma O.2. *Let G be a C_2C_4QD-group with a subdirect decomposition π such that $-1 \in G\pi$. Then, possibly after adjoining further components to π, $R[\pi]^* = G\pi$ ($\cong G$).*

Proof. Partition the projections in $bs\pi$ to give a subdirect decomposition

$$\pi = (\delta_1, \ldots, \delta_r, \gamma) : G \rightarrowtail_{sd} G_1 \times \cdots \times G_r \times G_*,$$

where the $\delta_i : G \twoheadrightarrow G_i$ ($1 \leq i \leq r$) are the D-blocks and γ contains the projections onto C_2, C_4, Q. By Theorem M.4 we may assume that extra D-components have been adjoined to each δ_i to guarantee that

$$R[\delta_i]^* = G\delta_i = G_i \ (1 \leq i \leq r).$$

Then by Lemma E.1 we certainly have

$$O_3 R[\pi]^* = \prod_1^r O_3 R[\delta_i]^* = \prod_1^r O_3 G_i = O_3 G,$$

where we are making our usual identification so that we also have, quite trivially

$$G = G\pi \leq R[\pi]^*.$$

Now we already know that the ring homomorphism $\varepsilon : RD \twoheadrightarrow \mathbb{F}_9$ maps D homomorphically into $(\mathbb{F}_9)^*$ with kernel $O_3(D)$, so that it maps the subgroup C_4 of D isomorphically onto $D\varepsilon$. If we now define

$$R^+ D = \{(x, y) \in RD \oplus RC_4 \mid x\varepsilon = y\varepsilon\} \leq_{\text{sd}} RD \oplus RC_4,$$

then

$$(R^+ D)^* = \{(x, y) \in D \times C_4 \mid x\varepsilon = y\varepsilon\} = D(1, \tau) \cong D,$$

where $\tau : D \twoheadrightarrow C_4$ is the retraction with kernel $O_3(D)$. Note that the projection $R^+ D \twoheadrightarrow RD$ maps $(R^+ D)^* = D(1, \tau)$ isomorphically onto $(RD)^* = D$, while the projection $R^+ D \twoheadrightarrow RC_4$ maps $(R^+ D)^*$ homomorphically onto C_4 with kernel $O_3(R^+ D)^* = O_3(D) \times \{1\}$.

For each D-component σ in π adjoin to γ the composite $G \xrightarrow{\sigma} D \xrightarrow{\tau} C_4$. Theorem L.2 then allows us to adjoin to γ further components so that

$$R[\gamma]^* = G\gamma.$$

Then the projection $\gamma^\natural : R[\pi] \twoheadrightarrow R[\gamma]$ maps $R[\pi]^*$ into $R[\gamma]^* = G\gamma$ with kernel $O_3 R[\pi]^* = O_3(G)$ and the subgroup G of $R[\pi]^*$ onto $G\gamma$ with kernel $O_3(G)$. It follows at once that $R[\pi]^* = G$. □

Lemma O.3. *Let G be a reduced B-group with r B-blocks. Then there exists a subdirect embedding*

$$\beta^+ : G \times \langle -1 \rangle \hookrightarrow_{\text{sd}} B^n \times C_6^r \times C_2$$

such that

$$R[\beta^+]^* = G \times \langle -1 \rangle.$$

Proof. Let the B-blocks be $\beta_i : G \twoheadrightarrow G_i$ $(1 \leq i \leq r)$. We may assume the G_i reduced, by Lemma N.3. If G_i contains its diagonal involution, then we know by Theorem K.5 that $R[\beta_i]^* = G\beta_i = G_i$; if not, and $G_i[2] = 2^{m_i}$, then by replacing β_i by a suitable set μ_i of m_i projections from β, we reduce to this case. So we may assume that

$$R[\beta_i]^* = G\beta_i = G_i \quad (1 \leq i \leq r).$$

For each i, let $\zeta_i : G_i \twoheadrightarrow \langle w \rangle = C_3 \leq C_6$ be a homomorphism with kernel $O_2(G_i)$ and let $\zeta_i^+ : G \times \langle -1 \rangle \twoheadrightarrow C_6$ be the homomorphism agreeing with $\beta_i \circ \zeta_i$ on G and with the injection $\langle -1 \rangle \rightarrowtail C_6$ on $\langle -1 \rangle$. Then

$$\beta_i^+ = (\beta_i, \zeta_i^+)$$

is a C_6B-block, therefore by Theorem K.5
$$R[\beta_i^+]^* = (G_i \times \langle -1 \rangle)(\beta_i^+, \zeta_i^+) = G(\beta_i, \zeta_i) \times \langle -1 \rangle \leq_{\text{sd}} G_i \times C_6.$$
Here the projection $R[\beta_i^+]^* \twoheadrightarrow RC_6$ has kernel $O_2(G_i) \times \{1\} = O_2(G_i)$.
Then
$$\boldsymbol{\beta}^+ = (\beta_1^+, \ldots, \beta_r^+) : G \rightarrowtail_{\text{sd}} R[\beta_1^+]^* \times \cdots \times R[\beta_r^+]^* \leq_{\text{sd}} G_1 \times \cdots \times G_r \times C_6^r$$
and the projection $R[\boldsymbol{\beta}^+] \twoheadrightarrow R[\boldsymbol{\zeta}^+]$ where $\boldsymbol{\zeta}^+ = (\zeta_1^+, \ldots, \zeta_r^+)$ maps $R[\boldsymbol{\beta}^+]^*$ into $R[\boldsymbol{\zeta}^+]^* = G\boldsymbol{\zeta}^+ \times \langle -1 \rangle$ with kernel $O_2(G_1) \times \cdots \times O_2(G_r) = O_2(G)$, where we have used the fundamental 3-realisation Lemma I.1; but the subgroup $G \times \langle -1 \rangle$ is mapped onto $G\boldsymbol{\zeta}^+ \times \langle -1 \rangle$ with kernel $O_2(G)$. Hence
$$R[\boldsymbol{\beta}^+]^* = G\boldsymbol{\beta}^+ \times \langle -1 \rangle.$$

It remains to bring in the factor C_2. Let $\zeta : G \times \langle -1 \rangle \twoheadrightarrow C_2$ be the homomorphism with $\operatorname{Ker} \zeta = G$, and take $\boldsymbol{\beta}^{++} = (\boldsymbol{\beta}^+, \zeta)$. It is easy to see that this works, using the auxiliary ring
$$R_0 = \{(x_1, \ldots, x_r, y) \in RC_6^r \oplus RC_2 \,|\, x_1\varepsilon = \cdots = x_r\varepsilon = y\varepsilon\},$$
where $\varepsilon : RC_6 \twoheadrightarrow \mathbb{F}_3$ is the homomorphism mapping C_6 onto $(\mathbb{F}_3)^* \cong C_2$ with kernel $O_3(C_6)$.

Proof of sufficiency in Theorem O.1. Let G satisfy the conditions of the theorem. We distinguish two cases.

(b) Here $G = G_0 \times G_1 \times \cdots \times G_r$ and it is enough to note that each of the direct factors is realisable, G_0 by Lemma O.2, and the B-blocks $G_1, \ldots, G_r$ as in the proof of Lemma O.3

(a) By Theorem N.5 we may consider a direct decomposition
$$G = C_2 \times G_1 \times G_2 \times G_3,$$
where $G_1 \in \mathfrak{A}_3$, G_2 is a finite C_2C_4QD-group (not necessarily reduced), and G_3 is a finite reduced B-group. We identify the first factor with $\langle -1 \rangle$ and correspondingly take $\zeta : G \twoheadrightarrow C_2$ to be the homomorphism with kernel $G_1 \times G_2 \times G_3$. Then by Lemma I.1 (with the remark at the end of the proof of Lemma O.3), Lemma O.2 and Lemma O.3 there exist families of components π_1, π_2, π_3 such that
$$R[(\zeta, \pi_i)]^* = (\langle -1 \rangle \times G_i)(\zeta, \pi_i) = `\langle -1 \rangle` \times G_i \ (i = 1, 2, 3).$$
Consider a unit $u \in R[(\zeta, \pi_1, \pi_2, \pi_3)]^*$. The ring homomorphism
$$(\zeta, \pi_i)^\natural : R[(\zeta, \pi_1, \pi_2, \pi_3)] \twoheadrightarrow R[(\zeta, \pi_i)]$$
maps u to an element of $(\langle -1 \rangle \times G_i)(\zeta, \pi_i)$, so
$$u(\zeta, \pi_i)^\natural = (`(-1)^{k_i}`, g_i)$$

for some $k_i \in \mathbb{Z}$, $g_i \in G$. But then $u\zeta = (-1)^{k_i}$, so these are all equal, and we have

$$u = ((-1)^{k_1}, g_1, g_2, g_3) \in \langle -1 \rangle \times G_1 \times G_2 \times G_3 = G.$$

Since in any case $G \leq R[(\zeta, \pi_1, \pi_2, \pi_3)]^*$, we have proved that

$$R[\boldsymbol{\pi}]^* = G\boldsymbol{\pi} = G$$

where $\boldsymbol{\pi} = (\zeta, \pi_1, \pi_2, \pi_3)$. □

Appendix by Federico Menegazzo

The following proposition provides a proof of the "more desirable, intrinsic requirement" referred to in the remark in Section O.

Proposition. *Suppose that G is a finite CQDB-group and either*

- *G has a direct factor of order 2, or*
- *G admits a direct decomposition*

$$G = G_0 \times G_1 \times \cdots \times G_r,$$

where $G_1, \ldots, G_r$ are B-blocks and G_0 is a C_4QD-group containing an element g_0 of order 4 whose centraliser in G_0 is a 2-group.

Then G may be embedded as a subdirect product of copies of C_2, C_4, C_6, Q, D, B in such a way that it contains the diagonal involution -1.

(Each of C_2, C_4, C_6, Q, D, B contains a unique involution that will be denoted by -1; the meaning of 'the diagonal involution -1' in a direct product of groups of this form should be clear.)

Suppose first that $G = \langle e \rangle \times L \leq_{\text{sd}} A_1 \times \cdots \times A_n$ where e is an involution and $A_i \in \{C_2, C_4, C_6, Q, D, B\}$ for every i. If $-1 \notin G$, then the subgroup $\langle -1 \rangle \times L$ of $A_1 \times \cdots \times A_n$ is isomorphic to G, and it is easily seen that it is a subdirect product. We denote by π_i the projection $A_1 \times \cdots \times A_n \to A_i$ according to the given decomposition: we need to show that $(\langle -1 \rangle \times L)\pi_i = A_i$ for every i. If $A_i \notin \{C_2, C_6\}$ then $e\pi_i$ is in the Frattini subgroup of A_i, so that $A_i = G\pi_i = L\pi_i = (\langle -1 \rangle \times L)\pi_i$. If $A_i = C_2$ then $\langle -1 \rangle \pi_i = A_i$. If $A_i = C_6$ there is some $l \in L$ such that $l\pi_i$ has order 3; but then $\langle -1 \cdot l \rangle \pi_i = A_i$.

To prove the second statement, we choose the notation as follows: $G \leq_{\text{sd}} A_1 \times \cdots \times A_n$ where $A_i \in \{C_4, Q, D, B\}$ for every i, π_i is the projection $A_1 \times \cdots \times A_n \to A_i$ according to the given decomposition, $A_i \in \{C_4, Q\}$ for $1 \leq i \leq r_0$, $A_i = D$ for $r_0 < i \leq s_0$, $A_i = B$ for $s_0 < i \leq n$, so that $G_0 \leq_{\text{sd}} A_1 \times \cdots \times A_{s_0}$, and the factors B are ordered according to the different B-blocks in such a way that $G_i \leq_{\text{sd}} A_{s_{i-1}+1} \times \cdots \times A_{s_i}$ for $i = 1, \ldots, r$, with $s_r = n$.

We will slightly modify the given subdirect product representation in order to achieve $-1 \in G_i$. We begin by studying G_0. The element g_0 acts by conjugation on $O_3(G_0)$ as a fixed-point-free automorphism of order 2, i.e it inverts every element of $O_3(G_0)$. For $r_0 < i \leq s_0$ let $x_i \in O_3(G_0)$ such that $x_i\pi_i \neq 1$: $(x_i\pi_i)^{(g_0\pi_i)} = (x_i\pi_i)^{-1}$ implies that the order of $g_0\pi_i$ is 4 and $g_0^2\pi_i = -1$. We put $\delta = (\pi_{r_0+1}, \ldots, \pi_{s_0})$: $A_1 \times \cdots \times A_n \to D^{s_0-r_0}$. We get $G_0\delta \leq_{\mathrm{sd}} D^{s_0-r_0}$ and $g_0^2\delta = -1$ (of $D^{s_0-r_0}$).

$G \cap \ker \delta$ is a 2-group; its elements of order ≤ 2 give a subgroup E that is elementary abelian of order 2^{m_0} with $m_0 \leq r_0$ since $E \hookrightarrow A_1 \times \cdots \times A_{r_0}$. As $\bigcap_{i=1}^{r_0}(E \cap \ker \pi_i) = 1$, we can find m_0 projections, say $\pi_1, \ldots, \pi_{m_0}$, such that $\bigcap_{i=1}^{m_0}(E \cap \ker \pi_i) = 1$; if we call $\mu = (\pi_1, \ldots, \pi_{m_0})$ we get $G_0\mu \leq_{\mathrm{sd}} A_1 \times \cdots \times A_{m_0}$. We similarly define $\sigma = (\mu, \delta) = (\pi_1, \ldots, \pi_{m_0}, \pi_{r_0+1}, \ldots, \pi_{s_0})$; the image $G_0\sigma \leq_{\mathrm{sd}} G_0\mu \times G_0\delta \leq_{\mathrm{sd}} A_1 \times \cdots \times A_{m_0} \times D^{s_0-r_0}$. Now $G_0 \cap \ker \sigma = G_0 \cap \ker \mu \cap \ker \delta$ is a 2-group (a subgroup of $\ker \delta$), and its involutions belong to $E \cap \ker \mu = 1$. It follows that σ gives a subdirect embedding of G_0 into $A_1 \times \cdots \times A_{m_0} \times A_{r_0+1} \times \cdots \times A_{s_0}$. In particular, $|E\sigma| = |E| = 2^{r_0}$ and $E\sigma \leq A_1 \times \cdots \times A_{m_0} \times \langle 1 \rangle$ say that all the elements of the form $(e_1, \ldots, e_{m_0}, 1, \ldots, 1)$ (here $e_j \in \langle -1 \rangle$) belong to $E\sigma \leq G_0\sigma$. But then $g_0^2\sigma = (g_0^2\mu, g_0^2\delta) = (g_0^2\mu, 1)(1, g_0^2\delta) \in G_0\sigma$, both $(-1, 1)$ and $(g_0^2\mu, 1)$ are in $G_0\sigma$, and finally $-1 \in G_0\sigma$.

To deal with the remaining G_i, $i \geq 0$ we proceed in a similar way. We recall that $G_i = O_2(G_i) \cdot X_i$ where X_i has order 3, and note that the kernel of every projection π_j for $s_{i-1} < j \leq s_i$ intersects G_i in a 2-group. The involutions in G_i generate an elementary abelian subgroup E of order 2^{m_i}, say, with $m_i \leq s_i - s_{i-1}$. It is possible to select a subset $S_i \subseteq \{\pi_{s_{i-1}+1}, \ldots, \pi_{s_i}\}$ consisting of exactly m_i projections such that $\bigcap_{\pi_j \in S_i}(E \cap \ker \pi_j) = 1$. As above, the map $\mu_i = (\pi_j \mid \pi_j \in S_i)$ gives a subdirect embedding $G_i \hookrightarrow G_i\mu_i \leq_{\mathrm{sd}} B^{m_i}$. In particular, the order $|E\mu_i| = |E| = 2^{m_i}$ is the order of the subgroup generated by the involutions of B^{m_i}, so that $-1 \in E\mu_i \leq G_i\mu_i$.

References

[1] A. L. S. Corner, *Every countable reduced torsion-free ring is an endomorphism ring*, Proc. London Math. Soc. **13** (1963), 687–710.

[2] A. L. S. Corner and R. Göbel, *Prescribing endomorphism algebras, a unified treatment*, Proc. London Math. Soc. **50** (1985), 447–479.

[3] L. Fuchs, *Infinite Abelian Groups, Vol. II,* Academic Press 1974.

[4] M. Hall, Jr., *Theory of Groups,* Chelsea Publishing Co., New York 1976.

[5] T. J. Hallett and K. A. Hirsch, *Torsion-free groups having finite automorphism groups*, J. Algebra **2** (1965), 287–298.

[6] T. J. Hallett and K. A. Hirsch, *Die Konstruktion von Gruppen mit vorgeschriebenen Automorphismengruppen*, J. Reine Angew. Math. **239/240** (1969), 32–46.

[7] T. J. Hallett and K. A. Hirsch, *Groups of exponent 4 as automorphism groups,* Math. Z. **117** (1970), 183–188.

[8] T. J. Hallett and K. A. Hirsch, *Finite groups of exponent 4 as automorphism groups II*, Math. Z. **131** (1973), 1–10.

[9] T. J. Hallett and K. A. Hirsch, *Finite groups of exponent 12 as automorphism groups,* Math. Z. **155** (1977), 43–53.

[10] K. A. Hirsch and H. Zassenhaus, *Finite automorphism groups of torsion-free groups*, J. London Math. Soc. **41** (1966), 545–549.

[11] B. Huppert, *Endliche Gruppen I*, Springer, Berlin 1967.

[12] G. Nöbeling, *Verallgemeinerung eines Satzes von Herrn E. Specker*, Invent. Math. **6** (1968), 41–55.

Author information

A. L. S. Corner, formerly of Worcester College, Oxford, England.

Basic subgroups and freeness, a counterexample

Andreas Blass and Saharon Shelah

Abstract. We construct a non-free but $\aleph_1$-separable, torsion-free abelian group G with a pure free subgroup B such that all subgroups of G disjoint from B are free and such that G/B is divisible. This answers a question of Irwin and shows that a theorem of Blass and Irwin cannot be strengthened so as to give an exact analog for torsion-free groups of a result proved for p-groups by Benabdallah and Irwin.

Key words. Abelian group, free, divisible, stationary set, Gamma invariant.

AMS classification. 20K20, 03E05.

1 Introduction

All groups in this paper are abelian and, except for some motivating remarks about p-groups in this introduction, all groups are torsion-free. A subgroup B of a group G is *basic* in G if

- B is a direct sum of cyclic groups,
- B is a pure subgroup of G, and
- G/B is divisible.

Of course in the torsion-free case, "a direct sum of cyclic groups" can be shortened to "free."

Benabdallah and Irwin proved in [1] the following result:

Theorem 1.1. *Suppose G is a p-group with no elements of infinite height. Suppose further that G has a basic subgroup B such that every subgroup of G disjoint from B is a direct sum of cyclic groups. Then G itself is a direct sum of cyclic groups.*

"Disjoint" means that the intersection is $\{0\}$, not $\varnothing$, as the latter is impossible for subgroups.

Later, Irwin asked whether an analogous theorem holds for torsion-free groups. The following partial affirmative answer was given in [2]. Note that, unlike p-groups, torsion-free groups need not have basic subgroups.

First author: Research partially supported by NSF grants DMS-0070723 and DMS-0653696.

Second author: Research partially supported by NSF grants DMS-0072560 and DMS-0600940 and German-Israeli Foundation for Scientific Research & Development Grant No. I-706-54.6/2001. Publication number 910.

Theorem 1.2. *Suppose G is a torsion-free group such that*

- *G has a basic subgroup of infinite rank, and*
- *for every basic subgroup B of G, all subgroups of G disjoint from B are free.*

Then G is free.

This result is weaker in two ways than the hoped-for analog of Theorem 1.1. First, not only must there be a basic subgroup, but it must have infinite rank. (It was shown in [3] that all basic subgroups of a torsion-free group have the same rank.) Second, the assumption that all subgroups disjoint from B are free is needed not just for one basic subgroup B but for all of them.

The assumption that a basic subgroup has infinite rank is needed. As was pointed out in [2], Fuchs and Loonstra constructed in [5] a torsion-free group of rank 2 such that every subgroup of rank 1 is free and every torsion-free quotient of rank 1 is divisible. In such a group G, every pure subgroup B of rank 1 is basic, every subgroup disjoint from B has rank at most 1 and is therefore free, yet G is certainly not free.

It has remained an open question until now whether the second weakness of Theorem 1.2 can be removed. Can "for every basic subgroup" be replaced with "for some basic subgroup" in the second hypothesis? In this paper, we answer this question negatively.

Theorem 1.3. *There exists an $\aleph_1$-separable torsion-free group G of size $\aleph_1$ with a basic subgroup B of rank $\aleph_1$ such that all subgroups of G disjoint from B are free but G itself is not free.*

The rest of this paper is devoted to the proof of this theorem. The group G and the subgroup B will be constructed in Section 2 and the claimed properties will be proved in Section 3.

The proof will show a little more than is stated in the theorem. We can arrange for the Gamma invariant $\Gamma(G)$ to be any prescribed non-zero element of the Boolean algebra $\mathcal{P}(\aleph_1)/NS$ of subsets of $\aleph_1$ modulo non-stationary subsets. (See [4, Section IV.1] for the definition and basic properties of Γ.)

2 Construction

Our construction is somewhat similar to the construction of $\aleph_1$-separable groups in [4, Section VIII.1]. We shall, however, present our result in detail, not presupposing familiarity with the cited construction from [4]. We begin by fixing notations for a set-theoretic ingredient and a group-theoretic ingredient of our construction.

Notation 2.1. Fix a set S of countable limit ordinals such that S is stationary in $\aleph_1$. Also fix, for each $\delta \in S$, a strictly increasing sequence $\langle \eta(\delta, n) : n \in \omega \rangle$ with limit δ.

The equivalence class of S in $\mathcal{P}(\aleph_1)/NS$ will be the Gamma invariant of the group G that we construct. Since the countable limit ordinals form a closed unbounded subset of $\aleph_1$, every non-zero element of $\mathcal{P}(\aleph_1)/NS$ is the equivalence class of an S as in Notation 2.1 and can therefore occur as $\Gamma(G)$ in Theorem 1.3.

Notation 2.2. Fix a torsion-free group E of rank 2 such that all rank 1 subgroups are free and all torsion-free rank-1 quotients are divisible. Such a group exists by [5, Lemma 2]. Also fix a pure subgroup of E of rank 1 and, since it is free, fix a generator a for it. Since $E/\langle a \rangle$ is a torsion-free rank-1 quotient of E, it is divisible and thus isomorphic to $\mathbb{Q}$. Fix an isomorphism φ from $\mathbb{Q}$ to $E/\langle a \rangle$ and fix, for each positive integer n, a representative $b_n \in E$ of $\varphi(1/n!)$. Since

$$\varphi\left(\frac{1}{n!}\right) = (n+1)\varphi\left(\frac{1}{(n+1)!}\right),$$

there are (unique) integers q_n such that

$$b_n = (n+1)b_{n+1} + q_n a$$

for all n. Fix this notation q_n for the rest of the paper.

Remark 2.3. We shall not need the full strength of the conditions on E. Specifically, we need divisibility only for $E/\langle a \rangle$, not for all the other torsion-free rank-1 quotients of E.

Lemma 2.4. *The generators a and b_n for $n \in \omega$ and the relations $b_n = (n+1)b_{n+1} + q_n a$ constitute a presentation of E.*

Proof. Since $\mathbb{Q}$ is generated by the elements $1/n!$, $E/\langle a \rangle$ is generated by the images $[b_n]$ of the elements b_n. Therefore E is generated by these elements together with a.

It remains to show that every relation between these generators that holds in E is a consequence of the specified relations $b_n = (n+1)b_{n+1} + q_n a$. Consider an arbitrary relation $ca + \sum_{n \in F} d_n b_n = 0$ that holds in E; here F is a finite subset of ω and c and the d_n's are integers.

The given relations $b_n = (n+1)b_{n+1} + q_n a$ allow us to eliminate any b_n in favor of b_{n+1} at the cost of changing the coefficient of a. So, at a similar cost, we can replace any b_n with a multiple of b_m for any desired $m > n$. Thus, we can arrange to have only a single b_n occurring; that is, the relation under consideration can, via the given relations, be converted to the form $c'a + d'b_n = 0$.

Since this relation holds in E, we have $d'[b_n] = 0$ in $E/\langle a \rangle$. But $E/\langle a \rangle$ is torsion-free and $[b_n] = \varphi(1/n!)$ is non-zero. So $d' = 0$ and our relation is simply $c'a = 0$. Since $\langle a \rangle$ is torsion-free, $c' = 0$. Thus, the given relations $b_n = (n+1)b_{n+1} + q_n a$ have reduced our original $ca + \sum_{n \in F} d_n b_n = 0$ to $0 = 0$. Equivalently, $ca + \sum_{n \in F} d_n b_n = 0$ is a consequence of the given relations. □

We are now ready to define the group G and subgroup B required in Theorem 1.3.

Definition 2.5. G is the group generated by symbols x_α for all $\alpha < \aleph_1$ and $y_{\delta,n}$ for all $\delta \in S$ and $n \in \omega$, subject to the defining relations, one for each $\delta \in S$ and $n \in \omega$,

$$y_{\delta,n} = (n+1)y_{\delta,n+1} + q_n x_\delta + x_{\eta(\delta,n)}.$$

B is the subgroup of G generated by all of the x_α's.

Since there are exactly $\aleph_1$ generators in this presentation, the inequality $|G| \leq \aleph_1$ is obvious. The reverse inequality also holds because, as we shall show in Subsection 3.1 below, B is free of rank $\aleph_1$.

We shall sometimes have to discuss formal words in the generators of G, i.e., elements of the free group on the x_α's and $y_{\delta,n}$'s without the defining relations above. We shall call such formal words *expressions* and we say that an expression *denotes* its image in G, i.e., its equivalence class modulo the defining relations. We call two expressions *equivalent* if they denote the same element, i.e., if one can be converted into the other by applying the defining relations.

We shall sometimes refer to the defining relation $y_{\delta,n} = (n+1)y_{\delta,n+1} + q_n x_\delta + x_{\eta(\delta,n)}$ as the defining relation for δ and n; when n varies but δ is fixed, we shall also refer to a defining relation for δ.

Given an expression that contains $y_{\delta,n}$ for a certain δ and n, we can eliminate this $y_{\delta,n}$ in favor of $y_{\delta,n+1}$ by applying the defining relation for δ and n. In the resulting equivalent expression, the coefficient of the newly produced $y_{\delta,n+1}$ will be $n+1$ times the original coefficient of $y_{\delta,n}$, and a couple of x terms, namely that original coefficient times $q_n x_\delta + x_{\eta(\delta,n)}$, are introduced as well. We shall refer to this manipulation of expressions as "raising the subscript n of $y_{\delta,n}$ to $n+1$", and we shall refer to the introduced x terms as being "spun off" in the raising process.

By repeating this process, we can raise the subscript n of $y_{\delta,n}$ to any desired $m > n$. If the original $y_{\delta,n}$ had coefficient c, then the newly produced $y_{\delta,m}$ will have coefficient $c \cdot m!/n!$. There will also be spun off terms, namely x_δ with coefficient $c \sum_{k=n}^{m-1} \frac{k!}{n!} q_k$, and $x_{\eta(\delta,k)}$ with coefficient $c \frac{k!}{n!}$ for each k in the range $n \leq k < m$.

We shall need the notion of a linear combination of defining relations, by which we mean the result of taking finitely many of the defining relations, multiplying each of these by an integer, and adding the resulting equations. It will sometimes be convenient to think of equations $t = u$ (particularly defining relations and their linear combinations) as normalized to the form $t - u = 0$. In particular, we shall say that a generator (x_α or $y_{\delta,n}$) *occurs* in $t = u$ if it occurs in the expression $t - u$, i.e., if its total coefficient in this expression is non-zero.

As a side effect of these conventions, we do not distinguish between two equations if their normalized forms are the same, i.e., if the equations differ only by adding the same expression to both sides, a special case of which is transposing terms from one side to the other.

Notice that an equation $t = u$ is (identified with) a linear combination of defining relations if and only if t and u denote the same element of G.

3 Proofs

In this section, we verify the properties of G and B claimed in Theorem 1.3.

3.1 B is free of rank $\aleph_1$

We show that the generators x_α of B are linearly independent, by showing that no nontrivial linear combination of the defining relations can involve only x's without any y's. In fact, we show somewhat more, because it will be useful later.

Lemma 3.1. *If x_α occurs in a linear combination of defining relations, then so does $y_{\delta,n}$ for some $\delta \geq \alpha$ and some n. Furthermore, if $y_{\delta,n}$ occurs in a linear combination of defining relations, then so does $y_{\delta,m}$ for at least one $m \neq n$ (and the same δ).*

Proof. For the first statement, consider a linear combination of defining relations in which x_α occurs, and consider one of the defining relations, say $y_{\delta,n} = (n+1)y_{\delta,n+1} + q_n x_\delta + x_{\eta(\delta,n)}$, used in this linear combination and containing x_α. So either $\alpha = \delta$ or $\alpha = \eta(\delta, n)$. In either case $\delta \geq \alpha$. Fix this δ and consider all the defining relations for this δ that are used in the given linear combination. If they are the defining relations for δ and $n_1 < \ldots < n_k$, then the y_{δ,n_1} from the first of these relations is not in any of the others, so it cannot be canceled and therefore occurs in the linear combination.

For the second statement, again suppose that the linear combination involves the defining relations for δ and $n_1 < \ldots < n_k$ (perhaps along with defining relations for other ordinals $\delta' \neq \delta$). As above, the y_{δ,n_1} from the first of these cannot be canceled. Neither can the y_{δ,n_k+1} from the last. So at least these two $y_{\delta,n}$'s occur in the linear combination. □

3.2 G/B is divisible and torsion-free

We get a presentation of G/B from the defining presentation of G by adjoining the relations $x_\alpha = 0$ for all the generators x_α of B. The resulting presentation amounts to having generators $y_{\delta,n}$ for all $\delta \in S$ and all $n \in \omega$ with relations

$$y_{\delta,n} = (n+1)y_{\delta,n+1}.$$

For any fixed $\delta \in S$, the generators and relations with δ in the subscripts are a presentation of $\mathbb{Q}$, with $y_{\delta,n}$ corresponding to $1/n!$. With δ varying over S, therefore, we have a presentation of $\bigoplus_{\delta \in S} \mathbb{Q}$, a torsion-free, divisible group.

Corollary 3.2. *G is a torsion-free group, and B is a basic subgroup.*

Proof. Since both the subgroup B and the quotient G/B are torsion-free, so is G. B is pure in G because G/B is torsion-free. Since B is free and G/B is divisible, B is basic. □

3.3 G is $\aleph_1$-free

To prove that G is $\aleph_1$-free, i.e., that all its countable subgroups are free, we use Pontryagin's criterion [4, Theorem IV.2.3]. We must show that every finite subset of G is included in a finitely generated pure subgroup of G.

Let F be an arbitrary finite subset of G, and provisionally choose, for each element of F, an expression denoting it. ("Provisionally" means that we shall modify these choices several times during the following argument. The first modification comes immediately.) Raising subscripts on the y's, we may assume that, for each δ, there is at most one m such that $y_{\delta,m}$ occurs in the chosen expressions. In fact, with further raising if necessary, we may and do assume that it is the same m, which we name m_1, for all δ. Notice that, although there is still some freedom in choosing the expressions (for example, we could raise the subscript m_1 further), there is no ambiguity as to the set Δ of δ's that occur as the first subscripts of y's in our expressions. Indeed, if ε occurs exactly once in one expression but doesn't occur in another expression, then, according to the second part of Lemma 3.1, these two expressions cannot be equivalent.

Let us say that an ordinal α is *used* in our (current) provisional expressions if either it is in Δ or x_α occurs in one of these expressions. (In other words, α occurs either as a subscript on an x or as the first subscript on a y.) Of course, only finitely many ordinals are used. So, by raising subscripts again from m_1 to a suitable m_2, we can assume that, if $\delta \in \Delta$ and if $\alpha < \delta$ was used (before the current raising), then $\alpha < \eta(\delta, m_2)$.

We would prefer to omit the phrase "before the current raising," but this needs some more work. The problem is that the raising process spins off x's whose subscripts may not have been used before but are used after the raising. We analyze this situation, with the intention of correcting it by a further raising of subscripts. The problem is that, in raising the subscript from m_1 to m_2 for y_{δ,m_1}, we spin off x_δ and $x_{\eta(\delta,k)}$ for certain k, namely those in the range $m_1 \leq k < m_2$, and the subscript used here (δ or $\eta(\delta, k)$) may be $< \delta'$ but $\geq \eta(\delta', m_2)$ for some $\delta' \in \Delta$.

The problem cannot arise from x_δ. That is, we will not have $\eta(\delta', m_2) \leq \delta < \delta'$. This is because m_2 was chosen so that (among other things), when $\delta, \delta' \in \Delta$ and $\delta < \delta'$, then $\delta < \eta(\delta', m_2)$.

So the problem can only be that $\eta(\delta', m_2) \leq \eta(\delta, k) < \delta'$. Here we cannot have $\delta = \delta'$ because $\eta(\delta, n)$ is a strictly increasing function of n and $k < m_2$. Nor can we have $\delta < \delta'$, for then we would have $\eta(\delta, k) < \delta < \eta(\delta', m_2)$ by our choice of m_2. So we must have $\delta' < \delta$.

Unfortunately, this situation cannot be excluded, so one further modification of our provisional expressions is needed. We raise the subscript from m_2 to an m_3 so large that, whenever $\eta(\delta, k) < \delta' < \delta$ with $k < m_2$ and $\delta, \delta' \in \Delta$, then $\eta(\delta', m_3) > \eta(\delta, k)$.

This raising from m_2 to m_3 solves the problem under consideration, but one might fear that it introduces a new problem, just like the old one but higher up. That is, the latest raising spins off new x's, so some new ordinals get used. Could they be below some $\delta' \in \Delta$ but $\geq \eta(\delta', m_3)$? Fortunately not. To see this, repeat the preceding discussion, now with m_3 in place of m_2, and notice in addition that the newly spun off $x_{\eta(\delta,k)}$ will have $m_2 \leq k < m_3$. As before, the problem can only be that $\eta(\delta', m_3) \leq \eta(\delta, k) < \delta'$ with $\delta' < \delta$. But now this is impossible, since $\delta' < \delta$ implies $\delta' < \eta(\delta, m_2) \leq \eta(\delta, k)$, thanks to our choice of m_2 and the monotonicity of η with respect to its second argument.

Rearranging the preceding argument slightly, we obtain the following additional information.

Lemma 3.3. *With notation as above, it never happens that $\delta, \delta' \in \Delta$ and $k < m_3$ and $\eta(\delta', m_3) \leq \eta(\delta, k) < \delta'$.*

Proof. Suppose we had δ, δ', and k violating the lemma. We consider several cases.

If $\delta = \delta'$ then the suppositions $\eta(\delta', m_3) \leq \eta(\delta, k)$ and $k < m_3$ violate the monotonicity of η with respect to the second argument.

If $\delta < \delta'$, then $\eta(\delta, k) < \delta < \eta(\delta', m_3)$ (in fact even with m_2 in place of m_3), contrary to the supposition.

If $\delta' < \delta$ and $k < m_2$ then our choice of m_3 ensures that $\eta(\delta', m_3) > \eta(\delta, k)$, contrary to the supposition.

Finally, if $\delta' < \delta$ and $k \geq m_2$ then $\delta' < \eta(\delta, m_2) \leq \eta(\delta, k)$, again contrary to the supposition. □

What we have achieved by all this raising of subscripts can be summarized as follows, where Δ and "used" refer to the final version of our expressions. (Actually, the raising process doesn't change Δ, but it generally changes what is used.) We have an expression for each element of F. There is a fixed integer m (previously called m_3) such that the only y's occurring in any of these expressions are $y_{\delta,m}$ for $\delta \in \Delta$. If $\delta \in \Delta$ and α is used and $\alpha < \delta$, then $\alpha < \eta(\delta, m)$. Furthermore, by the lemma, if $\delta, \delta' \in \Delta$ and $k < m$ and $\eta(\delta, k) < \delta'$ then $\eta(\delta, k) < \eta(\delta', m)$.

These expressions for the members of F will remain fixed from now on. Thus, the meanings of Δ and "used" will also remain unchanged. Also, m will no longer change.

Let M be the set of

- all the x's and y's occurring in the (final) expressions for elements of F,
- x_δ for all $\delta \in \Delta$, and
- $x_{\eta(\delta,k)}$ for all $\delta \in \Delta$ and all $k < m$.

Clearly, M is a finite subset of G and the subgroup $\langle M \rangle$ that it generates includes F. To finish verifying Pontryagin's criterion, we must show that $\langle M \rangle$ is pure in G.

We point out for future reference that the only y's in M are $y_{\delta,m}$ for the one fixed m and for $\delta \in \Delta$.

Suppose, toward a contradiction, that $\langle M \rangle$ is not pure, so there exist an integer $r \geq 2$ and an element $g \in G$ such that $rg \in \langle M \rangle$ but $g \notin \langle M \rangle$. Choose an expression $\hat{g}$ for g in which (by raising subscripts if necessary) no two y's occur with the same first subscript δ. In fact, arrange (by further raising) that the second subscript on all y's in $\hat{g}$ is the same n, independent of δ. Also choose an expression $\hat{d}$ for rg where $\hat{d}$ is a linear combination of elements of M. We may suppose that $\hat{d}$ is minimal in the sense that the number of elements of M occurring in $\hat{d}$ is as small as possible, for any r, g, and $\hat{d}$ as above.

Consider any $y_{\delta,n}$ that occurs in $\hat{g}$. According to Lemma 3.1, we must have $\delta \in \Delta$, because the equation $r\hat{g} = \hat{d}$ is a linear combination of defining relations.

If $n \leq m$, then we can raise the subscript n to m in $\hat{g}$, obtaining a new expression $\hat{g}'$ for the same element g. Since the equation $r\hat{g}' = \hat{d}$ is a linear combination of defining relations and since it no longer contains $y_{\delta,k}$ for any $k \neq m$ (and the same δ), we can apply Lemma 3.1 again to conclude that $y_{\delta,m}$ has the same coefficient in $r\hat{g}'$ and in $\hat{d}$. So, if we delete the terms involving $y_{\delta,m}$ from both $\hat{g}'$ and $\hat{d}$, we get another counterexample to purity with fewer elements of M occurring in $\hat{d}$. This contradicts the minimality of $\hat{d}$.

We therefore have $n > m$. Now consider what happens in $\hat{d}$ if we raise the subscripts of all the $y_{\delta,m}$ terms to n. Call the resulting expression $\hat{d}'$. (Note that $\hat{d}'$ will no longer be a combination of the generators listed for M.) The same argument as in the preceding paragraph shows that each $y_{\delta,n}$ has the same coefficient in $r\hat{g}$ and $\hat{d}'$. Therefore, if we remove all the y terms from both $\hat{g}$ and $\hat{d}'$, obtaining $\hat{g}^-$ and $\hat{d}^-$, then $r\hat{g}^-$ and $\hat{d}^-$ denote the same element in G. But we saw earlier that the x's are linearly independent in G, so $r\hat{g}^-$ and $\hat{d}^-$ must be the same expression. In particular, all the coefficients in $\hat{d}^-$ must be divisible by r. These are the same as the coefficients of the x terms in $\hat{d}'$.

Let δ be the largest ordinal such that $y_{\delta,m}$ occurred in $\hat{d}$. Let c be the coefficient of $y_{\delta,m}$ in $\hat{d}$.

When we raised the subscript of $y_{\delta,m}$ from m to n in going from $\hat{d}$ to $\hat{d}'$, the first step spun off (a multiple of x_δ and) $cx_{\eta(\delta,m)}$. The subscript $\eta(\delta,m)$ here is larger than all the other elements $\delta' \in \Delta$ that occur as subscripts of y's in $\hat{d}$, because of our choice of δ as largest and our choice of m. It is also, by choice of m, not among the α's for which $x_\alpha \in M$. As a result, no other occurrences of $x_{\eta(\delta,m)}$ were present in $\hat{d}$ or arose in the raising process leading to $\hat{d}'$. (Raising for smaller δ' spun off only x's whose subscripts are ordinals smaller than $\delta' < \eta(\delta,m)$, and later steps in the raising for δ spun off only x's with subscripts $> \eta(\delta,m)$.) This means that the coefficient of $x_{\eta(\delta,m)}$ in $\hat{d}'$ is c. Since we already showed that all coefficients of x's in $\hat{d}'$ are divisible by r, we conclude that r divides c.

Now we can delete the term $cy_{\delta,m}$ from $\hat{d}$ and subtract $\frac{c}{r}y_{\delta,m}$ from g to get a violation of purity with fewer terms in its $\hat{d}$. That contradicts our choice of $\hat{d}$ as minimal, and this contradiction completes the proof that $\langle M \rangle$ is pure in G. By Pontryagin's criterion, G is $\aleph_1$-free.

3.4 G is $\aleph_1$-separable

A group is κ-separable if every subset of size $< \kappa$ is included in a free direct summand of size $< \kappa$ (see [4, Section IV.2]). So we must prove in this subsection that every countable subset of G is included in a countable free direct summand of G. We begin by defining the natural filtration of G.

Definition 3.4. For any countable ordinal ν, let G_ν be the subgroup of G generated by the elements x_α for $\alpha < \nu$ and the elements $y_{\delta,n}$ for $\delta \in S \cap \nu$ and $n \in \omega$. (In writing $S \cap \nu$, we use the usual identification of an ordinal with the set of all smaller ordinals.)

Clearly, $G_\lambda = \bigcup_{\nu < \lambda} G_\nu$ for limit ordinals λ, the sequence $\langle G_\nu : \nu < \aleph_1 \rangle$ is increasing, and it covers G, so we have a filtration. Because G is $\aleph_1$-free and each G_ν is countable, each G_ν is free. Furthermore, every countable subset of G is included in some G_ν. So to complete the proof that G is $\aleph_1$-separable, we need only show that there are arbitrarily large $\nu < \aleph_1$ such that G_ν is a direct summand of G. In fact, we shall show that G_ν is a direct summand whenever $\nu \notin S$. Recall that the stationary S in Notation 2.1 was chosen to consist of limit ordinals, so, in particular, G_ν will be a direct summand for all successor ν.

Fix an arbitrary $\nu \notin S$. We shall show that G_ν is a direct summand of G by explicitly defining a projection homomorphism $p : G \to G_\nu$ that is the identity on G_ν. For this purpose, it suffices to define p on the generators x_α and $y_{\delta,n}$ of G and to show that the defining relations of G are preserved.

Of course, we define $p(x_\alpha) = x_\alpha$ for all $\alpha < \nu$ and $p(y_{\delta,n}) = y_{\delta,n}$ for all $\delta \in S \cap \nu$ and all $n \in \omega$, so that p is the identity on G_ν. For $\alpha \geq \nu$, we set $p(x_\alpha) = 0$. Finally, for $\delta \in S - \nu$ and $n \in \omega$, we set

$$p(y_{\delta,n}) = \sum_{k \geq n} \frac{k!}{n!} p(x_{\eta(\delta,k)}).$$

Although the sum appears to be over infinitely many k's, only finitely many of them give non-zero terms in the sum. Indeed, since $\nu \notin S$ and $\delta \in S - \nu$, we have $\nu < \delta$; therefore, for all sufficiently large $k \in \omega$, we have $\nu < \eta(\delta, k)$ and so $p(x_{\eta(\delta,k)}) = 0$.

It remains to check that p respects the defining relations of G, i.e., that, for all $\delta \in S$ and all $n \in \omega$,

$$p(y_{\delta,n}) = (n+1)p(y_{\delta,n+1}) + q_n p(x_\delta) + p(x_{\eta(\delta,n)}).$$

If $\delta < \nu$ this is trivial, since all four applications of p do nothing. $\delta = \nu$ is impossible as $\delta \in S$ and $\nu \notin S$. So we assume from now on that $\delta > \nu$. In this case, the term $q_n p(x_\delta)$ vanishes and what we must check is, in view of the definition of p,

$$\sum_{k \geq n} \frac{k!}{n!} p(x_{\eta(\delta,k)}) = (n+1) \sum_{k \geq n+1} \frac{k!}{(n+1)!} p(x_{\eta(\delta,k)}) + p(x_{\eta(\delta,n)}).$$

But this equation is obvious, and so the proof is complete.

3.5 G is not free

Using the filtration from the preceding subsection, we can easily show that G is not free because its Gamma invariant, $\Gamma(G)$, is at least (the equivalence class in $\mathcal{P}(\aleph_1)/NS$ of) S. (See [4, Section IV.1] for Gamma invariants and their connection with freeness.) Indeed, for any $\delta \in S$, the quotient group $G_{\delta+1}/G_\delta$ is generated by x_δ and the $y_{\delta,n}$ for $n \in \omega$, subject to the relations

$$y_{\delta,n} = (n+1)y_{\delta,n+1} + q_n x_\delta,$$

because the remaining term in the defining relation for G, namely $x_{\eta(\delta,n)}$, is zero in the quotient. But this presentation of $G_{\delta+1}/G_\delta$ is, except for the names of the generators, identical with the presentation of E in Lemma 2.4. Since E isn't free, G/G_δ isn't $\aleph_1$-free, and so $\delta \in \Gamma(G)$.

Although the preceding completes the verification that G isn't free, we point out that $\Gamma(G)$ is exactly (the equivalence class of) S. Indeed, we showed in the preceding subsection that, when $\nu \notin S$, then G_ν is a direct summand of G. Thus, the quotient G/G_ν is isomorphic to a subgroup of G and is therefore $\aleph_1$-free.

3.6 Subgroups of G disjoint from B are free

Suppose, toward a contradiction, that H is a non-free subgroup of G disjoint from B. So $\Gamma(H) \neq 0$. The Gamma invariant here can be computed using any filtration of H; we choose the one induced by the filtration of G already introduced. So we set $H_\nu = G_\nu \cap H$ and conclude that the set

$$A = \{\nu < \aleph_1 : H/H_\nu \text{ is not } \aleph_1\text{-free}\}$$
$$= \{\nu < \aleph_1 : \text{For some } \mu > \nu, \ H_\mu/H_\nu \text{ is not free}\}$$

must be stationary.

Thanks to our choice of the filtration $\langle H_\nu \rangle$, we have, for all $\nu < \mu < \aleph_1$,

$$\frac{H_\mu}{H_\nu} = \frac{H_\mu}{H_\mu \cap G_\nu} \cong \frac{H_\mu + G_\nu}{G_\nu} \subseteq \frac{G_\mu}{G_\nu},$$

the isomorphism being induced by the inclusion map of H_μ into $H_\mu + G_\nu$. We already saw that, when $\nu \notin S$, the groups G_μ/G_ν are free; therefore, so are the groups H_μ/H_ν. Thus, $A \subseteq S$.

Temporarily fix some $\nu \in A$. For any $\mu > \nu$, we have an exact sequence

$$0 \to \frac{H_{\nu+1}}{H_\nu} \to \frac{H_\mu}{H_\nu} \to \frac{H_\mu}{H_{\nu+1}} \to 0.$$

Since $\nu \in A$, the middle group here is not free for certain μ. The group on the right, $H_\mu/H_{\nu+1}$, on the other hand, is free because $\nu + 1 \notin S$. (Recall that S consists of limit ordinals.) So the exact sequence splits and therefore the group on the left, $H_{\nu+1}/H_\nu$, is not free.

Since $\nu \in S$, we know, from a calculation in the preceding subsection, that $G_{\nu+1}/G_\nu$ is isomorphic to E, and we saw above that $H_{\nu+1}/H_\nu$ is isomorphic to a subgroup of this (via the map induced by the inclusion of $H_{\nu+1}$ into $G_{\nu+1}$). Since all rank-1 subgroups of E are free but $H_{\nu+1}/H_\nu$ is not free, $H_{\nu+1}/H_\nu$ must have the same rank 2 as the whole group $G_{\nu+1}/G_\nu$. So the purification of $H_{\nu+1}/H_\nu$ in $G_{\nu+1}/G_\nu$ is all of $G_{\nu+1}/G_\nu$.

In particular, this purification must contain the coset of the element $x_\nu \in G_{\nu+1}$. That is, there must exist an integer $n \neq 0$ and an element $g \in G_\nu$ such that $nx_\nu - g \in H_{\nu+1}$.

Now un-fix ν. Of course the n and g obtained above can depend on ν, so we write them from now on with subscripts ν. Thus we have, for all $\nu \in A$, some $n_\nu \in \mathbb{Z} - \{0\}$ and some $g_\nu \in G_\nu$ such that

$$n_\nu x_\nu - g_\nu \in H_{\nu+1}.$$

Because A is stationary and all values of n_ν lie in a countable set, there is a stationary $A' \subseteq A$ such that n_ν has the same value n for all $\nu \in A'$. Furthermore, by Fodor's theorem, there is a stationary set $A'' \subseteq A'$ such that g_ν has the same value g for all $\nu \in A''$. (In more detail: For each $\nu \in A' \subseteq S$, we know that ν is a limit ordinal, so $G_\nu = \bigcup_{\alpha < \nu} G_\alpha$. Thus, $g_\nu \in G_{r(\nu)}$ for some $r(\nu) < \nu$. This r is a regressive function on A', so by Fodor's theorem it is constant, say with value ρ, on a stationary subset. For ν in this stationary set, g_ν has values in the countable set G_ρ and is therefore constant on a smaller stationary subset A''.)

Consider any two distinct elements ν and ξ of A''. Since $n_\nu = n_\xi = n$ and $g_\nu = g_\xi = g$, we have that H contains both $nx_\nu - g$ and $nx_\xi - g$. So it contains their difference $n(x_\nu - x_\xi)$. Since $n \neq 0$ and $\nu \neq \xi$, this contradicts the assumption that H is disjoint from the subgroup B generated by all the x_α's.

References

[1] K. Benabdallah and J. M. Irwin, *An application of B-high subgroups of abelian p-groups*, Journal of Algebra 34 (1975) pp. 213–216.

[2] A. Blass and J. M. Irwin, *Basic subgroups and a freeness criterion for torsion-free abelian groups*. Abelian Groups and Modules, International Conference in Dublin, August 10–14, 1998, ed. P. Eklof and R. Göbel, pp. 247–255. Birkhäuser-Verlag, 1999.

[3] M. Dugas and J. M. Irwin, *On basic subgroups of $\prod \mathbb{Z}$*, Communications in Algebra 19 (1991) pp. 2907–2921.

[4] P. C. Eklof and A. H. Mekler, *Almost Free Modules: Set-theoretic Methods*, North-Holland Mathematical Library 46. North-Holland, 1990.

[5] L. Fuchs and F. Loonstra, *On the cancellation of modules in direct sums over Dedekind domains*, Indagationes Mathematicae 33 (also Koninklijke Nederlandse Akademie van Wetenschappen. Proceedings, Series A, 74) (1971) pp. 163–169.

Author information

Andreas Blass, Mathematics Department, University of Michigan, Ann Arbor, MI 48109–1043, U.S.A.
E-mail: ablass@umich.edu

Saharon Shelah, Einstein Institute of Mathematics, Edmond J. Safra Campus, The Hebrew University of Jerusalem, Jerusalem 91904, Israel, and Mathematics Department, Rutgers University, New Brunswick, NJ 08854, U.S.A.
E-mail: shelah@math.huji.ac.il

An extension of Butler's theorem on endomorphism rings

Rüdiger Göbel and Jasmin Matz

Abstract. This note is based on three papers by A. L. S. Corner, H. Zassenhaus and M. C. R. Butler, respectively ([4, 8, 3]). Corner showed in his celebrated paper [4] that any countable ring R with torsion-free reduced additive group is endomorphism ring of some torsion-free abelian group M, thus $R \simeq \mathrm{End}_{\mathbb{Z}} M$. He also showed that the rank of M can be chosen to be $2\,\mathrm{rk}\,R$, and his examples in [4] demonstrate that this bound is actually minimal. A few years later Zassenhaus [8] improved the rank restriction by imposing that the additive group of R is free. Under this restriction we can choose $\mathrm{rk}\,M = \mathrm{rk}\,R$. A year later Butler replaced the freeness condition on R by local freeness, thus assuming that the additive group of R is only locally free. The minimal rank condition can also be expressed as $R \subseteq M \subseteq \mathbb{Q}R$, so a reference to finiteness of ranks can be avoided. Using $R \subseteq M \subseteq \mathbb{Q}R$ as the main point in the definition of "Zassenhaus-rings", it was recently shown in Dugas, Göbel [5], using counting arguments for partial endomorphisms, that all algebraic rings satisfying the conditions of Zassenhaus' theorem, but with possibly infinite countable rank, are Zassenhaus rings. It remained an open question if this is also true under Butler's weaker restrictions. In this note we will show that this is the case indeed; see Theorem 1.2.

Key words. Endomorphism rings, Butler–Zassenhaus' theorem.

AMS classification. Primary: 20K15, 20K20, 20K30. Secondary: 13F20, 16Gxx, 16S50.

1 Introduction

If σ is an endomorphism of a torsion-free abelian group M of finite rank n, then σ extends uniquely to an endomorphism $\hat{\sigma}$ of $\mathbb{Q}M$ ($= \mathbb{Q} \otimes_{\mathbb{Z}} M$ which is a vector space of dimension n over $\mathbb{Q}$). Thus a well-known argument from linear algebra applies. Since $\dim_{\mathbb{Q}} \mathrm{End}\,\mathbb{Q}M = n^2$ is also finite, there is a polynomial $0 \neq f(x) \in \mathbb{Q}[x]$ with $f(\hat{\sigma}) = 0$. Thus $\hat{\sigma}$ is algebraic over $\mathbb{Q}$. Multiplying f with a suitable natural number, we also find a polynomial $0 \neq F(x) \in \mathbb{Z}[x]$ with $F(\hat{\sigma}) = 0$; in particular $F(\sigma) = 0$. Thus σ is also algebraic over $\mathbb{Z}$.

Recall that an element r in a ring extension of R is *algebraic over* R if r is a root of a non-trivial polynomial from $R[x]$. We will also say that an element of a ring R is algebraic, if it is algebraic over the subring $1\mathbb{Z}$ generated by 1. If all elements of R are algebraic, then we say for short that R is algebraic. This shows the following simple

Observation 1. *The endomorphism ring of a torsion-free abelian group of finite rank is algebraic.*

If M is an arbitrary torsion-free abelian group and $\mathbb{Z}_{(p)}$ denotes the localization of $\mathbb{Z}$ at a prime p, then (as usual, [6]) we write $M_{(p)} = \mathbb{Z}_{(p)} \otimes_{\mathbb{Z}} M$ for the localization of M at p. The group M is called locally free, if $M_{(p)}$ is free as a $\mathbb{Z}_{(p)}$-module for all primes p.

If R is a ring with 1, then R^+ denotes its additive group and if $a \in R$, then we will write a_r for the endomorphism $a_r \in \operatorname{End} R^+$ which is scalar multiplication by a from the right, i.e. $a_r : R \longrightarrow R\ (x \mapsto xa)$. Moreover, let $R_r = \{a_r \mid a \in R\}$ which is a subring of $\operatorname{End} R^+$. We also adopt a definition from [5] about rings.

Definition 1.1. A ring R is a *Zassenhaus ring* if there is an R-module M with $R \subseteq M \subseteq \mathbb{Q}R$ (as R-modules) and $\operatorname{End} M = R_r$. We call M the *Zassenhaus module* for R.

The first condition on R allows us to circumvent the notion of rank, which now extends also naturally to infinite ranks.

Using ideas from [3, 8, 5] we want to show the following

Theorem 1.2. *Let R be any torsion-free ring with 1 of at most countable rank. If R is algebraic and R^+ is locally free, then R is a Zassenhaus ring. Moreover, the Zassenhaus module M is also a locally free R-module.*

Assuming that R^+ is free, we obtain a main result from [5]. If R has finite rank, then R is algebraic by Observation 1. Thus Theorem 1.2 applies without this assumption. We obtain as a

Corollary 1.3. *If R has finite rank and R^+ is locally free, then R is a Zassenhaus ring. Moreover, M above is also a locally free R-module.*

This is Butler's theorem [3]. If we also assume that R^+ is free, then we arrive at Zassenhaus' theorem from [8], see also [7, pp. 407–410]. There are many obvious examples showing that it is necessary to assume that R is algebraic, see also [1]. Examples due to Corner also show that the restriction to particular additive groups R^+ is also necessary for Corollary 1.3. He exhibited in [4] rings R of rank n such that any torsion-free abelian group M with $\operatorname{End} M \cong R$ has rank $\operatorname{rk} M \geq 2n$, where $2n$ is the minimal possible choice.

The main idea of the proof of Theorem 1.2 is a counting argument which allows us to control endomorphisms of R-modules of countable rank. Moreover we need a result which ensures that unwanted endomorphisms (not belonging to R_r) can be removed. This is the following version of a theorem by M. C. R. Butler [3] which will follow from Hensel's lemma.

2 Some preliminaries

Lemma 2.1. *Let $f \in \mathbb{Q}[x]\backslash\mathbb{Q}$. Then there are infinitely many primes such that f has a root in the ring of p-adic integers $\mathbb{Z}_p$.*

Proof. Firstly, assume $f \in \mathbb{Z}[x]\backslash\mathbb{Z}$ and f is monic. By Hensel's lemma it suffices to show that $f(x) \equiv 0 \bmod p$ is soluble for infinitely many primes. Since f is not equal to a constant, there is some $z \in \mathbb{Z}$ with $f(z) \notin \{1, -1\}$. Thus there is a prime p_1 with $p_1 | f(z)$, i.e. $f(x) \equiv 0 \bmod p_1$ is soluble.

Suppose the congruence is soluble for $p_1, \ldots, p_n$, $n \geq 1$, with solutions $\alpha_1, \ldots, \alpha_n$. Put
$$\alpha := \sum_{k=1}^{n} \Big(\prod_{\substack{i=1 \\ i \neq k}}^{n} p_i\Big)\alpha_k.$$

Then $f(\alpha) \equiv 0 \mod p_j$ for $j = 1, \ldots, n$ and therefore
$$f(\alpha) \equiv 0 \mod p_1 \cdot \ldots \cdot p_n.$$

If $f(\alpha) = 0$, we add some multiple of $p_1 \cdot \ldots \cdot p_n$ to α such that this new number is not a root of f. Such a choice is always possible since f has only finitely many roots. We call this new number α again. Choose some $\beta \in \mathbb{Z}$ such that $\beta \equiv \alpha \mod (f(\alpha))^2$ and $|f(\beta)| > |f(\alpha)|$. Since $f \in \mathbb{Z}[x]$, there is some $c \in \mathbb{Z}$ with $f(\beta) = f(\alpha) + cf(\alpha)^2 = f(\alpha)(1 + cf(\alpha))$ and $|1 + cf(\alpha)| > 1$. Since $p_1 \cdot \ldots \cdot p_n | f(\alpha)$, we get $\gcd(1 + cf(\alpha), p_1 \cdot \ldots \cdot p_n) = 1$. Therefore there exists a prime $p_{n+1} \notin \{p_1, \ldots, p_n\}$ with $p_{n+1} | f(\beta)$. Thus $f(x) \equiv 0 \mod p_{n+1}$ is soluble. Inductively we can find infinitely many primes p such that the congruence $f(x) \equiv 0 \mod p$ is soluble.

Now let $f \in \mathbb{Q}[x] \setminus \mathbb{Q}$ be arbitrary. There is a $q \in \mathbb{Z} \setminus \{0\}$ with $\tilde{f} := qf \in \mathbb{Z}[x] \setminus \mathbb{Z}$. Let $\tilde{f} = a_n x^n + \ldots + a_0$ with $a_n \neq 0$. We consider the (finite) set P of all primes dividing $a_n q$. Hensel's Lemma can be used, so that we still obtain an infinite set of primes not in P for which the congruence $\tilde{f}(x) \equiv 0 \mod p$ is soluble. Since all prime divisors of q are in P, it follows that $f(x) \equiv 0 \mod p$ is soluble for infinitely many primes p. □

The proof shows that $f(x) \equiv 0 \mod p$ is soluble even in $\mathbb{Z}$ for infinitely many primes.

In the following G denotes a free abelian group, $0 \neq e \in G$ some fixed element, $\varphi \in \text{End } G$ an algebraic endomorphism, $V = \sum \mathbb{Q} e \varphi^k$, and $V_p := V \cap G_{(p)}$. Since φ is algebraic, V is a finite dimensional vector space, $d := \dim V$, say. Put $W_p := \sum_{i=0}^{d-1} \mathbb{Z}_{(p)} e \varphi^i \subseteq G_{(p)}$. Then there is a minimal $n \in \mathbb{N}$ with $p^n V_p \subseteq W_p$.

Lemma 2.2. *Let $a, c \in \mathbb{Z}$. Suppose that c is not an eigenvalue of φ, and that $ae \in G_{(p)}\tau$ for $\tau := c_r - \varphi$. Then $\det(\tau \restriction W_p)$ divides $p^n a$ in $\mathbb{Z}_{(p)}$.*

Proof. Since φ is algebraic, so is τ and we have the relation $h(\tau) = 0$ for some non trivial polynomial $h \in \mathbb{Q}[x]$ of minimal degree. We may assume h to be monic and $h(x) \in \mathbb{Z}_{(p)}$ since τ is an endomorphism of $G_{(p)}$, which is a free $\mathbb{Z}_{(p)}$-module. Since h is of minimal degree and c is not an eigenvalue of φ, $h(0) \neq 0$, and we can write $h(x) = a_m x^m + \ldots + a_0$ with $m \leq d$, $a_m = 1$, and $a_0 \neq 0$. Clearly, τ is invertible as an element of $\text{End } V$. Moreover, it can be written as $a_0 \tau^{-1} = -\sum_{i=1}^{m} a_i \tau^{i-1}$. By our assumption $ae \in G_{(p)}\tau$, and therefore $ae \in V\tau$ and $ae \in V_p \tau$. So we get

$$V_p \ni ae\tau^{-1} = -\sum_{i=1}^{m} \frac{a_i a}{a_0} e \tau^{i-1}. \qquad (*)$$

By definition of our modules we have $V_p \subseteq \frac{1}{p^n} W_p = \frac{1}{p^n} \sum i = 0^{d-1} \mathbb{Z}_{(p)} e \phi^i$. By substituting $c_r - \phi$ for τ in $(*)$ we see that the coefficient of the highest power of ϕ is

$\frac{aa_m}{a_0} = \frac{a}{a_0}$. Since $\{e, \ldots, e\phi^{d-1}\}$ is linearly independent, we must have $\frac{a}{a_0} \in \frac{1}{p^n}\mathbb{Z}_{(p)}$, and therefore $\frac{ap^n}{a_0} \in \mathbb{Z}_{(p)}$. The determinant of τ restricted to W_p is exactly a_0, and thus we have shown our claim. □

Lemma 2.3. *Let R be a ring such that all elements are algebraic over $\mathbb{Q}$. Moreover, suppose that $\varphi \in \mathrm{End}_{\mathbb{Z}} R^+$ is algebraic and choose $0 \neq e \in R$, and $\Pi \subseteq \mathbb{N}$ a finite set of primes. Then there is a prime number $p \notin \Pi$ and some $c \in \mathbb{Z}$ such that $c_r - \varphi \in \mathrm{End}_{\mathbb{Q}} \mathbb{Q}R$ is a monomorphism and $e \notin R_{(p)}(c_r - \varphi)$.*

Proof. Since φ is algebraic, there is some monic polynomial $g(x) \in \mathbb{Q}[x]\setminus\{0\}$ with $g(\varphi) = 0$. Therefore $V := \sum \mathbb{Q}(e\varphi)$ is a vector space of finite dimension $m := \dim_{\mathbb{Q}} V \le \deg(g)$, and $\{1, e\varphi, \ldots, e\varphi^{m-1}\}$ is a basis of V over $\mathbb{Q}$. Let $f(x) := x^m + a_{m-1}x^{m-1} + \ldots + a_0 \in \mathbb{Q}[x]$ be the characteristic polynomial of $\varphi \in \mathrm{End}\, V$, which is at the same time the minimal polynomial μ of φ (note, if $\deg(\mu) < m$, then $\dim_{\mathbb{Q}} V < m$).

Put $f_k(x) := x^k + a_{m-1}x^{k-1} + \ldots + a_{m-k}$ $k = 1, \ldots, m$ and $f_0(x) := 1$. Then (using $a_m = 1$)

$$\left(e \sum_{i=0}^{m-1} f_i(x)\varphi^{m-i-1}\right)(x - \varphi) = e\left(\sum_{i=0}^{m-1} x f_i(x)\varphi^{m-i-1} - \sum_{i=0}^{m-1} f_i(x)\varphi^{m-i}\right)$$

$$= e\left(\sum_{i=0}^{m-1} (f_{i+1}(x) - a_{m-i-1})\varphi^{m-i-1} - \sum_{i=0}^{m-1} f_i(x)\varphi^{m-i}\right)$$

$$= e\left(\sum_{i=1}^{m} f_i(x)\varphi^{m-i} - \sum_{i=1}^{m} a_{m-i}\varphi^{m-i} - \sum_{i=0}^{m-1} f_i(x)\varphi^{m-i}\right)$$

$$= e\left(f_m(x) - f_0(x)\varphi^m - \sum_{i=1}^{m} a_{m-i}\varphi^{m-i}\right)$$

$$= e\left(f(x) - \varphi^m - \sum_{i=1}^{m} a_{m-i}\varphi^{m-i}\right)$$

$$= e\left(f(x) - \sum_{i=0}^{m} a_{m-i}\varphi^{m-i}\right) = e(f(x) - f(\varphi)) = ef(x).$$

By Lemma 2.1 there exists a prime number $p \notin \Pi$ such that f has a root in the ring of p-adic integers, i.e. $f(\sum_{i \in \omega} c_i p^i) = 0$ for suitable $c_i \in \{0, \ldots, p-1\}$. By our assumption $R_{(p)}$ is a free $\mathbb{Z}_{(p)}$-ring, thus, using $\dim_{\mathbb{Q}} V = m$, $V_p := R_{(p)} \cap V$ is a free $\mathbb{Z}_{(p)}$-module of rank m. Let $W_p := \bigoplus_{i=0}^{m-1} \mathbb{Z}_{(p)} e\varphi^i \subseteq V_p$. The elements $e, e\varphi, \ldots, e\varphi^{m-1}$ are linearly independent over $\mathbb{Z}_{(p)}$, thus W_p is also a free $\mathbb{Z}_{(p)}$-module of the same rank as V_p. Let $n \in \mathbb{N}$ be such that $p^n V_p \subseteq W_p$. For all $k > n$ the finite sum $\gamma_k := \sum_{0 \le j \le k} c_j p^j$ is a solution for $f(x) \equiv 0 \bmod p^{n+1}$. Since R is torsion-free and φ is algebraic, φ has only finitely many eigenvalues. Choose some $c \in \{\gamma_k | k > n\}$ which is not an eigenvalue of φ. Thus we have that $f(c) \equiv 0 \bmod p^{n+1}$, and that $\sigma := c_r - \varphi \in \mathrm{End}_{\mathbb{Q}} \mathbb{Q}R$

is a monomorphism, but $f(c) \neq 0$, because otherwise c was an eigenvalue of φ. It follows that $\frac{p^n}{f(c)} \notin \mathbb{Z}_{(p)}$. Suppose $e\sigma^{-1} \in R_{(p)}$. Then by Lemma 2.2 we have $f(c) = \det(\sigma \upharpoonright W_p) \mid p^n$ in $\mathbb{Z}_{(p)}$ contradicting $\frac{p^n}{f(c)} \notin \mathbb{Z}_{(p)}$. It follows $e \notin R_{(p)}\sigma$. □

Proposition 2.4. *Let Π be a finite set of primes and R a ring with torsion-free additive group of at most countable rank. If $0 \neq \sigma \in \mathrm{End}_\mathbb{Q} \mathbb{Q}R$ with $1\sigma = 0$, then there is a prime $p \notin \Pi$ and a free $\mathbb{Z}_{(p)}$-module M_p such that the following holds:*

(i) $R \subseteq M_p \subseteq \mathbb{Q}R$.

(ii) $M_p\sigma \not\subseteq M_p$.

(iii) *If $a \in \mathbb{Q}R$, then $a_r \in \mathrm{End}\, M_p$ if and only if $a \in R_{(p)}$.*

Proof. Since $\sigma \neq 0$, there is some $\tilde{t} \in \mathbb{Q}R$ such that $0 \neq \tilde{t}\sigma \in \mathbb{Q}R$; and since R is torsion-free, we also find $l \in \mathbb{Z}$ with $t := l\tilde{t} \in R$ and $0 \neq e := t\sigma = (l\tilde{t})\sigma \in R$. Let $\varphi : \mathbb{Q}R \longrightarrow \mathbb{Q}R (x \longmapsto tx)$. All elements of R are algebraic (by our assumption), so φ is algebraic as well. And $R\varphi \subseteq R$ also follows from $t \in R$. By Lemma 2.3 there is a prime $p \notin \Pi$, and some $c \in \mathbb{Z}$ such that $\tau := c_r - \varphi \in \mathrm{End}_\mathbb{Q} \mathbb{Q}R$ is a monomorphism and $e \notin R_{(p)}\tau$.

Since c and φ are algebraic, so is τ. Let $g(x) := x^n + b_{n-1}x^{n-1} + \ldots + b_0 \in \mathbb{Q}[x]$ be the minimal polynomial of τ, hence $g(\tau) = 0$. We have $b_0 \neq 0$, because τ is injective. Therefore we get $1 = -(\sum_{k=1}^n \frac{b_k}{b_0} \tau^{k-1})\tau$ (with $b_n := 1$). Let $m \in \mathbb{Z}$ be such that $p^m \frac{b_k}{b_0} \in \mathbb{Z}_{(p)}$ for $k = 1, \ldots, n$. Then

$$p^m R_{(p)} = R_{(p)} \Big(\sum_{k=1}^n p^m \frac{b_k}{b_0} \tau^{k-1} \Big) \tau \subseteq R_{(p)}\tau \subseteq R_{(p)}.$$

Put $M_p := p^{-m}R_{(p)}\tau$. Then we have $R \subseteq R_{(p)} \subseteq M_p \subseteq \mathbb{Q}R$, and (i) holds. Since $p^{-m}t \in R \subseteq M_p$ and $(p^{-m}t)\sigma = p^{-m}e \notin p^{-m}R_{(p)}\tau = M_p$, (ii) holds too. To show that (iii) is also true, let $a \in \mathbb{Q}R$ with $a_r \in \mathrm{End}\, M_p$. Then $a = 1a_r \in M_p$. Since all elements of R are algebraic over $\mathbb{Q}$, a is also algebraic over $\mathbb{Q}$, i.e. there are $\alpha_N, \ldots, \alpha_0 \in \mathbb{Z}$ such that $\alpha_N a^N = -(\alpha_{N-1}a^{N-1} + \ldots + \alpha_0)$. Therefore $a^N \in \frac{1}{p^K}\sum_{j=0}^{N-1} a^j \mathbb{Z}_{(p)}$ for some $K \geq 0$. Let $\tilde{F} := \sum_{j=0}^{N-1} a^j \mathbb{Z}_{(p)}$ and F be the purification of $\tilde{F}$ in M_p. Then F is a free $\mathbb{Z}_{(p)}$-module of finite rank, and by purity we have $p^K F = F \cap p^K M_p$. Thus $aF \subseteq p^{-K}F \cap M_p = F$, i.e. $a^n \in F$ for all $n \in \mathbb{N}$. But since F is a free $\mathbb{Z}_{(p)}$-module of finite rank, $1 \in F$ is divisible only a finite number of times by p. This implies that $a \in R_{(p)}$ and (iii) holds as well. □

3 Proof of Theorem 1.2

Now we are ready to prove our main result:

Theorem 3.1. *Let R be any torsion-free ring with 1 of at most countable rank. If R is algebraic and R^+ is locally free, then R is a Zassenhaus ring. Moreover, the Zassenhaus module M is also a locally free R-module.*

Proof. Let

$$E := \{\varphi \in \text{Hom}(V, \mathbb{Q}R) \mid 1 \in V \subseteq \mathbb{Q}R \text{ a subspace of finite dimension}\}$$

and $E_0 := \{\varphi \in E \mid 1\varphi = 0\}$.

Since $\text{rk}(R^+) \leq \aleph_0$, E_0 is countable, and we can enumerate the set $E_0 \setminus \{0\} = \{\Phi_1, \Phi_2, \ldots\}$. Let V_n denote the domain of Φ_n. Since V_n is a subspace of the vector space $\mathbb{Q}R$, the homomorphism Φ_n can be extended to $\mathbb{Q}R$ by the zero map on a fixed complement. Thus we can view Φ_n as an endomorphism of $\mathbb{Q}R$.

We now construct inductively a sequence of primes $(p_n)_{n \in \omega}$, and a sequence of free $\mathbb{Z}_{(p_n)}$-modules M_{p_n} with the following properties (for all $n \in \mathbb{N}$):

(i) $p_k \neq p_n$ for all $k \neq n$,

(ii) $R \subseteq M_{p_n} \subseteq \mathbb{Q}R$,

(iii) $M_{p_n} \Phi_n \not\subseteq M_{p_n}$,

(iv) for all $a \in \mathbb{Q}R$ we have $a_r \in \text{End}\, M_{p_n}$ iff $a \in R_{(p_n)}$.

By Proposition 2.4 there exist such sequences of primes (p_n) and free $\mathbb{Z}_{(p_n)}$-modules (M_{p_n}). For all primes with $p \notin \{p_n \mid n \in \mathbb{N}\}$ put $M_p := R_{(p)}$. By our assumptions about $R_{(p)}$, (ii) and (iv) also hold for such M_p. If Π denotes the set of all primes in $\mathbb{N}$, then put

$$M := \bigcap_{p \in \Pi} M_p.$$

Clearly, M is locally free with $M_{(p)} = M_p$ for all primes p, and $R \subseteq M \subseteq \mathbb{Q}R$ holds. Let $\varphi \in \text{End}\, M$, and set $\psi := \varphi - (1\varphi)_r \in \text{End}_\mathbb{Q}\, \mathbb{Q}R$ for the unique extension, so we have $1\psi = 0$. Suppose $\psi \neq 0$. Thus there is some pure subgroup $D \subseteq M$ of finite rank with $1 \in D$ and $\psi \restriction D \neq 0$. Since D is of finite rank and a pure subgroup of M there is some $k \in \mathbb{N}$ such that $kD\psi \subseteq M$. By definition of E there is some $n \in \mathbb{N}$ with $\Phi_n \restriction D = k\psi \restriction D$, and $M_{p_n} \Phi_n \not\subseteq M_{p_n}$. But by our choice of k we have $M\Phi_n = D(k\psi) \subseteq M$. Thus we get $M_{p_n} \Phi_n = M_{(p_n)} \Phi_n \subseteq M_{(p_n)} = M_{p_n}$ contradicting (iii). It follows that ψ is the zero-map, and $\varphi = (1\varphi)_r \in (\mathbb{Q}R)_r \cap \text{End}\, M = R_r$ by property (iv) ($\mathbb{Q}R_r$ and R_r denote the set of all right-multiplications by elements of $\mathbb{Q}R$ and R, respectively). Together with the properties of M already shown above, this yields $\text{End}\, M = R_r$. □

References

[1] J. Buckner, M. Dugas, *Quasi-localizations of* $\mathbb{Z}$, to appear in Israel J. Math. (2007).

[2] M. C. R. Butler, *A class of torsion-free rings of finite rank*, Proc. London Math. Soc. **15** (1965), 680–698.

[3] M. C. R. Butler, *On locally free torsion-free rings of finite rank*, J. London Math. Soc. **43** (1968), 297–300.

[4] A. L. S. Corner, *Every countable reduced torsion-free ring is an endomorphism ring*, Proc. London Math. Soc. (3) **13** (1963), 687–710.

[5] M. Dugas, R. Göbel *An Extension of Zassenhaus' Theorem on Endomorphism Rings*, Fundamenta Mathematicae, **194** (2007), 239–251.

[6] L. Fuchs, *Infinite Abelian Groups* – Vol. 1 & 2, Academic Press, New York (1970, 1973).

[7] R. Göbel, J. Trlifaj, *Approximation Theory and Endomorphism Algebras*, Walter de Gruyter, Berlin (2006).

[8] H. Zassenhaus, *Orders as endomorphism rings of modules of the same rank*, J. London Math. Soc. **42** (1967), 180–182.

Author information

Rüdiger Göbel, Fachbereich Mathematik, Universität Duisburg-Essen, 45117 Essen, Germany.
E-mail: `Ruediger.Goebel@uni-due.de`

Jasmin Matz, Fachbereich Mathematik, Universität Duisburg-Essen, 45117 Essen, Germany.
E-mail: `jasminmatz@web.de`

Locally free abelian groups of finite rank

David M. Arnold

Abstract. A torsion-free abelian group G of finite rank is called locally free if $G_{(p)}$ is a free $\mathbb{Z}_{(p)}$-module for each prime p. A locally free group G is determined, up to quasi-isomorphism, by its rank, Richman type, and an equivalence relation on a sequence of integer matrices. Given a prime p, there are correspondences from square p-adic matrices and torsion-free $\mathbb{Z}_{(p)}$-modules of finite rank to locally free groups. Included is the answer to an open question on the cancellation property for tensor products of locally free groups, as well as a variety of examples.

Key words. Locally free groups, torsion-free abelian groups of finite rank.

AMS classification. 20K15.

A torsion-free abelian group G of finite rank is called *locally free* if $G_{(p)}$ is a free $\mathbb{Z}_{(p)}$-module for each prime p, where $\mathbb{Z}_{(p)} = \{\frac{a}{b} \in \mathbb{Q} : \gcd(b,p) = 1\}$ is the localization of the integers $\mathbb{Z}$ and $G_{(p)} = \mathbb{Z}_{(p)} \otimes_{\mathbb{Z}} G$.

Locally free groups have been investigated in a variety of contexts. A duality for locally free groups, now called Warfield duality, is given in the seminal paper [19]. Warfield duality is generalized to other classes of abelian groups in [17], [7] and to modules over more general domains in [10], [12], [15]. An equivalence from the quasi-isomorphism category of locally free groups to a quasi-isomorphism category of mixed abelian groups with finite torsion-free rank is given in [20].

The diversity of locally free groups is demonstrated by the fact that if R is a ring with locally free additive group of rank n, then there is a locally free group G of rank n with $R = \mathrm{End}(G)$, [5]. This result is extended to modules over Dedekind domains in [13].

Examples of locally free groups include torsion-free images of a finite direct sum of locally free groups of rank 1, called locally free *Butler groups*. Various properties and examples of Butler groups are given in [4] and [2].

Each locally free group G of rank n can be constructed from a sequence $\Psi = (M_i)$ of $n \times n$ integer matrices and a torsion group (Theorem 2.3(a)). Moreover, G is a Butler group if and only if the matrices M_i can be chosen with only finitely many distinct rows (Theorem 2.4). Locally free groups with various properties are given in Examples 2.5, 2.7 and 2.8.

Given a prime p, there are injections **g** from $n \times n$ p-adic matrices to locally free groups of rank n and surjections **f** from locally free groups of rank n to $n \times n$ p-adic matrices with $\mathbf{fg}(M) = M$ for each matrix M (Theorem 3.1). A finite rank torsion-free abelian group G is a locally free Butler group if and only if G is quasi-isomorphic to a group in the image of some **g** (Corollary 3.2).

As a consequence of Theorem 3.1 and Corollary 3.2, given a prime p there are injections **h** from torsion-free $\mathbb{Z}_{(p)}$-modules of finite rank to locally free Butler groups (Corollary 4.1). Multi-functional generalizations **h*** of **h** yield locally free groups that are not Butler groups, as illustrated in Example 4.2. These correspondences are somewhat surprising in view of the conventional wisdom that "...locally free modules lie at the opposite end of the spectrum from quotient divisible modules.", [10], and the fact that $\mathbb{Z}_{(p)}$-modules are quotient divisible.

An uncountable class of rank-2 locally free groups is used to prove that the Borel complexity of the isomorphism problem (see [16]) for locally free groups of rank 2 is not hyperfinite, [11]. Each of these groups is of the form $\mathbf{h}^*(A)$ for some torsion-free $\mathbb{Z}_{(p)}$-module A of rank 2 (Example 4.3).

Techniques of this paper are applied to address the cancellation property for tensor products of locally free groups.

Corollary 4.4 ([10]). *Let X, G, and G' be locally free groups of finite rank and assume that X has rank 1. If $X \otimes_{\mathbb{Z}} G$ is quasi-isomorphic to $X \otimes_{\mathbb{Z}} G'$, then G is quasi-isomorphic to G'.*

The final example answers an open question of E. L. Lady, [10, p. 130].

Example 4.5. *There are locally free groups X, G, and G' of rank 2 such that $X \otimes_{\mathbb{Z}} G$ is quasi-isomorphic to $X \otimes_{\mathbb{Z}} G'$, but G is not quasi-isomorphic to G'.*

1 Preliminaries

In this section, some standard notation and terminology is given, following [9], [2], and [3]. The *rank* of a torsion-free abelian group G is the cardinality of a maximal $\mathbb{Z}$-independent subset of G.

A *type* is the equivalence class $[(h_i)]$ of a height sequence $(h_i)_{p_i \in \Pi}$, where $\Pi = \{p_1, \ldots, p_n, \ldots\}$ is the natural indexing of the set of primes, each h_i is a non-negative integer or ∞, and two height sequences (h_i) and (g_i) are *equivalent* if $h_i = g_i$ except for a finite number of i with both h_i and $g_i < \infty$. The set of all types is a lattice under the operations of meet and join.

There is a bijection from the set of isomorphism classes $[X]$ of torsion-free abelian groups X of rank 1 to the set of types given by $[X] \to \text{type}(X) = [(h_i(x))]$, where x is any fixed non-zero element of X and $h_i(x)$ is the p_i-height of x in X, [9, Theorem 85.1]. The *typeset* of a torsion-free abelian group G is the set of types of pure rank-1 subgroups of G and the *cotypeset of G* is the set of types of rank-1 torsion-free images of G.

Two abelian groups A and B are *quasi-isomorphic* if there is a homomorphism $f : A \to B$ such that both $\ker(f)$ and $\text{coker}(f) = B/f(A)$ are bounded groups, [18]. A *Richman type* is a quasi-isomorphism class $R(T) = [T]$ of a torsion group T. The *Richman type* of a torsion-free abelian group G is $RT(G) = [G/F]$, where F is a free subgroup of G with G/F a torsion group, and is independent of the choice of F, [14]. A relation $\leq$ is defined on Richman types by setting $R(T) \leq R(T')$ if T is isomorphic to a subgroup of T'.

Let $\hat{\mathbb{Z}}_{(p)}$ denote the completion of $\mathbb{Z}$ in the p-adic topology, called the *ring of p-adic integers*. Each element α of $\hat{\mathbb{Z}}_{(p)}$ can be written as the limit of a Cauchy sequence $(\alpha_1 + \alpha_2 p + \ldots + \alpha_i p^{i-1})_i$ with each $\alpha_i \in \mathbb{Z}$. The sequence (α_i) of coefficients is not unique but α can be written uniquely as the limit of a Cauchy sequence $(\alpha_1 + \alpha_2 p + \ldots + \alpha_i p^{i-1})_i$ with $0 \leq \alpha_i \leq p-1$ for each $i \geq 1$, [8, p. 18].

2 Locally free groups

The proof of the following proposition is straightforward, e.g. see [2, Theorem 1.14].

Proposition 2.1. *Let G be a torsion-free abelian group of finite rank n. The following statements are equivalent:*

(a) *G is a locally free group;*

(b) *The $\mathbb{Z}/p\mathbb{Z}$ dimension of G/pG is equal to n for each prime p;*

(c) *$RT(G) = [T]$ for some reduced subgroup T of $(\mathbb{Q}/\mathbb{Z})^n$;*

(d) *$RT(G) = [T]$ for a direct sum of finite p-groups $T \subset (\mathbb{Q}/\mathbb{Z})^n$;*

(e) *G is isomorphic to a subgroup of X^n for a locally free group X of rank 1.*

Let $\mathcal{LF}$ denote the category of locally free groups with morphism sets the usual group homomorphisms. The following proposition is easily proved from elementary properties of free $\mathbb{Z}_{(p)}$-modules and the fact that $\mathbb{Z}_{(p)}$ is a torsion-free, hence flat, abelian group. A type $\tau = \text{type}(X)$ is a *locally free type* if X is a locally free group of rank 1.

Proposition 2.2. *The category $\mathcal{LF}$ is an additive category closed under finite direct sums, pure subgroups, torsion-free quotients, and extensions. Moreover, each type in the typeset or cotypeset of a locally free group G is a locally free type.*

As a consequence of the next theorem, a locally free group G is determined, up to quasi-isomorphism, by its rank, Richman type $RT(G) = [G/F]$, and an equivalence relation on a sequence of integer matrices. The rows of these matrices correspond to preimages in G of generators of cyclic p-group summands of G/F. Henceforth, a cyclic group with p^e elements is denoted by $\mathbb{Z}(p^e)$.

Let $\Psi = (M_i)_{i \geq 1}$ be a sequence of $n \times n$ integer matrices and $T = \oplus_{i \geq 1} T_i$ a torsion group with

$$T_i = \mathbb{Z}(p_i^{e(i,1)}) \oplus \ldots \oplus \mathbb{Z}(p_i^{e(i,n)})$$

and $0 \leq e(i,1) \leq \ldots \leq e(i,n)$ for each i.

Define $G(\Psi, T)$ to be the subgroup of $\mathbb{Q}^n$ generated by $\mathbb{Z}^n$ and

$$\{(1/p_i^{e(i,j)})\text{row}_j(M_i) : 1 \leq j \leq n, i \geq 1\},$$

where $\text{row}_j(M_i) \in \mathbb{Z}^n$ denotes the jth row of M_i.

It is easily confirmed that $G(\Psi, T)$ is a locally free group of rank n with $RT(G) \leq R(T)$ and that if $R(T) = R(T')$, then $G(\Psi, T)$ is quasi-isomorphic to $G(\Psi, T')$.

In general, the Richman type of $G(\Psi, T)$ is not equal to $R(T)$. For example, let $\Psi = (M_i)$ and

$$T = \oplus_i (\mathbb{Z}(p_i^{e(i,1)}) \oplus \ldots \oplus \mathbb{Z}(p_i^{e(i,n)}))$$

such that $e(i,j) > 0$ and each element of $\text{row}_j(M_i)$ is divisible by $p_i^{e(i,j)}$ for some j and infinitely many i. As

$$(1/p_i^{e(i,j)})\text{row}_j(M_i) + \mathbb{Z}^n = 0 + \mathbb{Z}^n \in \mathbb{Z}(p_i^{e(i,j)})$$

for some j and infinitely many i with $e(i,j) > 0$, $RT(G(\Psi, T)) < R(T)$ because $G(\Psi, T)/\mathbb{Z}^n$ is isomorphic to a subgroup of T of unbounded index.

The *exponent* $\exp(T)$ of a finite p-group T is the least non-negative integer b with $p^b T = 0$. The first theorem is a mild variation of the classical Kurosh matrix invariants for torsion-free abelian groups of finite rank, [9, Theorem 93.5].

Theorem 2.3. (a) *A locally free group G of rank n is isomorphic to $G(\Psi, T)$ for some sequence $\Psi = (M_i)$ of $n \times n$ integer matrices and torsion group*

$$T = \oplus_i (\mathbb{Z}(p_i^{e(i,1)}) \oplus \ldots \oplus \mathbb{Z}(p_i^{e(i,n)}))$$

with $0 \leq e(i,1) \leq \ldots \leq e(i,n)$ for each i and $R(G(\Psi, T)) = R(T)$.

(b) *G is quasi-isomorphic to $G' = G(\Psi', T')$ if and only if*

(i) $\text{rank}(G) = \text{rank}(G') = n$;

(ii) $RT(G) = RT(G')$; *and*

(iii) *there are $n \times n$ integer matrices Φ, K_i, N_i with $\det(\Phi) \neq 0$ such that, for almost all i,*

$$M_i \Phi = K_i M_i' + \begin{bmatrix} p_i^{e(i,1)} & \ldots & 0 \\ . & \ldots & . \\ 0 & \ldots & p_i^{e(i,n)} \end{bmatrix} N_i.$$

Proof. (a) Let $F = \mathbb{Z}x_1 \oplus \ldots \oplus \mathbb{Z}x_n$ be a free subgroup of G with

$$T = G/F = \oplus_i (\mathbb{Z}(p_i^{e(i,1)}) \oplus \ldots \oplus \mathbb{Z}(p_i^{e(i,n)}))$$

and $0 \leq e(i,1) \leq \ldots \leq e(i,n)$ for each i. Define an $n \times n$ integer matrix M_i by setting

$$\text{row}_j(M_i) = (r_{i,j,1}, \ldots, r_{i,j,n}),$$

where

$$(1/p_i^{e(i,j)})(r_{i,j,1}x_1 + \ldots + r_{i,j,n}x_n) + F$$

is a generator of $\mathbb{Z}(p_i^{e(i,j)})$ for each $1 \leq j \leq n$, $i \geq 1$.

There is an isomorphism $f : F \to \mathbb{Z}^n$ defined by
$$f(a_1 x_1 + \ldots + a_n x_n) = (a_1, \ldots, a_n).$$
Extend f to an isomorphism $f : \mathbb{Q}F = \mathbb{Q}x_1 \oplus \ldots \oplus \mathbb{Q}x_n \to \mathbb{Q}^n$ and observe that
$$f(r_{i,j,1} x_1 + \ldots + r_{i,j,n} x_n) = (r_{i,j,1}, \ldots, r_{i,j,n})$$
for each $\text{row}_j(M_i) = (r_{i,j,1}, \ldots, r_{i,j,n})$. Because
$$\{(1/p_i^{e(i,j)})(r_{i,j,1} x_1 + \ldots + r_{i,j,n} x_n) + F : 1 \leq j \leq n, i \geq 1\}$$
is a set of generators for G/F and
$$\{(1/p_i^{e(i,j)})\text{row}_j(M_i) : 1 \leq j \leq n, i \geq 1\}$$
is a set of generators for $G(\Psi, T)/\mathbb{Z}^n$, it follows that f induces an isomorphism
$$G \to G(\Psi, T).$$
Since $T = G/F$, $R(T) = RT(G)$.

(b) Assume
$$f : G = G(\Psi, T) \to G' = G(\Psi', T')$$
is a quasi-isomorphism. Clearly, $\text{rank}(G) = \text{rank}(G')$. Write $RT(G) = [G/F]$ for some free subgroup F of G with G/F a torsion group. Then f induces a quasi-isomorphism $G/F \to G'/f(F)$ so that $RT(G) = RT(G')$.

Extend f to a vector space isomorphism $f : \mathbb{Q}^n \to \mathbb{Q}^n$ and represent f as an $n \times n$ rational matrix M relative to the standard basis of $\mathbb{Q}^n$. Since f is a quasi-isomorphism and M is a rational matrix, there is a non-zero integer m such that
$$mf : G \to G', (mf)(\mathbb{Z}^n) \subseteq \mathbb{Z}^n,$$
and $mM = \Phi$ is an integer matrix with non-zero determinant. Then mf induces a quasi-isomorphism $mf : G/\mathbb{Z}^n \to G'/\mathbb{Z}^n$ and so, for almost all i, there are integer matrices K_i, N_i with
$$\text{row}_j(M_i)\Phi = mf(\text{row}_j(M_i)) = \text{row}_j(K_i)M_i' + p_i^{e(i,j)} \text{row}_j(N_i),$$
whence
$$M_i \Phi = K_i M_i' + \begin{bmatrix} p_i^{e(i,1)} & \ldots & 0 \\ . & \ldots & . \\ 0 & \ldots & p_i^{e(i,n)} \end{bmatrix} N_i.$$

Conversely, given (i), (ii), and (iii), the integer matrix Φ induces an endomorphism of $\mathbb{Z}^n$ that extends to an isomorphism $f : \mathbb{Q}^n \to \mathbb{Q}^n$. As a consequence of the hypotheses, f induces a quasi-isomorphism from $G/\mathbb{Z}^n \to G'/\mathbb{Z}^n$, and so a quasi-isomorphism from G to G'. □

Theorem 2.4. *A locally free group G is a Butler group if and only if G is quasi-isomorphic to $G(\Psi, T)$ for some sequence $\Psi = (M_i)$ of integer matrices with only finitely many distinct row vectors and torsion group*

$$T = \oplus_i (\mathbb{Z}(p_i^{e(i,1)}) \oplus \ldots \oplus \mathbb{Z}(p_i^{e(i,n)}))$$

with $0 \leq e(i,1) \leq \ldots \leq e(i,n)$ for each i.

Proof. Assume that G is quasi-isomorphic to $G(\Psi, T)$ for some sequence $\Psi = (M_i)$ of $n \times n$ integer matrices with

$$\{\text{row}_j(M_i) : 1 \leq j \leq n, i \geq 1\} = \{\bar{x}_1, \ldots, \bar{x}_m\}$$

a finite subset of $\mathbb{Z}^n$. Let X_t be the pure rank-1 subgroup of $G(\Psi, T)$ generated by $\bar{x}_t$ and write

$$G(\Psi, T)/\mathbb{Z}^n = \oplus_i (\mathbb{Z}(p_i^{e(i,1)}) \oplus \ldots \oplus \mathbb{Z}(p_i^{e(i,n)}))$$

with $0 \leq e(i,1) \leq \ldots \leq e(i,n)$ for each i.

As

$$\{(1/p_i^{e(i,j)})\text{row}_j(M_i) + \mathbb{Z}^n : 1 \leq j \leq n, i \geq 1\}$$

is a set of generators for $G(\Psi, T)/\mathbb{Z}^n$ and each

$$(1/p_i^{e(i,j)})\text{row}_j(M_i) + \mathbb{Z}^n = (1/p_i^{e(i,j)})\bar{x}_t + \mathbb{Z}^n$$

for some t, the group $G(\Psi, T)$ is generated by $\mathbb{Z}^n$ and $X_1, \ldots, X_m$. Therefore, $G(\Psi, T)$ is a Butler group and, because Butler groups are closed under quasi-isomorphism, G is also a Butler group

Conversely, assume that G is a locally free Butler group of rank n generated by finitely many pure locally free rank-1 subgroups $X_1, \ldots, X_m$ with $m \geq n$. Choose a maximal linearly independent subset $\{X_1, \ldots, X_n\}$ of $\{X_1, \ldots, X_m\}$, $0 \neq x_t \in X_t$ for each $1 \leq t \leq n$, define $F = \mathbb{Z}x_1 \oplus \ldots \oplus \mathbb{Z}x_n$, and choose $0 \neq x_t \in X_t \cap F$ for each $n+1 \leq t \leq m$. Then F is a free subgroup of G with

$$G/F = \oplus_i (\mathbb{Z}(p_i^{e(i,1)}) \oplus \ldots \oplus \mathbb{Z}(p_i^{e(i,n)}))$$

and $0 \leq e(i,1) \leq \ldots \leq e(i,n)$.

Given $1 \leq t \leq m$, write $X_t/\mathbb{Z}x_t = \oplus_i \mathbb{Z}(p_i^{f(i,t)})$ so that $(1/p_i^{f(i,t)})x_t + \mathbb{Z}x_t$ is a generator of $\mathbb{Z}(p_i^{f(i,t)})$ for all i. Because there is an epimorphism

$$X_1/\mathbb{Z}x_1 \oplus \ldots \oplus X_m/\mathbb{Z}x_m \to G/F,$$

it follows that there is $M_i \in \text{Mat}_n(\mathbb{Z})$ with each

$$(1/p_i^{e(i,j)})\text{row}_j(M_i) + F$$

a generator of $\mathbb{Z}(p_i^{e(i,j)})$, and $\{\text{row}_j(M_i) : 1 \leq j \leq n, i \geq 1\}$ a finite set. As a consequence of Theorem 2.3(a), G is quasi-isomorphic to the group $G(\Psi, T)$ with $\Psi = (M_i)$, $T = G/F$, and only finitely many distinct row vectors among the rows of the M_i's. □

A torsion-free abelian group G is *strongly indecomposable* if

$$\mathbb{Q}\mathrm{End}(G) = \mathbb{Q} \otimes_{\mathbb{Z}} \mathrm{End}(G)$$

has no non-trivial idempotents, e.g. see [3]. Following is an example of a strongly indecomposable locally free Butler group.

Example 2.5. For each positive integer n, irreducible $f(x) \in \mathbb{Q}[x]$, and positive integer e with degree $f(x)^e = n$, there is a strongly indecomposable, locally free, Butler group G of rank $2n$ such that:

(a) $RT(G) = [\oplus_i (\mathbb{Z}(p_i)^n]$ and

(b) $\mathbb{Q}\mathrm{End}(G)$ is isomorphic to $\mathbb{Q}[x]/\langle f(x)^e \rangle$.

Proof. (a) Choose

$$N = (n_{ij}) \in \mathrm{Mat}_n(\mathbb{Z})$$

such that $f(x)^e$ is the minimal polynomial of N. For each $1 \leq j \leq 4$, let

$$P_j = \{p_{4k+j} : k \geq 0\}.$$

Then P_1, P_2, P_3, P_4 are infinite sets partitioning Π. Given $1 \leq j \leq 4$, define X_j to be the subgroup of $\mathbb{Q}$ containing 1 with $\mathrm{type}(X_j) = [(a_{ji})]$, where $a_{ji} = 1$ if $p_i \in P_j$ and $a_{ji} = 0$ otherwise.

Define integer matrices M_i for $i \geq 1$ by setting

$$M_{4k+4} = \begin{bmatrix} 0 & 0 \\ I_n & 0 \end{bmatrix}, \quad M_{4k+1} = \begin{bmatrix} 0 & 0 \\ 0 & I_n \end{bmatrix},$$

$$M_{4k+2} = \begin{bmatrix} 0 & 0 \\ I_n & I_n \end{bmatrix}, \quad M_{4k+3} = \begin{bmatrix} 0 & 0 \\ I_n & N \end{bmatrix}.$$

Define $G = G(\Psi, T)$, where $\Psi = (M_i)_{i \geq 0}$ and $T = \oplus_i \mathbb{Z}(p_i)^n$, a locally free group with rank $2n$. For each i, M_i has n rows containing a 1, from which it follows that $RT(G) = R(T)$.

(b) In view of Theorem 2.4 and the fact the M_i's have only finitely many distinct rows, G is a Butler group. It follows from the definitions that for each $1 \leq j \leq 4$, X_j^n is quasi-isomorphic to the pure subgroup of G generated by the finitely many non-zero rows of the M_{4k+j}'s. Consequently, $G = G(\Psi, T)$ is quasi-isomorphic to the subgroup G_N of $\mathbb{Q}^{2n}$ generated by $X_1^n \oplus X_2^n$,

$$X_3^n(1+1) = \{(y, y) : y \in X_3^n\},$$

and

$$X_4^n(1+N) = \{(y, Ny) : y \in X_4^n\}.$$

Then $\mathbb{Q}\mathrm{End}(G_N)$ is isomorphic to $\mathbb{Q}[x]/\langle f(x)^e \rangle$, [3, Example 3.3.10], whence G is strongly indecomposable because $\mathbb{Q}\mathrm{End}(G)$ is isomorphic to $\mathbb{Q}\mathrm{End}(G_N)$ and $\mathbb{Q}\mathrm{End}(G_N)$ has no non-trivial idempotents. □

Remark 2.6. The group G_N constructed in the previous example is a Butler group constructed from a classical $\mathbb{Q}$-representation

$$(\mathbb{Q}^{2n}, \mathbb{Q}^n \oplus 0, 0 \oplus \mathbb{Q}^n, \mathbb{Q}^n(1+1), \mathbb{Q}^n(1+N))$$

of an antichain with 4 elements. There is a well-known equivalence, due to M. C. R. Butler, [6], from the quasi-isomorphism category of Butler groups with typeset contained in a finite lattice T of types to $\mathbb{Q}$-representations of a finite partially ordered set S_T derived from T, e.g., see [3, Theorem 3.3.2]. If T is a lattice of locally free types, then the quasi-isomorphism category of locally free Butler groups with typeset contained in T is equivalent to the category of $\mathbb{Q}$-representations of S_T. As a consequence, techniques of Example 2.5 can be used to construct, up to quasi-isomorphism, a locally free Butler group from an arbitrary $\mathbb{Q}$-representation of a finite partially ordered set of locally free types.

The next two examples are examples of locally free groups that are not Butler groups.

Example 2.7. For each positive integer n, there is a locally free group G with rank $n+1$ such that:

(a) G has infinite typeset and so is not a Butler group,

(b) $RT(G) = [\oplus_i \mathbb{Z}(p_i)]$, and

(c) for each pure subgroup H of G, the endomorphism ring of H is isomorphic to $\mathbb{Z}$.

Proof. Let $T = \{t_1, \ldots, t_m, \ldots\}$ denote an indexing of the infinite set

$$\{(1, \alpha_1, \ldots, \alpha_n) \in \mathbb{Z}^{n+1} : 0 \leq \alpha_i\}.$$

Write $t_m = (1, t_{1,m}, \ldots, t_{n,m}) \in \mathbb{Z}^{n+1}$ for non-negative integers $t_{j,m}$ and choose a partition $P_1, \ldots, P_m, \ldots$ of infinite subsets of Π. For each $1 \leq j \leq n, i \geq 1$, define $\alpha_{j,i} = t_{j,m}$ if $p_i \in P_m$.

Let $\Psi = (M_i)$ be the sequence of $(n+1) \times (n+1)$ integer matrices given by

$$M_i = \begin{bmatrix} 0 & 0 & \ldots & 0 \\ \ldots & \ldots & \ldots & \ldots \\ 1 & \alpha_{1,i} & \ldots & \alpha_{n,i} \end{bmatrix}$$

and let $G = G(\Psi, T)$ with $T = \oplus_i \mathbb{Z}(p_i)$. Because each $(1, \alpha_{1,i}, \ldots, \alpha_{n,i}) = t_m$ for some m, G is the subgroup of $\mathbb{Q}^{n+1}$ generated by $\mathbb{Z}^n$ and $\{(1/p_i)t_m : p_i \in P_m, m \geq 1\}$.

The remainder of the proof is as given in [2, Example 2.8], i.e. typeset$(G) = \{\text{type}(t_m) : m \geq 1\}$ is an infinite set of pairwise incomparable types and each pure subgroup H of G is strongly indecomposable with $\text{End}(H) = \mathbb{Z}$. □

Example 2.8. For each pair of positive integers m and n, there is a locally free group G with rank $n+1$ and Richman type $[\oplus_i \mathbb{Z}(p_i)^n]$ such that G has a finite typeset with $\geq m$ elements but G is not a Butler group.

Proof. Choose a strictly ascending chain

$$0 < t_{1,1} < \ldots < t_{n,1} < t_{1,2} < \ldots < t_{1,i} < \ldots < t_{n,i} < \ldots$$

of integers with $\gcd(t_{ji}, p_i) = 1$ for each $1 \leq j \leq n, i \geq 1$ and define

$$P_s = \{p_{rm+s} : r \geq 0\}$$

for each $1 \leq s \leq m$. Then $P_1, \ldots, P_m$ is a partition of infinite subsets of Π.

Define $\alpha_{j,i} = t_{j,k}$ if $p_i \in P_k$ and let $\Psi = (M_i)$ be a sequence of $(n+1) \times (n+1)$ integer matrices with

$$M_i = \begin{bmatrix} 0 & 0 & \ldots & 0 \\ 1 & 0 & \ldots & \alpha_{1,i} \\ 0 & 1 & & a_{2,i} \\ \vdots & \vdots & \ldots & \vdots \\ 0 & 0 & \ldots & \alpha_{n,i} \end{bmatrix}.$$

Then $G = G(\Psi, T)$, with $T = \oplus_i \mathbb{Z}(p_i)^n$, is a locally free group with rank $n+1$ and Richman type $R(T)$.

Let $X_{j,i}$ be the pure subgroup of G generated by $(0, \ldots, 0, 1, 0, \ldots, \alpha_{j,i})$. Then

$$\{\text{type}(X_{j,i}) : 1 \leq j \leq n, i \geq 1\} = \{\tau_1, \ldots, \tau_m\}$$

is a subset of m pairwise incomparable elements of the typeset of G. By construction, G is generated by $G(\tau_1), \ldots, G(\tau_m)$, where $G(\tau_t)$ is the pure subgroup of G generated by elements of G of type τ_t. Hence, the typeset of G is contained in the finite sublattice of the set of all types generated by $\{\tau_1, \ldots, \tau_m\}$ under the operations of meet and join. In particular, G has a finite typeset with $\geq m$ elements. Because the $t_{j,i}$'s are an infinite strictly ascending chain of integers, G can not be generated by only finitely many pure rank-1 subgroups and so is not a Butler group. □

3 Locally free groups and p-adic matrices

The set of $n \times n$ matrices with entries in a ring R is denoted by $\text{Mat}_n(R)$. A matrix $M = (m_{ij}) \in \text{Mat}_n(\mathbb{Z})$ is *p-bounded* for a prime p if $0 \leq m_{ij} \leq p-1$ for each i, j. A sequence of $n \times n$ integer matrices $\Psi = (M_i)_{i \geq 1}$ is *p-bounded* if each M_i is p-bounded. The sequence $\Psi = (M_i)$ is *coefficient convergent* to $M \in \text{Mat}_n(\hat{\mathbb{Z}}_{(p)})$ if

$$M = \lim_{i \to \infty} (M_1 + pM_2 + \ldots + p^i M_{i+1}).$$

Theorem 3.1. *Let p be a prime, n a positive integer, and*

$$T = \oplus_i (\mathbb{Z}(p_i^{e(i,1)}) \oplus \ldots \oplus \mathbb{Z}(p_i^{e(i,n)}))$$

with $0 \leq e(i,1) \leq \ldots \leq e(i,n)$ for each i.

(a) *There is an injection* **g** *from* $\text{Mat}_n(\hat{\mathbb{Z}}_{(p)})$ *to locally free groups of rank n such that* $RT(\mathbf{g}(M)) \leq R(T)$ *for each M.*

(b) *There is a surjection* **f** *from locally free groups of the form* $G(\Psi, T)$ *to* $\text{Mat}_n(\hat{\mathbb{Z}}_{(p)})$ *such that* $\mathbf{fg}(M) = M$ *for each M.*

Proof. (a) Define $\mathbf{g}(M) = G(\Psi, T)$, where $\Psi = (M_i)$ is the unique p-bounded sequence of integer matrices coefficient convergent to M. It follows from Theorem 2.3(a) that $\mathbf{g}(M)$ is a locally free group of rank n with $RT(\mathbf{g}(M)) \leq R(T)$.

(b) Define a function **f** by $\mathbf{f}(G(\Psi, T)) = M$, where $\Psi = (M_i)$ coefficient converges to $M \in \text{Mat}_n(\hat{\mathbb{Z}}_{(p)})$. Here, Ψ need not be a p-bounded sequence. Since $\mathbf{fg}(M) = M$ for each M, **g** is an injection and **f** is a surjection. □

Corollary 3.2. *A torsion-free abelian group G of finite rank is a locally free Butler group if and only if G is quasi-isomorphic to a group in the image of* **g** *for some prime p and torsion group*

$$T = \oplus_i (\mathbb{Z}(p_i^{e(i,1)}) \oplus \ldots \oplus \mathbb{Z}(p_i^{e(i,n)}))$$

with $0 \leq e(i, 1) \leq \ldots \leq e(i, n)$ for each i.

Proof. Assume p is a prime and G is quasi-isomorphic to $\mathbf{g}(M) = G(\Psi, T)$. Since $\Psi = (M_i)$ is a p-bounded sequence,

$$\{\text{row}_j(M_i) : 1 \leq j \leq n, i \geq 1\}$$

is a finite subset of $\mathbb{Z}^n$. By Theorem 2.4, G is a locally free Butler group.

Conversely, if G is a locally free Butler group, then, by Theorem 2.4, G is quasi-isomorphic to $G(\Psi, T)$ for some torsion group T and sequence $\Psi = (M_i)$ of integer matrices $M_i = (\alpha_{j,r,i})_{j,r}$ with only finitely many distinct row vectors among the M_i's. It follows that there is a prime p with $p > a_{j,r,i}$ for each i, j, r. Then $\Psi = (M_i)$ is a p-bounded sequence and so G is quasi-isomorphic to $\mathbf{g}(M) = G(\Psi, T)$, where $\Psi = (M_i)$ coefficient converges to M. □

A slight modification of the proof of Theorem 3.1 yields the following corollary. The set of locally free types is denoted by LFT.

Corollary 3.3. *There is an injection* $\mathbf{g} : \hat{\mathbb{Z}}_{(p)}/\mathbb{Z} \to LFT$ *defined by* $\mathbf{g}(\alpha + \mathbb{Z}) = [(a_i)]$, *where each $0 \leq a_i < p$ and $\alpha = \lim_{i \to \infty}(\alpha_1 + \alpha_2 p + \ldots + \alpha_i p^{i-1})$.*

4 Locally free groups and $\mathbb{Z}_{(p)}$-modules

For a prime p and finite rank torsion-free $\mathbb{Z}_{(p)}$-module A, *p-rank(A)* is the $\mathbb{Z}/p\mathbb{Z}$-dimension of A/pA. Let $\Gamma_{k \times n}$ be a $k \times n$ $\hat{\mathbb{Z}}_{(p)}$-matrix and define $A(\Gamma)$ to be the pure $\mathbb{Z}_{(p)}$-submodule of $\hat{\mathbb{Z}}_{(p)}^n$ generated by $\mathbb{Z}_{(p)}^n$ and the k rows of Γ. Then $A(\Gamma)$ is a torsion-free $\mathbb{Z}_{(p)}$-module with p-rank n and rank $\leq n + k$.

Each torsion-free $\mathbb{Z}_{(p)}$-module A with p-rank n and rank $n + k$ is isomorphic to $A(\Gamma)$ for some $k \times n$ $\hat{\mathbb{Z}}_{(p)}$-matrix Γ, [1, Corollary 1.7].

Corollary 4.1. *Let p be a prime, k, n positive integers, and*
$$T = \oplus_i (\mathbb{Z}(p_i^{e(i,1)}) \oplus \ldots \oplus \mathbb{Z}(p_i^{e(i,n)}))$$
with $0 \leq e(i,1) \leq \ldots \leq e(i,n)$ for each i. There is an injection $\mathbf{h}$ from torsion-free $\mathbb{Z}_{(p)}$-modules $A = A(\Gamma_{k \times n})$ with p-rank n and rank $n+k$ to locally free Butler groups of rank $n+k$ such that $RT(\mathbf{h}(A)) \leq R(T)$.

Proof. Define $\mathbf{h}(A(\Gamma)) = G(\Psi, T)$, where $\Psi = (M_i)$ is a p-bounded sequence of integer matrices coefficient convergent to
$$M = \begin{bmatrix} 0_{n \times k} & 0_{n \times n} \\ I_{k \times k} & \Gamma_{k \times n} \end{bmatrix} \in \mathrm{Mat}_{n+k}(\hat{\mathbb{Z}}_{(p)}).$$

Apply Theorem 3.1 to complete the proof, observing that $\mathbf{h}(A(\Gamma)) = \mathbf{g}(M)$. □

Given a torsion group
$$T = \oplus_i (\mathbb{Z}(p_i^{e(i,1)}) \oplus \ldots \oplus \mathbb{Z}(p_i^{e(i,n)}))$$
with $0 \leq e(i,1) \leq \ldots \leq e(i,n)$ for each i and a prime p, define a relation $\mathbf{h}^*$ from $\mathbb{Z}_{(p)}$-modules $A = A(\Gamma_{k \times n})$ with p-rank n and rank $n+k$ to locally free Butler groups of rank $n+k$ by $\mathbf{h}^*(A(\Gamma)) = G(\Psi, T)$, where $\Psi = (M_i)$ is any sequence of integer matrices coefficient convergent to
$$M = \begin{bmatrix} 0_{n \times k} & 0_{n \times n} \\ I_{k \times k} & \Gamma_{k \times n} \end{bmatrix} \in \mathrm{Mat}_{n+k}(\hat{\mathbb{Z}}_{(p)}).$$

Notice that $\mathbf{h}^*$ is a generalization of $\mathbf{h}$ but is not a function because sequences of matrices coefficient convergent to M are not unique.

Example 4.2. Each of the locally free groups $G(\Psi, T)$ defined in Examples 2.7 and 2.8 is quasi-isomorphic to $\mathbf{h}^*(A(\Gamma))$ for some Γ. For Example 2.7, $A(\Gamma_{1 \times n})$ has p-rank n and rank $n+1$ and, for Example 2.8, $A(\Gamma_{n \times 1})$ has p-rank 1 and rank $n+1$.

Proof. Follows immediately from the definitions. □

Example 4.3. Let $T = \oplus_i \mathbb{Z}(p_i)$. For $\alpha \in \hat{\mathbb{Z}}_{(p)}$, write
$$\alpha = \lim_{i \to \infty} (\alpha_1 + \alpha_2 p + \ldots + \alpha_i p^{i-1})$$
with each $0 \leq \alpha_i < p_i$ and define $\mathbf{h}^*(A(\alpha)) = G(\Psi_\alpha, T)$, where $\Psi_\alpha = (M_{\alpha, i})$ with
$$M_{\alpha, i} = \begin{bmatrix} 0 & 0 \\ 1 & \alpha_i \end{bmatrix} \in \mathrm{Mat}_2(\hat{\mathbb{Z}}_{(p)}).$$

Then $\{\mathbf{h}^*(A(\alpha)) : \alpha \in \hat{\mathbb{Z}}_{(p)}\}$ is the uncountable collection $S_f(\mathbb{Q}^2)$ of locally free groups constructed in [11].

Proof. The set $S_f(\mathbb{Q}^2)$ is defined in [11] to be the set of subgroups A of $\mathbb{Q}^2$ containing $\mathbb{Z}^2$ "...which at each prime p have exactly one of $(\frac{1}{p}, 0), (\frac{1}{p}, \frac{1}{p}), \ldots, (\frac{1}{p}, \frac{p-1}{p})$ in A but at no (m, n) not divisible by p do we have $(\frac{m}{p^2}, \frac{n}{p^2}) \in A$." It follows that $\{h^*(A(\alpha)) : \alpha \in S\} = S_f(\mathbb{Q}^2)$. □

Corollary 4.4 ([10]). *Let X, G, and G' be locally free torsion-free abelian groups of finite rank with X a rank-1 group. If $X \otimes_\mathbb{Z} G$ is quasi-isomorphic to $X \otimes_\mathbb{Z} G'$, then G is quasi-isomorphic to G'.*

Proof. By Theorem 2.3(a), G is quasi-isomorphic to $G(\Psi, T)$ for some sequence $\Psi = (M_i)$ and torsion group T. Similarly, G' is quasi-isomorphic to some $G(\Psi', T')$ with $\Psi' = (M_i')$.

Write $RT(G) = [G/F] = [\oplus_i T_i]$, where each T_i is a finite p_i-group. Since X is a rank-1 group, $RT(X) = \oplus_i \mathbb{Z}(p^{f_i})$ for some $0 \leq f_i < \infty$. As $\mathbb{Z} \otimes_\mathbb{Z} F$ is a free subgroup of $X \otimes_\mathbb{Z} G$ with $(X \otimes_\mathbb{Z} G)/(\mathbb{Z} \otimes_\mathbb{Z} F)$ a torsion group, $RT(X \otimes_\mathbb{Z} G) = [\oplus_i U_i]$ with U_i isomorphic to $\mathbb{Z}(p_i^{f_i}) \otimes_\mathbb{Z} T_i$ for each i.

By Theorem 2.3(a), $X \otimes_\mathbb{Z} G$ is quasi-isomorphic to $G(\Delta, U)$ for some $U = \oplus_i U_i$ and sequence $\Delta = (L_i)$ of integer matrices with $p^{f_i} L_i = M_i$ for each i. Similarly, $X \otimes_\mathbb{Z} G'$ is quasi-isomorphic to $G(\Delta', RT')$ with $\Delta' = (L_i')$ and $p^{f_i} L_i' = M_i'$ for each i.

Since $G(\Delta, T)$ is quasi-isomorphic to $G(\Delta', T')$, by Theorem 2.3(b) there are integer matrices Φ, K_i, N_i with $\det(\Phi) \neq 0$ such that, for almost all i,

$$L_i \Phi = K_i L_i' + \begin{bmatrix} p_i^{e(i,1)} & \ldots & 0 \\ . & \ldots & . \\ 0 & \ldots & p_i^{e(i,n)} \end{bmatrix} N_i.$$

Multiplying this equation by $p_i^{f_i}$ yields

$$M_i \Phi = K_i M_i' + \begin{bmatrix} p_i^{e(i,1)+f_i} & \ldots & 0 \\ . & \ldots & . \\ 0 & \ldots & p_i^{e(i,n)+f_i} \end{bmatrix} N_i$$

since $p_i^{f_i} L_i = M_i$ and $p_i^{f_i} L_i' = M_i'$ for each i. By Theorem 2.3(b), $G(\Psi, T)$ is quasi-isomorphic to $G(\Psi', T')$ and so G is quasi-isomorphic to G'. □

Example 4.5. *There are locally free groups X, G, and G' of rank 2 such that $X \otimes_\mathbb{Z} G$ is quasi-isomorphic to $X \otimes_\mathbb{Z} G'$, but G is not quasi-isomorphic to G'.*

Proof. Define sequences (a_i) and (b_i) with $a_i = \pm 1, b_i = \pm 1$ such that for each integer matrix

$$\Phi = \begin{bmatrix} \phi_{11} & \phi_{12} \\ \phi_{21} & \phi_{22} \end{bmatrix}$$

with non-zero determinant,

$$\phi_{12} + a_i \phi_{22} \neq b_i(\phi_{11} a_i + \phi_{2,1}) \pmod{p_i}$$

for infinitely many i. For example, let $a_{2k} = b_{2k-1} = 1$ and $a_{2k-1} = b_{2k} = -1$ for each $k \geq 1$.

Let $X = G(\Gamma, T)$, where $\Gamma = (\Gamma_i)$ with

$$\Gamma_i = \begin{bmatrix} 0 & 0 \\ 1 & a_i \end{bmatrix}$$

and $T = \oplus_i \mathbb{Z}(p_i)$.

Define $G = G(\Psi, T)$, where $\Psi = (\Psi_i)$ with

$$\Psi_i = \begin{bmatrix} 0 & 0 \\ 1 & b_i \end{bmatrix}$$

and $G' = G(\Psi', T)$, where $\Psi' = (\Psi'_i)$ with

$$\Psi'_i = \begin{bmatrix} 0 & 0 \\ 1 & a_i b_i \end{bmatrix}.$$

Then X, G, and G' are locally free groups of rank 2.

Now $X \otimes_\mathbb{Z} G$ is quasi-isomorphic to $G(\Xi, T)$, where $\Xi = (\Gamma_i \otimes \Psi_i)$ and

$$\Gamma_i \otimes \Psi_i = \begin{bmatrix} 0 & 0 & 0 & 0 \\ 0 & 0 & 0 & 0 \\ 0 & 0 & 0 & 0 \\ 1 & b_i & a_i & a_i b_i \end{bmatrix}$$

is the Kronecker product of Γ_i and Ψ_i.

Similarly, $X \otimes_\mathbb{Z} G'$ is quasi-isomorphic to $G(\Xi', T)$, where $\Xi = (\Gamma_i \otimes \Psi'_i)$ and

$$\Gamma_i \otimes \Psi'_i = \begin{bmatrix} 0 & 0 & 0 & 0 \\ 0 & 0 & 0 & 0 \\ 0 & 0 & 0 & 0 \\ 1 & a_i b_i & a_i & a_i^2 b_i = b_i \end{bmatrix}.$$

It is now easy to see that $X \otimes_\mathbb{Z} G$ is quasi-isomorphic to $X \otimes_\mathbb{Z} G'$.

Assume that G is quasi-isomorphic to G'. The last row of the matrix equation of Theorem 2.3(b)(iii) yields equations

$$\phi_{11} + a_i \phi_{2,1} = k_{2,2,i} \pmod{p_i},$$
$$\phi_{12} + a_i \phi_{22} = k_{2,2,i} a_i b_i \pmod{p_i}$$

for almost all i. Substitute for $k_{2,2,i}$ in the second equation to get

$$\begin{aligned} \phi_{12} + a_i \phi_{22} &= (\phi_{11} + a_i \phi_{2,1}) a_i b_i \pmod{p_i} \\ &= b_i(\phi_{11} a_i + \phi_{2,1}) \pmod{p_i}, \end{aligned}$$

recalling that $a_i^2 = 1$. Then

$$\phi_{12} + a_i \phi_{22} = b_i(\phi_{11} a_i + \phi_{2,1}) \pmod{p_i}$$

for almost all i, a contradiction to the choice of the a_i's and b_i's. □

Acknowledgements. The author wishes to thank the referee for a careful reading of the original manuscript.

References

[1] D. M. Arnold, *A duality for quotient divisible abelian groups of finite rank*, Pacific Journal of Mathematics 42 (1972), pp. 11–15.

[2] D. M. Arnold, *Finite Rank Torsion-Free Abelian Groups and Rings,* Lecture Notes in Mathematics 931, Springer, New York, 1982.

[3] D. M. Arnold, *Abelian Groups and Representations of Finite Partially ordered Sets,* CMS Books in Mathematics, Springer, New York, 2000.

[4] M. C. R. Butler, *A class of torsion-free abelian groups of finite rank*, Journal of London Mathematical Society 15 (1965), pp. 680–698.

[5] M. C. R. Butler, *On locally free torsion-free rings of finite rank*, Journal of London Mathematical Society 43 (1968), pp. 297–300.

[6] M. C. R. Butler, *Torsion-free modules and diagrams of vector spaces*, Proceedings of London Mathematical Society 18 (1968), pp. 635–652.

[7] T. Faticoni, H. P. Goeters, and C. Vinsonhaler, *Torsion-free duality is Warfield*, Proceedings American Mathematical Society 125 (1997), pp. 961–969.

[8] L. Fuchs, *Infinite Abelian Groups*, Volume I, Academic Press, New York, 1970.

[9] L. Fuchs, *Infinite Abelian Groups*, Volume II, Academic Press, New York, 1973.

[10] E. L. Lady, *Warfield duality and rank-one quasi-summands of tensor products of finite rank locally free modules over Dedekind domains,* Journal of Algebra 121 (1989), pp. 129–138.

[11] G. Hjorth, *The locally free finite rank TFA groups*, preprint.

[12] J. D. Reid, *Warfield duality and irreducible groups*, Abelian Groups and Non-commutative Rings, Contemporary Mathematics 130, American Mathematical Society, Providence, 1992, pp. 361–370.

[13] J. D. Reid and C. Vinsonhaler, *A theorem of M. C. R. Butler for Dedekind domains*, Journal of Algebra 175 (1995), pp. 979–989.

[14] F. Richman, *A class of rank 2 torsion free groups*, Studies on Abelian Groups, Springer Dunod, Paris, 1968, pp. 327–333.

[15] L. Salce, *Warfield domains: module theory from linear algebra to commutative algebra through abelian groups*, Milan Journal of Mathematics 70 (2002), pp. 163–185.

[16] S. Thomas, *The classification problem for torsion-free abelian groups of finite rank*, Journal of American Mathematical Society 16 (2003), pp. 233–258.

[17] C. Vinsonhaler and W. Wickless, *Dualities for locally completely decomposable abelian groups,* Journal of Algebra 191 (1997), pp. 628–652.

[18] E. A. Walker, *Quotient categories and quasi-isomorphisms of abelian groups*, Proceedings of Colloquium on Abelian Groups, Budapest, 1964, pp. 147–162.

[19] R. B. Warfield, Jr., *Homomorphisms and duality for torsion-free groups*, Math Zeitschrift 107 (1968), pp. 189–200.

[20] W. J. Wickless, *A functor from mixed groups to torsion-free groups*, Abelian Group Theory and Related Topics, Contemporary Mathematics 171, American Mathematical Society, Providence, 1994, pp. 407–417.

Author information

David M. Arnold, Department of Mathematics, Baylor University, Waco, TX 76798-7328, U.S.A.
E-mail: David_Arnold@baylor.edu

Decompositions of global crq-groups

A. Mader, L. G. Nongxa and M. A. Ould-Beddi

Abstract. Almost completely decomposable groups with a cyclic regulating quotient, the crq-groups, are a reasonably accessible class of groups for arbitrary critical typesets and have been studied intensively. There is a satisfactory classification up to near-isomorphism and in the local case the possible direct decompositions are known. Local-global relations can be used in order to get information about direct decompositions of global groups. A description of the possible decompositions of global crq-groups in terms of invariants is derived in the special case when the group has a unique regulating subgroup. The general problem remains open.

Key words. Socle, radical, near-isomorphism, almost completely decomposable group, cyclic regulating quotient.

AMS classification. 20K15, 20K35.

1 Introduction

An *almost completely decomposable group* X is a finite (torsion-free abelian) extension of a completely decomposable group A of finite rank. In 1974 E. L. Lady [15] initiated a systematic theory of such groups based on the fundamental concept of *regulating subgroup*. The regulating subgroups can be defined as the completely decomposable subgroups of least index in an almost completely decomposable group X. This least index is the *regulating index* rgi X. An almost completely decomposable group is *local* if its regulating index is a prime power; otherwise it is a *global* group. Details on the beginnings and subsequent developments of the theory of almost completely decomposable groups can be found in the monograph [17]. Rather than citing the original sources we will quote [17] which contains an extensive bibliography of journal articles.

Almost completely decomposable groups initially served mainly as a source of examples of groups having drastically different direct decompositions with indecomposable summands. We will call decompositions with indecomposable summands *indecomposable decompositions* for short. After the theory had developed and tools became available attempts were made to describe the possible direct decompositions of special classes of almost completely decomposable groups. An early result in this direction stated that the indecomposable almost completely decomposable groups with two critical types have rank one or rank two. This result appeared first in [2] with a flawed

---

Third author: Thanks the University of the Western Cape for the kind hospitality and support during the preparation of this paper.

proof and has been proved many times since ([1], [11], [16], [18], [22]). A fundamental tool in studying direct decompositions of almost completely decomposable groups is the following theorem of David Arnold ([3, Corollary 12.9], [4, Corollary 2.2.11], [17, Theorem 10.2.5]).

Theorem 1.1 (Arnold Decomposition Theorem). *If X and Y are nearly isomorphic torsion-free abelian groups of finite rank and $X = X_1 \oplus X_2$, then $Y = Y_1 \oplus Y_2$ for some subgroups $Y_1 \cong_{\mathrm{nr}} X_1$ and $Y_2 \cong_{\mathrm{nr}} X_2$.*

Suppose a class $\mathcal{C}$ of almost completely decomposable groups is given whose members are classified up to near-isomorphism by numerical invariants. One can find decompositions of a given group X in the class by constructing a direct sum $Y = Y_1 \oplus Y_2$ in the class with the same invariants as X. Then $X \cong_{\mathrm{nr}} Y$ and by Arnold's theorem X decomposes in a way similar to the decomposition of Y. This strategy was used in [22], [16], and in [6].

The following very striking theorem was proved by Faticoni and Schultz ([13], [17, Corollary 10.4.6]).

Theorem 1.2 (Faticoni–Schultz Uniqueness Theorem). *Let X and Y be p-reduced nearly isomorphic almost completely decomposable groups with p-power regulating index. If $X = \bigoplus_{i=1}^{m} X_i$ and $Y = \bigoplus_{i=1}^{n} Y_i$ are indecomposable decompositions, then $m = n$ and after relabeling $X_i \cong_{\mathrm{nr}} Y_i$ for $1 \leq i \leq n$.*

This means that "pathological decompositions" arise only if the regulating index is composite. In the local case the question of decompositions (up to near isomorphism) reduces to finding the indecomposable groups.

In [9], [10] and [23], the relatively simple but important subclass of almost completely decomposable groups with a cyclic regulating quotient was studied. These crq-*groups* thus are almost completely decomposable groups X containing a regulating subgroup A such that X/A is cyclic. The direct decompositions of local crq-groups were determined, based on the concept of "sharp type", and shown to be unique up to isomorphism in [23] ([17, Theorem 6.5.11]). In [10] partial results were obtained for global crq-groups.

In [19] a definitive classification by numerical near-isomorphism invariants was obtained for global crq-groups but the question of direct decompositions was not considered. We will now exploit the invariants in order to study direct decompositions of global crq-groups.

Recently various authors ([5], [7], [8], [12], [20], [21]) studied generalizations of almost completely decomposable groups that allow infinite ranks. In particular, Blagoveshchenskaya and Göbel [5] generalized near-isomorphism to certain classes of torsion-free groups of infinite rank. It would be interesting to see whether the techniques of the present paper and local-global principles can be used for these more general classes of groups.

2 General background

The set of all prime numbers is denoted by $\mathbb{P}$. The purification of a subgroup H in a torsion-free group G is denoted by H_*^G. We take it for granted that the reader is familiar with the usual type subgroups, namely the *socles* $G(\tau)$, $G^*(\tau)$, $G^\sharp(\tau) = G^*(\tau)_*^G$, and the *radicals* $G[\tau]$, $G^\sharp[\tau] = \bigcap_{\rho < \tau} G[\rho]$. A type τ is *critical* for G if $G(\tau)/G^\sharp(\tau) \neq 0$. The *critical typeset* $\mathrm{T}_{\mathrm{cr}}(G)$ is the set of all critical types of G. If A is a completely decomposable group, then $A = \bigoplus_{\rho \in \mathrm{T}_{\mathrm{cr}}(A)} A_\rho$ is always assumed to be a decomposition of A into ρ-homogeneous components $A_\rho (\neq 0)$. The *typeset* of a group G is denoted by $\mathrm{Tst}(G)$. If G is an almost completely decomposable group, then its typeset is the meet closure of its critical typeset.

Every almost completely decomposable group X can be decomposed as $X = X_{cd} \oplus X_{cl}$ where X_{cd} is completely decomposable and X_{cl} is *clipped*, i.e., X_{cl} has no completely decomposable direct summands. The existence of such a *main decomposition* is immediate since X has finite rank but it is not unique. Lady showed that X_{cd} is unique up to isomorphism and consequently X_{cl} is unique up to near-isomorphism ([17, Theorem 9.2.7]).

The Purification Lemma [17, Lemma 11.4.1] and [17, Corollary 11.2.5] will be an important tool in calculating the invariants of purifications of direct summands of regulating subgroups of crq-groups. For future use we add a variant of part of the Purification Lemma. It involves a generalized greatest common divisor $\gcd^A(N, a^\downarrow)$ of a non-singular integral $k \times k$ matrix N and a column vector $a^\downarrow$ of k elements of A ([17, Chapter 11]), but for the purposes of this article $k = 1$ and the reader may think of N as a positive integer and $a^\downarrow$ as a single element of A.

Lemma 2.1 (Purification Lemma). *Let X be an almost completely decomposable group and A a subgroup of finite index in X. Suppose that $A = B \oplus C$ and $X = A + \vec{\mathbb{Z}} N^{-1} a^\downarrow$ with $a^\downarrow = b^\downarrow + c^\downarrow$ where $a^\downarrow \in A^\downarrow$, $b^\downarrow \in B^\downarrow$ and $c^\downarrow \in C^\downarrow$. Then $B_*^X = B + \vec{\mathbb{Z}} N_B^{-1} b^\downarrow$ where $N_B = \gcd^A(N, c^\downarrow)$. If A is a regulating subgroup of X, then B is regulating in B_*^X.*

Let A be a regulating subgroup of the almost completely decomposable group X. Lemma 2.1 implies the well-known facts that $A(\tau)$ is regulating in $X(\tau)$, $A^\sharp(\tau)$ is regulating in $X^\sharp(\tau)$, $A[\tau]$ is regulating in $X[\tau]$, $A^\sharp[\tau]$ is regulating in $X^\sharp[\tau]$ but also less evidently that $A[\tau](\sigma)$ is regulating in $X[\tau](\sigma)$, $A^\sharp(\tau)[\tau]$ is regulating in $X^\sharp(\tau)[\tau]$ and more. If X is a crq-group, then every one of these canonical subgroups are crq-groups also.

Recall that a subgroup $A = \bigoplus_{\rho \in \mathrm{T}_{\mathrm{cr}}(A)} A_\rho$ of an almost completely decomposable group X is regulating in X if and only if $X(\tau) = A_\tau \oplus X^\sharp(\tau)$ for every $\tau \in \mathrm{T}_{\mathrm{cr}}(X)$. In particular, the homogeneous components A_τ of a regulating subgroup are pure in X and therefore $\mathrm{hgt}_p^{A_\tau}(a) = \mathrm{hgt}_p^A(a) = \mathrm{hgt}_p^X(a)$ for any $a \in A_\tau$. We also mention in this context that for a subset T of $\mathrm{T}_{\mathrm{cr}}(A)$ and $a = \sum_{\rho \in T} a_\rho$, where $a_\tau \in A_\tau$, it is true that $\gcd^A(n, a) = \gcd\{\gcd^A(n, a_\rho) : \rho \in T\}$. This well-known and easily verified fact ([17, Lemma 11.2.9]) will be used freely and without reference in the sequel.

For general background on torsion-free abelian groups, and for almost completely decomposable groups in particular, we refer to [17]. See also [3], [14], and especially [4].

3 Local-global issues

We recall some basic results that can be found in [17, Chapter 5]. The *regulator* $R(X)$ of an almost completely decomposable group is the intersection of all regulating subgroups of X and was shown by Burkhardt to be itself a completely decomposable subgroup of finite index in X. If $A = \bigoplus_{\rho \in T_{cr}(A)} A_\rho$ is any regulating subgroup of X, then $R(X) = \bigoplus_{\rho \in T_{cr}(A)} \beta_\rho^X A_\rho$ where the *Burkhardt invariants* are given by $\beta_\tau^X = \exp(X^\sharp(\tau))/R(X^\sharp(\tau))$. If X has only one regulating subgroup that then equals the regulator, we say that X has *a regulating regulator*. By definition $X_{lp}/R(X) = (X/R(X))_p$, the p-primary component of the finite abelian group $X/R(X)$, and for an integer n, n_{lp} is the highest p-power dividing n. In the following we consider a fixed almost completely decomposable group X and various of its subgroups that are themselves almost completely decomposable groups and have localizations as such. The reader must therefore be cautioned that we consider exclusively localization with respect to the regulator $R(X)$ of the fixed group X. Explicitly, if H is a subgroup of X, its *p-local constituent* H_{lp} is given by

$$H_{lp} = H \cap \mathbb{Z}[p^{-1}] R(X), \qquad \frac{H_{lp} + R(X)}{R(X)} = \left(\frac{H + R(X)}{R(X)}\right)_p.$$

We recall some important facts ([17, Theorem 5.1.3]).

Lemma 3.1. *Let X be an almost completely decomposable group.*

(1) $R(X) = R(X_{lp})$.

(2) $\beta_\tau^{X_{lp}} = \left(\beta_\tau^X\right)_{lp}$.

(3) $\mathrm{rgi}(X_{lp}) = (\mathrm{rgi}\, X)_{lp}$.

(4) *If $A = \bigoplus_{\rho \in T_{cr}(A)} A_\rho \in \mathrm{Regg}(X)$, then $A_{lp} = \bigoplus_{\rho \in T_{cr}(A)} (A_\rho)_{lp} \in \mathrm{Regg}(X_{lp})$, where $(A_\tau)_{lp} = (\beta_\tau^X/(\beta_\tau^X)_{lp})A_\tau$.*

(5) $X(\tau)_{lp} = X_{lp}(\tau); X^\sharp(\tau)_{lp} = X_{lp}^\sharp(\tau); X[\tau]_{lp} = X_{lp}[\tau]; X^\sharp[\tau]_{lp} = X_{lp}^\sharp[\tau]$.

The next lemma is a general fact in local-global affairs.

Lemma 3.2 ([19]). *Let X be any almost completely decomposable group and suppose that $X = Y \oplus Z$. Then $Y_{lp} = Y \cap X_{lp}$, $Z_{lp} = Z \cap X_{lp}$, $X_{lp} = Y_{lp} \oplus Z_{lp}$, $Y = \sum_{p \in \mathbb{P}}(Y \cap X_{lp})$ and $Z = \sum_{p \in \mathbb{P}}(Z \cap X_{lp})$.*

Given an almost completely decomposable group X, one can pass to the localizations X_{lp} that are considerably easier to deal with than the original global group and then draw conclusions for X. It turns out that it is also easy to start with a family of local groups and construct a global group that has the original groups as localizations. The only restriction is that the given groups have a common regulator, a condition that is necessary by Lemma 3.1.1. Proposition 3.3 can be used to construct global groups from local groups with known properties. For example, if just one of the local groups is indecomposable, then so is the global group.

Proposition 3.3 ([19]). *Let P be a finite set of primes and let R be a completely decomposable group of finite rank. Suppose that for each $p \in P$, X_{lpl} is a group such that $R \leq X_{lpl} \leq \mathbb{Q}R$, X_{lpl}/R is a finite p-group and $\mathrm{R}(X_{lpl}) = R$. Then $X := \sum_{p \in P} X_{lpl}$ is an almost completely decomposable group such that $\mathrm{R}(X) = R$ and $X_{lp} = X_{lpl}$ for each $p \in P$ where X_{lp} is the p-local constituent of X (with respect to $\mathrm{R}(X)$).*

We also need to clarify what the localization of a crq-group looks like and relate top decompositions with main decompositions of localizations. We recall the definition of top decomposition for easy reference.

Definition 3.4. Let $X = A + \mathbb{Z}n^{-1}a$ be an almost completely decomposable group with completely decomposable subgroup $A = \bigoplus_{\rho \in \mathrm{T_{cr}}(A)} A_\rho$ of finite index. Write $a = \sum_{\rho \in \mathrm{T_{cr}}(A)} a_\rho$ with $a_\tau \in A_\tau$. A critical type τ is called *p-movable* if p is a prime divisor of n and either $\mathrm{hgt}_p^A(a_\tau) \geq \mathrm{hgt}_p^{\mathbb{Z}}(n)$ or there exists some critical type $\sigma < \tau$ with $\mathrm{hgt}_p^A(a_\sigma) \leq \mathrm{hgt}_p^A(a_\tau)$. The homogeneous decomposition $A = \bigoplus_{\rho \in \mathrm{T_{cr}}(A)} A_\rho$ is called a *top decomposition* of A, if, for all prime divisors p of n, $\mathrm{hgt}_p^A(a_\tau) \geq \mathrm{hgt}_p^{\mathbb{Z}}(n)$ for each p-movable type τ. If A is in top decomposition, then we say that we have a *top description* of X.

It was shown in [10] ([17, Section 6.3]) that a top decomposition can always be achieved, and after doing so the completely decomposable direct summands of the crq-group under consideration become visible. The latter is also true when passing to localizations.

Theorem 3.5. *Let $X = A + \mathbb{Z}n^{-1}a$ be a crq-group where $A \in \mathrm{Regg}(X)$, $a \in A$ and $\gcd^A(n, a) = 1$ (or equivalently, $n = \mathrm{rgi}\, X$). Let $A = \bigoplus_{\rho \in \mathrm{T_{cr}}(X)} A_\rho$ be a homogeneous decomposition of A. Then the following hold.*

(1) $X_{lp} = A_{lp} + \mathbb{Z}n_{lp}^{-1}a$ where $A_{lp} \in \mathrm{Regg}(X_{lp})$, $a \in A_{lp}$, and $\gcd^{A_{lp}}(n_{lp}, a) = 1$ (or equivalently, $n_{lp} = \mathrm{rgi}(X_{lp})$).

(2) $A = \bigoplus_{\rho \in \mathrm{T_{cr}}(A)} A_\rho$ is a top decomposition for X if and only if for all $p \in \mathbb{P}$,

$$X_{lp} = \left(\bigoplus_\rho \{(A_\rho)_{lp} : n_{lp} = \gcd^A(n, a_\rho)_{lp}\}\right)$$

$$\oplus \left(\bigoplus_\rho \{(A_\rho)_{lp} : n_{lp} > \gcd^A(n, a_\rho)_{lp}\} + \mathbb{Z}\tfrac{1}{n_{lp}}\sum_\rho \{a_\rho : n_{lp} > \gcd^A(n, a_\rho)_{lp}\}\right)$$

is a main decomposition of X_{lp}.

4 Invariants and classification of global crq-groups

The following proposition extends and generalizes [19, Proposition 4.4]. It was necessitated by the fact that a critical type of a group does not need to be a critical type of a direct summand of the group. Note that the sums ranging over the set of all types $\mathbb{T}$ are really finite sums. The least type $\mathbb{Z}$ requires special attention. We stipulate that $X^\sharp[\mathbb{Z}] = X$ which complies with the fact that $X^\sharp[\tau]$ becomes larger as τ becomes smaller, and with the isomorphism $X^\sharp[\tau]/X[\tau] \cong X(\tau)/X^\sharp(\tau)$.

Proposition 4.1. *Let* $X = A + \mathbb{Z}n^{-1}a$ *be a crq-group with* $\gcd^A(n, a) = 1$ *and* $A = \bigoplus_{\rho \in \mathbb{T}} A_\rho \in \operatorname{Regg} X$. *Write* $a = \sum_{\rho \in \mathbb{T}} a_\rho$ *such that* $a_\tau \in A_\tau$ *and set* $s_\tau = \gcd^A(n, a_\tau)$. *Further let* $X = X_{cd} \oplus X_{cl}$ *be a main decomposition of* X. *Then the following hold for any type* $\tau \in \mathbb{T}$.

(1) $\operatorname{rgi} X[\tau] = \gcd\{s_\sigma : \sigma \in \mathbb{T}, \sigma \leq \tau\}$.

(2) $\operatorname{rgi} X^\sharp[\mathbb{Z}] = \operatorname{rgi} X$, *and for* $\tau \not\cong \mathbb{Z}$, $\operatorname{rgi} X^\sharp[\tau] = \gcd\{s_\sigma : \sigma \in \mathbb{T}, \sigma < \tau\}$.

(3) $\operatorname{rgi} X[\tau] = \operatorname{rgi} X^\sharp[\tau]$ *if and only if* A_τ *is a summand of* X.

(4) X *is clipped if and only if it is slim and* $\operatorname{rgi} X[\tau] \neq \operatorname{rgi} X^\sharp[\tau]$ *for every* $\tau \in \operatorname{T_{cr}}(X)$.

(5) $\tau \in \operatorname{T_{cr}}(X_{cl})$ *if and only if* $\operatorname{rgi} X^\sharp[\tau] \neq \operatorname{rgi} X[\tau]$.

(6) $\forall \tau \in \mathbb{T}$, $\gcd\{\operatorname{rgi} X[\sigma] : \sigma \in \operatorname{T_{cr}}(X), \sigma \neq \tau\} = 1$.

(7) *For any two types* τ, σ, *if* $\sigma \leq \tau$, *then* $\operatorname{rgi} X[\tau]$ *divides* $\operatorname{rgi} X[\sigma]$.

(8) $\operatorname{rgi} X^\sharp[\tau] = \gcd\{\operatorname{rgi} X[\sigma] : \sigma < \tau\}$ *and by convention* $\operatorname{rgi} X^\sharp[\mathbb{Z}] = \operatorname{rgi} X$.

(9) $\operatorname{rgi} X(\tau) = \gcd\{\operatorname{rgi} X[\sigma] : \sigma \in \operatorname{T_{cr}}(X), \sigma \not\geq \tau\}$ *and* $\operatorname{rgi} X^\sharp(\tau) = \gcd\{\operatorname{rgi} X[\sigma] : \sigma \in \operatorname{T_{cr}}(X), \sigma \not> \tau\}$.

(10) $\beta^X_\tau = \gcd\{\operatorname{rgi} X[\sigma] : \sigma \in \operatorname{T_{cr}}(X), \sigma \not\geq \tau\}$
    $= \gcd\{\operatorname{rgi} X[\sigma] : \sigma \in \operatorname{T_{cr}}(X), \sigma \not> \tau\}$.

Proof. Note that $s_\sigma = n$ whenever $a_\sigma = 0$, and this is the case for almost all types σ.
(1) We distinguish two cases. Suppose that $\{\sigma \in \operatorname{T_{cr}}(X) : \sigma \leq \tau\} = \emptyset$. Then $X[\tau] = X$ and $\operatorname{rgi} X[\tau] = n$. Also $a_\sigma = 0$ for all $\sigma \leq \tau$, so $s_\sigma = n$, and $\gcd\{s_\sigma : \sigma \leq \tau\} = n$. The claim is established in this case. So suppose next that $\{\sigma \in \operatorname{T_{cr}}(X) : \sigma \leq \tau\} \neq \emptyset$. Then $\operatorname{rgi} X[\tau] = \gcd\{s_\sigma : \sigma \in \operatorname{T_{cr}}(X), \sigma \leq \tau\} = \gcd\{s_\sigma : \sigma \in \mathbb{T}, \sigma \leq \tau\}$.
(2) In addition to the case handled by conventions, there are again two cases and they are settled as in (1).
(3) First assume that $\tau \notin \operatorname{T_{cr}}(X)$. Then $A_\tau = 0$. Also $\{\sigma : \sigma \in \operatorname{T_{cr}}(X), \sigma \leq \tau\} = \{\sigma : \sigma \in \operatorname{T_{cr}}(X), \sigma < \tau\}$ and this implies that $X[\tau] = X^\sharp[\tau]$ and even more that $\operatorname{rgi} X[\tau] = \operatorname{rgi} X^\sharp[\tau]$. On the other hand the case $\tau \in \operatorname{T_{cr}}(X)$ is settled explicitly in [19, Proposition 4.4].
(4) Immediate from (3).

(5) Note that $\operatorname{rgi} X[\tau] = (\operatorname{rgi} X_{cd}[\tau])(\operatorname{rgi} X_{cl}[\tau]) = \operatorname{rgi} X_{cl}[\tau]$ and similarly $\operatorname{rgi} X^\sharp[\tau] = \operatorname{rgi} X^\sharp_{cl}[\tau]$. Suppose that $\tau \in \operatorname{T_{cr}}(X_{cl})$. By part (4) $\operatorname{rgi} X[\tau] = \operatorname{rgi} X_{cl}[\tau] \neq \operatorname{rgi} X^\sharp_{cl}[\tau] = \operatorname{rgi} X^\sharp[\tau]$. Conversely, assume $\operatorname{rgi} X[\tau] = \operatorname{rgi} X^\sharp[\tau]$. Then $\operatorname{rgi} X^\sharp_{cl}[\tau] = \operatorname{rgi} X^\sharp[\tau] = \operatorname{rgi} X[\tau] = \operatorname{rgi} X^\sharp_{cl}[\tau]$. By part (3) $\tau \notin \operatorname{T_{cr}}(X_{cl})$ else the clipped group X_{cl} would have a completely decomposable summand of type τ.

(6) $\gcd\{\operatorname{rgi} X[\sigma] : \sigma \in \operatorname{T_{cr}}(X), \sigma \neq \tau\}$ divides $\gcd\{s_\sigma : \sigma \in \operatorname{T_{cr}}(X), \sigma \neq \tau\} = 1$ which is true by the Purification Lemma and the fact that A_τ is pure in X, A being a regulating subgroup of X.

(7) Obvious from part (1).

(8) $\gcd\{\operatorname{rgi} X[\sigma] : \sigma < \tau\} = \gcd\{\gcd\{s_\rho : \rho \leq \sigma\} : \sigma < \tau\} = \gcd\{s_\sigma : \sigma < \tau\} = \operatorname{rgi} X^\sharp[\tau]$.

(9) Note that $\rho \not\geq \tau$ if $\rho \leq \sigma$ and $\sigma \not\geq \tau$. By the Purification Lemma $\operatorname{rgi} X(\tau) = \gcd\{s_\sigma : \sigma \not\geq \tau\}$. In the expression $\gcd\{s_\sigma : \sigma \not\geq \tau\}$, the values s_σ can be regrouped and duplicated without changing the greatest common divisor, and furthermore partial greatest common divisors can be formed. Therefore, using (1),

$$\begin{aligned} \operatorname{rgi} X(\tau) &= \gcd\{s_\sigma : \sigma \not\geq \tau\} \\ &= \gcd\{\gcd\{s_\rho : \rho \leq \sigma\} : \sigma \not\geq \tau\} \\ &= \gcd\{\operatorname{rgi} X[\sigma] : \sigma \not\geq \tau\}. \end{aligned}$$

The argument for $\operatorname{rgi} X^\sharp(\tau)$ is the same with $>$ replacing $\geq$.

(10) By [17, Theorem 6.2.2] $\beta^X_\tau = \operatorname{rgi} X(\tau) = \operatorname{rgi} X^\sharp(\tau)$. □

The near-classification theorem for global crq-groups is as follows ([19, Theorem 4.5]).

Theorem 4.2. *Two* crq-*groups X and Y are nearly isomorphic if and only if*

(1) $\operatorname{R}(X) \cong \operatorname{R}(Y)$,

(2) $\operatorname{rgi}(X) = \operatorname{rgi}(Y)$,

(3) $\operatorname{rgi} X[\tau] = \operatorname{rgi} Y[\tau]$ *for every* $\tau \in \operatorname{T_{cr}}(X) = \operatorname{T_{cr}}(Y)$.

We also recall an easy corollary ([19, Corollary 4.7]) which simplifies the description of crq-groups by passing to a nearly isomorphic group.

Corollary 4.3. *Let $A = \bigoplus_{\rho \in \operatorname{T_{cr}}(A)} A_\rho$ be a homogeneous decomposition of the completely decomposable group A, and let $a_\tau \in A_\tau$. Consider the two* crq-*groups $X = A + \mathbb{Z}\frac{1}{n}\sum_{\rho \in \operatorname{T_{cr}}(A)} t_\rho a_\rho$ and $Y = A + \mathbb{Z}\frac{1}{n}\sum_{\rho \in \operatorname{T_{cr}}(A)} a_\rho$ where the t_τ are integers. If $\gcd(n, t_\tau) = 1$ for every $\tau \in \operatorname{T_{cr}}(A)$, then $X \cong_{\operatorname{nr}} Y$.*

Finally we establish a general existence theorem.

Theorem 4.4. *Let T be a finite set of pairwise non-isomorphic rational groups, $\{s_\rho : \rho \in T\}$ a set of positive integers and n a positive integer that is a multiple of $\operatorname{lcm}\{s_\rho : \rho \in T\}$. Let $A = \bigoplus_{\rho \in T} \rho v_\rho$, $a = \sum_{\rho \in T} s_\rho v_\rho$, and $X = A + \mathbb{Z}n^{-1}a$. Then X is a slim* crq-*group with $\operatorname{T_{cr}}(X) = T$. Consider the properties* (GCD) *through* (NCRQ).

(GCD) *For every prime p dividing n, $(s_\tau)_{lp} \leq n_{lp}$, and if $(s_\tau)_{lp} < n_{lp}$, then $p^{-1} \notin \tau$.*

(IDX) $\gcd\{s_\rho : \rho \in T\} = 1$.

(RGG) *For every $\tau \in T$, $\gcd\{s_\rho : \rho \not\geq \tau\} = \gcd\{s_\rho : \rho \not> \tau\}$.*

(RGR) *For every $\tau \in T$, $\gcd\{s_\rho : \rho \not\geq \tau\} = 1$.*

(CLP) $\gcd\{s_\rho : \rho < \tau\}$ *does not divide* s_τ.

(NCRQ) *Whenever $\sigma \leq \tau$, then s_τ divides s_σ.*

The following is true.

(a) *If* (GCD), *then, for every $\tau \in T$, $\gcd^A(n, s_\tau v_\tau) = s_\tau$.*

Assuming (GCD), *the following hold.*

(b) (IDX) *if and only if $[X : A] = n$.*

(c) (RGG) *if and only if A is regulating in X.*

(d) (RGR) *if and only if A is the regulating regulator of X.*

(e) (CLP) *if and only if X is clipped.*

(f) (NCRQ) *if and only if $s_\tau = \mathrm{rgi}\, X[\tau]$ for every $\tau \in T$.*

Proof. (a) Assume (GCD). Suppose first that $n_{lp} = (s_\tau)_{lp}$. Then, by (GCD), $(\hat{s}_\tau)_{lp} = n_{lp} = \gcd^A(n, s_\tau v_\tau)$. On the other hand, if $(s_\tau)_{lp} < n_{lp}$, then, by (GCD), $p^{-1} \notin \tau$ and hence $\gcd^A(n, s_\tau v_\tau) = s_\tau$.

Note now that, by (a) and [17, Lemma 11.2.9], for any subset S of T,

$$\gcd^A\left(n, \sum_{\rho \in S} s_\rho v_\rho\right) = \gcd\{s_\rho : \rho \in S\}. \tag{4.5}$$

(b) (4.5) and [17, Lemma 6.1.2].

(c) (4.5) and [17, Theorem 6.2.2].

(d) The condition (RGR) implies both (IDX) and (RGG). Hence A is regulating in X. Furthermore, $\beta_\tau^X = \gcd^A(n, \sum\{s_\rho v_\rho : \rho \not\geq \tau\} = \{s_\rho : \rho \not\geq \tau\} = 1$ which means that the regulator is A.

(e) (4.5) and Proposition 4.1(2).

(f) Suppose that (NCRQ) holds. Then $\mathrm{rgi}\, X[\tau] = \gcd(\{s_\rho : \rho \leq \tau\}) = s_\tau$ by (NCRQ). Conversely, assume that $s_\tau = \mathrm{rgi}\, X[\tau]$ for every $\tau \in T$. Then (NCRQ) holds since $\mathrm{rgi}\, X[\tau]$ divides $\mathrm{rgi}\, X[\sigma]$ whenever $\sigma \leq \tau$. □

We need a relationship between the values s_τ of Theorem 4.4 and the invariants $\mathrm{rgi}\, X[\tau]$ of Theorem 4.2. Let $X = A + \mathbb{Z}n^{-1}a$ be a crq-group and A a regulating subgroup of X. Suppose that $n = \mathrm{rgi}\, X$, $A = \bigoplus_{\rho \in \mathrm{T_{cr}}(A)} A_\rho$, and $a = \sum_{\rho \in \mathrm{T_{cr}}(X)} a_\rho$ with $a_\tau \in A_\tau$. Then $\mathrm{rgi}\, X[\tau] = \gcd\{\gcd^A(n, a_\rho) : \rho \leq \tau\}$ always divides $\gcd^A(n, a_\tau)$. It would be nice to have the equality $\gcd^A(n, a_\tau) = \operatorname{rgi} X[\tau]$ but this cannot be achieved if $\tau \in \mathrm{T}_{\mathrm{cr}}(X)$ is p-divisible for some prime factor p of the regulating index $\operatorname{rgi} X$ and some critical types $\leq \tau$ are not p-divisible (see Theorem 4.4(f)). We wish to describe the factors t_τ with the property that $\gcd^A(n, a_\tau) = t_\tau \operatorname{rgi} X[\tau]$. Consider the troublesome case that $p\tau = \tau$ for a prime factor p of the regulating index. Then $\gcd^A(n, a_\tau)_{\wr p} = n_{\wr p}$ and hence to achieve the desired equality it is required that $(t_\tau)_{\wr p} = (\operatorname{rgi} X)_{\wr p}/(\operatorname{rgi} X[\tau])_{\wr p}$. In all other cases we set $(t_\tau)_{\wr p} = 1$, so that $t_\tau = \prod_{p \in \mathbb{P}}(t_\tau)_{\wr p}$ is given by specifying its prime factorization. We have shown that the following lemma is valid.

Lemma 4.6. *Let $X = A + \mathbb{Z}n^{-1}a$ be a crq-group and A a regulating subgroup of X. Suppose that $n = \operatorname{rgi} X$ and $A = \bigoplus_{\rho \in \mathrm{T}_{\mathrm{cr}}(A)} A_\rho$ is a top decomposition of X. Write $a = \sum_{\rho \in \mathrm{T}_{\mathrm{cr}}(X)} a_\rho$ with $a_\tau \in A_\tau$. Then*

$$\gcd^A(n, a_\tau) = t_\tau \operatorname{rgi} X[\tau],$$

where

$$t_\tau = \begin{cases} 1 & \text{if } \{p : p \mid \operatorname{rgi} X, p\tau = \tau\} = \emptyset, \\ \prod \left\{ \frac{(\operatorname{rgi} X)_{\wr p}}{(\operatorname{rgi} X[\tau])_{\wr p}} : p \mid \operatorname{rgi} X, p\tau = \tau \right\} & \text{otherwise.} \end{cases}$$

Corollary 4.7. *Let X be a crq-group. Then the following hold.*

(1) $\gcd\{t_\rho \operatorname{rgi} X[\rho] : \rho \in \mathrm{T}_{\mathrm{cr}}(X)\} = 1$.

(2) *For every $\tau \in \mathrm{T}_{\mathrm{cr}}(X)$, $\gcd\{t_\rho \operatorname{rgi} X[\rho] : \tau \neq \rho \in \mathrm{T}_{\mathrm{cr}}(X)\} = 1$.*

(3) *For every $\tau \in \mathrm{T}_{\mathrm{cr}}(X)$,*

$$\beta_\tau^X = \gcd\{t_\rho \operatorname{rgi} X[\rho] : \rho \not\geq \tau\} = \gcd\{t_\rho \operatorname{rgi} X[\rho] : \rho \not> \tau\}.$$

(4) *For every $\tau \in \mathrm{T}_{\mathrm{cr}}(X)$,*

$$\gcd\{t_\rho \operatorname{rgi} X[\rho] : \rho \leq \tau\} = \operatorname{rgi} X[\tau].$$

Proof. Choose a regulating subgroup $A = \bigoplus_{\rho \in \mathrm{T}_{\mathrm{cr}}(A)} A_\rho$ of X in top decomposition such that X/A is cyclic and apply Lemma 4.6. Then (1) reflects the fact that $\operatorname{rgi} X = [X : A]$, while (2) follows because A_τ is pure in X. For (3) we have

$$\begin{aligned} \beta_\tau^X &= \operatorname{rgi} X(\tau) = \operatorname{rgi} X^\sharp(\tau), \\ \operatorname{rgi} X(\tau) &= \gcd\{t_\rho \operatorname{rgi} X[\rho] : \rho \not\geq \tau\}, \\ \operatorname{rgi} X^\sharp(\tau) &= \gcd\{t_\rho \operatorname{rgi} X[\rho] : \rho \not> \tau\}. \end{aligned}$$

The claim (4) is obtained by another application of the Purification Lemma. □

5 Indecomposable crq-groups

The direct decompositions of local crq-groups were determined in [23] and shown to be unique up to isomorphism. We will utilize the localizations of global crq-groups in order to gain insight into decompositions of global crq-groups.

We observe first that a τ-Butler complement of a crq-group X is a direct summand of X if and only if the τ-Butler complements of all localizations X_{lp} are direct summands of X_{lp}.

Proposition 5.1. *Let X be an arbitrary* crq-*group. Let A_τ be a τ-Butler complement of X and for each prime p, let $A_{\tau p}$ be a τ-Butler complement of X_{lp}. Then A_τ is a direct summand of X if and only if $A_{\tau p}$ is a direct summand of X_{lp} for every prime p.*

Proof. By Lemma 3.1 rgi $X_{lp}[\tau] = (\text{rgi}\,X[\tau])_{lp}$ and rgi $X_{lp}^\sharp[\tau] = (\text{rgi}\,X^\sharp[\tau])_{lp}$. Hence rgi $X[\tau] = $ rgi $X^\sharp[\tau]$ if and only if rgi $X_{lp}[\tau] = $ rgi $X_{lp}^\sharp[\tau]$ for every p. The claim now follows from Proposition 4.1(3). □

The following is a useful general triviality.

Lemma 5.2. *Let X be an almost completely decomposable group and $X = X_1 \oplus \cdots \oplus X_m$ a direct decomposition of X. Then $X_{lp} = (X_1)_{lp} \oplus \cdots \oplus (X_m)_{lp}$ for any prime p. Consequently, if X_{lp} is indecomposable for some p, then X is indecomposable also. It can happen that X is indecomposable but every localization X_{lp} is decomposable.*

Proof. Lemma 3.2 and Example 5.6. □

Lemma 5.2 is particularly useful if the almost completely decomposable group X is slim.

Theorem 5.3. *Let X be a slim almost completely decomposable group.*

(1) *If $X = X_1 \oplus \cdots \oplus X_m$ is a direct decomposition of X, then $\text{T}_{\text{cr}}(X) = \text{T}_{\text{cr}}(X_1) \cup \cdots \cup \text{T}_{\text{cr}}(X_m)$ is a partition of $\text{T}_{\text{cr}}(X)$.*

(2) *Suppose that X is a local group. Then there is a unique partition $\text{T}_{\text{cr}}(X) = T_1 \cup \cdots \cup T_m$ such that for any indecomposable decomposition $X = X_1 \oplus \cdots \oplus X_k$ we have $k = m$ and, after relabeling if necessary, $\text{T}_{\text{cr}}(X_i) = T_i$.*

(3) *Let X be global and let $\text{T}_{\text{cr}}(X) = \text{T}_{\text{cr}}(X_{lp}) = T_{p1} \cup \cdots \cup T_{pm_p}$ be the unique partition determined in (2) for the local group X_{lp}. If $X = X_1 \oplus \cdots \oplus X_m$ is a direct decomposition of X, then for any prime p and any $i \in \{1, \ldots, m_p\}$ there is j such that $T_{pi} \subseteq \text{T}_{\text{cr}}(X_j)$.*

Proof. (1) is a direct consequence of slimness. (2) follows from the Unique Decomposition Theorem of Faticoni–Schultz (Theorem 1.2) and the fact that nearly isomorphic almost completely decomposable groups have equal critical typesets. Finally (3) is a consequence of (1), (2), and Lemma 5.2. □

Definition 5.4. (1) Let X be a slim local almost completely decomposable group. The unique partition of $\mathrm{T}_{\mathrm{cr}}(X)$ established in Theorem 5.3.2 will be called the *decomposition partition* of X.

(2) Let X be a slim global almost completely decomposable group and let $\mathrm{T}_{\mathrm{cr}}(X) = \mathrm{T}_{\mathrm{cr}}(X_{lp}) = T_{p1} \cup \cdots \cup T_{pm_p}$ be the decomposition partitions of its localizations. We define the *interlacing graph* of X as follows. The vertices of the graph are the elements of $\mathrm{T}_{\mathrm{cr}}(X)$. Two vertices σ, τ are joined by an edge if and only if there exist p, i such that $\sigma, \tau \in T_{pi}$.

The interlacing graph is similar to the "frame" in [17, Definition 13.1.9]. The following corollary follows immediately from Theorem 5.3.

Corollary 5.5. *Let X be a slim almost completely decomposable group. Then X is indecomposable if its interlacing graph is connected.*

We will now give an example of an indecomposable crq-group whose localizations are as close to being completely decomposable as is possible.

Example 5.6. There is an indecomposable crq-group of rank 4 all of whose localizations have completely decomposable summands of rank 2 and clipped parts of rank 2.

Proof. Let $\tau_1, \ldots, \tau_4$ be rational groups ordered as types as shown.

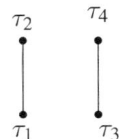 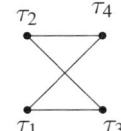

Poset of Types   Interlacing Graph

Assume that $p_1, \ldots, p_4$ are distinct prime numbers such that $p_i^{-1} \notin \tau_j$ for $i, j \in \{1, 2, 3, 4\}$. Let $A = \bigoplus_{i=1}^{4} \tau_i v_i$ and let

$$X = A + \mathbb{Z} \frac{1}{p_1 p_2 p_3 p_4} (p_1 p_2 v_1 + p_3 p_4 v_2 + p_1 p_3 v_3 + p_2 p_4 v_4).$$

It is easily checked that $X(\tau_1) = A(\tau_1)$ and $X(\tau_3) = A(\tau_3)$ which shows that A is regulating in X and since $\mathrm{T}_{\mathrm{cr}}(X)$ is an inverted forest, A is also the regulator of X. We use Theorem 3.5.1 to find the localizations and Corollary 4.3 to simplify the expressions.

$$X_{\ell p_1} = A + \mathbb{Z}\frac{1}{p_1}(p_3p_4v_2 + p_2p_4v_4) \cong_{\mathrm{nr}} A + \mathbb{Z}\frac{1}{p_1}(v_2 + v_4),$$

$$X_{\ell p_2} = A + \mathbb{Z}\frac{1}{p_2}(p_3p_4v_2 + p_1p_3v_3) \cong_{\mathrm{nr}} A + \mathbb{Z}\frac{1}{p_2}(v_2 + v_3),$$

$$X_{\ell p_3} = A + \mathbb{Z}\frac{1}{p_3}(p_1p_2v_1 + p_2p_4v_4) \cong_{\mathrm{nr}} A + \mathbb{Z}\frac{1}{p_3}(v_1 + v_4),$$

$$X_{\ell p_4} = A + \mathbb{Z}\frac{1}{p_4}(p_1p_2v_1 + p_1p_3v_3) \cong_{\mathrm{nr}} A + \mathbb{Z}\frac{1}{p_4}(v_1 + v_3).$$

The interlacing graph is obviously connected and shows that the group is indecomposable. □

On the basis of Proposition 3.3, the group X in Example 5.6 was actually obtained by adding the above localizations and then simplifying by passing to a near-isomorphic group.

In a special case we can show the converse of Corollary 5.5. We have no example of an indecomposable crq-group whose interlacing graph is not connected.

Proposition 5.7. *Let X be a slim crq-group such that for every prime p the clipped part $(X_{\ell p})_{cl}$ of the localization $X_{\ell p}$ is indecomposable. Then X is indecomposable if and only if the interlacing graph of X is connected.*

Proof. One direction of the claim being true in general, we only need to show that a disconnected interlacing graph leads to a direct decomposition of X. Suppose that the interlacing graph of X has two components without edges between them. Let T_1 and T_2 be the non-void sets of vertices of these two components. Assume that the group X is given in a top description $X = A + \mathbb{Z}n^{-1}a$ where $n = \mathrm{rgi}\, X$, $A = \bigoplus_{\rho \in \mathrm{T}_{\mathrm{cr}}(A)} A_\rho$ is a top decomposition, and $a = \sum_{\rho \in \mathrm{T}_{\mathrm{cr}}(A)} s_\rho a_\rho$ is such that $a_\tau \in A_\tau$ and $\gcd^A(n, s_\tau a_\tau) = s_\tau$. Recall the resulting main decompositions of the localizations given in Theorem 3.5(4). It follows from this and the assumption that $(X_{\ell p})_{cl}$ is indecomposable that $\sigma \in \mathrm{T}_{\mathrm{cr}}((X_{\ell p})_{cl}) \cap T_i$ implies $n_{\ell p} = (s_\tau)_{\ell p}$ for every $\tau \in T_j$ where $j \neq i$.

Let $X_i = (\bigoplus_{\rho \in T_i} A_\rho)_*^X$. To show that $X = X_1 \oplus X_2$ it is necessary and sufficient that $n = \mathrm{rgi}\, X = \mathrm{rgi}\, X_1 \cdot \mathrm{rgi}\, X_2$. The latter is true for the following reasons. Let p be a prime divisor of n. There is $\sigma \in \mathrm{T}_{\mathrm{cr}}(X)$ such that $(s_\sigma)_{\ell p} < n_{\ell p}$ (in fact there is a σ such that $(s_\sigma)_{\ell p} = 1$). The critical type σ either belongs to T_1 or T_2. Without loss of generality $\sigma \in T_1$. By the first paragraph of this proof $n_{\ell p}$ divides s_τ for every $\tau \in T_2$ and hence $n_{\ell p}$ divides $\mathrm{rgi}\, X_1 = \gcd\{s_\rho : \rho \in T_2\}$. Thus for every prime p the prime power factor $n_{\ell p}$ divides either $\mathrm{rgi}\, X_1$ or $\mathrm{rgi}\, X_2$ and it follows that $n = \mathrm{rgi}\, X = \mathrm{rgi}\, X_1 \cdot \mathrm{rgi}\, X_2$ as desired. □

6 Decompositions of crq-groups with regulating regulator

The idea is to generalize the results of Blagoveshchenskaya–Mader ([17, Section 13]) on block-rigid crq-groups to crq-groups with regulating regulator. The class of groups with regulating regulator is much larger than the class of block-rigid almost completely decomposable groups. For example, if the critical typeset of an almost completely decomposable group is an inverted forest, then the group has a regulating regulator ([17, Proposition 4.5.4]).

An indecomposability criterion is readily available.

Proposition 6.1. *A crq-group with regulating regulator is indecomposable if and only if it is slim and its interlacing graph is connected.*

Proof. Let X be a crq-group. Suppose first that X is indecomposable. Then it is clipped and hence slim ([17, Lemma 6.3.2]). Since X has a regulating regulator, every one of its localizations X_{lp} has a regulating regulator (Lemma 3.1, (1) and (4)). Therefore the clipped part of each localization X_{lp} is indecomposable ([17, Theorem 6.5.7]). Hence Proposition 5.7 applies and says that the interlacing graph is connected. Conversely, if X is slim and its interlacing graph is connected, then X is indecomposable by Corollary 5.5. □

We need indecomposability expressed in terms of the invariants rgi $X[\tau]$. By Proposition 4.1.8 we may use rgi $X^{\sharp}[\tau]$ as well.

Corollary 6.2. *A crq-group X with regulating regulator and $\operatorname{rk} X \geq 2$ is indecomposable if and only if X is slim and there is a labeling $p_1, \ldots, p_k$ of the prime factors of $\operatorname{rgi} X$ such that*

$$T_i \cap T_{i+1} \neq \emptyset \quad \text{where} \quad T_j = \left\{\tau \in \mathrm{T}_{\mathrm{cr}}(X) : (\operatorname{rgi} X[\tau])_{lp_j} \neq (\operatorname{rgi} X^{\sharp}[\tau])_{lp_j}\right\}.$$

Proof. It follows from Proposition 4.1.6 and Lemma 3.1.5 that $T_j = \mathrm{T}_{\mathrm{cr}}(X_{lp_j})_{cl}$. □

The following lemma is likely to be known to many, but it will be needed later.

Lemma 6.3. *Suppose that Y is a finite cyclic extension of B and that Z is a finite cyclic extension of C, and that the indices $[Y : B]$ and $[Z : C]$ are relatively prime. Then $X = Y \oplus Z$ is a cyclic extension of $A = B \oplus C$ with $[X : A] = [Y : B][Z : C]$.*

Proof. Set $n_Y = [Y : B], n_Z = [Z : C], n = n_Y n_Z$. By hypothesis there exist integers u, v such that $1 = un_Y + vn_Z$. Write $Y = B + \mathbb{Z} n_Y^{-1} b$ and $Z = C + \mathbb{Z} n_Z^{-1} c$. Set

$$X' = A + \mathbb{Z} \tfrac{1}{n} a, \quad \text{where} \quad a = un_Y c + vn_Z b.$$

We claim that $X = X'$. This follows from the identities

$$\frac{1}{n}a = u\frac{1}{n_Z}c + v\frac{1}{n_Y}b, \quad \frac{1}{n_Y}b = n_Z\frac{1}{n}a + ub - uc, \quad \frac{1}{n_Z}c = n_Y\frac{1}{n}a - vb + vc. \quad \square$$

The strategy is the following.

(1) A crq-group with regulating regulator is determined up to near-isomorphism by certain invariants (Theorem 4.2).

(2) Summands of crq-groups with regulating regulator are again crq-groups with regulating regulator (Lemma 6.5).

(3) The relationship of the invariants of a group and of the invariants of the summands of a decomposition is determined (Lemma 6.5).

(4) Given a set of invariants as in (3), a direct sum of crq-groups with regulating regulator is formed having the given invariants.

(5) By the classification theorem the direct sum is nearly isomorphic to the original group, and the latter has a decomposition corresponding to the invariants by Arnold's theorem.

Note that [17, Example 4.4.7] shows that the direct sum of two crq-groups with regulating regulators does not need to have a regulating regulator even if one summand is completely decomposable and the other a local crq-group.

Lemma 6.4. *Let X be an almost completely decomposable group with regulating subgroup A. Suppose that $A \leq Y \leq X$. Then A is a regulating subgroup of Y and any regulating subgroup of Y is a regulating subgroup of X. In particular, if X has a regulating regulator, then so does Y.*

Proof. That A is regulating in Y is well-known ([17, Lemma 4.1.9]). Suppose $B \in$ Regg(Y). Then
$$[X:B] = [X:Y][Y:B] = [X:Y][Y:A] = [X:A].$$
Hence B is regulating in X. □

We collect in a lemma the relationships between the invariants of a group and those of its summands.

Lemma 6.5. *Let X be a crq-group with regulating regulator $A = \mathrm{R}(X)$. Suppose that $X = Y \oplus Z$. Then the following hold.*

(1)   (a) $\mathrm{R}(X) = \mathrm{R}(Y) \oplus \mathrm{R}(Z)$,

(b) $X/\mathrm{R}(X) \cong Y/\mathrm{R}(Y) \oplus Z/\mathrm{R}(Z)$,

(c) $\mathrm{rgi}\, X = \mathrm{rgi}\, Y \,\mathrm{rgi}\, Z$,

(d) $\gcd(\mathrm{rgi}\, Y, \mathrm{rgi}\, Z) = 1$.

(2) *For every $\tau \in \mathrm{T}_{\mathrm{cr}}(X)$,*

(a) $X[\tau] = Y[\tau] \oplus Z[\tau]$,

(b) $X[\tau], Y[\tau],$ and $Z[\tau]$ are crq-groups with regulating regulators,

(c) $\mathrm{R}(X[\tau]) = \mathrm{R}(Y[\tau]) \oplus \mathrm{R}(Z[\tau])$,

(d) $X[\tau]/\mathrm{R}(X[\tau]) \cong Y[\tau]/\mathrm{R}(Y[\tau]) \oplus Z[\tau]/\mathrm{R}(Z[\tau])$,

(e) $\mathrm{rgi}\, X[\tau] = \mathrm{rgi}\, Y[\tau]\, \mathrm{rgi}\, Z[\tau]$,

(f) $\gcd(\mathrm{rgi}\, Y[\tau], \mathrm{rgi}\, Z[\tau]) = 1$.

Proof. (1) Summands of groups with regulating regulators have regulating regulators by [17, Corollary 4.5.3]. The quotient $X/\operatorname{R}(X)$ is cyclic by hypothesis and hence the summands $Y/\operatorname{R}(Y)$ and $Z/\operatorname{R}(Z)$ are also cyclic and have relatively prime orders. These orders are the regulating indices and also $[X:\operatorname{R}(X)] = [Y:\operatorname{R}(Y)] \cdot [Z:\operatorname{R}(Z)]$.
(2) By Lemma 6.4 (with $Y = B \oplus X[\tau]$ where $A = B \oplus A[\tau]$ is the regulating regulator of X) the subgroup $X[\tau]$ of X has a regulating regulator. The rest is an application of (1). □

We now formulate a decomposition criterion.

Theorem 6.6. *Let X be a crq-group that is not completely decomposable and has a regulating regulator; set $n = \operatorname{rgi} X (> 1)$ and $n_\tau = \operatorname{rgi} X[\tau]$ for $\tau \in \operatorname{T}_{\operatorname{cr}}(X)$. Let*

$$X = X_1 \oplus X_2$$

be a proper direct decomposition of X. Then there are factorizations $n = n_1 \cdot n_2$ and $n_\tau = n_{1\tau} \cdot n_{2\tau}$ for each $\tau \in \operatorname{T}_{\operatorname{cr}}(X)$ such that the following conditions are satisfied where $T_i = \{\tau \in \operatorname{T}_{\operatorname{cr}}(X) : n_{i\tau}^\sharp \neq n_{i\tau}\}$ and

$$n_{i\tau}^\sharp = \begin{cases} n_i & \text{if } \{\sigma \in \operatorname{T}_{\operatorname{cr}}(X) : \sigma < \tau\} = \emptyset, \\ \gcd\{n_{i\sigma} : \sigma < \tau\} & \text{if } \{\sigma \in \operatorname{T}_{\operatorname{cr}}(X) : \sigma < \tau\} \neq \emptyset. \end{cases}$$

(2Dec1) $\gcd(n_1, n_2) = 1$;

(2Dec2) $n_{i\tau}$ *divides* n_i;

(2Dec3) $n_{i\tau}$ *divides* $n_{i\sigma}$ *whenever* $\tau \geq \sigma$;

(2Dec4) *for each* $\tau \in \operatorname{T}_{\operatorname{cr}}(X)$, $\left|\{i : n_{i\tau} \neq n_{i\tau}^\sharp\}\right| \leq \operatorname{rk}\left(\frac{X(\tau)}{X^\sharp(\tau)}\right)$.

(2Dec5) *For $\tau \in T_i$ define an integer $t_{i\tau}$ by stipulating that $t_{i\tau} = \prod_{p | n_i} (t_{i\tau})_p$ where for each prime divisor p of n_i,*

$$(t_{i\tau})_p = \begin{cases} 1 & \text{if } p\tau \neq \tau, \\ (n_i)_p / (n_{i\tau})_p & \text{if } p\tau = \tau. \end{cases}$$

With these values the following hold.

(a) $\gcd\{t_{i\rho} n_{i\rho} : \rho \in T_i, \rho \not\geq \tau\} = 1$ *for every* $\tau \in T_i$.

(b) $\gcd\{t_{i\rho} n_{i\rho} : \rho \in T_i, \rho \leq \tau\} = n_{i\tau}$ *for every* $\tau \in T_i$.

Proof. Let $n_i = \operatorname{rgi} X_i$ and $n_{i\tau} = \operatorname{rgi} X_i[\tau]$. Then (2Dec1) holds by Lemma 6.5.1(d). (2Dec2) and (2Dec3) hold by the Purification Lemma. (2Dec4) requires more argument. The inequality $n_{i\tau} \neq n_{i\tau}^\sharp$ means that τ is a critical type of $(X_i)_{cl}$ by Proposition 4.1(5). The claim follows from the fact that $\operatorname{rk}((X_i)_{cl}(\tau)/(X_i)_{cl}^\sharp(\tau)) = 1$ if τ is a critical type of X_i and the formula

$$\frac{X(\tau)}{X^\sharp(\tau)} \cong \frac{X_1(\tau)}{X_1^\sharp(\tau)} \oplus \frac{X_2(\tau)}{X_2^\sharp(\tau)}.$$

Finally, (2Dec5) follows from Corollary 4.7 and the fact that X_i has a regulating regulator and hence Burkhardt invariants equal to 1. □

We show conversely that a factorization of the sort in Theorem 6.8 results in a corresponding decomposition of the group.

Theorem 6.7. *Let X be a crq-group with a regulating regulator, set $n = \operatorname{rgi} X$ and $n_\tau = \operatorname{rgi} X[\tau]$ for $\tau \in \operatorname{T_{cr}}(X)$. Suppose there are factorizations $n = n_1 n_2$ and $n_\tau = n_{1\tau} n_{2\tau}$ for each $\tau \in \operatorname{T_{cr}}(X)$ such that the conditions (2Dec1) through (2Dec5) are satisfied with $T_i = \{\tau \in \operatorname{T_{cr}}(X) : n_{i\tau}^\sharp \neq n_{i\tau}\}$ and*

$$n_{i\tau}^\sharp = \begin{cases} n_i & \text{if } \{\sigma \in \operatorname{T_{cr}}(X) : \sigma < \tau\} = \emptyset, \\ \gcd\{n_{i\sigma} : \sigma < \tau\} & \text{if } \{\sigma \in \operatorname{T_{cr}}(X) : \sigma < \tau\} \neq \emptyset. \end{cases}$$

Then there is a decomposition

$$X = X_1 \oplus X_2$$

such that X_1, X_2 are crq-groups with $\operatorname{rgi} X_i = n_i$, $\operatorname{rgi} X_i[\tau] = n_{i\tau}$, $\operatorname{rgi} X^\sharp[\tau] = n_{i\tau}^\sharp$.

Proof. Let $1 \leq i \leq 2$ and for each type in $T_i := \{\tau : n_{i\tau} \neq n_{i\tau}^\sharp\}$ choose a rational group having this type such that for every prime factor p of n either $1/p \notin \tau$ or $p\tau = \tau$. Let

$$A_i = \bigoplus_{\rho \in T_i} \rho v_{i\rho}, \quad X_i = A_i + \mathbb{Z}\tfrac{1}{n_i}\left(\sum_{\rho \in T_i} t_{i\rho} n_{i\rho} v_{i\rho}\right), \quad i = 1, 2,$$

where $t_{i\rho}$ is chosen as in (2Dec5). We then have that $\gcd^A(n_i, t_{i\rho} n_{i\rho} v_{i\rho}) = t_{i\rho} n_{i\rho}$ for $\rho \in T_i$. The existence theorem Theorem 4.4 can and will now be used without detailed reference. By (Dec5)(a) it follows that $n_i = \operatorname{rgi} X_i$ and hence $[X : A] = n$. By (2Dec5)(a) A_i is the regulating regulator of X_i. By the Purification Lemma and (2Dec5)(b), $\operatorname{rgi} X_i[\tau] = \gcd\{t_{i\rho} n_{i\rho} : \rho \in T_i, \rho \leq \tau\} = n_{i\tau}$. By (2Dec4) there is a completely decomposable group D such that $A_1 \oplus A_2 \oplus D \cong A$. Let $X' = X_1 \oplus X_2 \oplus D$. Then X' is a crq-group by Lemma 6.3. Thus X and X' are crq-groups with $\operatorname{R}(X) \cong \operatorname{R}(X')$, $\operatorname{rgi} X = \operatorname{rgi} X'$, and $\operatorname{rgi} X[\tau] = \operatorname{rgi} X'[\tau]$. Hence X and X' are nearly isomorphic and by Arnold's theorem X has a decomposition of the desired kind. □

Iterating the previous result and adding the indecomposability criterion we get the following.

Theorem 6.8. *Let X be a proper crq-group with a regulating regulator, set $n = \operatorname{rgi} X$ (> 1) and $n_\tau = \operatorname{rgi} X[\tau]$ for $\tau \in \operatorname{T_{cr}}(X)$. Let*

$$X = X_1 \oplus \cdots \oplus X_k \oplus X_{k+1} \oplus \cdots \oplus X_K$$

be an indecomposable decomposition of X arranged such that $X_1, \ldots, X_k$ are clipped groups and $X_{k+1}, \ldots, X_K$ are rank-one groups. Then there are factorizations $n = \prod_{i=1}^k n_i$ and $n_\tau = \prod_{i=1}^k n_{i\tau}$, for each $\tau \in \operatorname{T_{cr}}(X)$, such that the following conditions are satisfied with $T_i = \{\tau \in \operatorname{T_{cr}}(X) : n_{i\tau}^\sharp \neq n_{i\tau}\}$ and

$$n_{i\tau}^\sharp = \begin{cases} n_i & \text{if } \{\sigma \in \operatorname{T_{cr}}(X) : \sigma < \tau\} = \emptyset, \\ \gcd\{n_{i\sigma} : \sigma < \tau\} & \text{if } \{\sigma \in \operatorname{T_{cr}}(X) : \sigma < \tau\} \neq \emptyset. \end{cases}$$

(Dec0) $n_1, \ldots, n_k > 1$;

(Dec1) $\gcd(n_i, n_j) = 1$ whenever $i \neq j$;

(Dec2) $n_{i\tau}$ divides n_i;

(Dec3) $n_{i\tau}$ divides $n_{i\sigma}$ whenever $\tau \geq \sigma$;

(Dec4) for each $\tau \in \mathrm{T}_{\mathrm{cr}}(X)$, $\left|\left\{i : n_{i\tau} \neq n_{i\tau}^\sharp\right\}\right| \leq \mathrm{rk}\left(\frac{X(\tau)}{X^\sharp(\tau)}\right)$;

(Dec5) For $\tau \in T_i$ define an integer $t_{i\tau}$ by stipulating that $t_{i\tau} = \prod_{p \mid n_i} (t_{i\tau})_{\wr p}$ where for each prime divisor p of n_i,

$$(t_{i\tau})_{\wr p} = \begin{cases} 1 & \text{if } p\tau \neq \tau, \\ (n_i)_{\wr p} / (n_{i\tau})_{\wr p} & \text{if } p\tau = \tau. \end{cases}$$

With these values the following hold.

(a) $\gcd\{t_{i\rho} n_{i\rho} : \rho \in T_i, \rho \not\geq \tau\} = 1$ for every $\tau \in T_i$.

(b) $\gcd\{t_{i\rho} n_{i\rho} : \rho \in T_i, \rho \leq \tau\} = n_{i\tau}$ for every $\tau \in T_i$.

(Dec6) there is a labeling $p_1, \ldots, p_\ell$ of the prime factors of n_i such that

$$T_{ij} \cap T_{i,j+1} \neq \emptyset \text{ where } T_{ij} = \left\{\tau \in \mathrm{T}_{\mathrm{cr}}(X) : (n_{i\tau})_{\wr p_j} \neq \left(n_{i\tau}^\sharp\right)_{\wr p_j}\right\}.$$

Conversely, a factorization of the sort in Theorem 6.8 results in a corresponding decomposition of the group.

Theorem 6.9. *Let X be a crq-group with a regulating regulator, set $n = \mathrm{rgi}\, X$ and $n_\tau = \mathrm{rgi}\, X[\tau]$ for $\tau \in \mathrm{T}_{\mathrm{cr}}(X)$. Suppose there are factorizations $n = \prod_{i=1}^k n_i$ and $n_\tau = \prod_{i=1}^k n_{i\tau}$ for each $\tau \in \mathrm{T}_{\mathrm{cr}}(X)$ such that the conditions* (Dec0) *through* (Dec6) *are satisfied with $T_i = \{\tau \in \mathrm{T}_{\mathrm{cr}}(X) : n_{i\tau}^\sharp \neq n_{i\tau}\}$ and*

$$n_{i\tau}^\sharp = \begin{cases} n_i & \text{if } \{\sigma \in \mathrm{T}_{\mathrm{cr}}(X) : \sigma < \tau\} = \emptyset, \\ \gcd\{n_{i\sigma} : \sigma < \tau\} & \text{if } \{\sigma \in \mathrm{T}_{\mathrm{cr}}(X) : \sigma < \tau\} \neq \emptyset. \end{cases}$$

Then there is an indecomposable decomposition

$$X = X_1 \oplus \cdots \oplus X_k \oplus X_{k+1} \oplus \cdots \oplus X_K$$

such that $X_1, \ldots, X_k$ are indecomposable crq-groups with $\mathrm{rgi}\, X_i = n_i$, $\mathrm{rgi}\, X_i[\tau] = n_{i\tau}$, $\mathrm{rgi}\, X_i^\sharp[\tau] = n_{i\tau}^\sharp$, and $X_{k+1}, \ldots, X_K$ are rank-one groups.

We illustrate our results with examples starting with some general observations.

Let X be any almost completely decomposable group whose critical typeset $\{\tau_1, \tau_2, \tau_3\}$ has the Hasse diagram drawn. Then X has a regulating regulator since $T_{cr}(X)$ is V-free ([17, Proposition 4.5.4]). Also $X[\tau_1] = X(\tau_2)$ is completely decomposable since the critical typeset of this Butler subgroup is linearly ordered ([17, Corollary 4.1.4]). Hence $\operatorname{rgi} X[\tau_1] = 1$. Also $X[\tau_3] = X(\tau_1)$ is completely decomposable since $X(\tau_1)$ is τ_1-homogeneous. Suppose in addition that $X = A + \mathbb{Z}n^{-1}a$ is a crq-group, $\gcd^A(n, a) = 1$, $A = A_{\tau_1} \oplus A_{\tau_2} \oplus A_{\tau_3}$ is its regulating regulator, and $a = a_1 + a_2 + a_3$ with $a_i \in A_{\tau_i}$. Then the last invariant is $\operatorname{rgi} X[\tau_2] = \gcd^A(n, a_2)$. The invariants of such a group X are

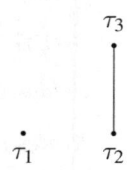

$$\operatorname{rgi} X = n,$$
$$\operatorname{rgi} X[\tau_1] = 1, \qquad \operatorname{rgi} X^\sharp[\tau_1] = n,$$
$$\operatorname{rgi} X[\tau_2] = \gcd^A(n, a_2), \quad \operatorname{rgi} X^\sharp[\tau_2] = n,$$
$$\operatorname{rgi} X[\tau_3] = 1, \qquad \operatorname{rgi} X^\sharp[\tau_3] = \gcd^A(n, a_2).$$

This permits the following conclusions. We assume that $n > 1$ so that X is not completely decomposable.

$$\begin{aligned}
&\text{In any case } \tau_1 \in T_{cr}(X_{cl}),\\
&\tau_2 \in T_{cr}(X_{cl}) \text{ if and only if } \gcd^A(n, a_2) < n,\\
&\tau_3 \in T_{cr}(X_{cl}) \text{ if and only if } \gcd^A(n, a_2) > 1,\\
&\text{if } 1 < \gcd^A(n, a_2) < n, \text{ then } T_{cr}(X_{cl}) = \{\tau_1, \tau_2, \tau_3\}.
\end{aligned} \quad (6.10)$$

Since τ_1 is a critical type of every clipped summand, there are at most $\operatorname{rk} \frac{X(\tau_1)}{X^\sharp(\tau_1)}$ such summands. The clipped summands of X are slim crq-groups and therefore have rank at most 3 with critical typesets that are subsets of $\{\tau_1, \tau_2, \tau_3\}$. The main decomposition of X has a clipped summand of rank 2 or of rank 3, and by the uniqueness properties of main decompositions ([17, Theorem 9.2.7]), exactly one of the two occurs for a given group. We now consider specific examples.

Example 6.11. Let $r_1 = 3$, $r_2 = 2$, $r_3 = 2$, and let

$$A = A_{\tau_1} \oplus A_{\tau_2} \oplus A_{\tau_3}, \quad \text{where} \quad A_{\tau_i} = \bigoplus_{1 \leq j \leq r_i} \tau_i v_{ij}$$

and assume that none of the rational groups τ_1, τ_2, τ_3, contains $\frac{1}{2}$, $\frac{1}{3}$, $\frac{1}{5}$, or $\frac{1}{7}$. Let $n = 2^2 3^3 5^2 7^2$. Our assumptions assure that

$$\gcd^A(n, s_{ij} v_{ij}) = s_{ij} \quad \text{provided that} \quad s_{ij} \mid n.$$

Let X be a crq-group with $R(X) = A$. Possible ranks of summands in an indecomposable decomposition are

(1) $2 + 1 + 1 + 1 + 1 + 1 = 7$,

(2) $3 + 1 + 1 + 1 + 1 = 7$,

(3) $3 + 3 + 1 = 7$,

(4) $3 + 2 + 1 + 1 = 7$,

(5) $2 + 2 + 1 + 1 + 1 = 7$,

(6) $3 + 2 + 2 = 7$,

(7) $2 + 2 + 2 + 1 = 7$.

Cases (1) and (2) correspond to the main decomposition and only one of the two can occur. It is reasonable to write an example in its main decomposition and then investigate its possible direct decompositions.

- Let $X_{cl} = (\tau_1 v_{11} \oplus \tau_2 v_{21} \oplus \tau_3 v_{31}) + \mathbb{Z}\frac{1}{n}(v_{11} + 2 \cdot 3^2 \cdot 5 \cdot 7 v_{21} + v_{31})$. It is routine to check that A is the regulating regulator of $X = \tau_1 v_{22} \oplus \tau_1 v_{23} \oplus \tau_2 v_{22} \oplus \tau_3 v_{32} \oplus X_{cl}$ and that X_{cl} is in fact indecomposable. The invariants of X are

$$\begin{aligned}
&\text{rgi } X = 2^2 3^3 5^2 7^2, \\
&\text{rgi } X[\tau_1] = 1, &&\text{rgi } X^\sharp[\tau_1] = 2^2 3^3 5^2 7^2, \\
&\text{rgi } X[\tau_2] = 2 \cdot 3^2 \cdot 5 \cdot 7, &&\text{rgi } X^\sharp[\tau_2] = 2^2 3^3 5^2 7^2, \\
&\text{rgi } X[\tau_3] = 1, &&\text{rgi } X^\sharp[\tau_3] = 2 \cdot 3^2 \cdot 5 \cdot 7.
\end{aligned}$$

It is apparent that every factorization of rgi X into relatively prime factors results in the last case of (6.10) and therefore X has no indecomposable summand of rank 2. However, decompositions of X into a sum of two rank 3 groups and one rank 1 group is possible in $\binom{4}{1} + \binom{4}{2} = 10$ ways corresponding to the factorizations $n = (2^2)(3^3 5^2 7^2) = (3^3)(2^2 5^2 7^2) = (5^2)(2^2 3^3 7^2) = (7^2)(2^2 3^3 5^2) = (2^2 3^3)(5^2 7^2) = (2^2 5^2)(3^3 7^2) = (2^2 7^2)(3^3 5^2) = (3^3 5^2)(2^2 7^2) = (3^3 7^2)(2^2 5^2) = (5^2 7^2)(2^2 3^3)$. Note that the factorization of rgi X determines the factorization of rgi $X[\tau_2]$.

- Let $Y_{cl} = (\tau_1 v_{11} \oplus \tau_2 v_{21} \oplus \tau_3 v_{31}) + \mathbb{Z}\frac{1}{n}(v_{11} + 5 \cdot 7 v_{21} + v_{31})$. Again it is routine to check that A is the regulating regulator of $Y = \tau_1 v_{22} \oplus \tau_1 v_{23} \oplus \tau_2 v_{22} \oplus \tau_3 v_{32} \oplus Y_{cl}$ and that Y_{cl} is in fact indecomposable. The invariants of Y are

$$\begin{aligned}
&\text{rgi } Y = 2^2 3^3 5^2 7^2, \\
&\text{rgi } Y[\tau_1] = 1, &&\text{rgi } Y^\sharp[\tau_1] = 2^2 3^3 5^2 7^2, \\
&\text{rgi } Y[\tau_2] = 5 \cdot 7, &&\text{rgi } Y^\sharp[\tau_2] = 2^2 3^3 5^2 7^2, \\
&\text{rgi } Y[\tau_3] = 1, &&\text{rgi } Y^\sharp[\tau_3] = 5 \cdot 7.
\end{aligned}$$

There is more flexibility in this case and the factorization rgi $Y = (2^2)(3^2)(5^2 7^2)$ and rgi $Y[\tau_2] = (1)(1)(5 \cdot 7)$ shows that Y is the direct sum of an indecomposable group of rank 2, an indecomposable group of rank 3, and two rank-1 groups.

We introduce further invariants in order to connect with earlier results.

Definition 6.12. Let X be an almost completely decomposable group. For each $\tau \in \mathrm{Tst}(X)$ define
$$\nu_\tau(X) = \frac{\mathrm{rgi}\, X}{\mathrm{rgi}\, X[\tau]}.$$
If X is a block-rigid crq-group, then
$$\nu_\tau(X) = \mu_\tau(X)$$
where the latter is the μ-*invariant* of X introduced in [17, Definition 12.6.2] and generalised by Vinsonhaler ([17, Section 16.5]). In fact, let X be a block-rigid crq-group and $A = \mathrm{R}(X)$ its regulator. Let $e = \mathrm{rgi}\, X$. Then $eX \subset A$, A is a completely decomposable group, and A has a unique decomposition into homogeneous components $A = \bigoplus_{\rho \in \mathbb{T}} A_\rho$ since $\mathrm{T}_{\mathrm{cr}}(A) = \mathrm{T}_{\mathrm{cr}}(X)$ is an antichain. In fact, $A_\tau = X(\tau)$. Now X has a description
$$X = A + \mathbb{Z} e^{-1} a, \text{ where } a = \sum_{\rho \in \mathbb{T}} a_\rho \text{ for some } a_\tau \in A_\tau.$$
We find
$$\mu_\tau(X) := \mathrm{ord}(a_\tau + eA) = \tfrac{e}{\gcd^A(e, a_\tau)} = \tfrac{e}{\mathrm{rgi}\, X[\tau]} = \nu_\tau(X).$$
It is now routine to check that the results of [6] are obtained as a special case of the above.

References

[1] D. Arnold and M. Dugas. Locally free finite rank Butler groups and near isomorphism. In *Abelian Groups and Modules, Proceedings of the 1994 Padova Conference*, pages 41–48. Kluwer Academic Publishers, 1995.

[2] D. M. Arnold. A class of pure subgroups of completely decomposable abelian groups. *Proc. Amer. Math. Soc.*, 41:37–44, 1973.

[3] D. M. Arnold. *Finite Rank Torsion Free Abelian Groups and Rings*, volume 931 of *Lecture Notes in Mathematics*. Springer Verlag, 1982.

[4] D. M. Arnold. *Abelian Groups and Representations of Finite Partially Ordered Sets*, CMS Books in Mathematics, Springer Verlag, 2000.

[5] E. Blagoveshchenskaya and R. Göbel. Classification and direct decompositions of some Butler groups of countable rank. *Communications in Algebra*, 30:3403–3427, 2002.

[6] E. A. Blagoveshchenskaya and A. Mader. Decompositions of almost completely decomposable groups. *Contemporary Mathematics*, 171:21–36, 1994.

[7] E. Blagoveshchenskaya and L. Strüngmann. Near isomorphism for a class of infinite rank torsion-free abelian groups. *Communications in Algebra*, to appear, 2007.

[8] E. Blagoveshchenskaya. Determination of a class of countable rank torsion-free abelian groups by their endomorphism rings. Preprint, 2007.

[9] R. Burkhardt and O. Mutzbauer. Almost completely decomposable groups with primary cyclic regulating quotient. *Rend. Sem. Mat. Univ. Padova*, 95:81–93, 1996.

[10] U. Dittmann, A. Mader, and O. Mutzbauer. Almost completely decomposable groups with a cyclic regulating quotient. *Communications in Algebra*, 25:769–784, 1997.

[11] M. Dugas. Butler groups with two types. *Unpublished manuscript*, 10 pp., 1990.

[12] A. Elter. *Torsionsfreie Gruppen mit linear geordneter Typenmenge*. PhD thesis, Universität Essen, 1996.

[13] T. Faticoni and P. Schultz. Direct decompositions of acd groups with primary regulating index. In *Abelian Groups and Modules, Proceedings of the 1995 Colorado Springs Conference*, pages 233–241. Marcel Dekker, Inc., 1996.

[14] L. Fuchs. *Infinite Abelian Groups, Vol. I, II*. Academic Press, 1970 and 1973.

[15] E. L. Lady. Almost completely decomposable torsion-free abelian groups. *Proc. Amer. Math. Soc.*, 45:41–47, 1974.

[16] W. S. Lewis. Almost completely decomposable groups with two critical types. *Communications in Algebra*, 21:607–614, 1993.

[17] A. Mader. *Almost Completely Decomposable Groups*, volume 13 of *Algebra, Logic and Applications*. Gordon and Breach Science Publishers, 2000.

[18] A. Mader, O. Mutzbauer, and L. Nongxa. Representing matrices of almost completely decomposable groups. *J. Pure Appl. Alg.*, 158:247–265, 2001.

[19] A. Mader, L. G. Nongxa, and M. Ould-Beddi. Invariants of global crq-groups. *Contemporary Mathematics*, 273:209–222, 2001. Abelian Groups and Modules, International Conference in Perth, July 2000.

[20] A. Mader and L. Strüngmann. Bounded essential extensions of completely decomposable groups. *J. Algebra*, 229:205–233, 2000.

[21] A. Mader and L. Strüngmann. Generalized almost completely decomposable groups. *Rend. Sem. Mat. Univ. Padova*, 113:47–69, 2005.

[22] A. Mader and C. Vinsonhaler. Classifying almost completely decomposable groups. *J. Algebra*, 170:754–780, 1994.

[23] A. Mader and C. Vinsonhaler. Almost completely decomposable groups with a cyclic regulating quotient. *J. Algebra*, 177:463–492, 1995.

Author information

A. Mader, Department of Mathematics, University of Hawaii, 2565 McCarthy Mall, Honolulu, HI 96822, USA.
E-mail: adolf@math.hawaii.edu

L. G. Nongxa, University of the Witwatersrand, Private Bag 3, Wits 2050, Republic of South Africa.
E-mail: nongxal@vco.wits.ac.za

M. A. Ould-Beddi, Faculté des Sciences et Techniques, Université de Nouakchott, B. P. 5026, Nouakchott, Mauritania.

Endomorphisms and automorphisms of squares of abelian groups

Grigore Călugăreanu and Phill Schultz

Abstract. A major theme in Tony Corner's work is the interaction between abelian groups and their rings of endomorphisms and groups of automorphisms. Here we study the properties of an abelian group A which are reflected in $\mathrm{Aut}(A \oplus A)$.

Key words. Automorphism groups, decompositions, involutions.

AMS classification. Primary: 20K30. Secondary: 20K15, 20K20.

1 Introduction and notation

Let A be an abelian group. We say that A is determined in a category $\mathcal{A}$ of abelian groups by its endomorphism ring if whenever $B \in \mathcal{A}$ has the property that $\mathcal{E}(A) \cong \mathcal{E}(B)$ as rings, then $A \cong B$ as abelian groups. Similarly, we say that A is determined in $\mathcal{A}$ by its automorphism group if whenever $B \in \mathcal{A}$ has the property that $\mathrm{Aut}(A) \cong \mathrm{Aut}(B)$ as groups, then $A \cong B$ as abelian groups.

Usually a group is determined neither by its automorphism group nor its endomorphism ring in a non-trivial category of torsion-free abelian groups closed under isomorphisms. However, if A is a rational group of idempotent type, that is, a subring of the rational numbers, then A is determined by its endomorphism ring in the category of rational groups of idempotent type, [MS00]. In fact, since A is determined by the set of primes p for which $pA = A$, A is even determined by its additive endomorphism group $\mathrm{End}(A)$ in this category. On the other hand, A is determined by its automorphism group in the same category only if $A \cong \mathbb{Z}$, whereas $\mathbb{Q}$ is determined by $\mathrm{Aut}(\mathbb{Q})$ in the category of torsion-free divisible groups, as well as by $\mathrm{End}(\mathbb{Q})$ in the same category. Of course, if A is determined in a given category by $\mathrm{Aut}(A)$ or $\mathrm{End}(A)$, then it is also determined by $\mathcal{E}(A)$.

A new phenomenon was considered by the first author in [Cal06]. He shows that more information about A is sometimes provided by $\mathrm{Aut}(A^2) = \mathrm{Aut}(A \oplus A)$. For example, if A and B are rational groups of idempotent type and both are divisible by exactly n primes, where n is a positive integer or infinity, then always $\mathrm{Aut}(A) \cong \mathrm{Aut}(B)$ [MS00, Proposition 4.3], and the following are equivalent:

(1) $\mathrm{Aut}(A^2) \cong \mathrm{Aut}(B^2)$.

(2) $\mathrm{End}(A) \cong \mathrm{End}(B)$.

(3) A and B are both divisible by the same sets of primes.

(4) $A \cong B$.

We say that A is determined in a category $\mathcal{A}$ by its square automorphism group if whenever $H \in \mathcal{A}$ has the property that $\operatorname{Aut}(A^2) \cong \operatorname{Aut}(H)$ as groups, then $H = B^2$ with $A \cong B$ as abelian groups. Similarly, A is determined in a category $\mathcal{A}$ by its square endomorphism ring, if whenever $H \in \mathcal{A}$ has the property that $\mathcal{E}(A^2) \cong \mathcal{E}(H)$ as rings, then $H = B^2$ with $A \cong B$ as abelian groups. Determination by the square endomorphism group $\operatorname{End}(A^2)$ can be similarly defined.

It is the purpose of this paper to investigate this phenomenon. There are well known examples in the literature [F70, Theorem 90.3] of non-isomorphic indecomposable torsion-free groups A and B for which $A^2 \cong B^2$, so in general $\operatorname{Aut}(A^2) \cong \operatorname{Aut}(B^2)$ does not imply that $A \cong B$.

One could also consider groups determined by the automorphism group of higher powers; for example, Hahn and O'Meara [HO'M89, 3.3.8 and 3.3.11] showed that A is determined by $\operatorname{Aut}(A^n)$ for all $n \geq 3$ in the category of abelian groups whose endomorphism ring is a principal ideal domain, and Krylov et al. [KMT03] showed that A is determined by $\operatorname{Aut}(A^n)$ for all $n \geq 4$ in the category of abelian groups whose endomorphism ring is commutative. However, in this paper we only consider the square case.

Although we have not obtained definitive results on the structure of groups determined by their square automorphism groups, we have found several properties of pairs of groups whose square automorphism groups are isomorphic. The main result is that if A and H are torsion-free abelian groups divisible by 2 such that $\operatorname{Aut}(A^2) \cong \operatorname{Aut}(H)$, then $H = B^2$ for some group B such that $\operatorname{End}(A) \cong \operatorname{End}(B)$. Furthermore, we show that in this case, $\operatorname{Aut}(A) \times \operatorname{Aut}(A) \cong \operatorname{Aut}(B) \times \operatorname{Aut}(B)$.

Consequently, A is determined by its square automorphism group in the category of torsion-free abelian groups if A is determined by its endomorphism group in that category. We also pose some intriguing questions which will be the subject of further research.

Of course, the fact that $\operatorname{End}(A) \cong \operatorname{End}(B)$ does not imply that $\mathcal{E}(A) \cong \mathcal{E}(B)$ is well known. As long ago as 1959, Sasiada [S59] found examples of rank 2 torsion-free groups A and B, A being completely decomposable, satisfying the former but not the latter. Nowadays, such examples are easy to construct using the methods of Mader [M00, Section 15.2].

The notation is mostly standard as in [F70]. In particular, $\mathbb{Z}$ denotes the group or ring of integers and $\mathbb{Q}$ the group or ring of rationals. Unless specifically excepted, the word group will denote a torsion-free abelian group. We use the common symbol $\cong$ for group, abelian group or ring isomorphism, the meaning being specified when it is ambiguous. Similarly, $Z(X)$ denotes the center of the group or ring X and $X \ominus Y$ is the direct sum or direct product of groups or rings X and Y.

Let $G = A \oplus B$ be a direct sum of groups A and B. Then $\mathcal{E}(G)$ can be represented by the ring of 2×2 matrices $\begin{bmatrix} \operatorname{End}(A) & \operatorname{Hom}(A,B) \\ \operatorname{Hom}(B,A) & \operatorname{End}(B) \end{bmatrix}$, the action being given by

$$(a,b) \begin{bmatrix} r & s \\ t & u \end{bmatrix} = (ar + bt, as + bu) \text{ for all } (a,b) \in G.$$

In particular, if $G = A^2$, then we can represent $\mathcal{E}(G)$ by $\mathcal{M} = M(2, \mathcal{E}(A))$, the ring of 2×2 matrices with entries from $\mathcal{E}(A)$, where the slight abuse of notation causes no harm. It is straightforward to verify that $\mathrm{GL}(2, \mathcal{E}(A)) = \mathrm{Aut}(A^2)$ corresponds to the group of invertible matrices in $\mathcal{M}$.

In general, we denote abstract automorphisms of G by Greek lower case letters, but their representation by matrices by capital Latin letters, so for example an automorphism α is represented by $A = \begin{bmatrix} r & s \\ t & u \end{bmatrix}$.

In accordance with common practice in group theory, function names are written on the right of their arguments but functor names are written on the left. For example, $Z(\mathrm{Aut}(A))$ denotes the center of $\mathrm{Aut}(A)$ and $Z(\mathcal{E}(A))$ the center of $\mathcal{E}(A)$. In particular, if F is a sub-functor of the identity in the category under consideration and $f \in \mathrm{Hom}(A, B)$ then $F(A)f \subseteq F(B)$.

When X and Y are groups, abelian or not, the notation $X \leq Y$ means that X is a subgroup of Y and $X \trianglelefteq Y$ means that X is a normal subgroup of Y. When $X \leq Y$,

$$\mathcal{C}_Y(X) = \{y \in Y : yx = xy \text{ for all } x \in X\}$$

is the centralizer of X in Y and

$$\mathcal{N}_Y(X) = \{y \in Y : y^{-1}xy \in X \text{ for all } x \in X\}$$

is the normalizer of X in Y.

For any unital ring R, R^+ denotes the additive group of R and $U(R)$ the unit group of R so that $U(M(2, \mathcal{E}(A))) = \mathrm{GL}(2, \mathcal{E}(A))$.

2 Properties of automorphism groups preserved by an isomorphism

In this section, we pose several naturally arising questions concerning these properties. Throughout the section, A is a group divisible by 2, $G = A^2$ and $\theta : \mathrm{Aut}(G) \to \mathrm{Aut}(H)$ is an isomorphism. Note that in general, if A and B are groups such that $2A = A$, $\mathrm{Aut}(A) \cong \mathrm{Aut}(B)$ does not imply that $2B = B$, which prompts our first question:

Question 2.1. For which $G = 2G$ does it follow that θ maps multiplication by 2 onto multiplication by 2?

Because of the result of the first author cited above, the answer is affirmative for A a rank 1 group of idempotent type. Incidentally, 2-divisibility is an *additive* property of groups, but in our situation, it can be determined multiplicatively.

Proposition 2.2. *2 is an automorphism of G if and only if* $Y = \begin{bmatrix} 1 & 1 \\ 0 & 1 \end{bmatrix} = \begin{bmatrix} 1 & s \\ 0 & 1 \end{bmatrix}^2$ *for some $s \in \mathcal{E}(A)$.*

Proof. If 2 is an automorphism of G then A is also 2-divisible so there exists $s \in \mathcal{E}(A)$ such that $2s = 1$. Hence $\begin{bmatrix} 1 & s \\ 0 & 1 \end{bmatrix}^2 = \begin{bmatrix} 1 & 2s \\ 0 & 1 \end{bmatrix} = Y$.

Conversely, if $\begin{bmatrix} 1 & s \\ 0 & 1 \end{bmatrix}^2 = \begin{bmatrix} 1 & 2s \\ 0 & 1 \end{bmatrix} = Y$, then $1 = 2s$ in $\mathrm{Aut}(A)$ so A and hence G are 2-divisible. □

There is a similar criterion using Y^T (Y transpose), but these results are not strong enough to settle Question 2.1.

Since orders and centralizers of automorphisms are preserved by isomorphisms, it is important to consider special elements of G for which these characteristics are known. For example, -1 is a central automorphism of G of order 2, so $(-1)\theta$ is a central automorphism of H of order 2. Hence one may ask:

Question 2.3. When is $(-1)\theta = -1$?

This question will be discussed further in Section 3. Other special automorphisms of G whose algebraic properties are reflected in $\mathrm{Aut}(H)$ include

(1) $X = \begin{bmatrix} 0 & 1 \\ 1 & 0 \end{bmatrix}$ and its additive inverse $-X$ of order 2,

(2) $Z = \begin{bmatrix} 0 & 1 \\ -1 & 0 \end{bmatrix}$ of order 4. Its additive and multiplicative inverse coincide.

Note that conjugation by X and by Z map upper triangular elements and subgroups to lower triangular elements and subgroups, while conjugation by Y preserves upper and lower triangularity.

3 Involutions

In this section, G is a 2-divisible group and H is a group such that $\mathrm{Aut}(G) \cong \mathrm{Aut}(H)$.

An involution μ of G is an automorphism of order 2. As shown in [F73, Section 113], μ determines a decomposition $G = G^\mu \oplus G^{-\mu}$ where $G^\mu = \{x \in G : x\mu = x\}$ and $G^{-\mu} = \{x \in G : x\mu = -x\}$. Conversely, every decomposition $G = A \oplus B$ determines an involution μ for which $G^\mu = A$ and $G^{-\mu} = B$, namely $\mu = 2\pi_A - 1$ where π_A is the projection of G along B onto A.

If G is any group, then the involution -1 of G is called the *trivial involution* since the corresponding decomposition is $0 \oplus G$. The importance of involutions for our problem is that if $\theta : \mathrm{Aut}(G) \to \mathrm{Aut}(H)$ is an isomorphism, then θ maps involutions of G onto involutions of H. This in itself does not imply that non-trivial decompositions of G induce non-trivial decompositions of H or *vice versa*, firstly because H may not be 2-divisible and secondly because $(-1)\theta$ may not equal -1.

We now assume in the rest of this section that H is 2-divisible and study the image of the central involution -1 under θ. Let $\beta = (-1)\theta$. We call θ *good* if $\beta = -1$, otherwise *bad*.

Proposition 3.1. *Let G and H be 2-divisible groups and let $\theta : \mathrm{Aut}(G) \to \mathrm{Aut}(H)$ be an isomorphism. If θ is bad, then there are decompositions $G = A \oplus B$, $H = C \oplus D$ such that A and B are fully invariant in G and C and D are fully invariant in H.*

Proof. Since θ is bad, β is a non-trivial involution and hence H has a decomposition $H = H^{\beta} \oplus H^{-\beta}$ with respect to which β has the matrix form $\begin{bmatrix} 1 & 0 \\ 0 & -1 \end{bmatrix}$. Let $C = H^{\beta}$ and $D = H^{-\beta}$.

Since β is central in $\mathrm{Aut}(H)$, β commutes with all automorphisms of the form $\begin{bmatrix} 1 & s \\ 0 & 1 \end{bmatrix}$ with $s \in \mathrm{Hom}(C, D)$. But this implies that $s = -s$ for all $s \in \mathrm{Hom}(C, D)$ and hence $\mathrm{Hom}(C, D) = 0$ so C is fully invariant in H. Similarly, by considering automorphisms of the form $\begin{bmatrix} 1 & 0 \\ t & 1 \end{bmatrix}$ with $t \in \mathrm{Hom}(D, C)$ we conclude that D is fully invariant in H.

Now consider $-1 \in \mathrm{Aut}(H)$ and $\alpha = (-1)\theta^{-1}$. Since $\alpha \neq -1$, α is a non-trivial central involution in $\mathrm{Aut}(G)$, so by the same argument, we conclude that $G = A \oplus B$ with A and B fully invariant in G. $\square$

Now for the good news. Even though θ may not map -1 to -1, it can always be perturbed to do so.

Proposition 3.2. *If $\mathrm{Aut}(G) \cong \mathrm{Aut}(H)$ then there exists a good isomorphism $\phi : \mathrm{Aut}(G) \to \mathrm{Aut}(H)$.*

Proof. Let $\theta : \mathrm{Aut}(G) \to \mathrm{Aut}(H)$ be an isomorphism. Let $(-1)\theta = \beta$. If $\beta \neq -1$ then by Proposition 3.1, $H = C \oplus D$ such that $\beta|C = 1$ and $\beta|D = -1$. Let $\gamma \in \mathrm{Aut}(H)$ act as -1 on C and the identity on D. Then $\phi = \beta \circ \gamma$ is an isomorphism of $\mathrm{Aut}(G)$ onto $\mathrm{Aut}(H)$ which maps -1 to -1. $\square$

Proposition 3.2 has some useful consequences.

Corollary 3.3. *If $\mathrm{Aut}(G) \cong \mathrm{Aut}(H)$ then there is an isomorphism $\phi : \mathrm{Aut}(G) \to \mathrm{Aut}(H)$ such that for all $\alpha \in \mathrm{Aut}(G)$, $(-\alpha)\phi = -(\alpha\phi)$.* $\square$

Corollary 3.4. *Let $\theta : \mathrm{Aut}(G) \to \mathrm{Aut}(H)$ be a good isomorphism, and let $G = A \oplus B$ be any non-trivial decomposition of G. Let α with matrix $M = \begin{bmatrix} 1 & 0 \\ 0 & -1 \end{bmatrix}$ be the involution corresponding to this decomposition of G. Then there is a non-trivial decomposition $H = C \oplus D$ such that $\alpha\theta$ has the same matrix M with respect to this decomposition.* $\square$

Given a non-trivial decomposition $G = A \oplus B$ of G, we are interested in involutions ν for which $G^{\nu} = A$ and $G^{-\nu} = C$ for some different complementary summand C. A pair $(r, s) \in \mathrm{Hom}(A, B) \times \mathcal{E}(B)$ is called a *decomposition pair* if $(1 + s)r = 0$ and s is an involution in $\mathrm{Aut}(B)$. Decomposition pairs are used to classify the decompositions of G for which A is fixed as the first summand.

Proposition 3.5. *Let $G = A \oplus B$. Then any involution fixing A corresponds to a matrix of the form $M = \begin{bmatrix} 1 & 0 \\ r & s \end{bmatrix}$ for some decomposition pair (r, s). Conversely, if $M = \begin{bmatrix} 1 & 0 \\ r & s \end{bmatrix}$ for some decomposition pair (r, s) then M is the matrix corresponding to an involution fixing A.*

Proof. Let M be an involution fixing $\{(a, 0) : a \in A\}$, so its first row has the form $(1, 0)$. Hence $M = \begin{bmatrix} 1 & 0 \\ r & s \end{bmatrix}$ for some pair $(r, s) \in \mathrm{Hom}(B, A) \times \mathcal{E}(B)$. Then $M^2 = 1$ implies that $(1 + s)r = 0$ and $s^2 = 1$. $M \neq I$ implies that $s \neq 1$. Hence (r, s) is a decomposition pair.

Conversely, any such M is an involution and $G = M^+ \oplus M^-$ where $M^+ = \{(a, 0) \in G : a \in A\}$ and $M^- = \{(a, b) \in G : br = -2a \text{ and } bs = -b\}$. □

Remarks 3.6. (1) $s = -1$ for all decomposition pairs if and only if B is directly indecomposable.

(2) $(r, -1)$ is a decomposition pair for any $r \in \mathrm{Hom}(B, A)$.

(3) The given decomposition $G = A^2$ corresponds to the decomposition pair $(0, -1)$.

(4) We refer to the involution defined in Proposition 3.5 as $\mu(r, s)$, the corresponding matrix M as $M(r, s)$, and the complementary summand C as $B(r, s)$.

Corollary 3.7. *For any fixed involution s of B, the mapping $r \mapsto B(r, s)$ is a bijection from $\mathrm{Hom}(B, A)$ onto complementary summands of A in G and the mapping $r \mapsto M(r, s)$ is a bijection of $\mathrm{Hom}(B, A)$ onto involutions of G fixing A.* □

Now let $\theta : \mathrm{Aut}(G) \to \mathrm{Aut}(H)$ be a good isomorphism and let $\beta \in \mathrm{Aut}(H)$ be the image under θ of $\mu(0, -1)$. Then β is a non-trivial involution of H. Let $C = H^\beta$ and $D = H^{-\beta}$ so that $H = C \oplus D$ and, with respect to this decomposition, β is represented by $M' = \begin{bmatrix} 1 & 0 \\ 0 & -1 \end{bmatrix}$. By our remarks above, every complementary summand to C in H has the form $D(r', s')$ for some decomposition pair $(r', s') \in \mathrm{Hom}(D, C) \times \mathcal{E}(D)$. Furthermore, every involution in $\mathrm{Aut}(H)$ fixing C has the form $M'(r', s')$ for such a decomposition pair.

Proposition 3.8. *With the notation above,*

(1) *for each decomposition pair $(r, s) \in \mathrm{Hom}(B, A) \times \mathcal{E}(B)$, $M(r, s)\theta = M'(r', s')$ for some decomposition pair $(r', s') \in \mathrm{Hom}(D, C) \times \mathcal{E}(D)$.*

(2) *the mapping $(r, s) \mapsto (r', s')$ is a bijection.*

(3) *if $s = -1$ then $s' = -1$. In particular, if B is indecomposable, then B' is indecomposable.*

Proof. (1) Let $G = A \oplus B$ and (r, s) be a decomposition pair, so that $M(r, s)$ is an involution of G which fixes A. Let $M(r, s)\theta = \beta \in \mathrm{Aut}(H)$. We have seen that H has a decomposition $C \oplus D$ for which $\beta = M'(r', s')$ for some decomposition pair (r', s').

(2) Since (r', s') is determined uniquely by (r, s) and *vice versa*, the correspondence is a bijection.

(3) We have seen that the correspondence maps $(r, -1)$ to $(r', -1)$. □

4 Automorphisms of squares of groups

We now consider the case in which $G = A^2$. Throughout this section, whose two theorems are the main results of this paper, G and H are 2-divisible groups and $\theta : \text{Aut}(G) \to \text{Aut}(H)$ is a good isomorphism. Furthermore, $G = A \oplus A$ is a fixed decomposition of G, and $\mu \in \text{Aut}(G)$ has matrix $M = \begin{bmatrix} 1 & 0 \\ 0 & -1 \end{bmatrix}$ with respect to this decomposition. Let $\nu = \mu\theta$ and let $H = C \oplus D$ be the corresponding decomposition of H so that by Corollary 3.4 the matrix of ν with respect to this decomposition is also M.

Now $\text{Aut}(G)$ contains an involution of a different type, namely χ which maps the first copy of A identically onto the second and *vice versa*. The corresponding matrix is $X = \begin{bmatrix} 0 & 1 \\ 1 & 0 \end{bmatrix}$, which has the following properties:

(1) For all $(a, b) \in G$, $(a, b)X = (b, a)$.

(2) For all decomposition pairs (r, s), $XM(r,s)X$ is an involution fixing the second copy of A.

(3) $XM = -MX$.

Theorem 4.1. *Let $G = A \oplus A$, $H = C \oplus D$ and χ be as described in the two paragraphs above and let $\lambda = \chi\theta$. Then the matrix of λ with respect to this decomposition of H is $L = \begin{bmatrix} 0 & b \\ b^{-1} & 0 \end{bmatrix}$ for some isomorphism $b : C \to D$.*

Proof. Let λ have matrix $L = \begin{bmatrix} a & b \\ c & d \end{bmatrix}$ with respect to the decomposition $H = C \oplus D$. Since θ is a group isomorphism, by (3) above, $ML = -LM$, so

$$\begin{bmatrix} a & b \\ -c & -d \end{bmatrix} = \begin{bmatrix} -a & b \\ -c & d \end{bmatrix}.$$

Since $\text{End}(C)$ and $\text{End}(D)$ are torsion-free, this implies that $a = 0 = d$. Since L represents an involution, it follows that $c \in \text{Hom}(D, C)$ is an isomorphism with inverse $b \in \text{Hom}(C, D)$. □

Theorem 4.1 shows that if $G = A^2$ and H are 2-divisible groups with $\text{Aut}(G) \cong \text{Aut}(H)$, then $H = B^2$ for some group B and there is a good isomorphism $\theta : \text{Aut}(G) \to \text{Aut}(H)$ which maps the involution corresponding to a fixed decomposition $G = A \oplus A$ to the involution corresponding to a fixed decomposition $H = B \oplus B$. But as yet, we have found no relationship between A and B. We now construct an additive isomorphism from $\text{End}(A)$ to $\text{End}(B)$.

Theorem 4.2. *Let $G = A^2$ and $H = B^2$ be 2-divisible groups with $\mathrm{Aut}(G) \cong \mathrm{Aut}(H)$. Then there exists an isomorphism $\theta : \mathrm{Aut}(G)$ onto $\mathrm{Aut}(H)$ inducing an additive isomorphism $\phi : \mathrm{End}(A) \to \mathrm{End}(B)$.*

Proof. By Proposition 3.8 there is a good isomorphism $\theta : \mathrm{Aut}(G) \to \mathrm{Aut}(H)$ which induces the matrix bijection $\begin{bmatrix} 1 & 0 \\ r & -1 \end{bmatrix} \mapsto \begin{bmatrix} 1 & 0 \\ r' & -1 \end{bmatrix}$, and hence by Corollary 3.4 the matrix bijection $\begin{bmatrix} 1 & 0 \\ r & 1 \end{bmatrix} \mapsto \begin{bmatrix} 1 & 0 \\ r' & 1 \end{bmatrix}$.

Define $\phi : \mathrm{End}(A) \to \mathrm{End}(B)$ by $r \mapsto r'$. Since

$$\begin{bmatrix} 1 & 0 \\ r+s & 1 \end{bmatrix} = \begin{bmatrix} 1 & 0 \\ r & 1 \end{bmatrix} \begin{bmatrix} 1 & 0 \\ s & 1 \end{bmatrix}$$

ϕ is an additive homomorphism which is clearly bijective. □

5 The action of θ on subgroups of $\mathrm{Aut}(G)$

In the final two sections of this paper, we no longer need the hypotheses that G and H are torsion-free and 2-divisible so we replace them by the weaker conditions that G and H are abelian groups satisfying $G = A^2$, $H = B^2$ and $\theta : \mathrm{Aut}(G) \to \mathrm{Aut}(H)$ is an isomorphism mapping $\begin{bmatrix} 1 & 0 \\ 0 & -1 \end{bmatrix}$ in $\mathrm{Aut}(G)$ to $\begin{bmatrix} 1 & 0 \\ 0 & -1 \end{bmatrix}$ in $\mathrm{Aut}(H)$. All matrices are with respect to fixed decompositions $G = A \oplus A$ and $H = B \oplus B$, so we can regard θ as mapping matrices to matrices.

By the fundamental homomorphism theorems, θ maps the lattice of subgroups and the lattice of normal subgroups of $\mathrm{Aut}(G)$ to isomorphic lattices of subgroups and normal subgroups in $\mathrm{Aut}(H)$. Call a subgroup X of a subgroup Y of $\mathrm{Aut}(G)$ *functorial* if there is a sub-functor of the identity F on the category of groups such that $X = F(Y)$. Then $X\theta = F(Y\theta)$. Examples include $F(Y) = Z(Y)$, the center of Y, $\mathcal{C}_{\mathrm{Aut}(G)}(Y)$, the centralizer of Y and $\mathcal{N}_{\mathrm{Aut}(G)}(Y)$ the normalizer of Y. More generally, for all $X \leq Y$ in $\mathrm{Aut}(G)$, $(\mathcal{C}_Y(X))\theta = (\mathcal{C}_{Y\theta}X)\theta$ and similarly for normalizers.

Unfortunately the special subgroups familiar in Linear Algebra and Geometry may not be defined in the present context, and need not be preserved by θ when they are. Nevertheless, some are defined and preserved.

We first consider centralizers of involutions. Recall that $\alpha \in \mathrm{Aut}(G)$ is the involution with matrix $M = \begin{bmatrix} 1 & 0 \\ 0 & -1 \end{bmatrix}$ and that θ maps M to a matrix of the same form as M. Hence θ maps $\mathcal{C}_{\mathrm{Aut}(G)}(M) \to \mathcal{C}_{\mathrm{Aut}(H)}(M)$. It is easy to see that $\mathcal{C}_{\mathrm{Aut}(G)}(M) = \begin{bmatrix} \mathrm{Aut}(A) & 0 \\ 0 & \mathrm{Aut}(A) \end{bmatrix} \cong \mathrm{Aut}(A) \times \mathrm{Aut}(A)$ and similarly, $\mathcal{C}_{\mathrm{Aut}(H)}(M) \cong \mathrm{Aut}(B) \times \mathrm{Aut}(B)$, so that $\mathrm{Aut}(A) \times \mathrm{Aut}(A) \cong \mathrm{Aut}(B) \times \mathrm{Aut}(B)$.

Question 5.1. Does this imply $\mathrm{Aut}(A) \cong \mathrm{Aut}(B)$?

Now consider the corresponding result for the involutions $N = \begin{bmatrix} 0 & 1 \\ 1 & 0 \end{bmatrix}$ and $K = \begin{bmatrix} 0 & b \\ b^{-1} & 0 \end{bmatrix} \in \text{Aut}(H)$. A simple calculation shows

Proposition 5.2. *With the notation above, θ maps*

$$\mathcal{C}_{\text{Aut}(G)}(N) = \left\{ \begin{bmatrix} a & b \\ b & a \end{bmatrix} \in \text{Aut}(G) \right\} \text{ to}$$

$$\mathcal{C}_{\text{Aut}(H)}(K) = \left\{ \begin{bmatrix} x & y \\ b^{-1}yb^{-1} & b^{-1}xb \end{bmatrix} \in \text{Aut}(H) \right\}.$$

□

We now consider the isomorphism of the centres of the automorphism groups $\text{Aut}(G)$ and $\text{Aut}(H)$.

Proposition 5.3. $Z(\text{Aut}(G)) = \left\{ \begin{bmatrix} a & 0 \\ 0 & a \end{bmatrix} : a \in Z(\mathcal{E}(A)) \cap \text{Aut}(A) \right\}$ *and there exists an isomorphism* $\phi : Z(\mathcal{E}(A)) \cap \text{Aut}(A) \to Z(\mathcal{E}(B)) \cap \text{Aut}(B)$ *such that* $\begin{bmatrix} a & 0 \\ 0 & a \end{bmatrix} \theta = \begin{bmatrix} a\phi & 0 \\ 0 & a\phi \end{bmatrix}$.

Proof. The description of the center of $\text{Aut}(G)$ follows from routine calculations.

Since the center is a functorial subgroup, $Z(\text{Aut}(G))\theta = Z\text{Aut}(H)$. Define $\phi : Z(\mathcal{E}(A)) \cap \text{Aut}(A) \to Z(\mathcal{E}(B)) \cap \text{Aut}(B)$ by $a\phi = b$ if $\begin{bmatrix} a & 0 \\ 0 & a \end{bmatrix} \theta = \begin{bmatrix} b & 0 \\ 0 & b \end{bmatrix}$. Then ϕ is a multiplication preserving bijection.

□

Corollary 5.4. *Let A and B be abelian groups with commutative endomorphism rings. Then $\text{Aut}(A^2) \cong \text{Aut}(B^2)$ implies $\text{Aut}(A) \cong \text{Aut}(B)$.*

□

By considering the example of rational groups of idempotent type, one can see that the converse of Corollary 5.4 generally fails.

6 Triangular subgroups

We now consider the properties of triangular subgroups of $\text{Aut}(G)$, defined as follows. An element $X \in \text{Aut}(G)$ is called *upper triangular* if it has the form $X = \begin{bmatrix} r & s \\ 0 & t \end{bmatrix}$. Note that if X is upper triangular, then necessarily r and $t \in \text{Aut}(A)$. A subgroup is upper triangular if all its elements are. The set of all such matrices with r and t in $\text{Aut}(A)$ and $s \in \mathcal{E}(A)$ is a subgroup of $\text{Aut}(G)$ called the *full upper triangular subgroup*. Similarly, we can define lower triangular elements and subgroups. All the following examples have lower triangular counterparts.

It is not true that the triangular property of automorphisms is preserved by θ, but several useful properties of triangular automorphisms are.

Let $\mathcal{T} = \mathcal{T}(G)$ denote the group of upper triangular matrices in $\text{Aut}(G)$. We shall describe the lattices of subgroups and normal subgroups of $\mathcal{T}$ in terms of subgroups of $\text{Aut}(A) \times \text{Aut}(A)$ acting on $\mathcal{E}(A)$. Let $(B,C) \leq \text{Aut}(A) \times \text{Aut}(A)$. Then (B,C) is a double operator on $\mathcal{E}(A)$ in the sense that for all $(b,c) \in (B,C)$ and for all $s \in \mathcal{E}(A)$, $bsc \in \mathcal{E}(A)$ is defined by $x(bsc) = ((xb^{-1})s)c \in A$. A (B,C) module J in $\mathcal{E}(A)$ is an additive subgroup of $\mathcal{E}(A)$ closed under the double operator (B,C).

This seemingly obscure structure is simply the multiplicative version of a bimodule; in fact J is a (B,C) module if and only if J is a left $\mathbb{Z}[B]$-right $\mathbb{Z}[C]$-bimodule over the group rings $\mathbb{Z}[B]$ and $\mathbb{Z}[C]$. For example, every ideal of $\mathcal{E}(A)$ is an $(\text{Aut}(A), \text{Aut}(A))$ module. However, not every (B,C) module is an ideal; for example, let J be the cyclic additive subgroup of $\mathcal{E}(A)$ generated by the identity map. Then J is a $\{1\} \times \{1\}$ module. The importance of this concept is that (B,C) modules can be used to classify all subgroups and normal subgroups of $\mathcal{T}$.

Proposition 6.1. *Let $U \leq \mathcal{T}$. Then there exists a unique subgroup $(B,C) \leq \text{Aut}(A) \times \text{Aut}(A)$ and a unique (B,C) module J of $\mathcal{E}(A)$ such that*

$$U = \left\{ \begin{bmatrix} b & s \\ 0 & c \end{bmatrix} : (b,c) \in (B,C) \text{ and } s \in J \right\}.$$

Conversely, each such U is a subgroup of $\mathcal{T}$.

Furthermore, $U \trianglelefteq \mathcal{T}$ if and only if $B \trianglelefteq \text{Aut}(A)$ and $C \trianglelefteq \text{Aut}(A)$.

Proof. Let $\begin{bmatrix} b & s \\ 0 & c \end{bmatrix}$ and $\begin{bmatrix} u & v \\ 0 & w \end{bmatrix} \in U$. Then

$$\begin{bmatrix} b & s \\ 0 & c \end{bmatrix} \begin{bmatrix} u & v \\ 0 & w \end{bmatrix} = \begin{bmatrix} bu & bv+sw \\ 0 & cw \end{bmatrix} \text{ and } \begin{bmatrix} r & s \\ 0 & t \end{bmatrix}^{-1} = \begin{bmatrix} r^{-1} & -r^{-1}st^{-1} \\ 0 & t^{-1} \end{bmatrix}.$$

Since $U \leq \mathcal{T}$, the set of diagonals of elements of U is closed under multiplication and inverses, so form a subgroup (B,C) of $\text{Aut}(A) \times \text{Aut}(A)$. Furthermore, the set of entries in the north-east corners of elements of U is closed under addition and the action of (B,C), so form a (B,C) module. Conversely, if U satisfies these conditions, then U is a subgroup of $\mathcal{T}$.

By the description of multiplication above, U is normal if and only if B and C are. □

By conjugation with the matrix $\begin{bmatrix} 0 & 1 \\ 1 & 0 \end{bmatrix}$, we obtain an equivalent characterization of lower triangular subgroups of $\text{Aut}(G)$.

The following examples will play a significant rôle in the sequel.

Examples 6.2. (1) If $(B,C) = \text{Aut}(A) \times \text{Aut}(A)$ and $J = \mathcal{E}(A)$, we recover $\mathcal{T}$.

(2) If $(B,C) = (\{1\},\{1\})$ and $J = \mathcal{E}(A)$, we have the group $E(G)$ of upper transvections in $\text{Aut}(G)$.

(3) Let $Z = Z(\mathcal{E}(A)) \cap \text{Aut}(G)$, and let (B,C) be the diagonal of $Z \times Z$ and $J = \mathcal{E}(A)$. Then we have the group

$$F(G) = \left\{ \begin{bmatrix} a & b \\ 0 & a \end{bmatrix} : a \in Z, b \in \mathcal{E}(A) \right\}.$$

Note that for fixed (B,C) the set of (B,C) submodules of $\mathcal{E}(A)$ is a lattice under inclusion, and we have the following description of subgroups and normal subgroups of $\mathcal{T}$. To simplify the notation, denote by $\mathcal{T}((B,C),J)$ the subgroup of $\mathcal{T}$ described in Proposition 6.1.

Proposition 6.3. *Let $B \leq B'$ and $C \leq C'$ be subgroups of $\text{Aut}(A) \times \text{Aut}(A)$ and let J be a (B,C) submodule of the (B',C') module J'. Then $\mathcal{T}((B,C),J) \leq \mathcal{T}((B',C'),J')$.*

Moreover, if $B \trianglelefteq B'$ and $C \trianglelefteq C'$ then $\mathcal{T}((B,C),J) \trianglelefteq \mathcal{T}((B',C'),J')$.

Proof. This is an immediate consequence of the definitions. □

Corollary 6.4. *For fixed (B,C), the lattice $\mathcal{J}$ of (B,C) modules determines a lattice of subgroups $\mathcal{T}((B,C),\mathcal{J}) = \{\mathcal{T}((B,C),J) : J \in \mathcal{J}\}$ of subgroups of $\text{Aut}(G)$. If B and C are normal in $\text{Aut}(A)$, then $\mathcal{T}((B,C),\mathcal{J})$ is a lattice of normal subgroups.* □

Examples 6.5. (1) Taking $(B,C) = (\{1\},\{1\})$ and $\mathcal{J}$ the lattice of additive subgroups of $\mathcal{E}(A)$ we obtain a lattice of normal subgroups of $\text{Aut}(G)$ isomorphic to the lattice of additive subgroups of $\mathcal{E}(A)$.

(2) Taking $(B,C) = \text{Aut}(A) \times \text{Aut}(A)$ and $\mathcal{J}$ the lattice of ideals of $\mathcal{E}(A)$, we obtain a sublattice of (1) isomorphic to the lattice of ideals of $\mathcal{E}(A)$.

In the following proposition, $F(G)$ and $E(G)$ are the subgroups of $\text{Aut}(G)$ defined in Examples 6.2(2) and (3).

Proposition 6.6. (1) *$F(G)$ is an abelian subgroup of $\text{Aut}(G)$, normal in $\mathcal{T}(G)$.*

(2) *$F(G)$ is the direct product $Z(\text{Aut}(G)) \times E(G)$.*

Proof. (1) It is clear that $F(G)$ is commutative. By Proposition 6.1, $F(G) \trianglelefteq \mathcal{T}(G)$.

(2) $Z(\text{Aut}(G)) \cap E(G) = \{I_2\}$ and $\begin{bmatrix} r & s \\ 0 & r \end{bmatrix} = \begin{bmatrix} r & 0 \\ 0 & r \end{bmatrix} \begin{bmatrix} 1 & r^{-1}s \\ 0 & 1 \end{bmatrix}$ with $\begin{bmatrix} r & 0 \\ 0 & r \end{bmatrix} \in Z(\text{Aut}(G))$ and $\begin{bmatrix} 1 & r^{-1}s \\ 0 & 1 \end{bmatrix} \in E(G)$.

□

Matrices in $F(G)$ can be characterized as follows.

Lemma 6.7. *Let $M \in \operatorname{Aut}(G)$. Then $M \in F(G)$ if and only if $E(G) \leq \mathcal{C}_{\operatorname{Aut}(G)}(M)$, the centralizer of M in $\operatorname{Aut}(G)$.*

Proof. Note that $E(G) \leq \mathcal{C}_{\operatorname{Aut}(G)}(M)$ if and only if for all $\begin{bmatrix} a & b \\ c & d \end{bmatrix} \in \operatorname{Aut}(G)$ and all $s \in \mathcal{E}(A)$, $\begin{bmatrix} 1 & s \\ 0 & 1 \end{bmatrix} \begin{bmatrix} a & b \\ c & d \end{bmatrix} = \begin{bmatrix} a & b \\ c & d \end{bmatrix} \begin{bmatrix} 1 & s \\ 0 & 1 \end{bmatrix}$.

But this is true if and only if $c = 0$ and $a = d \in Z(\mathcal{E}(A))$ or equivalently, if and only if
$$M \in F(G) = \left\{ \begin{bmatrix} a & b \\ 0 & a \end{bmatrix} : a \in Z(\mathcal{E}(A)) \cap \operatorname{Aut}(A), b \in \mathcal{E}(A) \right\}.$$
□

Corollary 6.8. *Let $F(G) < H \leq \operatorname{Aut}(G)$. Then H is not commutative.*

Proof. Indeed, if $M \in H \setminus F(G)$ there is transvection $X \in E(G) \leq F(G) < H$ such that $MX \neq XM$. □

Recall that a subgroup is called *maximal abelian* if it is maximal among all the abelian subgroups and *abelian maximal* if it is abelian and maximal among all subgroups. Of course, every abelian maximal subgroup is maximal abelian but the converse may fail.

Corollary 6.9. *$F(G)$ is a maximal abelian subgroup in $\operatorname{Aut}(G)$.* □

The next result is folklore.

Lemma 6.10. *Let $H \leq G$. Then $H = \mathcal{C}_G(H)$ if and only if H is a maximal abelian subgroup of G.* □

Corollary 6.11. *$F(G) = \mathcal{C}_{\operatorname{Aut}(G)}(F(G))$.* □

At last we have a property of $\operatorname{Aut}(G)$ that θ transfers to $\operatorname{Aut}(H)$.

Corollary 6.12. *The image $F(G)\theta$ is a maximal abelian subgroup in $\operatorname{Aut}(H)$.* □

Of course this does not yet imply that $F(G)\theta = F(H)$, just that they are both maximal abelian subgroups of $\operatorname{Aut}(H)$. However, we may conclude:

Theorem 6.13. *Let $\mathcal{A}$ be the class of all abelian groups A such that $\operatorname{Aut}(A \oplus A)$ has a unique maximal abelian subgroup. If $A, B \in \mathcal{A}$, then $\operatorname{Aut}(A \oplus A) \cong \operatorname{Aut}(B \oplus B)$ implies $\operatorname{End}(A) \cong \operatorname{End}(B)$.*

Proof. By Corollary 6.12 and the hypothesis, $F(G)\theta = F(H)$. By Proposition 6.6, $F(G) = Z(\operatorname{Aut}(G)) \times E(G)$ implies $F(H) = Z(\operatorname{Aut}(H)) \times E(G)\theta$ and since also $F(H) = Z(\operatorname{Aut}(H)) \times E(H)$, these direct complements are isomorphic. Hence finally $E(G) \cong E(H)$. It remains only to use the additive embedding

$$f_A : \operatorname{End}(A) \longrightarrow E(G), \quad s \mapsto \begin{bmatrix} 1 & s \\ 0 & 1 \end{bmatrix},$$

yielding the isomorphism $\operatorname{End}(A) \xrightarrow{f_A} E(G) \longrightarrow E(H) \xrightarrow{(f_B)^{-1}} \operatorname{End}(B)$. □

Remark 6.14. Let $\mathcal{A}'$ be any class of abelian groups such that for every group $G \in \mathcal{A}'$, all maximal abelian subgroups in $\text{Aut}(G)$ are isomorphic and we can cancel $Z(\text{Aut}(G))$ from direct products. Then again, by the proof above, $\text{Aut}(A \oplus A) \cong \text{Aut}(B \oplus B)$ implies $\text{End}(A) \cong \text{End}(B)$.

In another direction, notice that Lemma 6.7 above gives a little more than Corollary 6.9:

Proposition 6.15. *Among all the abelian subgroups of* $\text{Aut}(G)$ *containing* $E(G)$, $F(G)$ *is the greatest (i.e.,* $E(G) \leq L \leq \text{Aut}(G)$, L *abelian, implies* $L \leq F(G)$).

Proof. By contradiction: if $L \nsubseteq F(G)$, there is a matrix $M \in L$ such that $M \notin F(G)$. By Lemma 6.7, there is a transvection $X \in E(G)$ with $MX \neq XM$ and so H is not commutative. □

References

[Cal06]   G. Călugăreanu, *Determination of torsion-free Abelian groups by their direct powers*, http://www.math.ubbcluj.ro/~calu/ Preprint, 2006.

[F70]   L. Fuchs, *Infinite abelian groups*. Vol. I. Pure and Applied Mathematics. Vol. 36-I. Academic Press, New York, London, 1970.

[F73]   L. Fuchs, *Infinite abelian groups*. Vol. II. Pure and Applied Mathematics. Vol. 36-II. Academic Press, New York, London, 1973.

[HO'M89] A. J. Hahn and O. T. O'Meara, *The classical groups and K-theory*. Grundlehren der Mathematischen Wissenschaften, 291. Springer-Verlag, Berlin, 1989.

[KMT03] P. A. Krylov, A. V. Mikhalev and A. A. Tuganbaev, *Endomorphism rings of abelian groups*. Algebras and Applications, 2. Kluwer Academic Publishers, Dordrecht, 2003.

[M00]   A. Mader, *Almost comletely decomposable groups*. Algebra, Logic and Applications Series, Vol. 13, Gordon and Breach, 2000.

[MS00]   A. Mader and P. Schultz, *Endomorphism rings and automorphism groups of almost completely decomposable groups* Comm. Algebra, 28 (1), (2000), 51–68.

[S59]   E. Sasiada, *On two problems concerning endomorphism groups* Ann. Univ. Sci. Budapest, 2 (1959), 65–66.

Author information

Grigore Călugăreanu, Department of Mathematics and Computer Science, Faculty of Science, Kuwait University, 13060 Kuwait.
E-mail: calu@sci.kuniv.edu.kw

Phill Schultz, School of Mathematics and Statistics, The University of Western Australia, Nedlands, 6009 Australia.
E-mail: schultz@maths.uwa.edu.au

Classification of a class of finite rank Butler groups

Ekaterina Blagoveshchenskaya

Abstract. A new class of torsion-free abelian groups of finite rank is introduced. Groups of this class are defined as epimorphic images of some almost completely decomposable rigid groups and naturally included in the wider class of Butler groups. Their classification up to near isomorphism is obtained.

Key words. Torsion-free abelian groups of finite rank, almost completely decomposable groups, Butler groups.

AMS classification. 20K15.

1 Introduction

Torsion-free abelian groups of finite rank have been investigated in a number of papers, including [1–11]. We consider Butler groups which are epimorphic images of certain almost completely decomposable groups. They inherit a lot of properties from acd-groups and can be studied by similar methods. So, on the one hand, it is a generalization of the results on acd-groups of Blagoveshchenskaya and Mader [10]. On the other hand, the subject derives from the class of Butler groups, which have been introduced and investigated by Arnold, Fuchs, Metelli, Vinsonhaler in [2], [3]. This class of groups can be also considered as a generalization of those studied by A. Corner in [6].

We adopt the standard notation of Fuchs [4] and Mader [5]. For a group generated by some system of elements we use the symbol $\langle \ldots \rangle$, the rank of a torsion-free group X is denoted by $\operatorname{rk} X$. A torsion-free abelian group is completely decomposable, a *cd-group*, if it is a direct sum of subgroups of the rationals $\mathbb{Q}$. An *acd-group* X (almost completely decomposable group) is a torsion-free abelian group of finite rank, that contains a completely decomposable group V for which X/V is a finite group. If in addition X/V is a cyclic group, then X is called a *crq-group* (i.e. an acd-group with cyclic regulator quotient).

We concentrate on a *rigid* crq-group X of *ring* type. This means that the critical typeset $T_{cr}(X)$ is an antichain and consists only of idempotent types (i.e. those which can be represented by characteristics consisting only of 0's and ∞'s), moreover, $\operatorname{rk} X(\tau) = 1$ for each $\tau \in T_{cr}(X)$, see [5, pp. 13, 34, 37].

Such a group X has a distinguished completely decomposable subgroup $R(X)$, its *regulator*. The group $V =: R(X)$ decomposes uniquely into rank-one τ-homogeneous components $V_\tau = V(\tau) = X(\tau)$ of pairwise incomparable types, which are pure in X for any $\tau \in T_{cr}(X)$. Then $R(X)$ is a fully invariant subgroup of X.

We construct a new class of Butler groups, which are epimorphic images of rigid crq-groups. These Butler groups turn out to be extensions of finite direct sums of strongly indecomposable groups by finite groups. So we call them **almost strongly decomposable** groups (*asd-groups*) accepting that a direct sum of strongly indecomposable groups can be naturally called a **strongly decomposable** group (*sd-group*). In particular, a completely decomposable group is strongly decomposable. We restrict ourselves to sd-groups which are uniquely decomposable up to isomorphism (in general this is not true, see [2, Ex. 3.2]).

We need the necessary and sufficient condition for some Butler groups (the so-called $\mathfrak{B}^{(1)}$-groups) to be strongly indecomposable, see Arnold and Vinsonhaler [2, 1.5] or Fuchs and Metelli [3, 3.3].

Our notation is standard and as in [2, 3, 4, 5, 10]. If a group X is isomorphic to Y, we write $X \cong Y$, near-isomorphism of them is denoted by $X \cong_n Y$. As usual $\mathbb{Z}$ is the group of all integers, $\mathbb{N}$ is the set of all natural numbers, $\mathbb{Q}$ is the additive group of rational numbers.

If integer q is divisible by integer p, we will write $p|q$. The symbol $|c|$ denotes the order of a group element $c \in X$, $|C|$ serves as cardinality of a group C and it will be used only for finite groups and sets.

For any element a of a torsion-free group X and a natural number q the only one element $b \in X$ can satisfy $qb = a$. If such an element b exists we may denote it by $\frac{a}{q}$ and say that a is divisible by q in X, recorded as $q|a$. In general, denote the group Y such that $qY = X$ by $\frac{X}{q}$.

We write $f \in \text{Mon}(G, F)$ if $f : G \longmapsto F$ is an injective homomorphism.

The author is very grateful to Claudia Metelli for her attention to these results. The interesting discussions with Laszlo Fuchs and Chuck Vinsonhaler were also very useful for the paper preparation, which took quite a long time.

2 Almost strongly decomposable abelian groups

First we consider an epimorphic image of a completely decomposable rigid group $A = \bigoplus_{\tau \in T_{cr}(A)} \tau a_\tau$ defined by rank-one rational groups τ, containing $\mathbb{Z}$, of pair-wise incomparable idempotent types. We denote by $T_{cr}(A)$ the set of all critical types of A, see [5, p. 37, Definition 2.4.6], [4, Section 85]. We write $\tau(p) = \infty$ if $1/p^n$ belongs to τ for any natural number n (p is a prime).

Definition 2.1. Let $A = \bigoplus_{\tau \in T_{cr}(A)} \tau a_\tau$ be a rigid completely decomposable group of finite rank and let $\{\alpha_\tau : \tau \in T_{cr}(A)\}$ be integers, for which the following hold,

(i) $\text{rk } A = |T_{cr}(A)| \geqslant 3$,

(ii) for any $\tau \in T_{cr}(A)$ there is a prime p with $\tau(p) = \infty$ and $\sigma(p) \neq \infty$ for all $\sigma \neq \tau$, $\sigma \in T_{cr}(A)$,

(iii) $\bigcap_{\tau \neq \sigma} \tau = \mathbb{Z}$ for each $\sigma \in T_{cr}(A)$,

(iv) $\gcd(\{\alpha_\tau | \tau \neq \sigma, \tau \in T_{cr}(A)\}) = 1$ for any $\sigma \in T_{cr}(A)$,

(v) each α_τ is not p-divisible if $\sigma(p) = \infty$ for some $\sigma \in T_{cr}(A)$.

Let $K(A) \cong \mathbb{Z}$ be a group generated by the element $\sum_{\tau \in T_{cr}(A)} \alpha_\tau a_\tau$. Then the group $B(A) = A/K(A)$ is called a *proper* $\mathfrak{B}^{(1)}$-*group*.

Note that $K(A)$ is pure in A because the elements $\{\alpha_\tau a_\tau : \tau \in T_{cr}(A)\}$ have no non-trivial common divisors in A, see [5, p. 29, Definition 2.1.6]. In the Definition 2.1 we summarize and generalize the definitions and notations of [2], [3]. As we can see, a $\mathfrak{B}^{(1)}$-group is completely determined by $T_{cr}(A)$ and the numbers $\{\alpha_\tau\}$. A proper group $B(A)$ is cotrimmed (that is the image of each τa_τ is pure in $B(A)$ under the canonical mapping $A \longrightarrow B(A)$), see [2, p. 19] and strongly indecomposable, see [4, Section 92], [3, 3.3].

The condition (i) of Definition 2.1 is necessary to make $B(A)$ cotrimmed if the condition (ii) is satisfied.

Remark 2.2. It was shown in [3, 5.1] that any two strongly indecomposable groups $B(A)$ and $B(A')$ are isomorphic if and only if the cd-groups A and A' are isomorphic and the coefficients α_τ in the definitions of $B(A)$ and $B(A')$ coincide for any $\tau \in T_{cr}(A) = T_{cr}(A')$, see Definition 2.1. Furthermore, we have for the endomorphism ring of a proper $\mathfrak{B}^{(1)}$-group that $\operatorname{End}(B(A)) \cong \bigcap_{\tau \in T_{cr}(A)} \tau = \mathbb{Z}$ by [3, 3.5].

Now we introduce a new class of torsion-free abelian groups. Let $\{A_1, \ldots, A_t\}, t \geqslant 2$, be a finite set of completely decomposable rigid groups of finite ranks $k_1, \ldots, k_t$. Let $T = T_{cr}(\bigoplus_{i=1}^t A_i), T_i = T_{cr}(A_i)$ and let T be an antichain with $T_i \cap T_j = \varnothing$ if $i \neq j$. We consider proper $\mathfrak{B}^{(1)}$-groups $B(A_i)$ (if rk $A_i = 1$ we put $B(A_i) = A_i$ in order to have rank-one $\mathfrak{B}^{(1)}$-groups cotrimmed).

Define

$$A_0 = \bigoplus_{i=1}^{t} B(A_i). \tag{2.1}$$

We naturally say that A_0 is a *strongly decomposable* rigid group (*sd-group*) as it is a direct sum of strongly indecomposable groups and $T = T_{cr}(A_0)$ is an antichain.

Definition 2.3. If a group B contains a strongly decomposable group as a subgroup of finite index, then B will be called an almost strongly decomposable group (*asd-group*).

A direct decomposition of A_0 into indecomposable summands is uniquely determined because its endomorphism ring satisfies $\operatorname{End}(A_0) \cong \bigoplus_{i=1,\ldots,t} \operatorname{End}(B(A_i))$.

We are particularly interested in asd-groups with a cyclic quotient over a strongly decomposable subgroup, which are epimorphic images of rigid crq-groups. Take $V = A_1 \oplus \cdots \oplus A_t$ as the regulator $R(X) = \bigoplus_\tau X(\tau)$ of a rigid crq-group X with cyclic X/V of order Q, see [5, p. 88, Theorem 4.4.4], [10, Preliminaries]. Recall that the regulator homogeneous components $V_\tau = X(\tau)$ are pure subgroups of X.

Let $e \in \mathbb{N}$ be a number divisible by Q ($e = Q$ is admissible) and let

$$\overline{} : V \longmapsto V/eV = \overline{V} \tag{2.2}$$

denote the natural epimorphism. Choose a generator $u + V$ of X/V and write $Qu = \sum_{\tau \in T} v_\tau, v_\tau \in V(\tau)$. We will use the following numerical invariants, introduced in

[10, p. 22, Definition 2.1], [5, p. 265, Definition 12.6.2],
$$m_\tau(X) = |\overline{v_\tau}| = |v_\tau + QV|, \qquad (2.3)$$

so that $Q = \text{lcm}_\tau\, m_\tau(X)$. It was shown in [10, 2.2] that $m_\tau(X)$ are invariants of a crq-group X indeed, because they do not depend on the choices of an element u and number e.

Definition 2.4. Let X be a rigid crq-group of ring type and let α_τ, $\tau \in T$, be integers such that the following conditions hold,

(i) $V = A_1 \oplus \cdots \oplus A_t$ is the regulator of X with $A_i = \bigoplus_{\tau \in T_i} \tau a_\tau$, $T = T_{cr}(X)$, $T_i = T_{cr}(A_i)$ and $\text{rk}\, A_i = |T_i| \neq 2$, for any $i = 1, \ldots, t$;

(ii) for any $\tau \in T$ there exists a prime p such that $\tau(p) = \infty$ and $\sigma(p) \neq \infty$ for all $\sigma \neq \tau$;

(iii) $T_i \cap T_j = \varnothing$ if $i \neq j$;

(iv) $\bigcap_{\tau \neq \sigma, \tau \in T_i} \tau = \text{tp}\,\mathbb{Z}$ for any $\sigma \in T_i$, $i = 1, \ldots, t$, if $|T_i| \neq 1$;

(v) each α_τ, $\tau \in T_i$, is not p-divisible if $\sigma(p) = \infty$ for some $\sigma \in T_i$, if $|T_i| \neq 1$; $\alpha_\tau = 0$ if $\tau \in T_i$ with $|T_i| = 1$;

(vi) $\gcd(\{\alpha_\tau | \tau \neq \sigma, \tau \in T_i\}) = 1$ for any $\sigma \in T_i$, $i = 1, \ldots, t$ with $|T_i| \neq 1$;

(vii) $\gcd(m_\tau(X), m_\sigma(X)) = 1$ if $\tau \neq \sigma$ and there exists i with $\tau \in T_i$, $\sigma \in T_i$;

(viii) α_τ, $\tau \in T_i$, are relatively prime with $Q = |X/V|$ if $|T_i| \neq 1$;

(ix) $\tau(p) \neq \infty$ for any prime divisor p of Q and any $\tau \in T$.

Let $K = \bigoplus_{i \leqslant t} K(A_i) \subset X$ with $K(A_i) = \langle \sum_{\tau \in T_i} \alpha_\tau a_\tau \rangle$.
The group $B = X/K$ is called a $\mathfrak{B}^{(1)}$crq-group.

Definition 2.5. The canonical epimorphism $\phi : X \longmapsto B = X/K$ with a rigid crq-group X will be called a regular representation of a $\mathfrak{B}^{(1)}$crq-group B with the partition $T = \bigcup_{i \leqslant t} T_i$ and coefficients α_τ, $\tau \in T_{cr}(X)$, if the groups X and K satisfy the above conditions (i)–(ix).

Example. Let $\tau_1 = \langle \frac{1}{3^k} : k \in \mathbb{N}\rangle$, $\tau_2 = \langle \frac{1}{5^k} : k \in \mathbb{N}\rangle$, $\tau_3 = \langle \frac{1}{7^k} : k \in \mathbb{N}\rangle$, $\tau_4 = \langle \frac{1}{11^k} : k \in \mathbb{N}\rangle$ and let

$$X = (\tau_1 a_1 \oplus \tau_2 a_2 \oplus \tau_3 a_3 \oplus \tau_4 a_4) + \mathbb{Z}\frac{1}{26}(13a_1 + 2a_2 + 26a_3 + a_4).$$

Take $\phi : X \longmapsto B$, a regular representation of a $\mathfrak{B}^{(1)}$crq-group B with the partition $T_{cr}(X) = \{\tau_1, \tau_2, \tau_3\} \cup \{\tau_4\}$ and coefficients $\alpha_{\tau_1} = \alpha_{\tau_2} = \alpha_{\tau_3} = 1$, $\alpha_{\tau_4} = 0$. We have a $\mathfrak{B}^{(1)}$crq-group

$$B \cong (\tau_1 a_1 \oplus \tau_2 a_2 \oplus \tau_3 a_3)/(a_1 + a_2 + a_3) \oplus \alpha_{\tau_4} a_4 + \mathbb{Z}\frac{1}{13}(2a_2 + a_4) + \mathbb{Z}\frac{1}{2}(a_1 + a_4),$$

which is not a crq-group, because it does not contain a completely decomposable group as a subgroup of finite index. The invariants $m_{\tau_1}(X) = 2$, $m_{\tau_2}(X) = 13$, $m_{\tau_3}(X) = 1$, $m_{\tau_4}(X) = 26$ satisfy condition (vii).

With the notation of Definitions 2.4 and 2.5, note that $X/V \cong (X/K)/(V/K) = B/A_0$ is a cyclic group of order $Q = |X/V|$, where $A_0 = V/K$ is a strongly decomposable group with proper $\mathfrak{B}^{(1)}$-groups $A_i/K(A_i)$. We allow $\mathrm{rk}(A_i) = 1$ and $K(A_i) = 0$ for some i, only the case of $\mathrm{rk}(A_i) = 2$ is excluded by Definition 2.1.

A few more observations concerning crq-group X are needed later. Fix a representation of its regulator $R(X) = V = \bigoplus_{\tau \in T_{cr}(X)} \tau a_\tau$ with pure subgroups $X(\tau) = \tau a_\tau$, see [10, Preliminaries]. Let

$$Q = \prod_{p \in P} r_p \text{ with } r_p = p^{\gamma_p} \qquad (2.4)$$

be the prime factor decomposition of Q for a finite set P of primes p, $\gamma_p \in \mathbb{N}$. It is well known, see [5, p. 105, Proposition 5.1.5], that

$$X = \sum_{p \in P} X_p$$

with crq-groups X_p, which have the regulator V and $r_p X_p \subset V$. Note that $X_p = X \cap \frac{V}{p^{\gamma_p}}$ are fully invariant subgroups of X because this is true for V, and, evidently,

$$B = \sum_{p \in P} B_p$$

with $\mathfrak{B}^{(1)}$crq-groups $B_p = \phi(X_p) = X_p/K$, moreover, $B/A_0 = \bigoplus_{p \in P} B_p/A_0$ and $r_p(B_p/A_0) = 0$. Evidently, the restriction of ϕ to X_p induces the regular representation $\phi : X_p \longmapsto B_p = X_p/K$ of the $\mathfrak{B}^{(1)}$crq-group B_p with the partition $T = \bigcup_{i \leqslant t} T_i$ and coefficients $\alpha_\tau, \tau \in T_{cr}(V)$.

Suppose that the generators $\{v_p + V | p \in P\}$ of X/V are chosen to be of the p-primary orders $r_p = p^{\gamma_p}$, $\prod_{p \in P} r_p = Q$. Then $X_p = \langle V, v_p \rangle$ and $X/V = \langle \sum_{p \in P} v_p + V \rangle$ with $r_p v_p = \sum_{\tau \in T} v_{\tau p}$, $v_{\tau p} \in \tau a_\tau$. It follows from (2.3) that $m_\tau = m_\tau(X) = \prod_{p \in P} m_\tau(X_p)$.

Remark 2.6. There exist at least two critical types τ_1, τ_2 with $m_{\tau_1}(X_p) = m_{\tau_2}(X_p) = p^{\gamma_p}$ (otherwise the homogeneous subgroup $V(\tau) = V_\tau = \tau a_\tau$ with $m_\tau(X_p) = p^{\gamma_p}$ is not pure in X_p which contradicts the property of V to be the regulator of X_p and X).

Denote

$$m_\tau^p = m_\tau(X_p) = \gcd(m_\tau, r_p) \text{ and } d_\tau^p = \frac{r_p}{m_\tau^p} \text{ for any } \tau \in T, p \in P. \qquad (2.5)$$

If $d_\tau^p = p^{\beta_\tau}$, $\beta_\tau \in \mathbb{N} \cup \{0\}$, then $v_{\tau p}$ is divisible by p^{β_τ} in V and not divisible by $p^{\beta_\tau + 1}$, see (2.3).

Without loss of generality we may assume that $v_{\tau p} = 0$, if $\gcd(m_\tau, p) = 1$. This immediately leads to the system of elements v_p such that $v_{\tau p} \neq 0$ if and only if $m_\tau^p = m_\tau(X_p) \neq 1$. We define the subsets R_p of T by taking

$$R_p = \{\tau \in T : m_\tau^p \neq 1\}, \quad p \in P, \tag{2.6}$$

then

$$r_p v_p = \sum_{\tau \in R_p} v_{\tau p}, \tag{2.7}$$

and for each i the set $R_p \cap T_i$ is either empty set or a singleton by condition (vii) (Definition 2.4).

Remark 2.7. It follows from (2.6) that $R_p \subset T$ is the minimal set such that the factor-group $(X \cap \frac{\bigoplus_{\tau \in R_p} V(\tau)}{p^{\gamma_p}} / \bigoplus_{\tau \in R_p} V(\tau))$ is cyclic of order p^{γ_p}. The condition $|R_p \cap T_i| \leq 1$ is equivalent to the condition (vii) of Definition 2.4 by (2.5).

Recall that
$$X_p = \langle V, v_p \rangle \text{ and } R(X) = V = \bigoplus_{\tau \in T_{cr}(X)} \tau a_\tau \tag{2.8}$$

is a representation of X satisfying (2.7) with rank-one homogeneous groups $V_\tau = V(\tau) = \tau a_\tau$. Let

$$b_\tau = \phi(a_\tau), \tau \in T = T_{cr}(X), \tag{2.9}$$

then $A_0(\tau) = \tau b_\tau, \tau \in T = T_{cr}(X)$.

For any prime p we denote by $\chi_X^p(x)$ the p-entry of the height sequence, which is the *characteristic* $\chi_X(x)$ of an element $x \neq 0$ in a group X, see [4, Section 85].

Lemma 2.8. *Let $\phi : X \longmapsto B = X/K$ be a regular representation of a rigid $\mathfrak{B}^{(1)}$crq-group B and $R(X) = V = \bigoplus_{\tau \in T_{cr}(X)} \tau a_\tau$. Then for any prime divisor p of $|X/R(X)|$ the following holds,*

$$0 = \chi_V^p(a_\tau) = \chi_{X_p}^p(a_\tau) = \chi_{A_0}^p(\phi(a_\tau)) = \chi_{B_p}^p(\phi(a_\tau)).$$

Proof. Fix a prime $p \in P$. Only the last one of the required equalities has not yet been proved. Since $X_p = \langle V, v_p \rangle$, we have $B_p = \langle A_0, \phi(v_p) \rangle$ with $v_p \in \bigoplus_{\tau \in R_p} \tau a_\tau$ and $\phi(v_p) \in \bigoplus_{\tau \in R_p} \tau b_\tau$, then we can restrict ourselves to $\tau \in R_p$, see (2.7). As $r_p \phi(v_p) = \sum_{\tau \in R_p} \phi(v_{\tau p})$ and

$$v_{\tau p} = p^{\beta_\tau} \frac{s_\tau}{l_\tau} a_\tau \tag{2.10}$$

with reduced fractions $\frac{s_\tau}{l_\tau} \in \tau$ satisfying $\gcd(s_\tau, p) = 1$, we keep in mind the equality $\chi_V^p(v_{\tau p}) = \chi_{A_0}^p(\phi(v_{\tau p}))$ and claim that $\chi_{B_p}^p(\phi(v_{\tau p})) = \chi_{A_0}^p(\phi(v_{\tau p}))$ for each τ. The converse leads to the divisibility of the linearly independent elements $\phi(v_{\tau p})$ and $\sum_{\sigma \in R_p, \sigma \neq \tau} \phi(v_{\sigma p})$ by $p^{\beta_\tau + 1}$ in B_p and not in A_0, because there exists $\sigma \in R_p$, $\sigma \neq \tau$ such that $m_\tau(X_p) \leq m_\sigma(X_p)$, see Remark 2.6. This means that the element $\phi(v_{\sigma p})$ is not divisible by $p^{\beta_\tau + 1}$ in A_0, see (2.5). This contradicts the fact that the cyclic group $X_p/V \cong B_p/A_0$ does not contain a direct sum of p-primary groups. Therefore, $\chi^p_{B_p}(\phi(v_{\tau p})) = \chi^p_{A_0}(\phi(v_{\tau p}))$ which implies by (2.10), $0 = \chi^p_V(a_\tau) = \chi^p_{X_p}(a_\tau) = \chi^p_{A_0}(\phi(a_\tau)) = \chi^p_{B_p}(\phi(a_\tau))$.

The proof is complete. □

We have $A_0 = \sum_{\tau \in T} A_0(\tau)$ with homogeneous summands $A_0(\tau) = \tau b_\tau$, which are pure in each $B_p = B \cap \frac{A_0}{p^{\gamma_p}}$ and in B.

Corollary 2.9. *For any $\tau \in T$ the group $\phi(V_\tau)$ is pure in B_p and the sd-group $A_0 = \phi(V)$ is a fully invariant subgroup of the $\mathfrak{B}^{(1)}$crq-group B. The groups B_p are also fully invariant subgroups of B.*

Definition 2.10. The strongly decomposable group A_0 of a $\mathfrak{B}^{(1)}$crq-group B, which is the image of the regulator V of X under the regular representation ϕ, will be called the *regulator* of B, that is $A_0 = R(B)$.

Using (2.6)–(2.8) and Lemma 2.8, which says that $0 = \chi^p_V(a_\tau) = \chi^p_{A_0}(\phi(a_\tau))$, for each $p \in P$ we can construct the following commutative diagram ($\times p^{\gamma_p}$ means multiplication by $r_p = p^{\gamma_p}$),

$$\begin{array}{ccccc} X_p & \xrightarrow{\times p^{\gamma_p}} & V & \longrightarrow & \overline{V} = V/p^{\gamma_p} V \\ \downarrow \phi & & \downarrow \phi|_V & & \downarrow \overline{\phi} \\ B_p & \xrightarrow{\times p^{\gamma_p}} & A_0 & \longrightarrow & \overline{A_0} = A_0/p^{\gamma_p} A_0 \end{array} \quad (2.11)$$

with the natural epimorphism $\overline{\phi} : \overline{V} \longrightarrow \overline{A_0} \cong \overline{V}/\overline{K}$ defined on $\overline{V} = \bigoplus_{\tau \in T} \langle \overline{a_\tau} \rangle$ by $\overline{K} = \bigoplus_{i \leqslant t} \overline{K(A_i)}$ and $\overline{K(A_i)} = \langle \sum_{\tau \in T_i} \alpha_\tau \overline{a_\tau} \rangle$, see Definition 2.4.

For any w_p satisfying $X_p = \langle V, w_p \rangle$ with

$$p^{\gamma_p} w_p = \sum_{\tau \in T} w_{\tau p} \quad \text{and} \quad w_{\tau p} \in V_\tau, \quad (2.12)$$

we obtain from the above diagram, using Lemma 2.8 and (2.3), that $m_\tau(X_p) = |\overline{w_{\tau p}}| = |\overline{v_{\tau p}}| = |\overline{\phi(v_{\tau p})}| = |\overline{\phi}\,(\overline{v_{\tau p}})|$ with $\overline{\phi}\,(\overline{v_{\tau p}}) \in \overline{A_0(\tau)}$, $\tau \in T$ (by construction, $v_{\tau p} = 0$ and $|\overline{\phi(v_{\tau p})}| = 0$ if and only if $\tau \in T \setminus R_p$, see (2.7)).

Since $\phi : X_p \longrightarrow B_p$ is an epimorphism, then $B_p = \langle \phi(V), u_p \rangle$ implies $u_p = \phi(w_p)$ for some w_p satisfying (2.12), such that $X_p/V = \langle w_p + V \rangle$. We can introduce the following numerical invariants of a $\mathfrak{B}^{(1)}$crq-group B_p with a primary regulator quotient B_p/A_0.

Let $B_p = \langle A_0, u_p \rangle$ and $p^{\gamma_p} u_p = \sum_{\tau \in T} u_{\tau p}$ with $u_{\tau p} = \phi(w_{\tau p}) \in A_0(\tau)$ arising from $p^{\gamma_p} w_p = \sum_{\tau \in T} w_{\tau p}$ such that $X_p = \langle V, w_p \rangle$ and $u_p = \phi(w_p)$. Define

$$m_\tau(B_p) = |\overline{u_{\tau p}}| = |u_{\tau p} + p^{\gamma_p} A_0|. \quad (2.13)$$

We see from the above that $m_\tau(B_p) = m_\tau(X_p)$ for any τ and it is equal to 1 if $\tau \notin R_p$.

Furthermore, let
$$m_\tau(B) = \prod_{p \in P} m_\tau(B_p) = m_\tau(X). \tag{2.14}$$

We need the following

Proposition 2.11. *Let $\phi : X \longmapsto B = X/K$ with a rigid crq-group X be a regular representation of a $\mathfrak{B}^{(1)}$ crq-group B (see Definition 2.4). Then the following hold:*

(i) τa_τ, $\tau \in T_{cr}(X)$, *is a homogeneous component of the regulator $V = R(X)$ if and only if $\phi(\tau a_\tau)$ is pure in B;*

(ii) $\gcd(m_\tau(X), m_\sigma(X)) = 1$ *whenever $\tau \neq \sigma$ and $\tau \in T_i$, $\sigma \in T_i$ for some i, if and only if there exist the minimal sets $R_p \subset T$, such that the factor-group*

$$\left(B \cap \frac{\bigoplus_{\tau \in R_p} A_0(\tau)}{p^{\gamma_p}} \right) \Big/ \bigoplus_{\tau \in R_p} A_0(\tau)$$

is cyclic of order p^{γ_p} and $|R_p \cap T_i| \leqslant 1$ for any $p \in P$.

Proof. The statement (i) is given by Lemma 2.8. The statement (ii) is based on Remark 2.7 and Lemma 2.8. □

We are now able to give an equivalent definition of a rigid $\mathfrak{B}^{(1)}$ crq-group B without reference to its regular representation and to see that the invariants introduced don't depend on it. The following system of axioms coincides with that of Definition 2.4 by Proposition 2.11, which concerns items (i) and (vii).

Definition 2.12. Let B be a group with $T = T_{cr}(B) = \bigcup_{i=1,\ldots,t} T_i$ for pair-wise disjoint sets T_i and $A_0 \cong \bigoplus_{i=1,\ldots,t} B(A_i)$ be its strongly decomposable subgroup with $B(A_i) = A_i/K(A_i)$ defined by rigid completely decomposable groups of ring type $A_i = \bigoplus_{\tau \in T_i} \tau a_\tau$ and $K(A_i) = \sum_{\tau \in T_i} \alpha_\tau a_\tau$, $\alpha_\tau \in \mathbb{Z}$, $\tau \in T$. Suppose that B/A_0 is a finite cyclic group of order $Q = \prod_{p \in P} p^{\gamma_p}$, where P is a finite set of primes, $\gamma_p \in \mathbb{N}$.

Let the following conditions hold:

(i) for any $i = 1, \ldots, t$ and $\tau \in T_i$ the group $\tau a_\tau + K(A_i)$ is pure in B and $\operatorname{rk} A_i = |T_i| \neq 2$;

(ii) for any $\tau \in T$ there exists a prime p such that $\tau(p) = \infty$ and $\sigma(p) \neq \infty$ for all $\sigma \neq \tau$;

(iii) $T_i \cap T_j = \emptyset$ if $i \neq j$;

(iv) $\bigcap_{\tau \neq \sigma, \tau \in T_i} \tau = \operatorname{tp} \mathbb{Z}$ for any $\sigma \in T_i$, $i = 1, \ldots, t$, if $|T_i| \neq 1$;

(v) each α_τ, $\tau \in T_i$, is not p-divisible if $\sigma(p) = \infty$ for some $\sigma \in T_i$, if $|T_i| \neq 1$; $\alpha_\tau = 0$ if $\tau \in T_i$ with $|T_i| = 1$;

(vi) $\gcd(\{\alpha_\tau | \tau \neq \sigma, \tau \in T_i\}) = 1$ for any $\sigma \in T_i$, $i = 1, \ldots, t$ with $|T_i| \neq 1$;

(vii) there exist the minimal sets $R_p \subset T$, such that the factor-group $(B \cap \frac{\bigoplus_{\tau \in R_p} A_0(\tau)}{p^{\gamma_p}}) / \bigoplus_{\tau \in R_p} A_0(\tau))$ is cyclic of order p^{γ_p} and $|R_p \cap T_i| \leq 1$ for any $p \in P$;

(viii) α_τ, $\tau \in T_i$, are relatively prime with $Q = |X/V|$ if $|T_i| \neq 1$;

(ix) $\tau(p) \neq \infty$ for any prime divisor p of Q and any $\tau \in T$.

Then B is a $\mathfrak{B}^{(1)}$crq-group.

It is easy to see that a group $B = \sum_{p \in P} B_p$ defined by $B_p = B \cap \frac{A_0}{p^{\gamma_p}}$ (see Definition 2.12) possesses a regular representation 2.5, $\phi : X \longmapsto B = X/\bigoplus_{i=1,\ldots,t} K(A_i)$, with a crq-group $X = \sum_{p \in P} X_p$ having the regulator $V = \bigoplus_{i=1,\ldots,t} A_i = \bigoplus_{\tau \in T} \tau a_\tau$. This implies the regular representations $\phi : X_p \longmapsto B_p = X_p / \bigoplus_{i=1,\ldots,t} K(A_i)$ for any $p \in P$, where $B_p = \langle A_0, u_p \rangle$ with $p^{\gamma_p} u_p = \sum_{\tau \in R_p} u_{\tau p}$, $u_{\tau p} \in A_0(\tau)$, and the group $X_p = \langle \bigoplus_{\tau \in T} \tau a_\tau, v_p \rangle$ satisfies $p^{\gamma_p} v_p = \sum_{\tau \in R_p} v_{\tau p}$, $v_{\tau p} \in \tau a_\tau$ with $\phi(v_{\tau p}) = u_{\tau p}$.

Theorem 2.13. *Let B be a $\mathfrak{B}^{(1)}$crq-group (Definition 2.12) and let $\phi : X \longmapsto B$ be a regular representation of B (Definition 2.5). Then a crq-group X is uniquely determined up to isomorphism.*

Proof. As above, $B = \sum_{p \in P} B_p$ with $B_p = B \cap \frac{A_0}{p^{\gamma_p}}$, so $B = \langle A_0, \{u_p : p \in P\} \rangle$ for some $u_p \in B$ with $p^{\gamma_p} u_p \in A_0$.

It is evident that the regulator of X, $V = \bigoplus_{i=1,\ldots,t} A_i$, a set of primes P, a partition $T_{cr}(X) = \bigcup_{i=1,\ldots,t} T_{cr}(A_i)$ and coefficients α_τ, $\tau \in T_{cr}(X)$, are uniquely determined by Definition 2.12. So, there exists only one map $\phi : V \longmapsto A_0$ satisfying (i)–(vi), (viii) and (ix) of Definition 2.4.

It remains to show that a preimage in V of each $p^{\gamma_p} u_p \in A_0 \cap p^{\gamma_p} B_p$, $p \in P$, is also unique for the regular representation ϕ of a $\mathfrak{B}^{(1)}$crq-group B (see Definition 2.5). It follows from $V = \bigoplus_{\tau \in T} \tau a_\tau$ and condition (vii) that $p^{\gamma_p} u_p = \sum_{\tau \in R_p} u_{\tau p}$ with $u_{\tau p} \in A_0(\tau) = \tau \phi(a_\tau)$ (see Definition 2.12). If $\tau \in T_i$ with $|T_i| \geq 3$, for some rational β we have

$$u_{\tau p} = \phi(\beta_\tau \alpha_\tau a_\tau + d_1) = \phi\left(-\beta_\tau \sum_{\sigma \in T_i,\, \sigma \neq \tau} \alpha_\sigma a_\sigma + d_2\right)$$

with any $d_1, d_2 \in K(A_i)$. We easily see that only $\beta_\tau \alpha_\tau a_\tau$ as a preimage of $u_{\tau p}$ in A_i makes the group X satisfy condition (vii) of Definition 2.4.

In case of $\tau \in T_i$ with $|T_i| = 1$ the uniqueness of such a preimage of $u_{\tau p}$ in A_i is evident. Hence, a crq-group X with regulator V is uniquely determined and the proof is complete. □

Corollary 2.14. *The numbers $m_\tau(B)$ defined by (2.13) and (2.14) are invariants of a $\mathfrak{B}^{(1)}$crq-group B indeed.*

3 Classification of rigid $\mathfrak{B}^{(1)}$crq-groups

We use the symmetric form of near-isomorphism definition.

Definition 3.1. Let G and H be torsion-free abelian groups of finite rank. Then G and H are called nearly isomorphic (in symbols $G \cong_{nr} H$) if and only if for any prime q there are monomorphisms $\eta_q : G \longrightarrow H$ and $\xi_q : H \longrightarrow G$ such that $H/\eta_q(G)$ and $G/\xi_q(H)$ are finite groups and $|H/\eta_q(G)|$ and q as well as $|G/\xi_q(H)|$ and q are relatively prime.

Let B and C be nearly isomorphic rigid $\mathfrak{B}^{(1)}$crq-groups, so there exist monomorphisms $\eta_q : B \longrightarrow C$ and $\xi_q : C \longrightarrow B$ such that $C/\eta_q(B)$ and $B/\xi_q(C)$ are finite groups. This means that $\operatorname{rk} B = \operatorname{rk} C$ and the critical typesets of the groups coincide, $T_{cr}(B) = T_{cr}(C)$, because they consist of the maximal group element types and, clearly, $\eta_q(R(B)_\tau) \subset R(C)_\tau$, $\xi_q(R(C)_\tau) \subset R(B)_\tau$ for τ-homogeneous rank-one groups $R(B)_\tau = R(B)(\tau)$, $R(C)_\tau = R(C)(\tau)$. Furthermore, if $\phi_q(x) = a \in R(C)_\tau$ for some $x \in C$, then, by construction, $x \in R(B)_\tau$, therefore, $R(C)/\eta_q(R(B)) \subset C/\eta_q(B)$ and, by symmetry, $R(B)/\xi_q(R(C)) \subset B/\xi_q(C)$ for any prime q, hence $R(B) \cong_{nr} R(C)$. It is immediate from [1] and [2, Remark, p. 26] that near-isomorphism implies isomorphism $R(B) \cong R(C)$, since the regulators of rigid $\mathfrak{B}^{(1)}$crq-groups are direct sums of strongly indecomposable fully invariant Butler subgroups with endomorphism rings principal ideal domains.

Note that in the case of $\mathfrak{B}^{(1)}$crq-groups the monomorphisms $\eta_q : B \longrightarrow C$ and $\xi_q : C \longrightarrow B$ trivially exist if the primes q divide neither $|C/\eta_q(B)|$ nor $|B/\xi_q(C)|$.

Remark 3.2. It is well known that two finite rank torsion-free abelian groups G and H are nearly isomorphic if there exist only one-sided monomorphisms $\eta_q : G \longrightarrow H$ or $\xi_q : H \longrightarrow G$ with the required properties, see [1, pp. 80, 151, 153].

We will use the **Near-Isomorphism Criterion for crq-groups** ([10, Theorem 2.4], [5, Theorem 12.6.5]): Let X and Y be block-rigid (in particular, rigid) crq-groups. Then $X \cong_{nr} Y$ if and only if and only if their regulators $R(X)$ and $R(Y)$ are isomorphic and $m_\tau(X) = m_\tau(Y)$ for all types τ.

We need the following

Lemma 3.3. Let X and Y be nearly isomorphic rigid crq-groups of ring type with $T = T_{cr}(X) = T_{cr}(Y)$ and $m_\tau = m_\tau(X) = m_\tau(Y)$, $\phi : X \longmapsto B$ and $\psi : Y \longmapsto C$ be regular representations of $\mathfrak{B}^{(1)}$crq-groups B and C with the partition $T = \bigcup_{i \leqslant t} T_i$ and the same coefficients α_τ, $\tau \in T$. Then B and C are nearly isomorphic.

Proof. Since $X \cong_{nr} Y$, there is a monomorphism $\chi : X \longmapsto Y$ such that $\gcd([Y : \chi(X)], Q) = 1$, where $Q = \operatorname{lcm}_\tau m_\tau$. Note that $[Y : \chi(X)] = [R(Y) : \chi(R(X))] = \prod_{\tau \in T}[Y(\tau) : \chi(X(\tau))]$ by [5, 9.2.2].

Let $s_\tau = [Y(\tau) : \chi(X(\tau))]$. Choose $a_\tau \in X$ and $a'_\tau \in Y$ with $\tau \in T$ so that $K_1 = \operatorname{Ker}\phi = \bigoplus_{i=1,\ldots,t} \langle \sum_{\tau \in T_i} \alpha_\tau a_\tau \rangle$ and $K_2 = \operatorname{Ker}\psi = \bigoplus_{i=1,\ldots,t} \langle \sum_{\tau \in T_i} \alpha_\tau a'_\tau \rangle$. Then $X(\tau) = \tau a_\tau$, $Y(\tau) = \tau a'_\tau$ and $\chi(a_\tau) = d_\tau a'_\tau$, where $d_\tau = \frac{l_\tau s_\tau}{n_\tau}$ are reduced fractions with $\tau(p) = \infty$ for every prime divisor p of n_τ or l_τ and $\tau(q) \neq \infty$ if $q|s_\tau$. Let $l = \text{lcm}_\tau n_\tau$ and $\chi_1 : X \longmapsto Y$ be a monomorphism defined by $\chi_1(a_\tau) = ld_\tau a'_\tau$, so $[Y(\tau) : \chi_1(X(\tau))]$ divides ls_τ, that is relatively prime with Q, because l and s_τ have the same property, see Definition 2.12 (ix). By the Chinese Remainder Theorem there exist positive integers x_i such that $x_i \equiv ld_\tau \pmod{m_\tau}$ for any $\tau \in T_i$ and $\gcd(x_i, Q) = 1$, see [11, Lemma 3.4, proof]. Consider $\chi_0 \in \text{Mon}(X, Y)$ defined by $\chi_0(a_\tau) = x_i a'_\tau, \tau \in T_i, i = 1, \ldots, t$.

Denote $Y' = \chi_0(X) \cong X$. Then the representation $\psi : Y \longmapsto C$ implies the representation $\psi|_{Y'} : Y' \longmapsto \psi(Y')$ with the same coefficients as ϕ and ψ. Therefore, $\psi(Y') \subset \psi(Y) = C$ is a subgroup of the same index as Y' in Y because every $\psi(Y(\tau))$ is pure in C.

Since $B \cong \psi(Y')$, near-isomorphism of B and C has been established by Remark 3.2. □

Let B and C be nearly isomorphic $\mathfrak{B}^{(1)}$crq-groups and let the monomorphisms $\eta_q : B \longrightarrow C$ and $\xi_q : C \longrightarrow B$ satisfy the required conditions for all primes q. Let $\phi : X \longmapsto B$ and $\psi : Y \longmapsto C$ be the regular representations of the groups.

It is immediate from Remark 2.2 that $R(B) \cong R(C)$, so $R(X) \cong R(Y)$ and the two regular representations have the same partitions of $T = T_{cr}(X) = T_{cr}(Y)$ and the same coefficients α_τ.

Furthermore, the crq-groups X and Y are nearly isomorphic because $|C/\eta_q(B)| = |Y/\eta_q(X)|$ is relatively prime with q for any prime q.

This leads us to the final

Theorem 3.4 (Near-Isomorphism Criterion for rigid $\mathfrak{B}^{(1)}$crq-groups). *Rigid $\mathfrak{B}^{(1)}$crq-groups B and C of ring type are nearly isomorphic if and only if their regulators are isomorphic, $R(B) \cong R(C)$, and for any $\tau \in T_{cr}(B) = T_{cr}(C)$ their invariants coincide, $m_\tau(B) = m_\tau(C)$.*

Proof. It was shown in Lemma 3.3 and below it that $\mathfrak{B}^{(1)}$crq-groups B and C are nearly isomorphic if and only if their regular representations $\phi : X \longmapsto B$ and $\psi : Y \longmapsto C$ satisfy the conditions:

(i) $X \cong_{nr} Y$;

(ii) the regular representations of $\mathfrak{B}^{(1)}$crq-groups B and C have the same partitions of $T = T_{cr}(X) = T_{cr}(Y)$ and the same coefficients $\alpha_\tau, \tau \in T$, see Definition 2.12.

We apply the above near-isomorphism criterion for crq-groups, Remark 2.2 and the equalities (2.14), $m_\tau(X) = m_\tau(B), m_\tau(Y) = m_\tau(C)$, for all τ, to complete the proof. □

References

[1] D. Arnold. *Finite Rank Torsion Free Abelian Groups and Rings*, Lecture Notes in Mathematics, vol. 931, Springer-Verlag, 1982.

[2] D. Arnold and C. Vinsonhaler. *Finite Rank Butler Groups: A Survey of Recent Results*, Lecture Notes in Pure and Applied Mathematics, vol. 146, pp. 17–41, 1993.

[3] L. Fuchs and C. Metelli. *On a class of Butler groups*, Manuscripta Math., 71, pp. 1–28, 1991.

[4] L. Fuchs. *Infinite Abelian Groups*, vol. 1, 2, Academic Press 1970, 1973.

[5] A. Mader. *Almost completely decomposable abelian groups*, Gordon and Breach, *Algebra, Logic and Applications* Vol. **13**, Amsterdam, 1999.

[6] A. L. S. Corner, *A note on rank and decomposition of torsion-free abelian groups*, Proceedings Cambridge Philos. Soc. **57** (1961), 230–233, and **66** (1969) 239–240.

[7] E. L. Lady, *Almost completely decomposable groups torsion-free abelian groups*. Proc. Amer. Math. Soc. vol. 45, pp. 41–47, 1974.

[8] E. L. Lady, *Summands of finite rank torsion-free abelian groups*. J. Algebra, vol. 32, pp. 51–52, 1974.

[9] E. Blagoveshchenskaya, G. Ivanov and P. Schultz, *The Baer-Kaplansky theorem for almost completely decomposable groups*, Contemporary Mathematics 273, pp. 85–93, 2001.

[10] E. Blagoveshchenskaya and A. Mader. *Decompositions of almost completely decomposable abelian groups*, Contemporary Mathematics, vol. 171, pp. 21–36, 1994.

[11] E. Blagoveshchenskaya. *Classification of a class of almost completely decomposable groups*. Rings, Modules, Algebras and Abelian Groups (Lecture Notes in Pure and Applied Mathematics Series 236), pp. 45–54, 2004.

Author information

Ekaterina Blagoveshchenskaya, Department of Mathematics, St. Petersburg State Polytechnical University, Polytechnicheskaya 29, St. Petersburg 195251, Russia.
E-mail: `kate@robotek.ru` and `kblag2002@yahoo.com`

Quotient divisible and almost completely decomposable groups

Alexander A. Fomin

Abstract. The quotient divisible abelian groups, which are dual to the almost completely decomposable torsion-free abelian groups, are investigated. In particular, the well known example of anomalous direct decompositions by A. L. S. Corner is considered on dual quotient divisible groups.

Key words. Abelian group, module.

AMS classification. 20K15, 20K21.

1 Introduction

The notion of quotient divisible group has been introduced in [1] as a generalization of two classes of group. The first one is the class $\mathcal{G}$ of honestly mixed groups introduced earlier by S. Glaz and W. Wickless [2], and the second class is the well known class of torsion free quotient divisible groups by R. Beaumont and R. Pierce [3]. The mixed quotient divisible groups are considered also in [4–11]. The main result motivating the introduction of the quotient divisible mixed groups is the duality between the quotient divisible groups and the torsion free groups of finite rank introduced in [1] as well.

The almost completely decomposable groups have been researched by many authors for a long time. We mention contributions of D. Arnold, K. Benabdallah, E. A. Blagoveshenskaya, R. Burkhardt, A. L. S. Corner, M. Dugas, T. Faticoni, L. Fuchs, R. Goebel, B. Jonsson, S. F. Kozhukhov, A. Mader, O. Mutzbauer, E. Lee Lady, F. Loonstra, J. Reid, P. Schultz, C. Vinsonhaler, A. V. Yakovlev, the list is obviously far from being complete.

The main goal of the present paper is an application of the mentioned above duality for investigation of the almost completely decomposable groups. Since the original duality is a duality of categories with quasi-homomorphisms, a direct application is impossible. There is no difference between the almost completely decomposable groups and the completely decomposable groups in such a category. Thus we use a new approach developed in [10].

Every pair consisting of a torsion free group A and a basis (a maximal linearly independent set of elements) $x_1, \ldots, x_n$ of A gives a dual pair consisting of a quotient divisible group A^* and a basis $x_1^*, \ldots, x_n^*$ of A^* and conversely. It is proved (Corollary 10) that for every almost completely decomposable group A it is possible to choose a basis $x_1, \ldots, x_n$ of A such that the quotient divisible group A^* is decomposed into a direct sum of rank-1 quotient divisible subgroups. This is a simplification. Considering a finite extension A of a completely decomposable group B and a common basis $x_1, \ldots, x_n$ for two groups, we obtain two dual bases $x_{1B}^*, \ldots, x_{nB}^*$ and $x_{1A}^*, \ldots, x_{nA}^*$ according to B and to A. They differ by torsion elements $t_1, \ldots, t_n$ such that $x_{1A}^* = x_{1B}^* + t_1, \ldots, x_{nA}^* = x_{nB}^* + t_n$. The basis $x_{1A}^*, \ldots, x_{nA}^*$ and therefore the sequence $t_1, \ldots, t_n$ determines completely the group A in this configuration with the fixed basis $x_1, \ldots, x_n$. In such a way we obtain a description of the almost completely decomposable groups (Theorem 15) in terms of the sequences $(t_1, \ldots, t_n)$ of torsion elements. Note that all quotient divisible groups considered in Theorem 15 are decomposed into a direct sum of rank-1 quotient divisible subgroups, while their dual almost completely decomposable groups can be indecomposable (Theorem 11 and Corollary 16) or they can have anomalous direct decompositions as in an example below.

In the final section we apply this description for a dualization of the famous masterpiece by A. L. S. Corner [12]. For every pair $0 < k \leq n$ of integers, there exists a torsion free group C of rank n such that for every decomposition of the number $n = n_1 + \ldots + n_k$ into a sum of k positive integers, the group C can be decomposed into a direct sum of k indecomposable subgroups of ranks $n_1, \ldots, n_k$, respectively.

2 Preliminaries

All groups will be additive abelian groups. Let n be a positive integer and p a prime number, $Z, Q, Z_n = Z/nZ, \widehat{Z}_p$ denote the ring of integers, the field of rational numbers, the ring of residue classes modulo n, the ring of p-adic integers, respectively. $Q_p = \{\frac{m}{n} \in Q \,|\, g.c.d.\,(n,p) = 1\}$ and $Q^{(p)} = \{\frac{m}{p^n} \in Q \,|\, m, n \in Z\}$. The ring $\widehat{Z} = \prod_p \widehat{Z}_p$ is the Z-adic completion of Z, it is called the ring of *universal integers*. The additive groups of the rings have the same notations.

If $x_1, \ldots, x_n$ are elements of an abelian group A, then $\langle x_1, \ldots, x_n \rangle$ is the subgroup of A generated by these elements, $\langle x_1, \ldots, x_n \rangle_*$ is the pure hull of these elements, that is $a \in \langle x_1, \ldots, x_n \rangle_* \Leftrightarrow$ there exists a nonzero integer m such that $ma \in \langle x_1, \ldots, x_n \rangle$. In particular, all torsion elements of A belong to $\langle x_1, \ldots, x_n \rangle_*$. At last, $\langle x_1, \ldots, x_n \rangle_R$ denotes the submodule of an R-module generated by these elements.

A set of elements $x_1, \ldots, x_n$ of an abelian group (of an R-module) is called *linearly independent over Z*, if every equality $m_1 x_1 + \ldots + m_n x_n = 0$ with integer coefficients implies $m_1 = \ldots = m_n = 0$. A set of elements $x_1, \ldots, x_n$ of a $\widehat{Z}$-module is called *linearly independent over $\widehat{Z}$*, if every equality $\alpha_1 x_1 + \ldots + \alpha_n x_n = 0$ with universal integer coefficients implies $\alpha_1 x_1 = \ldots = \alpha_n x_n = 0$. In particular, the set $0, \ldots, 0$ is linearly independent over $\widehat{Z}$.

We use the characteristics (m_p) and the types $\tau = [(m_p)]$ in the same manner as in [13] denoting the zero characteristic and the zero type by 0. As usual $(m_p) \geq (k_p)$ if $m_p \geq k_p$ for all prime numbers p. In this case we define $(m_p) - (k_p) = (m_p - k_p)$ setting $\infty - \infty = 0$.

If $\alpha = (\alpha_p) \in \widehat{Z}$, we define the *characteristic of α* as $\text{char}\,(\alpha) = (m_p)$, where α_p is divisible by p^{m_p} in $\widehat{Z}_p$ and m_p is the maximal power. If $\alpha_p = 0$ then $m_p = \infty$. Every finitely generated ideal I of the ring $\widehat{Z}$ is of the form $I = I_\chi = \{\alpha \in$

$\widehat{Z} \,|\, \mathrm{char}\,(\alpha) \geq \chi\}$ for a characteristic χ. Let $Z_\chi = \widehat{Z}/I_\chi$. As a $\widehat{Z}$-module, Z_χ is cyclic and finitely presented.

A $\widehat{Z}$-module M is called *finitely presented*, if there exists an exact sequence of $\widehat{Z}$-module homomorphisms $\widehat{Z}^m \to \widehat{Z}^n \to M \to 0$ for positive integers m and n. Every finitely presented $\widehat{Z}$-module M is of the form $M \cong Z_{\chi_1} \oplus \ldots \oplus Z_{\chi_n}$. The decomposition is not unique in general, even the number of summands is not an invariant. But it is uniquely definite at the additional condition on the characteristics $\chi_1 \leq \ldots \leq \chi_n$. Every finitely generated submodule N of a finitely presented $\widehat{Z}$-module M is finitely presented and the quotient M/N is finitely presented as well, see [14].

For an element x of a finitely presented $\widehat{Z}$-module M and a prime p, we define: m_p is the greatest nonnegative integer such that p^{m_p} divides x in M and k_p is the least nonnegative integer such that the element $p^{k_p} x$ is divisible by all powers of p. If such a number m_p or k_p doesn't exist then $m_p = \infty$ or $k_p = \infty$, respectively. The characteristics $\mathrm{char}\,(x) = (m_p)$ and $\mathrm{cochar}\,(x) = (k_p)$ are called the *characteristic* and the *co-characteristic* of the element x in the module M. The type $[\mathrm{cochar}\,(x)]$ is called the *co-type* of the element x. The co-characteristic is an analog of the order of an element. If $x \in Z_\chi$ then $\mathrm{cochar}\,(x) = \chi - \mathrm{char}\,(x)$ and $\mathrm{char}\,(x) \geq \chi - \mathrm{cochar}\,(x)$, the inequality can be strict.

The ring $R = \langle 1, \bigoplus_p \widehat{Z}_p \rangle_* \subset \widehat{Z}$ is called the ring of *pseudo-rational* numbers. See [8] for basic properties of this ring, where the concept has been introduced. The mentioned class of mixed groups $\mathcal{G}$ has the following characterization. The category of groups $\mathcal{G}$ coincides with the category of all finitely generated R-modules such that their p-components are torsion for all prime numbers p (Theorem 5.2 in [8]). A. V. Tsarev is developing an interesting theory of modules over the ring of pseudo-rational numbers in [11,15,16] which is very close to the quotient divisible group theory.

For every characteristic $\chi = (m_p)$ we define the ideal $J_\chi = \bigoplus_p p^{m_p} \widehat{Z}_p$ of the ring R, assuming $p^\infty = 0$, and the ring $R(\chi) = R/J_\chi$. An inequality $\chi \geq \kappa$ for two characteristics implies the inclusion $J_\chi \subset J_\kappa$ which determines in turn the natural homomorphism of rings

$$g_\kappa^\chi : R(\chi) \to R(\kappa) \text{ for } \chi \geq \kappa.$$

In this paper we are interested in the subrings $R^\chi = \langle 1 \rangle_* \subset R(\chi)$ of the rings $R(\chi)$ keeping the same notation for their additive groups as well. Restrictions of g_κ^χ on R^χ induce the following spectrum of homomorphisms of rings (abelian groups)

$$g_\kappa^\chi : R^\chi \to R^\kappa \text{ for } \chi \geq \kappa. \qquad (2.1)$$

Note that the homomorphisms (2.1) are not necessarily surjective, for example, the natural embedding $Z \to Q$ is exactly the homomorphism g_κ^χ for the pair of characteristics $\chi = (\infty, \infty, \ldots) \geq \kappa = (0, 0, \ldots)$. If a characteristic $\chi = (m_p)$ belongs to a nonzero type, then the ring R^χ coincides with the subring $\langle 1 \rangle_* \subset Z_\chi$. In this case, the co-characteristic of 1 in Z_χ coincides with χ. This is one of the reasons why we'll call the characteristic χ as the *co-characteristic* of the group R^χ. If a characteristic $\chi = (m_p)$ belongs to the zero type, then $R^\chi = Z_m \oplus Q$, where $m = \prod_p p^{m_p}$.

3 Quotient divisible groups

Definition ([1]). An abelian group A without nonzero torsion divisible subgroups is called *quotient divisible* if it contains a free subgroup F of finite rank such that the quotient group A/F is torsion divisible. Every free basis $x_1, \ldots, x_n$ of the group F is called a *basis* of the quotient divisible group A, the number n is the *rank* of A.

The groups R^χ serve as examples of the quotient divisible groups. The rank of R^χ is equal to 1 and a basis of R^χ is the unity element $1 \in R^\chi$ considering R^χ as a ring. And what is more, every quotient divisible group of rank 1 is isomorphic to a group R^χ for some characteristic χ, and $R^\chi \cong R^\kappa \iff \chi = \kappa$ (see [7]). Therefore an arbitrary quotient divisible group A of rank 1 with a basis x may be denoted as $A = xR^\chi$ and the characteristic χ is the *co-characteristic* of the quotient divisible rank-1 group A.

As it is shown in [1] and [10], every reduced quotient divisible group A can be presented as a pure hull $A \cong \langle x_1, \ldots, x_n \rangle_* \subset M$ of a linearly independent over Z set of elements $x_1, \ldots, x_n$ of a finitely presented $\widehat{Z}$-module M such that $M = \langle x_1, \ldots, x_n \rangle_{\widehat{Z}}$. Namely, $M = \widehat{A}$ is the Z-adic completion of A and the set $x_1, \ldots, x_n$ is the image of a basis of A in $\widehat{A}$.

The divisible part of a quotient divisible group is a divisible torsion free group of finite rank. A reduced complement of the divisible part is not necessarily quotient divisible, for example it is true for a group R^χ, where χ is a nonzero characteristic of the zero type. In general, this complement is a direct sum of a finite group and a quotient divisible reduced group.

The following lemma is useful for us. Let p be a prime number, we understand under p-rank of a group A the dimension of the vector space A/pA over the field Z_p.

Lemma 1 (of complement). *Let A be a quotient divisible group and $\langle t \rangle$ a cyclic p-primary group for a prime number p. If the p-rank of A is strictly less than the rank of A, then the group $A \oplus \langle t \rangle$ is quotient divisible as well.*

Proof. Let $x_1, \ldots, x_n$ be a basis of A and $r = \mathrm{rank}_p A, r < n$. The vector space A/pA over the field Z_p is generated by the set of vectors $\overline{x}_1 = x_1 + pA, \ldots, \overline{x}_n = x_n + pA$. This set of vectors contains a basis, say $\overline{x}_1, \ldots, \overline{x}_r$, of A/pA. Then the set of elements $x_1, \ldots, x_r, x_{r+1} + t, x_{r+2}, \ldots, x_n$ is a basis of the quotient divisible group $B = A \oplus \langle t \rangle$. □

4 Duality

The duality [1] between the quotient divisible groups and the torsion free groups of finite rank can be considered as a part of a more general construction as it has been done in [10]. Namely, we have a commutative diagram of the following category functors.

$$\begin{array}{ccc} & \mathcal{RM} & \\ c' \nearrow \swarrow c & & b \searrow \nwarrow b' \\ & d & \\ & \rightleftarrows & \\ \mathcal{QD} & d' & \mathcal{QTF} \end{array} \qquad (4.1)$$

It is convenient to consider three objects simultaneously. Thus we prefer to call the situation "the triplicity" such that the duality d and d' is a part of it. We briefly introduce now all three categories and the functors referring a reader to [10] for details.

(i) An object of the category $\mathcal{RM}$ is an arbitrary sequence of elements $x_1^0, \ldots, x_n^0$ of an arbitrary finitely presented $\widehat{Z}$-module. Note that, choosing a basis $y_1, \ldots, y_m$ of the module $M = \langle x_1^0, \ldots, x_n^0 \rangle_{\widehat{Z}} = y_1 Z_{\chi_1} \oplus \ldots \oplus y_m Z_{\chi_m}$, we obtain a matrix

$$\begin{pmatrix} \alpha_{11} & \cdots & \alpha_{1n} \\ \cdots & \cdots & \cdots \\ \alpha_{m1} & \cdots & \alpha_{mn} \end{pmatrix}$$

with $\alpha_{ij} \in Z_{\chi_i}$ at the equalities $x_i^0 = \alpha_{1i} y_1 + \ldots + \alpha_{mi} y_m$, $i = 1, \ldots, n$. The matrix is reduced, i.e. its columns generate the $\widehat{Z}$-module M. This point of view has been employed in [10]. That is why the category $\mathcal{RM}$ is called the category of reduced matrices.

(ii) An object of the category $\mathcal{QTF}$ is a pair consisting of a torsion-free finite-rank group A and its basis $x_1, \ldots, x_n$, that is a maximal linearly independent set of elements. For an object $x_1^0, \ldots, x_n^0$ of the category $\mathcal{RM}$, the object $b\left(x_1^0, \ldots, x_n^0\right)$ of the category $\mathcal{QTF}$ is defined in the following way. We define A as a group located between a free group F and a divisible group V

$$F = x_1 Z \oplus \ldots \oplus x_n Z \subset A \subset x_1 Q \oplus \ldots \oplus x_n Q = V.$$

For elements $\gamma_1 = \frac{a_1}{k} + Z, \ldots, \gamma_n = \frac{a_n}{k} + Z$ of the group Q/Z, we define $\gamma_1 x_1 + \ldots + \gamma_n x_n = k^{-1} (a_1 x_1 + \ldots + a_n x_n) \in V$. Then

$$A = \left\langle f\left(x_1^0\right) x_1 + \ldots + f\left(x_n^0\right) x_n \, \big| \, f \in \mathrm{Hom}_{\widehat{Z}}(M, Q/Z) \right\rangle,$$

reminding $M = \langle x_1^0, \ldots, x_n^0 \rangle_{\widehat{Z}}$. Conversely (the functor b'), we define a function $x_i^0 : A/F \to Q/Z$ in the following way. Let $z = k^{-1}(a_1 x_1 + \ldots + a_n x_n) + F \in A/F$. Then $x_i^0(z) = \frac{a_i}{k} + Z \in Q/Z, i = 1, \ldots, n$. Thus the elements $x_1^0, \ldots, x_n^0$ belong to the finitely presented $\widehat{Z}$-module $M = \mathrm{Hom}_{\widehat{Z}}(A/F, Q/Z)$. Note that the group $M = \mathrm{Hom}_{\widehat{Z}}(A/F, Q/Z)$ with the Z-adic topology coincides with the group of Pontryagin's characters ([17]) for the discrete group A/F.

(iii) An object of the category $\mathcal{QD}$ is a pair consisting of a quotient divisible group A^* and its basis $x_1^*, \ldots, x_n^*$. The object $c\left(x_1^0, \ldots, x_n^0\right)$ of the category $\mathcal{QD}$ is defined in the following way. Let $d_1, \ldots, d_n$ be a linearly independent set of elements of a torsion-free divisible group D. Then

$$x_1^* = x_1^0 + d_1, \ldots, x_n^* = x_n^0 + d_n \tag{4.2}$$

is a linearly independent set of the group $M \oplus D$, where M is here the additive group of the $\widehat{Z}$-module $M = \langle x_1^0, \ldots, x_n^0 \rangle_{\widehat{Z}}$. And at last, $A^* = \langle x_1^*, \ldots, x_n^* \rangle_*$

is the pure hull of the elements $x_1^*, \ldots, x_n^*$ in the group $M \oplus D$. It is clear that this definition of the quotient divisible group A^* doesn't depend up to isomorphism on the choice of the elements $d_1, \ldots, d_n$. Nevertheless we can use further the freedom of choice for the elements $d_1, \ldots, d_n$ considering inclusions in the Theorem 3. Conversely (the functor c'), let $\mu : A^* \to \widehat{A^*}$ be the Z-adic completion of a quotient divisible group A^* (see [13], Chapter 39). Then $x_1^0 = \mu(x_1^*), \ldots, x_n^0 = \mu(x_n^*)$.

The functors d and d' are defined like this $d = bc'$ and $d' = cb'$.

The morphisms from an object $(x_1^0, \ldots, x_n^0)$ to an object $(z_1^0, \ldots, z_k^0)$ of the category $\mathcal{RM}$ are pairs (φ, T), where $\varphi : \langle x_1^0, \ldots, x_n^0 \rangle_{\widehat{Z}} \to \langle z_1^0, \ldots, z_k^0 \rangle_{\widehat{Z}}$ is a quasi-homomorphism of the $\widehat{Z}$-modules and T is a matrix with rational entries of dimension $k \times n$, such that the equality $(\varphi x_1^0, \ldots, \varphi x_n^0) = (z_1^0, \ldots, z_k^0) T$ takes place in the module $Q \otimes \langle z_1^0, \ldots, z_k^0 \rangle_{\widehat{Z}}$. The morphisms of the categories $\mathcal{QD}$ and $\mathcal{QTF}$ are the quasi-homomorphisms of groups. The functors b and c transform a morphism (φ, T) of the category $\mathcal{RM}$ to the morphisms $f : B \to A$ in $\mathcal{QTF}$ and $f^* : A^* \to B^*$ in $\mathcal{QD}$, where $B, z_1, \ldots, z_k$ and $B^*, z_1^*, \ldots, z_k^*$ correspond to the object $z_1^0, \ldots, z_k^0$, the quasi-homomorphisms f and f^* are defined by the matrix equalities

$$\begin{pmatrix} f(z_1) \\ \ldots \\ f(z_k) \end{pmatrix} = T \begin{pmatrix} x_1 \\ \ldots \\ x_n \end{pmatrix} \quad \text{and} \quad (f^*(x_1^*), \ldots, f^*(x_n^*)) = (z_1^*, \ldots, z_k^*) T.$$

It is shown in [10] that the mutually inverse functors c and c' present a category equivalence. The functors b and b' present a category duality, which can be considered as a modern version of the description by Kurosh–Malcev–Derry. The functors d and d' present the category duality [1].

Note that our definitions of the categories $\mathcal{QD}$ and $\mathcal{QTF}$ differ a little bit from the original definitions in [1] and [10], where the objects are groups and the morphisms are quasi-homomorphisms. But evidently our definitions (with fixed bases) give the equivalent categories and we may keep the same notations for them. The basis fixing gives an advantage for the investigations of almost completely decomposable groups. It allows to introduce the following definition.

Definition. A *triple* is a set of three objects of the categories $\mathcal{RM}$, $\mathcal{QD}$ and $\mathcal{QTF}$ such that each of them corresponds to each other at the functors of the diagram (4.1). Namely, it is:

- A set of elements $x_1^0, \ldots, x_n^0$ of a finitely presented $\widehat{Z}$-module,
- A torsion-free finite-rank group A with a basis $x_1, \ldots, x_n$,
- A quotient divisible group A^* with a basis $x_1^*, \ldots, x_n^*$.

We underline that every element of the triple determines uniquely the remaining two objects of the triple. Moreover, we consolidate further this notation for a triple to simplify formulations without an additional explanation.

Theorem 2 ([10]). *The following statements are equivalent for a triple:*

(i) $A = B \oplus C$, where $B = \langle x_1, \ldots, x_k \rangle_*$ and $C = \langle x_{k+1}, \ldots, x_n \rangle_*$,

(ii) $\langle x_1^0, \ldots, x_n^0 \rangle_{\widehat{Z}} = \langle x_1^0, \ldots, x_k^0 \rangle_{\widehat{Z}} \oplus \langle x_{k+1}^0, \ldots, x_n^0 \rangle_{\widehat{Z}}$,

(iii) $A^* = B^* \oplus C^*$.

In this theorem, the groups B, B^* and the set $x_1^0, \ldots, x_k^0$ form a separate triple as well as the groups C, C^* with the set $x_{k+1}^0, \ldots, x_n^0$. In particular, we use the pure hull $\langle x_1^0, \ldots, x_k^0 \rangle_*$ in $\langle x_1^0, \ldots, x_k^0 \rangle_{\widehat{Z}}$, but not in $\langle x_1^0, \ldots, x_n^0 \rangle_{\widehat{Z}}$, by the construction of the reduced part of the group B^* at the functor c.

5 Change of bases

Two different bases $x_1, \ldots, x_n$ and $y_1, \ldots, y_n$ of a torsion free group A give us two different triples. One of them contains also a set $x_1^0, \ldots, x_n^0$ of a finitely presented $\widehat{Z}$-module and a quotient divisible group A_X^* with a basis $x_1^*, \ldots, x_n^*$. The second one contains a set $y_1^0, \ldots, y_n^0$ and a quotient divisible group A_Y^* with a basis $y_1^*, \ldots, y_n^*$. The following theorem considers relations between them.

Theorem 3. *Let two bases $x_1, \ldots, x_n$ and $y_1, \ldots, y_n$ of a torsion free group A be written in the form of columns X and Y. If $X = SY$ for a nonsingular matrix S with integer entries, then:*

(i) *The $\widehat{Z}$-module $\langle y_1^0, \ldots, y_n^0 \rangle_{\widehat{Z}}$ is a submodule of index $|\det S|$ of the module $\langle x_1^0, \ldots, x_n^0 \rangle_{\widehat{Z}}$ and the following matrix equality takes place*

$$(y_1^0, \ldots, y_n^0) = (x_1^0, \ldots, x_n^0) S. \qquad (5.1)$$

(ii) *Defining the bases of the groups A_X^* and A_Y^* by the equalities (4.2), we can choose elements $d_1, \ldots, d_n$ in these equalities such that $A_Y^* \subset A_X^*$ and the following equality takes place*

$$(y_1^*, \ldots, y_n^*) = (x_1^*, \ldots, x_n^*) S. \qquad (5.2)$$

Moreover $|A_X^*/A_Y^*| = |\det S|$, where $|\det S|$ is the absolute value of the determinant.

Proof. Since $X = SY$, the following inclusion takes place $F = \langle x_1, \ldots, x_n \rangle \subset G = \langle y_1, \ldots, y_n \rangle \subset A$. Applying the functor $\mathrm{Hom}\,(-, Q/Z)$ to the short exact sequence $0 \to G/F \xrightarrow{i} A/F \xrightarrow{j} A/G \to 0$, we obtain the exact sequence of $\widehat{Z}$-modules $0 \to M_1 \xrightarrow{j^*} M \xrightarrow{i^*} \mathrm{Hom}\,(G/F, Q/Z) \to 0$, where $M_1 = \langle y_1^0, \ldots, y_n^0 \rangle_{\widehat{Z}}$, $M = \langle x_1^0, \ldots, x_n^0 \rangle_{\widehat{Z}}$ and the $\widehat{Z}$-module on the right is a finite abelian group which is isomorphic to the group G/F.

An arbitrary element z of the group A/F is of the form $z = m^{-1}\left(\sum_{i=1}^n a_i x_i\right) + F$, where $a_i \in Z, 0 \neq m \in Z$. Substituting $x_i = \sum_{k=1}^n s_{ik} y_k$, where $S = \|s_{ik}\|$ and $X = SY$, we obtain $z = m^{-1}\left(\sum_{i=1}^n a_i \sum_{k=1}^n s_{ik} y_k\right) + F = m^{-1}\left(\sum_{k=1}^n \left(\sum_{i=1}^n a_i s_{ik}\right) y_k\right) + F$. By the definition of the elements x_i^0 and y_i^0 in Section 4, the function $z \longmapsto m^{-1}\left(\sum_{i=1}^n a_i s_{ik}\right) + Z \in Q/Z$ is exactly the function $j^*\left(y_k^0\right) : A/F \to Q/Z$ and $x_i^0(z) = m^{-1} a_i + Z \in Q/Z$. Identifying $j^*\left(y_k^0\right) = y_k^0$, we obtain finally $y_k^0(z) = m^{-1}\left(\sum_{i=1}^n a_i s_{ik}\right) + Z = \sum_{i=1}^n \left(m^{-1} a_i + Z\right) s_{ik} = \sum_{i=1}^n x_i^0(z) s_{ik}$. Since the values of two functions $y_k^0(z)$ and $\sum_{i=1}^n x_i^0(z) s_{ik}$ coincide for every $z \in A/F$, the functions coincide as well and the equality (5.1) is proved.

The index of M_1 in M is equal to $|G/F|$. The matrix S can be presented in the form $S = T_1 T T_2$, where T_1 and T_2 are invertible, T is diagonal and they are all with integer entries. The matrix equality $X = SY = (T_1 T T_2) Y$ implies the equality $T_1^{-1} X = T(T_2 Y)$. Thus the basis $T_1^{-1} X$ of the free group F is expressed over the basis $T_2 Y$ of the free group G with help of the diagonal matrix

$$T = \begin{pmatrix} t_1 & 0 & \cdots & 0 \\ \cdots & & \ddots & \cdots \\ 0 & 0 & \cdots & t_n \end{pmatrix}.$$

Hence $G/F = C_1 \oplus \ldots \oplus C_n$, where the direct summands $C_1, \ldots, C_n$ are cyclic of order $|t_1|, \ldots, |t_n|$, respectively. Therefore $|G/F| = |t_1| \cdot \ldots \cdot |t_n| = |\det S|$. It accomplishes the proof of the first part of the theorem.

If we just have, say, $x_1^* = x_1^0 + d_1, \ldots, x_n^* = x_n^0 + d_n$, then we choose elements $d_1', \ldots, d_n'$ of the divisible torsion free group $D = \langle d_1, \ldots, d_n\rangle_*$ in such a way that $(d_1', \ldots, d_n') = (d_1, \ldots, d_n) S$. Defining $y_1^* = y_1^0 + d_1', \ldots, y_n^* = y_n^0 + d_n'$, we obtain immediately $A_Y^* \subset A_X^*$ and the equality (5.2). The only thing to do is to prove $|A_X^*/A_Y^*| = |\det S|$. Without loss of generality assume $A_X^* = \langle x_1^0, \ldots, x_n^0\rangle_*$, that is the set $x_1^0, \ldots, x_n^0$ is linearly independent over Z. The natural homomorphism $\theta : A_X^* \to M/M_1$ is surjective, because the images of elements $x_1^0, \ldots, x_n^0$ generate the finite group M/M_1. The kernel of θ is equal to $A_X^* \cap M_1$ and the intersection $A_X^* \cap \langle y_1^0, \ldots, y_n^0\rangle_{\widehat{Z}}$ coincides in turn with A_Y^*. We obtain finally $A_X^*/A_Y^* \cong M/M_1 \cong G/F$. □

We distinguish a particular case of Theorem 3.

Corollary 4. *Let* $x_1 = my_1, \ldots, x_n = my_n$ *for an integer* $m \neq 0$. *Then* $\langle y_1^0, \ldots, y_n^0\rangle_{\widehat{Z}} \subset \langle x_1^0, \ldots, x_n^0\rangle_{\widehat{Z}}$ *and* $y_1^0 = mx_1^0, \ldots, y_n^0 = mx_n^0$.

Corollary 5. *Let two sequences* $x_1^0, \ldots, x_n^0$ *and* $y_1^0, \ldots, y_n^0$ *of elements be given in a finitely presented* $\widehat{Z}$-*module. If a matrix equality* $(y_1^0, \ldots, y_n^0) = (x_1^0, \ldots, x_n^0) S$ *takes place for an integer matrix* S *with* $\det S = \pm 1$, *then the torsion free groups coincide and the quotient divisible groups coincide in two triples corresponding to the given sequences. In particular, the groups A and A^* do not depend on the order of elements in the sequence* $x_1^0, \ldots, x_n^0$ *of a triple.*

Corollary 6. *The dual quotient divisible (torsion free) group with respect to a basis $x_1, \ldots, x_n$ $(x_1^*, \ldots, x_n^*)$ doesn't depend on the choice of the basis in the free group $F = \langle x_1, \ldots, x_n \rangle$ $(F^* = \langle x_1^*, \ldots, x_n^* \rangle)$. Therefore, it depends only on the choice of the free subgroup $F(F^*)$. Moreover, it doesn't depend even on the choice of the free subgroup up to quasi-equality.*

The following example shows that an indecomposable quotient divisible group can be dual to a completely decomposable torsion free group.

Example 1. We consider a torsion free group $A = x_1 Q_2 \oplus x_2 Q_5$ with the basis x_1, x_2. The dual quotient divisible group A_X^* with respect to this basis is of the form $A_X^* = x_1^* Q^{(2)} \oplus x_2^* Q^{(5)}$. Let us consider now two new bases of the group A: $y_1 = \frac{1}{3}(x_1 - x_2)$, $y_2 = x_2$ and $z_1 = \frac{1}{3} x_1, z_2 = \frac{1}{3} x_2$. We have

$$\begin{pmatrix} x_1 \\ x_2 \end{pmatrix} = \begin{pmatrix} 3 & 1 \\ 0 & 1 \end{pmatrix} \begin{pmatrix} y_1 \\ y_2 \end{pmatrix} \quad \text{and} \quad \begin{pmatrix} y_1 \\ y_2 \end{pmatrix} = \begin{pmatrix} 1 & -1 \\ 0 & 3 \end{pmatrix} \begin{pmatrix} z_1 \\ z_2 \end{pmatrix}$$

in the group A. By Theorem 3 and Corollary 4, we obtain the inclusions for dual quotient divisible groups $A_Z^* \subset A_Y^* \subset A_X^*$ and the relations $z_1^* = 3x_1^*, z_2^* = 3x_2^*$ and $z_1^* = y_1^*, z_2^* = -y_1^* + 3y_2^*$ and $y_1^* = 3x_1^*, y_2^* = x_1^* + x_2^*$. Note that $A_Z^* = z_1^* Q^{(2)} \oplus z_2^* Q^{(5)} = 3 A_X^* \cong A_X^*$ and $A_Y^* = \langle A_Z^*, \frac{z_1^* + z_2^*}{3} \rangle$. The group A_Y^* is indecomposable (see [13], Example 88.2).

6 Almost completely decomposable groups

For a characteristic χ, we denote by R_χ the subgroup of Q such that $1 \in R_\chi$ and the characteristic of 1 in R_χ is equal to χ. The lattice of characteristics gives a spectrum of the natural embeddings

$$f_\kappa^\chi : R_\kappa \to R_\chi \quad \text{for } \kappa \leq \chi, \tag{6.1}$$

where $f_\kappa^\chi(1) = 1$. The quotient divisible group R^χ is dual to R_χ with respect to the natural basis $1 \in R_\chi$ and the dual basis is $1^* = 1 \in R^\chi$. The homomorphisms (2.1) g_κ^χ are dual to f_κ^χ and the following spectrum of the homomorphisms of the quotient divisible groups is dual to (6.1)

$$g_\kappa^\chi : R^\chi \to R^\kappa \quad \text{for } \kappa \leq \chi, \tag{6.2}$$

where $g_\kappa^\chi(1) = 1$. It is interesting to note that every group of the spectrum (6.2) is naturally a ring and then g_κ^χ are homomorphisms of rings, while the groups of the spectrum (6.1) are subrings of Q if and only if they are quotient divisible. $R^\chi = R_\kappa \iff \chi \vee \kappa = (\infty, \infty, \ldots)$ and $\chi \wedge \kappa = (0, 0, \ldots)$.

An arbitrary torsion-free rank-1 group is of the form $A = xR_\chi$, where x is its basis and χ is the characteristic of x. The dual to A quotient divisible group is $A^* = x^* R^\chi$. The last group can be considered sometimes as a free rank-1 module over the ring R^χ. The bases x and x^* are mutually dual. The following theorem describes triples for the completely decomposable groups.

Theorem 7 ([10]). *The following statements are equivalent for a triple:*

(i) *The set of elements $x_1^0, \ldots, x_n^0$ is linearly independent over $\widehat{Z}$ and the co-characteristics of these elements are $\chi_1, \ldots, \chi_n$, respectively.*

(ii) $A^0 = x_1^0 Z_{\chi_1} \oplus \ldots \oplus x_n^0 Z_{\chi_n}$, *where* $A^0 = \langle x_1^0, \ldots, x_n^0 \rangle_{\widehat{Z}}$.

(iii) $A = x_1 R_{\chi_1} \oplus \ldots \oplus x_n R_{\chi_n}$.

(iv) $A^* = x_1^* R^{\chi_1} \oplus \ldots \oplus x_n^* R^{\chi_n}$.

We are generalizing this theorem on the almost completely decomposable groups in the present section.

Definition. A set of elements $y_1, \ldots, y_n$ of a finitely presented $\widehat{Z}$-module is called *almost linearly independent over* $\widehat{Z}$ if the equality $\alpha_1 y_1 + \ldots + \alpha_n y_n = 0$ with universal integer coefficients implies that all the elements $\alpha_1 y_1, \ldots, \alpha_n y_n$ have finite order, that is $m\alpha_1 y_1 = \ldots = m\alpha_n y_n = 0$ for some non-zero integer m.

Lemma 8. *Let $y_1, \ldots, y_n$ be an almost linearly independent set of elements of a finitely presented $\widehat{Z}$-module $M = \langle y_1, \ldots, y_n \rangle_{\widehat{Z}}$. Then there exist elements of finite order $t_1, \ldots, t_n \in M$ such that $M = \langle y_1 + t_1, \ldots, y_n + t_n \rangle_{\widehat{Z}}$ and the set of the elements $y_1 + t_1, \ldots, y_n + t_n$ is linearly independent over $\widehat{Z}$.*

Proof. For every $i = 1, \ldots, n$, the $\widehat{Z}$-module $T_i = \langle y_i \rangle_{\widehat{Z}} \cap \langle y_1, \ldots, y_{i-1}, y_{i+1}, \ldots, y_n \rangle_{\widehat{Z}}$ is finitely presented and cyclic, that is it is isomorphic to Z_χ for a characteristic χ. Since the set $y_1, \ldots, y_n$ is almost linearly independent, it follows that χ belongs to the zero type and hence T_i is a cyclic group. Consider the set $P = \{p_1, \ldots, p_s\}$ of prime divisors of the orders of the groups $T_1, \ldots, T_n$ and carry out the following operation for a prime number $p \in P$.

We remind that as every finitely presented $\widehat{Z}$-module the module M is of the form $M = \prod_p M_p$, where $M_p = \langle a_1 \rangle_{\widehat{Z}_p} \oplus \ldots \oplus \langle a_n \rangle_{\widehat{Z}_p}$. The first r direct summands are p-primary cyclic groups and the remaining summands are isomorphic to $\widehat{Z}_p, 0 \leq r \leq n$. We obtain a direct decomposition $M = \langle a_1 \rangle_{\widehat{Z}} \oplus \ldots \oplus \langle a_r \rangle_{\widehat{Z}} \oplus N$, where the $\widehat{Z}$-module N has no p-torsion, and also we obtain the equalities $y_1 = s_1 + y_1', \ldots, y_n = s_n + y_n'$ with respect to this decomposition, where $s_1, \ldots, s_n \in \langle a_1 \rangle_{\widehat{Z}} \oplus \ldots \oplus \langle a_r \rangle_{\widehat{Z}}$ and $y_1', \ldots, y_n' \in N$. Let ε_p be the universal integer such that all its components are zeros except for the p-component which is equal to 1. The elements $\varepsilon_p y_1', \ldots, \varepsilon_p y_n'$ generate the free p-adic module $N_p = \varepsilon_p N$ of rank $n - r$. Exactly $n - r$ elements in the sequence $\varepsilon_p y_1', \ldots, \varepsilon_p y_n'$ are different from 0, otherwise we obtain a contradiction with the property of the almost linear independence. Therefore, r elements, say $\varepsilon_p y_1', \ldots, \varepsilon_p y_r'$, are equal to 0. We define now $z_1 = a_1 + y_1', \ldots, z_r = a_r + y_r', z_{r+1} = y_{r+1}', \ldots, z_n = y_n'$. It is easy to see that $z_1 = y_1 + t_1, \ldots, z_n = y_n + t_n$, where $t_1, \ldots, t_n$ are p-primary elements of the module M, $M = \langle z_1, \ldots, z_n \rangle_{\widehat{Z}}$, and the set of elements $z_1 = y_1 + t_1, \ldots, z_n = y_n + t_n$ is almost linearly independent.

The corresponding set of prime numbers for the almost linearly independent set of elements $z_1, \ldots, z_n$ is equal to $P \setminus \{p\}$. By the hypothesis of induction the statement of lemma takes place for the elements $z_1, \ldots, z_n$, therefore it takes place for the set $y_1, \ldots, y_n$ as well. □

Theorem 9. *The following statements are equivalent for a triple.*

(i) *The group A contains a subgroup of finite index of the form $x_1 R_{\chi_1} \oplus \ldots \oplus x_n R_{\chi_n}$ for some characteristics $\chi_1, \ldots, \chi_n$.*

(ii) *The set of elements $x_1^0, \ldots, x_n^0$ is almost linearly independent over $\widehat{Z}$.*

(iii) *There exist torsion elements $t_1, \ldots, t_n \in A^*$ such that $A^* = (x_1^* + t_1) R^{\kappa_1} \oplus \ldots \oplus (x_n^* + t_n) R^{\kappa_n}$. Moreover, $[\kappa_1] = [\chi_1], \ldots, [\kappa_n] = [\chi_n]$.*

Proof. (i) $\Rightarrow$ (ii). Denote $B = x_1 R_{\chi_1} \oplus \ldots \oplus x_n R_{\chi_n}$ and $F = \langle x_1, \ldots, x_n \rangle$. The exact sequence $0 \to B \to A \to C \to 0$ induces the exact sequence $0 \to B/F \to A/F \to C \to 0$ with a finite group $C \cong A/B$. Applying the functor $\operatorname{Hom}(-, Q/Z)$ to the last sequence, we obtain an exact sequence $0 \to C^0 \to A^0 \to B^0 \to 0$ of $\widehat{Z}$-module homomorphisms, where the group $C^0 = \operatorname{Hom}(C, Q/Z) \cong C$ is finite. The elements $x_1^0, \ldots, x_n^0$ of the triple are located in the $\widehat{Z}$-module A^0. Suppose $\alpha_1 x_1^0 + \ldots + \alpha_n x_n^0 = 0$ with universal integer coefficients. Passing to the $\widehat{Z}$-module $B^0 = x_1^0 Z_{\chi_1} \oplus \ldots \oplus x_n^0 Z_{\chi_n}$ we obtain the equalities $\alpha_1 x_1^0 = \ldots = \alpha_n x_n^0 = 0$ in B^0. Therefore, the elements $\alpha_1 x_1^0, \ldots, \alpha_n x_n^0$ belong to the image of C^0 in A^0, that is they are periodic. Hence the set of elements $x_1^0, \ldots, x_n^0 \in A^0$ is almost linearly independent over $\widehat{Z}$.

(ii) $\Rightarrow$ (iii). Let the set $x_1^0, \ldots, x_n^0$ be almost linearly independent over $\widehat{Z}$. By Lemma 8, $A^0 = \langle x_1^0, \ldots, x_n^0 \rangle_{\widehat{Z}} = \langle x_1^0 + t_1, \ldots, x_n^0 + t_n \rangle_{\widehat{Z}} = \langle x_1^0 + t_1 \rangle_{\widehat{Z}} \oplus \ldots \oplus \langle x_n^0 + t_n \rangle_{\widehat{Z}}$ for some torsion elements $t_1, \ldots, t_n \in A^0$. Since the torsion parts of the groups A^0 and A^* coincide, $t_1, \ldots, t_n \in A^*$. Moreover, it is easy to see that $\langle x_1^0, \ldots, x_n^0 \rangle_* = \langle x_1^0 + t_1, \ldots, x_n^0 + t_n \rangle_* = \langle x_1^0 + t_1 \rangle_* \oplus \ldots \oplus \langle x_n^0 + t_n \rangle_*$. The last pure hulls are considered in the modules $\langle x_1^0 + t_1 \rangle_{\widehat{Z}}, \ldots, \langle x_n^0 + t_n \rangle_{\widehat{Z}}$, respectively. Thus we obtain $A^* = (x_1^* + t_1) R^{\kappa_1} \oplus \ldots \oplus (x_n^* + t_n) R^{\kappa_n}$, where $\kappa_1, \ldots, \kappa_n$ are the co-characteristics of the elements $x_1^0 + t_1, \ldots, x_n^0 + t_n$ in the module A^0, which are equivalent to the co-characteristics of the elements $x_1^0, \ldots, x_n^0$ in the module A^0, which are equivalent in turn to the characteristics $\chi_1, \ldots, \chi_n$, respectively.

(iii) $\Rightarrow$ (i). We have two bases in the quotient divisible group A^*, namely $x_1^*, \ldots, x_n^*$ and $y_1^* = x_1^* + t_1, \ldots, y_n^* = x_n^* + t_n$. The dual torsion free group with respect to the first basis coincides with the group A of the given triple, the fixed basis of A is $x_1, \ldots, x_n$. The dual group with respect to the second basis $y_1^*, \ldots, y_n^*$ belongs to other triple. We denote it as A_Y, its basis $y_1, \ldots, y_n$ is dual to the basis $y_1^*, \ldots, y_n^* \in A^*$. By Theorem 7, $A_Y = y_1 R_{\kappa_1} \oplus \ldots \oplus y_n R_{\kappa_n}$, where $\kappa_1, \ldots, \kappa_n$ are co-characteristics of the elements $x_1^0 + t_1, \ldots, x_n^0 + t_n$ in the module A^0.

There exists a non-zero integer m such that $mx_1^* = my_1^*, \ldots, mx_n^* = my_n^*$. The homomorphism $f : A^* \to A^*$ with $f(z) = mz$ induces two dual quasi-homomorphisms $f_1^* : A \to A_Y$ and $f_2^* : A_Y \to A$ according two different triples. By the definitions of Section 4, $f_1^*(x_1) = my_1, \ldots, f_1^*(x_n) = my_n$ and $f_2^*(y_1) = mx_1, \ldots, f_2^*(y_n) = mx_n$. For a non-zero integer k, two morphisms kf_1 and kf_2 are not only homomorphisms, but monomorphisms as well. Identifying along the monomorphisms kf_1 and kf_2, we obtain the inclusions

$$\left(k^2 m^2 y_1\right) R_{\kappa_1} \oplus \ldots \oplus \left(k^2 m^2 y_n\right) R_{\kappa_n} \subset A \subset y_1 R_{\kappa_1} \oplus \ldots \oplus y_n R_{\kappa_n}.$$

Since $kmy_i = x_i, i = 1, \ldots, n$, under the identification, it follows that the first inclusion is of the form $(kmx_1) R_{\kappa_1} \oplus \ldots \oplus (kmx_n) R_{\kappa_n} \subset A$. Thus we obtain $(kmx_1) R_{\kappa_1} \oplus \ldots \oplus (kmx_n) R_{\kappa_n} \subset x_1 R_{\chi_1} \oplus \ldots \oplus x_n R_{\chi_n} \subset A$. The index of the subgroup is not greater than $(mk)^{2n}$, and the characteristics $\chi_1, \ldots, \chi_n$ are equivalent to the characteristics $\kappa_1, \ldots, \kappa_n$, respectively. □

Example 1 shows in particular that the quotient divisible group dual to an almost completely decomposable group is not necessarily decomposed into a direct sum of subgroups. But nevertheless the following corollary of Theorem 9 takes place.

Corollary 10. *Every almost completely decomposable group contains a basis such that the dual quotient divisible group with respect to it is decomposed into a direct sum of quotient divisible groups of rank 1.*

7 Dualization of a lemma by L. Fuchs

Lemma by L. Fuchs [18] gives a sufficient condition of indecomposability for a torsion-free finite-rank group, see Lemma 88.3 in [13]. The following theorem is a dualization of this lemma.

Theorem 11. *Let a set $x_1^0, \ldots, x_n^0$ of elements of a finitely presented $\widehat{Z}$-module determine a triple with a torsion free group A. If:*

(i) *The set $x_1^0, \ldots, x_n^0$ is almost linearly independent,*

(ii) *The co-characteristics $\chi_1, \ldots, \chi_n$ of $x_1^0, \ldots, x_n^0$ belong to pairwise incomparable types,*

(iii) $\langle x_1^0 \rangle_{\widehat{Z}} \cap \langle x_i^0 \rangle_{\widehat{Z}} \neq 0$ *for each $i = 2, \ldots, n$,*

then the group A is not decomposable into a direct sum of nonzero subgroups.

Proof. We show first that the set $x_1^0, \ldots, x_n^0$ is linearly independent over Z. Let $m_1 x_1^0 + \ldots + m_n x_n^0 = 0$, $m_1, \ldots, m_n \in Z$. If, say, $m_1 \neq 0$, then the element $m_1 x_1^0$ is periodic because of the first condition. Therefore $[\chi_1] = 0$ and this is a contradiction with the second condition. Thus $m_1 = \ldots = m_n = 0$. By Section 4, we obtain that $x_1^* = x_1^0, \ldots, x_n^* = x_n^0$ and $A^* = \langle x_1^0, \ldots, x_n^0 \rangle_*$.

Let $x \in A^*$ be an arbitrary element of infinite order and χ be its co-characteristic in $\langle x_1^0, \ldots, x_n^0 \rangle_{\widehat{Z}}$. Then $mx = m_1 x_1^0 + \ldots + m_n x_n^0$ for some integer coefficients with $m \neq 0$. Multiplying the equality by an arbitrary universal number α of characteristic χ, we obtain $m_1 \alpha x_1^0 + \ldots + m_n \alpha x_n^0 = 0$. Since all the summands must be periodic, we obtain that $[\chi] \geq [\chi_i]$ for every i with $m_i \neq 0$. Thus the co-type of x is greater than or equal to at least one of the co-types of elements $x_1^0, \ldots, x_n^0$. Suppose now that $[\chi] \leq [\chi_j]$ for some j, then $[\chi_i] \leq [\chi] \leq [\chi_j]$, and by the second condition we obtain $i = j$ and $[\chi] = [\chi_j]$. In this case, only one coefficient m_j is different from zero in the equality $mx = m_1 x_1^0 + \ldots + m_n x_n^0$ on the right. Hence $mx = m_j x_j^0$ for some non-zero integers m and m_j. The torsion elements of A^* have the zero co-type. Thus it is proved that if cotype $(x) \leq$ cotype (x_i^0) for $x \in A^*$ and some $i = 1, \ldots, n$, then the elements x and x_i^0 are colinear or x is torsion.

Let us suppose now that the torsion free group A with the basis $x_1, \ldots, x_n$ is decomposed into a direct sum of non-zero subgroups. Then there exists a basis $y_1, \ldots, y_n \in \langle x_1, \ldots, x_n \rangle \subset A$ such that $\langle y_1, \ldots, y_k \rangle_* \oplus \langle y_{k+1}, \ldots, y_n \rangle_* = A, 0 < k < n$. Moreover,

$$\begin{pmatrix} y_1 \\ \cdots \\ y_n \end{pmatrix} = S \begin{pmatrix} x_1 \\ \cdots \\ x_n \end{pmatrix},$$

where S is a nonsingular matrix with integer entries. Applying Theorem 3, we obtain that $A^* = A_X^* \subset A_Y^*$ and $(x_1^*, \ldots, x_n^*) = (y_1^*, \ldots, y_n^*) S$. By Theorem 2, $A_Y^* = B \oplus C$, where

$$B = \langle y_1^0, \ldots, y_k^0 \rangle_* \text{ in } \langle y_1^0, \ldots, y_k^0 \rangle_{\widehat{Z}} \quad \text{and} \quad C = \langle y_{k+1}^0, \ldots, y_n^0 \rangle_* \text{ in } \langle y_{k+1}^0, \ldots, y_n^0 \rangle_{\widehat{Z}}.$$

By the projections $A_Y^* \to B$ and $A_Y^* \to C$, the co-characteristics of elements are decreasing as it takes place for any homomorphism of quotient divisible groups. Since A^* and A_Y^* are quasi-equal, the sets of their co-types coincide. Therefore one of the projections of the element x_i^* must have the co-type $[\chi_i]$ and the other projection has the co-type 0 for every $i = 1, \ldots, n$. Thus $mx_i^* \in B$ or $mx_i^* \in C$ for a suitable integer $m \neq 0$. On the other hand, $x_i^0 = s_{1i}y_1^0 + \ldots + s_{ni}y_n^0$, where s_{ki} are the entries of the matrix S. If $mx_i^0 \in B$, then necessarily $s_{k+1\,i} = \ldots = s_{ni} = 0$, hence $x_i^* = x_i^0 = s_{1i}y_1^0 + \ldots + s_{ki}y_k^0 \in B$. We obtain $x_i^* = x_i^0 \in B$ or $x_i^* = x_i^0 \in C$ for every $i = 1, \ldots, n$. Let $x_1^0 \in B$ and $x_j^0 \in C$ for some j. Then for some element $0 \neq t \in \langle x_1^0 \rangle_{\widehat{Z}} \cap \langle x_j^0 \rangle_{\widehat{Z}}$, we obtain $t \in B \cap C$ and it is a contradiction. Thus the group A is indecomposable. □

8 Completely decomposable homogeneous groups

Theorem 12. *Let $y_1^0, \ldots, y_n^0$ be a linearly independent over $\widehat{Z}$ set of elements of a finitely presented $\widehat{Z}$-module, such that the co-characteristics $\chi_1, \ldots, \chi_n$ of $y_1^0, \ldots, y_n^0$ are equal $\chi_1 = \ldots = \chi_n = \chi$. Let a set of elements $x_1^0, \ldots, x_n^0$ of the same module be defined by a matrix equality $(x_1^0, \ldots, x_n^0) = (y_1^0, \ldots, y_n^0) S$, where S is an integer matrix of dimension $n \times n$ with $\det S = \pm 1$. Then the triple corresponding to the set $x_1^0, \ldots, x_n^0$ has the following properties:*

(i) *The set $x_1^0, \ldots, x_n^0$ is linearly independent over $\widehat{Z}$ and all the co-characteristics of the elements are equal to χ,*

(ii) $A = x_1 R_\chi \oplus \ldots \oplus x_n R_\chi$,

(iii) $A^* = x_1^* R^\chi \oplus \ldots \oplus x_n^* R^\chi$.

Proof. The module $M = \langle y_1^0, \ldots, y_n^0 \rangle_{\widehat{Z}} = y_1^0 Z_\chi \oplus \ldots \oplus y_n^0 Z_\chi$ is a free module over the ring Z_χ as well. The correspondence $y_1^0 \longmapsto x_1^0, \ldots, y_n^0 \longmapsto x_n^0$ determines an automorphism of the Z_χ-module M which maps the free basis $y_1^0, \ldots, y_n^0$ to the free basis $x_1^0, \ldots, x_n^0$. It proves the first statement of the theorem. Applying Theorem 7, we finish the proof. □

Corollary 13. *Let a quotient divisible group $B = y_1 R^\chi \oplus \ldots \oplus y_n R^\chi$ be a direct sum of copies isomorphic to R^χ. For every integer matrix S of dimension $n \times n$ with $\det S = \pm 1$, the set $x_1, \ldots, x_n \in B$, defined by the matrix equality $(x_1, \ldots, x_n) = (y_1, \ldots, y_n) S$, is a basis of the quotient divisible group B as well. The dual to B torsion free group B^* is the same considering it with respect to each of two bases. Moreover, the following two decompositions of B^* take place with respect to the dual bases: $B^* = y_1^* R_\chi \oplus \ldots \oplus y_n^* R_\chi = x_1^* R_\chi \oplus \ldots \oplus x_n^* R_\chi$.*

Theorem 12 would not be true if we replace the equality of the co-characteristics $\chi_1 = \ldots = \chi_n = \chi$ by the equivalence of the co-characteristics $\chi_1 \sim \ldots \sim \chi_n \sim \chi$. It is shown in the following example.

Example 2. First we define a triple. We consider three characteristics $\chi_1 = (0, 1, 0, 0, \ldots)$, $\chi_2 = (1, 0, 0, 0, \ldots)$, $\chi_3 = \chi_1 + \chi_2 = (1, 1, 0, 0, \ldots)$ and a finitely presented $\widehat{Z}$-module $Z_6 = \{\overline{0}, \overline{1}, \overline{2}, \overline{3}, \overline{4}, \overline{5}\}$. Let $y_1^0 = \overline{2}$ and $y_2^0 = \overline{3}$. Then cochar $(y_1^0) = \chi_1$ and cochar $(y_2^0) = \chi_2$. According to Section 4, $y_1^* = \overline{2} + d_1, y_2^* = \overline{3} + d_2, y_1^* R^{\chi_1} = \langle \overline{2} \rangle \oplus d_1 Q, y_2^* R^{\chi_2} = \langle \overline{3} \rangle \oplus d_2 Q$. Since the set y_1^0, y_2^0 is linearly independent over $\widehat{Z}$, we have the following direct decompositions according to Theorem 7:

(i) $A^* = y_1^* R^{\chi_1} \oplus y_2^* R^{\chi_2} = d_1 Q \oplus d_2 Q \oplus Z_6$,

(ii) $A = y_1 R_{\chi_1} \oplus y_2 R_{\chi_2}$. The rank-1 group $y_1 R_{\chi_1}$ contains an element $v_1 = \frac{1}{3} y_1$ and $y_1 R_{\chi_1} = \langle v_1 \rangle$. Analogously, $y_2 R_{\chi_2} = \langle v_2 \rangle$, where $v_2 = \frac{1}{2} y_2$. Thus $A = v_1 Z \oplus v_2 Z$ is a free group of rank 2 and the fixed basis is $y_1 = 3v_1, y_2 = 2v_2$.

We consider now another triple which is corresponding to the set x_1^0, x_2^0 defined by the matrix equality $(x_1^0, x_2^0) = (y_1^0, y_2^0) \begin{pmatrix} 1 & 2 \\ 2 & 3 \end{pmatrix}$. We note immediately that the groups A and A^* of the new triple are the same, because the matrix is invertible. Only the pair of mutually dual bases is different. Namely, we have $x_1^0 = \overline{2}$, $x_2^0 = \overline{1}$, the co-characteristics of x_1^0 and x_2^0 are χ_1 and χ_3, respectively. Theorem 7 can not be used, because the set x_1^0, x_2^0 is not linearly independent over $\widehat{Z}$, it is only almost linearly independent over $\widehat{Z}$. And we can not obtain a direct decomposition "along" the bases x_1, x_2 and x_1^*, x_2^*. According to Theorem 3, $x_1^* = \overline{2} + (d_1 + 2d_2), x_2^* = \overline{1} + (2d_1 + 3d_2)$ and $x_1 = -9v_1 + 4v_2, x_2 = 6v_1 - 2v_2$. The quotient divisible group A^* contains the quotient divisible rank-1 subgroups $x_1^* R^{\chi_1} = \langle \overline{2} \rangle + (d_1 + 2d_2) Q$ and $x_2^* R^{\chi_3} = \langle \overline{1} \rangle + (2d_1 + 3d_2) Q$. Moreover, $A^* = x_1^* R^{\chi_1} + x_2^* R^{\chi_3}$, but $x_1^* R^{\chi_1} \cap x_2^* R^{\chi_3} = \langle \overline{2} \rangle$ and the sum is not direct. On the other hand, $\langle x_1 \rangle_* = x_1 Z \subset A, \langle x_2 \rangle_* = x_2 R_{\chi_2} \subset A$. Of course, $\langle x_1 \rangle_* \cap \langle x_2 \rangle_* = 0$, but the direct sum $\langle x_1 \rangle_* \oplus \langle x_2 \rangle_*$ doesn't coincide with the group A, it is of the index 3.

9 Lattice of admissible almost completely decomposable groups

Definition. An element t of an arbitrary group is called *admissible* with respect to a characteristic χ if it is torsion and the p-component of the characteristic χ is equal to zero for every prime divisor p of the order of the element t.

The next proposition follows easily from Lemma 1.

Proposition 14. *Let xR^χ be a rank-1 quotient divisible group of a co-characteristic χ with a basis x and $\langle t \rangle$ be a cyclic group. The group $xR^\chi \oplus \langle t \rangle$ is quotient divisible if and only if the element t is admissible with respect to the characteristic χ. Moreover, if $xR^\chi \oplus \langle t \rangle$ is quotient divisible, then its rank is 1 and the element $x + t$ is its basis.*

We fix now an arbitrary sequence of characteristics $\Xi = (\chi_1, \ldots, \chi_n)$ and a basis $x_1, \ldots, x_n$ of a vector space V over Q. The group $B = x_1 R_{\chi_1} \oplus \ldots \oplus x_n R_{\chi_n} \subset V$ is completely decomposable torsion free. The group $B^* = x_1^* R^{\chi_1} \oplus \ldots \oplus x_n^* R^{\chi_n}$ is dual to B quotient divisible. In this section, we consider some finite extensions A of the group B with the same common fixed basis $x_1, \ldots, x_n$ for all them. Every such group A determines a pair: the dual quotient divisible group A^* and the dual basis (to the fixed basis $x_1, \ldots, x_n$). For different groups A those dual bases are different, the dual groups A^* are different of course as well, though they all are quasi-equal. Thus we obtain a fan of different quotient divisible groups and their bases. Connections between them are described in Theorem 15. The different dual bases differ by torsion elements. So the sequences of the torsion elements are terms of this description.

Definition. A sequence of elements $T = (t_1, \ldots, t_n)$ of a group $G_T = \langle t_1, \ldots, t_n \rangle$ is called *admissible* with respect to the sequence of characteristics $\Xi = (\chi_1, \ldots, \chi_n)$ if each element t_i is admissible with respect to the characteristic $\chi_i, i = 1, \ldots, n$.

Theorem 15. *Let $B = x_1 R_{\chi_1} \oplus \ldots \oplus x_n R_{\chi_n}$ and $B^* = x_1^* R^{\chi_1} \oplus \ldots \oplus x_n^* R^{\chi_n}$ be mutually dual groups as it is defined above. For every admissible sequence of torsion elements $T = (t_1, \ldots, t_n)$ with respect to the sequence of the characteristics $\Xi = (\chi_1, \ldots, \chi_n)$ the following statements take place:*

(i) *The group $B^* \oplus G_T$ is quotient divisible. Moreover, it is a direct sum of quotient divisible rank-1 subgroups. The set $x_1^* + t_1, \ldots, x_n^* + t_n$ is a basis of the group $B^* \oplus G_T$.*

(ii) *The dual to $B^* \oplus G_T$ torsion free group A_T with respect to the basis $x_1^* + t_1, \ldots, x_n^* + t_n$ is an almost completely decomposable group with the basis $x_1, \ldots, x_n$. Moreover, $B \subset A_T$ and $A_T/B \cong G_T$. Thus every admissible sequence T gives an almost completely decomposable group A_T.*

(iii) *Let A_S be an almost completely decomposable group corresponding to another admissible sequence $S = (s_1, \ldots, s_n)$ for the same sequence of the characteristics $\Xi = (\chi_1, \ldots, \chi_n)$.*
The inclusion $A_T \subset A_S$ takes place if and only if there exists a homomorphism $\eta : G_S \to G_T$ such that $\eta(s_1) = t_1, \ldots, \eta(s_n) = t_n$.

Proof. The first statement of the theorem is a direct consequence of definitions and Lemma 1.

The Z-adic completion M of the group $B^* \oplus G_T$ is of the form $M = B^0 \oplus G_T$, where $B^0 = x_1^0 Z_{\chi_1} \oplus \ldots \oplus x_n^0 Z_{\chi_n}$. The set of elements $x_1^0 + t_1, \ldots, x_n^0 + t_n$ generates the module M over the ring $\widehat{Z}$ and it is a part of the triple corresponding to the quotient divisible group $B^* \oplus G_T$ with the basis $x_1^* + t_1, \ldots, x_n^* + t_n$. The set $x_1^0 + t_1, \ldots, x_n^0 + t_n$ is not necessarily linearly independent over $\widehat{Z}$, for example t_1 can be equal to t_2, but it is surely almost linearly independent over $\widehat{Z}$.

The group A_T is generated in V by all elements of the form

$$f\left(x_1^0 + t_1\right) x_1 + \ldots + f\left(x_n^0 + t_n\right) x_n, \tag{9.1}$$

where f runs through the group

$$\mathrm{Hom}_{\widehat{Z}}\left(B^0 \oplus G_T, Q/Z\right) = \mathrm{Hom}_{\widehat{Z}}\left(B^0, Q/Z\right) \oplus \mathrm{Hom}_{\widehat{Z}}\left(G_T, Q/Z\right).$$

If the function f is running only through the first direct summand $\mathrm{Hom}_{\widehat{Z}}\left(B^0, Q/Z\right)$, then the elements (9.1) generate in total the group $B = x_1 R_{\chi_1} \oplus \ldots \oplus x_n R_{\chi_n}$. Thus the group A_T is generated by B and the finite set of elements (9.1), where f is running through $\mathrm{Hom}_{\widehat{Z}}\left(G_T, Q/Z\right)$.

We denote $G_T^* = \mathrm{Hom}_{\widehat{Z}}\left(G_T, Q/Z\right)$ and identify $G_T^{**} = G_T$. If $t \in G_T$ and $f \in G_T^*$, then $t : G_T^* \to Q/Z$ is defined as $t(f) = f(t)$. It is easy to see that the function $\theta : G_T^* \to V/B$, where $\theta(f) = (f(t_1) x_1 + \ldots + f(t_n) x_n) + B$, is a homomorphism. Let us prove that it is injective. Suppose $\theta(f) = 0$. It means that $f(t_1) x_1 + \ldots + f(t_n) x_n \in B$. Let $f(t_i) = \frac{k_i}{m_i} + Z$, $\gcd(k_i, m_i) = 1$. Every prime divisor p of m_i is a divisor of the order of the element t_i. Since t_i is admissible with respect to χ_i, the p-component of χ_i is zero and hence $\frac{1}{p} x_i \notin B$. This contradiction shows that $m_i = 1$ for every i and therefore $f = 0$.

Identifying along the monomorphism θ, we obtain $G_T^* \subset V/B$. The preimage of G_T^* at the natural homomorphism $V \to V/B$ is exactly the group A_T. Thus $B \subset A_T$ and $A_T/B = G_T^*$. The observation $G_T^* \cong G_T$ completes the second statement of the theorem.

Let $S = (s_1, \ldots, s_n)$ be another admissible sequence which similarly determines a group $A_S \subset V$. It is clear that $A_T \subset A_S \Leftrightarrow A_T/B \subset A_S/B$. Let $A_T \subset A_S$. Taking into account all the identifications, the embedding $\mathrm{id} : A_T/B \to A_S/B$ can be described in the following way.

Let $f \in A_T/B = G_T^* = \mathrm{Hom}_{\widehat{Z}}(G_T, Q/Z)$. Then $\mathrm{id}(f)$ is such a homomorphism $g \in A_S/B = G_S^* = \mathrm{Hom}_{\widehat{Z}}(G_S, Q/Z)$ that $(f(t_1) x_1 + \ldots + f(t_n) x_n) + B = (g(s_1) x_1 + \ldots + g(s_n) x_n) + B$, that is $(f(t_1) - g(s_1)) x_1 + \ldots + (f(t_n) - g(s_n)) x_n \in B$. Similarly to the injectivity of θ, it follows that $f(t_1) = g(s_1), \ldots, f(t_n) = g(s_n)$. Since $\mathrm{id} : G_T^* \to G_S^*$ is injective, the dual homomorphism $\mathrm{id}^* : G_S^{**} \to G_T^{**}$ is surjective. Here $\mathrm{id}^*(h) = h \circ \mathrm{id}$ for an element $h : G_S^* \to Q/Z$ of the group G_S^{**}. Considering $s_i \in G_S = G_S^{**}$, we have $(\mathrm{id}^*(s_i))(f) = s_i(\mathrm{id}(f)) = s_i(g) = g(s_i) = f(t_i) = t_i(f)$ for every $f \in G_T^*$. Thus $\mathrm{id}^*(s_i) = t_i$ for all i and $\mathrm{id}^* : G_S \to G_T$ is the desired homomorphism.

Conversely, if $\eta : G_S \to G_T$ is a homomorphism such that $\eta(s_1) = t_1, \ldots, \eta(s_n) = t_n$, then every generator of the form $f(t_1)x_1 + \ldots + f(t_n)x_n$ of the group A_T can be represented in the form $(f\eta)(s_1)x_1 + \ldots + (f\eta)(s_n)x_n$ as a generator of the group A_S. Therefore $A_T \subset A_S$. □

This theorem together with Theorems 2 and 11 leads to the following corollary.

Corollary 16. *Let a sequence of elements $T = (t_1, \ldots, t_n)$ be admissible with respect to a sequence of characteristics $\Xi = (\chi_1, \ldots, \chi_n)$. Then the following statements hold for the group A_T and the basis $x_1, \ldots x_n$ defined in Theorem 15:*

(i) $A_T = \langle x_1, \ldots, x_k \rangle_* \oplus \langle x_{k+1}, \ldots, x_n \rangle_*$ *if and only if* $G_T = \langle t_1, \ldots, t_k \rangle \oplus \langle t_{k+1}, \ldots, t_n \rangle$.

(ii) *The group A_T is indecomposable if the characteristics $\chi_1, \ldots, \chi_n$ belong to pairwise incomparable types and the intersections of the cyclic group $\langle t_1 \rangle$ with each of the cyclic groups $\langle t_2 \rangle, \ldots, \langle t_n \rangle$ are different from zero.*

For every two admissible sequences $T = (t_1, \ldots, t_n)$ and $S = (s_1, \ldots, s_n)$ with respect to the same sequence of characteristics $\Xi = (\chi_1, \ldots, \chi_n)$, we define:

- $T \leq S$ if there exists a homomorphism $\eta : G_S \to G_T$ such that $\eta(s_1) = t_1, \ldots, \eta(s_n) = t_n$.

- $T \sim S$ if there exists an isomorphism $\eta : G_S \to G_T$ such that $\eta(s_1) = t_1, \ldots, \eta(s_n) = t_n$.

The second relation is an equivalence. The first relation is an order on the set of equivalence classes of admissible sequences. Thus we obtain a lattice of admissible sequences L_Ξ. We call a group of the form A_T as *admissible* with respect to Ξ.

Corollary 17. *Let $B = x_1 R_{\chi_1} \oplus \ldots \oplus x_n R_{\chi_n}$ be a completely decomposable torsion free group. The lattice by inclusion of the admissible extensions of B is isomorphic to the lattice L_Ξ.*

The restriction of admissibility is not very hard as it is shown in the following theorem.

Theorem 18. *For every almost completely decomposable group A there exist a sequence of characteristics $\Xi = (\chi_1, \ldots, \chi_n)$ and an admissible sequence of elements $T = (t_1, \ldots, t_n)$ such that $A = A_T$.*

Proof. The group A contains a completely decomposable subgroup $B = x_1 R_{\kappa_1} \oplus \ldots \oplus x_n R_{\kappa_n}$ of finite index. Let $P = \{p_1, \ldots, p_m\}$ be the set of all prime divisors of this index. Replacing the finite p-components of the characteristics $\kappa_1, \ldots, \kappa_n$ by zeros for all $p \in P$, we obtain a new sequence of characteristics $\chi_1 \leq \kappa_1, \ldots, \chi_n \leq \kappa_n$ of the same types. Then A is a finite extension of the group $B_1 = x_1 R_{\chi_1} \oplus \ldots \oplus x_n R_{\chi_n}$ and the set of prime divisors of the index is the same $P = \{p_1, \ldots, p_m\}$. Now it is easy to see from Theorem 9 that there exists an admissible sequence T for $\Xi = (\chi_1, \ldots, \chi_n)$ such that $A = A_T$. □

10 The group of A. L. S. Corner

10.1 Partitions

If $k = n$ in Corner's theorem, then the Corner's group is $C = \bigoplus_k Q^{(p)}$ for a prime number p. Excluding this trivial case, we consider different prime numbers $q_1, \ldots, q_{n-k}$, where $0 < k < n$, and a group $G = \langle t_1 \rangle \oplus \ldots \oplus \langle t_{n-k} \rangle$, where the order of t_i is q_i for $i = 1, \ldots, n-k$. The group G is cyclic itself of the order $m = q_1 \cdot \ldots \cdot q_{n-k}$, $G = \langle t \rangle$, where $t = t_1 + \ldots + t_{n-k}$.

Since the greatest common divisor of the integers $\frac{m}{q_1}, \frac{m}{q_2}, \ldots, \frac{m}{q_{n-k}}$ is equal to 1, there exist integers $r_1, r_2, \ldots, r_{n-k}$ such that $r_1 \frac{m}{q_1} + r_2 \frac{m}{q_2} + \ldots + r_{n-k} \frac{m}{q_{n-k}} = 1$. We are interested in the last equality just as in the sum of the integers which is equal to 1.

$$m_1 + m_2 + \ldots + m_{n-k} = 1 \qquad (10.1)$$

The numbers $m_1, m_2, \ldots, m_{n-k}$ have the following property

$$m_i t_i = t_i \text{ for all } i \text{ and } m_i t_j = 0 \text{ for } i \neq j.$$

A sequence of natural numbers $(P_1, \ldots, P_{k-1})$ satisfying $0 \leq P_1 \leq \ldots \leq P_{k-1} \leq n-k$, is called a *subdivision* of the interval $[1, n-k]$ into k parts by $k-1$ *partitions* $P_1, \ldots, P_{k-1}$. We define the following four sequences for a given subdivision $(P_1, \ldots, P_{k-1})$.

(i) The sequence of non-negative integers:

$$n_1 = P_1, \ldots, n_i = P_i - P_{i-1}, \ldots, n_k = (n-k) - P_{k-1}.$$

It is easy to see that $n - k = n_1 + \ldots + n_k$ and the number of different subdivisions is $\binom{n-1}{k-1}$. It is equal to the number of different representations of $n - k$ as a sum of k non-negative integers $n - k = n_1 + \ldots + n_k$.

(ii) The sequence of integers obtained from (10.1):

$$\begin{aligned} s_1 &= m_1 + \ldots + m_{P_1}, \ldots, s_i = m_{P_{i-1}+1} + \ldots + m_{P_i}, \ldots, s_k \\ &= m_{P_{k-1}+1} + \ldots + m_{n-k}. \end{aligned}$$

Note that if $n_i = 0$ then the sum s_i is empty and we define $s_i = 0$. Obviously, $s_1 + \ldots + s_k = 1$.

(iii) The sequence of elements of the group G:

$$g_1 = t_1 + \ldots + t_{P_1}, \ldots, g_i = t_{P_{i-1}+1} + \ldots + t_{P_i}, \ldots, g_k = t_{P_{k-1}+1} + \ldots + t_{n-k}.$$

If $n_i = 0$ we define $t_i = 0$. Obviously, $g_1 + \ldots + g_k = t$.

(iv) The sequence of subsets of a linearly independent set $X = \{x_1, \ldots, x_{n-k}\}$, where $x_1, \ldots, x_{n-k}$ are vectors of a rational vector space:

$$X_1 = \{x_j \mid j \leq P_1\}, \ldots, X_i = \{x_j \mid P_{i-1} < j \leq P_i\}, \ldots,$$
$$X_k = \{x_j \mid P_{k-1} < j \leq n-k\}.$$

If $n_i = 0$ then $X_i = \emptyset$.

The following properties take place for a subdivision $(P_1, \ldots, P_{k-1})$:

(i) $g_1 = s_1 t, \ldots, g_k = s_k t$,

(ii) $G = \langle g_1 \rangle \oplus \ldots \oplus \langle g_k \rangle$,

(iii) the set X_i consists of n_i elements, $X = \{x_1, \ldots, x_{n-k}\} = X_1 \cup \ldots \cup X_k$ and $X_i \cap X_j = \emptyset$ for $i \neq j$,

(iv) $\langle g_i \rangle \cap \langle t_j \rangle \neq 0 \iff x_j \in X_i$.

10.2 The Corner's group

Now we are able to define the Corner's group. Besides the set of the prime numbers $q_1, \ldots, q_{n-k}$, we fix prime numbers $p, p_1, \ldots, p_{n-k}$ such that all they are different. We also define $n - k + 1$ characteristics in the following way. The p-component of a characteristic χ is ∞ and all other components are equal to 0. The p_i-component of a characteristic χ_i is ∞ and all other components are equal to 0, $i = 1, \ldots, n - k$. Then $R_\chi = Q^{(p)}, R_{\chi_i} = Q^{(p_i)}$ and $R^\chi = Q_p, R^{\chi_i} = Q_{p_i}$.

The group $B = \left(u_1 Q^{(p)} \oplus \ldots \oplus u_k Q^{(p)}\right) \oplus \left(x_1 Q^{(p_1)} \oplus \ldots \oplus x_{n-k} Q^{(p_{n-k})}\right)$ is torsion free completely decomposable with a basis $u_1, \ldots, u_k, x_1, \ldots, x_{n-k}$ of rank n, we keep here the original notation of the book [13]. The dual to B quotient divisible group is of the form $B^* = \left(u_1^* Q_p \oplus \ldots \oplus u_k^* Q_p\right) \oplus \left(x_1^* Q_{p_1} \oplus \ldots \oplus x_{n-k}^* Q_{p_{n-k}}\right)$. Thus the sequence of characteristics is $\Xi = (\chi, \ldots, \chi, \chi_1, \ldots, \chi_{n-k})$. The sequence of torsion elements $T = (t, 0, \ldots, 0, t_1, \ldots, t_{n-k})$ is admissible with respect to Ξ.

We can apply Theorem 15 and define now the group of A. L. S. Corner as $C = A_T$. That is the torsion free group dual to the group $B^* \oplus G$ with respect to the basis $u_1^* + t, u_2^*, \ldots, u_k^*, x_1^* + t_1, \ldots, x_{n-k}^* + t_{n-k}$. The basis of C dual to this one is $u_1, u_2, \ldots, u_k, x_1, \ldots, x_{n-k}$.

First of all, we can see immediately by Corollary 16 that $A_T = \langle u_1, x_1, \ldots, x_{n-k} \rangle_* \oplus \langle u_2 \rangle_* \oplus \langle u_3 \rangle_* \oplus \ldots \oplus \langle u_k \rangle_*$ is the decomposition of the Corner's group into a direct sum of k indecomposable groups, because the types $[\chi], [\chi_1], \ldots, [\chi_{n-k}]$ are pairwise incomparable.

For an arbitrary representation $n - k = n_1 + \ldots + n_k$, we change only the part $u_1, u_2, \ldots, u_k$ of the common basis $u_1, u_2, \ldots, u_k, x_1, \ldots, x_{n-k}$ of the groups B and C with help of the matrix

$$L = \begin{pmatrix} s_1 & s_2 & s_3 & \ldots & s_k \\ -1 & 1 & 0 & \ldots & 0 \\ \ldots & \ldots & \ldots & \ldots & \ldots \\ -1 & 0 & 0 & \ldots & 1 \end{pmatrix}$$

such that

$$\begin{pmatrix} u_1 \\ \ldots \\ u_k \end{pmatrix} = L \begin{pmatrix} v_1 \\ \ldots \\ v_k \end{pmatrix}.$$

Thus we obtain a new common basis $v_1, \ldots, v_k, x_1, \ldots, x_{n-k}$ of the groups B and C. Since $\det L = s_1 + \ldots + s_k = 1$, it follows from Corollary 6 that the dual to B and C quotient divisible groups are the same groups B^* and $C^* = B^* \oplus G$, respectively. Due to Theorem 3, the dual to $v_1, \ldots, v_k, x_1, \ldots, x_{n-k}$ bases for these two groups are $v_{1B}^*, \ldots, v_{kB}^*, x_1^*, \ldots, x_{n-k}^*$ in B^* and $v_{1C}^*, \ldots, v_{kC}^*, x_1^* + t_1, \ldots, x_{n-k}^* + t_{n-k}$ in $B^* \oplus G$, where

$$\begin{aligned}(v_{1B}^*, v_{2B}^*, \ldots, v_{kB}^*) &= (u_1^*, u_2^*, \ldots, u_k^*) L, \\ (v_{1C}^*, \ldots, v_{kC}^*) &= (u_1^* + t, u_2^*, \ldots, u_k^*) L.\end{aligned} \qquad (10.2)$$

Note that by Corollary 13 we have two direct decompositions

$$B = \left(v_1 Q^{(p)} \oplus \ldots \oplus v_k Q^{(p)}\right) \oplus \left(x_1 Q^{(p_1)} \oplus \ldots \oplus x_{n-k} Q^{(p_{n-k})}\right)$$

and

$$B^* = (v_1^* Q_p \oplus \ldots \oplus v_k^* Q_p) \oplus (x_1^* Q_{p_1} \oplus \ldots \oplus x_{n-k}^* Q_{p_{n-k}}),$$

we denote here and further $v_1^* = v_{1B}^*, \ldots, v_k^* = v_{kB}^*$. It means that we can apply Theorem 15 once more for the fixed common basis $v_1, \ldots, v_k, x_1, \ldots, x_{n-k}$ of the groups B and C. Subtracting from the second equality (10.2) the first one, we obtain $(v_{1C}^*, \ldots, v_{kC}^*) - (v_{1B}^*, v_{2B}^*, \ldots, v_{kB}^*) = (t, 0, \ldots, 0) L = (s_1 t, s_2 t, \ldots, s_k t) = (g_1, \ldots, g_k)$. Thus the sequence of torsion elements is $S = (g_1, \ldots, g_k, t_1, \ldots, t_{n-k})$ which is obviously admissible with respect to the same sequence of characteristics $\Xi = (\chi, \ldots, \chi, \chi_1, \ldots, \chi_{n-k})$. The group C (with the basis $v_1, \ldots, v_k, x_1, \ldots, x_{n-k}$) is dual to $B^* \oplus G$ with respect to the basis $v_1^* + g_1, v_2^* + g_2, \ldots, v_k^* + g_k, x_1^* + t_1, \ldots, x_{n-k}^* + t_{n-k}$. In other words, $C = A_S$ in terms of Theorem 15. Due to Corollary 16, the equality $G = \langle g_1 \rangle \oplus \ldots \oplus \langle g_k \rangle$ implies a direct decomposition

$$C = A_S = \langle v_1, X_1 \rangle_* \oplus \langle v_2, X_2 \rangle_* \oplus \ldots \oplus \langle v_k, X_k \rangle_*. \qquad (10.3)$$

Every direct summand $\langle v_i, X_i \rangle_*$ is an extension of the group

$$B_i = v_i Q^{(p)} \oplus \left(\bigoplus_{x_j \in X_i} x_j Q^{(p_j)}\right)$$

with help of the group $\langle g_i \rangle$. Applying Corollary 16, we can conclude that every group $\langle v_i, X_i \rangle_*$ is indecomposable, because $\langle g_i \rangle \cap \langle t_j \rangle \neq 0$ for all $x_j \in X_i$ and the types $[\chi], [\chi_1], \ldots, [\chi_{n-k}]$ are pairwise incomparable. The indecomposable summands of the decomposition (10.3) have ranks $n_1 + 1, n_2 + 1, \ldots, n_k + 1$, respectively.

Thus the Corner's group $C_{nk} = C = A_T = A_S$ depends on a pair $0 < k < n$ of integers, it has rank n and the following property. For every decomposition of the number $n = n_1 + \ldots + n_k$ into a sum of k positive integers, the group C_{nk} can be decomposed into a direct sum of k indecomposable subgroups of ranks $n_1, \ldots, n_k$, respectively.

References

[1] A. A. Fomin and W. Wickless, *Quotient divisible abelian groups,* Proc. Amer. Math. Soc. 126 (1998), pp. 45–52.

[2] S. Glaz and W. Wickless, *Regular and principal projective endomorphism rings of mixed abelian groups,* Comm. in Alg. 22 (1994), pp. 1161–1176.

[3] R. Beaumont and R. Pierce, *Torsion-free rings,* Ill. J. Math. 5 (1961), pp. 61–98.

[4] 4 U. Albrecht and W. Wickless, *Finitely generated and cogenerated qd groups,* Lecture Notes in Pure and Applied Math., 236, Dekker, NY, pp. 13–26.

[5] W. Wickless, *Multi-isomorphism for quotient divisible groups,* Houston J. Math., 31 (2006), pp. 1–19.

[6] U. Albrecht, S. Breaz, C. Vinsonhaler and W. Wickless, *Cancellation properties for quotient divisible groups,* J. Algebra, 70 (2007), pp. 1–11.

[7] O.I. Davidova, *Rank-1 quotient divisible groups,* Fundam. Prikl. Mat., 13 (2007), no. 3, pp. 25–33.

[8] A. A. Fomin, *Some mixed abelian groups as modules over the ring of pseudo-rational numbers, in Abelian Groups and Modules,* Trends in Mathematics (1999), Birkhäuser Verlag Basel, pp. 87–100.

[9] A. A. Fomin, *Quotient divisible mixed groups, Abelian Groups, Rings and Modules,* Contemp. Math., vol. 273 (2001), pp. 117–128.

[10] A. A. Fomin, *A category of matrices representing two categories of abelian groups,* Fundam. Prikl. Mat., 13 (2007), no. 3, pp. 223–244.

[11] A. V. Tsarev, *Modules over the ring of pseudo-rational numbers and quotient divisible groups,* Sanct Petersburg Math.J., 18 (2006), no. 4, pp. 198–214.

[12] A. L. S. Corner, *A note on rank and direct decompositions of torsion free abelian groups,* Proc. Cambridge Philos. Soc., 57 (1961), pp. 230–233; 66 (1969), pp. 239-240.

[13] L. Fuchs, *Infinite Abelian Groups I+II,* Academic Press (1970, 1973).

[14] A. A. Fomin, *Finitely presented modules over the ring of universal numbers,* Contemp. Math. 171 (1995), pp. 109–120.

[15] A. V. Tsarev, *Pseudo-rational rank of an abelian group,* Sib. Math.J., 46 (2005), no. 1, pp. 312–325.

[16] A. V. Tsarev, *Projective and generating modules over the ring of pseudo-rational numbers,* Math. Notes, 80 (2006), no. 3, pp. 437–448.

[17] L. S. Pontryagin, *The theory of topological commutative groups,* Ann. of Math., 35 (1934), pp. 361–388.

[18] L. Fuchs, *The existence of indecomposable abelian groups of arbitrary power,* Acta Math. Acad. Sci. Hungar., 10 (1959), pp. 453–457.

Author information

Alexander A. Fomin, Department of Mathematics, Moscow Pedagogical State University, Russia.
E-mail: `alexander.fomin@mail.ru`

Pure subgroups of completely decomposable groups – an algorithmic approach

Daniel Herden and Lutz Strüngmann

Abstract. In [1] Arnold gave an example of a torsion-free abelian group G having the property that every finite rank pure subgroup H of G is a pure subgroup of some completely decomposable group. These groups are called B_0-(Butler)groups. However, Arnold proved that G itself cannot be purely embedded into any completely decomposable group. Starting from this classical result, we want to discuss this embedding problem for a large class of abelian groups including the class of bounded extensions of completely decomposable groups. On the one hand, using a linear algebra approach, we provide an algorithmic way to embed a given torsion-free group from our class into a completely decomposable group C with an explicit construction formula for C. On the other hand we give new examples of groups that have the same property as Arnold's example.

Key words. Almost completely decomposable groups, pure embeddings, infinite rank Butler groups.

AMS classification. 20K15, 20K20, 20K25, 15A36.

1 Introduction

It is well known that a torsion-free abelian group of finite rank is a pure subgroup of some completely decomposable group if and only if it is the epimorphic image of a completely decomposable group of finite rank. This result goes back to M. C. R. Butler [7] and consequently these groups were called *Butler groups* (of finite rank). Naturally, scientists asked for an extension of the notion of Butler groups to the infinite rank case. It turned out that various such notions exist defining classes of abelian groups with different properties. Probably the most natural attempt is to consider the class of pure subgroups of completely decomposable groups of arbitrary rank. Alternatively, one can require that the groups are *locally Butler*, meaning that every pure subgroup of finite rank is a Butler group. We call these groups B_0-groups in this article. Bican and Salce [4] introduced two further generalizations, namely B_2-groups and B_1-groups. B_2-groups are unions of smooth ascending chains of pure subgroups such that each level can be obtained from the previous one by adding a finite rank Butler group. B_1-groups are defined by means of the homological functor $\text{Bext}(-, -)$. All these classes of B_i-groups coincide in the countable case [4] while in the uncountable case it is undecidable if B_1-groups are the same as B_2-groups [10], [15]. The situation is different if one looks at the pure subgroups of completely decomposable groups. An example by

---

The first author was supported by a grant from the Minerva Fellowship, Max Planck Foundation.
The second author was supported by a grant from the German Research Foundation DFG.

Arnold [1] shows that there exist countable B_0-groups that cannot be purely embedded into any completely decomposable group. This leads to the question of which groups can be embedded as pure subgroups into completely decomposable groups.

We first consider Arnold's example from [1] and correct a little gap that was contained in the proof. We then consider torsion-free abelian groups and express their defining relations as a matrix. Using linear algebra we give an algorithmic way to embed certain groups into completely decomposable groups.

Our notation is standard as can be found for instance in [8] and we write maps on the left.

2 Preliminaries

In this section we recall some basic definitions and terminology that will be used frequently below. Let Π be the set of primes. A group X is called *rational* if it is isomorphic to a subgroup of the group of rationals $\mathbb{Q}$, i.e. X is a torsion-free group of rank 1. If $[X]$ is the class of all groups isomorphic to X we may define an equivalence relation by setting $[X] \leq [X']$ if $\text{Hom}\,(X, X') \neq 0$. The class $[X]$ is called the *type* of X and if $x \in G$ is an element of the torsion-free group G, then we denote by $\text{type}_G(x)$ the *type* of x in G which is defined as $\text{type}_G(x) = [\langle x \rangle_*^G]$. Here, $\langle x \rangle_*^G$ is the pure subgroup of G generated by x. It is well-known that types can also be described by equivalence classes of infinite sequences $(n_p : p \in \Pi)$ of natural numbers and the symbol ∞. For instance, one can choose $(n_p : p \in \Pi)$ such that $X = \langle 1/p^{m_p} : m_p \leq n_p \rangle$. Two such sequences $(n_p : p \in \Pi)$ and $(n'_p : p \in \Pi)$ describe the same type if and only if they differ in finitely many finite entries only, i.e. if $\sum_{p \in \Pi} |n_p - n'_p|$ is finite. For further details on types see [8]. Since there is no danger of confusion we will assume that $\mathbb{Z} \subseteq X$ for every rational group X and use X both for the group X and its type $[X]$.

A torsion-free group G is called *completely decomposable* if G is a direct sum of rational groups. A *Butler group* is a pure subgroup of a finite rank completely decomposable group or equivalently an epimorphic image of a completely decomposable group of finite rank. Examples of Butler groups are the so-called *almost completely decomposable groups* (acd-groups) which are finite extensions of completely decomposable groups of finite rank. More generally, a torsion-free group G is a B_0-group if every pure finite rank subgroup of G is a Butler group. Examples of B_0-groups are countable bcd-groups: recall that a *bounded completely decomposable group* (bcd-group) is a bounded extension of a completely decomposable group of arbitrary rank. For further terminology and details on Butler groups, acd-groups and bcd-groups see [2], [8], [11] and [13].

3 Countable B_0-groups

In [1] the following result was proved which shows that the class of pure subgroups of completely decomposable groups is a proper subclass of the class of B_0-groups.

Theorem 3.1. *There exists a countable B_0-group that is not a pure subgroup of a completely decomposable group.*

The proof given in [1] contains a minor combinatorial gap which we fix in this section. A correction is also contained in [3] by Arnold and Rangaswamy but our approach is different and will be used later on to provide new examples. Therefore we include it here and start with a sketch of the basic construction from [1].

Let $S := \{s : n \to \{0,1\} \mid n \in \omega\}$ be the set of all words in the alphabet $\{0,1\}$ where we consider $n \in \omega$ as the set $n = \{0, 1, \ldots, n-1\}$ in the definition of functions $s : n \to \{0,1\}$. We choose an infinite tree of proper subgroups $\mathfrak{X} = \{X_s \mid s \in S\}$ of $\mathbb{Q}$ with the following property:

(1*) If $s, t, u \in S$ with $t = s0$, $u = s1$, then $X_s \subseteq X_t \cap X_u$.

Examples of trees $\mathfrak{X}$ having the property **(1*)** are easily constructed (see Lemma 3.4 or [1]). We now put

$$G_n := \bigoplus \{X_s \mid s \in S, \mathrm{Dom}\,(s) = n\}$$

for every $n \in \omega$ and define the injections $\varphi_n : G_n \to G_{n+1}$ by mapping every $x_s \in X_s$ onto $(x_s, x_s) \in X_{s0} \oplus X_{s1}$. Here $\mathrm{Dom}\,(s) = n = \{0, 1, \ldots, n-1\}$ is the domain of s. The maps φ_n ($n \in \omega$) are well-defined by property **(1*)**. The desired group is obtained as the direct limit

$$G := G_{\mathfrak{X}} := \varinjlim \{G_m, \varphi_m \mid m \in \omega\},$$

i.e. $G = A/B$ with $A = \bigoplus_{m \in \omega} G_m$ and $B = \langle g_m - g_m \varphi_m \mid m \in \omega, g_m \in G_m \rangle$. In [1] a proof of Theorem 3.1 was given assuming further conditions on $\mathfrak{X}$, namely

(1) If $s, t, u \in S$ with $t = s0$, $u = s1$, then $X_s = X_t \cap X_u$.

(2*) If τ is a type and for each positive integer n there is a word $s(n) \in S$ of length greater than or equal to n with $\tau \geq X_{s(n)}$, then $\tau = \mathbb{Q}$.

However, it turns out that these two conditions **(1)** and **(2*)** are inconsistent.

Lemma 3.2. *There exists no tree $\mathfrak{X}$ of proper subgroups of $\mathbb{Q}$ such that **(1)** and **(2*)** are satisfied.*

Proof. Assume $\mathfrak{X} = \{X_s \mid s \in S\}$ is an infinite tree of proper subgroups of $\mathbb{Q}$ with properties **(1)** and **(2*)**. First we prove that for every prime $p \in \Pi$ there exists a natural number $N(p) \in \omega$ such that X_s is p-divisible whenever $s \in S$ with $\mathrm{Dom}\,(s) \geq N(p)$. By way of contradiction assume that there exist arbitrarily large words $s \in S$ such that X_s is not p-divisible. Then $\mathbb{Z}_{(p)} = \langle 1/q^n : n \in \omega, p \neq q \in \Pi \rangle < \mathbb{Q}$ is an upper bound for all the X_s and hence we obtain a contradiction to condition **(2*)**.

However, on the other hand condition **(1)** implies that there exists some $p \in \Pi$ such that for all $n \in \omega$ there is $s_n \in S$ with $\mathrm{Dom}\,(s_n) = n$ and X_{s_n} not p-divisible. To see this recall that the group $X_\emptyset$ is reduced, thus there is some $p \in \Pi$ such that $X_\emptyset$ is p-reduced. With property **(1)** we now construct recursively an ascending chain of words s_n such that $\mathrm{Dom}\,(s_n) = n$ and X_{s_n} is p-reduced. Here we use that $X_s = X_{s0} \cap X_{s1}$

and X_s p-reduced implies that not both groups X_{s0} and X_{s1} can be p-divisible. We obtain a contradiction and thus such trees do not exist. □

One attempt to fix this problem is to assume conditions (**1***) and (**2***) which are consistent. Indeed, there is a tree $\mathfrak{X}$ satisfying (**1***) and (**2***), however, the corresponding group $G_{\mathfrak{X}}$ is divisible, hence does not provide an example as stated in Theorem 3.1.

The problem can be fixed replacing Arnold's property (**2***) by:

(**2**) Assume that $k \in \omega$ and τ_i ($0 \leq i \leq k$) are types such that for each positive integer n there are words $s_i^n \in S$ of length greater than or equal to n with $\tau_i \geq X_{s_i^n}$. Then we have $\bigcap_{i=0}^{k} \tau_i > X_\emptyset$.

We obtain the following theorem.

Theorem 3.3. *Let $\mathfrak{X} = \{X_s | s \in S\}$ be an infinite tree of proper subgroups of $\mathbb{Q}$ with properties* (**1**) *and* (**2**). *Then the group $G_{\mathfrak{X}}$ is a countable B_0-group that is not a pure subgroup of a completely decomposable group.*

Proof. The proof is similar to the one in [1], hence we only state the main steps. First we have to prove that G is a countable reduced B_0-group. This follows as in [1] using condition (**1**). The required filtration turns out to be

$$G = \bigcup_{n \in \omega} G'_n \text{ where } G'_n = (A_n + B)/B \cong A_n/B_n$$

with $A_n = \bigoplus_{m=0}^{n} G_m$ and $B_n = \langle g_m - g_m \varphi_m | m \in \omega, m < n, g_m \in G_m \rangle$.

In order to complete the proof we have to show that G is not a pure subgroup of a completely decomposable group. By way of contradiction assume that $G \subseteq_* C = \bigoplus_{i \in \omega} C_i$ for some rational groups $C_i \subseteq \mathbb{Q}$. Without loss of generality we may assume that C_i is reduced for all $i \in \omega$. In fact, if we let $I := \{i | C_i < \mathbb{Q}\} \subseteq \omega$, then it easily follows that G embeds into $\bigoplus_{i \in I} C_i$ via the canonical projection $\pi : G \subseteq C \to \bigoplus_{i \in I} C_i$.

Now observe that $B_0 = \{0\}$ and $G'_0 \cong A_0/B_0 = A_0 = X_\emptyset$. We choose $0 \neq x \in G'_0 = (X_\emptyset + B)/B \subseteq_* G$ and let $x = \sum_{j \in J} c_j \in \bigoplus_{j \in J} C_j$ with $c_j \neq 0$ for all $j \in J$, where $J \subseteq \omega$ is finite. Then $X_\emptyset \cong G'_0 \subseteq_* \bigoplus_{j \in J} C_j$, and we obtain that $X_\emptyset = \text{type}_G(x) = \bigcap_{j \in J} C_j$. For any $n \geq 0$ there exists some representative $\bigoplus \{x_s | \text{Dom}(s) = n\} \in G_n$ of x in G, where $x_s \in X_s \subseteq G_n$. We write

$$x_s + B = c(1, s) \oplus \ldots \oplus c(k, s) \quad \text{with} \quad c(i, s) \in C_i \text{ for some } k \geq \max J.$$

From $c_j \neq 0$ ($j \in J$) it follows that there exists some $s'_n \in S$ with $\text{Dom}(s'_n) = n$ and $c(j, s'_n) \neq 0$ (here s'_n depends on j and n), while

$$X_{s'_n} \leq \text{type}_G(x_{s'_n} + B) \leq \text{type}_C(c(j, s'_n)) = C_j.$$

Since this holds for every $n \in \omega$ we get that $\tau_j := C_j \geq X_{s'_n}$ ($j \in J$) and property (**2**) yields $\bigcap_{j \in J} C_j > X_\emptyset$ contradicting the fact that $\bigcap_{j \in J} C_j = X_\emptyset$. □

Thus to complete the proof of Theorem 3.1 using Theorem 3.3 we must show the consistency of properties **(1)** and **(2)**.

Lemma 3.4. *There exists a tree* $\mathfrak{X} = \{X_s | s \in S\}$ *of proper subgroups of* $\mathbb{Q}$ *satisfying conditions* **(1)** *and* **(2)**.

Proof. For every $s \in S$ let $\mathrm{inv}(s)$ be the *reversed word* of s, i.e. $(\mathrm{inv}(s))(i) = s(n - i - 1)$ for all $i \in n := \mathrm{Dom}(s) = \mathrm{Dom}(\mathrm{inv}(s))$. Furthermore, let $(p_n)_{n \in \omega}$ be the ascending sequence of all primes. Recall that the binary representation of a natural number n is a finite sequence of 0s and 1s such that the first digit is 1. For instance the binary representation of 9 is $(1, 0, 0, 1) = 2^3 + 2^0$. Put

$$\Pi_s := \{p_n | \text{ The binary representation of } n \text{ ends in } \mathrm{inv}(s).\}$$

and define X_s as

$$X_s := \langle 1, \frac{1}{p^k} | k \in \omega, p \in \Pi \setminus \Pi_s \rangle \subseteq \mathbb{Q}$$

for all $s \in S$. Obviously, $\Pi_\emptyset = \Pi$ and $X_\emptyset = \mathbb{Z}$. Then the rational groups X_s ($s \in S$) are reduced because Π_s is non-empty for all $s \in S$. Since $\Pi_s \subseteq \Pi_{s'}$ whenever $s, s' \in S$ such that $s' \subseteq s$ we conclude that $X_{s'} \subseteq X_s$ for $s' \subseteq s$, thus $\mathfrak{X} = \{X_s | s \in S\}$ is an infinite tree of proper subgroups of $\mathbb{Q}$. Furthermore, $\Pi_s = \Pi_{s0} \cup \Pi_{s1}$ for every $s \in S$ gives $X_s = X_{s0} \cap X_{s1}$ and hence property **(1)** is satisfied. Observe also that $\Pi_s \cap \Pi_{s'} = \emptyset$ for all $s, s' \in S$ if neither $s' \not\subseteq s$ nor $s \not\subseteq s'$. Thus $X_s \cup X_{s'} = \mathbb{Q}$ for incomparable $s, s' \in S$. Now assume that τ is a type such that for each positive integer n there is some word $s_n \in S$ of length $\geq n$ with $\tau \geq X_{s_n}$. Then

$$\tau \geq \bigcup_{n \in \omega} X_{s_n}$$

where all groups X_s are idempotent. If there exist $n, n' \in \omega$ with $s_n, s_{n'}$ incomparable, then $\tau = \mathbb{Q}$ follows immediately. Otherwise the words s_n are linearly ordered by inclusion and the supremum $\bar{s} = \bigcup_{n \in \omega} s_n$ exists as an infinite word, i.e. an infinite sequence $\bar{s} : \omega \to \{0, 1\}$. Then τ is divisible by any prime p_n such that the binary representation of n does not coincide with $\mathrm{inv}(\bar{s} \upharpoonright_k)$ for any $k \in \omega$, i.e. the infinite word $\bar{s}$ does not have the reversed binary representation of n as an initial segment. Thus, given finitely many types τ_i ($0 \leq i \leq k$) as in property **(2)**, every τ_i corresponds to an infinite word $\bar{s}_i$ ($0 \leq i \leq k$). Since there are 2^{m-1} different binary representations of length m for any $m \in \omega$ it follows that there is some $l \in \omega$ such that no $\mathrm{inv}(\bar{s}_i \upharpoonright_n)$ coincides with the binary representation of l for any $0 \leq i \leq k$ and $n \in \omega$. Thus each τ_i must be divisible by p_l and therefore $\bigcap_{i=0}^k \tau_i > \mathbb{Z} = X_\emptyset$ follows which shows that property **(2)** holds. □

4 Extensions of completely decomposable groups by torsion groups

In this section we consider extensions of completely decomposable groups by torsion groups. We relate these groups to certain matrices with rational entries which will in certain cases be helpful to define pure embeddings into completely decomposable groups in the next sections.

In the sequel let $C = \bigoplus_{i \in \kappa} R_i e_i$ be a completely decomposable group of rank κ for some cardinal κ and rational groups R_i $(i \in \kappa)$. Moreover, let $D = \mathbb{Q} \otimes C$ be the divisible hull of C.

By $\mathcal{G}(C)$ we denote the class of all (torsion) extensions of C inside D. This is to say
$$\mathcal{G}(C) := \{G : C \subseteq G \subseteq D\}.$$
Note that any $G \in \mathcal{G}(C)$ satisfies that G/C is torsion.

At this point we give some examples of groups in $\mathcal{G}(C)$ which will be of importance later on.

Example 4.1.
- If C is of finite rank, then $\mathcal{G}(C)$ contains all finite extensions of C, i.e all almost completely decomposable groups with regulating subgroup C (see [11] for details on regulating subgroups).
- If C is of infinite rank, then $\mathcal{G}(C)$ contains all bounded essential extensions of C, i.e. all groups $C \subseteq G \subseteq D$ such that $nG \subseteq C$ for some $n \in \mathbb{N}$. These groups are called n-bcd-groups (see [13] for details).
- Let C be of infinite countable rank and $\{p_i : i \in \omega\}$ an increasing sequence of primes, $p \in \Pi$, then $\mathcal{G}(C)$ contains the groups $G_C^1 := \langle C, \frac{e_i + e_{i+1}}{p_i} : i \in \omega \rangle$, $G_C^2 := \langle C, \frac{e_i + e_i}{p_i} : i \in \omega \rangle$, $G_C^3 := \langle C, \frac{e_i + e_i}{p^i} : i \in \omega \rangle$ and $G_C^4 := \langle C, \frac{e_i + e_{i+1}}{p^i} : i \in \omega \rangle$.

We now aim for a workable description of the groups in $\mathcal{G}(C)$. For the convenience of the reader we will discuss an example after every step. Although the example group will be free it serves as a good example for the main changes of the group presentation which in the end will induce a pure embedding into some completely decomposable group.

Definition 4.2. We say that a group $G \in \mathcal{G}(C)$ is given in standard form if there are elements $d_j \in D$, primes $p_j \in \Pi$ (not necessarily distinct) and integers $n_j \in \mathbb{N}_0$ for $j \in \kappa$ such that

(i) $G = \langle C, d_j : j \in \kappa \rangle$ and

(ii) $d_j = \sum_{i \in \kappa} \frac{d_{i,j}}{p_j^{n_j}} e_i$ with $d_{i,j} \in \mathbb{Z}$ for all $i, j \in \kappa$.

Note that in Definition 4.2, for all $j \in \kappa$ the elements $d_{i,j}$ are almost all equal to zero since $d_j \in D$ and that the standard form of G is by no means unique.

Example 4.3 (G^*). Let $C := \bigoplus_{n \in \omega} \frac{1}{p_n p_{n-1}} \mathbb{Z} e_n$ where $\Pi = \{p_n : n \in \omega\}$ is in increasing order and $p_{-1} := 1$. Then $G^* = \langle C, \frac{e_n + e_{n+1}}{p_n^2} : n \in \omega \rangle \in \mathcal{G}(C)$ is given in standard form.

We have a first easy lemma.

Lemma 4.4. *Let $G \in \mathcal{G}(C)$. Then there is a free group F such that $G \oplus F \in \mathcal{G}(C \oplus F)$ can be given in standard form and $(G \oplus F)/(C \oplus F) \cong G/C$.*

Proof. Let $G \in \mathcal{G}(C)$ be given. Certainly we can find a cardinal δ and elements $d_j \in D$ ($j \in \delta$) such that $G = \langle C, d_j : j \in \delta \rangle$. If $\kappa < \delta$ (this can only happen if κ is finite), then choose a free group F with $\delta = \operatorname{rk}(C \oplus F)$ and hence Definition 4.2 (i) is satisfied for $G \oplus F, C \oplus F$ and δ. If $\delta < \kappa$, then choose δ' disjoint to δ with $|\delta \cup \delta'| = \kappa$ and put $d_j = c$ for some $0 \neq c \in C$ and $j \in \delta'$. Thus $G = \langle C, d_j : j \in \delta \cup \delta' \rangle$ and Definition 4.2 (i) is satisfied with $F = \{0\}$. It remains to ensure Definition 4.2 (ii). But this follows easily using Euclid's algorithm. Without loss of generality we may assume that each d_j is of the form $d_j = \sum_{i \in \kappa} \frac{r_{i,j}}{p_j^{n_j}} e_i$ with $r_{i,j} \in R_i$. Let n be such that $nr_{i,j} \in \mathbb{Z}$ for all $i \in \kappa$. Choose n' such that $n = n' p_j^t$ for some $t \in \mathbb{N}_0$ and $\gcd(n', p_j) = 1$. Representing $1 = kn' + lp_j^{n_j}$ for some $l, k \in \mathbb{Z}$ we obtain that $d_j \equiv \sum_{i \in \kappa} \frac{knr_{i,j}}{p_j^{n_j+t}} e_i$ modulo C which shows (ii). Obviously, $(G \oplus F)/(C \oplus F) \cong G/C$ holds. □

We now associate to each group G in $\mathcal{G}(C)$ which is given in standard form a matrix $B(G)$ with rational entries. $B(G)$ describes the relations d_j ($j \in \kappa$) that define the quotient G/C.

Definition 4.5. Let $G = \langle C, d_j : j \in \kappa \rangle \in \mathcal{G}(C)$ be given in standard form. If $d_j = \sum_{i \in \kappa} \frac{d_{i,j}}{p_j^{n_j}} e_i$ with $d_{i,j} \in \mathbb{Z}$ put $B(G) = \left(\frac{d_{i,j}}{p_j^{n_j}} \right)_{i,j \in \kappa}$, the $\kappa \times \kappa$ matrix with $\frac{d_{i,j}}{p_j^{n_j}}$ as entry in the ith row and jth column. We call $B(G)$ the matrix associated to G.

Note that the matrix $B(G)$ from Definition 4.5 is column finite and hence an endomorphism of $D = \mathbb{Q}^{(\kappa)}$ mapping e_i to d_i, i.e. $B(G)e_i = d_i \in D$. From this it is clear that $B(G)$ acts on D as the usual matrix operation on a vector space. However, notice that $B(G)$ does not define a homomorphism from C to G since e_i is of type R_i but d_i is not necessarily of this type inside G. We revisit the example G^*.

Example 4.6 (G^*). Let G^* be defined as in Example 4.3, then its associated matrix $B = B(G^*)$ is

$$B = \begin{pmatrix} \frac{1}{4} & 0 & 0 & 0 & 0 & \cdots \\ \frac{1}{4} & \frac{1}{9} & 0 & 0 & 0 & \cdots \\ 0 & \frac{1}{9} & \frac{1}{25} & 0 & 0 & \cdots \\ 0 & 0 & \frac{1}{25} & \frac{1}{49} & 0 & \cdots \\ 0 & 0 & 0 & \frac{1}{49} & \frac{1}{121} & \cdots \\ \vdots & \vdots & \vdots & \vdots & \vdots & \ddots \end{pmatrix}.$$

Our next lemma shows that we can simplify the representation of G even more depending on the properties of $B(G)$. Recall that for an element $d = \sum_{i \in \kappa} q_i e_i \in D$ the *support* of d is defined as $[d] = \{i \in \kappa : q_i \neq 0\}$.

Definition 4.7. Let $G \in \mathcal{G}(C)$. We say that G is given in normal form if there are elements $d_j \in D$, primes $p_j \in \Pi$ (not necessarily distinct) and integers $n_j \in \mathbb{N}_0$ for $j \in \kappa$ such that

(i) $G = \langle C, d_j : j \in \kappa \rangle$;

(ii) $d_j = \sum_{i \in \kappa} \frac{d_{i,j}}{p_j^{n_j}} e_i$ with $d_{i,j} \in \mathbb{Z}$ for all $i, j \in \kappa$ and

(iii) $h_{p_j}^{R_i}(1) = 0$ for all $i \in [d_j]$ and $j \in \kappa$.

This means that G is in standard form and satisfies (iii).

Here $h_{p_j}^{R_i}(1)$ denotes as usual the p_j-height of 1 in R_i.

Lemma 4.8. Let $G = \langle C, d_j : j \in \kappa \rangle \in \mathcal{G}(C)$ be given in standard form and $\mathcal{B}(G)$ its associated matrix. If either

(i) $\{p_j : j \in \kappa\}$ is finite or

(ii) for every row $\bar{b}$ of $B(G)$ there is a natural number $n \in \mathbb{N}$ such that $n\bar{b} \in \mathbb{Z}^\kappa$ (e.g. $B(G)$ is row finite),

then there is a completely decomposable group $C' = \bigoplus_{i \in \kappa} R'_i e'_i \cong C$ and a torsion-free group $G' \cong G$ such that $G' \in \mathcal{G}(C')$ is given in normal form. Moreover, the isomorphism $G \cong G'$ is induced by the isomorphism $C \cong C'$.

Proof. Let $G = \langle C, d_j : j \in \kappa \rangle \in \mathcal{G}(C)$ be given. Fix $i \in \kappa$. By our assumptions the set $I_i = \{p_j : i \in [d_j]\}$ is finite. If $p_j \in I_i$ and $p_j R_i = R_i$, then we may replace d_j by $d_j - \frac{d_{i,j}}{p_j^{n_j}} e_i$ since $\frac{d_{i,j}}{p_j^{n_j}} e_i \in C$. Hence without loss of generality we may assume that $p_j R_i \neq R_i$ for all $p_j \in I_i$. By the finiteness of I_i there is a rational group $R'_i \cong R_i$ such that $h_{p_j}^{R_i}(1) = 0$ for all $i \in [d_j]$, $j \in \kappa$. The induced isomorphism $\alpha : C \to C' = \bigoplus_{i \in \kappa} R'_i e'_i$ extends to $\alpha' : \mathbb{Q}^{(\kappa)} \to \mathbb{Q}^{(\kappa)}$ which, restricted to G, yields a group $G' \in \mathcal{G}(C')$ having the desired properties. □

Again we take a look at our example G^*.

Example 4.9 (G^*). Let G^* be defined as in Example 4.3, then its associated matrix $B = B(G^*)$ is row finite, hence Lemma 4.8 yields $I_0 = \{p_0\}$, $I_n = \{p_{n-1}, p_n\}$ ($n \geq 1$) and $R'_n = \mathbb{Z}$, $C' = \bigoplus_{n \in \omega} \mathbb{Z} e'_n$. Moreover, $G' = \langle C', \frac{p_{n-1} e'_n + p_{n+1} e'_{n+1}}{p_n} : n \in \omega \rangle$ and the matrix $B' = B(G')$ associated to G' is

$$B' = \begin{pmatrix} \frac{1}{2} & 0 & 0 & 0 & 0 & \cdots \\ \frac{3}{2} & \frac{2}{3} & 0 & 0 & 0 & \cdots \\ 0 & \frac{5}{3} & \frac{3}{5} & 0 & 0 & \cdots \\ 0 & 0 & \frac{7}{5} & \frac{5}{7} & 0 & \cdots \\ 0 & 0 & 0 & \frac{11}{7} & \frac{7}{11} & \cdots \\ \vdots & \vdots & \vdots & \vdots & \vdots & \ddots \end{pmatrix}.$$

5 bcd-groups

In this section we show that every bcd-group is a pure subgroup of some completely decomposable group. We first reduce the problem to quotients of exponent $p \in \Pi$, i.e. p-bcd-groups.

Lemma 5.1. *Assume that every p-bcd group is a pure subgroup of a completely decomposable group for every $p \in \Pi$. Then every bcd-group is a pure subgroup of a completely decomposable group.*

Proof. Let C be a completely decomposable group and $G \in \mathcal{G}(C)$ an n-bcd-group, i.e. $nG \subseteq C$. We induct on $n \in \mathbb{N}$. If $n = 1$, then there is nothing to show. Thus assume $n > 1$ and let $n = mp$ for some $p \in \Pi$ and $m \in \mathbb{N}$. We consider the following short exact sequence
$$0 \to C + pG \to G \to G/(C + pG) \to 0.$$
Since $G \in \mathcal{G}(C)$ it follows that $C + pG \in \mathcal{G}(C)$ is an m-bcd group. By induction hypothesis there exists a completely decomposable group C' and a pure embedding $\alpha : C + pG \to C'$. Let G' be the pushout of G and C'; thus G is a pure subgroup of G'.

$$\begin{array}{ccccc}
0 \to C + pG & \longrightarrow & G \to G/(C + pG) & \to 0 \\
{\scriptstyle *}\downarrow & & {\scriptstyle *}\downarrow \quad \Big\| & & \\
0 \to C' & \longrightarrow & G' \to G/(C + pG) & \to 0 \\
& & {\scriptstyle *}\downarrow & & \\
& & C'' & &
\end{array}$$

Then $G'/C' \cong G/(C + pG)$ is p-bounded and hence G' is a p-bcd group. By assumption G' and hence also G embeds purely into some completely decomposable group C''. □

We now return to matrices. If α, β are ordinals and $R \subseteq \mathbb{Q}$, then let $\mathrm{Mat}_{(\alpha \times \beta)}(R)$ consist of all column finite matrices with α rows and β columns and with entries in R. Note that any $B \in \mathrm{Mat}_{(\alpha \times \beta)}(R)$ defines a homomorphism from $\mathbb{Q}^{(\beta)}$ to $\mathbb{Q}^{(\alpha)}$ in the obvious way.

Definition 5.2. Let $G \in \mathcal{G}(C)$ be a p-bcd group in normal form for some prime $p \in \Pi$ and $B = B(G) \in \mathrm{Mat}_{(\kappa \times \kappa)}(\frac{1}{p}\mathbb{Z})$ its associated matrix. A matrix
$$A = (a_{i,j})_{i \in \kappa + \kappa, j \in \kappa} \in \mathrm{Mat}_{((\kappa + \kappa) \times \kappa)}(\mathbb{Z})$$
is B-good if the following two conditions are satisfied:

(i) $a_{\kappa + \alpha, \alpha} = p$ and $a_{\kappa + \alpha, \beta} = 0$ for all $\alpha \neq \beta \in \kappa$;

(ii) $\mathrm{Im}\,(pB) = \mathrm{Ker}\,(A)$ modulo $(p\mathbb{Z})^{(\kappa)}$ if one regards pB and A as homomorphisms from $\mathbb{Z}^{(\kappa)}$ to $\mathbb{Z}^{(\kappa)}$ respectively to $\mathbb{Z}^{(\kappa + \kappa)}$.

We have an immediate lemma.

Lemma 5.3. *Let $G \in \mathcal{G}(C)$ be a p-bcd group given in normal form for some prime $p \in \Pi$ and $B = B(G) \in \text{Mat}_{(\kappa \times \kappa)}(\frac{1}{p}\mathbb{Z})$ its associated matrix. If $A = (a_{i,j})_{i \in \kappa+\kappa, j \in \kappa} \in \text{Mat}_{((\kappa+\kappa) \times \kappa)}(\mathbb{Z})$ is B-good, then the following holds:*

(i) $AB \in \text{Mat}_{((\kappa+\kappa) \times \kappa)}(\mathbb{Z})$;

(ii) $gcd(a_{i,j} : i \in \kappa + \kappa) = 1$ for all $j \in \kappa$ *(without loss of generality).*

Proof. Follows immediately from the definition of goodness. □

We now prove that p-bcd-groups produce good matrices.

Lemma 5.4. *Let $G \in \mathcal{G}(C)$ be a p-bcd group for some prime $p \in \Pi$ and $B = B(G) \in \text{Mat}_{(\kappa \times \kappa)}(\frac{1}{p}\mathbb{Z})$ its associated matrix. Then there exists a B-good matrix $A \in \text{Mat}_{((\kappa+\kappa) \times \kappa)}(\mathbb{Z})$.*

Proof. The matrix pB can be regarded as a homomorphism from $\mathbb{Z}^{(\kappa)}$ to $\mathbb{Z}^{(\kappa)}$. The induced mapping $p\bar{B}$ modulo $(p\mathbb{Z})^{(\kappa)}$ is therefore an endomorphism of the vector space $(\mathbb{Z}/p\mathbb{Z})^{(\kappa)}$. By linear algebra it follows that there is a column finite matrix $A' \in \text{Mat}_{(\kappa \times \kappa)}(\mathbb{Z})$ such that $\text{Im}(pB) = \text{Ker}(A')$ modulo $(p\mathbb{Z})^{(\kappa)}$. If $A' = (a'_{i,j})_{i,j \in \kappa}$ we put $A = (a_{i,j})_{i \in \kappa+\kappa, j \in \kappa}$ with $a_{i,j} = a'_{i,j}$ for $i, j \in \kappa$ and $a_{\kappa+i,j} = p\delta_{i,j}$ for $i, j \in \kappa$. It is easy to see that A is as required, i.e. B-good. □

We are now in the position to prove our main theorem of this section.

Theorem 5.5 (Main Theorem). *Let $G \in \mathcal{G}(C)$ be a p-bcd group for some prime $p \in \Pi$. Then G is a pure subgroup of a completely decomposable group.*

Proof. Let $G \in \mathcal{G}(C)$ be given with $C = \bigoplus_{j \in \kappa} R_j e_j$. We may assume by Lemma 4.4 that B is given in standard form and by Lemma 4.8 that B is given in normal form. Note that without loss of generality $h_p^{R_j}(1) = 0$ for all $j \in \kappa$. Let $B = B(G)$ be the matrix associated to G and let $A = (a_{i,j})_{i \in \kappa+\kappa, j \in \kappa}$ be B-good. For $i \in \kappa + \kappa$ we put $R'_i = \mathbb{Z} + \sum_{a_{i,j} \neq 0, j \in \kappa} R_j \subseteq \mathbb{Q}$ and note that $R_{\kappa+\alpha} = R_\alpha$ for all $\alpha \in \kappa$. Let $C' = \bigoplus_{i \in \kappa+\kappa} R'_i e'_i$ and define

$$\varphi : C \to C', \quad e_j \mapsto \sum_{i \in \kappa+\kappa} a_{i,j} e'_i.$$

By the choice of the R'_i the mapping φ is well-defined and we have that $\kappa+i \in [\varphi(e_j)]$ if and only if $i = j$. We claim that φ extends to $\varphi : G \to C'$ and is a pure embedding. Let $\tilde{\varphi}$ be the extension of φ to $D = \mathbb{Q} \otimes G = \mathbb{Q}^{(\kappa)}$. We have to prove first that $\tilde{\varphi}(d_k) \in C'$ for all $k \in \kappa$. Fix $k \in \kappa$. Then $pd_k \in \bigoplus_{j \in \kappa} \mathbb{Z} e_j$ and $pd_k = pBe_k \in \text{Im}(pB)$. Thus $\tilde{\varphi}(pd_k) = A(pd_k) = 0$ modulo $(p \bigoplus_{i \in \kappa+\kappa} \mathbb{Z} e'_i)$ since A is B-good. We conclude that $p\tilde{\varphi}(d_k) \in p\bigoplus_{i \in \kappa+\kappa} \mathbb{Z} e'_i$. Thus $\tilde{\varphi}(d_k) \in \bigoplus_{i \in \kappa+\kappa} \mathbb{Z} e'_i \subseteq C'$. Therefore φ extends to $\varphi : G \to C'$.

In order to see that φ is a monomorphism assume that $\varphi(g) = 0$ for some $g \in G$ and represent g as $g = \sum_{j \in \kappa} g_j e_j$ with $g_j \in \mathbb{Q}$. If $i \in [g]$ then $\kappa + i \in [\varphi(g)]$, in fact $\varphi(g) \lceil_{R'_{\kappa+i} e'_{\kappa+i}} = pg_i e'_{\kappa+i}$ and hence $g_i = 0$ follows. Therefore $g = 0$ and φ is a monomorphism.

Finally, we have to show that $\text{Im}(\varphi)$ is pure in C'. Let q be a prime and $g = \sum_{j \in \kappa} r_j e_j \in G$ with $r_j \in \frac{1}{p} R_j$. Assume that

$$\varphi(g) = q \sum_{i \in \kappa + \kappa} r'_i e'_i \in C'$$

for some $r'_i \in R'_i$. We distinguish two cases.

Case 1: $q \neq p$.
We consider the $R'_{\kappa+j} e'_{\kappa+j}$ component of C' and deduce $qr'_{\kappa+j} = pr_j \in R'_{\kappa+j} = R_j$ for all $j \in \kappa$. Since $\gcd(p, q) = 1$ it is easy to see that q divides pg inside C and hence g is divisible by q inside G. Thus $\varphi(\frac{1}{q}g) = \sum_{i \in \kappa + \kappa} r'_i e'_i \in \text{Im}(\varphi)$.

Case 2: $q = p$.
As in Case 1 we obtain that $pr'_{\kappa+j} = pr_j$ for all $j \in \kappa$. Hence $r_j = r'_{\kappa+j} \in R'_{\kappa+j} = R_j$ and therefore $g \in C$. Let h be the least integer such that $hr_j \in \mathbb{Z}$ and $hr'_i \in \mathbb{Z}$ for all $j \in \kappa$ and $i \in \kappa + \kappa$. It follows that $phg' = \varphi(hg) = A(hg)$. Since both elements, $hg' \in \bigoplus_{i \in \kappa + \kappa} \mathbb{Z} e'_i$ and $hg \in \bigoplus_{j \in \kappa} \mathbb{Z} e_j$ we deduce that $A(hg) = 0$ modulo $\bigoplus_{i \in \kappa + \kappa} p\mathbb{Z} e_i$ and hence $hg \in \text{Im}(pB)$ modulo $\bigoplus_{j \in \kappa} p\mathbb{Z} e_j$. Therefore $hg = \sum_{k \in \kappa} n_k p d_k + pc = pd$ for some integers $n_k \in \mathbb{Z}$ ($k \in \kappa$) and $c \in \bigoplus_{j \in \kappa} \mathbb{Z} e_j \subseteq C$, $d \in G$. Now recall that $h_p^{R'_i}(1) = h_p^{R_j}(1) = 0$ for all $i \in \kappa + \kappa, j \in \kappa$. Hence $\gcd(h, p) = 1$ because $r'_i \in R'_i, r_j \in R_j$ for all $i \in \kappa + \kappa, j \in \kappa$. Again using Euclid we obtain that g is divisible by p inside G and thus $g' = \varphi(\frac{1}{p}g) \in \text{Im}(\varphi)$. □

Let us demonstrate the construction used in the proof of Theorem 5.5 in an example.

Example 5.6. Let $G = \langle \bigoplus_{n \in \omega} R_n, \frac{e_n + e_{n+1}}{p} : n \in \omega \rangle$ for some prime p where $R_n = \langle \frac{1}{p_i^n} : p_i \neq p, i \in \omega \rangle$ and $\Pi = \{p_i : i \in \omega\}$ in increasing order. Then G is already in normal form and the associated matrix $B = B(G)$ is

$$B = \begin{pmatrix} \frac{1}{p} & 0 & 0 & 0 & 0 & \cdots \\ \frac{1}{p} & \frac{1}{p} & 0 & 0 & 0 & \cdots \\ 0 & \frac{1}{p} & \frac{1}{p} & 0 & 0 & \cdots \\ 0 & 0 & \frac{1}{p} & \frac{1}{p} & 0 & \cdots \\ 0 & 0 & 0 & \frac{1}{p} & \frac{1}{p} & \cdots \\ \vdots & \vdots & \vdots & \vdots & \vdots & \ddots \end{pmatrix}.$$

Thus
$$pB = \begin{pmatrix} 1 & 0 & 0 & 0 & 0 & \cdots \\ 1 & 1 & 0 & 0 & 0 & \cdots \\ 0 & 1 & 1 & 0 & 0 & \cdots \\ 0 & 0 & 1 & 1 & 0 & \cdots \\ 0 & 0 & 0 & 1 & 1 & \cdots \\ \vdots & \vdots & \vdots & \vdots & \vdots & \ddots \end{pmatrix}$$

and it is easy to see that the following matrix A is B-good,

$$A' = \begin{pmatrix} 1 & -1 & 1 & -1 & 1 & \cdots \\ 0 & 0 & 0 & 0 & 0 & \cdots \\ 0 & 0 & 0 & 0 & 0 & \cdots \\ 0 & 0 & 0 & 0 & 0 & \cdots \\ 0 & 0 & 0 & 0 & 0 & \cdots \\ \vdots & \vdots & \vdots & \vdots & \vdots & \ddots \end{pmatrix}$$

and
$$A = \begin{pmatrix} A' \\ p\,\mathrm{Id.} \end{pmatrix}.$$

Thus we obtain $R'_0 = \sum_{n \in \omega} R_n = \mathbb{Q}_{(p)}$ and $R'_n = \mathbb{Z}$ for $n \geq 1$ as well as $R'_{\omega+n} = R_n$ for all $n \in \omega$. Hence $C' = \mathbb{Q}_{(p)} e'_0 \oplus \bigoplus_{1 \leq n \in \omega} \mathbb{Z} e'_n \oplus \bigoplus_{n \in \omega} R_n e'_{\omega+n}$. The pure embedding $\varphi : G \to C'$ we obtain does the following :

$$e_n \mapsto (-1)^n e'_0 + p e'_{\omega+n}$$

for all $n \in \omega$.

Corollary 5.7. *Every bcd-group is a pure subgroup of a completely decomposable group.*

Proof. Follows from Main Theorem 5.5 and Lemma 5.1. □

Corollary 5.8. *Every pure subgroup of a direct sum of bcd-groups is a pure subgroup of a completely decomposable group.*

Proof. Clear. □

Our Theorem 5.5 (and later on Theorem 6.2) shows that many groups in our class $\mathcal{G}(C)$ can be embedded into completely decomposable groups; in fact, all bcd-groups can be. Moreover, the proof gives an algorithmic way to obtain those embeddings and the corresponding completely decomposable groups. All the calculations can be done by a computer software except for the homological argument in Lemma 5.1 that is used to restrict ourselves to p-bcd-groups. However, we would like to remark that one can easily modify the definitions and proofs in this section to obtain an algorithmic proof for general bcd-groups. In order to do so one has to replace Definition 5.2 by

Definition 5.9. Let $G \in \mathcal{G}(C)$ be an n-bcd group for some natural number n and $B = B(G) \in \text{Mat}_{(\kappa \times \kappa)}(\frac{1}{n}\mathbb{Z})$ its associated matrix. A matrix $A = (a_{i,j})_{i \in \kappa+\kappa, j \in \kappa} \in \text{Mat}_{((\kappa+\kappa) \times \kappa)}(\mathbb{Z})$ is B-good if the following two conditions are satisfied:

(i) $a_{\kappa+\alpha,\alpha} = n$ and $a_{\kappa+\alpha,\beta} = 0$ for all $\alpha \neq \beta \in \kappa$;

(ii) $\text{Im}(nB) = \text{Ker}(A)$ modulo $p\mathbb{Z}^{(\kappa)}$ for all prime divisors p of n if one regards nB and A as homomorphisms from $\mathbb{Z}^{(\kappa)}$ to $\mathbb{Z}^{(\kappa)}$ respectively to $\mathbb{Z}^{(\kappa+\kappa)}$.

In the proof of Lemma 5.4 one obtains matrices A_p satisfying Definition 5.9 (ii) for each prime divisor p of n. Now use the Chinese Remainder Theorem to get the desired matrix A. The main difference is that $h_p^{R_j}(1) = \infty$ for some prime p dividing n may happen. Finally, one uses induction to show that φ in Theorem 5.5 extends and the rest of the proof is straight forward.

Here is our algorithm that works in particular in the case of acd-groups.

Algorithm 5.10. Let $G \in \mathcal{G}(C)$ be a bcd-group. Then do the following.

- Calculate a standard form of G.
- Calculate a normal form of G.
- Compute the matrix $B(G)$ associated to G.
- Compute a B-good matrix A as in Lemma 5.4.
- Construct the completely decomposable group C' and the pure embedding $\varphi : G \to C'$ as in Theorem 5.5.

6 Pure subgroups of direct sums of bcd-groups

By Corollary 5.8 every pure subgroup of a direct sum of bcd-groups can be embedded purely into some completely decomposable group. Therefore we show in this section that a large class of so-called *local bcd-groups* serves as examples of pure subgroups of direct sums of bcd-groups. We start with the definition. As before let $C = \bigoplus_{i \in \kappa} R_i e_i$ be a completely decomposable group of rank κ and let $G \in \mathcal{G}(C)$ be given in standard from, i.e.
$$G = \langle C, d_j : j \in \kappa \rangle$$
where $d_j = \sum_{i \in \kappa} \frac{d_{i,j}}{p_j^{n_j}} e_i$ for some primes $p_j \in \Pi$, natural numbers $n_j \in \mathbb{N}_0$ and integers $d_{i,j} \in \mathbb{Z}$ ($i,j \in \kappa$).

Definition 6.1. Let $G = \langle C, d_j : j \in \kappa \rangle \in \mathcal{G}(C)$ be given in standard form. Then G is called a local bcd-group if for all $i \in \kappa$ we have $\text{lcm}(p_j^{n_j} : i \in [d_j]) < \infty$ and for every $p \in \Pi$ the set $\{n_j : j \in \kappa, p = p_j\}$ is bounded.

For instance the group $G_C^1 = \langle C, \frac{e_i + e_{i+1}}{p_i} : i \in \omega \rangle$ from Example 4.1 is a local bcd-group while the group $G_C^2 = \langle C, \frac{e_1 + e_i}{p_i} : i \in \omega \rangle$ from the same Example 4.1 is not a local bcd-group ($\kappa = \omega$, $\{p_i : i \in \omega\}$ an increasing sequence of primes). If $B = B(G)$ is the matrix associated to a local bcd-group G, then for every row $\bar{b}$ of B there is a natural number $n \in \mathbb{N}$ such that $n\bar{b} \in \mathbb{Z}^\kappa$. However, note that B does not have to be row finite.

By Lemma 4.8 we may assume without loss of generality that any local bcd-group is given in normal form, hence $h_{p_j}^{R_i}(1) = 0$ for all $i \in [d_j]$, $j \in \kappa$.

Theorem 6.2. *Let $G \in \mathcal{G}(C)$ be a local bcd-group. Then G is a pure subgroup of a direct sum of bcd-groups.*

Proof. Let $G = \langle C, d_j : j \in \kappa \rangle \in \mathcal{G}(C)$ be given with $C = \bigoplus_{i \in \kappa} R_i e_i$. Without loss of generality we may assume that G is given in normal form. We will define homomorphisms $\pi_p : G \to D_p$ for some bcd-groups D_p where $p \in \Pi_0 = \Pi \cup \{0\}$ such that in the end the summation $\pi = \bigoplus_{p \in \Pi_0} \pi_p : G \to \bigoplus_{p \in \Pi_0} D_p$ is a well-defined pure embedding. Let us start with $p = 0$ and put $D_0 = \bigoplus_{i \in \kappa} R_i e'_i$. Let $\pi_0 : G \to D_0$ be given by $\pi_0(e_i) = e'_i$ if $i \notin [d_j]$ for all $j \in \kappa$ and $\pi_0(e_i) = 0$ else. Clearly, π_0 is well-defined and D_0 is a bcd-group. Now let $p \in \Pi$. For $i \in \kappa$ let $m_{p,i} \in \mathbb{N}_0$ such that $\mathrm{lcm}\,(p_j^{n_j} : i \in [d_j]) = p^{m_{p,i}} u_{p,i}$ with $\gcd(u_{p,i}, p) = 1$. Define

$$\pi_p : G \to D_p := \langle \bigoplus_{i \in \kappa} R_i e'_i, \mathrm{Im}\,\pi_p \rangle$$

via

$$e_i \mapsto \begin{cases} \frac{1}{p^{m_{p,i}}} \mathrm{lcm}\,(p_j^{n_j} : i \in [d_j]) e'_i & \text{if there exists } j \in \kappa \text{ such that } p = p_j \text{ and } i \in [d_j], \\ 0 & \text{else.} \end{cases}$$

Then π_p is a well-defined homomorphism and it is easy to check that D_p is a p^{l_p}-bcd-group where $l_p = \max\{m_{p,i} : i \in \kappa\}$. Note that l_p exists by Definition 6.1.

Moreover, since for $i \in \kappa$ we have $\mathrm{lcm}\,(p_j^{n_j} : i \in [d_j])$ is finite it follows that $\pi_p(e_i) = 0$ for almost all $p \in \Pi_0$. Hence $\pi = \bigoplus_{p \in \Pi_0} \pi_p : G \to \bigoplus_{p \in \Pi_0} D_p$ is a well-defined homomorphism. We leave it to the reader to verify that π is even a monomorphism.

Finally, it remains to show that π is a pure embedding which is the hardest part of the proof. Therefore assume that

$$s = \sum_{i \in \kappa} r'_i e_i + \sum_{j \in \kappa} z_j d_j \in G$$

with $r'_i \in R_i$ and $z_j \in \mathbb{Z}$ ($i, j \in \kappa$). Furthermore let $q \in \Pi$ and suppose that $t = \sum_{p \in \Pi_0} t_p \in \bigoplus_{p \in \Pi_0} D_p$ such that

$$\pi(s) = qt = q \sum_{p \in \Pi_0} t_p. \tag{$*$}$$

Using the euclidean algorithm it is readily seen that we may write $z_j d_j$ as $z_j d_j = q z'_j d_j + c_j$ for some $z'_j \in \mathbb{Z}$ and $c_j \in C$ if $q \neq p_j$. Combining $\sum_{i \in \kappa} r'_i e_i$ and

$\sum_{j\in\kappa,q\neq p_j} c_j$ we obtain that

$$s = \sum_{i\in\kappa} r_i e_i + \sum_{j\in\kappa,q=p_j} z_j d_j + q \sum_{j\in\kappa,q\neq p_j} z'_j d_j$$

for some $r_i \in R_i$ and $z_j, z'_j \in \mathbb{Z}$. Thus

$$\pi\Big(\sum_{i\in\kappa} r_i e_i + \sum_{j\in\kappa,q=p_j} z_j d_j\Big) = q(t - \pi\Big(\sum_{j\in\kappa,q\neq p_j} z'_j d_j\Big))$$

and we may assume without loss of generality that s is of the form

$$s = \sum_{i\in\kappa} r_i e_i + \sum_{j\in\kappa,q=p_j} z_j d_j. \tag{**}$$

We now consider the ith component of s and distinguish two main cases. Fix $i \in [s]$.

<u>Case I:</u> $q = p_j$ implies $i \notin [d_j]$ for all $j \in \kappa$.
We have to distinguish two subcases:

<u>Case Ia:</u> $i \notin [d_j]$ for all $j \in \kappa$.
In this case it follows from (*) that q divides $\pi_0(s)$ inside D_0. By definition of π_0 we have that $\pi_0(e_i) = e_i$, hence $q|r_i e_i \in R_i e_i$ and therefore $q|r_i \in R_i$.

<u>Case Ib:</u> There exists $p_j \in \Pi$ and $i \in [d_j]$ for some $j \in \kappa$.
In this case we conclude by our main assumption that $p_j \neq q$, while (*) implies that q divides $\pi_{p_j}(s)$ in D_{p_j}. Thus q divides $\frac{1}{m_{p_j,i}} \text{lcm}\,(p_l^{n_l} : i \in [d_l]) r_i e'_i$ inside $\frac{1}{m_{p_j,i}} R_i e'_i$ by the definition of π_{p_j}. This implies that $q|\text{lcm}\,(p_l^{n_l} : i \in [d_l]) r_i e'_i$ inside $R_i e'_i$. Therefore $q|r_i$ inside R_i. Note that q does not divide $\text{lcm}\,(p_l^{n_l} : i \in [d_l])$ by our main assumption.

We have seen that r_i is divisible by q inside R_i whenever Case I is satisfied, i.e. whenever $i \notin [d_j]$ for all $q = p_j$. Thus it remains to consider the following case.

<u>Case II:</u> For all $i \in [s]$ there exists $j \in \kappa$ such that $q = p_j$ and $i \in [d_j]$.
First note that our assumption implies that $h_q^{R_i}(1) = 0$ for all $i \in [s]$ since G was assumed to be in normal form. By (*) we have that q divides $\pi_q(s)$ inside D_q. More precisely we have that

$$\pi_q(s) = qt_q = q\sum_{i\in\kappa} r'_i e'_i + q\pi_q(s') \text{ for some } s' \in G.$$

We may even assume without loss of generality that $s' \in \langle d_j : p_j = q\rangle$ since all other generators e_i and d_j, $p_j \neq q$ of G are mapped into $\bigoplus_{i\in\kappa} R_i e'_i$ by π_q. Now

$$\pi_q(s - qs') = q\sum_{i\in\kappa} r'_i e'_i.$$

We claim that $s - qs' = qc$ for some $c \in C$ which follows from the fact that π_q is injective restricted to $\bigoplus_{i\in[s]\cup[s']} R_i e_i$ by the definition of π_q and the main assumption of Case II. In detail write $s - qs'$ as

$$s - qs' = \sum_{i \in [s] \cup [s']} \frac{t_i}{q^m} e_i$$

for some $t_i \in R_i$ and some $m \in \mathbb{N}$. By the definition of π_q we obtain

$$\pi_q\left(\frac{t_i}{q^m} e_i\right) = \frac{t_i}{q^m} u_{q,i} e'_i = qr'_i e'_i \in qR_i$$

for all $i \in [s] \cup [s']$. Hence $t_i u_{q,i} = q^{m+1} r'_i$ inside R_i. Since $h_q^{R_i}(1) = 0$ and $\gcd(q, u_{q,i}) = 1$ for all $i \in [s] \cup [s']$ we conclude that $q^{m+1} | t_i$ inside R_i for all $i \in [s] \cup [s']$ and thus $s - qs' = \sum_{i \in [s] \cup [s']} \frac{t_i}{q^m} e_i \in q \bigoplus_{i \in [s] \cup [s']} R_i e_i \subseteq qC$. It follows that

$$s - qs' = qc$$

for some $c \in C$ and hence $s = qs' + qc \in qG$ which finishes the proof. □

We discuss an example which is again trivial because it is free but it demonstrates nicely how our algorithm works.

Example 6.3. Let $G = \langle \bigoplus_{i \in \omega} R_i e_i, \frac{e_i + e_{i+1}}{p_i} : i \in \omega \rangle$ where $R_i = \mathbb{Z}$ and $\Pi = \{p_i : i \in \omega\}$ is in increasing order. Then G is given in normal form and a local bcd-group. By construction we get $\pi_0 = 0$ and

$$\pi_{p_i} : G \to \langle \mathbb{Z} e'_i \oplus \mathbb{Z} e'_{i+1}, \frac{p_{i-1} e'_i + p_{i+1} e'_{i+1}}{p_i} \rangle = D_{p_i}$$

via

$$e_i \mapsto p_{i-1} e'_i, \quad e_{i+1} \mapsto p_{i+1} e'_{i+1} \quad \text{and} \quad e_j \mapsto 0 \text{ for } j \neq i, i+1.$$

Thus G is a pure subgroup of a direct sum of acd-groups $\bigoplus_{i \in \omega} D_{p_i}$ using $\pi : G \to \bigoplus_{i \in \omega} D_{p_i}, g \mapsto \sum_{i \in \omega} \pi_{p_i}(g)$.

We finish with a corollary.

Corollary 6.4. *Any local bcd-group is a pure subgroup of some completely decomposable group.*

Proof. Follows immediately from Theorem 6.2 and Corollary 5.8. □

Again we state an algorithm that works for local bcd-groups.

Algorithm 6.5. Let $G \in \mathcal{G}(C)$ be a local bcd-group. Then do the following.

- Compute a normal form of G.
- Construct a direct sum of bcd-groups $H = \bigoplus_{p \in \Pi_0} D_p$ and a pure embedding $\varphi : G \to H$ as in Theorem 6.2.
- Use Algorithm 5.10 to obtain a completely decomposable group C' and a pure embedding of G into C'.

7 An example

The following example shows that Theorem 6.2 cannot be improved canonically. Moreover, it provides a new example for a countable B_0-group that cannot be embedded into any completely decomposable group (see Theorem 3.1).

Lemma 7.1. *Let* $D := \mathbb{Z}e \oplus \bigoplus_{i \in \omega} \mathbb{Z}_{(p_i)} e_i$, *where* $\Pi = \{p_i | i \in \omega\}$ *is in increasing order and* $\mathbb{Z}_{(p_i)} := \langle \frac{1}{p_j^m} | j, m \in \omega, j \neq i \rangle$. *Then* $G := \langle D, \frac{e+e_i}{p_i} : i \in \omega \rangle$ *is a countable B_0-group that is not a pure subgroup of a completely decomposable group.*

Proof. An easy calculation shows that for

$$G_n := \langle \mathbb{Z}e \oplus \bigoplus_{i=0}^{n} \mathbb{Z}_{(p_i)} e_i, \frac{e+e_i}{p_i} : i \in \omega, i \leq n \rangle$$

we have a pure ascending chain $G = \bigcup_{n \in \omega} G_n$ of Butler groups, and thus G is a reduced B_0-group. In order to complete the proof we again have to show that G is not a pure subgroup of a completely decomposable group (see Theorem 3.3):

By way of contradiction assume that $G \subseteq_* C = \bigoplus_{k \in \omega} C_k$ for some rational groups $C_k \subseteq \mathbb{Q}$. Without loss of generality we may assume that C_k is reduced for all $k \in \omega$.

Let $e = \sum_{k \in K} c_k \in \bigoplus_{k \in K} C_k$ with $c_k \neq 0$ for all $k \in K$ be the representation of e in C, where $K \subseteq \omega$ is finite. Then for every $i \in \omega$ there exists some $k(i) \in K$ such that the $C_{k(i)}$-component of e_i in C is nonzero, as $e + e_i$ is divisible by p_i while e is p_i-reduced by purity. In particular, $\text{type}_G(e_i) \leq C_{k(i)}$ holds. Now using a pigeonhole principle there exist $i \neq i' \in \omega$ and some $k \in K$ with $k = k(i) = k(i')$. Hence

$$\mathbb{Q} = \text{type}_G(e_i) \cup \text{type}_G(e_{i'}) \leq C_k,$$

which implies $C_k = \mathbb{Q}$. This contradicts the fact that C_k is reduced. □

As mentioned earlier there are two cases that we cannot handle with our techniques since they violate one of the conditions in the definition of local bcd-groups (Definition 6.2). In fact we do not know if this kind of groups are pure subgroups of completely decomposable groups or not. Therefore we finish with an open question.

Question 7.2. *Are the groups* $G_C^3 := \langle C, \frac{e_1+e_i}{p^i} : i \in \omega \rangle$ *and* $G_C^4 := \langle C, \frac{e_i+e_{i+1}}{p^i} : i \in \omega \rangle$ *from Example 4.1 pure subgroups of completely decomposable groups? For which completely decomposable groups C is $G_C^2 := \langle C, \frac{e_1+e_i}{p_i} : i \in \omega \rangle$ from Example 4.1 a pure subgroup of some completely decomposable group (see Lemma 7.1)?*

References

[1] D. M. Arnold, *Notes on Butler groups and balanced extensions*, Bull. U.M.I. **6** 5-A (1986), pp. 175–184.

[2] D. M. Arnold, *Abelian Groups and Representations of Finite Partially Ordered Sets*, CMS Books in Mathematics, Springer-Verlag, New York (2000).

[3] D. M. Arnold and K. M. Rangaswamy, *A note on countable Butler groups*, submitted.

[4] L. Bican and L. Salce, *Butler groups of infinite rank*, Abelian Group Theory, Lecture Notes in Math. **1006**, Springer-Verlag (1983), pp. 171–189.

[5] E. Blagoveshchenskaya and R. Göbel, *Classification and direct decompositions of some Butler groups of countable rank*, Comm. Algebra **30** (2002), pp. 3403–3427.

[6] E. Blagoveshchenskaya and L. Strüngmann, *Near isomorphism for a class of torsion-free abelian groups*, Comm. Algebra **35** (2007), pp. 1055–1072.

[7] M. C. R. Butler, *A class of torsion-free abelian groups of finite rank*, Proc. London Math. Soc. **15** (1965), pp. 680–698.

[8] L. Fuchs, *Infinite Abelian Groups* - Vol. 1&2, Academic Press, New York (1970,1973).

[9] L. Fuchs, *A survey on Butler groups of infinite rank*, Contemp. Math. **171** (1994), pp. 121–139.

[10] L. Fuchs and M. Magidor, *Butler groups of arbitrary cardinality*, Israel J. Math. **84** (1993), pp. 239–263.

[11] A. Mader, *Almost completely decomposable groups*, Algebra, Logic and Applications **13**, Gordon and Breach Science Publishers (2000).

[12] A. Mader, O. Mutzbauer and K. M. Rangaswamy *A generalization of Butler groups*, Contemp. Math. **171** (1994), pp. 257–275.

[13] A. Mader and L. Strüngmann, *Bounded essential extensions*, J. Algebra **229** (2000), pp. 205–233.

[14] A. Mader and L. Strüngmann, *A class of Butler groups and their endomorphism rings*, to appear in Hokkaido J. Math. (2008).

[15] S. Shelah and L. Strüngmann, *It is consistent with ZFC that B_1 groups are not B_2 groups*, Forum Math. **15** (2003), pp. 507–524.

Author information

Daniel Herden, Fachbereich Mathematik, Universität Duisburg-Essen, Campus Essen, 45117 Essen, Germany. Current address: Einstein Institute of Mathematics, The Hebrew University of Jerusalem, Edmond Safra Campus, Givat Ram, Jerusalem 91904, Israel.
E-mail: daniel.herden@uni-due.de

Lutz Strüngmann, Fachbereich Mathematik, Universität Duisburg-Essen, Campus Essen, 45117 Essen, Germany.
E-mail: lutz.struengmann@uni-due.de

Strong subgroup chains and the Baer–Specker group

O. Kolman

Abstract. Examples are given of non-elementary properties that are preserved under $\mathcal{C}$-filtrations for various classes $\mathcal{C}$ of Abelian groups. The Baer–Specker group $\mathbb{Z}^\omega$ is never the union of a chain $\{A_\alpha : \alpha < \delta\}$ of proper subgroups such that $\mathbb{Z}^\omega/A_\alpha$ is cotorsion-free. cotorsion-free groups form an abstract elementary class (AEC). The Kaplansky invariants of $\mathbb{Z}^\omega/\mathbb{Z}^{(\omega)}$ are used to determine the AECs $^\perp(\mathbb{Z}^\omega/\mathbb{Z}^{(\omega)})$ and $^\perp(B/A)$, where B/A is obtained by factoring the Baer–Specker group B of a ZFC extension by the Baer–Specker group A of the ground model, under various hypotheses, yielding information about its stability spectrum.

Key words. Cotorsion, Baer–Specker group, Kaplansky invariants, abstract elementary class, infinitary logic.

AMS classification. Primary: 03C45, 03C52, 03C75. Secondary: 20K25, 20K20, 03C55.

1 Introduction

Unions of chains of subgroups of the Baer–Specker group $\mathbb{Z}^\omega$, the product of ω many copies of the infinite cyclic group $\mathbb{Z}$, have been studied in recent work of Blass and Irwin [7] and Fuchs and Göbel [15]. In [15], the authors show that if $\{A_\alpha : \alpha < \delta\}$ is a continuous ascending chain of Abelian groups such that $A_0 = 0$ and for all $\alpha < \delta$, $A_{\alpha+1}/A_\alpha$ is slender, then the union $\bigcup_{\alpha<\delta} A_\alpha$ is slender; they deduce that the Baer–Specker group $\mathbb{Z}^\omega$ is not the union of any continuous ascending chain $\{A_\alpha : \alpha < \delta\}$ of proper subgroups such that $\mathbb{Z}^\omega/A_\alpha$ is cotorsion-free, provided that $\delta <$ cov($\mathbf{B}$). Recall that the covering number cov($\mathbf{B}$) is the least cardinal κ such that κ meagre sets cover the real line. Fuchs and Göbel also prove that $\mathbb{Z}^\omega$ is not the union of any countable ascending chain $\{A_n : n < \omega\}$ of slender pure subgroups. Blass and Irwin [7] demonstrate that $\mathbb{Z}^\omega$ is never the union of a chain of proper subgroups each isomorphic to $\mathbb{Z}^\omega$, if the chain has length less than cov($\mathbf{B}$).

Eliminating the hypothesis that $\tau <$ cov($\mathbf{B}$) from the Fuchs–Göbel corollary, we prove that $\mathbb{Z}^\omega$ is never the union of a chain $\{A_\alpha : \alpha < \delta\}$ of proper subgroups such that $\mathbb{Z}^\omega/A_\alpha$ is cotorsion-free. The analogous question concerning the higher Baer–Specker group $\mathbb{Z}^\kappa$ for an uncountable cardinal κ is examined briefly under additional set-theoretic axioms. Using results of Dugas and Göbel [11], we deduce for example that if the axiom of constructibility $V = L$ holds, then $\mathbb{Z}^\kappa$ is not the union of any continuous ascending chain $\{A_\alpha : \alpha < \delta\}$ of subgroups such that for all $\alpha < \delta$, (1) A_α is not isomorphic to $\mathbb{Z}^\kappa$ and (2) $\mathbb{Z}^\kappa/A_\alpha$ is a product.

Throughout the paper, it is tacitly assumed that all groups under consideration are Abelian.

Let $\mathcal{C}$ be a class of groups that is closed under isomorphism. A $\mathcal{C}$-filtration of a group G is a continuous increasing chain $\{A_\alpha : \alpha \leq \delta\}$ of its subgroups such that $A_0 = 0, A_\delta = G$, and for all $\alpha < \delta, A_{\alpha+1}/A_\alpha$ is isomorphic to a member of $\mathcal{C}$. In the terminology of $\mathcal{C}$-filtrations, the Fuchs-Göbel theorem states that if $\mathcal{S}$ is the class of slender groups and $\{A_\alpha : \alpha \leq \delta\}$ is an $\mathcal{S}$-filtration, then A_δ is slender. If a class $\mathcal{C}$ contains 0 and is closed under extensions, then any class that is also closed under $\mathcal{C}$-filtrations satisfies some of the characteristic axioms of an *abstract elementary class* (AEC) ([28] and [18]). In the area of module theory, AECs have been used recently as a unifying framework for classes of modules that do not necessarily possess a first-order axiomatization (see [6] and references therein). Baldwin [2] asks for examples of AECs that are not given syntactically. We show that the class $\mathcal{K}$ of cotorsion-free groups is an AEC under a suitable strong submodel relation. However, $\mathcal{K}$ is never of the form $^\perp\mathcal{C} = \{G : \text{Ext}(G, X) = 0 \text{ for all } X \in \mathcal{C}\}$, for any class $\mathcal{C}$.

The paper concludes with some brief remarks on the algebraic structure of the quotient groups obtained by factoring the Baer–Specker group B of an extension (in the set-theoretic sense) by the Baer–Specker group A of the ground model. If M and N are transitive models of ZFC (Zermelo Fraenkel set theory with the axiom of choice), and N is an extension of M, then $A = (\mathbb{Z}^\omega)^M$, the Baer–Specker group in M, is a subgroup of $B = (\mathbb{Z}^\omega)^N$, the Baer–Specker group in N. The Baer–Specker quotient group, B/A, is a torsion-free cotorsion group (in N) and hence is pure-injective (algebraically compact). The Kaplansky invariants of B/A determine its algebraic structure completely. The class $^\perp(B/A)$ is an example of an AEC whose stability properties can be altered by forcing. We compute $^\perp(B/A)$ under various hypotheses. Applications include $^\perp(\mathbb{Z}^\omega/(\mathbb{Z}^\omega \cap L))$ when there are only countably many constructible reals, and when $V = L$. In the case where N is a forcing extension of M, the algebraic properties of the quotient provide a measure of the distance between the reals in the ground model and the reals in the generic extension. Profound works in this area abound: see [4], [26] and the papers of many other authors in this tradition. We make just the simple remark that if a forcing **P** adds reals, then in $M[G]$, there is no $\mathcal{K}$-filtration of B from A, where $\mathcal{K}$ is the class of cotorsion-free groups in $M[G]$.

The notation is standard and follows [12], [21] and [6].

Let us recall some classical results to which repeated appeals will be made throughout the paper. We shall call a group G *a product* if G is isomorphic to $\mathbb{Z}^\lambda$ for some cardinal λ. A group G is a *countable product* if G is isomorphic to $\mathbb{Z}^X$ for some $X \in \omega \cup \{\omega\}$. The following important results, due to Nunke, Dugas and Göbel, and Łoś, will prove invaluable; they are mostly collected in [12, 13, 14].

Theorem (Nunke [25]). *Every epimorphic image of the Baer–Specker group $\mathbb{Z}^\omega$ is a direct sum of a cotorsion group and a countable product.*

Theorem (Nunke [24]; Dugas, Göbel [11]). *Suppose that κ is not ω-measurable and A is a subgroup of $\mathbb{Z}^\kappa$ such that $\mathbb{Z}^\kappa/A$ is a product. Then A is a direct summand of $\mathbb{Z}^\kappa$ if any of the following conditions holds:*

(1) $\kappa = \omega$;

(2) *every Whitehead group of cardinality at most κ is free;*

(3) $\mathbb{Z}^\kappa/A$ is a countable product.

Note that condition (2) subsumes condition (1) since countable Whitehead groups are free.

Theorem (Łoś; see [14]). *Suppose that κ is an infinite cardinal that is not ω-measurable. No proper direct summand of $\mathbb{Z}^\kappa$ contains $\mathbb{Z}^{(\kappa)}$.*

2 Filtrations by cotorsion-free groups

Definition. Let $\mathcal{C}$ be a class of groups that is closed under isomorphism, and suppose that A is a subgroup of B.

(1) A $\mathcal{C}$-*filtration* of B from A is a continuous increasing chain $\{A_\alpha : \alpha \leq \delta\}$ of subgroups of B such that $A_0 = A, A_\delta = B$, and for all $\alpha < \delta$, $A_{\alpha+1}/A_\alpha$ is isomorphic to a member of $\mathcal{C}$.

(2) B is $\mathcal{C}$-*filtrable* from A if there is a $\mathcal{C}$-filtration of B from A.

Thus, a $\mathcal{C}$-filtration of B is a $\mathcal{C}$-filtration of B from 0. Also if $\{A_\alpha : \alpha \leq \delta\}$ is a $\mathcal{C}$-filtration of B from A, then $\{A_\alpha/A : \alpha \leq \delta\}$ is a $\mathcal{C}$-filtration of B/A.

Definition. A class $\mathcal{C}$ of groups is closed under extensions if $G/A \in \mathcal{C}$ and $A \in \mathcal{C}$ imply $G \in \mathcal{C}$.

Proposition 2.1. *Suppose $\mathcal{C}$ is closed under extensions. The following are equivalent:*

(1) *For all A and B, for every $\mathcal{C}$-filtration $\{A_\alpha : \alpha \leq \delta\}$ of B from A, $A \in \mathcal{C}$ implies $B \in \mathcal{C}$.*

(2) *For all A and B, for every $\mathcal{C}$-filtration $\{A_\alpha : \alpha \leq \delta\}$ of B from A, $A \in \mathcal{C}$ and $\alpha < \delta$ imply $B/A_\alpha \in \mathcal{C}$.*

Proof. For (1) $\Rightarrow$ (2), apply (1) to the $\mathcal{C}$-filtration $\{A_\beta/A_\alpha : \alpha \leq \beta < \delta\}$ of B/A_α; for (2) $\Rightarrow$ (1), use (2) to deduce that $B/A = B/A_0 \in \mathcal{C}$, and then note that $\mathcal{C}$ is closed under extensions. □

Recall that the class of cotorsion-free groups is closed under subgroups, extensions and direct products. In [15], the authors observe and implicitly prove that if $\mathcal{K}$ is the class of cotorsion-free groups and $\{A_\alpha : \alpha \leq \delta\}$ is a $\mathcal{K}$-filtration of B from $A \in K$, then $B \in K$. We therefore just sketch the argument, making explicit the role that closure of the class $\mathcal{K}$ under extensions plays, since this will be of interest in the discussion of abstract elementary classes.

Theorem 2.2. *Let $\mathcal{K}$ be the class of cotorsion-free groups. If there exists a $\mathcal{K}$-filtration $\{A_\alpha : \alpha \leq \delta\}$ of B from $A \in \mathcal{K}$, then $B \in \mathcal{K}$.*

Proof. We prove simultaneously by induction on δ the following two claims:

(1) $B \in \mathcal{K}$, and (2) for all $\alpha < \delta$, $B/A_\alpha \in \mathcal{K}$.

If $\delta = 0$, then the result simply restates the hypothesis that A is cotorsion-free. If $\delta = \gamma + 1$, then B is cotorsion-free, being an extension of A_γ by a cotorsion-free group and B/A_α is cotorsion-free since the quotient $(B/A_\alpha)/(A_\gamma/A_\alpha)$ is cotorsion-free and $\mathcal{K}$ is closed under extensions.

If δ is a limit ordinal, then for all $\alpha < \delta$, A_α is a cotorsion-free pure subgroup of B. It is easy to see that neither B nor B/A_α can contain $\mathbb{Q}$ or $\mathbb{Z}(p)$ for any prime p; if, for a contradiction, B contains J_p for some prime p, note that since $A_{\alpha+1}/A_\alpha$ is cotorsion-free, then as in [15] there exists $\alpha < \beta < \delta$ such that A_β/A_α is not cotorsion-free, contradicting the induction hypothesis for (2) applied to A_β/A_α. Therefore B cannot contain J_p; by Proposition 2.1 B/A_α is also cotorsion-free. □

It will be convenient to have a separate statement of (2) for later.

Corollary 2.3. *Let $\mathcal{K}$ be the class of cotorsion-free groups. Suppose that $\{A_\alpha : \alpha \le \delta\}$ is a $\mathcal{K}$-filtration of B from $A \in \mathcal{K}$. Then for all $\alpha < \delta$, $B/A_\alpha \in K$.*

Theorem 2.4. *The Baer–Specker group $\mathbb{Z}^\omega$ is never the union of a chain $\{A_\alpha : \alpha \le \delta\}$ of proper subgroups such that $\mathbb{Z}^\omega/A_\alpha$ is cotorsion-free.*

Proof. Suppose that $\{A_\alpha : \alpha < \delta\}$ is a chain of proper subgroups such that $\mathbb{Z}^\omega/A_\alpha$ is cotorsion-free. If $\mathrm{cf}(\delta) = \omega$, let $\{\alpha_n : n < \omega\}$ be a cofinal sequence in δ. Then by [15], since $\omega < \mathrm{cov}(\mathbf{B})$, $\mathbb{Z}^\omega \ne \bigcup_{n<\omega} A_{\alpha_n} = \bigcup_{\alpha<\delta} A_\alpha$. If $\mathrm{cf}(\delta) > \omega$, then for each n, there exists α_n such that $e_n \in A_{\alpha_n}$ (recall e_n is the unit vector with 1 at the nth place, 0 elsewhere); so $\alpha^* = \sup\{\alpha_n : n < \omega\} < \delta$, and $\mathbb{Z}^{(\omega)}$ is contained in A_{α^*}. But $\mathbb{Z}^\omega/\mathbb{Z}^{(\omega)}$ is cotorsion and therefore so is its quotient $\mathbb{Z}^\omega/A_{\alpha^*}$, contradicting the hypothesis that $\mathbb{Z}^\omega/A_{\alpha^*}$ is cotorsion-free. □

In the above result, the chain is not assumed continuous.

Corollary 2.5. *Suppose that $\{A_\alpha : \alpha < \delta\}$ is a chain of proper direct summands of $\mathbb{Z}^\omega$. Then $\mathbb{Z}^\omega \ne \bigcup_{\alpha<\delta} A_\alpha$.*

Corollary 2.6. *The Baer–Specker group $\mathbb{Z}^\omega$ is never the union of a chain $\{A_\alpha : \alpha \le \delta\}$ of slender direct summands.*

Note however that $\mathbb{Z}^\omega$ is the union of a continuous ascending chain of length $\mathrm{cf}(2^{\aleph_0})$ of slender subgroups; also, by Kulikov's theorem, $\mathbb{Z}^\omega$ is the union of a ascending chain of countable length of subgroups that are free, hence slender (see [13]).

We consider next whether these corollaries about $\mathbb{Z}^\omega$ have provable analogues in higher cardinalities. This will require information on the quotients of the higher Baer–Specker group $\mathbb{Z}^\kappa$ for an uncountable cardinal κ. Recall the well-known fact that the axiom of constructibility $V = L$ implies that measurable (and hence ω-measurable) cardinals do not exist.

Theorem 2.7 (Dugas, Göbel [11]). *If $V = L$, then for all cardinals κ and all subgroups A of the higher Baer–Specker group $\mathbb{Z}^\kappa$, A is a direct summand of $\mathbb{Z}^\kappa$ if and only if $\mathbb{Z}^\kappa/A$ is a product.*

Theorem 2.8. *Assume that $V = L$ holds. Suppose that $\{A_\alpha : \alpha < \delta\}$ is a continuous ascending chain of subgroups of $\mathbb{Z}^\kappa$ such that for all $\alpha < \tau$ (1) A_α is not isomorphic to $\mathbb{Z}^\kappa$ and (2) $\mathbb{Z}^\kappa/A_\alpha$ is a product. Then $\mathbb{Z}^\kappa \neq \bigcup_{\alpha<\delta} A_\alpha$.*

Proof. Case 1: $\kappa < \mathrm{cf}(\delta)$. Then for some $\alpha^* < \delta$, $\mathbb{Z}^{(\kappa)}$ is contained in A_{α^*}. By the results of Dugas and Göbel and Łoś, it follows that $A_{\alpha^*} = \mathbb{Z}^\kappa$, a contradiction. Case 2: $\mathrm{cf}(\delta) \leq \kappa$. Passing to a continuous cofinal subchain if necessary, we may assume that $\mathrm{cf}(\delta) = \delta$. Since $\delta \leq \kappa < \mathrm{cf}(|\mathbb{Z}^\kappa|)$, for some $\alpha^* < \delta$, A_{α^*} has cardinality 2^κ and is a direct summand of $\mathbb{Z}^\kappa$; referring to [12, Theorem 1.4, pp. 294–295] and remembering that GCH is a consequence of $V = L$, it follows that A_{α^*} must be isomorphic to $\mathbb{Z}^\kappa$, contradicting (1). □

The referee generously provided a direct proof of the last assertion and for convenience it is reproduced here with thanks.

Claim: if κ is not ω-measurable, then any direct summand of $\mathbb{Z}^\kappa$ must be isomorphic to $\mathbb{Z}^\lambda$ for some $\lambda \leq \kappa$. Why? Suppose $\mathbb{Z}^\kappa$ is isomorphic to $A \oplus B$. The non-ω-measurability hypothesis implies that $\mathbb{Z}^\kappa$ is reflexive, and therefore so are the summands A and B. Non-ω-measurability also implies that the dual group $(\mathbb{Z}^\kappa)^*$ is free of rank κ, and therefore its summands A^* and B^* are also free. Let λ be the rank of A^*. Then $\lambda \leq \kappa$ and

$$A \cong A^{**} \cong (\mathbb{Z}^{(\lambda)})^* \cong \mathbb{Z}^\lambda,$$

as claimed.

Since GCH follows from $V = L$, if $\lambda < \kappa$, then $|A_{\alpha^*}| = |\mathbb{Z}^\lambda| = \lambda^+ \leq \kappa < 2^\kappa$. In case (2) of the proof above, $|A_{\alpha^*}| = 2^\kappa$, so we can conclude that $\lambda = \kappa$.

Theorem 2.9 (Dugas, Göbel [11]). *Suppose that κ and λ are cardinals which are not ω-measurable and A is a subgroup of the higher Baer–Specker group $\mathbb{Z}^\kappa$ which is a homomorphic image of $\mathbb{Z}^\lambda$. Then $\mathbb{Z}^\kappa/A$ is the direct sum of a cotorsion group and a product.*

Corollary 2.10. *Assume that $V = L$ holds. Suppose that for all $\alpha < \delta, \lambda_\alpha < \kappa$ and $\{A_\alpha : \alpha < \delta\}$ is a continuous ascending chain of subgroups of $\mathbb{Z}^\kappa$ such that (1) A_α is a homomorphic image of $\mathbb{Z}^{\lambda_\alpha}$ and (2) $\mathbb{Z}^\kappa/A_\alpha$ is cotorsion-free. Then $\mathbb{Z}^\kappa \neq \bigcup_{\alpha<\delta} A_\alpha$.*

Proof. Since GCH holds in L and $\lambda_\alpha < \kappa$, A_α cannot contain an isomorphic copy of $\mathbb{Z}^\kappa$ and by (2) $\mathbb{Z}^\kappa/A_\alpha$ is a product; an appeal to Theorem 2.8 completes the proof. □

3 Abstract elementary classes and cotorsion-freeness

Let us return to the class $\mathcal{K}$ of cotorsion-free groups and the natural question whether there exists a partial order $<_\mathcal{K}$ on $\mathcal{K}$ such that $(\mathcal{K}, <_\mathcal{K})$ is an abstract elementary class.

First, recall the concept of an elementary class. Let $\mathcal{C}$ be a class of structures all of the same similarity type L($\mathcal{C}$). The class $\mathcal{C}$ is an *elementary class* if there exists a first-order theory T in L($\mathcal{C}$) such that $\mathcal{C} = \text{Mod}(T)$, the class of models of T.

For definiteness, the similarity type of groups is $\{+,-,0\}$; for $\mathbb{Z}$-modules (or generally R-modules), add a unary function symbol for each element r of the ring to express scalar multiplication by r. Many natural classes of Abelian groups are non-elementary. For example, since every countable elementary submodel of the group J_p of the p-adic integers is cotorsion-free, it follows that the class $\mathcal{K}$ of cotorsion-free groups is non-elementary. Similarly, by the Łoś Lemma, $\mathbb{Z}$ is elementarily equivalent to the ultrapower $\mathbb{Z}^\omega/\mathcal{U}$ by a non-principal ultrafilter $\mathcal{U}$, and so the class of slender groups is non-elementary (in fact it is not pseudo-elementary).

We shall write $M \subseteq N$ to mean that M is a substructure of N.

Definition. Let $\mathcal{C}$ be a class of structures all of the same similarity type L($\mathcal{C}$). The ordered pair $(\mathcal{C}, <_\mathcal{C})$ is an *abstract elementary class* (AEC) if the following axioms are satisfied:

(A0) Both $\mathcal{C}$ and the binary relation $<_\mathcal{C}$ on $\mathcal{C}$ are closed under isomorphism, i.e.

  (i) if f is an isomorphism of $M \in \mathcal{C}$ onto N, then $N \in \mathcal{C}$, and

  (ii) if $f_i (i = 0, 1)$ is an isomorphism from M_i onto N_i, $M_0 <_\mathcal{C} M_1$, and f_1 extends f_0 (i.e., $f_0 \subseteq f_1$), then $N_0 <_\mathcal{C} N_1$.

(A1) $<_\mathcal{C}$ is a partial order on $\mathcal{C}$.

(A2) If $M <_\mathcal{C} N$, then $M \subseteq N$.

(A3) If $\{A_\alpha : \alpha \leq \delta\}$ is a continuous $<_\mathcal{C}$-increasing chain, then:

  (1) $\bigcup_{\alpha<\delta} A_\alpha \in \mathcal{C}$;

  (2) for each $\beta < \delta$, $A_\beta <_\mathcal{C} \bigcup_{\alpha<\delta} A_\alpha$;

  (3) if for each $\beta < \delta$, $A_\beta <_\mathcal{C} M \in \mathcal{C}$, then $\bigcup_{\alpha<\delta} A_\alpha <_\mathcal{C} M$.

(A4) If $A, B, C \in \mathcal{C}$, $A <_\mathcal{C} C$, $B <_\mathcal{C} C$, and $A \subseteq B$, then $A <_\mathcal{C} B$.

(A5) There is a Löwenheim–Skolem number LS($\mathcal{C}$) such that if $A \subseteq B \in \mathcal{C}$, there exists $A' \in \mathcal{C}$ such that $A \subseteq A'$, $A' <_\mathcal{C} B$, and $|A'| \leq |A| + \text{LS}(\mathcal{C})$.

We say that $\{A_\alpha : \alpha \leq \delta\}$ is a continuous $<_\mathcal{C}$-increasing chain if for all $\alpha < \delta$, $A_\alpha \in \mathcal{C}$, $A_\alpha <_\mathcal{C} A_{\alpha+1}$, and if ξ is a limit ordinal then $A_\xi = \bigcup_{\zeta<\xi} A_\zeta$. If $M <_\mathcal{C} N$, then M is called a *strong submodel* of N.

Numerous examples of AECs are given in [18], [3] and [6]. See [18] for an explanation of AECs in the context of classification theory for non-elementary classes. We recall from [6] the following definition and notation. For a class $\mathcal{C}$ of groups, let $(^\perp \mathcal{C}, <_\mathcal{C})$ be defined as follows: $^\perp \mathcal{C} = \{G : \text{Ext}(G, X) = 0 \text{ for all } X \in \mathcal{C}\}$, and the partial order $<_\mathcal{C}$ is defined by $G <_\mathcal{C} H$ if G is a subgroup of H and G and H/G belong to $^\perp \mathcal{C}$. Baldwin, Eklof and Trlifaj ([6, Theorem 1.20]) prove that $(^\perp \mathcal{C}, <_\mathcal{C})$ is an AEC if and only if every member of $\mathcal{C}$ is cotorsion. We seek a strong subgroup relation $<_\mathcal{K}$

on the class $\mathcal{K}$ of cotorsion-free groups under which $(\mathcal{K}, <_\mathcal{K})$ is an abstract elementary class. Theorem 2.1 and its corollary express the fact that $\mathcal{K}$ satisfies the axioms (A3)(1) and (A3)(2) of an abstract elementary class under the partial order defined as follows: $G < H$ if G is a subgroup of H and G and H/G belong to $\mathcal{K}$. The axioms (A0), (A1), (A2) and (A4) are also evident for $(\mathcal{K}, <_\mathcal{K})$. The first problematic axiom is (A3)(3): if we take $M = \mathbb{Z}^\omega \in \mathcal{K}$, and identify $A_n = \mathbb{Z}^n$ with its inclusion in $\mathbb{Z}^\omega$, then $A_n < A_{n+1}$, $\bigcup_{n<\omega} A_n = \mathbb{Z}^{(\omega)}$, but $\mathbb{Z}^\omega/\mathbb{Z}^{(\omega)}$ is a cotorsion group.

Theorem ([22]). *There exists a sentence ψ in the infinitary language $L_{\lambda,\omega}$ with $\lambda = (2^{\aleph_0})^+$ such that $\mathcal{K} = \mathrm{Mod}(\psi)$.*

Fix a fragment A of $L_{\lambda,\omega}$ containing the sentence ψ and define $G <_\mathcal{K} H$ if and only if G is an L_A-elementary substructure of H.

Theorem 3.1. *The class $(\mathcal{K}, <_\mathcal{K})$ of cotorsion-free groups is an abstract elementary class, where $G <_\mathcal{K} H$ if and only if G is an L_A-elementary substructure of H. The Löwenheim–Skolem number $\mathrm{LS}(\mathcal{K})$ is $|A|$.*

Proof. Immediate from the quoted theorem and well-known model theory of the infinitary language $L_{\lambda,\omega}$. □

Since every Whitehead group is slender and hence cotorsion-free, but the cotorsion-free group $\mathbb{Z}^\omega$ is not Whitehead, it follows that $^\perp\mathbb{Z}$ is a proper subclass of $\mathcal{K}$. However, in answer to a question of the referee, $\mathcal{K}$ cannot be represented as $^\perp\mathcal{C}$ for any class $\mathcal{C}$ of Abelian groups.

Corollary 3.2. *The class $\mathcal{K}$ of cotorsion-free groups is never of the form $^\perp\mathcal{C}$ for any class $\mathcal{C}$ of Abelian groups.*

Proof. The class $\mathcal{K}$ of cotorsion-free groups is closed under subgroups and products, and it contains $\mathbb{Z}$, but not $\mathbb{Z}^\omega/\mathbb{Z}^{<\omega}$. If $\mathcal{K} = {^\perp\mathcal{C}}$, then by [17, Lemma 4.3.17], $\mathbb{Z}^\omega/\mathbb{Z}^{<\omega} \in {^\perp\mathcal{C}}$. □

4 Baer–Specker quotient groups

We wish to exhibit some examples of AECs that arise from the quotient groups obtained by factoring the Baer–Specker group B of an extension (in the set-theoretic sense) by the Baer–Specker group A of the ground model. The Baer–Specker quotient group, B/A, is a torsion-free cotorsion group (in N) and hence is pure-injective. By [6], the class $^\perp(B/A)$ is an AEC in N. We use the Kaplansky invariants of B/A to compute $^\perp(B/A)$ under various hypotheses, yielding information about the stability properties of the class. The following facts will be the cornerstone of the calculations:

Theorem ([19, 1]). (1) *The group $\mathbb{Z}^\omega/\mathbb{Z}^{(\omega)}$ is pure-injective (algebraically compact).*

(2) *The invariants of the group $\mathbb{Z}^\omega/\mathbb{Z}^{(\omega)}$ are $\alpha_{p,n} = 0, \beta_p = 2^{\aleph_0}, \gamma_p = 0, \delta = 2^{\aleph_0}$, and $\mathbb{Z}^\omega/\mathbb{Z}^{(\omega)}$ is isomorphic to $\prod_{p \in P} A_p \oplus \mathbb{Q}^{(\delta)}$ where for each prime p, A_p is the p-adic completion of $J_p^{(\beta_p)}$.*

Theorem 4.1. (1) $^\perp\mathbb{Q}$ is the class of all Abelian groups.

(2) $^\perp(\mathbb{Z}^\omega/\mathbb{Z}^{(\omega)}) = \bigcap_{p\in P} {^\perp J_p}$.

(3) $^\perp(\mathbb{Z}^\kappa/B^{<\infty}) = {^\perp\mathbb{Q}}$, where $B^{<\infty}$ is the subgroup of bounded functions.

Proof. (1) The group $\mathbb{Q}$ is divisible, hence $\mathrm{Ext}(G,\mathbb{Q}) = 0$ for every Abelian group G.
(2) The p-adic completion of $J_p^{(\beta_p)}$ is J_p^ω since $\beta_p = 2^{\aleph_0}$, and $\mathbb{Q}^{(\delta)}$ is isomorphic to $\mathbb{Q}^\omega$ since $\delta = 2^{\aleph_0}$. The result follows on applying the isomorphism $\mathrm{Ext}(G, \prod_{i\in I} C_i) \cong \prod_{i\in I} \mathrm{Ext}(G, C_i)$.
(3) It is a remark of P. Hill (see [16]) that the group $\mathbb{Z}^\kappa/B^{<\infty}$ is divisible; hence it is of form $\mathbb{Q}^{(\lambda)}$ which is divisible. □

Recall from [6] that for a set P of maximal ideals of $\mathbb{Z}$, $\mathcal{K}(P)$ is defined as the class of all Abelian groups that are d-torsion-free for all $d \in P$.

Corollary 4.2. (1) If CH holds, then $^\perp(\mathbb{Z}^\omega/\mathbb{Z}^{(\omega)})$ is stable in $\aleph_1$ (in fact in all $\aleph_n, 0 < n < \omega$), but not stable in $\aleph_\omega$.

(2) If CH fails, then $^\perp(\mathbb{Z}^\omega/\mathbb{Z}^{(\omega)})$ is not stable in $\aleph_1$.

Proof. By [6], the AEC $^\perp(\mathbb{Z}^\omega/\mathbb{Z}^{(\omega)})$ is $(\mathcal{K}(P), <_{\mathcal{K}(P)})$ for some set P of maximal ideals, and the latter is stable in an infinite cardinal λ if and only if $\lambda^{\aleph_0} = \lambda$. □

Suppose that $M = (M, \epsilon^M)$ and $N = (N, \epsilon^N)$ are models of ZF and that N is an extension of M, i.e., every element of M belongs to N and ϵ^N of N agrees on $M \times M$ with ϵ^M. There are thus two Baer–Specker groups, $A = (\mathbb{Z}^\omega)^M$, the Baer–Specker group in M, and $B = (\mathbb{Z}^\omega)^N$, the Baer–Specker group in N. We shall also suppose that M and N are transitive models of ZF. Note that if M is a transitive model of ZF, then $(\mathbb{Z}^\omega)^M = \mathbb{Z}^\omega \cap M$, where $\mathbb{Z}^\omega$ is the real Baer–Specker group, i.e. $(\mathbb{Z}^\omega)^V$. By absoluteness, $(\mathbb{Z}^{(\omega)})^M = (\mathbb{Z}^{(\omega)})^N = \mathbb{Z}^{(\omega)}$, and hence $\mathbb{Z}^{(\omega)}$ is always a subgroup of A. Let us call B/A the *Baer–Specker quotient group*.

To determine the structure of B/A, we shall appeal to some well-known classical results.

Theorem (Chase [9]). *Suppose that H is a countable pure subgroup of $\mathbb{Z}^\omega$ and H is dense in the product topology, where $\mathbb{Z}$ is equipped with the discrete topology. Then there exists an automorphism $\alpha \in \mathrm{Aut}(\mathbb{Z}^\omega)$ such that α maps H onto $\mathbb{Z}^{(\omega)}$.*

Corollary 4.3. *Suppose that H is a countable pure subgroup of $\mathbb{Z}^\omega$ and H is dense in the product topology. Then:*

(1) $\mathbb{Z}^\omega/H$ is isomorphic to $\prod_{p\in P} A_p \oplus \mathbb{Q}^{(\delta)}$ where for each prime p, A_p is the completion of $J_p^{(\beta_p)}$, with $\beta_p = 2^{\aleph_0}$ and $\delta = 2^{\aleph_0}$.

(2) $^\perp(\mathbb{Z}^\omega/H) = \bigcap_{p\in P} {^\perp J_p}$.

Proof. By the theorem of Chase, $\mathbb{Z}^\omega/H$ is isomorphic to the cotorsion group $\mathbb{Z}^\omega/\mathbb{Z}^{(\omega)}$; now apply the result of Balcerzyk and Theorem 4.1. □

Observe that A is a subgroup of B, and hence in N, B/A is a cotorsion group. Since $A = \mathbb{Z}^\omega \cap M$, it follows that A is pure in B and hence B/A is torsion-free, so that B/A is algebraically compact (pure-injective). Kaplansky's classification theorem for algebraically compact groups (see [12] or [13]) applied in N says that in N, B/A has the form $\prod_{p \in P} A_p \oplus \mathbb{Q}^{(\delta)}$, where for each prime p, A_p is the completion of $J_p^{(\beta_p)}$, $0 \leq \beta_p \leq 2^{\aleph_0}$ for each prime p, and $0 \leq \delta \leq 2^{\aleph_0}$.

Proposition 4.4. *Suppose that M is a transitive model of ZF. Then the Baer–Specker quotient group B/A is pure-injective in N. If A is countable, then the cardinal invariants of B/A are $\alpha_{p,n} = 0, \beta_p = 2^{\aleph_0}, \gamma_p = 0, \delta = 2^{\aleph_0}$ and $^\perp(B/A) = \bigcap_{p \in P} {}^\perp J_p$.*

Proof. $A/\mathbb{Z}^{(\omega)}$ is a pure subgroup of the cotorsion group $B/\mathbb{Z}^{(\omega)}$ in N. □

Now consider what happens when N is obtained as a generic extension of M by some forcing **P**.

Corollary 4.5. *Suppose that M is a transitive model of (ZF + CH). Suppose that the forcing **P** collapses $\aleph_1$. Then in $M[G]$, $^\perp(B/A) = \bigcap_{p \in P} {}^\perp J_p$. The stability spectrum of $^\perp(B/A)$ can be calculated from the cardinal arithmetic of $M[G]$.*

If the forcing **P** adds reals, then the Baer–Specker group acquires new elements in the generic extension $M[G]$, and the Baer–Specker group A of the ground model may become slender. If the forcing is iterated, then new reals may appear at the stages of the iteration, giving rise to an ascending chain of subgroups of B in the generic extension. A final observation about B/A shows that if **P** adds reals, then in $M[G]$, B is never $\mathcal{K}$-filtrable from A, where $\mathcal{K}$ is the class of cotorsion-free groups in $M[G]$.

Proposition 4.6. *Suppose that the forcing **P** adds reals. Then in $M[G]$, there is no $\mathcal{K}$-filtration of B from A, where $\mathcal{K}$ is the class of cotorsion-free groups in $M[G]$.*

Proof. Suppose towards a contradiction that in $M[G]$ $\{A_\alpha : \alpha \leq \delta\}$ is a $\mathcal{K}$-filtration of B from A. Then $A \in \mathcal{K}$, and hence by Corollary 2.3, $B/A \in \mathcal{K}$. But B/A is cotorsion, hence 0. □

Let us next suppose that $M = L$ and $N = V$, i.e. M is the universe of constructible sets and N is the real universe. We shall call the group $\mathbb{Z}^\omega \cap L$ the *constructible Baer–Specker group*. Recall that reduced torsion-free groups of cardinality less than continuum are slender (Sąsiada [27]), as are direct sums of slender groups (Fuchs [14]); also cotorsion-free groups are reduced and torsion-free; see [12]. The following proposition records some basic facts about the constructible Baer–Specker group. For more detailed information on L and $0^\sharp$, see [10, 12, 21, 23] or [20]; $\aleph_1^L$ denotes the first uncountable cardinal in L.

Proposition 4.7. (1) *The constructible Baer–Specker group is a pure subgroup of $\mathbb{Z}^\omega$ properly containing $\mathbb{Z}^{(\omega)}$.*

(2) *If $V = L$, then the constructible Baer–Specker group is not slender, and hence not free.*

(3) If $\aleph_1^L = \aleph_1$ and $2^{\aleph_0} > \aleph_1$, then the constructible Baer–Specker group is slender, but not free. If $\aleph_1^L < \aleph_1$, then the constructible Baer–Specker group is free.

(4) If there exists a non-constructible real, then $\mathbb{Z}^\omega \cap L$ is not a direct summand of $\mathbb{Z}^\omega$.

(5) If $0^\sharp$ exists, then the constructible Baer–Specker group is free.

(6) If there exists a measurable cardinal, then the constructible Baer–Specker group is free.

Proof. Most of these are either trivial or evident from well-known results concerning L and $0^\sharp$. For (3), recall that $\mathbb{Z}^\omega \cap L$ is cotorsion-free and, having cardinality $\aleph_1 < 2^{\aleph_0}$, is therefore slender. The subgroup H of $\mathbb{Z}^\omega \cap L$ consisting of elements whose tails are divisible by arbitrarily large powers of a fixed prime p is uncountable and non-free. The second part of (3) is immediate on recalling that for $\alpha = \aleph_1^L$, $\mathbb{Z}^\omega \cap L \subseteq L_\alpha$, so if $\alpha = \aleph_1^L < \aleph_1$, then $\mathbb{Z}^\omega \cap L$ is countable and free, since $\mathbb{Z}^\omega$ is $\aleph_1$-free. The claim (4) is immediate from (1) and the theorem of Łoś quoted in the introduction. Since $\mathbb{Z}^\omega$ is $\aleph_1$-free and the hypotheses of (5) and (6) separately imply that $\mathbb{Z}^\omega \cap L$ is countable (in V), it follows that $\mathbb{Z}^\omega \cap L$ is free. □

The invariants of the quotient $\mathbb{Z}^\omega/(\mathbb{Z}^\omega \cap L)$ are easily seen to be independent of ordinary set theory. If $V = L$, then all the invariants are zero and by a result of Baldwin *et al.* [5] quoted in [6], $\mathcal{K}(\emptyset) = {}^\perp(\mathbb{Z}^\omega/(\mathbb{Z}^\omega \cap L))$ is the class of all Abelian groups and is stable. If $\aleph_1^L < \aleph_1$, then the cardinal invariants of $\mathbb{Z}^\omega/(\mathbb{Z}^\omega \cap L)$ are $\alpha_{p,n} = 0, \beta_p = 2^{\aleph_0}, \gamma_p = 0, \delta = 2^{\aleph_0}$; ${}^\perp(\mathbb{Z}^\omega/(\mathbb{Z}^\omega \cap L))$ is $\bigcap_{p \in P} {}^\perp J_p$, and the stability spectrum depends on the cardinal arithmetic of V.

Acknowledgements. I thank the referee for two detailed reports containing fundamental corrections and invaluable comments that led to a total reworking of this paper and a fuller exploitation of the pure injectivity of $\mathbb{Z}^\omega/\mathbb{Z}^{(\omega)}$.

References

[1] Balcerzyk, S., *On factor groups of some subgroups of a complete direct sum of infinite cyclic groups*, Bull. Acad. Polon. Sci., Sér. Sci. Math. Astron. Phys. 7 (1959), pp. 141–142.

[2] Baldwin, J. T., *Abstract Elementary Classes: Some Answers, More Questions*, preprint, July 7, 2006.

[3] Baldwin, J. T. and Lessmann, O., *Uncountable categoricity of local abstract elementary classes with amalgamation*, Ann. Pure Appl. Logic 143, 1-3, (2006), pp. 29–42.

[4] Bartoszyński, T., *Combinatorial aspects of measure and category*, Fund. Math. 127 (1987), pp. 225–239.

[5] Baldwin, J. T., Calvert, W., Goodrick, J., Villaveces, A. and Walczak-Typke, A., *Abelian groups as AECs*, preprint, 2007.

[6] Baldwin, J. T., Eklof, P. C. and Trlifaj, J., *${}^\perp N$ as an abstract elementary class*, preprint, July 5, 2007.

[7] Blass, A. and Irwin, J., *Baer meets Baire: Applications of category arguments and descriptive set theory to* $\mathbb{Z}^{\aleph_0}$, in: Arnold, D. M. et al. (eds.), Abelian Groups and Modules. Proceedings of the international conference at Colorado Springs, CO, USA, August 7-12, 1995; New York, NY, Marcel Dekker, Lecture Notes Pure Applied Mathematics, Vol. 182, 1996, pp. 193–202.

[8] Chang, C. C. and Keisler, H. J., Model Theory, Amsterdam, North-Holland, 1973; revised third edition, second impression, 1991.

[9] Chase, S. U., *Function topologies on abelian groups*, Illinois J. Math. 7 (1963), pp. 593–608.

[10] Devlin, K. J., Constructibility, Perspectives in Mathematical Logic, Berlin, Heidelberg, New York, and Tokyo, Springer-Verlag, 1984.

[11] Dugas, M. and Göbel, R., *Die Struktur kartesischer Produkte ganzer Zahlen modulo kartesische Produkte ganzer Zahlen*, Math. Z. 168 (1979), pp. 15–21.

[12] Eklof, P. C. and Mekler, A. H., Almost Free Modules: Set-Theoretic Methods, Rev. ed., North-Holland Math. Library, Vol. 65, Amsterdam, 2002.

[13] Fuchs, L., Infinite Abelian Groups, Vol. 1, Academic Press, 1970.

[14] Fuchs, L., Infinite Abelian Groups, Vol. 2, Academic Press, 1973.

[15] Fuchs, L. and Göbel, R., *Unions of slender groups*, Arch. Math. 87, No. 1 (2006), pp. 6–14.

[16] Gerstner, O., *Algebraische Kompaktheit bei Faktorgruppen von Gruppen ganzzahliger Abbildungen*, Manuscr. Math. 11 (1974), pp. 103–109.

[17] Göbel, R. and Trlifaj, J., Approximations and Endomorphism Algebras of Modules, Expositions in Mathematics, 41, de Gruyter, Berlin, New York, 2006.

[18] Grossberg, R., *Classification theory for abstract elementary classes*, In: Logic and Algebra, Yi Zhang(ed.), Contemporary Mathematics, Vol. 302, AMS, 2002, pp. 165–204.

[19] Golema, K. and Hulanicki, A., *The structure of the factor group of the unrestricted sum by the restricted sum of Abelian groups. II*, Fund. Math. 53ă(1963), pp. 177–185.

[20] Jech, T., Set Theory, 2nd. ed., Heidelberg, Springer-Verlag, 1997.

[21] Kunen, K., Set Theory, An Introduction to Independence Proofs, Studies in Logic and the Foundations of Mathematics, Vol. 102, North-Holland, 1993.

[22] Kolman, O. and Shelah, S., *Infinitary axiomatizability of slender and cotorsion-free groups*, Bull. Belg. Math. Soc. 7 (2000), pp. 623–629.

[23] Moschovakis, Y. N., Descriptive Set Theory, Studies in Logic and the Foundations of Mathematics, Vol. 100, North-Holland, 1980.

[24] Nunke, R., *On direct products of infinite cyclic groups*, Proc. Amer. Math. Soc. 13 (1962), pp. 66–71.

[25] Nunke, R., *Slender groups*, Acta Sci. Math. (Szeged) 23 (1963), pp. 67–73.

[26] Rosłanowski, A. and Shelah, S., *Localizations of infinite subsets of* ω, Arch. Math. Logic 35 (1996), pp. 315–339.

[27] Sąsiada, E., *Proof that every countable and reduced torsion-free abelian group is slender*, Bull. Acad. Polon. Sci. Ser. Sci. Math. Astronom. Phys. 7 (1959), pp. 143–144.

[28] Shelah, S., *Classification of non-elementary classes II, Abstract elementary classes*, In: Classification theory (Chicago, IL, 1985), Proceedings of the USA-Israel Conference on Classification Theory, Chicago, December 1985, Baldwin, J. T. (ed.), Lecture Notes in Mathematics, Vol. 1292, Springer, Berlin, 1987, pp. 419–497.

Author information

O. Kolman, Laboratoire de Mathématiques Nicolas Oresme – CNRS, Université de Caen BP 5186, 14032 Caen Cedex, France.
E-mail: `okolman@member.ams.org`

On direct decompositions of Butler $B(2)$-groups

Clorinda De Vivo and Claudia Metelli

Abstract. A $B(2)$-group is a sum of finitely many torsionfree Abelian groups of rank 1, subject to two independent relations. We study here a particular kind of decomposition, that mimics the general case for $B(1)$-groups; we give necessary and sufficient conditions for it in two out of three cases, and a counterexample in the third.

Key words. Abelian group, torsion-free, finite rank, Butler group, $B(1)$-group, $B(2)$-group, type, tent, base change, direct decomposition.

AMS classification. 20K15, 06B99, 06F99.

Introduction

In this paper *group* means *torsionfree Abelian group of finite rank*. A Butler $B(n)$-*group* is a finite sum of rank 1 groups, $G = \langle g_1 \rangle_* + \cdots + \langle g_m \rangle_*$, subject to n ($\leq m$) independent relations; the elements g_i are called *base elements* of G. The definition of G is then two-faced: linear (the relations linking the base elements) and order-theoretical (the types [= isomorphism classes] of the rank one groups $\langle g_i \rangle_*$ [where $*$ stands for pure subgroup]). Since a $B(n-1)$-group can be represented as a $B(n)$-group, the classes $B(n)$ form a chain: $B(0)$ is the class of completely decomposable groups [FII]; $B(1)$-groups have been amply studied (for history, see [A]). $B(2)$-groups that are a direct sum of two $B(1)$-groups have been characterized in [DVM8]; the only other result on $B(2)$ is in [VWW], giving a sufficient condition for indecomposability in a subclass of $B(2)$. General results on $B(n)$-groups can be found in [DVM10], [DVM11]. Throughout, as is usual in this subject, we use as a basic equivalence *quasi-isomorphism* (i.e. isomorphism up to finite index, [FII]) instead of isomorphism; we write "*isomorphic, indecomposable, direct decomposition, ...*" instead of "quasi-isomorphic, strongly indecomposable, quasi-direct decomposition, ...".

In our study of $B(2)$-groups, we use the basic tools for $B(n)$-groups introduced in [DVM11]: in particular, the extension to $B(n)$-groups of the *tent* of G, a lattice-theoretical structure determined by the types of the base elements, that was essentially responsible for the structure of $B(1)$-groups. Of the two main problems on $B(n)$-groups - direct decompositions and base changes - we deal here with the first, and use the results on $B(1)$ as guidelines; the presence of a second relation produces significant changes, adding linear conditions to the order-theoretical ones that dominate the field in $B(1)$; still, in more cases than expected, it is the tent that determines the result.

A decomposable $B(1)$-group has two main properties: its direct summands are $B(1)$-groups, and it always has a decomposition "over *one* base element" (in $B(n)$, a decomposition of our group G into $G = G' \oplus G''$ occurs *over d base elements* if all but d of its base elements reappear whole either in G' or in G''). In [DVM10] we determined the condition for a $B(n)$-group to split over one base element; here, besides giving an operative equivalent of that condition for $B(2)$-groups (Corollary 1.11), we investigate decompositions of a $B(2)$-group over *two* base elements. We show in particular that such a decomposition ensures that the summands are $B(2)$-groups (Theorem 4.2), thus giving a first answer to the problem of whether a decomposable $B(2)$-group will always split into a direct sum of $B(2)$-groups.

In Section 3 (Theorem 3.5) we give necessary and sufficient conditions for G to split over two base elements which are *not adjacent* (that is, they are separated by the relations of G); these conditions turn out to be type-theoretical, determined by the tent of G. In Section 4 we analyse the *adjacent* case, where the decomposition may occur in two different ways: in one (Case A) we give necessary and sufficient conditions for the splitting, which is again determined by the tent (Theorem 4.6); while in Case B the condition is not just type-theoretical: in Example 4.8 we show two $B(2)$-groups (with the same tent but different linear relations), one of which splits over two base elements while the other is indecomposable. Section 5 recalls the main open problems.

Let us add a word about our point of view on this research. We firmly believe that, when an elementary class of objects (e.g. finite sums of subgroups of $\mathbb{Q}$) is first approached, the most elegant results are those that use basic methods: it was our luck that for $B(1)$-groups, using the quasi-isomorphism approximation and our representation by types and primes (i.e. tents), it was possible to obtain satisfactory results using both simple algorithms for the solution of the problems and basic tools for their proofs. In the $B(2)$ setting, statements as the ones of Theorems 3.5 and 4.6, that might look cumbersome, are instead of the desired quality, because they offer simple recipes (both to check decomposability and to describe the summands) that can be easily checked by hand on reasonably small ranks. Once the class shows its structure, it will suggest meaningful generalizations and abstractions.

1 Notation and first results

Lower case Greek letters, with the exception of σ and τ, denote rational numbers. We will keep notation and tools introduced in our previous papers (e.g. [DVM4], [DVM11]) on $B(1)$- and $B(2)$-groups. In particular, $\mathbb{T}(\wedge, \vee)$ denotes the lattice of all types (a *type* is the isomorphism class of a rank 1 group), with the added maximum ∞ for the type of the 0 group; if g is an element of a group G, $t_G(g)$ denotes the type in G of the pure subgroup $\langle g \rangle_*$; typeset$(G) = \{t_G(g) \mid g \in G\}$ is a finite sub-$\wedge$-semilattice of $\mathbb{T}$, hence (having ∞ as a maximum) a lattice. $\mathcal{P}(I)$ denotes the set of parts of I; $\mathbb{P}(I)(\vee, \wedge)$ is the lattice of partitions $\mathcal{A} = \{A_1, \ldots, A_k\}$ of I under the ordering "bigger = coarser"; blocks A_i of partitions are nonempty by definition; their complements $I \setminus A_i$ are called *coblocks*.

Throughout, $I = \{1, \ldots, m\}$; if $E \subseteq I$, set

$$E^{-1} = I \setminus E.$$

For a group $W = \langle w_1 \rangle_* + \cdots + \langle w_m \rangle_*$, and $E \subseteq I$, let
$$w_E = \sum\{w_i \mid i \in E\}, \quad W_E = \langle w_i \mid i \in E \rangle_*.$$
Given the element $w = \gamma_1 w_{C_1} + \cdots + \gamma_h w_{C_h}$ of W, if $\gamma_i \neq \gamma_j$ for $i \neq j$ then $\mathcal{C} = \{C_1, \ldots, C_h\}$ is called *a partition of I into equal-coefficient blocks* for w, or shortly *a partition of w*, with respect to the elements $w_1, \ldots, w_m$; when these elements are fixed, we set $\mathcal{C} = \text{part}_W(w)$.

Definition 1.1 ([DVM11]). If $(t_1, \ldots, t_m)$ is a fixed m-tuple of types and $E \subseteq I$, set
$$\tau(E) = \wedge\{t_i \mid i \in E\},$$
in particular, $\tau(\emptyset) = \infty$; if $\mathcal{E}$ is a set of subsets of I define
$$t(\mathcal{E}) = \vee\{\tau(E^{-1}) \mid E \in \mathcal{E}\};$$
the thus defined map $t : \mathcal{P}(\mathcal{P}(I)) \to \mathbb{T}$ is called *tent* (details in [DVM11]), and $(t_1, \ldots, t_m)$ is its *base*; we will often call tent the base itself. □

In the following, our B(2)-group $G = \langle g_1 \rangle_* + \cdots + \langle g_m \rangle_*$ will be *regular*, i.e. the *base elements* $g_1, \ldots, g_m$ will satisfy the two *basic relations*:
$$g_I = g_1 + \cdots + g_m = 0 \text{ (the diagonal relation)}$$
and
$$g_0 = \alpha_1 g_{A_1} + \cdots + \alpha_k g_{A_k} = 0 \text{ (the second relation)};$$
here $k > 1$, and for $j', j'' \in J = \{1, \ldots, k\}$ we have $\alpha_{j'} \neq \alpha_{j''}$ iff $j' \neq j''$. The partition
$$\mathcal{A} = \{A_1, \ldots, A_k\}$$
of I is called the *basic partition* of G, and its blocks A_j are called *sections*. Base elements indexed in the same section are called *adjacent*; they are not separated by the relations, and often this requires a different proof from the non-adjacent case.

Clearly, an element $g \in G$ can be written in many ways as a linear combination of base elements; we will call each such linear combination a *representative* of g. For representatives of 0 we have

Lemma 1.2. *If $0 = \gamma_1 g_{C_1} + \cdots + \gamma_h g_{C_h}$ with $\gamma_i \neq \gamma_j$ if $i \neq j$ then either $h = 1$ or $\mathcal{C} = \mathcal{A}$.* □

Lemma 1.3. *If $k = 2$, G is a degenerate $B(2)$-group (a direct sum of two $B(1)$-groups, [DVM8]). If the second relation is of the form $\alpha g_i + \beta g_j = 0$, G reduces trivially to a $B(1)$-group.* □

In view of the above result *we will generally assume $k \geq 3, \mathcal{A} \neq \{\{i\}, \{j\}, \{i,j\}^{-1}\}$*.

For all $i \in I$, $t_i = t_G(g_i)$ is a *base type* of G, and $(t_1, \ldots, t_m)$ the *type-base* of G. We adopt the way to compute types introduced in previous papers:

Definition 1.4. For $g = \beta_1 g_1 + \cdots + \beta_m g_m \in G$, call $Z(g) = \{i \in I \mid \beta_i = 0\}$ a *zero-block* of g.

Call $\mathrm{maxfam}_G(g)$ the set of maximal zero-blocks of g in its various representations ($\mathrm{maxfam}_G(g)$ coincides with the partition $\mathrm{part}_G(g)$ if G is $B(1)$).

Following [DVM11, 1.1], for $\mathcal{E}, \mathcal{F} \in \mathcal{P}(\mathcal{P}(I))$, set $\mathcal{E} \leq \mathcal{F}$ ($\mathcal{E}$ is *finer* than $\mathcal{F}$) if for each $E \in \mathcal{E}$ there is an $F \in \mathcal{F}$ such that $E \subseteq F$. □

Lemma 1.5 ([DVM11]). *We have*

$$t_G(g) = t(\mathrm{maxfam}_G(g)) = \vee\{\tau(Z^{-1}) \mid Z \text{ a zero-block of } g\};$$

in particular, the types of G are suprema of infima of base types.

Moreover, just as for part_G *in $B(1)$-groups*, $\mathrm{maxfam}_G(g) \leq \mathrm{maxfam}_G(g')$ *implies* $t_G(g) \leq t_G(g')$. □

As an application, let us compute $t_G(g_E)$ for $\emptyset \neq E \subseteq A_1$. Recall first from [DVM11, 2.10] that

$$\mathrm{maxfam}_G(g_E) = \{E^{-1}, E \cup A_2, \ldots, E \cup A_k\} \tag{0}$$

with $(E \cup A_j)^{-1} = E^{-1} \cap A_j^{-1} = E^{-1} \cap (\cup \{A_{j'} \mid j' \neq j\}) = (E^{-1} \cap A_1) \cup (\cup \{A_{j'} \mid j' \neq 1, j\}) = (A_1 \backslash E) \cup (\cup \{A_{j'} \mid j' \neq 1, j\})$, since $E^{-1} \supseteq A_{j'}$ if $j' \neq 1$. Then

$$\begin{aligned}
t_G(g_E) &= \tau(E) \vee \tau((E \cup A_2)^{-1}) \vee \cdots \vee \tau((E \cup A_k)^{-1}) \\
&= \tau(E) \vee \tau((E^{-1} \cap A_2^{-1})) \vee \cdots \vee \tau((E^{-1} \cap A_k^{-1})) \\
&= \tau(E) \vee [\tau(A_1 \backslash E) \wedge \{\vee\{\tau(\cup A_{j'}) \mid j' \neq 1, j\} \mid j = 2, \ldots, k\}].
\end{aligned}$$

More generally, setting $S = \{1, \ldots, r\}$, we have

Lemma 1.6. *Let $g \in G$ have a representative indexed inside a section, say $g = \gamma_1 g_{E_1} + \cdots + \gamma_r g_{E_r}$ with $\gamma_i \neq \gamma_j$ if $i \neq j$ and $E = \cup \{E_s \mid s \in S\} \subseteq A_1$. Then*

a) *if $\gamma_s \neq 0$ for each $s \in S$,*

$$\mathrm{maxfam}_G(g) = \{E^{-1}, E_s \cup A_j \mid s \in S; j = 2, \ldots, k\};$$

b) *if for an $s' \in S$ - w. l. o. g. , $s' = 1$ - we have $\gamma_1 = 0$, then*

$$\mathrm{maxfam}_G(g) = \{E^{-1} \cup E_1, E_s \cup A_j \mid s \in S \backslash \{1\}; j = 2, \ldots, k\}.$$

Proof. From

$$\begin{aligned}
g &= \gamma_1 g_{E_1} + \cdots + \gamma_r g_{E_r} + \mu_1 g_I + \mu_2 g_0 \\
&= \mu_1 g_{A_1 \backslash E} + \sum \{(\mu_1 + \gamma_s) g_{E_s} \mid s \in S\} + \sum \{(\mu_1 + \mu_2 \alpha_j) g_{A_j} \mid j = 2, \ldots, k\},
\end{aligned}$$

we have:

in case a), the maximal zero-block E^{-1} for $\mu_1 = \mu_2 = 0$; while for $\mu_1 \neq 0$ and $\mu_2 = -\mu_1/\alpha_j$, the maximal zero-blocks $E_s \cup A_j$;

in case b), the maximal zero-block $E^{-1} \cup E_1$ for $\mu_1 = \mu_2 = 0$; the maximal zero-blocks $E_s \cup A_j$ otherwise. □

Following [DVM4] and [DVM11], without loss of generality *we will take the base types to consist of all zeros but a finite number of infinities*; they form a finite table (associated to the tent) where sections are marked; its *columns* are also called *primes*, and we consider each type as a product of its primes, e.g.

A_1	t_1	$=$	∞	∞	0	∞	zeros...	$=$	2	3	$\cdot$	7
A_2	t_2	$=$	∞	0	0	∞	"	$=$	2	$\cdot$	$\cdot$	7
	t_3	$=$	∞	0	0	0	"	$=$	2	$\cdot$	$\cdot$	$\cdot$
A_3	t_4	$=$	∞	∞	∞	0	"	$=$	2	3	5	$\cdot$
	t_5	$=$	∞	∞	0	0	"	$=$	2	3	$\cdot$	$\cdot$

If the type σ is a product of primes among which there is p, we say p *divides* σ ($p \mid \sigma$), or σ *has the prime* p; p *is a prime of* g, or *divides* g, if p divides $t_G(g)$. Each prime p has a zero-block $Z(p)$, with $\mathrm{supp}(p) = Z(p)^{-1}$: e.g., for $p = 5$ resp. 7 above, $Z(5) = \{1, 2, 3, 5\}$, $\mathrm{supp}(5) = \{4\}$, $Z(7) = \{3, 4, 5\}$, $\mathrm{supp}(7) = \{1, 2\}$. For a type σ, the set of all zero-blocks of primes dividing σ is

$$ZB_t(\sigma) = \{Z(p) \mid p \text{ divides } \sigma\}.$$

$ZB_t(\sigma)$ takes the place of what was in $B(1)$-groups $\mathrm{part}_t(\sigma)$ (the partition of I obtained by uniting zero-blocks of primes of σ which intersect properly); we will recover partitions when dealing inside a section.

A prime p with $Z(p) = I \setminus \{i\}$ (like 5 above) is called a *locking prime*; by Lemma 1.5 the base type t_i where p occurs (a *locked type*, e.g. type t_4 above) is bound to belong to every type-base of G.

Lemma 1.7 (Regularity, [DVM11]). *In the tent of a regular $B(2)$-group there are no primes with only one zero in a section, and all other zeros in another section.*

Proof. In [DVM11] this is shown to be a consequence of the initial setting: $t_i = t_G(g_i)$. (For instance, a prime with only one hole cannot occur: if it divides all base elements but one, then by the diagonal relation it divides also the remaining base element.) □

Note that the above tent is not the tent of a regular $B(2)$-goup, due to the prime 7.

Lemma 1.8. *The subset E of I contains the zero-block $Z(p)$ if and only if p divides $\tau(E^{-1})$; p is a prime of g if and only if some zero-block E of g contains $Z(p)$; a type σ divides g if and only if $ZB_t(\sigma) \leq \mathrm{maxfam}_G(g)$.*

Proof. E contains $Z(p)$ if and only if p divides all base types t_i (all base elements g_i) with $i \in E^{-1}$. □

Recall from [DVM4] that in a $B(1)$-group G the type σ divides $t_G(g)$ if and only if $\mathrm{part}_t(\sigma) \Leftarrow \mathrm{part}_G(g)$.

Definition 1.9. Since w. l. o. g. the second relation of G skips one of the sections – say, A_j – we can define, for any subset E of A_j and any $g \in G$, the partition $\underline{\mathrm{part}_{G|E}(g)}$ of E

into equal-coefficient blocks for g: this partition is independent from the representation of g.

Analogously, for a type σ, define $\text{part}_{t|E}(\sigma)$ to be the partition of σ in the tent obtained from t by restricting it to the types t_i with $i \in E$. □

Proposition 1.10. *If $E \subseteq A_j$, $\text{part}_{G|E}(g)$ is the partition of $g + G_{E^{-1}}$ in the $B(1)$-group $G/G_{E^{-1}} = \sum\{\langle g_i + G_{E^{-1}}\rangle_* | \ i \in E\}$ (with the diagonal relation). The tent of this $B(1)$-group consists of the types t_i ($i \in E$), regularized.*

Proof. W.l.o.g. let $E \subseteq A_1$. For $i \in E$ and $h \in G_{E^{-1}}$, consider the representative $g_i + h + \mu_1(g_E + g_{E^{-1}}) + \mu_2 g_0$ of $g_i + G_{E^{-1}}$. The zero-block $I \setminus \{i\}$ (obtained for $h = 0$, $\mu_1 = \mu_2 = 0$) is clearly the maximum zero-block not containing $\{i\}$, and supplies t_i to the supremum constituting the type t'_i of $g_i + G_{E^{-1}}$ in $G' = G/G_{E^{-1}}$; while $\{i\} \cup E^{-1}$ (obtained for $\mu_1 = -1$, $\mu_2 = 0$ and $h = g_{E^{-1}}$) is the maximum zero-block containing $\{i\}$, and supplies $\tau(E \setminus \{i\})$ to t'_i (filling any prime having only one hole placed on t_i, and thus regularizing the tent of G'). □

As an application, recall that, in [DVM10], the necessary and sufficient condition for a $B(n)$-group to split over one base element (say, g_1), specified for $B(2)$-groups, was the existence of a partition $\{\{1\}, E, F\}$ of I such that E (say) is contained in A_1 and $t_1 \leq t_G(g_E)$; then $G = G_E \oplus G_F$. Set $F_1 = F \cap A_1$, $A = A_2 \cup \cdots \cup A_k$. The previous proposition and (0) help making this condition operative:

Corollary 1.11. *The $B(2)$-group G splits over its base element $g_1 \in A_1$ if and only if $p \mid t_1$ (i.e. $1 \in Z(p)$) implies that p divides one of $\tau(E)$, $\tau((E \cup A_2)^{-1}), \ldots,$ $\tau((E \cup A_k)^{-1})$; i.e. that either $Z(p) \subseteq E^{-1}$ or $Z(p) \subseteq E \cup A_j$ for some $j \neq 1$; that is, either p fills E or p fills $F_1 \cup (A \setminus A_j)$ for some $j \neq 1$.*

An easily checked necessary condition is then that $G/G_{A_1^{-1}}$ split over $g_1 + G_{A_1^{-1}}$; that is (see [DVM4]) that $\text{part}_{t|A_1}(t_1) \neq \{\{1\}, A_1^{-1}\}$. Once this condition is satisfied, it will provide all possible choices of E, F_1 to be checked for the remaining part of the sufficient condition. □

Exercise. Let t_1 be the minimum type of G. Then $G = (\oplus\{\langle g_i\rangle_* | \ i \in A_1 \setminus \{1\}\}) \oplus G_A$, a (non regular) $B(1)$-group. If moreover a non-adjacent type t_j is $\leq \tau(A)$, then G is completely decomposable. [Solution: here we may take $E = \emptyset$, $G = 0 \oplus G_{I \setminus \{1\}}$, a non-regular $B(1)$-group with the second relation, which (having set as usual $\alpha_1 = 0$) involves only the base elements indexed in A. G_A becomes regular by turning the second relation (which, there, has only nonzero coefficients) into the diagonal. The last claim is well known for $B(1)$-groups.] □

We prove now a very useful proposition, describing the pure subgroups G_E of a $B(2)$-group G as $B(n)$-groups, when E or E^{-1} is contained in a section. In both cases, the description does not depend on the tent of G; Example 1.13 illustrates the behavior in the general case.

Proposition 1.12. *Let G be a $B(2)$-group, $E \subseteq A_j$ for some $j \in J$. Then*

i) $G_E = \sum\{\langle g_i \rangle_* | \, i \in E\} + \langle g_{E^{-1}} \rangle_*$ *is a $B(1)$-group of rank $|E|$, with the diagonal relation, and*

ii) $G_{E^{-1}} = \sum\{\langle g_i \rangle_* | \, i \in E^{-1}\} + \langle g_E \rangle_*$ *is a $B(2)$-group of rank $|E^{-1}| - 1$, with the same relations as G.*

Proof. W. l. o. g. let $j = 1$, $\alpha_1 = 0$.

i) We only need to prove that, for any

$$g \in H = \sum\{\langle g_i \rangle_* | \, i \in E\} + \langle g_{E^{-1}} \rangle_*,$$

we have $t_G(g) \leq t_H(g)$; then H will be pure, hence will coincide with G_E. Recalling that $g_E = -g_{E^{-1}}$ (from the diagonal relation), set

$$t_0 = t_G(g_E),$$

so the index set of the $B(1)$-group H is $E \cup \{0\}$, and $\text{maxfam}_H(g) = \text{part}_H(g)$; τ_H plays in H the role of τ.

Let $g \in H$, $g = \sum\{\delta_i g_i \, | \, i \in E \cup \{E^{-1}\}\} = \beta_1 g_{E_1} + \cdots + \beta_r g_{E_r} + \beta g_{E^{-1}}$, where $A_1 \supseteq E = E_1 \cup \cdots \cup E_r$ (disjoint union) and $\beta_s \neq \beta_{s'}$ if $s \neq s' \in S = \{1, \ldots, r\}$. We have

- $(E_s \cup A_j)^{-1} = (E \backslash E_s) \cup (A_1 \backslash E) \cup (\cup\{A_{j'} \, | \, j' \neq 1, j\}) \supseteq E \backslash E_s$, for $j \neq 1$;
- $(E^{-1} \cup E_1)^{-1} = E \backslash E_1$; moreover, by (0), for $j \neq 1$, $\quad (\ddagger)$
  $E_s \cup A_j$ is contained in a member of $\text{maxfam}_G(g_E)$, hence
- $\tau((E_s \cup A_j)^{-1}) \leq t_0$.

We have two cases.

a) Let $\beta_s \neq \beta$ for all $s \in S$. Then $\text{maxfam}_H(g) = \text{part}_H(g) = \{\{0\}, E_1, \ldots, E_r\}$; $\tau_H(\{0\}) = t_0$; $\tau_H(E) = \tau(E) \leq t_0$; $\tau(E) \leq \tau(E \backslash E_s)$ for all $s \in S$; thus, computing types in the $B(1)$-group H as suprema of τ_H's of coblocks in $E \cup \{0\}$, we get

$$t_H(g) = \tau_H(E) \vee \tau_H(\{0\} \cup (E \backslash E_1)) \vee \cdots \vee \tau_H(\{0\} \cup (E \backslash E_r))$$
$$= t_0 \wedge [\tau(E \backslash E_1) \vee \cdots \vee \tau(E \backslash E_r)].$$

Computing now the type of g in G_E, that is in the $B(2)$-group G, from Lemma 1.6a) for $\gamma_s = \beta_s - \beta$ we have

$$\text{maxfam}_G(g) = \{E^{-1}, E_s \cup A_j \, | \, s \in S; j = 2, \ldots, k\}.$$

The complements in I of the elements of $\text{maxfam}_G(g)$ yield: $\tau(E)$, which is $\leq t_0$ and $\leq \tau(E \backslash E_s)$ for all $s \in S$; and $\tau((E_s \cup A_j)^{-1})$, which by $(\ddagger)$ is $\leq t_0$, and $\leq \tau(E \backslash E_s)$ for all $s \in S$; therefore $t_G(g) \leq t_H(g)$, as desired.

b) Let $\beta = \beta_{s'}$ for an $s' \in S$, w.l.o.g. $s' = 1$. Then $\text{maxfam}_H(g) = \text{part}_H(g) = \{\{0\} \cup E_1, E_s \mid s \in S \setminus \{1\}\}$, hence

$$\begin{aligned} t_H(g) &= \tau_H(E \setminus E_1) \vee [\vee \{\tau_H(\{0\} \cup (E \setminus E_s)) \mid s \in S \setminus \{1\}\}] \\ &= \tau(E \setminus E_1) \vee (t_0 \wedge [\vee \{\tau(E \setminus E_s) \mid s \in S \setminus \{1\}\}]). \end{aligned}$$

Here $\text{maxfam}_G(g) = \{E^{-1} \cup E_1, E_s \cup A_j \mid s \in S \setminus \{1\}; j = 2, \ldots, k\}$, thus

$$t_G(g) = \tau(E \setminus E_1) \vee (\vee \{\tau((E_s \cup A_j)^{-1}) \mid s \in S \setminus \{1\}; j = 2, \ldots, k\}.$$

Then the above comparisons apply, to yield $t_G(g) \le t_H(g)$.

ii) Again, we need to show that for any

$$g \in H' = \sum \{\langle g_i \rangle_* \mid i \in E^{-1}\} + \langle g_E \rangle_*$$

we have $t_G(g) \le t_{H'}(g)$. Since $E \subseteq A_1$, base elements indexed in E are not separated by the relations; thus a block $B \in \text{maxfam}_{H'}(g)$ either contains $\{0\}$ or is contained in E^{-1}. In the first case, $\tau_H(B^{-1}) = \tau_G(B^{-1})$; in the second, the place of t_0 in $\tau_H(B^{-1})$ is taken by $\tau(E)$ in $\tau_G(B^{-1})$; but $t_0 \ge \tau(E)$; hence $t_H(g) \ge t_G(g)$. □

We show in the next example that if, for $E \subseteq I$, neither E nor E^{-1} is contained in a section, obtaining a linear description of G_E as a $B(n)$-group looks hopeless.

Example 1.13. Let $G = \langle g_1 \rangle_* + \cdots + \langle g_4 \rangle_*$ be the rank 2 $B(2)$-group with relations $g_I = 0 = \alpha_2 g_2 + \alpha_3 g_3 + \alpha_4 g_4$, where the α_i are pairwise distinct and $\ne 0$, hence $\mathcal{A} = \{\{1\}, \{2\}, \{3\}, \{4\}\}$. Set

A_1	t_1	=	p
A_2	t_2	=	·
A_3	t_3	=	·
A_4	t_4	=	·

Let $E = \{3, 4\}$; then $G_E = G$, but since t_1 is locked, the only way to recover G as a sum of rank 1 subgroups is to put $\langle g_1 \rangle_*$ among them; and we do not see any formal way to obtain this from information on E, on $\mathcal{A}$, and on the relations of G. □

2 $B(2)$-groups decomposing over two base elements

Given the $B(2)$-group $G = G' \oplus G''$, decompose accordingly each base element: $g_i = g'_i + g''_i$. Set $E = \{i \in I \mid g''_i = 0\}$, $F = \{i \in I \mid g'_i = 0\}$. Then $G_E \le G'$, $G_F \le G''$; while the set $D = I \setminus (E \cup F)$ indexes the base elements of G not belonging to $G' \cup G''$. Note that if the splitting is trivial – say, $G'' = 0$ – then $I = E$ and $D = \emptyset$.

Definition 2.1. In the above situation, if $|D| = d \ne 0$ we say <u>G splits over d base elements</u>. □

From above we have: $g_i = g'_i$ for $i \in E$; $g_i = g''_i$ for $i \in F$. If $i \in D$, since $g_i = g'_i + g''_i$ implies $t_i = t_G(g'_i) \wedge t_G(g''_i)$, we have $\langle g_i \rangle_* \leq \langle g'_i \rangle_* + \langle g''_i \rangle_*$; then $G = \sum\{\langle g'_i \rangle_* + \langle g''_i \rangle_* |\ i \in D\} + \sum\{\langle g_i \rangle_* |\ i \in E \cup F\}$ hence, setting $I' = E \cup D$, $I'' = F \cup D$,

$$G' = \sum\{\langle g'_i \rangle_* |\ i \in I'\}, \qquad G'' = \sum\{\langle g''_i \rangle_* |\ i \in I''\}. \tag{1}$$

Then $g_I = \sum\{g'_i \mid i \in I'\} + \sum\{g''_i \mid i \in I''\} = 0$ yields $g'_{I'} = \sum\{g'_i \mid i \in I'\} = -g''_{I''} = -\sum\{g''_i \mid i \in I''\} \in G' \cap G'' = 0$: from this we get $g'_{I'} = 0$ and $g''_{I''} = 0$, the diagonal relations on G' resp. G''. Analogously, $g'_0 = 0$ and $g''_0 = 0$ are relations of G' resp. G''.

Set now $|I'| = m'$, $|I''| = m''$; then

$$m = m' + m'' - d.$$

Here rank $G = m - 2$; rank $G' = m' - n'$, where n' is the number of independent relations of G', which is thus a $B(n')$-group; rank $G'' = m'' - n''$, hence G'' is a $B(n'')$-group. Since rank G' + rank G'' = rank G, we must have $m - 2 = m' + m'' - d - 2 = m' - n' + m'' - n''$, that is

$$n' + n'' = d + 2. \tag{2}$$

Lemma 2.2. *If a $B(2)$-group splits over d base elements then it splits into the sum of a $B(n')$- and a $B(n'')$-group satisfying (2).* □

In any case a splitting yields a base-change for $G = G' \oplus G''$ from a $B(2)$-group to a $B(n' + n'')$-group.

Let us evidence the two extreme cases. If $D = I$ (i.e. $E = F = \emptyset$, no base element survives whole), that is $m' = m'' = m$, we have $d = m$, $n' + n'' = m + 2$. If $D = \emptyset$ (i.e. each base element belongs to G' or to G''), hence $n' + n'' = 2$, we have the degenerate case, which is the standard outcome when we perform the external direct sum either of two $B(1)$-groups or of a $B(0)$- and a $B(2)$-group, by just uniting generators and relations. This case contains the trivial splitting (e.g. $G'' = 0$, obtained for $n' = 2$, $d = n'' = 0$). The case $d = 1$ has been treated in general in [DVM10].

Lemma 2.3. *If $k = |\mathcal{A}| \geq 3$, the following are equivalent for $i, j \in I$, $i \neq j$:*

i) $G = G_{\{i,j\}^{-1}}$,

ii) *i and j are not adjacent.*

Moreover, if i) *or* ii) *holds, and $\{i,j\}^{-1} = E \cup F$ (disjoint union), then*

iii) $G_E \cap G_F = 0$.

Proof. i) $\Leftrightarrow$ ii). $G = G_{\{i,j\}^{-1}}$ contains g_i and g_j; this means that the system of the two (independent) basic relations $g_I = 0$, $g_0 = 0$ is solvable with respect to g_i and g_j, but this is possible if and only if they occur with different coefficients in g_0, i.e. i and j belong to different sections.

ii) ⇒ iii). If $0 \neq g \in G_E \cap G_F$ then some multiple rg of g satisfies $rg = \sum\{\beta_i g_i \mid i \in E\} = -\sum\{\beta_j g_j \mid j \in F\}$. Then $g' = \sum\{\beta_s g_s \mid s \in E \cup F\} = 0$; there g_i and g_j have the same coefficient (zero), hence i and j belong to the same block of the partition $\mathcal{C}$ of g'; then, by Lemma 1.2, i and j are adjacent. □

Lemma 2.4. *If the $B(2)$-group G splits nontrivially over two base elements and, in the above notation, one of E, F is empty, then G is the direct sum of two $B(2)$-groups.*

Proof. Say $E = \emptyset$, hence $G_E = 0$. From the above lemma, in the non-adjacent case $G = G_{\{i,j\}^{-1}} = G_F$, and the splitting becomes trivial. If instead g_i and g_j are adjacent, the two relations coincide on G', yielding $g'_i + g'_j = 0$, so $G' = \langle g'_i \rangle_*$. We have $|F| = |I \setminus \{i,j\}| = m - 2 = \text{rank } G$; since F^{-1} is contained in a section, by Proposition 1.12 G_F is a $B(2)$-group of rank $|F| - 1$; thus $G_F = G''$. □

To solve an apparent contradiction, since here $n' = 1$, observe that the group $G_F = G''$, viewed as G'' where $n'' = 3$, is in fact a $B(3)$-group on the base $\{g''_1, g''_2, g_i \mid i \in F\}$, with relations $g''_I = g''_0 = 0$, and the relation $g''_1 \in G_F$; while, viewed as G_F, it is a $B(2)$-group on the base $\{g_{F^{-1}}, g_i \mid i \in F\}$.

3 $B(2)$-groups splitting over two non-adjacent base elements

Lemma 2.4 shows that a split over two base elements behaves differently depending on whether the two base elements are or aren't adjacent. Let $\{\{1, m\}, E, F\}$ be a partition of I, with $1, m$ not adjacent; in this Section we give necessary and sufficient conditions *on the tent of G, to have $G = G_E \oplus G_F$*. Example 3.6 illustrates the situation.

Let $k \geq 3$, and w. l. o. g. $1 \in A_1, m \in A_k$. By Lemma 2.3 we have $G = G_{\{1,m\}^{-1}} = (G_E + G_F)_*$, $G_E \cap G_F = 0$; thus, in order to get $G = G_E \oplus G_F$, we need necessary and sufficient conditions ensuring that g_1 and g_m belong to $G_E + G_F$ *with their whole type*. We start with the necessary conditions.

If $G = G_E \oplus G_F$ we have $g_1 = g'_1 + g''_1$, with $g'_1 \in G_E$, $g''_1 \in G_F$, and analogously $g_m = g'_m + g''_m$. Moreover

$$t_1 = t_G(g_1) \leq t_{G_E}(g'_1) \wedge t_{G_F}(g''_1), \quad t_m = t_G(g_m) = t_{G_E}(g'_m) \wedge t_{G_F}(g''_m)$$

(in both cases $\geq$ always holds).

Set $E_j = E \cap A_j$, $F_j = F \cap A_j$ for all $j \in J$. The two relations $g_I = g_0 = 0$ of G decompose each into two elements of $G_E \cap G_F$:

$$g'_I = g'_1 + \sum\{g_{E_j} \mid j \in J\} + g'_m = -(g''_1 + \sum\{g_{F_j} \mid j \in J\} + g''_m) = -g''_I$$

from the diagonal relation, and

$$\begin{aligned} g'_0 &= \alpha_1 g'_1 + \sum\{\alpha_j g_{E_j} \mid j \in J\} + \alpha_k g'_m \\ &= -(\alpha_1 g''_1 + \sum\{\alpha_j g_{F_j} \mid j \in J\} + \alpha_k g''_m) = -g''_0 \end{aligned}$$

from the second relation. $G_E \cap G_F = 0$ then yields $g'_1 = g'_0 = g''_1 = g''_0 = 0$, providing the diagonal and second relations of G_E resp. G_F.

Set $t_{1'} = t_{G_E}(g'_1)$, $t_{1''} = t_{G_F}(g''_1)$, $t_{m'} = t_{G_E}(g'_m)$, $t_{m''} = t_{G_F}(g''_m)$; the types are the same if computed in G. The index sets of the two summands are $I' = \{1'\} \cup E \cup \{m'\}$ for G_E, resp. $I'' = \{1''\} \cup F \cup \{m''\}$ for G_F.

Let us compute $t_{1'}$. We have, from $g'_0 = 0$,

$$\alpha_1 g'_1 = -\sum \{\alpha_j g_{E_j} \mid j \in J\} - \alpha_k g'_m,$$

and from $g'_I = 0$,

$$\begin{aligned}
\alpha_1 g'_1 &= \alpha_1 g'_1 + \alpha_k g'_I = -\sum \{\alpha_j g_{E_j} \mid j \in J\} - \alpha_k g'_m + \alpha_k g'_I \\
&= \sum \{(-\alpha_j + \alpha_k) g_{E_j} \mid j \in J\} + \alpha_k g'_1,
\end{aligned}$$

from which

$$\begin{aligned}
(\alpha_1 - \alpha_k) g'_1 &= \sum \{(\alpha_k - \alpha_j) g_{E_j} \mid j \in J\} + \mu_0 g_0 + \mu_1 g_I \quad \text{(for arbitrary } \mu_0, \mu_1\text{)} \\
&= (\alpha_k - \alpha_1 + \mu_0 \alpha_1 + \mu_1) g_{E_1} + (\mu_0 \alpha_1 + \mu_1) g_{F_1 \cup \{1\}} \\
&\quad + (\alpha_k - \alpha_2 + \mu_0 \alpha_2 + \mu_1) g_{E_2} + (\mu_0 \alpha_2 + \mu_1) g_{F_2} \\
&\quad + \cdots \\
&\quad + (\alpha_k - \alpha_{k-1} + \mu_0 \alpha_{k-1} + \mu_1) g_{E_{k-1}} + (\mu_0 \alpha_{k-1} + \mu_1) g_{F_{k-1}} \\
&\quad + (\mu_0 \alpha_k + \mu_1) g_{A_k}.
\end{aligned}$$

By equating to zero two coefficients at a time we get all the maximal zero-blocks of $(\alpha_1 - \alpha_k) g'_1$ (i.e. of g'_1):

for $\mu_0 = \mu_1 = 0$, we get $\{1\} \cup F \cup A_k$;

for $\mu_0 = 1$, $\mu_1 = -\alpha_k$, we get $E \cup A_k$;

for $\mu_0 = (\alpha_k - \alpha_j)/(\alpha_1 - \alpha_j)$, $\mu_1 = -\mu_0 \alpha_1$, we get $E_j \cup \{1\} \cup F_1$ ($j = 2, \ldots, k-1$);

for $\mu_0 = (\alpha_k - \alpha_{j'})/(\alpha_j - \alpha_{j'})$, $\mu_1 = -\mu_0 \alpha_j$, we get $E_{j'} \cup F_j$ ($j = 2, \ldots, k-1$; $j' = 1, \ldots, k-1$, $j \neq j'$). Therefore

$$\begin{aligned}
t_{1'} &= \tau((\{1\} \cup F \cup A_k)^{-1}) \vee \tau((E \cup A_k)^{-1}) \\
&\quad \vee [\vee \{\tau((E_j \cup \{1\} \cup F_1)^{-1}) \mid j = 2, \ldots, k-1\}] \\
&\quad \vee [\vee \{\tau((E_{j'} \cup F_j)^{-1}) \mid j = 2, \ldots, k-1, j' = 1, \ldots, k-1, j \neq j'\}].
\end{aligned}$$

Having $t_1 \leq t_{1'}$ means that any prime dividing t_1 must occur among the primes of at least one term of the $\vee$.

- A prime p divides $\tau((\{1\} \cup F \cup A_k)^{-1})$ if and only if $Z(p) \subseteq \{1\} \cup F \cup A_k$ (Lemma 1.8);
- p divides $\tau((E \cup A_k)^{-1})$ if and only if $Z(p) \subseteq E \cup A_k$;
- p divides $\tau((E_j \cup \{1\} \cup F_1)^{-1})$ if and only if $Z(p) \subseteq E_j \cup \{1\} \cup F_1$;
- p divides $\tau((E_{j'} \cup F_j)^{-1})$ if and only if $Z(p) \subseteq E_{j'} \cup F_j$; etc.

Lemma 3.1. $t_1 \leq t_{1'}$ if and only if the primes p dividing t_1 have $Z(p) \subseteq \{1\} \cup F \cup A_k$; or $Z(p) \subseteq E \cup A_k$; or $Z(p) \subseteq E_j \cup \{1\} \cup F_1$ for some $j = 2, \ldots, k-1$; or $Z(p) \subseteq E_{j'} \cup F_j$ for some $j = 2, \ldots, k-1$, $j' = 1, \ldots, k-1$, $j \neq j'$. □

An analogous computation yields for $t_1 \leq t_{1''}$:

$$t_{1''} = \tau((\{1\} \cup E \cup A_k)^{-1}) \vee \tau((F \cup A_k)^{-1})$$
$$\vee [\vee \{\tau((F_j \cup \{1\} \cup E_1)^{-1}) \mid j = 2, \ldots, k-1\}]$$
$$\vee [\vee \{\tau((F_{j'} \cup E_j)^{-1}) \mid j = 2, \ldots, k-1, j' = 1, \ldots, k-1, j \neq j'\}];$$

Lemma 3.2. $t_1 \leq t_{1''}$ if and only if the primes p dividing t_1 have $Z(p) \subseteq \{1\} \cup E \cup A_k$; or $Z(p) \subseteq F \cup A_k$; or $Z(p) \subseteq F_j \cup \{1\} \cup E_1$ for some $j = 2, \ldots, k-1$; or $Z(p) \subseteq F_{j'} \cup E_j$ for some $j = 2, \ldots, k-1$, $j' = 1, \ldots, k-1$. □

Performing all intersections, and enforcing regularity, we get

Proposition 3.3. $t_1 \leq t_{1'} \wedge t_{1''}$ if and only if the primes p dividing t_1 have
$Z(p) \subseteq E \cup A_k$, or
$Z(p) \subseteq F \cup A_k$, or
$Z(p) \subseteq F_{j'} \cup E_j$ for some $j \neq j' \in J \setminus \{k\}$. □

Analogous computations yield:
$t_{m'} = \tau((\{m\} \cup F \cup A_1)^{-1}) \vee \tau((E \cup A_1)^{-1}) \vee [\vee \{\tau((E_j \cup \{m\} \cup F_k)^{-1}) \mid j = 2, \ldots, k-1\}] \vee [\vee \{\tau((E_{j'} \cup F_j)^{-1}) \mid j = 2, \ldots, k-1, j' = 2, \ldots, k\}];$
$t_{m''} = \tau((\{m\} \cup E \cup A_1)^{-1}) \vee \tau((F \cup A_1)^{-1}) \vee [\vee \{\tau((F_j \cup \{m\} \cup E_k)^{-1}) \mid j = 2, \ldots, k-1\}] \vee [\vee \{\tau((F_{j'} \cup E_j)^{-1}) \mid j = 2, \ldots, k-1, j' = 2, \ldots, k\}].$

Proposition 3.4. $t_m \leq t_{m'} \wedge t_{m''}$ if and only if the primes p dividing t_m have
$Z(p) \subseteq E \cup A_1$, or
$Z(p) \subseteq F \cup A_1$, or
$Z(p) \subseteq F_{j'} \cup E_j$ for some $j \neq j' \in J \setminus \{1\}$. □

We can conclude with

Theorem 3.5. Let $\{\{1, m\}, E, F\}$ be a partition of I with $1 \in A_1$, $m \in A_k$; set $E_j = E \cap A_j$, $F_j = F \cap A_j$ for all $j \in J$. Then $G = G_E \oplus G_F$ if and only if the primes dividing t_1 satisfy Proposition 3.3 and the primes dividing t_m satisfy Proposition 3.4; that is,
$$ZB_t(t_1) \leq \{E \cup A_k, F \cup A_k, F_{j'} \cup E_j \mid j \neq j' \in J \setminus \{k\}\}, \quad \text{and}$$
$$ZB_t(t_m) \leq \{E \cup A_1, F \cup A_1, F_{j'} \cup E_j \mid j \neq j' \in J \setminus \{1\}\}.$$

Proof. Necessity has just been proved. For sufficiency, observe that $G = (G_E + G_F)_*$; hence, for a suitable integer $r \neq 0$ we have $rg_1, rg_m \in G_E + G_F$. Replacing each base element g_i with rg_i and redenominating, we end up with $g_1 = g_1' + g_1''$, $g_m = g_m' + g_m''$, to which the above Propositions 3.3 and 3.4 apply, ensuring that $t_G(g_1) \leq t_{G_E}(g_1') \wedge t_{G_F}(g_1'')$, $t_G(g_m) \leq t_{G_E}(g_m') \wedge t_{G_F}(g_m'')$, i.e. that G is (quasi!-) equal to $G_E \oplus G_F$. □

Note that the above characterization is purely order-theoretical.

Here is an example of a $B(2)$-group splitting over two non-adjacent base elements.

Example 3.6. Let $G = \langle g_1 \rangle_* + \cdots + \langle g_7 \rangle_*$ have relations $g_I = 0 = \alpha_1 g_1 + \cdots + \alpha_7 g_7$ with the α_i pairwise distinct (and w.l.o.g. $\alpha_1 = 0$), hence $\mathcal{A} = \{\{1\}, \{2\}, \{3\}, \{4\}, \{5\}, \{6\}, \{7\}\} = \min$; set

A_1	t_1	$=$	p_1	q_1	$\cdot$	$\cdot$	p
A_2	t_2	$=$	p_1	$\cdot$	p_7	$\cdot$	p
A_3	t_3	$=$	p_1	$\cdot$	p_7	$\cdot$	p
A_4	t_4	$=$	$\cdot$	q_1	$\cdot$	q_7	$\cdot$
A_5	t_5	$=$	$\cdot$	q_1	$\cdot$	q_7	$\cdot$
A_6	t_6	$=$	$\cdot$	q_1	$\cdot$	q_7	$\cdot$
A_7	t_7	$=$	$\cdot$	$\cdot$	p_7	q_7	p

For $E = \{2, 3\}$, $F = \{4, 5, 6\}$, we have $G = G' \oplus G''$ for

$$G' = G_{\{2,3\}} = \langle g_1' \rangle_* + \sum \{\langle g_i \rangle_* \mid i \in E\} + \langle g_7' \rangle_*,$$

$$G'' = G_{\{4,5,6\}} = \langle g_1'' \rangle_* + \sum \{\langle g_i \rangle_* \mid i \in F\} + \langle g_7'' \rangle_*;$$

$$g_1' = \alpha_7^{-1}((\alpha_2 - \alpha_7)g_2 + (\alpha_3 - \alpha_7)g_3),$$

$$g_1'' = \alpha_7^{-1}((\alpha_4 - \alpha_7)g_4 + (\alpha_5 - \alpha_7)g_5 + (\alpha_6 - \alpha_7)g_6),$$

$$g_7' = -\alpha_7^{-1}(\alpha_2 g_2 + \alpha_3 g_3), \quad g_7'' = -\alpha_7^{-1}(\alpha_4 g_4 + \alpha_5 g_5 + \alpha_6 g_6).$$

Here p_1, q_1, p satisfy Proposition 3.3; p_7, q_7, p satisfy Proposition 3.4. Note that cutting down on types would violate regularity. □

4 $B(2)$-groups splitting over two adjacent base elements

W.l.o.g. consider the case where G splits over its base elements g_1, g_2, with $1, 2 \in A_1$.

Lemma 4.1. *If $\{\{1, 2\}, E, F\}$ is a partition of I with $\{1, 2\} \cup E$ contained in a section, then $G_E \cap G_F = 0$.*

Proof. Let $g \in G_E \cap G_F$, with $0 \neq g = \sum\{\beta_i g_i \mid i \in E\} = \sum\{\beta_i g_i \mid i \in F\}$. The partition of $0 = g - g$ into equal coefficient blocks differs both from $\{I\}$ and $\mathcal{A}$, against Lemma 1.2. □

We start with necessary conditions. Let $G = G' \oplus G''$, E and F with the usual meaning; $G' \geq G_E$, $G'' \geq G_F$, with $G > (G_E + G_F)_*$; the last containment is proper by Lemma 2.3. Since $\mathrm{rk}(G) = |E| + |F|$, w.l.o.g. we suppose $\mathrm{rk}(G_F) < |F|$, thus the g_i indexed in F are not independent; their relation is not the diagonal relation of G (otherwise $F = I$ and $E = \emptyset$), hence it can be taken to be the second relation $g_0 = 0$ of G; and since there is no room for a third relation, we have $\mathrm{rk}(G_F) = |F| - 1$,

$\text{rk}(G_E) = |E|$. This implies that $F^{-1} = \{1, 2\} \cup E$ is contained in the section A_1, hence, by Proposition 1.12, G_E is a $B(1)$-group of rank $|E|$ and G_F is a $B(2)$-group of rank $|F| - 1$.

For $i = 1, 2$ set $g_i = g'_i + g''_i$. We have two cases:

Case A) $G' = G_E$ and $G'' > G_F$. Here $G' = \langle g'_1 \rangle_* + \langle g'_2 \rangle_* + \sum \{\langle g_i \rangle_* | \ i \in E\}$ is a $B(2)$-group of rank $|E|$, with relations $g'_{I'} = 0$, $g'_1 \in G_E$ (while, in its form as $G_E = \sum \{\langle g_i \rangle_* | \ i \in E\} + \langle g_{E^{-1}} \rangle_*$, it is a $B(1)$-group with the diagonal relation). Thus rank $G'' = |F|$; but by (1) G'' is the sum of $|F| + 2$ rank one groups, hence it is a $B(2)$-group, strictly containing the $B(2)$-group G_F.

Case B) $G' > G_E$ and $G'' = G_F$. Then G'' – in its form as G_F – is a $B(2)$-group of rank $|F| - 1$, hence rank $G' = |E| + 1$; but by (1) G' is the sum of $|E| + 2$ rank 1 groups, hence is a $B(1)$-group.

At this point, adding these considerations to Lemma 2.4 we can in our setting narrow down the condition $n' + n'' = 4$ into

Theorem 4.2. *A $B(2)$-group splitting over two base elements is always the direct sum of two $B(2)$-groups.* □

We will now treat separately cases A and B, showing that in case A the splitting of $G = G' \oplus G''$ over two base elements can be read on the tent, just as in the non-adjacent case; while in case B Example 4.8 will demonstrate that this does not happen in general.

Case A. From $G' \cap G'' = 0$ we have:
$G' = \langle g'_1 \rangle_* + \langle g'_2 \rangle_* + \sum \{\langle g_i \rangle_* | \ i \in E\}$, with relations:
$$g'_{I'} = g'_1 + g'_2 + g_E = 0, \quad g'_1 \in G_E;$$
$G'' = \langle g''_1 \rangle_* + \langle g''_2 \rangle_* + \sum \{\langle g_i \rangle_* | \ i \in F\}$, with relations:
$$g''_{I''} = g''_1 + g''_2 + g_F = 0, \quad g''_0 = g_0 = \sum \{\alpha_j g_{A_j} | \ j \in J\} = 0,$$

where w. l. o. g. $\alpha_1 = 0$.

Lemma 4.3. *In the above setting, we have $G \leq G' + G''$ if and only if $t_i \leq t_G(g'_i)$ for $i = 1, 2$.*

Proof. Since $g''_i = g_i - g'_i$, $t_i \leq t_G(g'_i)$ is equivalent to $t_i \leq t_G(g''_i)$, hence to $\langle g_i \rangle_* \leq \langle g'_i \rangle_* + \langle g''_i \rangle_*$. □

We look now for conditions on the primes of the tent to obtain $t_i \leq t_G(g'_i)$ for $i = 1, 2$ (the other primes can be arbitrary).

To compute $t_G(g'_i)$, set $F_1 = F \cap A_1$, $A = A_2 \cup \cdots \cup A_k$, $g'_1 = -\sum \{\gamma_i g_i | \ i \in E\}$, and let $\mathcal{E} = \text{part}_{G|E}(g'_1) = \{E_s | \ s \in S\}$; then, since $g'_2 = -g'_1 - g_E$, also $\text{part}_{G|E}(g'_2) = \mathcal{E}$. We have

$$\begin{aligned} g'_1 &= -\sum \{\beta_s g_{E_s} | \ s \in S\} + \mu_1 g_I + \mu_2 g_0 \ (\text{with } \beta_s \neq \beta_{s'} \text{ if } s \neq s') \\ &= \mu_1 g_{\{1,2\} \cup F_1} + \sum \{(\mu_1 - \beta_s) g_{E_s} | \ s \in S\} \\ &\quad + \sum \{(\mu_1 + \mu_2 \alpha_j) g_{A_j} | \ j = 2, \ldots, k\}. \end{aligned}$$

Let $s_1 \in S$ be such that $\beta_{s_1} = 0$ (*if there is no such s_1, set $S\setminus\{s_1\} = S$, $E_{s_1} = \emptyset$*); then (computing as before) we have

$$\mathrm{maxfam}_G(g_1') = \{\{1,2\} \cup F_1 \cup E_{s_1} \cup A, E_s \cup A_j \mid s \in S\setminus\{s_1\}; j = 2, \ldots, k\}.$$

For

$$\begin{aligned}
g_2' &= -g_1' - g_E = \sum\{(\beta_s - 1)g_{E_s} \mid s \in S\} + \nu_1 g_I + \nu_2 g_0 \\
&= \nu_1 g_{\{1,2\} \cup F_1} + \sum\{(\beta_s - 1 + \nu_1)g_{E_s} \mid s \in S\} \\
&\quad + \sum\{(\nu_1 + \nu_2 \alpha_j)g_{A_j} \mid j = 2, \ldots, k\},
\end{aligned}$$

let $s_2 \in S$ be such that $\beta_{s_2} = 1$ (*otherwise set $S\setminus\{s_2\} = S$, $E_{s_2} = \emptyset$*); then we have

$$\mathrm{maxfam}_G(g_2') = \{\{1,2\} \cup F_1 \cup E_{s_2} \cup A, E_s \cup A_j \mid s \in S\setminus\{s_2\}; j = 2, \ldots, k\}.$$

The condition $t_1 \leq t_G(g_1')$ is equivalent to requiring, on the tent, that if the prime p divides t_1 (i.e. $1 \in \mathrm{supp}\{p\}$) then p divide $t_G(g_1')$, that is $\mathrm{supp}(p)$ contain the inverse of some block of $\mathrm{maxfam}_G(g_1')$. That is, using the most convenient version,

either $\mathrm{supp}(p) \supseteq E \setminus E_{s_1}$, or $Z(p) \subseteq E_s \cup A_j$ for some $s \in S\setminus\{s_1\}, j = 2, \ldots, k$. (3)

Analogously, $t_2 \leq t_G(g_2')$ requires, for all primes q dividing t_2 (i.e. with 2 in their support):

either $\mathrm{supp}(q) \supseteq E \setminus E_{s_2}$, or $Z(q) \subseteq E_s \cup A_j$ for some $s \in S\setminus\{s_2\}, j = 2, \ldots, k$. (4)

We summarize:

Lemma 4.4. *In the above setting, we have $t_i = t_G(g_i') \wedge t_G(g_i'')$ ($i = 1, 2$) if and only if, for the primes p of the tent, $1 \in \mathrm{supp}\{p\}$ implies (3), $2 \in \mathrm{supp}\{p\}$ implies (4).* □

Observe that if, instead of $\mathrm{part}_{G|E}(g_1')$, we want $\mathrm{part}_{G|A_1}(g_1')$, we only have to replace E_{s_1} with the block $\{1,2\} \cup F_1 \cup E_{s_1}$, while for $\mathrm{part}_{G|A_1}(g_2')$ we replace E_{s_2} with the block $\{1,2\} \cup F_1 \cup E_{s_2}$: now the two partitions may differ. Looking at the tent of the $B(1)$-group $G/G_{A_1^{-1}}$ (see Proposition 1.10) the condition on the primes translates (following [DVM4]) as follows:

Proposition 4.5. *A necessary condition for the $B(2)$-group G to split into $G' \oplus G''$ over g_1 and g_2, with $G' = G_E$ for $\{1,2\} \cup E \subseteq A_1$ and the usual meaning for F, is that there be a partition $\{\{1\}, \{2\}, F_1, E_s \mid s \in S\}$ of A_1, and possibly indices $s_1 \neq s_2 \in S$, such that*

$$\begin{aligned}
\mathrm{part}_{t|A_1}(t_1) &\leq \{\{1\}, \{2\} \cup F_1 \cup E_{s_1}, E_s \mid s \in S\setminus\{s_1\}\}, \\
\mathrm{part}_{t|A_1}(t_2) &\leq \{\{2\}, \{1\} \cup F_1 \cup E_{s_2}, E_s \mid s \in S\setminus\{s_2\}\}.
\end{aligned} \quad (5)$$

□

Note that when the above partitions are different, they always differ on the blocks containing 1 and 2; condition (5) requires that they differ at most on another block.

We are now ready for the complete condition; the operative result is better than Theorem 3.5, in that it requires only a check on all pairs of base-types, rather than on all partitions $\{\{i,j\}, E, F\}$ of I.

Theorem 4.6. *The $B(2)$-group G splits over g_1 and g_2 with $\{1,2\} \cup E \subseteq A_1$ and $G' = G_E$ if and only if the following three conditions are satisfied:*

i) *the partitions of t_1 and t_2 over A_1 differ at most on three blocks, that is*
$$\text{part}_{t|A_1}(t_1) \leq \{\{1\}, \{2\} \cup C, E_2, E_3, \ldots, E_r\},$$
$$\text{part}_{t|A_1}(t_2) \leq \{\{2\}, \{1\} \cup D, E_1, E_3, \ldots, E_r\};$$

ii) *if $p \mid t_1$ (i.e. $1 \notin Z(p)$) then:*
- $Z(p) \cap A_1 \subseteq \{2\} \cup C$ *implies* $Z(p) \subseteq \{2\} \cup C \cup A$;
- $Z(p) \cap A_1 \subseteq E_s$ *for $s \in S \setminus \{1\}$ implies $Z(p) \subseteq E_s \cup A_j$ for some $j \in J \setminus \{1\}$;*

iii) *if $p \mid t_2$ (i.e. $2 \notin Z(p)$) then:*
- $Z(p) \cap A_1 \subseteq \{1\} \cup D$ *implies* $Z(p) \subseteq \{1\} \cup D \cup A$;
- $Z(p) \cap A_1 \subseteq E_s$ *for $s \in S \setminus \{2\}$ implies $Z(p) \subseteq E_s \cup A_j$ for some $j \in J \setminus \{1\}$.*

Proof. Necessity is clear from Lemma 4.4 and Proposition 4.5. For sufficiency, note that, setting $C \cap D = F_1$, we have $C = F_1 \cup E_1$, $D = F_1 \cup E_2$. We show that for $E = E_1 \cup E_2 \cup \cdots \cup E_r$ (disjoint union, by i)) and $F = F_1 \cup A$ we have $G = G_E \oplus G''$, with $G'' > G_F$.

Choose pairwise different coefficients $\gamma_3, \ldots, \gamma_r$ all $\neq 0, 1$, and consider the two elements of G_E:
$$g'_1 = -g_{E_2} - \sum \{\gamma_s g_{E_s} \mid s \in S \setminus \{1,2\}\},$$
$$g'_2 = -g'_1 - g_E = -g_{E_1} + \sum \{(\gamma_s - 1)g_{E_s} \mid s \in S \setminus \{1,2\}\}.$$
By construction,
$$\text{part}_{G|A_1}(g'_1) = \{\{1,2\} \cup F_1 \cup E_1, E_s \mid s \in S \setminus \{1\}\} \geq \text{part}_{t|A_1}(t_1),$$
$$\text{part}_{G|A_1}(g'_2) = \{\{1,2\} \cup F_1 \cup E_2, E_s \mid s \in S \setminus \{2\}\} \geq \text{part}_{t|A_1}(t_2).$$
From Lemma 1.6b),
$$\text{maxfam}_G(g'_1) = \{\{1,2\} \cup F_1 \cup E_1 \cup A, E_s \cup A_j \mid s \in S \setminus \{1\}; j = 2, \ldots, k\},$$
$$\text{maxfam}_G(g'_2) = \{\{1,2\} \cup F_1 \cup E_2 \cup A, E_s \cup A_j \mid s \in S \setminus \{2\}; j = 2, \ldots, k\}.$$
The hypotheses yield $ZB_t(t_1) \leq \text{maxfam}_G(g'_1)$, $ZB_t(t_2) \leq \text{maxfam}_G(g'_2)$, thus $t_G(g'_1) \geq t_1$, $t_G(g'_2) \geq t_2$. Setting for $i = 1, 2$, $g''_i = g_i - g'_i$, we have $t_i \leq t_G(g'_i) \wedge t_G(g''_i)$; defining $G'' = \langle G_F, g''_1 \rangle_*$ (which then contains $g''_2 = g_2 - g'_2 = g_2 + g_E + g'_1 = -g_1 - g_F + g'_1 = -g''_1 - g_F$), we have $G \leq G_E + G''$.

We are left to show that $G_E \cap G'' = 0$. Let $g \in G_E \cap G''$, $g = \sum \{\beta_i g_i \mid i \in E\} = -\sum \{\beta_i g_i \mid i \in F\} - \beta g''_1$. From $g''_1 = g_1 - g'_1$ with $g'_1 \in G_E$ we obtain a relation whose partition equals $\mathcal{A}$ only if $g = 0$. □

Here are two examples of $B(2)$-groups splitting over two adjacent base elements under condition A.

Example 4.7. (1) Let $G = \langle g_1 \rangle_* + \cdots + \langle g_{11} \rangle_*$ have relations $g_I = 0 = \alpha_2 g_{A_2} + \alpha_3 g_{A_3}$ with $\alpha_2 \neq \alpha_3 \neq 0$, $\mathcal{A} = \{A_1, A_2, A_3\} = \{\{1,\ldots,7\},\{8,9\},\{10,11\}\}$; set $E = \{3,4,5,6\}$ with $E_1 = \{3\}$, $E_2 = \{4\}$, $E_3 = \{5,6\}$; $F_1 = \{7\}$, $F = \{7,\ldots,11\}$; and let the tent be

A_1	E_1	t_1	$=$	p_1	$\cdot$	p	q	$\cdot$
		t_2	$=$	$\cdot$	p_2	p	q	$\cdot$
		t_3	$=$	$\cdot$	p_2	p	q	r
	E_2	t_4	$=$	p_1	$\cdot$	p	q	r
	E_3	t_5	$=$	p_1	p_2	$\cdot$	$\cdot$	r
		t_6	$=$	p_1	p_2	$\cdot$	$\cdot$	r
	F_1	t_7	$=$	p_1	$\cdot$	p	q	r
A_2		t_8	$=$	$\cdot$	$\cdot$	$\cdot$	q	$\cdot$
		t_9	$=$	$\cdot$	$\cdot$	$\cdot$	q	$\cdot$
A_3		t_{10}	$=$	$\cdot$	$\cdot$	p	$\cdot$	$\cdot$
		t_{11}	$=$	$\cdot$	$\cdot$	p	$\cdot$	$\cdot$

Then if $g_1' = -g_4 + 2g_{\{5,6\}}$ (and $g_1'' = -g_1' + g_1, \ldots$) we get $G = G_E \oplus G''$, for $G'' = \langle g_1'', G_F \rangle_*$; here both G_E and G'' are further decomposable. □

(2) Let $G = \langle g_1 \rangle_* + \cdots + \langle g_8 \rangle_*$ have relations $g_I = 0 = \alpha_2 g_{A_2} + \alpha_3 g_{A_3}$ with $\alpha_2 \neq \alpha_3 \neq 0$, $\mathcal{A} = \{A_1, A_2, A_3\} = \{\{1,\ldots,4\},\{5,6\},\{7,8\}\}$; set $E = \{3,4\}$, $F = \{5,\ldots,8\}$, with tent

A_1		t_1	$=$	q	p_1	$\cdot$	$\cdot$	$\cdot$	$\cdot$	$\cdot$	$\cdot$
		t_2	$=$	q	$\cdot$	p_2	$\cdot$	$\cdot$	$\cdot$	$\cdot$	$\cdot$
	E	t_3	$=$	$\cdot$	p_1	p_2	s_3	$\cdot$	$\cdot$	$\cdot$	$\cdot$
		t_4	$=$	$\cdot$	p_1	p_2	$\cdot$	s_4	$\cdot$	$\cdot$	$\cdot$
A_2		t_5	$=$	q	$\cdot$	$\cdot$	$\cdot$	$\cdot$	s_5	$\cdot$	$\cdot$
		t_6	$=$	q	$\cdot$	$\cdot$	$\cdot$	$\cdot$	$\cdot$	s_6	$\cdot$
A_3		t_7	$=$	q	$\cdot$	$\cdot$	$\cdot$	$\cdot$	$\cdot$	$\cdot$	s_7	$\cdot$
		t_8	$=$	q	$\cdot$	$\cdot$	$\cdot$	$\cdot$	$\cdot$	$\cdot$	$\cdot$	s_8

Here $\text{part}_{t|A_1}(t_1) = \{\{1\},\{2\},E\} = \text{part}_{t|A_1}(t_2)$. Setting $g_1' = g_{\{3,4\}}, g_2' = -2g_{\{3,4\}}$, we have $t_{\{3,4\}} = qp_1p_2$, thus the type-base $(t_3, t_4, t_{\{3,4\}})$ of G_E is locked by q, s_3 and s_4; G'' has type-base $(t_1'' = t_1, t_2'' = t_2, t_5, \ldots, t_8)$, locked by $p_1, p_2, s_5, \ldots, s_8$; hence both summands are indecomposable. □

Case B. We give an example of a $B(2)$-group G (with $E = \emptyset$, hence $G' (> G_E)$ of rank 1 and $G'' = G_F$) whose splitting over two base elements does not depend on the tent alone. In fact, a change in the second linear relation will yield two groups with the same tent, one of which splits over two base elements, while the other is indecomposable.

Example 4.8. Consider the rank 4 group $G = \langle g_1 \rangle_* + \cdots + \langle g_6 \rangle_*$ with relations $g_I = 0 = g_0$ and $g_0 = \alpha_3 g_3 + \cdots + \alpha_6 g_6$, with the α_i pairwise distinct and $\neq 0$, thus $\mathcal{A} = \{\{1,2\},\{3\},\{4\},\{5\},\{6\}\}$. Let the tent of G be

A_1	t_1	=	p_1	q_1	·	·	·	·	·
	t_2	=	·	·	p_2	q_2	·	·	·
A_2	t_3	=	·	q_1	·	q_2	s_3	·	·
A_3	t_4	=	·	q_1	p_2	·	·	s_4	·
A_4	t_5	=	p_1	·	·	q_2	·	·	s_5
A_5	t_6	=	p_1	·	p_2	·	·	·	s_6

Observe first that, since the types t_3, t_4, t_5, t_6 are locked by the primes s_i, their respective base elements cannot decompose properly in any decomposition of G; moreover, being related by the second relation, they must belong to the same summand, which then must have rank ≥ 3; therefore if G splits nontrivially, it does so over g_1 and g_2. In this case then $G = \langle h \rangle_* \oplus G_{\{3,4,5,6\}}$, with w.l.o.g. $h = g_1 + g'$ for $g' = \sum \{\beta_i g_i \mid i = 3, \ldots, 6\}$; since t_1 and t_2 are incomparable, h must be of type $t_1 \vee t_2$, hence must be divisible by p_1, p_2, q_1, q_2; that is, h must have representatives with zero-blocks containing $\{2,3,4\}$ resp. $\{2,5,6\}$ resp. $\{1,3,5\}$ resp. $\{1,4,6\}$.

Let now the second relation be $g_0 = 3g_3 - 3g_4 + 6g_5 - 2g_6 = 0$. If we choose $g' = -g_5 + 1/3 g_6$, we have

$$\begin{aligned} h &= g_1 + g' + \mu_1 g_I + \mu_2 g_0 \\ &= (1+\mu_1)g_1 + \mu_1 g_2 + (\mu_1 + 3\mu_2)g_3 + (\mu_1 - 3\mu_2)g_4 \\ &\quad + (-1 + \mu_1 + 6\mu_2)g_5 + (1/3 + \mu_1 - 2\mu_2)g_6, \end{aligned}$$

whose representative for $\mu_1 = 0$ and $\mu_2 = 0$ has the zero-block $\{2,3,4\}$; for $\mu_1 = 0$ and $\mu_2 = -1/6$, $\{2,5,6\}$; for $\mu_1 = -1$ and $\mu_2 = 1/3$, $\{1,3,5\}$; for $\mu_1 = -1$ and $\mu_2 = -1/3$, $\{1,4,6\}$.

If instead the second relation is $g_0 = g_3 - g_4 + 2g_5 - 2g_6 = 0$, it is not difficult to see that for no choice of the coefficients β_i of g' the element $h = g_1 + g' + \mu_1 g_I + \mu_2 g_0$ will have the assigned zero-blocks, hence the necessary type to ensure the splitting. □

5 Problems

Here are some open problems on a decomposable $B(2)$-group G; 2. and 3. arise naturally from the comparison with the situation of $B(1)$-groups. Note that since the results of Sections 3 and 4 depend on the representation of G, the solutions to 2. and 3. might involve base changes.

1. Give necessary and sufficient (linear and order-theoretical) conditions for the splitting in Case B.
2. Is a decomposable $B(2)$-group always a proper direct sum of $B(2)$-groups?
3. If G splits, does it always have a decomposition over ≤ 2 base elements? (A positive answer to this would solve Problem 2 in the positive, given Theorem 4.2.)

References

[A]   D. M. Arnold, *Abelian groups and representations of finite partially ordered sets*, CMS books in Mathematics 2. Springer-Verlag, New York, 2000.

[VWW]   S. L. Wallutis, C. Vinsonhaler and W. J. Wickless, *A class of $B(2)$-groups*, Comm. Algebra 33 No. 6 (2005), pp. 2025–2034.

[FII]   L. Fuchs, *Infinite Abelian Groups*, II. Academic Press, London - New York, 1973.

[DVM4]   C. De Vivo and C. Metelli, *Decomposing $B(1)$-groups: an algorithm*, Comm. Algebra 30 No. 12 (2002), pp. 5621–5637.

[DVM8]   C. De Vivo and C. Metelli, *On degenerate $B(2)$-groups*, Houston Math.J. 32 No. 3 (2006), pp. 633–649.

[DVM11]   C. De Vivo and C. Metelli, *Settings for a study of finite rank Butler groups*, J. Algebra 318 No. 1 (2007), pp. 456–483.

[DVM10]   C. De Vivo and C. Metelli, *Butler groups splitting over a base element*, Colloquium Mathematicum 109 No. 2 (2007), pp. 297–305.

Author information

Clorinda De Vivo, Dipartimento di Matematica e Applicazioni, Universita' Federico II di Napoli, 80100 Napoli, Italy.
E-mail: clorinda.devivo@dma.unina.it

Claudia Metelli, Dipartimento di Matematica e Applicazioni, Universita' Federico II di Napoli, 80100 Napoli, Italy.
E-mail: cmetelli@math.unipd.it

Diagonal equivalence of matrices

A. Mader and O. Mutzbauer

Abstract. Two matrices A and B are diagonally equivalent if there exist invertible diagonal matrices U and V such that $B = UAV^{-1}$. Diagonal equivalence is a crucial factor in classifying rigid, local, almost completely decomposable groups, i.e., uniform groups. We establish a criterion for diagonal equivalence of matrices over a commutative ring (Theorem 5.3) and use it to classify a certain class of uniform groups up to near-isomorphism (Theorem 6.2).

Key words. Normal form, permutation matrix, decomposition, group action, uniform, rigid, local, almost completely decomposable group, nearly isomorphic.

AMS classification. 15A99, 15A21.

1 Introduction

Saunders and Schneider [5], in 1978, considered diagonal equivalence of matrices. They translated the problem into a graph theoretic problem. Here, we deal with matrices over commutative rings with 1. Two $m \times n$ matrices A and B are diagonally equivalent if there exist invertible diagonal matrices D and E such that $B = DAE^{-1}$. For matrices A, B over a commutative ring with 1 and such that each line has a non-zero-divisor, we present an algorithmic decision procedure for diagonal equivalence. For integer matrices and for matrices which have a unit in each line we determine a normal form relative to diagonal equivalence. Finally, we apply our results to a classification problem in abelian groups.

2 Equivalence of matrices

Let $\mathbf{M}_{m \times n}(R)$ be the set of $m \times n$ matrices with entries in R. The group of all invertible matrices of size n over a commutative ring R with 1 is the *linear group* and is denoted by $\mathrm{GL}_n(R)$. Two $m \times n$ matrices A and B are *equivalent* if there are invertible matrices $U \in \mathrm{GL}_m(R)$ and $V \in \mathrm{GL}_n(R)$ such that

$$B = UAV^{-1}. \tag{2.1}$$

We deal with special equivalences restricting U and V to certain subgroups of the full linear groups, e.g. the subgroup of all permutation matrices. If both U and V in (2.1)

Second author: Supported by grant MU 628/8-1 of the German Research Foundation DFG. Moreover, he would like to thank the Department of Mathematics at the University of Hawaii for the kind hospitality afforded him.

are permutation matrices, then we call A and B *permutation equivalent*. A matrix B that is permutation equivalent to A is obtained from A by permuting rows and columns thus the entries are only rearranged but not changed. If U and V in (2.1) are diagonal matrices, then A and B are *diagonally equivalent*.

3 Direct decompositions of matrices

A *line* of a matrix is either a row or a column. A matrix B that can be obtained by striking out rows and columns of a matrix A is a *submatrix of A*. An $m \times n$ matrix $A = (a_{ij})$ is the *direct sum of its submatrices* $A_1, A_2, \ldots, A_k$, denoted $A = A_1 \dotplus \cdots \dotplus A_k$, if there exist sequences of natural numbers

$$1 \leq n_1 < \cdots < n_k = n, \quad 1 \leq m_1 < \cdots < m_k = m,$$

such that, setting $m_0 = n_0 = 0$, for $1 \leq i \leq k$,

$$A_i = (a_{rs})_{n_{i-1} < r \leq n_i, m_{i-1} < s \leq m_i}$$

and $a_{ij} = 0$ if (i, j) is a position that does not appear in any of the submatrices A_i. The last condition may be expressed loosely by saying that A is 0 outside the submatrices A_i.

If $k \geq 2$, then we call A *decomposed*. A matrix is *indecomposed* if it is not decomposed. A matrix is *decomposable* if there exists a matrix that is permutation equivalent to A and decomposed. Accordingly we call a matrix *indecomposable* if no permutation equivalent matrix is decomposed. A matrix is *fully decomposed* if it decomposed and its direct summands are indecomposable. The following two lemmata are obvious.

Lemma 3.1. *Every matrix is permutation equivalent to a fully decomposed matrix.*

Given a matrix A, we wish to determine whether it is indecomposable or not. To do so we simplify matters. The *pattern* of a matrix A is the matrix $\mathrm{pat}(A)$ that has the same dimensions as A and the entry in position (i, j) of $\mathrm{pat}(A)$ is 1 if and only if the entry in position (i, j) of A is $\neq 0$, and all other entries of $\mathrm{pat}(A)$ are 0. Thus $\mathrm{pat}(A)$ is an example of a $(0, 1)$-*matrix*, i.e., a matrix whose entries are either 0 or 1.

Lemma 3.2. *An $m \times n$ matrix A is indecomposable, decomposed, and fully decomposed if and only if its pattern* $\mathrm{pat}(A)$ *is indecomposable, decomposed, and fully decomposed, respectively.*

4 Frames

Decomposed matrices have a special structure. We will formalize "structure" of a matrix in this section.

Definition 4.1. Data

$$\mathrm{FD} = \left[(r_h)_1^k, (n_h)_1^k, (s_f)_1^l, (m_f)_1^l\right]$$

are $m \times n$ *frame data* if k, l are natural numbers, $(r_h)_1^k$, $(n_h)_1^k$ are sequences of natural numbers of length k, and $(s_f)_1^l$, $(m_f)_1^l$ are sequences of natural numbers of length l such that

(1) $1 = r_1 < r_2 < \cdots < r_k \leq m$ and $1 \leq n_1 < n_2 < \cdots < n_k = n$,

(2) $1 = s_1 < s_2 < \cdots < s_l \leq m$ and $1 \leq m_1 < m_2 < \cdots < m_l = m$,

(3) for all $h \leq k$ and all $f \leq l$, agreeing that $n_0 = m_0 = 0$,

$$\{(r_h, j) : n_{h-1} < j \leq n_h\} \cap \{(i, s_f) : m_{f-1} < i \leq m_f\} \neq \emptyset$$

$$\Leftrightarrow r_h = m_{f-1} + 1 \wedge s_f = n_{h-1} + 1.$$

Since for given $m \times n$ frame data FD the parameters m, n are built in by $n_k = n$ and $m_l = m$, we often omit the specification $m \times n$ for frame data. For given $m \times n$ frame data FD, we single out sets of positions (i, j) in the Cartesian product $\{1, \ldots, m\} \times \{1, \ldots, n\}$ as follows.

(1) The *horizontal frame* consists of all pairs (r_h, j) such that $h \leq k$ and $n_{h-1} < j \leq n_h$,

(2) the *vertical frame* consists of all pairs (i, s_f) such that $f \leq l$ and $m_{f-1} < i \leq m_f$,

(3) the *frame* is the union of the horizontal and the vertical frame,

(4) the *outer region* consists of all pairs (i, j) such that $i < r_h$ and $j > n_{h-1}$, or $j < s_f$ and $i > m_{f-1}$,

(5) the *inner region* consists of all remaining positions.

Figure 1 illustrates the frame with boxes and the various regions as positions in a matrix. The horizontal and the vertical frames overlap precisely at the positions specified by item (3) of the above definition of frame data, i.e., $(r_h, s_f) = (m_{f-1} + 1, n_{h-1} + 1)$. The position $(1, 1)$ is always included. The horizontal and the vertical frame always touch the right and the lower margin, respectively, of the matrix.

Definition 4.2. Let $\mathrm{FD} = \left[(r_h)_1^k, (n_h)_1^k, (s_f)_1^l, (m_f)_1^l\right]$ be $m \times n$ frame data. An $m \times n$ matrix $A = (a_{ij})$ is a *framed matrix with frame data* FD if the entries of A on the frame are all non-zero and the entries on the outer region of A are all equal to 0. The entries of the inner region may take any value. Note that a framed matrix A has no 0-lines, so in particular $A \neq 0$.

Let $A = (a_{ij})$ be an $m \times n$ matrix and FD $m \times n$ frame data. The $m \times n$ matrix $F = (f_{ij})$ with entries $f_{ij} = a_{ij}$ on the frame given by FD and all entries off the frame equal to 0, is called the FD-*template of A* and denoted by $F := A[\mathrm{FD}]$. An FD-template of A is framed if and only if all entries of A on the frame are non-zero. A matrix F is a *template* if it is framed with data FD and $F = F[FD]$, i.e., the matrix has a frame, non-zero entries on the frame, and zero entries off the frame.

If A is framed with data FD, then the FD-template $A[\mathrm{FD}]$ of A is also framed.

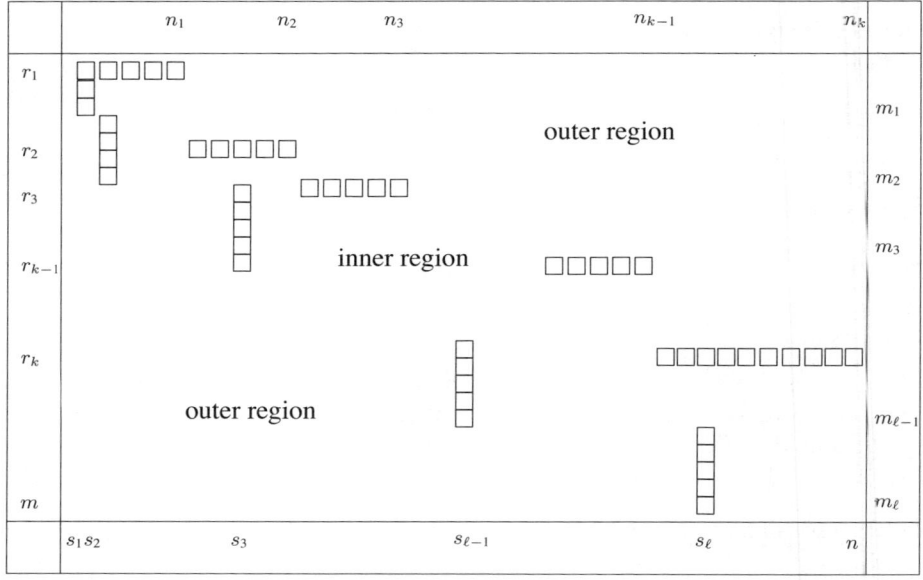

Figure 1. Frame data, frame, and regions of a matrix

We show next how framed matrices are associated with given matrices.

A $(0,1)$-matrix is *doubly ordered* if the rows are in lexicographic order from top to bottom and simultaneously the columns are in lexicographic order from left to right where we take $1 > 0$. Every $(0,1)$-matrix can be doubly ordered with a simple algorithm, that consists in interchanging rows and columns step by step in such a way that at each interchange the lexicographic order is improved. It was shown in [2, Theorem 2] that this leads, in a finite number of steps, to a doubly ordered $(0,1)$-matrix that is permutation equivalent to the original matrix.

Lemma 4.3 ([2, Theorem 2]). *A $(0,1)$-matrix without 0-lines is permutation equivalent to a doubly ordered matrix without 0-lines. A doubly ordered $(0,1)$-matrix without 0-lines is framed.*

We point out that there may be several different doubly ordered matrices permutation equivalent to a given $(0,1)$-matrix. Also there are algorithms transforming a $(0,1)$-matrix into a framed matrix that are simpler than double ordering and are easy to discover.

A matrix A has no 0-lines if and only if its pattern has no 0-lines and only matrices without 0-lines can be framed.

Proposition 4.4. *A framed matrix is fully decomposed and the direct summands are framed. The dimensions of the indecomposable direct summands are unique relative to permutation equivalence up to rearrangement.*

Proof. Let FD be the frame data of the framed matrix A. The horizontal and the vertical frames overlap at the positions specified by item (3) of the definition of frame data. Since the outer region of a framed matrix is filled with 0's, there is a corresponding direct decomposition of $A = \dotplus_{i=1}^{k} A_i$. The summands A_i are framed and indecomposed by their frames. We show that they are indecomposable. For this we define intersections of rows and columns of A as follows. Two rows $(a_{ij} \mid j)$ and $(a_{lj} \mid j)$ intersect if there is a column index j_0 with $a_{i,j_0} \neq 0$ and $a_{l,j_0} \neq 0$. Two columns $(a_{ij} \mid i)$ and $(a_{il} \mid i)$ intersect if there is a row index i_0 with $a_{i_0,j} \neq 0$ and $a_{i_0,l} \neq 0$. A row $(a_{i,j} \mid j)$ and a column $(a_{l,j} \mid l)$ intersect if $a_{i,j} \neq 0$. Note that those intersections will not be changed by permutation equivalence, and rows and columns that intersect belong to the same indecomposable direct summand.

Now for indecomposed framed matrices all rows and columns intersect or are connected by successively intersecting pairs of rows and columns. Thus all A_i are indecomposable. Moreover, again by the invariance of the above intersections relative to permutation equivalence, the dimensions of the indecomposable direct summands A_i are unique up to rearrangement. □

5 Diagonal equivalence

Let R be a commutative ring with 1. All matrices are matrices over R. The entries of a matrix diagonally equivalent to A differ from the entries of A only by unit factors, i.e., entries in the same position are *associate* in the sense of ring theory. Two $m \times n$ matrices $A = (a_{ij})$ and $B = (b_{ij})$ are *associate* if a_{ij} and b_{ij} are associate for all i, j. In particular, $a_{ij} = 0$ if and only if $b_{ij} = 0$. Note that diagonal equivalence over a ring with only one unit is equality.

If two matrices A, A' are diagonally equivalent, and A is fully decomposed, then A' is fully decomposed with the same direct sum structure. Moreover, two matrices A and A' that are fully decomposed are diagonally equivalent if and only if their corresponding direct summands are diagonally equivalent.

Let $\mathcal{D}_k \subset \mathrm{GL}_k(R)$ be the subgroup of (invertible) diagonal matrices. The equivalence classes of diagonally equivalent matrices are the orbits under the action of the group $\Delta = \mathcal{D}_m \times \mathcal{D}_n$ on $\mathbf{M}_{m \times n}(R)$ given by

$$\delta : \Delta \ni (D, E) \mapsto \begin{pmatrix} M \\ DME^{-1} \end{pmatrix} \in \mathrm{Sym}(\mathbf{M}_{m \times n}(R)),$$

where $\begin{pmatrix} M \\ DME^{-1} \end{pmatrix}$ denotes the permutation that maps each $M \in \mathbf{M}_{m \times n}(R)$ to $DME^{-1} \in \mathbf{M}_{m \times n}(R)$. The *kernel* of the action δ is easily seen to be

$$\Delta_0 := \{(\alpha I_m, \alpha I_n) : \alpha \in \mathrm{U}(R)\},$$

where I_k denotes the identity matrix of size k, and $\mathrm{U}(R)$ is the group of units of R. Thus the group Δ/Δ_0 acts faithfully on $\mathbf{M}_{m \times n}(R)$.

Let $S = (u_i \times v_i \mid 1 \leq i \leq k)$ be a sequence of dimensions of matrices with $\sum_{i=1}^{k} u_i = m$ and $\sum_{i=1}^{k} v_i = n$. Let $\mathbf{M}_S(R) = \mathbf{M}_{u_1 \times v_1}(R) \dotplus \cdots \dotplus \mathbf{M}_{u_k \times v_k}(R) \subset$

$M_{m \times n}(R)$ be the set of $m \times n$ matrices over R with a fixed direct sum structure. Then the restriction of δ to $M_S(R)$ is a well-defined action δ_S of Δ on $M_S(R)$. The *kernel* of this action δ_S is easily seen to be

$$\Delta_0(S) := \{(\dotplus_{i=1}^k \alpha_i I_{u_i}, \dotplus_{i=1}^k \alpha_i I_{v_i}) : \alpha_i \in U(R)\}.$$

Thus the quotient group $\Delta/\Delta_0(S)$ acts faithfully on $M_S(R)$.

The subgroup of invertible matrices generated by permutation and diagonal matrices is a semi-direct product with the diagonal matrices forming the normal subgroup. This has the following consequence.

Lemma 5.1. *Let A and B be matrices over R and let P, Q be permutation matrices. Then A and B are diagonally equivalent if and only if PAQ^{-1} and PBQ^{-1} are diagonally equivalent.*

Proof. Let A and B be diagonally equivalent, say $B = DAE^{-1}$ for invertible diagonal matrices D and E. Then

$$PBQ^{-1} = PDAE^{-1}Q^{-1} = (PDP^{-1})(PAQ^{-1})(QE^{-1}Q^{-1}),$$

where PDP^{-1} and $QE^{-1}Q^{-1}$ are invertible diagonal matrices since permutations matrices normalize the subgroup of all diagonal matrices. □

Three facts are crucial. First, Lemma 4.3 saying that matrices without 0-lines are permutation equivalent to framed matrices. Second, Lemma 5.1, stating that two matrices A, A' are diagonally equivalent if and only if also PAQ^{-1} and $PA'Q^{-1}$ are diagonally equivalent. Third, the following Lemma 5.2, saying that a template F is diagonally equivalent to any associate matrix F', and if $F' = DFE^{-1}$, then the pair (D, E) is uniquely determined modulo $\Delta_0(S)$ where S indicates the direct decomposition of F.

Lemma 5.2. *Let F be a template over a commutative ring with 1 and with only non-zero-divisors on the frame. Let $F = F_1 \dotplus \cdots \dotplus F_k$ be the direct sum of indecomposed framed submatrices F_i and let $S = (u_i \times v_i \mid 1 \leq i \leq k)$ be the sequence of the dimensions of the direct summands F_i. Let F' be a matrix associate to F. Then there is a pair of invertible diagonal matrices $(D, E) \in \Delta$ such that $F' = DFE^{-1}$ and this pair is unique modulo $\Delta_0(S)$.*

Proof. Let $\mathrm{FD} = [(r_h)_1^k, (n_h)_1^k, (s_f)_1^l, (m_f)_1^l]$ be the frame data of $F = (f_{ij})$. First we deal with the case of an indecomposed framed template F. Let $F' = (f'_{ij})$ be associate to F. Then F and F' have the same frame data, hence $f_{ij} = \rho_{ij} f'_{ij}$ on the frame, with units ρ_{ij}.

We need to find diagonal matrices D, E such that $DF = F'E$. To do so we assume that $DF = F'E$ where

$$D = \mathrm{diag}(d_1 = 1, d_2, \ldots, d_m), \quad E = \mathrm{diag}(e_1, e_2, \ldots, e_n).$$

We will show that there exist uniquely determined solutions d_i and e_j.

The matrix equation $DF = F'E$ is equivalent to the set of equations $d_i f_{ij} = d_i \rho_{ij} f'_{ij} = e_j f'_{ij}$ for (i, j) on the common frame of F and F'. Since the matrices F, F' are indecomposed framed templates, the only overlap of the horizontal and the vertical frame is at position $(1, 1)$, and therefore we have precisely $m + n - 1$ equations. Thus we have, as $d_1 = 1$, as many equations as unknowns.

Since the entries f_{ij}, f'_{ij} on the frame are non-zero-divisors, we obtain $d_i \rho_{ij} = e_j$ for all positions (i, j) on the frame of the matrices. Note that the units ρ_{ij} are given, hence d_i determines e_j uniquely and vice versa.

We proceed by induction on the row index. Note first that for fixed i there is a unique natural number w such that $r_w \leq i < r_{w+1}$. For instance if $i = 1$, then $w = 1$ since $1 = r_1 \leq 1 < r_2$.

We start the induction with $i = 1$. Since for the first row $d_1 = 1$ we get that $e_j = \rho_{1j}$ is uniquely determined for $1 \leq j \leq n_w$ and here $w = 1$. If $m = 1$, hence $n_1 = n$, and the proof is finished.

The induction hypothesis for i with $1 < i < m$ is that $d_1, \ldots, d_i$ and $e_1, \ldots, e_{n_w}$ are uniquely determined, where w is given by the inequality $r_w \leq i < r_{w+1}$.

Now we consider the row with index $i + 1$. Since the vertical frame reaches from row 1 to row m there exists a (unique) index f such that f_{i+1,s_f} is a non-zero-divisor.

If $s_f > n_w$, then F is decomposed because all entries in positions (x, y) with $x > i$ and $y < s_f$ are 0 and all entries in position (x, y) with $x \leq i$ and $y > n_w$ are also 0. Thus $s_f \leq n_w$ and $d_{i+1} = \rho_{i+1,s_f}^{-1} e_{s_f}$, i.e., d_{i+1} is uniquely determined. Note that the e_j are already determined for $j \leq n_w$ by the induction hypothesis. If $n_w = n$ or if $i + 1 < r_{w+1}$, then there are no more e_j to determine. Otherwise $r_{w+1} = i + 1$ and we need to check the values e_j for $n_w < j \leq n_{w+1}$. But now $e_j = d_{i+1} \rho_{i+1,j}$ is uniquely determined for $n_w < j \leq n_{w+1}$. This concludes the induction proof.

Since the matrices F, F' are templates, we actually have $DF = F'E$.

In the general case, if F is the direct sum of the F_i the frame data FD_i of F_i are segments of the frame data FD of F. The associate matrix F' has a corresponding direct decomposition with summands F'_i associate to F_i, respectively. By Proposition 4.4 the F_i are indecomposed framed templates. By the above there is for each pair F_i, F'_i a pair (D_i, E_i) of diagonal matrices such that $F'_i = D_i F_i E_i^{-1}$. This pair (D_i, E_i) is unique modulo the corresponding kernel $\Delta_{0,i} = \{(\alpha I_{u_i}, \alpha I_{v_i}) : \alpha \in U(R)\}$. The matrices $D = \dotplus_i D_i$ and $E = \dotplus_i E_i$ solve the equation $F' = DFE^{-1}$ and are unique modulo $\Delta_0(S)$. □

Let A, A' be associate framed matrices with entries on the frame that are non-zero-divisors. Then A and A' have common frame data FD, and a common decomposition structure $S = (u_i \times v_i \mid 1 \leq i \leq k)$. Thus the FD-templates $F := A[\text{FD}]$ and $F' := A'[\text{FD}]$ of A and A', respectively, are framed templates whose entries on the frame are non-zero-divisors. Hence, by Lemma 5.2, there is a pair (D, E) of diagonal matrices such that $F' = DFE^{-1}$, and this pair is uniquely determined modulo $\Delta_0(S)$. We call such a pair (D, E) *forced by A, A'* or *forced by F, F'* or *a forced pair for A, A'* or *for F, F'*. Note that a forced pair is determined by the entries on the frames of A, A' alone.

For a matrix A over a commutative ring with 1, the $(0, 1)$-matrix that is obtained by replacing the entries of A that are non-zero-divisors by 1 and all other entries by 0 is the *essential pattern* of A. If the essential pattern M of a matrix A has no 0-lines, or in other words if there is a non-zero-divisor in each line of A, then there are permutation matrices P, Q such that PMQ^{-1} is a framed matrix. The matrix A is then called (P, Q)-*essential*. Suppose that A is (P, Q)-*essential*. Let FD be the frame data of PMQ^{-1}. Clearly, PAQ^{-1} is in general not framed but the FD-template $F = (PAQ^{-1})[\text{FD}]$ of PAQ^{-1} is a framed template, and all entries of PAQ^{-1} in the outer region of the frame given by FD are zero divisors. Note that there might be other pairs (P', Q') of permutation matrices such that A is also (P', Q')-essential. If the matrix A is framed and its entries on the frame are all non-zero-divisors then we say briefly that A is *essential*. Associate matrices A, A' have the same essential pattern, thus A is (P, Q)-essential if and only if A' is (P, Q)-essential. Let A, A' be associate essential (framed) matrices with (common) frame data FD. Then the FD-templates $A[\text{FD}]$ and $A'[\text{FD}]$ are essential framed templates and by Lemma 5.2 there is a forced pair (D, E) of diagonal matrices, i.e., $A'[\text{FD}] = D(A[\text{FD}])E^{-1}$, but $A' \neq DAE^{-1}$ in general. Now A, A' are diagonally equivalent if and only if $A' = DAE^{-1}$ for this forced pair (D, E).

The following theorem is the main result on diagonal equivalence.

Theorem 5.3. *Let A be a matrix over a commutative ring with 1 and assume that the essential pattern M of A has no 0-lines. Then there exist permutation matrices P, Q such that PMQ^{-1} is framed with frame data* FD, *and the* FD*-template $(PAQ^{-1})[\text{FD}]$ of PAQ^{-1} is an essential template.*

Then A, A' are diagonally equivalent if and only if the following hold.

(1) *A, A' are associate.*

(2) *For any forced pair (D, E) for $(PAQ^{-1})[\text{FD}]$ and $(PA'Q^{-1})[\text{FD}]$*

$$A' = (P^{-1}DP)\,A\,(Q^{-1}E^{-1}Q).$$

Proof. Since the essential pattern M of A has no 0-lines, there are permutation matrices P, Q such that PMQ^{-1} is framed by Lemma 4.3, with frame data, say FD. Hence the FD-template $F = (PAQ^{-1})[\text{FD}]$ of PAQ^{-1} is essential.

If A' is associate to A, then the FD-template $F' = (PA'Q^{-1})[\text{FD}]$ of $PA'Q^{-1}$ is associate to F, hence also essential. By Lemma 5.2 there is a forced pair (D, E) for F, F', i.e., (2) makes sense and establishes the diagonal equivalence of A and A'.

Conversely, suppose that A, A' are diagonally equivalent, in particular, associate, i.e., (1) holds trivially, and it suffices to verify (2). Since A, A' have non-zero-divisors in each line and are associate the FD-templates

$$F = (PAQ^{-1})[\text{FD}] \quad \text{and} \quad F' = (PA'Q^{-1})[\text{FD}]$$

of PAQ^{-1} and of $PA'Q^{-1}$ are essential and associate, and there is a forced pair (D, E) for F, F', i.e., $F' = DFE^{-1}$. Since A, A' are diagonally equivalent, also PAQ^{-1} and $PA'Q^{-1}$ are diagonally equivalent by Lemma 5.1, and $PA'Q^{-1} = D(PAQ^{-1})E^{-1}$ as desired. □

Theorem 5.3 constitutes an algorithmic decision procedure for diagonal equivalence. The determination whether two matrices A, A' over a commutative ring with 1, and with at least two units, are diagonally equivalent, i.e., whether $A' = DAE^{-1}$ for some diagonal matrices D, E, is achieved in steps.

First of all A, A' must be associate. Thus they have the same essential pattern M. It is assumed that M has no 0-lines. Then M is permutation equivalent to a framed matrix. This framed form PMQ^{-1} of M, together with the permutation matrices P, Q, can be found by double ordering the $(0, 1)$-matrix M, a simple algorithm, cf. [2, Theorem 2].

Thus PAQ^{-1} and $PA'Q^{-1}$ have associate FD-templates

$$F = (PAQ^{-1})[\text{FD}] \text{ and } F' = (PA'Q^{-1})[\text{FD}].$$

These templates are essential. Hence there is a forced pair (D, E) for F, F' and the proof of Lemma 5.2 provides an algorithm to determine such a forced pair.

Then deciding whether A, A' are diagonally equivalent is reduced to checking whether the equation

$$A' = (P^{-1}DP) A (Q^{-1}E^{-1}Q)$$

holds or not.

If there are frame data FD such that the FD-template $A[\text{FD}]$ is essential, then checking diagonal equivalence can be done without line permutations.

For domains Theorem 5.3 simplifies since there are no zero-divisors.

Corollary 5.4. *Let A be a matrix without 0-lines over a domain. Then there exist permutation matrices P, Q such that PAQ^{-1} is framed. Let A' be another matrix. Then A, A' are diagonally equivalent if and only if the following hold.*

(1) A, A' *are associate.*

(2) *For any forced pair (D, E) for $PAQ^{-1}, PA'Q^{-1}$*

$$A' = (P^{-1}DP) A (Q^{-1}E^{-1}Q).$$

Proof. The pattern of a matrix over a domain is automatically essential. Thus Theorem 5.3 applies. □

Since invertible integer diagonal matrices have entries ± 1 on the diagonal, integer matrices that are diagonally equivalent are absolutely equal, i.e., entries at the same positions have equal absolute values.

Proposition 5.5. *Let A be an integer matrix without 0-lines. Then there exist permutation matrices P, Q such that PAQ^{-1} is framed. Let P, Q be matrices such that PAQ^{-1} is framed and let A' be another matrix. Then A, A' are diagonally equivalent if and only if the following hold.*

(1) A, A' *are absolutely equal.*

(2) *For any forced pair* (D, E) *for* PAQ^{-1} *and* $PA'Q^{-1}$

$$A' = (P^{-1}DP)\, A\, (Q^{-1}E^{-1}Q).$$

Moreover, there is a matrix B diagonally equivalent to A such that all entries on the frame of PBQ^{-1} *are positive.*

Proof. The main part of the statement follows from Corollary 5.4.

Moreover, the FD-template $F = (PAQ^{-1})[\text{FD}] = (f_{ij})$ is framed and absolutely equal to the matrix $F_{\text{abs}} = (|f_{ij}|)$, where the entries are replaced by its absolute values. By Lemma 5.2 there is a forced pair (D, E) for F, F_{abs} such that $F_{\text{abs}} = DFE^{-1}$. Then

$$B = (P^{-1}DP)A(Q^{-1}E^{-1}Q)$$

is diagonally equivalent to A, and the FD-template $(PBQ^{-1})[\text{FD}] = F_{\text{abs}}$ as desired. □

For a pair (P, Q) of permutation matrices an integer matrix A is called (P, Q)-*positive*, if PAQ^{-1} is a framed matrix with only positive integers on the frame. By Proposition 5.5 a class of diagonally equivalent integer matrices without 0-lines contains a (P, Q)-positive representative for a suitable pair (P, Q) of permutation matrices. The following corollary indicates that for integer matrices without 0-lines the (P, Q)-positive matrices are a kind of normal form for diagonal equivalence.

Corollary 5.6. *Two integer matrices* A, A' *without 0-lines are diagonally equivalent if and only if there is a pair* (P, Q) *of permutation matrices such that* A, A' *are both diagonally equivalent to a* (P, Q)-*positive matrix* B.

Furthermore, two (P, Q)-*positive integer matrices* A, A' *are diagonally equivalent if and only if* $A = A'$.

Proof. If A and A' are both diagonally equivalent to B, then they are diagonally equivalent to one another. Suppose that A and A' are diagonally equivalent. It is to show that A, A' are both diagonally equivalent to a (P, Q)-positive matrix B for some pair (P, Q). By Proposition 5.5 the matrix A is diagonally equivalent to a certain (P, Q)-positive matrix and the matrix A' is diagonally equivalent to a certain (P', Q')-positive matrix. Diagonally equivalent matrices A, A' are absolutely equal, hence they are diagonally equivalent to (P, Q)-positive matrices B, B' for the same pairs (P, Q), i.e., PBQ^{-1}, and $PB'Q^{-1}$ are framed with the same frame data FD and positive entries on the frame. Thus the FD-templates $(PBQ^{-1})[\text{FD}]$ and $(PB'Q^{-1})[\text{FD}]$ are framed and absolutely equal with only positive entries on the frame, hence they are equal. But then (I, I) is a forced pair and $B = B'$ by Proposition 5.5, part (2). This shows also that if A, A' are diagonally equivalent and already (P, Q)-positive, then they are equal. □

While there is only one (P, Q)-positive matrix B diagonally equivalent to A, A might also be diagonally equivalent to a (P', Q')-positive matrix $B' \neq B$ for another pair (P', Q').

For a matrix A over a commutative ring with 1, the $(0, 1)$-matrix that is obtained by replacing the unit entries of A by 1 and all other entries by 0, is the *unit pattern* of A.

Theorem 5.7. *Let A be a matrix over a commutative ring with 1 and assume that the unit pattern M of A has no 0-lines. Then there exist permutation matrices P, Q such that PMQ^{-1} is framed with frame data, say FD, and the FD-template (PAQ^{-1})[FD] has only units on the frame. Then A, A' are diagonally equivalent if and only if the following hold.*

(1) *A, A' are associate.*

(2) *For any forced pair (D, E) for the FD-templates (PAQ^{-1})[FD] and $(PA'Q^{-1})$[FD]*
$$A' = (P^{-1}DP) \, A \, (Q^{-1}E^{-1}Q).$$

Moreover, there is a matrix B diagonally equivalent to A such that all entries on the frame of the FD-template (PBQ^{-1})[FD] are 1.

Proof. Since units are not zero divisors, the main part of the statement follows from Theorem 5.3.

Moreover, the FD-template $F = (PAQ^{-1})$[FD] is essential and associate to the matrix $F_1 = (f'_{ij})$, where $f'_{ij} = 1$ on the frame and 0 elsewhere. By Lemma 5.2 there is a forced pair (D, E) for F, F_1 such that $F_1 = DFE^{-1}$. Then
$$B = (P^{-1}DP)A(Q^{-1}E^{-1}Q)$$
is diagonally equivalent to A, and $F_1 = (PBQ^{-1})$[FD]. □

Let A be a matrix over a commutative ring with 1 and with units in each line. Let M be the unit pattern of A. Let (P, Q) be a pair of permutation matrices such that PMQ^{-1} is framed with frame data, say FD. Then A is called (P, Q)-*normed* if all entries on the frame of the FD-template (PAQ^{-1})[FD] are 1. Note that the matrix PAQ^{-1} is not framed in general but all entries in the outer region of the frame with frame data FD are non-units. By Theorem 5.7 an equivalence class of diagonally equivalent matrices over a commutative ring with 1 and with units in each line has (P, Q)-normed representatives for suitable pairs (P, Q) of permutation matrices. The following corollary indicates that for those matrices the (P, Q)-normed matrices are a kind of normal form for diagonal equivalence.

Corollary 5.8. *Two matrices A, A' over a commutative ring with 1 and with units in each line are diagonally equivalent if and only if there is a pair (P, Q) of permutation matrices such that A, A' are both diagonally equivalent to a (P, Q)-normed matrix B.*

Furthermore, two (P, Q)-normed matrices A, A' are diagonally equivalent if and only if $A = A'$.

Proof. If A and A' are both diagonally equivalent to B, then they are diagonally equivalent to one another. Suppose that A and A' are diagonally equivalent. It is to show that A, A' are both diagonally equivalent to a (P, Q)-normed matrix B for some pair (P, Q). By Theorem 5.7 the matrix A is diagonally equivalent to a certain (P, Q)-normed matrix and the matrix A' is diagonally equivalent to a certain (P', Q')-normed matrix. Diagonally equivalent matrices A, A' are associate, hence they are diagonally equivalent to (P,Q)-normed matrices B, B' for the same pair (P,Q). Thus B, B' are also diagonally equivalent and the FD-templates $(PBQ^{-1})[FD]$ and $(PB'Q^{-1})[FD]$ both have all entries equal to 1 on the frame, hence they are equal. But then (I, I) is a forced pair and $B = B'$ by Theorem 5.7, part (2). This shows also that if A, A' are diagonally equivalent and already (P,Q)-normed for the same pair (P,Q) then they are equal. □

While there is only one (P,Q)-normed matrix B diagonally equivalent to A, A might also be diagonally equivalent to a (P',Q')-normed matrix $B' \neq B$ for another pair (P',Q').

6 Applications to uniform groups

Uniform groups are rigid, local, and almost completely decomposable groups G with a homocyclic p-primary regulator quotient G/R. Let $\mathcal{C}(T, p, r, e)$ be the class of uniform groups with critical typeset $T = (\tau_1, \ldots, \tau_n)$ of cardinality n and a homocyclic p-group of exponent p^e and rank r as regulator quotient. A uniform group $U \in \mathcal{C}(T, p, r, e)$ has relative to a decomposition basis $(v_1, \ldots, v_n)$ of the regulator and a basis of the regulator quotient a representing matrix with entries in $\mathbb{Z}/p^e\mathbb{Z}$. Note that there is a bijection between the critical typeset T and the decomposition basis of the regulator, i.e., between T and the columns of the representing matrix. The rows of this representing matrix $M = (\alpha_{ij}) \in \mathrm{M}_{m \times n}(\mathbb{Z}/p^e\mathbb{Z})$ correspond to representatives

$$g_i = p^{-e} \sum_{j=1}^{n} \alpha'_{ij} v_j, \quad 1 \leq i \leq r,$$

of generators of the regulator quotient U/R, where the α'_{ij}'s are integers such that $\alpha_{ij} = \alpha'_{ij} + p^e \mathbb{Z}$. Dugas and Oxford proved in [1] the following theorem.

Theorem (Dugas–Oxford). *There is an ordering of the critical typeset of a uniform group $U \in \mathcal{C}(T, p, r, e)$ that allows a representing matrix of the form (I, A), where I is the identity matrix. Two groups U, U', given by the representing matrices (I, A) and (I, A'), respectively, relative to the same ordering of the critical typeset are nearly isomorphic if and only if A, A' are diagonally equivalent.*

Orderings of the critical typeset of a group in the class $\mathcal{C}(T, p, r, e)$ that allow a representing matrix of the form (I, A) are called *pivot orderings*. The critical types corresponding to the columns of I form the *pivot set*. It is well-known that nearly isomorphic groups in $\mathcal{C}(T, p, r, e)$ have the same critical typeset and the same pivot orderings. Note that there are in general different pivot orderings for a group.

By the regulator criterion, the matrix A has a unit in each row. Now let $\mathcal{C}^1(T, p, r, e)$ be the subclass of groups U in $\mathcal{C}(T, p, r, e)$ such that for the regulator R the subgroup $p^{-1}R \cap U$ of U is *clipped*, i.e., without a direct summand of rank 1. Then the matrix A has also a unit in each column.

Proposition 6.1. *If (I, A) is a representing matrix of a uniform group $U \in \mathcal{C}^1(T, p, r, e)$ relative to a pivot ordering of the critical typeset, then there is a pair (P, Q) of permutation matrices such that A is diagonally equivalent to a (P, Q)-normed matrix.*

Furthermore, let the groups $U, U' \in \mathcal{C}^1(T, p, r, e)$ be given by representing matrices (I, A) and (I, A') relative to the same pivot ordering of the critical typeset. Then U, U' are nearly isomorphic if and only if there is a pair (P, Q) such that A, A' are diagonally equivalent to the same (P, Q)-normed matrix.

In particular, a near isomorphism class of groups in $\mathcal{C}^1(T, p, r, e)$ is characterized by a pivot ordering of the critical typeset, a pair (P, Q) of permutation matrices and a representing matrix (I, A) where A is (P, Q)-normed.

Proof. By the theorem of Dugas–Oxford the critical typeset of a group $U \in \mathcal{C}(T, p, r, e)$ has a pivot ordering such that U has a representing matrix of the form (I, A). Since even $U \in \mathcal{C}^1(T, p, r, e)$, the matrix A has a unit in each line, and by Theorem 5.7 the matrix A is diagonally equivalent to a (P, Q)-normed matrix for some pair (P, Q) of permutation matrices.

Furthermore, by the theorem of Dugas–Oxford the groups U, U' are nearly isomorphic if and only if A, A' are diagonally equivalent, hence by Corollary 5.8 there is a pair (P, Q) such that A, A' are diagonally equivalent to the same (P, Q)-normed matrix.

In particular, by the theorem of Dugas–Oxford, a near isomorphism class of uniform groups in the class $\mathcal{C}^1(T, p, r, e)$ is characterized in the indicated way. □

An (I_m, I_n)-normed matrix is called *normed*. Note that a normed matrix A has a framed unit pattern with frame data, say FD, and the FD-template $A[\text{FD}]$ has all entries on the frame equal to 1. Moreover, all entries on the outer region of A relative to the given frame are not units. A pivot ordering of the critical typeset of a group $U \in \mathcal{C}^1(T, p, r, e)$ is called *admissible* if the matrix A in a corresponding representing matrix (I, A) has a framed unit pattern. In particular, if (I, A) is the representing matrix of a group $U \in \mathcal{C}^1(T, p, r, e)$ with A normed, then the corresponding ordering of the critical typeset, i.e., the ordering of the columns of (I, A), is admissible for U. Note that there may be not only different pivot orderings but also different admissible orderings for U that belong to the same pivot ordering.

The following theorem presents a solution of the classification problem in the class $\mathcal{C}^1(T, p, r, e)$ up to near isomorphism.

Theorem 6.2. *Groups in the class $\mathcal{C}^1(T, p, r, e)$ have admissible orderings of the critical typeset. In every near isomorphism class of groups in $\mathcal{C}^1(T, p, r, e)$ there is a group that has relative to an admissible ordering of its critical typeset a representing matrix of the form (I, A) where A is normed.*

Let two groups $U, U' \in \mathcal{C}^1(T, p, r, e)$ have representing matrices (I, A) and (I, A'), respectively, relative to the same admissible ordering, where A, A' are normed. Then U, U' are nearly isomorphic groups if and only if $A = A'$.

In particular, a near isomorphism class of a group in $\mathcal{C}^1(T, p, r, e)$ has as invariants an admissible ordering of T and a normed matrix A with frame data FD.

Proof. By the theorem of Dugas–Oxford there is a pivot ordering of the critical typeset of a group $U \in \mathcal{C}^1(T, p, r, e)$. Hence, again by the theorem of Dugas–Oxford and by Proposition 6.1 there is a group U_0 nearly isomorphic to U such that the part B of the representing matrix (I, B) of U_0 is (P, Q)-normed for a suitable pair (P, Q) of permutation matrices. The following matrix equation holds:

$$P(I, B)(P^{-1}, Q^{-1}) = (I, PBQ^{-1}).$$

This equation can be understood as a reordering of the columns and the rows of the representing matrix (I, B) of U_0. The group U_0 is not changed. But now PBQ^{-1} is normed, i.e., the ordering of the columns, respectively the critical typeset, is admissible. In other words, relative to this admissible ordering of the critical typeset T there is a representing matrix (I, A) of U_0 with A normed.

Finally, let two groups $U, U' \in \mathcal{C}^1(T, p, r, e)$ have equal admissible orderings of the critical typeset and corresponding representing matrices (I, A) and (I, A'), respectively, where A, A' are normed. Then, by Corollary 5.8, the groups U, U' are nearly isomorphic if and only if $A = A'$.

In particular, the final equation $A = A'$ proves that the admissible ordering of T and a normed matrix A with frame data FD are the complete invariants of a near isomorphism class of groups in the class $\mathcal{C}^1(T, p, r, e)$. □

Let $U \in \mathcal{C}^1(T, p, r, e)$ have the representing matrix (I, A) relative to the regulator and the ordering of the critical typeset given by the indexing of the columns. If the matrix A is normed we call (I, A) a *normal form* of U or *a normal form of the near isomorphism class* of U relative to the given admissible ordering of T.

Theorem 6.2 establishes a real normal form for uniform groups in the subclass $\mathcal{C}^1(T, p, r, e)$ up to near isomorphism. Each group has, up to near isomorphism, a representing matrix (I, A), with A normed, relative to an admissible ordering, and two such representations describe nearly isomorphic groups if and only if the representations are equal.

Moreover, there are simple algorithms that allow to determine step by step, such a normal form for a group $U \in \mathcal{C}^1(T, p, r, e)$. Namely, if U is given by a representing matrix M modulo the regulator R, then M is an $r \times |T|$ matrix over the ring $\mathbb{Z}/p^e\mathbb{Z}$ and has rank r by the regulator criterion, i.e., there are invertible $r \times r$ submatrices. The columns of M indicate a certain ordering of the critical typeset T.

(1) **Determine a pivot ordering.** This is done by sorting to the left r columns of M that form an invertible $r \times r$ submatrix J. We obtain the representing matrix $(J, B) = MS$ relative to this ordering of T, where S is a permutation matrix.

(2) **Determine the pivot form.** By the Gauß algorithm, or in other words by choice of a suitable basis of the regulator quotient U/R, the new representing matrix $(I, C) = J^{-1}MS = J^{-1}(J, B)$, i.e., $C = J^{-1}B$ is obtained. This algorithm works for all groups in the bigger class $\mathcal{C}(T, p, r, e)$.

(3) **Determine an admissible ordering.** Now we use that the group U is in the smaller class $\mathcal{C}^1(T, p, r, e)$. Then C has units in every line, i.e., there is a pair (P, Q) of permutation matrices such that the unit pattern of PCQ^{-1} is framed. Clearly, the matrix

$$(I, PCQ^{-1}) = P(I, C)(P^{-1}, Q^{-1})$$

is obtained by permuting rows and columns of the representing matrix (I, C) of the group U. Thus the group U is unchanged, but has now the new representing matrix (I, PCQ^{-1}) where the unit pattern of PCQ^{-1} is framed, i.e., we have an admissible ordering for U. The permutation matrices P, Q can be found by double ordering, cf. [2, Theorem 2].

(4) **Determine a normal form.** The frame data FD of the unit frame of PCQ^{-1} can be read off. The FD-template $F = (PCQ^{-1})[\text{FD}]$ is framed and has only units on the frame. The framed template F_1 with frame data FD, that has all entries equal to 1 on the frame, is associate to F. By Lemma 5.2 there is a forced pair (D, E) of diagonal matrices for F, F'. Thus $F_1 = DFE^{-1}$, hence $A = DPCQ^{-1}E^{-1}$ is normed. The proof of Lemma 5.2 describes an algorithm to obtain such a forced pair (D, E). Since diagonal equivalence of the second part of a representing matrix leads to a nearly isomorphic group, there is a group U_0 nearly isomorphic to the original group U with a representing matrix $(I, A) = (I, DPCQ^{-1}E^{-1})$ in normal form. Thus if M is the original representing matrix, we get the normal form

$$(I, A) = DPJ^{-1}MS(P^{-1}, Q^{-1})(D^{-1}, E^{-1}).$$

(5) **Comparing groups.** Let $U, U' \in \mathcal{C}^1(T, p, r, e)$ be given by representing matrices, (I, A) for U and M for U'. Let (I, A) be a normal form, i.e., A is normed with frame data FD and the columns of (I, A) reflect an admissible ordering for U. The ordering of the columns of M reflect possibly another ordering of T. There is a permutation matrix S such that the ordering of the columns of $M' = MS$, is now the same than the ordering given by (I, A). This must be a pivot ordering for U' otherwise U, U' are not nearly isomorphic. Then $M' = (J, B)$, where J is invertible and $(I, C) = J^{-1}M'$ where $C = J^{-1}B$ is a representing matrix for U' up to near isomorphism. Clearly, since C has a unit in each line there is a pair (P, Q) of permutation matrices such that the unit pattern PCQ^{-1} is framed. The pivot set is permuted by P, i.e., the critical types corresponding to the columns of I, and the complement of the pivot set is permuted by Q. The given ordering of the columns of M' is admissible for U if and only if P, Q can be chosen to be identity matrices. So the unit pattern of C is already framed, otherwise U, U' are not nearly isomorphic. Then, the frame data of C are FD, otherwise U, U' are not nearly isomorphic. Thus, there is a forced pair (D, E) for the FD-templates $A[\text{FD}]$ and $C[\text{FD}]$. Then $A' = DCE^{-1}$ is also normed. Finally, U, U' are nearly isomorphic if and only if $A = A'$.

Note that there are in general several admissible orderings of the critical typeset. Any admissible ordering allows the comparison of two groups. But comparison always happens modulo the same admissible ordering.

We mention in this context that in [4] there is a reduction of the classification in the more general class $\mathcal{C}(T, p, r, e)$ up to near isomorphism, by choosing admissible orderings that are differently defined. The reduction consists in specializing the diagonal equivalence in the theorem of Dugas–Oxford to a so called *modified diagonal similarity*. Moreover, there were determined certain numbers of near isomorphism classes in special cases.

The direct decompositions of a group in the general class $C(T, p, r, e)$ can be described, cf. [3].

Corollary 6.3 ([3]). *Let $U \in C(T, p, r, e)$ be given by the representing matrix (I, A). Then U is indecomposable if and only if A is indecomposable.*

Note that there is a certain consequence for groups in $C^1(T, p, r, e)$ given in normal form (I, A) with frame data FD. The direct decompositions of A reflect decompositions of the group. Since the unit pattern of A is framed, it is fully decomposed by Lemma 4.4. Thus direct decompositions of A induce decompositions of the unit pattern of A. But the direct decomposition of the unit pattern of A is in general a refinement of the decomposition of A, because in the outer region of the frame given by the frame data FD there might be entries of A that are not units but also not zero. In other words, the decomposition of the unit pattern of A, i.e., the frame data FD are always a refinement of the direct decomposition of A.

It should be mentioned that it is desirable to obtain a similarly explicit classification in the bigger class $C(T, p, r, e)$ instead of the special class $C^1(T, p, r, e)$. Moreover, there is at the moment no group theoretic interpretation of the invariants of a near isomorphism class of groups in the class $C^1(T, p, r, e)$, namely the admissible ordering of T and a normed matrix A with frame data FD. But there is some hope.

References

[1] M. Dugas and E. Oxford, *Near isomorphism invariants for a class of almost completely decomposable groups*, Abelian Groups, Proceedings of the 1991 Curaçao Conference, Marcel Dekker, Inc., 1993, pp. 129–150.

[2] A. Mader and O. Mutzbauer, *Double ordering of $(0, 1)$-matrices*, Ars Combinatoria **61** (2001), pp. 81–95.

[3] A. Mader and O. Mutzbauer, *Decompositions of uniform groups*, Coll. Math. **87** (2001), pp. 211–226.

[4] O. Mutzbauer, *Normal form of matrices with application to almost completely decomposable groups*, Abelian Groups and Modules, Trends in Mathematics, (1998), pp. 121–134.

[5] B. D. Saunders and H. Schneider, *Flows on graphs applied to diagonal similarity and diagonal equivalence for matrices*, Discrete Mathematics **24** (1978), pp. 205–220.

Author information

A. Mader, Department of Mathematics, University of Hawaii, 2565 McCarthy Mall, Honolulu, HI 96822, USA.
E-mail: `adolf@math.hawaii.edu`

O. Mutzbauer, Mathematisches Institut, Universität Würzburg, Am Hubland, 97074 Würzburg, Germany.
E-mail: `mutzbauer@mathematik.uni-wuerzburg.de`

Root bases of polynomials over integral domains

E. F. Cornelius Jr. and Phill Schultz

Abstract. The authors extend results on quotient groups derived from the evaluation of integral polynomials and power series to modules over integral domains. They demonstrate that valuation points which form arithmetic, geometric, and hypergeometric series produce stacked bases and hence tractable quotient modules.

Key words. Integral root basis, stacked bases, polynomials, power series, matrices.

AMS classification. Primary: 13G05, 13C13, 13P05, 13C05. Secondary: 15A99, 20K25.

1 Background

The Vandermonde matrix of degree n

$$V = \begin{bmatrix} 1 & s_0 & s_0^2 & \cdots & s_0^{n-1} \\ \vdots & \vdots & \vdots & \vdots & \vdots \\ 1 & s_{n-1} & s_{n-1}^2 & \cdots & s_{n-1}^{n-1} \end{bmatrix} = (s_i^j : i, j = 0, \ldots, n-1),$$

where $\mathbf{s} = (s_0, s_1, \ldots, s_{n-1})$ is a sequence of elements of an integral domain D, can be viewed as the matrix whose columns are the values of the polynomials $1, x, \ldots, x^{n-1}$ evaluated at the entries of $\mathbf{s}$. Although a formula for the determinant of V is well known,

$$\det(V) = \prod_{0 \leq i < j \leq n-1} (s_j - s_i),$$

it is hardly obvious when the matrix is expressed in the foregoing format.

However, when the standard basis of $D[x]_n$, the free D-module of polynomials of degree $< n$, is replaced by the basis

$$\{1, x - s_0, (x - s_0)(x - s_1), \ldots, (x - s_0)(x - s_1) \cdots (x - s_{n-2})\},$$

the corresponding matrix of values is

$$V' = \begin{bmatrix} 1 & 0 & 0 & \cdots & 0 \\ 1 & s_1 - s_0 & 0 & \cdots & 0 \\ 1 & s_2 - s_0 & (s_2 - s_0)(s_2 - s_1) & \cdots & 0 \\ \vdots & \vdots & \vdots & \vdots & \vdots \\ 1 & s_{n-1} - s_0 & (s_{n-1} - s_0)(s_{n-1} - s_1) & \cdots & (s_{n-1} - s_0) \cdots (s_{n-1} - s_{n-2}) \end{bmatrix}$$

whose determinant is immediately obvious.

Since V and V' represent the same valuation function with respect to different bases of $D[x]_n$ and the transition matrix between these bases has determinant 1, V and V' have equal determinants, although manifestly $\det(V')$ is easier to compute, being the product of the diagonal elements. As we shall see in Section 3 below, V and V' are in fact similar matrices.

Using bases of $D[x]_n$ tailored to solve particular problems is a well-known technique in algebra. In [CS08], the authors used what we have termed the *integral root basis* of $\mathbb{Z}[x]_n$, the basis just used for V' with D the integers and $s_i = i$, to determine the structure of the factor group $\mathbb{Z}^n/P_n$, where $\mathbb{Z}^n$ denotes the group of n-tuples of integers and P_n denotes the image of $\mathbb{Z}[x]_n$ in $\mathbb{Z}^n$ when the polynomials are evaluated at $0, 1, \ldots, n-1$. Evaluation of the integral root basis of $\mathbb{Z}[x]_n$ produces strongly stacked bases (definitions below) of $\mathbb{Z}^n$ and P_n. In [CS07], the evaluation of polynomials was extended to the evaluation of power series with respect to the integral root basis of $\mathbb{Z}[x]$, and in [CS08A] to polynomials and power series in several indeterminates.

Other examples include [B74], in which Biggs employs the integral root basis (without denominating it as such) to express complex chromatic polynomials for determining the number of vertex colorings of a finite graph; [GMc69], in which Gunji and McQuillan also employ the integral root basis of $\mathbb{Z}[x]$ (again without the terminology); and [N95], in which Narkiewicz characterizes the polynomials over the rationals $\mathbb{Q}$, which map $\mathbb{Z}$ into $\mathbb{Z}$. Other illustrative works include [CC97] and the numerous references therein, as well as [E06].

In order to produce meaningful factor groups, our evaluation images are rather sparse in the sense that the factor group $\mathbb{Z}_n/P_n$ has order $0!1!\cdots(n-1)!$ (or $(n-1)$-superfactorial; see [S07, Sequence A000178]). In the infinite case, the extension of P_n is nowhere dense in the Baer–Specker group [CS07, CS08], yielding a factor group of continuum cardinality. In contrast, the factor groups in most of the papers cited would prove to be trivial.

The purpose of this paper is to extend the methods of [CS07, CS08] to an arbitrary integral domain D through use of a generalization of the integral root basis to $D[x]$. The main difficulties in applying these methods with D in place of $\mathbb{Z}$ lie in the facts that there is no natural ordering on D enabling us to attack the problem using the weapons of combinatorial theory, and that D may lack the arithmetic properties of $\mathbb{Z}$ which make the tools of number theory useful. Also absent are the simple patterns which appear when integral polynomials are evaluated at fixed intervals.

Nevertheless, some basic algebraic properties of $\mathbb{Z}[x]$, in particular the Lagrange interpolation theorem, do extend to $D[x]$, enabling us to apply an extension of the integral root basis to the study of bases of the free D-module D^n and particular submodules.

2 Notation and terminology

D will denote an integral domain with 1, and K its field of fractions. This paper concerns properties of the images under evaluation, at a set S of elements of D, of the polynomial module $D[x]$ and of the power series module $D[[x]]_R$ (defined below). In particular, we are interested in how a choice of basis for the free D-module $D[x]$ affects the evaluations of polynomials and power series. Since our results are trivial for fields, we shall assume also that D is infinite.

Let $\mathbb{N}$ denote the natural numbers, starting with zero, and $\mathbb{N}^+$ the positive integers. For $n \in \mathbb{N}^+$, denote by $[n]$ the set $\{0, \ldots, n-1\}$. Let $S = \{s_i : i \in \mathbb{N}\}$ be a countable subset of distinct elements from D and let $S_n = \{s_i : i \in [n]\}$ denote the first n elements of S.

Let $D^n \subseteq K^n$ denote the sets of n-tuples from D and K, respectively, the elements of which we view as column vectors to facilitate matrix-vector multiplication, in which the matrix acts from the left on the column vector to the right. K^n is a vector space over K and a module over D, and D^n is a D-submodule of K^n. Elements of D^n or K^n are denoted by boldface lower case letters and their components by the corresponding subscripted plaintext characters, for example $\mathbf{b} = (b_i : i \in [n])$. In particular, the first n elements S_n of S form the n-tuple $(s_0, s_1, \ldots, s_{n-1}) \in D^n$, which may be denoted by $\mathbf{s}_n$.

Let $K[x]$ and $D[x]$ denote the rings of polynomials in the indeterminate x over K and D, respectively. For $n \in \mathbb{N}^+$, $K[x]_n$ and $D[x]_n$ denote the polynomials of degree $< n$. We shall use a number of linear transformations between K^n and $K[x]_n$ and their restrictions to module homomorphisms between D^n and $D[x]_n$.

1. Given $f \in K[x]_n$ and a sequence $\mathbf{b}$ in K^n, the *evaluation of f at $\mathbf{b} \in K^n$*, denoted by $f(\mathbf{b})$, is the sequence $(f(b_i) : i \in [n]) \in K^n$. The map $\mathrm{eval}_\mathbf{b} : f \mapsto f(\mathbf{b})$ is a linear transformation of $K[x]_n$ into K^n. If $\mathbf{b} \in D^n$, the restriction of $\mathrm{eval}_\mathbf{b}$ to $D[x]_n$ is a module homomorphism into D^n.

2. Given $\mathbf{a} \in K^n$ and a basis $B = \{b_0(x), b_1(x), \ldots, b_{n-1}(x)\}$ of $K[x]_n$, let $f_\mathbf{a} \in K[x]_n$ be the polynomial $\sum_{i=0}^{n-1} a_i b_i(x)$. The map $\mathrm{poly}_B : \mathbf{a} \mapsto f_\mathbf{a}$ is a linear transformation of K^n into $K[x]_n$. If $B \subset D[x]_n$, the restriction of poly_B to D^n is a module homomorphism into $D[x]_n$.

3. Given a basis B for $K[x]_n$ and $f \in K[x]_n$, the sequence of coefficients of f with respect to B is in K^n and the map, $\mathrm{coef}_B(f) : f \mapsto$ the sequence of coefficients of f with respect to B, is a linear transformation of $K[x]_n$ into K^n. In fact, the very definition shows that coef_B and poly_B are mutual inverses. If $B \subset D[x]_n$, then coef_B restricted to $D[x]_n$ is a module homomorphism into D^n, the inverse of the restriction of poly_B.

We now address the question of the inverse of $\mathrm{eval}_\mathbf{b}$. Let $\mathbf{b} \in K^n$ have n distinct entries. By Lagrange's interpolation theorem, for each $\mathbf{a} \in K^n$, there is a unique $f \in K[x]_n$ such that $\mathrm{eval}_\mathbf{b}(f) = \mathbf{a}$. Denote this f by $\ell_\mathbf{a}^\mathbf{b}$, so that $\mathbf{a} \mapsto \ell_\mathbf{a}^\mathbf{b}$ is also a linear transformation $K^n \to K[x]_n$, in fact the inverse of $\mathrm{eval}_\mathbf{b}$. This result, however, does not have a counterpart in $D[x]_n$, because if $\mathbf{a}, \mathbf{b} \in D^n$, $\ell_\mathbf{a}^\mathbf{b}$ is in $K[x]_n$ but not necessarily in $D[x]_n$. This means that for all $\mathbf{b} \in D^n$, the homomorphism $\mathrm{eval}_\mathbf{b} : D[x]_n \to D^n$ is monic but not in general epic.

We also need notation for matrices. For m a positive integer, we denote the rings of $m \times m$ matrices over D and K by $M(m, D)$ and $M(m, K)$, respectively. Similarly we denote the modules of $\omega \times \omega$ matrices over D and K by $M(\omega, D)$ and $M(\omega, K)$ and their respective rings of row-finite matrices by $FM(\omega, D)$ and $FM(\omega, K)$.

3 The S-root basis of $D[x]$

Let $S = \{s_i : i \in \mathbb{N}\}$ be a countable subset of distinct elements of D. Define the set $R = \{\rho_i(x) : i \in \mathbb{N}\}$ of monic polynomials in $D[x]$ by $\rho_0(x) \equiv 1$ and $\rho_{i+1}(x) = \rho_i(x)(x - s_i)$ for $i \in \mathbb{N}$. Let $R_n = \{\rho_i : i \in [n]\}$ denote the first n elements of R. Since R_n is a set of n monic polynomials of degrees $0, \ldots, n-1$, it is a basis for the free D-module $D[x]_n$ and for the vector space $K[x]_n$, so that R is a basis for the free D-module $D[x]$ and for the vector space $K[x]$.

We call R and R_n, which depend upon S, the S-*root bases* of $D[x]$ and $D[x]_n$, respectively. In the papers [CS07, CS08], in which we considered the case $D = \mathbb{Z}$ and $S = \mathbb{N}$, R took the particularly simple form $\rho_0 \equiv 1$ and $\rho_i = x(x-1)\cdots(x-i+1)$ for $i \geq 1$.

We now verify the claim in Section 1 that the matrices V and V' are similar. For any basis B of $D[x]_n$, consider the endomorphism $\sigma = \text{poly}_B \circ \text{eval}_{s_n} : D[x]_n \to D[x]_n$. When B is the standard basis of $D[x]_n$, σ maps a polynomial $f(x)$ to the polynomial $f(s_0) + f(s_1)x + \cdots + f(s_{n-1})x^{n-1}$. In particular, the standard basis element x^i is mapped to $s_0^i + s_1^i x + \cdots + s_{n-1}^i x^{n-1}$. Hence with respect to the standard basis of $D[x]_n$, σ has matrix V.

Similarly, when B is the S-root basis, σ maps a polynomial $f(x)$ to the polynomial $f(s_0)\rho_0(x) + f(s_1)\rho_1(x) + \cdots + f(s_{n-1})\rho_{n-1}(x)$ and in particular, the basis element $\rho_i(x)$ is mapped to $\rho_i(s_0)\rho_0(x) + \rho_i(s_1)\rho_1(x) + \cdots + \rho_i(s_{n-1})\rho_{n-1}(x)$. Since $\rho_j(s_i) = 0$ for $i < j$, with respect to the S-root basis of $D[x]_n$, σ has matrix V'. Since V and V' are $n \times n$ matrices over D representing the same endomorphism of $D[x]_n$ with respect to different bases, they are similar.

Since D is an integral domain, we may consider $\mathbb{Z}$ or $\mathbb{Z}_p$ (the field of order p for some prime p) as a subring of D, and in the characteristic zero case, we may also consider $\mathbb{Q}$ as a subfield of K. If D has characteristic zero, we can consider the results of [CS07, CS08] as special cases in which $S = \mathbb{N}$ and the subring $\mathbb{Z} \subseteq D$ is the operative integral domain. We refer to this case as the *type example* from now on.

Henceforth in this section we consider exclusively the S-root basis R_n for $D[x]_n$ and evaluation at s_n, so we simplify the notation by omitting the subscripts from eval_{s_n}, coef_{R_n}, poly_{R_n} and the superscript from $\ell_a^{s_n}$. We denote by P_n the image $\text{eval}(D[x]_n)$, a free submodule of rank n of D^n, which we call the module of *polynomial points*.

In Section 1 of this paper we exhibited the matrix V' of the transformation eval with respect to the S-root basis of $D[x]_n$. We now redesignate this matrix as $C_n = (c_{ij})$, where

$$c_{ij} = \rho_j(s_i) = \begin{cases} 0 & \text{if } i < j, \\ 1 & \text{if } j = 0, \\ \prod_{k=0}^{j-1}(s_i - s_k) & \text{if } i \geq j \geq 1. \end{cases}$$

In the type example, $c_{ij} = (i)_j$, the *falling factorial* $j!\binom{i}{j}$.

Note that C_n is lower triangular with i-th diagonal element $c_{ii} = \rho_i(s_i) = (s_i - s_0)\cdots(s_i - s_{i-1}) \neq 0$, so that C_n is invertible in $M(n, K)$ with lower triangular inverse.

Furthermore, if $1 \leq m \leq n$, then C_m is the $m \times m$ major submatrix of C_n, i.e., its top left corner.

We now construct the inverse of C_n. We note that the inverse of the Vandermonde matrix has been presented in the literature (e.g. [PS54, p. 99] or [K73, p. 36, Ex. 40]), but not to our knowledge in the setting of a general integral domain. Moreover, $C_n = V' \neq V$ in the notation of Section 1.

Define
$$b_{ij} = \begin{cases} 0 & \text{if } i < j, \\ c_{jj}^{-1} & \text{if } i = j, \\ (-1)^{i+j}(c_{jj} \prod_{k=j+1}^{i}(s_k - s_j))^{-1} & \text{if } i > j \end{cases}$$

and let $B_n = (b_{ij}) \in M(n, K)$. Note that B_n is lower triangular with diagonal elements the inverses of the corresponding diagonal elements of C_n. The definition reveals a recursive formula for B_n, namely for all $j \in [n]$, $b_{jj} = [(s_j - s_0) \cdots (s_j - s_{j-1})]^{-1}$ and for all $i > j$, $b_{ij} = -\frac{b_{i-1,j}}{s_i - s_j}$. In the type example, $b_{ij} = \frac{(-1)^{i+j}}{i!}\binom{i}{j}$.

The following lemma appears to be an extension to an arbitrary integral domain of the combinatorial identity $\sum_{i=1}^{n}(-1)^i \binom{n}{i} = -1$, to which it reduces in $\mathbb{Z}$ for $1, 2, \ldots, n$.

Lemma 3.1. *Let $d_1, \ldots, d_n$ be any $n \geq 1$ non-zero elements from an integral domain, and set $d_0 = 0$. Then*
$$\sum_{i=1}^{n}(-1)^i \frac{(d_n - d_0)(d_n - d_1) \cdots (d_n - d_{i-1})}{d_1 \cdots d_i} = -1.$$

Proof. The lemma is obviously true for any 1 or 2 elements, so assume its truth for any $n \geq 2$. Then
$$\sum_{i=1}^{n+1}(-1)^i \frac{(d_{n+1} - d_0) \cdots (d_{n+1} - d_{i-1})}{d_1 \cdots d_i}$$
$$= \sum_{i=1}^{n-1}(-1)^i \frac{(d_{n+1} - d_0) \cdots (d_{n+1} - d_{i-1})}{d_1 \cdots d_i}$$
$$+ (-1)^n \frac{(d_{n+1} - d_0) \cdots (d_{n+1} - d_{n-1})}{d_1 \cdots d_n}$$
$$+ (-1)^{n+1} \frac{(d_{n+1} - d_0) \cdots (d_{n+1} - d_{n-1})(d_{n+1} - d_n)}{d_1 \cdots d_n d_{n+1}}$$
$$= \sum_{i=1}^{n-1}(-1)^i \frac{(d_{n+1} - d_0) \cdots (d_{n+1} - d_{i-1})}{d_1 \cdots d_i}$$
$$+ (-1)^n \frac{(d_{n+1} - d_0) \cdots (d_{n+1} - d_{n-1})}{d_1 \cdots d_{n-1} d_n} \left(1 - \frac{d_{n+1} - d_n}{d_{n+1}}\right)$$

$$= \sum_{i=1}^{n-1}(-1)^i \frac{(d_{n+1}-d_0)\cdots(d_{n+1}-d_{i-1})}{d_1\cdots d_i}$$

$$+ (-1)^n \frac{(d_{n+1}-d_0)\cdots(d_{n+1}-d_{n-1})}{d_1\cdots d_n} \frac{d_n}{d_{n+1}}$$

$$= \sum_{i=1}^{n-1}(-1)^i \frac{(d_{n+1}-d_0)\cdots(d_{n+1}-d_{i-1})}{d_1\cdots d_i}$$

$$+ (-1)^n \frac{(d_{n+1}-d_0)\cdots(d_{n+1}-d_{n-1})}{d_1\cdots d_{n-1}d_{n+1}}.$$

Now this last expression is just the given expression using the n elements $d_1, \ldots, d_{n-1}, d_{n+1}$, so by induction, it equals -1. □

Theorem 3.2. (1) *Let B_n and C_n be the matrices defined above. Then $B_n = C_n^{-1}$.*

(2) *Let $\mathbf{a} \in K^n$ and $B_n\mathbf{a} = \mathbf{b}$. Then $C_n\mathbf{a} = \text{eval } f_\mathbf{a}$ and $\ell_\mathbf{a} = f_\mathbf{b}$.*

Proof. (1) Let $C_n B_n = E_n = (e_{ij})$. Then E_n is a lower triangular $n \times n$ matrix with 1's on the diagonal, so it remains to show that for all $0 \leq j < i$, $e_{ij} = \sum_{k=0}^{n-1} c_{ik}b_{kj} = 0$. Since $c_{ik} = 0$ if $k > i$ and $b_{kj} = 0$ if $k < j$, we may restrict the summation limits to $j \leq k \leq i$.

Because of the definition of c_{i0} we must deal with the special case $j = 0$ first. We have for all $1 \leq i < n$,

$$e_{i0} = c_{i0}b_{00} + \sum_{k=1}^{i} c_{ik}b_{k0} = 1 + \sum_{k=1}^{i}(-1)^k \frac{\prod_{\ell=0}^{k-1}(s_i - s_\ell)}{\prod_{\ell=1}^{k}(s_\ell - s_0)} = 0$$

by Lemma 3.1, with $n = i$ and $d_\ell = s_\ell - s_0$ for all $\ell = 1, \ldots, k$.

Now suppose that $j \geq 1$. Then

$$e_{ij} = \sum_{k=j}^{i} c_{ik}b_{kj} = c_{ij}b_{jj} + \sum_{k=j+1}^{i}(-1)^{k+j}\frac{(s_i-s_0)\cdots(s_i-s_{k-1})}{c_{jj}(s_{j+1}-s_j)\cdots(s_k-s_j)}$$

$$= \frac{c_{ij}}{c_{jj}} + \frac{1}{c_{jj}}\sum_{k=j+1}^{i}(-1)^{k+j}\frac{(s_i-s_0)\cdots(s_i-s_{k-1})}{(s_{j+1}-s_j)\cdots(s_k-s_j)}$$

$$(*) \quad = \frac{1}{c_{jj}}\left[c_{ij} + (s_i-s_0)\cdots(s_i-s_{j-1})\sum_{k=j+1}^{i}(-1)^{k+j}\frac{(s_i-s_j)\cdots(s_i-s_{k-1})}{(s_{j+1}-s_j)\cdots(s_k-s_j)}\right].$$

In Lemma 3.1, take 'i'$= k - j$, 'n'$= i - j$ and 'd_i'$= s_{i+j} - s_j$. Then

$$\sum_{k=j+1}^{i}(-1)^{k+j}\frac{(s_i-s_j)\cdots(s_i-s_{k-1})}{(s_{j+1}-s_j)\cdots(s_k-s_j)} = \sum_{i=1}^{n}(-1)^i\frac{(d_n-d_0)\cdots(d_n-d_{i-1})}{d_1\cdots d_i} = -1.$$

Hence the expression (∗) reduces to $c_{ij} - \rho_j(s_i)$ so that $e_{ij} = 0$.

(2) Since as remarked above, C_n is the matrix over K representing the linear transformation eval with respect to the S-root basis of $D[x]_n$, the first statement is a matter of definition. The second holds because from (1) and the first statement, $\mathbf{a} = C_n\mathbf{b} = \text{eval}(f_\mathbf{b})$. Now also $\mathbf{a} = \text{eval}(\ell_\mathbf{a})$ so that $f_\mathbf{b} = \ell_\mathbf{a}$. □

4 Stacked bases

We have shown above that any choice of the countably infinite set S of distinct elements of D determines, for each positive integer n, a matrix C_n whose columns are in D^n and are linearly independent over K and hence form a basis for a free submodule P_n of D^n, namely the image of R_n under evaluation at $\mathbf{s}_n$. We call this basis the S-*Gamma basis* of P_n, denoted $\mathcal{G}_n$.

Let M be a free submodule of D^n, with basis $B = \{b_i : i \in [n]\}$, and let $A = \{a_i : i \in [n]\}$ be a basis of D^n. We say A and B are *stacked* if there exist elements $\{k_i : i \in [n]\}$ of D such that $b_i = k_i a_i$ for all $i \in [n]$, and *strongly stacked* if further $k_i | k_{i+1}$ for all $i \in [n-1]$. In either case we call the set $\{k_i : i \in [n]\}$ the *stacking factors* of A and B. It is well known that if D is a principal ideal domain, then D^n and its rank n submodules always have strongly stacked bases; see for example [J74, Section 3.8]. A similar result does not hold for infinite products, even when $D = \mathbb{Z}$; see putative proof [F73, Lemma 95.1] and counterexample [G81].

Although we impose no restriction on the infinite integral domain D, we do posit the existence of a basis $\mathcal{A}_n$ of D^n such that $\mathcal{A}_n$ and $\mathcal{G}_n$ are stacked. This is equivalent to the fact that each column of C_n can be divided in D by its diagonal entry c_{jj} to produce a lower triangular matrix A_n whose diagonal entries are 1's. Since A_n has determinant 1, it is invertible in $M(n, D)$ so its columns form a basis $\mathcal{A}_n$ of D^n which is stacked with the S-Gamma basis $\mathcal{G}_n$ of P_n. In accordance with the terminology of [CS07], we call $\mathcal{A}_n$ the S-*Alpha basis* of D^n.

When C_n has this divisibility property, we say that S_n is *stackable*. Further, we say that S is *stackable* if S_n is stackable for every n. If S_n is stackable, and Δ_n denotes the diagonal matrix $\text{diag}(c_{jj} : j \in [n])$, then $C_n = A_n \Delta_n$ and the set $\{c_{jj} : j \in [n]\}$ forms the stacking factors of $\mathcal{A}_n$ and $\mathcal{G}_n$, to which we refer simply as the stacking factors of S_n.

It is important to note that if S is stackable, then for all $m \le n \in \mathbb{N}^+$, the S-Alpha bases $\mathcal{A}_m$ and $\mathcal{A}_n$ are compatible in the sense that the $m \times m$ major submatrix of A_n is A_m. Similarly the S-Gamma bases $\mathcal{G}_m$ and $\mathcal{G}_n$ are compatible. It follows that for all $m \le n$, the stacking factors of S_m are the initial m stacking factors of S_n.

In the type example, we showed [CS08] that A_n is Pascal's matrix $\left(\binom{i}{j}\right)$ with inverse $\left((-1)^{i+j}\binom{i}{j}\right)$. For the general case, a routine calculation, using $A_n^{-1} = \Delta_n B_n$ and Theorem 3.2(1), establishes the following proposition, in which we interpret products of the form $\prod_{k=0}^{-1}(s_j - s_k)$ and of the form $\prod_{k=j}^{j}(s_k - s_j)$ as 1.

Proposition 4.1. *Let S_n be stackable and let $A_n = C_n \Delta_n^{-1}$. Then $A_n = (a_{ij})$ and its inverse $A_n^{-1} = (\tilde{a}_{ij})$ are lower triangular with 1's on the diagonal and for all $0 \leq j \leq i \leq n-1$,*

$$a_{ij} = \frac{\rho_j(s_i)}{\rho_j(s_j)} = \prod_{k=0}^{j-1} \frac{s_i - s_k}{s_j - s_k}$$

and

$$\tilde{a}_{ij} = (-1)^{i+j} \frac{\rho_i(s_i)}{\rho_j(s_j) \prod_{k=j+1}^{i}(s_k - s_j)} = (-1)^{i+j} \frac{\prod_{k=0}^{i-1}(s_i - s_k)}{\prod_{k=0}^{j-1}(s_j - s_k) \prod_{k=j+1}^{i}(s_k - s_j)}.$$

□

If the columns of C_n and A_n determine strongly stacked bases, we say that S_n is *strongly stackable*. Accordingly, S is *strongly stackable* if S_n is strongly stackable for every n.

Note that if S_n is strongly stackable, then for all $j \in [n]$, the j-th column of C_n is $c_{jj} \times$ the j-th column of A_n and $c_{jj} | c_{j+1,j+1}$ for all $j \in [n-1]$. In this case, our stacking factors coincide with the well-known Smith invariants for modules and matrices over a principal ideal domain, except for a reversal of order.

We now show that in our context, stackable and strongly stackable are equivalent.

Proposition 4.2. *Let $S_n = \{s_i : i \in [n]\}$ be a set of distinct elements of D. If S_n is stackable, then S_n is strongly stackable.*

Proof. In the notation of Section 3, we must show that for all $j \in [n-1]$, $c_{jj} | c_{j+1,j+1}$. Since S_n is stackable, $c_{jj} | c_{j+1,j}$. But by definition, $c_{j+1,j+1} = c_{j+1,j}(s_{j+1} - s_j)$ so that $c_{jj} | c_{j+1,j+1}$ as required. □

Example 4.3. (i) For any D and S, S_2 is stackable with stacking factors $c_{00} = 1$ and $c_{11} = s_1 - s_0$. This follows from the fact that $C_2 = \begin{bmatrix} 1 & 0 \\ 1 & s_1 - s_0 \end{bmatrix}$.

(ii) If D is a field, then $P_n = D^n$ for all n so that any set of distinct elements S_n is stackable with stacking factors all equal to 1.

(iii) In [CS08, Theorem 3.2], we showed that if $D = \mathbb{Z}$ and $S = \mathbb{N}$, then S is stackable with stacking factors $\{c_{ii} = i! : i \in \mathbb{N}\}$.

(iv) Nevertheless, not every choice of S_n in $\mathbb{Z}$ is stackable. For example, let $S_3 = \{0, 2, 5\}$ in that order. Then

$$C_3 = \begin{bmatrix} 1 & 0 & 0 \\ 1 & 2 & 0 \\ 1 & 5 & 15 \end{bmatrix}.$$

Since the central element 2 does not divide the entry 5 below it, the corresponding basis of P_3 is not stacked with any basis of $\mathbb{Z}^3$. The usual row and column operations on C_3 produce a basis

$$\left\{ \mathbf{u}_0 = \begin{bmatrix} 1 \\ -2 \\ 3 \end{bmatrix}, \mathbf{u}_1 = \begin{bmatrix} 0 \\ 3 \\ -5 \end{bmatrix}, \mathbf{u}_2 = \begin{bmatrix} 0 \\ 44 \\ -73 \end{bmatrix} \right\}$$

of $\mathbb{Z}^3$ which is strongly stacked with the basis $\{\mathbf{u}_0, \mathbf{u}_1, 30\mathbf{u}_2\}$ of P_3. However, the resulting bases and the corresponding factor modules bear no discernible relationship to the given set S_3.

We now consider closure properties of stackable sequences.

Proposition 4.4. *Let α be an automorphism of D and $S_n = \{s_i : i \in [n]\}$ a stackable set. Then $\alpha(S_n) = \{\alpha(s_i) : i \in [n]\}$ is stackable.*

Proof. It is obvious that αS_n is a set of n distinct elements and the entries of the resulting matrix are the images under α of the corresponding entries of C_n. Since α preserves divisibility, αS_n is stackable. □

Let $a, b \in D$, $a \neq 0$. An *affine transformation* $T_{a,b}$ of D is one of the form $T_{a,b}(d) = ad + b$ for all $d \in D$. Note that $T_{a,b}$ is *not* an automorphism of D unless a is a unit. Nevertheless, we have

Proposition 4.5. *Let $S_n = \{s_i : i \in [n]\}$ be stackable and let $T_{a,b}$ be an affine transformation of D. Then $T_{a,b}(S_n) = \{as_0 + b, \ldots, as_{n-1} + b\}$ is stackable. In particular, $S_n + b = \{s_i + b : i \in [n]\}$ and $aS_n = \{as_i : i \in [n]\}$ are stackable.*

Proof. First note that for all relevant i and j, $a(s_i - s_j) = (as_i + b) - (as_j + b)$ so there is no loss in generality if we assume $b = 0$. Let C_n be the matrix corresponding to S_n and C'_n the matrix corresponding to aS_n. Then for all $j \in [n]$, the j-th column of C'_n is a^j times the j-th column of C_n. It follows that each entry of column j is divisible by the diagonal entry so that aS_n is stackable. □

Example 4.6. Proposition 4.5 confirms what observation of the C_n matrices reveals: Such matrices depend only upon the differences $s_i - s_0$, because $s_i - s_j = (s_i - s_0) - (s_j - s_0)$; i.e., the entire matrix is determined by its column 1. Thus sets S_n and $S'_n \subset D$ yield the same C_n matrix if and only if S_n and S'_n are translations of each other.

In order to present examples of stacked bases, we need a result which is interesting in its own right.

Theorem 4.7. *Let D be an integral domain. Then*

$$\frac{(x^k - 1)}{(x - 1)} \frac{(x^{k+1} - 1)}{(x^2 - 1)} \cdots \frac{(x^{k+p-1} - 1)}{(x^p - 1)} \in D[x] \quad \text{for all } k, p \in \mathbb{N}^+.$$

Proof. The theorem is certainly true for $k = 1$ and for all $p \in \mathbb{N}^+$, and for all $k \in \mathbb{N}^+$ and $p = 1, 2$. Assume it is true for some $k \in \mathbb{N}^+$ and for some $p \geq 2$.

Let
$$\begin{bmatrix} k+p-1 \\ p \end{bmatrix} = \frac{(x^k-1)}{(x-1)} \frac{(x^{k+1}-1)}{(x^2-1)} \cdots \frac{(x^{k+p-1}-1)}{(x^p-1)}.$$

Then for all $p \geq 2$,
$$\begin{bmatrix} k+p-1 \\ p \end{bmatrix} = \begin{bmatrix} k+p-2 \\ p \end{bmatrix} + \begin{bmatrix} k+p-2 \\ p-1 \end{bmatrix} x^{k-1} = \begin{bmatrix} k+p-2 \\ p \end{bmatrix} x^p + \begin{bmatrix} k+p-2 \\ p-1 \end{bmatrix}.$$

Hence by the inductive hypothesis, the theorem is true for $k + 1$ and for all $p \geq 3$. □

Remark 4.8. The polynomials in Theorem 4.7 are related to hypergeometric series and q-nomial coefficients [K73, p. 64, (37); p. 72, Ex. 58; p. 488, #58] and to Gaussian binomial coefficients [CC97, p. 46, Ex. 15]. The discussion here provides an illustration of how such concepts arise naturally in algebraic contexts.

The following examples show that if S is a non-constant arithmetic, geometric or hypergeometric sequence in D, then S is strongly stackable.

Example 4.9. (1) We have shown in [CS07] that the type example, $s_i = i$ for all $i \in \mathbb{N}$, is stackable.

(2) Hence by Proposition 4.5, any non-constant arithmetic sequence is stackable.

(3) Let d be a non-zero element of D, which is not a root of unity, and let $s_i = d^i$. As always, $c_{i0} = 1$ and for all $j \geq 1$,
$$c_{jj} = (d^j - 1)(d^j - d) \cdots (d^j - d^{j-1}) = d^{\binom{j}{2}}(d^j - 1)(d^{j-1} - 1) \cdots (d - 1),$$
whereas for all $i > j \geq 1$,
$$c_{ij} = (d^i - 1)(d^i - d) \cdots (d^i - d^{j-1}) = d^{\binom{j}{2}}(d^i - 1)(d^{i-1} - 1) \cdots (d^{i-j+1} - 1).$$

Thus $c_{jj} | c_{ij}$ if and only if $\frac{(d^i-1)}{(d^j-1)} \frac{(d^{i-1}-1)}{(d^{j-1}-1)} \cdots \frac{(d^{i-j+1}-1)}{(d-1)}$ is an element of D, which is true by Theorem 4.7. Hence $\{s_i = d^i : i \in [n]\}$ is stackable. When d is a root of unity, the process obviously stops at step $k - 1$ as soon as $d^k = 1$.

(4) Let d be a non-zero element of D, which is not a root of unity, and let $s_i = \frac{d^i-1}{d-1}$ so that $s_i = \sum_{k=0}^{i-1} d^k$ for $i \geq 1$. As always, $c_{i0} = 1$. Then for all $i \geq j \geq 1$, $c_{ij} = \prod_{m=1}^{j} \sum_{\ell=1}^{i-j+m} d^{i-\ell}$. It follows that for all $i > j \geq 1$,
$$c_{ij} = c_{jj} \frac{(d^{j+1}-1)}{(d-1)} \frac{(d^{j+2}-1)}{(d^2-1)} \cdots \frac{(d^i-1)}{(d^{i-j}-1)}.$$

By Theorem 4.7, the multiplier of c_{jj} is in D, so that $\{s_i\}$ is stackable. As in (3), when d is a root of unity, the process necessarily terminates.

5 Extension to $D[x]_R$

There are at least two approaches to extending the preceding results to modules of infinite rank: One may consider the union of the inclusions $D[x]_m \subseteq D[x]_n$ with bases $R_m \subseteq R_n$ for all $m \leq n$, which leads to the polynomial module $D[x]$ with basis R; or one may consider the module of formal power series with respect to the S-root basis, $D[[x]]_R = \{\sum_{i=0}^{\infty} a_i \rho_i(x) : a_i \in D\}$. In this section, we take the first approach.

Let D^ω be the module of countable sequences in D and $D^{(\omega)}$ its submodule of sequences which are 0 except at finitely many places. Algebraically D^ω is the direct product $\prod_\omega D$ and $D^{(\omega)}$ is the direct sum $\oplus_\omega D$. Extend the valuation maps eval : $D[x]_n \to D^n$ to a map, also called eval, from $D[x]$ to D^ω by defining eval$(f) = (f(s_i) : i \in \mathbb{N})$. Denote eval$(D[x])$ by P_ω. Extend the S-Gamma bases $\mathcal{G}_n$ of the P_n to a subset $\mathcal{G}_\omega = \{\gamma_j : j \in \mathbb{N}\}$ of D^ω, by defining $\gamma_j(i) = \rho_j(s_i)$, $i \in \mathbb{N}$.

Proposition 5.1. *The map eval is an isomorphism of $D[x]$ onto P_ω, which is a countably generated free submodule of D^ω with basis $\mathcal{G}_\omega$.*

Proof. The map eval is epic by definition and monic because only the zero polynomial has infinitely many zeros. P_ω is a countably generated and free because $D[x]$ is. $\mathcal{G}_\omega$ is a basis of P_ω because it is the image under eval of the basis R of $D[x]$. □

Suppose now that S is stackable so that for all $n \in \mathbb{N}^+$, D^n and P_n have strongly stacked bases. Extend the S-Alpha bases $\mathcal{A}_n$ to a subset $\mathcal{A}_\omega = \{\alpha_j : j \in \mathbb{N}\}$ of D^ω by defining $\alpha_j = \gamma_j/\rho_j(s_j)$. Note that $\mathcal{A}_\omega$ is a linearly independent set because $\mathcal{G}_\omega$ is. In an obvious extension of the finite-dimensional definition, we say that $\mathcal{A}_\omega$ and $\mathcal{G}_\omega$ are *stacked* with *stacking factors* $\rho_j(s_j)$.

If π_n is the canonical projection of D^ω onto D^n, then $\pi_n P_\omega = P_n$, $\mathcal{G}_n = \{\pi_n(\gamma_j) : j \in [n]\}$ and $\mathcal{A}_n = \{\pi_n(\alpha_j) : j \in [n]\}$.

Let P_* denote the pure submodule of D^ω generated by P_ω. Then we have

Proposition 5.2. *For stackable S, P_* is a countably generated free submodule of D^ω, with basis $\mathcal{A}_\omega$, and*

$$P_*/P_\omega \cong \bigoplus_{j \in \mathbb{N}} \frac{D}{\rho_j(s_j)D}.$$

Proof. Since it is clear that $\mathcal{A}_\omega$ is independent, it suffices to prove that it spans P_*. Because $\rho_j(s_j)\alpha_j = \gamma_j$, for all $j \in \mathbb{N}$, $\alpha_j \in P_*$ since P_* is pure. Let $0 \neq x \in P_*$. By purity, $dx \in P_\omega$ for some $0 \neq d \in D$ so that $dx = \sum_{j=0}^{n-1} d_j \gamma_j$ for some $n \in \mathbb{N}^+$, with $d_j \in D$, $j = 0, \ldots, n-1$. Thus $dx = \sum_{j=0}^{n-1} d_j \rho_j(s_j) \alpha_j$.

Application of the canonical projection π_n of D^ω onto D^n yields

$$d\pi_n(x) = \sum_{j=0}^{n-1} d_j \rho_j(s_j) \pi_n(\alpha_j).$$

Now $\{\pi_n(\alpha_j) : j \in [n]\} = \mathcal{A}_n$ is a basis of D^n so that $d | d_j \rho_j(s_j)$, $j = 0, \ldots, n-1$; i.e., x is a linear combination of the α_j. That P_*/P_ω has the claimed structure is clear. □

6 Extension to $D[[x]]_R$

Let $S = \{s_i : i \in \mathbb{N}\}$ be a set of distinct elements from D and let $\mathbf{a} = (a_i : i \in \mathbb{N}) \in D^\omega$. Define the (formal) power series with respect to the S-root basis with coefficient sequence $\mathbf{a}$ to be the expression $f_\mathbf{a}(x) = \sum_{i \in \mathbb{N}} a_i \rho_i(x)$. Let $D[[x]]_R$ denote the module under addition of all such power series. Similarly let $K[[x]]_R$ denote the module of power series $f_\mathbf{a}(x) = \sum_{i \in \mathbb{N}} a_i \rho_i(x)$ with $\mathbf{a} \in K^\omega$.

The linear transformations described in Section 2 have the following counterparts:
Given $f \in K[[x]]_R$ and the sequence in D^ω determined by S, $\mathbf{s} = (s_i : i \in \mathbb{N}) \in D^\omega$, the *evaluation of f at* $\mathbf{s}$, denoted $\text{eval}_\mathbf{s}(f)$, or simply $\text{eval}(f)$, is the sequence $(f(s_i) : i \in \mathbb{N}) \in K^\omega$. Note that since $\rho_j(s_i) = 0$ for all $j > i$, $f(s_i)$ is always a well defined element of K. If $f \in D[[x]]_R$ then $\text{eval}(f) \in D^\omega$ and eval restricted to $D[[x]]_R$ is an isomorphism of $D[[x]]_R$ into D^ω. We call the image $\text{eval}(D[[x]]_R)$ the module of *power points*, denoted $\mathbf{P}$. Since each $f(x) \in D[x]$ can be expressed uniquely as a power series in $D[[x]]_R$, eval maps $D[x]$ isomorphically onto $\mathbf{P}_\omega \subset \mathbf{P}$.

The mapping $\mathbf{a} \mapsto f_\mathbf{a}$ is an isomorphism of K^ω with $K[[x]]_R$, denoted pow, and the restriction of pow to D^ω is an isomorphism of D^ω with $D[[x]]_R$.

Given $f \in K[[x]]_R$, the sequence of coefficients of f with respect R, denoted $\text{coef}(f)$, is in K^ω and the map $f \mapsto \text{coef}(f)$ is an isomorphism of $K[[x]]_R$ with K^ω. The definition shows that coef and pow are mutual inverses in both cases, K^ω and D^ω.

We deal now with the analog for power series of the Lagrange interpolation polynomial. For all $\mathbf{a} \in K^\omega$, let $\ell_\mathbf{a}(x) \in K[[x]]_R$ be the unique power series whose initial n terms, for every $n \in \mathbb{N}$, is the polynomial $\ell_{\mathbf{a}_n}$, i.e., the Lagrange interpolation polynomial for the initial segment $\mathbf{a}_n$ of $\mathbf{a}$. This definition makes sense since we showed in [CS07, Lemma 4.1] that for $m \leq n$, $\ell_{\mathbf{a}_m}$ is the polynomial consisting of the initial m terms of the polynomial $\ell_{\mathbf{a}_n}$, and the proof there carries over readily to integral domains. We call $\ell_\mathbf{a}$ the *Lagrange interpolation power series for* $\mathbf{a}$.

As in the finite rank case, the functions eval and $\mathbf{a} \mapsto \ell_\mathbf{a}$ are inverses, so that eval is an isomorphism of $K[[x]]_R$ with K^ω.

The $\omega \times \omega$ lower triangular matrix C whose columns are the S-Gamma basis $\mathcal{G}_\omega$ has a lower triangular inverse $B \in FM(\omega, K)$, the top left $n \times n$ corner of which, for all $n \in \mathbb{N}^+$, is B_n. If S is stackable, then the $\omega \times \omega$ matrix A whose columns are the S-Alpha basis $\mathcal{A}_\omega$ has a lower triangular inverse $A^{-1} \in FM(\omega, D)$, since A is lower triangular with 1's down the diagonal. The top left $n \times n$ corner of A^{-1} is just A_n^{-1}. A, B and C act by left multiplication on the elements of K^ω when they are viewed as column vectors.

Proposition 6.1. *Let* $\mathbf{a} \in K^\omega$ *and* $B\mathbf{a} = \mathbf{b}$. *Then* $C\mathbf{a} = \text{eval } f_\mathbf{a}$ *and* $\ell_\mathbf{a} = f_\mathbf{b}$.

Proof. Both statements follow immediately from Theorem 3.2 and the properties of the sequences (C_n), $(f_{\mathbf{a}_n})$ and $(\ell_{\mathbf{a}_n})$. □

The relationship among power series, coefficients, and power points is given by

Theorem 6.2. *Let* $\mathbf{a} \in D^\omega$. *The following are equivalent:*

(1) $\mathbf{a} \in \mathbf{P}$.

(2) $\ell_{\mathbf{a}}(x) \in D[[x]]_R$.

(3) $B\mathbf{a} \in D^\omega$.

Proof. The equivalence of (1) and (3) is just Proposition 6.1. The equivalence of (2) and (3) is a consequence of the fact that the columns of B are the coefficients with respect to the S-root basis of the Lagrange interpolation power series of the standard basis elements of $D^{(\omega)}$. □

While the equivalence of (1) and (2) follows from the proof above, it is worth noting that since $\mathbf{a} \in \mathbf{P}$, each initial segment $\mathbf{a}_n = (a_0, a_1, \ldots, a_{n-1}) \in P_n$, so by Theorem 3.2(2), $\ell_{\mathbf{a}_n}(x) \in D[x]_n$. Hence by the construction of $\ell_{\mathbf{a}}(x)$, all its coefficients are in D.

Suppose now that S is stackable, so that we have a matrix A whose columns form the S-Alpha basis $\mathcal{A}_\omega$ of P_*, which is strongly stacked with the S-Gamma basis $\mathcal{G}_\omega$ of P_ω as in Section 5. Under these hypotheses, the following holds:

Theorem 6.3. $D^\omega/\mathbf{P} \cong \prod_{j \in \mathbb{N}} D/\rho_j(s_j)D$.

Proof. Consider the S-Alpha basis $\mathcal{A}_\omega = \{\alpha_j : j \in \mathbb{N}\}$. Because the matrix A with columns formed by the α_j's is lower triangular with 1's down the diagonal, each element $\mathbf{x} \in D^\omega$ can be expressed uniquely as $\mathbf{x} = \sum_{j \in \mathbb{N}} b_j \alpha_j$ for appropriate $b_j \in D$. Thus $D^\omega = \prod_{j \in \mathbb{N}} \langle \alpha_j \rangle$, where $\langle \alpha_j \rangle$ denotes the submodule of D^ω generated by α_j. By Proposition 6.1, $\mathbf{P} = CD^\omega$ so that each element $\mathbf{y} \in \mathbf{P}$ can be expressed as $\mathbf{y} = \sum_{j \in \mathbb{N}} a_j \gamma_j$, $a_j \in D$. Since C is invertible, the expression is unique and $\mathbf{P} = \prod_{j \in \mathbb{N}} \langle \gamma_j \rangle$. Finally, inasmuch as $\gamma_j = \rho_j(s_j)\alpha_j$ for all $j \in \mathbb{N}$, $D^\omega/\mathbf{P} \cong \prod_{j \in \mathbb{N}} D/\rho_j(s_j)D$. □

Example 6.4. (i) In the type example, the quotient is a non-splitting mixed group [CS07, Theorems 4.3 and 5.2].

(ii) Let $D = \mathbb{Z}[x]$ and let $S = \{x^n : n \in \mathbb{N}\}$. Then it is not difficult to see that, as a D-module, $D^\omega/\mathbf{P}$ is mixed and, as an abelian group, is isomorphic to the Baer–Specker group; [CS07, Lemma 4.3] may be of assistance with the latter.

Finally, we introduce the product analog of stacked bases. Let U be a submodule of D^ω. We say that U is a *product* in D^ω with *product basis* $B = \{\beta_n : n \in \mathbb{N}\}$, denoted $U = \prod_{n \in \mathbb{N}} \langle \beta_n \rangle$, if $B \subset U$ and each sum of the form $\sum_{n \in \mathbb{N}} d_n \beta_n$, $d_n \in D$, is a well defined element of U, and each element of U has a unique representation of that form, the sums being calculated in D^ω. We say that product bases $\{\beta_n : n \in \mathbb{N}\}$ of U and $\{\delta_n : n \in \mathbb{N}\}$ of D^ω are *stacked* if there is a sequence $\{t_n : n \in \mathbb{N}\}$ of elements of D such that for all $n \in \mathbb{N}$, $\beta_n = t_n \delta_n$, and *strongly stacked* if in addition, $t_n | t_{n+1}$ for all $n \in \mathbb{N}$. In either case we call $\{t_n\}$ the *stacking factors*.

In this terminology, Theorem 6.3 states that D^ω is a product with product basis the S-Alpha basis, $\mathbf{P}$ is a product with product basis the S-Gamma basis, and these product bases are strongly stacked with stacking factors $\{\rho_j(s_j)\}$.

References

[B74]   N. L. Biggs, *Algebraic Graph Theory*, Cambridge Tracts in Mathematics, 67, 1974

[CC97]   P-J. Cahen and J-L. Chabert, *Integer-Valued Polynomials*, Mathematical Surveys and Monographs, Vol. 48, Amer. Math. Soc., 1997

[CS07]   E. F. Cornelius Jr. and P. Schultz, *Polynomial Points*, Journal of Infinite Sequences, Vol. 10 (2007), Article 07.3.6

[CS08]   E. F. Cornelius Jr. and P. Schultz, *Sequences Generated by Polynomials*, Amer. Math. Monthly, 115 (2008), 154–158

[CS08A] E. F. Cornelius Jr. and P. Schultz, *Multinomial Points*, Houston J. Math., 34 (2008), 661–676

[E06]   J. Elliott, *Binomial rings, integer-valued polynomials and λ-rings*, J. Pure and Appl. Algebra, 207 (2006), 165–185

[F73]   L. Fuchs, *Infinite Abelian Groups*, Vol. 2, Academic Press, 1973

[G81]   B. Goldsmith, *A note on products of infinite cyclic groups*, Rend. Semin. Mat. Univ. Padova, 64 (1981), 243–246

[GMc69] H. Gunji and D. L. McQuillan, *On polynomials with integer coefficients*, J. Number Theory (1969), 486–493

[J74]   N. Jacobson, *Basic Algebra I*, W. H. Freeman and Co., 1974

[K73]   D. E. Knuth, *Fundamental Algorithms*, Vol. 1 of The Art of Computer Programming, Addison-Wesley, 2nd Ed. 1973

[N95]   W. Narkiewicz, *Polynomial Mappings*, Lecture Notes in Mathematics, Vol. 1600, Springer, 1995

[PS54]   G. Polya and G. Szegö, *Aufgaben und Lehrsätze aus der Analysis, 2*, Springer, 1954

[S07]   N. J. A. Sloane, *The On-Line Encyclopedia of Integer Sequences*, 2007. Available at http://www.research.att.com/~njas/sequences/

Author information

E. F. Cornelius Jr., College of Engineering and Science, University of Detroit Mercy, Detroit, MI 48221-3038, USA.
E-mail: efcornelius@comcast.net

Phill Schultz, School of Mathematics and Statistics, The University of Western Australia, Nedlands, 6009, Australia.
E-mail: schultz@maths.uwa.edu.au

A solution to a problem on lattice isomorphic Abelian groups

Grigore Călugăreanu and Kulumani M. Rangaswamy

Abstract. This paper completes the solution to the problem of deciding when two Abelian groups will have the lattices of their subgroups isomorphic.

Key words. Abelian groups, lattice of subgroups, torsion-free rank one.

AMS classification. 20K21, 20K27, 06C99.

1 Introduction

This paper completes the last step in the investigation initiated by R. Baer [1] to answer the question: When two Abelian groups G and H have their subgroup lattices $L(G)$ and $L(H)$ isomorphic? In this case, following Baer [1], we say that G and H are *projective groups*. Note that if G is projective with H, then G need not be isomorphic to H as is clear when G and H are cyclic groups of order p^n and q^n respectively, where p and q are different prime numbers.

R. Baer made substantial progress in solving this problem and proved, among other things, the following major results:

Theorem 1.1 ([1], [2]). *Let G and H be two Abelian groups and $L(G)$ and $L(H)$ be their subgroup lattices.*

(a) *If G has torsion-free rank > 1, then $L(G) \cong L(H)$ if and only if $G \cong H$.*

(b) *If G is a torsion group, then $L(G) \cong L(H)$ if and only if there is a bijection between the primary components of G and H such that the corresponding primary components P and Q are isomorphic whenever rank $P > 1$, and if P has rank 1, say $P \cong \mathbf{Z}(p^n)$ for some prime p with $n > 0$ or $n = \infty$, then the corresponding primary component Q is isomorphic to $\mathbf{Z}(q^n)$ for some (perhaps different) prime q.*

(c) *If G is torsion-free and has rank 1, then $L(G) \cong L(H)$ implies that H is a torsion-free group of rank 1. Moreover, $G \cong H$, if G is, in addition, infinite cyclic.*

L. Fuchs ([2]) extended Theorem 1.1(c) by proving the following:

This work was done when the second author visited Kuwait University during December 2006. He gratefully acknowledges the kind hospitality of the faculty and staff of the mathematics department.

Theorem 1.2 ([3]). *Let G be a rank 1 torsion-free Abelian group of type*

$$(k_1, \ldots, k_n, \ldots).$$

If G is projective with an Abelian group H, then H is a rank 1 torsion-free group with type $(l_1, \ldots, l_n, \ldots)$ where the l_i's are obtained from the k_i's by a permutation π of the indexing set of primes.

Extracting some of the ideas from the papers by R. Baer ([1]) and L. Fuchs ([2]) and making corrections, K. Mahdavi and J. Poland ([5]) obtained the following necessary condition for two mixed groups of torsion-free rank 1 to be projective:

Theorem 1.3 ([5]). *Let G and H be two mixed Abelian groups of torsion-free rank 1. If $f : L(G) \to L(H)$ is a lattice isomorphism, then the torsion part $T(G) \cong T(H)$, the height matrix $U(H)$ results from $U(G)$ by a permutation π of the primes (the rows of $U(G)$) which fixes the primes p occurring as orders of elements in G and π and f are related by the property that for any subgroup S of G and prime p, $f(pS) = \pi(p)f(S)$.*

The permutation π in Theorem 1.3 is actually *multiplicative*, that is, it extends to an endomorphism of the multiplicative semigroup of positive integers. Further, as pointed out in ([5]), the equation $f(nS) = \pi(n)f(S)$ holds for all positive integers n and for all subgroups S of G.

An example of Megibben (see [7], [6]) shows that the converse of Theorem 1.3 is false. This raises the problem of describing the mixed Abelian groups G and H with torsion-free rank 1 for which $L(G) \cong L(H)$.

K. Mahdavi and J. Poland ([5], [6]), and independently, U. Ostendorf ([8]), partially answered this by considering the special case when G is a splitting mixed Abelian group.

Our main theorem answers this problem completely.

Theorem 1.4. *Let G and H be mixed Abelian groups of torsion-free rank 1. Then $L(G) \cong L(H)$ if and only if*

(i) *G and H have isomorphic torsion parts: $T(G) \cong T(H)$;*

(ii) *there exist elements of infinite order $a \in G$, $b \in H$ such that the height matrix $\mathcal{H}(b) = U(H)$ arises from $\mathcal{H}(a) = U(G)$ by a permutation π of primes which fixes those primes occurring as orders of elements in G and is multiplicative;*

(iii) *there is a bijection between the p-components of $G/\langle a \rangle$ and $H/\langle b \rangle$ such that the corresponding components are isomorphic if they have rank > 1 and, if for a prime p, $(G/\langle a \rangle)_p$ has rank 1 and corresponds to $(H/\langle b \rangle)_q$ for some (not necessarily distinct) prime q, then they are of the form $\mathbf{Z}(p^n)$ and $\mathbf{Z}(q^n)$ respectively, where n is a positive integer or ∞.*

Observe that condition (iii) simply states that $G/\langle a \rangle$ and $H/\langle b \rangle$ are projective.

Our methods involve using a key idea of constructing a projectivity of a divisible group by Mahdavi and Poland ([5]) and, following Warfield, viewing a mixed Abelian group as an extension of a torsion-free group by a torsion group.

Thus Theorem 1.4 together with Theorem 1.1 provides a complete solution to the problem of describing when two Abelian groups have isomorphic subgroup lattices.

2 Preliminaries

Let **P** be the set of all prime integers. All groups that we consider are additively written Abelian groups and we generally follow the notation and terminology of the books by L. Fuchs ([3]) and R. Schmidt ([9]). For an Abelian group G, $L(G)$ denotes the lattice of subgroups of G under set inclusion. $T(G)$ denotes the torsion part of G. For a prime p, G_p denotes the p-component of $T(G)$ and $G[p]$ denotes the subgroup $\{a \in G \mid pa = 0\}$. For a subset S of G, $\langle S \rangle$ denotes the subgroup generated by S. If $g \in G$, $\langle g \rangle$ denotes $\langle \{g\} \rangle$, the cyclic subgroup generated by g. For all ordinals α, the subgroups $p^\alpha G$ are defined inductively: $pG = \{pa \mid a \in G\}$; if $\alpha = \gamma + 1$ then $p^{\gamma+1}G = p(p^\gamma G)$ and if α is a limit ordinal, then $p^\alpha G = \bigcap_{\gamma < \alpha} p^\gamma G$.

If $a \in p^\alpha G$ and $a \notin p^{\alpha+1}G$, we say a has p-*height* α and write $h_p(a) = \alpha$. If $a \in p^\alpha G$ for all α, then a has infinite height and we write $h_p(a) = \infty$.

The *height matrix* of an element a in a group G is a doubly infinite matrix $\mathcal{H}(a)$ indexed by the primes p and non-negative integers n whose (p, n) entry is $h_p(p^n a)$. If G is a mixed group of torsion-free rank one and if a, b are two infinite order elements in G, then $\mathcal{H}(a)$ and $\mathcal{H}(b)$ are *equivalent*, that is, almost all rows are equal and if the pth rows of $\mathcal{H}(a)$ and $\mathcal{H}(b)$ are not equal, then there exist integers n and m such that $h_p(p^{n+i}a) = h_p(p^{m+i}b)$ for every $i \in \{0, 1, 2, \ldots\}$. Thus to each mixed Abelian group G of torsion-free rank one we can assign uniquely an equivalence class $U(G)$ of the height matrix of any infinite order element in G. The pth row of $\mathcal{H}(a)$ is called the p-*indicator* of a.

An indicator $(\sigma_0, \sigma_1, \ldots)$ is said to have a *gap* if, for some $k \geq 0$, $\sigma_k + 1 < \sigma_{k+1}$; in this case the gap is said to *follow* σ_k or *precede* σ_{k+1}. For any prime p, the αth Ulm–Kaplansky invariant of G is denoted by $f_\alpha^p(G)$ and is defined as the dimension $\dim(\frac{p^\alpha G[p]}{p^{\alpha+1} G[p]})$. We shall be using the well-known result (see [3]), that two countable Abelian p-groups with the same Ulm–Kaplansky invariants are isomorphic.

3 The main result

We begin by mentioning two lemmas by R. Hunter [4] about mixed groups of torsion-free rank 1.

Lemma 3.1 ([4]). *Let G be a mixed Abelian group of torsion-free rank one and $a \in G$ an element of infinite order and p-indicator $(\sigma_0, \sigma_1, \ldots)$. Then, for any ordinal σ,*

$$f_\sigma^p(G) = \begin{cases} f_\sigma^p(G/\langle a \rangle) + 1 & \text{if } \sigma = \sigma_n \text{ and a gap follows } \sigma_n, \\ f_\sigma^p(G/\langle a \rangle) - 1 & \text{if } \sigma + 1 = \sigma_n \text{ and a gap precedes } \sigma_n, \\ f_\sigma^p(G/\langle a \rangle) & \text{otherwise.} \end{cases}$$

Lemma 3.2 ([4]). *If G and H are mixed Abelian groups of torsion-free rank 1, then $G \cong H$ if and only if there exist infinite order elements $a \in G$, $b \in H$ such that $\mathcal{H}(a) = \mathcal{H}(b)$ and $G/\langle a \rangle \cong H/\langle b \rangle$.*

The next result (which is Theorem 1.3.2 of [9]) states that certain bijections between cyclic subgroups induce projectivities.

Proposition 3.3 ([9]). *Let G and H be two Abelian groups. If g is a bijection from the set of cyclic subgroups of G to the set of cyclic subgroups of H, then g extends to a projectivity f from G to H if g satisfies the property*

$$A \leq B + C \Leftrightarrow g(A) \leq g(B) + g(C) \tag{3.1}$$

for all cyclic subgroups A, B, C of G.

We now consider the construction of autoprojectivities of a divisible Abelian group of torsion-free rank one corresponding to specific permutations of the set **P** of primes. The following result is a special case of Theorem E of [5]. For the sake of completeness, we shall give a (slightly simpler) proof which is more group-theoretical.

Following L. Fuchs ([3]), a prime p is said to be *relevant* to a group G, if G has an element of order p. A positive integer n is said to be *relevant* to a group G if each prime factor of n is relevant to G.

Proposition 3.4 ([5]). *Let D be a divisible mixed Abelian group of torsion-free rank one and let π be a permutation of the set **P** of primes which is multiplicative and fixes those primes that are relevant to D. Then there is a lattice isomorphism $f : L(D) \to L(D)$ such that $f(nS) = \pi(n) f(S)$ for all subgroups S of D and all positive integers n.*

Proof. Let $D = T \oplus \mathbf{Q}$ with T torsion divisible and $\mathbf{Q}$ the additive group of rational numbers. Let P' be the set of primes relevant to T. Let K be the localization of $\mathbf{Z}$ at the primes not relevant to T, that is, K is the subgroup of $\mathbf{Q}$ generated by $\{\frac{1}{p^s} : p \in P', k \in \mathbf{Z}\}$ i.e., $K = \langle\{1/m : m \text{ positive integer relevant to } T\}\rangle$, so that $\mathbf{Z} < K$ and $K/\mathbf{Z} \cong \bigoplus\{\mathbf{Z}(p^\infty) : p \in P'\}$. Clearly K is the union of cyclic subgroups $C_m = \langle x_m \rangle$, where $x_m = 1/m$ for various positive integers m relevant to T. Note that if $m = rs$, then $rx_m = x_s$ and x_m is in $\langle x_n \rangle$ if and only if $m|n$.

Let $D = T \oplus Q'$ with $Q' \cong \mathbf{Q}$ be another decomposition of D with $\langle y \rangle = Z' < Q'$ and $Z' \cong \mathbf{Z}$. Choose $K' > Z'$ analogous to the subgroup K, so that $K'/Z' \cong \bigoplus\{\mathbf{Z}(p^\infty) : p \in P'\}$ and we realize $Q'/K' \cong \{\mathbf{Z}(p^\infty) : p \in \pi(P)\setminus P'\}$. As before, K' is the union of cyclic subgroups $C'_m = \langle y_m \rangle = \langle 1/m \rangle$ for various positive integers m relevant to T with y_m satisfying $my_m = y$ and if $m = rs$, then $ry_m = y_s$.

Let $H = T \oplus K$ and $H' = T \oplus K'$. If $a \in D$ is an element of infinite order with $o(a + H) = e$, then there is a smallest integer n such that $ea \in T \oplus C_n$ so that

$$ea = u + dx_n = d(v + x_n) \text{ with } u = dv \in dT = T. \tag{3.2}$$

Similar equation (3.2) holds with respect to $T + H'$.

Now given such an a satisfying (3.2), choose b in D such that

$$\pi(e)b = \pi(d)(v + y_n). \tag{3.3}$$

Such an element b exists and is unique since D is divisible and $\pi(e)$ is not relevant to $T \oplus H'$. Note that if a is not in H, then b is not in H' as $\pi(e)$ is not relevant to $H'/(T \oplus Z')$. Also the minimality of e and n implies that $\gcd(d, e) = 1 = \gcd(d, n)$.

We now wish to define a bijection f from the set of cyclic subgroups of D to itself satisfying the property (3.1) of Proposition 3.3.

To this end, we use the natural isomorphism $K \to K'$ and the identity map $T \to T$, and define the function g by

$$g(\langle a \rangle) = \begin{cases} \langle a \rangle & \text{if } a \in T, \\ \langle ry_n \rangle & \text{if } a = rx_n \in H, \\ \langle b \rangle & \text{if } a \text{ has infinite order satisfying (3.2), and } b \text{ satisfies (3.3).} \end{cases}$$

Using the uniqueness of e, d, m and the inverse permutation π^{-1}, it is easy to see that g is a bijection between cyclic subgroups of $T \oplus \mathbf{Q}$ and $T \oplus Q'$ respectively.

To finish off the proof, we need only verify the condition (3.1) of Proposition 3.3 for the non-trivial case when $\langle z \rangle < \langle t \rangle \oplus \langle w \rangle$, where $w \in T$ and z, t have infinite order and then apply Lemma 3.1 of [5]. Specifically, if $e_1 z = d_1(v_1 + x_m)$ and $e_2 t = d_2(v_2 + x_n)$ as per equation (3.2) and $z = at + bw$ with a, b in $\mathbf{Z}$, then a direct calculation as done in the proof of Lemma 3.1 of [5] shows that m, e_1, d_2 are divisors respectively of n, e_2 and d_1 so that $n = a_1 m$, $e_2 = a_2 e_1$, and $d_1 = a_3 d_2$, $a = a_1 a_2 a_3$ and $d_1(v_1 - a_1 v_2) = e_1 bw$ and conversely. Since π is multiplicative, the corresponding elements z', t' where $g(\langle z \rangle) = \langle z' \rangle$ and $g(\langle t \rangle) = \langle t' \rangle$, satisfy the similar equation needed to reach the conclusion that $\langle z' \rangle$ is in $\langle t' \rangle \oplus \langle w \rangle$ and conversely. Hence the condition (3.1) of Proposition 3.3 holds and so g induces a projectivity f from D to D that satisfies $f(nS) = \pi(n) f(S)$ for all cyclic (and hence any) subgroup S of D and any positive integer n. □

Proof of Theorem 1.4. *The conditions are necessary*: Suppose $f : L(G) \to L(H)$ is a lattice isomorphism and let $a \in G$ be an infinite order element with $\mathcal{H}(a) = U(G)$. If for some $b \in H$, $f(\langle a \rangle) = \langle b \rangle$, then by Theorem 1.3, conditions (i) and (ii) are satisfied. Moreover, f induces a lattice isomorphism $\overline{f} : L(G/\langle a \rangle) \to L(H/\langle b \rangle)$. Then, by Theorem 1.1(b), condition (iii) holds.

The conditions are sufficient: Suppose G and H satisfy conditions (i)–(iii). Identifying the divisible hulls of $T(G)$ and $T(H)$, we may assume, without loss of generality, that G and H are (essential) subgroups of $D \oplus \mathbf{Q}$, where D is a torsion divisible Abelian group and $\mathbf{Q}$ is the additive group of rational numbers.

Now apply Proposition 3.4 to construct a projectivity $f : D \oplus \mathbf{Q} \to D \oplus \mathbf{Q}$ corresponding to the permutation π with the property that for all subgroups S of $D \oplus \mathbf{Q}$ and primes p, $f(pS) = \pi(p)(f(S))$.

So if $K = f(G)$ and $\langle c \rangle = f(\langle a \rangle)$, then G is projective with K, $\mathcal{H}(c)$ arises from $\mathcal{H}(a)$ by permuting the rows of $\mathcal{H}(a)$ by π and $G/\langle a \rangle$ is projective with $K/\langle c \rangle$. By Theorem 1.3, $T(K) \cong T(G)$. In view of condition (iii) of our hypothesis, Theorem 1.1(b), shows that $G/\langle a \rangle$ is projective with $H/\langle b \rangle$. Putting these facts together, we conclude that $T(K) \cong T(H)$, $\mathcal{H}^K(c) = \mathcal{H}^H(b)$ and $K/\langle c \rangle$ is projective with $H/\langle b \rangle$. By Theorem 1.1(b), there is a bijection between the p-components of the torsion groups $K/\langle c \rangle$ and $H/\langle b \rangle$ such that the corresponding components are projective and those of rank >1 are isomorphic.

Now $T(K) \cong T(H)$ implies the equality of the Ulm–Kaplansky invariants $f^p_\sigma(K) = f^p_\sigma(H)$ for all primes p and ordinals σ. Since $\mathcal{H}^K(c) = \mathcal{H}^H(b)$, Lemma 3.1 implies that $K/\langle c \rangle$ and $H/\langle b \rangle$ have the same Ulm–Kaplansky invariants, that is $f^p_\sigma(K/\langle c \rangle) = f^p_\sigma(H/\langle b \rangle)$.

Let $P' = \{p \text{ prime} \mid \text{rank}(K/\langle c \rangle)_p = 1\}$. If, for some $p \in P'$, $(K/\langle c \rangle)_p \cong \mathbf{Z}(p^n)$ with n a positive integer or ∞, then from the equality of the Ulm–Kaplansky invariants $f_\sigma^p((K/\langle c \rangle)_p) = f_\sigma^p(K/\langle c \rangle) = f_\sigma^p(H/\langle b \rangle)$, which holds for all ordinals σ, we derive $(H/\langle b \rangle)_p \neq 0$ and cannot have rank > 1. Hence $(H/\langle b \rangle)_p$ has rank 1 and is therefore a countable p-group with the same Ulm–Kaplansky invariants as $(K/\langle c \rangle)_p$ and by [3], $(K/\langle c \rangle)_p \cong (H/\langle b \rangle)_p$. Thus $K/\langle c \rangle \cong H/\langle b \rangle$. Finally, since $\mathcal{H}^K(c) = \mathcal{H}^H(b)$, we conclude, by Lemma 3.2, that $K \cong H$. Hence G is projective with H and the proof is complete. □

4 Corollaries

A *simply presented* group is an Abelian group G defined by generators and defining relations such that all the relations are induced by relations involving two generators, having the form $px = y$ or $px = 0$ where x, y belong to the given generating set and p varies over primes. A direct summand of a simply presented group is called a *Warfield* group.

First we specialize Theorem 1.4 to Warfield groups.

Corollary 1. *If G and H are Warfield groups of torsion-free rank one, then $L(G) \cong L(H)$ if and only if G and H satisfy conditions* (i) *and* (ii) *of Theorem 1.4.*

Proof. All one needs is to observe that in the proof of Theorem 1.4, $K/\langle c \rangle$ and $H/\langle b \rangle$ are now simply presented torsion groups with the same Ulm–Kaplansky invariants and hence are isomorphic. □

Corollary 2. *Suppose G and H are mixed Abelian groups of torsion-free rank one and, for every prime p, G has an element of order p. Then $L(G) \cong L(H)$ if and only if $G \cong H$.*

Proof. Let $f : L(G) \to L(H)$ be the lattice isomorphism. By Theorem 1.4(i), (ii), $T(G) \cong T(H)$ and, since π now fixes every prime p, there exist $a \in G, b \in H$ such that $\mathcal{H}(a) = U(G) = U(H) = \mathcal{H}(b)$, where $f(\langle a \rangle) = \langle b \rangle$. Since $T(G) \cong T(H)$, Lemma 3.1 implies that the torsion groups $G/\langle a \rangle$ and $H/\langle b \rangle$ have the same Ulm–Kaplansky invariants.

Now proceed as in the proof of Theorem 1.4 replacing the role of K by G, to conclude $G/\langle a \rangle \cong H/\langle b \rangle$. Then Lemma 3.2 yields that $G \cong H$. □

Corollary 3. *Suppose G and H are mixed Abelian groups of torsion-free rank one and $T(G)$ is a p-group for some prime p. Then $L(G) \cong L(H)$ if and only if conditions* (i) *and* (ii) *of Theorem 1.4 hold and $G/\langle a \rangle \cong H/\langle b \rangle$, whenever $T(G)$ has rank > 1.* □

References

[1] Baer, R., *The significance of the system of subgroups for the structure of a group*, Amer. Journ. Math., 71 (1939), 1–44.

[2] Fuchs, L., *Abelian Groups*, Publishing House of Hungarian Academy of Sci., Budapest (1958).

[3] Fuchs, L., *Infinite Abelian Groups*, Academic Press, vol. 1 & 2, 1970, 1973.

[4] Hunter, R., *Balanced subgroups of Abelian groups,* Trans. Amer. Math. Soc., 215 (1976), 81–98.

[5] Mahdavi, K., Poland, J., *On lattice-isomorphic abelian groups*. Arch. Math. (Basel) 58 (1992), no. 3, 220–230.

[6] Mahdavi, K., Poland, J., *On lattice isomorphic of mixed abelian groups*. Arch. Math. (Basel) 60 (1993), no. 4, 327–329.

[7] Megibben, C. K., *On mixed groups of torsion-free rank 1*, Illinois J. of Math., 11, (1967), 134–144.

[8] Ostendorf, U., *Projektivitätstypen torsionsfreier abelscher Gruppen vom Rang 1*. (German) [Projectivity types of torsion-free abelian groups of rank 1] Rend. Sem. Mat. Univ. Padova 86 (1991), 183–191.

[9] Schmidt, R., *Subgroup lattices of groups*, de Gruyter Expositions in Mathematics 14, de Gruyter, Berlin, 1994.

Author information

Grigore Călugăreanu, Department of Mathematics and Computer Science, Kuwait University, P.O. Box 5969, Safat, 13060, Kuwait.
E-mail: `calu@math.ubbcluj.ro`

Kulumani M. Rangaswamy, Department of Mathematics, University of Colorado, P.O. Box 7150, Colorado Springs, CO. 80933-7150, U.S.A.
E-mail: `krangasw@uccs.edu`

A note on Axiom 3 and its dual for abelian groups

Paul Hill and William Ullery

Abstract. In this paper, we initiate a study of the dual of Axiom 3, called Axiom 3*, and consider the problem of which abelian groups satisfy Axiom 3* with respect to summands. Among our main results is the demonstration that arbitrary direct sums of cyclic groups (including free groups) satisfy Axiom 3* with respect to summands. On the other hand, it is shown that a divisible group satisfies Axiom 3* with respect to summands if and only if it is countable. We also obtain some partial results regarding primary groups; in particular, a reduced d.s.c. that satisfies Axiom 3* with respect to summands must have countable length.

Key words. Axiom 3, Axiom 3* with respect to summands.

AMS classification. 20K10, 20K20, 20K21.

1 Introduction

Throughout, G denotes an additively written abelian group. In this note, we discuss the role of Axiom 3 and for the first time its dual, Axiom 3*, for such groups. The notion of Axiom 3 for abelian p-groups was introduced by one of the authors in [2]. Recall that a distinguished collection of subgroups $\mathcal{C}$ of G is an *Axiom 3 system* for G if the following three conditions are satisfied.

(0) $\mathcal{C}$ contains the trivial subgroup 0.

(1) If $\{N_\alpha\}_{\alpha \in A} \subseteq \mathcal{C}$, then $\sum_{\alpha \in A} N_\alpha \in \mathcal{C}$.

(2) If H is a countable subgroup of G, there exists a countable subgroup $N \in \mathcal{C}$ such that $H \subseteq N$.

It was shown in [2], the introductory paper of Axiom 3, that a p-primary abelian group is totally projective if and only if it has an Axiom 3 system consisting entirely of *nice* subgroups. This characterization was then used to extend (the Kaplansky–Mackey formulation of) Ulm's Theorem from countable p-groups and their direct sums to the much larger class of totally projective groups. Further applications of Axiom 3 in the structure theory of p-groups abound. They include the following. In [3] the isotype subgroups of totally projective p-groups that are themselves totally projective were characterized in terms of Axiom 3 systems of *separable* subgroups. A little later L. Fuchs and P. Hill showed in [1] that the dimension theory of p-local torsion groups can be described in terms of generalized Axiom 3 systems of *separable* subgroups. However, the usefulness of Axiom 3 is not limited to the structure theory of primary groups. For example, in [4] and [5], Hill and Megibben introduced *knice* subgroups in the contexts of p-local and torsion free groups, respectively. In [4], it was shown that a p-local group is a Warfield group if and only if it has an Axiom 3 system of knice subgroups, and in [5] the latter condition was shown to characterize completely decomposable torsion free groups. These characterizations led to new insights into the structure of isotype subgroups in the p-local case, and pure subgroups in the torsion free case.

In the sequence of papers [6] and [7], Hill and Megibben seemed to have reached the pinnacle of Axiom 3 characterizations: [6] provided the appropriate definition for the notion of a *knice* subgroup in the context of arbitrary mixed groups, and [7] demonstrated that global Warfield groups are precisely those groups that have an Axiom 3 system of knice subgroups. Megibben and Ullery [8, 9, 10] used this characterization of global Warfield groups to obtain for general (mixed) groups versions of the results that we have mentioned in [3] and [1] for p-groups. More precisely, in [10], the isotype subgroups of global Warfield groups that are again Warfield groups were described in terms of Axiom 3 systems of *almost strongly separable k-subgroups*, and in [8], the (sequentially pure projective) dimension of a mixed group with a decomposition basis was computed in terms of generalized Axiom 3 systems of *almost strongly κ-separable subgroups*. In [9], the main result of [8] was refined to compute the dimension of an arbitrary global k-group in terms of generalized Axiom 3 systems of *almost strongly κ-separable k-subgroups*.

The results cited above are by no means a complete account of the applications of Axiom 3 in the literature. Rather, they are intended to represent only some of the highlights. Notwithstanding the fact that the collection of all subgroups of any group G is an Axiom 3 system for G, we should emphasize that Axiom 3 systems are useful only when linked with special subgroup properties. There is little need here to present the precise definitions of the various types of subgroups mentioned above. For our purpose, it suffices to say that direct summands enjoy all of these subgroup properties. To set the stage for the problem considered herein, we observe the following.

Proposition 1.1. *An abelian group G satisfies Axiom 3 with respect to summands if and only if it is a direct sum of countable groups.*

In this paper we begin an initial study of the problem of characterizing those groups G that satisfy the hypothesis of the preceding proposition with Axiom 3 replaced by its dual. Since Axiom 3 has proved to be so fruitful in the study of the structure theory of abelian groups, it seems likely that its dual might also be a useful concept.

Definition 1.2. A collection of subgroups $\mathcal{C}$ of G is called an *Axiom 3* system* if the following three conditions are satisfied.

(0*) $G \in \mathcal{C}$.

(1*) If $\{N_\alpha\}_{\alpha \in A} \subseteq \mathcal{C}$, then $\bigcap_{\alpha \in A} N_\alpha \in \mathcal{C}$.

(2*) If H is a subgroup of countable index in G, there exists a subgroup $N \in \mathcal{C}$ of countable index in G such that $N \subseteq H$.

As we have mentioned, we want to study here groups G that satisfy Axiom 3* with respect to summands; that is, those G that have an Axiom 3* system consisting entirely of summands of G. To see that Axiom 3* is a nontrivial condition on a group G, we begin with the following observation.

Proposition 1.3. *If p is a prime and if $G = J_p$ is the additive group of p-adic integers, then G does not satisfy Axiom 3* with respect to summands.*

Proof. If we set $H = pG$, then G/H is cyclic of order p. Hence H has countable index in G. However, H does not even contain a nontrivial pure subgroup of $G = J_p$. □

It is clear why the group G in Proposition 1.3 has uncountable cardinality. Indeed, it follows immediately from the definition that any countable group satisfies Axiom 3* with respect to summands. The following two simple but useful propositions also follow easily from Definition 1.2.

Proposition 1.4. *If $\{G_n\}_{n<\omega_0}$ is a countable collection of groups, each of which satisfies Axiom 3* with respect to summands, then $G = \bigoplus_{n<\omega_0} G_n$ satisfies Axiom 3* with respect to summands.*

Proof. For each n, let $\mathcal{C}_n$ be an Axiom 3* system of summands for G_n. Then, it is routine to verify that
$$\mathcal{C} = \Big\{ \bigoplus_{n<\omega_0} N_n : N_n \in \mathcal{C}_n \Big\}$$
is an Axiom 3* system of summands for G. □

Proposition 1.5. *If G satisfies Axiom 3* with respect to summands, then any subgroup of countable index in G satisfies Axiom 3* with respect to summands.*

Proof. If $\mathcal{C}$ is an Axiom 3* system of summands for G and if H is a subgroup of G with $|G/H| \leq \aleph_0$, then
$$\mathcal{C}_H = \{N \in \mathcal{C} : N \subseteq H\} \cup \{H\}$$
is an Axiom 3* system of summands for H. □

We close this introductory section with our first application of Propositions 1.4 and 1.5.

Theorem 1.6. *Let C be a countable subgroup of G. If G/C satisfies Axiom 3* with respect to summands, then so does G.*

Proof. Let E denote the divisible hull of C, and let G^* be the pushout of C given by the inclusion maps $C \to E$ and $C \to G$. Thus, we may regard E and G as subgroups of G^* with $G^* = E + G$ and $E \cap G = C$. Since E is countable, G has countable index in G^*. So in view of Proposition 1.5, it suffices to show that G^* satisfies Axiom 3* with respect to summands.

To see that G^* satisfies Axiom 3* with respect to summands, we use the fact that E is divisible to write
$$G^* = E + G = E \oplus H$$

for some subgroup H of G^*. Hence, because $C \cap H$ is trivial,

$$G^*/C = E/C \oplus G/C \cong (E/C) \oplus H.$$

Since E/C is countable and G/C satisfies Axiom 3* with respect to summands by hypothesis, Proposition 1.4 implies that G^*/C satisfies Axiom 3* with respect to summands. Another application of Proposition 1.5 now shows that H satisfies Axiom 3* with respect to summands. Recalling that E is countable, $G^* = E \oplus H$ satisfies Axiom 3* with respect to summands by Proposition 1.4. □

2 Direct sums of finite cyclic groups

We begin this section by considering homogeneous p-groups of the form

$$G = \bigoplus_{i \in I} \langle x_i \rangle.$$

By *homogeneous*, we mean that each x_i has the same finite order p^n for some fixed prime p and positive integer n. We intend to show that G satisfies Axiom 3* with respect to summands. Without loss, we may assume that $n \geq 2$ because any subgroup of an elementary p-group is a summand. Moreover, we can assume that $|I| \geq \aleph_1$, for otherwise G is countable.

To construct an Axiom 3* system $\mathcal{C}$ for G, first let $\mathcal{J}$ be a collection of subsets of I. That is $\mathcal{J}$ is a subset of the power set $P(I)$ of I. Then, for each $\mathcal{J} \subseteq P(I)$, define

$$G(\mathcal{J}) = \left\{ \sum_{i \in I} t_i x_i : t_i \in \mathbb{Z},\ t_i = 0 \text{ for almost all } i, \text{ and } \sum_{i \in J} t_i = 0 \text{ for all } J \in \mathcal{J} \right\}.$$

If we now take

$$\mathcal{C} = \{G(\mathcal{J})\}_{\mathcal{J} \subseteq P(I)},$$

we obtain the following.

Lemma 2.1. *In the above notation, $\mathcal{C}$ is an Axiom 3* system of summands for the homogeneous group G.*

Proof. Since $G = G(\emptyset)$ where $\emptyset$ is the empty subset of $P(I)$, $G \in \mathcal{C}$. Moreover, $\mathcal{C}$ is closed under intersections because if $\{G(\mathcal{J}_\alpha)\}_{\alpha < \lambda}$ is a subset of $\mathcal{C}$ indexed by some ordinal λ with each $\mathcal{J}_\alpha \subseteq P(I)$, then

$$\bigcap_{\alpha < \lambda} G(\mathcal{J}_\alpha) = G\Big(\bigcup_{\alpha < \lambda} \mathcal{J}_\alpha\Big).$$

Thus, to complete the proof, it suffices to show that each member of $\mathcal{C}$ is a summand of G and that condition (2*) of Definition 1.2 is satisfied by $\mathcal{C}$.

To see that each member of $\mathcal{C}$ is a summand, suppose $N \in \mathcal{C}$ and that $N \neq G$. Thus, $N = G(\mathcal{J})$ for some nonempty $\mathcal{J} \subseteq P(I)$. Since N is bounded, it is enough to show that N is pure in G. This will be accomplished by showing that each member of the socle $N[p]$ has as much height in N as it has in G. In fact, we show that $N[p] \subseteq p^{n-1}N$. So suppose that $x \in N[p]$ and write $x = \sum_{i \in I} t_i x_i$ with each $t_i \in \mathbb{Z}$, and $\sum_{i \in J} t_i = 0$ for all $J \in \mathcal{J}$. Then, for each $i \in I$, $pt_i x_i = 0$ so that $p^{n-1} \mid t_i$ for all i. Hence $t_i = p^{n-1} s_i$ for some $s_i \in \mathbb{Z}$. It is now easily checked that $y = \sum_{i \in I} s_i x_i \in N$ and that $x = p^{n-1} y$ as desired.

Finally, to see that condition (2*) holds for $\mathcal{C}$, suppose that H is a subgroup of countable index in G. Select a (countable) set of representatives $\{y_j\}$ for the distinct cosets of H in G. For each j, we construct a subset I_j of I by decreeing that $i \in I_j$ if and only if $x_i - y_j \in H$. Now set $J = \{j : I_j \neq \emptyset\}$ and take $\mathcal{J} = \{I_j\}_{j \in J}$. Set $J_0 = \{j : |I_j| \geq 2\}$, and for each $j \in J_0$, select and fix a single $i(j) \in I_j$. Then

$$G(\mathcal{J}) = \bigoplus_{j \in J_0} \left(\bigoplus_{i \in I_j} \langle x_i - x_{i(j)} \rangle \right).$$

Clearly $G(\mathcal{J}) \subseteq H$. Moreover, $G(\mathcal{J})$ has countable index in G since its complement in G is the countable subgroup

$$\left(\bigoplus_{j \in J_0} \langle x_{i(j)} \rangle \right) \oplus \left(\bigoplus_{j \in J_1} \langle x_j \rangle \right),$$

where $J_1 = \{j : |I_j| = 1\}$. $\square$

Since any torsion group is a direct sum of its primary components, and a direct sum of cyclic p-groups is a direct sum of countably many homogeneous p-groups, the following is a direct consequence of Proposition 1.4 and Lemma 2.1.

Theorem 2.2. *If G is a direct sum of finite cyclic groups, then G satisfies Axiom 3* with respect to summands.*

3 Free groups

Suppose that G is a free abelian group and write $G = \bigoplus_{i \in I} \langle x_i \rangle$, where each x_i has infinite order. In order to show that G satisfies Axiom 3* with respect to summands, clearly we may assume that $|I| \geq \aleph_1$. Let $\mathcal{P}$ denote the set of all pairs (J, f), where J is a co-countable subset of I (that is, $|I \setminus J| \leq \aleph_0$), and f is a function $f : J \to I \setminus J$. For each $(J, f) \in \mathcal{P}$ define

$$G(J, f) = \bigoplus_{j \in J} \langle x_j - x_{f(j)} \rangle.$$

Observe that each $G(J, f)$ is a summand of countable index in G. Indeed,

$$G = G(J, f) \oplus \left(\bigoplus_{i \in I \setminus J} \langle x_i \rangle \right).$$

Now let
$$\mathcal{C} = \left\{ \bigcap_{(J,f)\in S} G(J,f) \right\}_{S \subseteq \mathcal{P}}$$
where S ranges over all subsets of $\mathcal{P}$, with the usual convention that the empty intersection is G.

Theorem 3.1. *In the above notation, $\mathcal{C}$ is an Axiom 3* system of summands for the free group G.*

Proof. By construction, $G \in \mathcal{C}$ and $\mathcal{C}$ is closed under arbitrary intersections. Thus, it remains to show that $\mathcal{C}$ satisfies condition (2*) of Definition 1.2 and that each member of $\mathcal{C}$ is a summand of G.

To verify condition (2*) for $\mathcal{C}$, suppose that a subgroup H of countable index in G is given and let $\{c_n\}$ be a complete set of representatives for the distinct cosets of H in G. Partition I into a countable set of subsets $\{I_n\}$ by decreeing that $i \in I_n$ if and only if $x_i - c_n \in H$. Discarding any I_n with $|I_n| \leq 1$, enumerate those that remain as $\{J_n\}_{n<\alpha}$, for some countable ordinal $\alpha \leq \omega_0$. For each $n < \alpha$, select and fix a single $i(n) \in J_n$ and set $J = (\bigcup_{n<\alpha} J_n) \setminus \{i(n)\}_{n<\alpha}$. Certainly J is a co-countable subset of I. Now define $f : J \to I \setminus J$ as follows: for a given $j \in J$, select the unique $n < \alpha$ with $j \in J_n$ and take $f(j) = i(n)$. It is clear that $G(J,f) \subseteq H$ and, as observed above, $G(J,f)$ has countable index in G.

To see that each member of $\mathcal{C}$ is a summand of G, suppose that S is a nonempty subset of $\mathcal{P}$, and for convenience of notation, set $K = \bigcap_{(J,f)\in S} G(J,f)$. Observe that the proof will be complete once we have shown that G/K is free. We accomplish this by showing that G/K is isomorphic to a group of bounded (possibly transfinite) sequences of integers. Consider the natural map $G/K \to \prod_{(J,f)\in S} G/G(J,f)$, in conjunction with the decompositions $G = G(J,f) \oplus (\bigoplus_{i \in I\setminus J} \langle x_i \rangle)$, to obtain an embedding
$$\varphi : G/K \to \prod_{(J,f)\in S} \prod_{i \in I\setminus J} \langle x_i \rangle.$$
Clearly it suffices to show that the sequence of coefficients of $\varphi(x_k + K)$ is bounded for all $k \in I$. But for given $k \in I$, $(J,f) \in S$ and $i \in I \setminus J$, the only possible values of the i-th component of $\varphi(x_k + K)$ are 0, or x_k if $k = i$, or $x_{f(k)}$ if $k \in J$ with $f(k) = i$. Thus, the sequence of coefficients is all 0's and 1's, and is therefore bounded. □

Using Proposition 1.4 and Theorem 3.1, we can now eliminate the finiteness condition from Theorem 2.2.

Theorem 3.2. *If G is a direct sum of cyclic groups, then G satisfies Axiom 3* with respect to summands.*

4 Divisible groups

In contrast to our results in the previous two sections, in this section we show that divisible groups of uncountable cardinality cannot satisfy Axiom 3* with respect to summands.

We require some preliminary results regarding torsion completions of direct sums of cyclic p-groups (where p is an arbitrary, but fixed, prime). If B is a direct sum of cyclic p-groups, we denote its torsion completion by $\overline{B}$, and write c for the cardinality of the continuum $2^{\aleph_0}$. Observe that if B is countable and unbounded, then $\overline{B}$ has cardinality c. We will use the notation $|x|_p^G$ for the p-height of x as computed in the containing group G.

Lemma 4.1. *If B is a countable unbounded direct sum of cyclic p-groups and if K is an uncountable pure subgroup of $\overline{B}$ with $B \subseteq K$, then $G = K \oplus A$ fails to satisfy Axiom 3^* with respect to summands for any group A.*

Proof. Suppose to the contrary that $G = K \oplus A$ satisfies Axiom 3^* with respect to summands. Since $\overline{B}/B$ is divisible and K/B is a pure subgroup of $\overline{B}/B$, K/B is divisible. Thus, $K = pK + B$ and we conclude that K/pK is countable. Hence, $G/(pK \oplus A)$ is countable. As a result, there exist, due to Axiom 3^*, subgroups N and C of G such that $G = N \oplus C$, with $N \subseteq pK \oplus A$ and $|C| \leq \aleph_0$.

We claim that $N \cap K = 0$. Indeed, if this is not the case, select a nonzero $x \in N \cap K$. Observe that $|x|_p^G = |x|_p^N = n$ is finite because $|x|_p^G = |x|_p^K \leq |x|_p^{\overline{B}}$ and $\overline{B}$ has no (nonzero) elements of infinite p-height. In particular, $x = p^n x_1$ for some $x_1 \in N$. Since $x_1 \in pK \oplus A$, $x_1 = pb + a$ for some $b \in K$ and $a \in A$. Then,

$$p^n x_1 = p^{n+1} b + p^n a = x$$

implies that $p^n a = 0$, and we obtain the contradiction $|x|_p^G \geq n + 1$. Therefore, $N \cap K = 0$ as claimed. But this is impossible because $|(N \oplus K)/N| \leq |G/N| = |C|$ is countable, while $|(N \oplus K)/N| = |K| \geq \aleph_1$. □

Lemma 4.2. *If B is a countable direct sum of cyclic p-groups and if K is a pure subgroup of $\overline{B}$ with $B \subseteq K$, then K has countable index in its divisible hull.*

Proof. Let E denote the divisible hull of B and observe that E/B is countable. Moreover, K/B is divisible because it is a pure subgroup of the divisible group $\overline{B}/B$. If $f : B \to E$ and $g : B \to K$ are the inclusion maps, let G be the pushout of B obtained from f and g. Regarding both E and K as subgroups of G, we have that $G/K \cong E/B$ and $G/E \cong K/B$. These last two isomorphisms imply, respectively, that G/K is countable and that G is divisible. Since G must contain the divisible hull of K, the proof is complete. □

Theorem 4.3. *An uncountable divisible group cannot satisfy Axiom 3^* with respect to summands.*

Proof. If G is an uncountable divisible group, we can write $G = D \oplus A$ where either $D = \bigoplus_{\aleph_1} Z(p^\infty)$ for some prime p, or else $D = \bigoplus_{\aleph_1} \mathbb{Q}$. We consider each of the two cases in turn.

Case 1. $D = \bigoplus_{\aleph_1} Z(p^\infty)$ for some prime p. Let B be a countable unbounded direct sum of cyclic p-groups. Since $|\overline{B}| = c \geq \aleph_1$ and B is countable, there exists a (reduced) pure subgroup K of $\overline{B}$ such that $B \subseteq K$ and $|K| = \aleph_1$. Since K and D have the same p-rank $\aleph_1$, we may regard D as the divisible hull of K (with $K \subseteq D$). By Lemma 4.2, K has countable index in D, so that $K \oplus A$ has countable index in G. Moreover, $K \oplus A$ does not satisfy Axiom 3* with respect to summands by Lemma 4.1. Therefore, Proposition 1.5 implies that G itself does not satisfy Axiom 3* with respect to summands.

Case 2. $D = \bigoplus_{\aleph_1} \mathbb{Q}$. Write D as $D = \bigoplus_{i \in I} Q(i)$ where $|I| = \aleph_1$ and $Q(i) = \mathbb{Q}$ for all i. For an arbitrary but fixed prime p, select a subgroup E of $\mathbb{Q}$ such that $\mathbb{Q}/E \cong Z(p^\infty)$ and set $F = \bigoplus_{i \in I} E(i)$ where, for all $i \in I$, $E(i) = E$ is regarded as a subgroup of $Q(i)$. Therefore, D/F is a divisible p-group of cardinality $\aleph_1$. As in Case 1, there is a reduced subgroup K of countable index in D/F. Select the subgroup H of D such that $F \subseteq H$ and $H/F = K$. Then, H has countable index in D, so it follows that $H \oplus A$ has countable index in G.

Now suppose to the contrary that G satisfies Axiom 3* with respect to summands. Then G has a summand N of countable index with $N \subseteq H \oplus A$. Observe that N is not contained in $F \oplus A$. Indeed, G/N is countable, while $G/(F \oplus A) \cong D/F$ has cardinality $\aleph_1$. Thus, $(N + (F \oplus A))/(F \oplus A)$ is a nonzero divisible subgroup of $G/(F \oplus A)$. However, this is impossible because

$$(N+(F\oplus A))/(F\oplus A) \subseteq ((H\oplus A)+(F\oplus A))/(F\oplus A) = (H\oplus A)/(F\oplus A) \cong H/F = K$$

implies that $(N + (F \oplus A))/(F \oplus A)$ is isomorphic to a subgroup of the reduced group K. □

Recalling that any countable group satisfies Axiom 3* with respect to summands, Theorem 4.3 can be reformulated as follows.

Corollary 4.4. *A divisible group satisfies Axiom* 3* *with respect to summands if and only if it is countable.*

Actually, the proofs of the preceding results yield

Theorem 4.5. *In order for a group to satisfy Axiom* 3* *with respect to summands, its divisible part must be countable.*

5 Reduced p-groups

In this final section we show that, for a reduced p-group G (p a prime), the condition that G satisfies Axiom 3* with respect to summands bounds its length by ω_1, where the length of a reduced p-group G is the smallest ordinal λ with $p^\lambda G = 0$. Moreover, if G is a reduced d.s.c. that satisfies Axiom 3* with respect to summands, we show that the length of G is countable. Recall that a p-group G is a d.s.c. if it is a direct sum of countable p-groups. Thus, an arbitrary reduced d.s.c. has length not exceeding ω_1.

The next lemma is implicitly known; however, we include a short proof for the sake of completeness.

Lemma 5.1. *Let C be a countable subgroup of a p-group G. If M is a subgroup of G maximal with respect to $M \cap C = 0$, then G/M is countable.*

Proof. Since C is countable, observe that the proof will be complete if we can show that $G/(M \oplus C)$ is countable. Thus, since a p-group with countable socle is countable, we need only show that $(G/(M \oplus C))[p]$ is countable. So, suppose to the contrary that $(G/(M \oplus C))[p]$ is uncountable. Then, because C is countable and there are an uncountable number of cosets $g + (M \oplus C)$ with $pg \in M \oplus C$, there must be x and y in G that represent distinct cosets in $(G/(M \oplus C))[p]$ such that the components in C of px and py are the same. In other words, there are distinct cosets $x + (M \oplus C)$ and $y + (M \oplus C)$ with $p(x - y) \in M$. However, if $z \in G$ has the property that $pz \in M$, then z must be in $M \oplus C$. Indeed, if this were not the case, we would have that $\langle M, z \rangle \cap C = 0$, violating the maximality of M. Therefore, $x - y \in M \oplus C$, contradicting that x and y represent distinct cosets modulo $M \oplus C$. □

Proposition 5.2. *If G is a reduced p-group that satisfies Axiom 3^* with respect to summands, then the length of G does not exceed ω_1.*

Proof. Suppose $0 \neq x \in G$ and select a subgroup H maximal with respect to $H \cap \langle x \rangle = 0$. By Lemma 5.1, $|G/H| \leq \aleph_0$. So, there exist subgroups N and C of G such that $G = N \oplus C$, $N \subseteq H$, and $|C| \leq \aleph_0$. Then $x = y + c$ for some $y \in N$ and $c \in C$. Observe that $c \neq 0$ because $x \notin N$. Therefore, $|x|_p^G = \min\{|y|_p^N, |c|_p^C\} \leq |c|_p^C < \omega_1$. □

If G is a d.s.c., Proposition 5.2 can be improved as follows.

Theorem 5.3. *A reduced d.s.c. p-group G that satisfies Axiom 3^* with respect to summands has countable length.*

Proof. Suppose to the contrary that the length of G is ω_1 and let G^* be an elongation of G by a cyclic group of order p. That is

(i) $p^{\omega_1} G^* = \langle z \rangle \cong \mathbb{Z}/p\mathbb{Z}$, and

(ii) $G^*/p^{\omega_1} G^* \cong G$.

Observe that G^* is not a d.s.c. because its length is $\omega_1 + 1$.

Select a subgroup M^* of G^* maximal with respect to $M^* \cap \langle z \rangle = 0$. Then M^* is not a d.s.c. because $|G^*/M^*| \leq \aleph_0$ by Lemma 5.1, and a countable extension of a d.s.c. is again a d.s.c. Thus, $K = (M^* \oplus \langle z \rangle)/\langle z \rangle \cong M^*$ is not a d.s.c. Regarding K as a subgroup of G by means of the composition $K \subseteq G^*/\langle z \rangle \cong G$, we have that $|G/K| \leq \aleph_0$. Since G satisfies Axiom 3^* with respect to summands, there are subgroups N and C of G such that $G = N \oplus C$, $N \subseteq K$ and $|C| \leq \aleph_0$. Then N is a d.s.c. by virtue of the fact that it is a summand of a d.s.c. However, this yields the contradiction that K is a d.s.c. because $K/N \subseteq G/N \cong C$ implies that $|K/N| \leq \aleph_0$. □

We conclude with a result regarding the structure of totally projective p-groups that satisfy Axiom 3^* with respect to summands. Here we do not require totally projective groups to be reduced.

Corollary 5.4. *If G is a totally projective p-group that satisfies Axiom 3^* with respect to summands, then $G = D \oplus H$ where D is a countable divisible group and H is a reduced d.s.c. of countable length.*

Proof. In general, $G = D \oplus H$ where D is divisible and H is a reduced totally projective p-group. By Theorem 4.5, D is countable, and so Proposition 1.5 implies that H satisfies Axiom 3^* with respect to summands. From Proposition 5.2, we have that H is a reduced totally projective p-group of length not exceeding ω_1. Therefore, H is a d.s.c. and an application of Theorem 5.3 completes the proof. □

References

[1] L. Fuchs and P. Hill, *The balanced projective dimension of abelian p-groups*, Trans. Amer. Math. Soc. 293 (1986), pp. 99–112.

[2] P. Hill, *On the classification of abelian groups*, photocopied manuscript, 1967.

[3] P. Hill, *Isotype subgroups of totally projective groups*, Lecture Notes in Math. 874, Springer, New York, 1981, pp. 305–321.

[4] P. Hill and C. Megibben, *Axiom 3 modules*, Trans. Amer. Math. Soc. 295 (1986), pp. 715–734.

[5] P. Hill and C. Megibben, *Torsion free groups*, Trans. Amer. Math. Soc. 295 (1986), pp. 735–751.

[6] P. Hill and C. Megibben, *Knice subgroups of mixed groups*, Abelian Group Theory, Gordon-Breach, New York, 1987, pp. 89–109.

[7] P. Hill and C. Megibben, *Mixed groups*, Trans. Amer. Math. Soc. 334 (1991), pp. 121–142.

[8] C. Megibben and W. Ullery, *The sequentially pure projective dimension of global groups with decomposition bases*, J. Pure and Applied Algebra 187 (2004), pp. 183–205.

[9] C. Megibben and W. Ullery, *On global abelian k-groups*, Houston J. Math. 31 (2005), pp. 675–692.

[10] C. Megibben and W. Ullery, *Isotype knice subgroups of global Warfield groups*, Czech. Math. J. 56 (2006), pp. 109–132.

Author information

Paul Hill, Department of Mathematics and Statistics, Auburn University, AL 36849-5310, USA.
E-mail: hillpad@auburn.edu

William Ullery, Department of Mathematics and Statistics, Auburn University, AL 36849-5310, USA.
E-mail: ullerwd@auburn.edu

Units of modular p-mixed abelian group algebras

Warren May

Abstract. Let F be a perfect field of characteristic p, G an abelian group, and $V(G)$ the group of normalized units in the group algebra FG. The question of when the torsion subgroup of $V(G)$ is a totally projective p-group is answered for G of countable torsion-free rank.

Key words. Abelian group algebras, modular, p-mixed, units.

AMS classification. 20K10, 20C07, 20K25.

1 Introduction

Let F be a perfect field of prime characteristic p and G an abelian group written multiplicatively. We let $V(G)$ denote the group of normalized units in the group algebra FG. If G is a p-primary group, then $V(G)$ is also p-primary, and it was shown in [4] that $V(G)$ is totally projective if and only if G is totally projective. Moreover, in that case, G is a direct factor of $V(G)$ with totally projective complement.

Now consider a p-mixed group G, that is, the torsion subgroup T of G is p-primary. Then $V(G)$ is also p-mixed, and Hill and Ullery [2] showed that for the torsion subgroup $S(G)$ of $V(G)$ to be totally projective, it is necessary that T be totally projective.

To prove a partial converse, Hill and Ullery needed a length condition on T that depended on the torsion-free rank of G. The p-length of T had to be either countable, or, in the case that G had countable torsion-free rank, less than $\omega_1 + \omega_0$. The proof used a new characterization of totally projective p-groups of countable length, and rested upon having a composition series for T whose terms had finite height spectra, since such subgroups would consequently be nice subgroups of G.

In this paper, we shall eliminate the length restriction on T so long as the torsion-free rank of G is countable. We do not require totally projective p-groups to be reduced.

Theorem. *Let G be a p-mixed abelian group of countable torsion-free rank and totally projective torsion subgroup T. Then $S(G)$ is totally projective. Moreover, $S(T)$ is a direct factor of $S(G)$ with totally projective complement.*

Corollary 1. *T is a direct factor of $S(G)$ with totally projective complement and G is a direct factor of $V(G)$ with the same complement.*

Corollary 2. *If $FG \cong FH$ for some group H, then the torsion subgroup of H is isomorphic to T, and $G \times T' \cong H \times T'$ for some totally projective p-group T'.*

Corollary 3. *If G has torsion-free rank 1, and $FG \cong FH$ for some H, then $H \cong G$.*

Our approach will be to utilize subgroups of G which are finitely generated extensions of T and to construct composition series for T that are relevant to these subgroups. Thus we make use of infinitely many specially constructed composition series for T. Moreover, a truncated exponential map is employed to carry out a crucial height argument.

2 Torsion units

Let $H \subseteq G$ and let $\{g_i | i < \mu\}$ represent the distinct cosets of H in G, taking $g_0 = 1$. If $\alpha \in FG$, then $\alpha = 1 + \sum_{i<\mu} g_i \beta_i$ for unique $\beta_i \in FH$.

Lemma 1. *Assume that $T \subseteq H$ and let $\alpha \in FG$ be as above.*

(i) *We have $\alpha \in S(G)$ if and only if β_i is nilpotent for every i.*

(ii) *$S(G)$ is p-primary and $V(G) = GS(G)$.*

Proof. Let $\alpha \in FG$ and suppose that every β_i is nilpotent. Then there exists n such that $\beta_i^{p^n} = 0$ for every i, hence $\alpha^{p^n} = 1$. Thus, α is in the p-torsion of $S(G)$. Now let $\alpha \in V(G)$ and let $H = T$. Suppose that α is in the kernel of the natural map $V(G) \to V(G/T)$. Each β_i goes to its augmentation, which therefore must be 0. But $\beta_i \in FT$ of augmentation 0 is nilpotent, consequently from the above, α is in the p-component of $S(G)$. It is well known that $V(G/T) = G/T$, hence $S(G)$ is the kernel and is therefore p-primary. It now also follows that $V(G) = GS(G)$.

To finish proving (i), we may assume that $\alpha^{p^n} = 1$, thus $\sum_{i<\mu} g_i^{p^n} \beta_i^{p^n} = 0$. It suffices to show that the supports of the summands are disjoint. Suppose for some $i < j < \mu$ there exist $h_i, h_j \in H$ such that $g_i^{p^n} h_i^{p^n} = g_j^{p^n} h_j^{p^n}$. Then $g_i g_j^{-1} h_i h_j^{-1} \in T$, hence $g_i g_j^{-1} \in H$, contrary to $i \neq j$. $\square$

Let $B \subseteq H \subseteq G$. Then the natural map $H \to H/B$ induces a surjective algebra homomorphism $\varphi: FH \to F(H/B)$ whose kernel we denote by $I_H(B)$. If $\{h_i | i < \mu\}$ represent the distinct cosets of B in H, then $\alpha \in I_H(B)$ if and only if $\alpha = \sum_{i<\mu} h_i \beta_i$, where $\beta_i \in FB$ has augmentation 0 for every i. Moreover, we have the direct sum representation $I_H(B) = \bigoplus_{i,b} Fh_i(b-1)$ ($i < \mu, b \in B, b \neq 1$). By restriction, φ induces a homomorphism $V(H) \to V(H/B)$ whose kernel we denote by $K_H(B)$.

Lemma 2. *Let $B \subseteq T \subseteq H \subseteq G$. Then $S(H) = K_H(T)$, the induced map $S(H) \to S(H/B)$ is onto, and $K_H(B) = 1 + I_H(B)$.*

Proof. The elements of FH which go to 1 under φ are $1 + I_H(B)$, thus $K_H(B) \subseteq 1 + I_H(B)$. Since $I_H(B)$ is generated by the nilpotent elements $\{b - 1 | b \in B\}$, we have $1 + I_H(B) \subseteq K_H(B)$, hence equality. Apply Lemma 1 with H and G replaced by T and H respectively. Nilpotent elements of FT are of augmentation 0, thus they belong to $I_H(T)$. We conclude that $S(H) \subseteq 1 + I_H(T) = K_H(T)$, consequently, $S(H) = K_H(T)$.

We have that $S(H) = 1 + I_H(T)$ and $S(H/B) = 1 + I_{H/B}(T/B)$. Since $I_H(T)$ is generated by elements of form $t - 1$ ($t \in T$), and $1 + I_{H/B}(T/B)$ is generated by elements of form $tB - 1$, it follows that $S(H) \to S(H/B)$ is surjective. $\square$

3 Heights and the truncated exponential map

We shall refer to p-heights simply as heights, and if σ is an ordinal, we shall write G^σ to mean G^{p^σ}. Define subalgebras of FG inductively by $(FG)^0 = FG, (FG)^{\sigma+1} = ((FG)^\sigma)^p$, and taking intersection at limit ordinals. Let $(FG)^\infty = (FG)^\sigma$, where σ is the smallest ordinal such that $(FG)^\sigma = (FG)^{\sigma+1}$. If $\alpha \in (FG)^\sigma \setminus (FG)^{\sigma+1}$, we denote the height of α by $|\alpha| = \sigma$. Since F is perfect, it is easy to show that $(FG)^\sigma = F(G^\sigma)$. Thus, the direct sum decomposition $FG = \bigoplus_{g \in G} Fg$ is actually a valuated direct sum. All heights will be understood to be computed in FG, or if we are dealing with a quotient group $\overline{G}$ of G, then heights are relative to $F\overline{G}$.

Define two maps ε and λ of FG into itself as follows. If $\alpha, \beta \in FG$, put $\varepsilon(\beta) = \sum_{0 \leq i < p}(1/i!)\beta^i$ and $\lambda(\alpha) = \sum_{1 \leq i < p}(-1)^{i+1}(1/i)(\alpha-1)^i$. Note that for any subgroup $H \subseteq G$, ε and λ map FH into itself, thus heights are not decreased under ε or λ. By Lemma 2, if $B \subseteq T \subseteq H \subseteq G$, then $K_H(B) = 1 + I_H(B)$. Since $I_H(B)$ is an ideal of FH, we see that ε maps $I_H(B)$ into $K_H(B)$ and λ maps $K_H(B)$ into $I_H(B)$.

If B has order p, then we have the following.

Lemma 3. *Assume that B has order p and that $T \subseteq H \subseteq G$. Then $\varepsilon : I_H(B) \to K_H(B)$ is a height-preserving isomorphism.*

Proof. By [6, Lemma 2], $\varepsilon : I_G(B) \to K_G(B)$ is a height-preserving isomorphism with inverse λ. Restrict to $I_H(B)$ and $K_H(B)$. □

For other subgroups B, ε may not be a homomorphism or λ its inverse. However, we shall prove that ε is always a height-preserving bijection if B is torsion. First a lemma on polynomials.

Lemma 4. *Let $f \in F[X]$ be either $f = X + X^n g$ or $f = X + (X-1)^n g$ for some $n > 1$ and $g \in F[X]$. For $m \geq 1$, let f^m denote the composition of f with itself m times. Then for every $N \geq 2$, there exists m such that either $f^m = X + X^N g_m$ or $f^m = X + (X-1)^N g_m$ respectively, for some $g_m \in F[X]$.*

Proof. Write $f = X + aX^n + X^{n+1}h$. By a simple induction, $f^k = X + kaX^n + X^{n+1}h_k$. Since $pa = 0$, $m = p^{N-2}$ will work. The proof of the other case is similar after expressing g as a polynomial in powers of $X - 1$. □

Proposition 1. *Let $B \subseteq T \subseteq H \subseteq G$. Then $\varepsilon : I_H(B) \to K_H(B)$ is a height-preserving bijection.*

Proof. Put $\theta = \lambda \circ \varepsilon : I_H(B) \to I_H(B)$. For $\beta \in I_H(B)$, one sees that $\theta(\beta) = \beta + \beta^2 g(\beta)$ for some $g \in F[X]$. Let $\beta_1, \beta_2 \in I_H(B)$. Since these are nilpotent elements, we may choose k such that $\beta_i^k = 0$ ($i = 1, 2$). By Lemma 4 there exists m such that $\theta^m(\beta_i) = \beta_i$ ($i = 1, 2$). It follows that ε is injective. Moreover, since both ε and λ do not decrease height, ε must be height-preserving. For the surjectivity of ε, put $\nu = \varepsilon \circ \lambda : K_H(B) \to K_H(B)$. If $\alpha \in K_H(B)$, then $\nu(\alpha) = \alpha + (\alpha - 1)^2 h(\alpha)$ for some $h \in F[X]$. Since $\alpha - 1$ is nilpotent, there exists m such that $\nu^m(\beta) = \beta$, hence ε is surjective. □

We shall use the truncated exponential map ε in Lemma 10 to translate height arguments from the additive to the multiplicative.

4 Some nice subgroups

Lemma 5. *Let $T \subseteq H \subseteq G$ and $\alpha \in S(G)$. If $\alpha = 1 + \beta_0 + \sum_{0<i<\mu} g_i \beta_i$ as in Lemma 1, then $\beta = 1 + \beta_0 \in S(H)$, and $\beta^{-1}\alpha = 1 + \sum_{0<i<\mu} g_i \beta^{-1}\beta_i$ has maximal height in its coset modulo $S(H)$.*

Proof. Since β_0 is nilpotent, $\beta = 1 + \beta_0 \in S(H)$ and $\beta^{-1}\alpha$ has the prescribed form. Suppose that $\beta' \in S(H)$ and that $|\beta'| = |\beta^{-1}\alpha|$. Then $\beta'\beta^{-1}\alpha = \beta' + \sum_{0<i<\mu} g_i \beta' \beta^{-1}\beta_i$, where the summands have disjoint supports. Thus $|\beta'\beta^{-1}\alpha| \leq |\beta'| = |\beta^{-1}\alpha|$, showing that $\beta^{-1}\alpha$ has maximal height. □

Corollary. *$S(H)$ is a nice subgroup of $S(G)$.*

If $\alpha \in FG, \alpha = \sum_{g \in G} r_g g$, we denote the support of α by $\mathrm{supp}(\alpha) = \{g \in G \mid r_g \neq 0\}$.

Lemma 6. *Let $B \subseteq T \subseteq H \subseteq G$. If B is a nice subgroup of H, then $K_H(B)$ is a nice subgroup of $S(H)$.*

Proof. The natural map $G \to G/B$ induces $\varphi \colon FG \to F(G/B)$, restricting to $S(H) \to S(H/B)$ with kernel $K_H(B)$. Let $\alpha \in S(H) \setminus K_H(B)$. We must show that there is a preimage of $\varphi(\alpha)$ in $S(H)$ having maximal height. Let us call a preimage α' of $\varphi(\alpha)$ minimal if φ establishes a bijection of $\mathrm{supp}(\alpha')$ with $\mathrm{supp}(\varphi(\alpha))$. If we can find a minimal preimage α' such that every element of $\mathrm{supp}(\alpha')$ has maximal height in its coset modulo B, then α' will have maximal height in its coset modulo $K_H(B)$.

By Lemma 1, $\varphi(\alpha) = 1 + \sum_i \bar{h}_i \bar{\beta}_i$ where the $\bar{h}_i$ represent distinct cosets of H/B modulo T/B, and the $\bar{\beta}_i \in F(T/B)$ are nonzero nilpotent elements. Fix i and choose $h_i \in H$ to be a preimage of $\bar{h}_i$. Then $\bar{\beta}_i = \sum_j r_j \bar{t}_j$, with $r_j \in F$, $r_j \neq 0$, and $\bar{t}_j \in T/B$ distinct. Since $\bar{\beta}_i$ is nilpotent, it has augmentation 0. Since B is nice in H, for every j we may choose a preimage $t_j \in T$ of $\bar{t}_j$ such that $t_j h_i$ has maximal height in its coset modulo B. Then $\beta_i = \sum_j r_j t_j$ is nilpotent in FH. Obtaining β_i for every i, we then have $\alpha' = 1 + \sum_i h_i \beta_i \in S(H)$. Since the supports of the $\bar{h}_i \bar{\beta}_i$ are disjoint, α' is the desired minimal preimage of $\varphi(\alpha)$. □

Note that since we are dealing with valuated groups, we do not claim that the maximal height attained in a coset modulo $K_H(B)$ is the same as the height of the image in $S(H/B)$.

Lemma 7. *Let $B \subseteq T \subseteq H' \subseteq H$ and $\alpha \in S(H)$ such that $\alpha = 1 + \beta$, where $\mathrm{supp}(\beta) \cap H' = \emptyset$. Assume that $\gamma \in K_H(B)$ such that $\gamma\alpha = \beta_0 + \beta'$, where $\mathrm{supp}(\beta_0) \subseteq H'$ and $\mathrm{supp}(\beta') \cap H' = \emptyset$. Then $\beta_0 \in K_{H'}(B)$.*

Proof. Let $\varphi\colon FH \to F(H/B)$ be induced by the natural map. Then $\varphi(\alpha) = 1 + \varphi(\beta)$, where $\operatorname{supp}(\varphi(\beta)) \cap (H'/B) = \emptyset$. Since $\gamma \in K_H(B)$, $\varphi(\alpha) = \varphi(\gamma\alpha) = \varphi(\beta_0) + \varphi(\beta')$, where $\operatorname{supp}(\varphi(\beta_0)) \subseteq H'/B$ and $\operatorname{supp}(\varphi(\beta')) \cap (H'/B) = \emptyset$. Therefore, $\varphi(\beta_0) = 1$. But $\gamma\alpha = \beta_0 + \beta' \in S(H)$, hence Lemma 1 (with H and H' replacing G and H) implies that $\beta_0 \in S(H)$, thus $\beta_0 \in K_H(B)$. □

Lemma 8. *Let $B \subseteq T \subseteq H' \subseteq H$. If N is a nice subgroup of $S(H)$ such that $S(H') \cap K_H(B) \subseteq N \subseteq K_H(B)$, then $S(H')N$ is nice in $S(H)$.*

Proof. Let $\alpha \in S(H)$. Applying Lemma 5 to $S(H')$, we may assume that $\alpha = 1 + \sum_{0<i<\mu} h_i \beta_i$, where $\{h_i | i < \mu\}$ represent the distinct cosets of H' in H with $h_0 = 1$, and $\beta_i \in FH'$. Let $\beta \in S(H')$ and $\gamma \in N$. Again by Lemma 5, we may assume without decreasing height that $\beta\gamma\alpha = 1 + \sum_{0<i<\mu} h_i \beta'_i \beta'_i \in FH'$. Let $\gamma\alpha = 1 + \beta''_0 + \sum_{0<i<\mu} h_i \beta''_i$ ($\beta''_i \in FH'$). Since $\gamma \in K_H(B)$, by Lemma 7, $1 + \beta''_0 \in S(H') \cap K_H(B) \subseteq N$. But we must have $\beta = (1 + \beta''_0)^{-1}$, thus $\beta\gamma \in N$. Thus an element of maximal height in the coset $N\alpha$ will be of maximal height in $S(H')N\alpha$. □

For the remainder of this section, let $C \subseteq B \subseteq T \subseteq H' \subseteq H$, where C is nice in H and B/C has order p. Let $\overline{B}, \overline{H'}$, and $\overline{H}$ be the quotients of B, H' and H respectively, modulo C. Let $\varphi\colon FG \to F(G/C)$ have restriction $\varphi_I\colon I_H(B) \to I_{\overline{H}}(\overline{B})$. Then φ_I is surjective with kernel $I_H(C)$. Let $\overline{B} = \langle \overline{b} \rangle$, where $\overline{b}$ has order p, and let $\{\overline{h}_i | i < \mu\}$ represent the cosets of $\overline{B}$ in $\overline{H}$ that lie outside of $\overline{H'}$. Then

$$I_{\overline{H}}(\overline{B}) = I_{\overline{H'}}(\overline{B}) \oplus \left(\bigoplus_{i,j} F\overline{h}_i(\overline{b}^j - 1)\right), \quad i < \mu, 1 \leq j < p.$$

Choose an additive composition series $\{F_k | k < \rho\}$ for F. We wish to allow partial sums from the above direct sum and also to incorporate terms from the composition series for F. Let $W \subseteq \rho \times \mu \times \{1, \ldots, p-1\}$ such that whenever (k, i, j) and (k', i, j) are in W, then $k = k'$. Put

$$\overline{N} = I_{\overline{H'}}(\overline{B}) \oplus \left(\bigoplus_W F_k \overline{h}_i(\overline{b}^j - 1)\right).$$

We now consider heights in the additive structure.

Lemma 9. *If $N = \varphi_I^{-1}(\overline{N})$, then N is nice in $I_H(B)$.*

Proof. Let $\overline{\beta} \in I_{\overline{H}}(\overline{B}) \setminus \overline{N}$. We must consider heights of all preimages in $I_H(B)$ of all $\overline{\beta} + \overline{\gamma}$ for all $\overline{\gamma} \in \overline{N}$ and show that a maximum exists. Since $I_{\overline{H'}}(\overline{B}) \subseteq \overline{N}$, we may assume that $\operatorname{supp}(\overline{\beta}) \subseteq \bigcup_{1 \leq i \leq n} \overline{h}_i \overline{B}$. In maximizing the height of preimages of $\overline{\beta} + \overline{\gamma}$, we may assume that $\operatorname{supp}(\overline{\gamma}) \subseteq \bigcup_{1 \leq i \leq n} \overline{h}_i \overline{B}$ since allowing preimages of support elements outside this set cannot increase height. This is a finite set, thus there are only finitely many support sets to consider for $\overline{\beta} + \overline{\gamma}$. Hence we may assume that $\overline{\beta} + \overline{\gamma}$ has a fixed support. As in the proof of Lemma 6, we need only consider minimal preimages, say of a set $\{\overline{h}_i(\overline{b}^j - 1) | 1 \leq i \leq n, 1 \leq j < p\}$. In fact, since the $\overline{h}_i$

represent distinct cosets of $\overline{B}$, we are reduced to considering $\{\overline{h}(\overline{b}_j - 1)|1 \leq j \leq k\}$, where $\overline{h}$ represents a coset of $\overline{B}$ and $\overline{b}_1, \ldots, \overline{b}_k$ are distinct nonidentity elements of $\overline{B}$. Since C is nice in H, we may choose a preimage h of $\overline{h}$ to be of maximal height. Then we can choose preimages b_j for each $\overline{b}_j$ such that hb_j has maximal height. Note that this height is independent of the choice of h, therefore we have maximized the height of the preimage $\sum_{1 \leq j \leq n} r_j h(b_j - 1)$. □

Lemma 10. *In the setting above, there exists a composition series from $K_{H'}(B)K_H(C)$ to $K_H(B)$, all terms of which are nice in $S(H)$.*

Proof. We may choose sets W above so that the various resulting $\overline{N}$ will form a composition series from $I_{\overline{H'}}(\overline{B})$ to $I_{\overline{H}}(\overline{B})$. The inverse images of this series under φ_I will give a composition series $\mathcal{S}_I$ from $I_{H'}(B) + I_H(C)$ to $I_H(B)$ with terms nice in $I_H(B)$.

Let $\varphi_K \colon K_H(B) \to K_{\overline{H}}(\overline{B})$ be induced by φ. By Lemma 3 and Proposition 1, we have a diagram

$$\begin{array}{ccc} I_H(B) & \xrightarrow{\varphi_I} & I_{\overline{H}}(\overline{B}) \\ \downarrow \varepsilon & & \downarrow \overline{\varepsilon} \\ K_H(B) & \xrightarrow{\varphi_K} & K_{\overline{H}}(\overline{B}) \end{array}$$

where ε is a height-preserving bijection and $\overline{\varepsilon}$, the truncated exponential map for the quotient groups, is a height-preserving isomorphism. Since φ_I and φ_K are restrictions of the algebra homomorphism φ and ε is polynomially defined, the diagram is commutative. By Lemma 3, $\overline{\varepsilon}(I_{\overline{H'}}(\overline{B})) = K_{\overline{H'}}(\overline{B})$. Apply $\overline{\varepsilon}$ to the nice composition series in $I_{\overline{H}}(\overline{B})$ to get a nice composition series from $K_{\overline{H'}}(\overline{B})$ to $K_{\overline{H}}(\overline{B})$. Take inverse images under φ_K to get a composition series $\mathcal{S}_K$ from $K_{H'}(B)K_H(C)$ to $K_H(B)$. Since the diagram is commutative and ε is bijective, we can conclude that $\varepsilon \colon I_H(B) \to K_H(B)$ maps cosets of terms in the composition series $\mathcal{S}_I$ bijectively to cosets of terms in the composition series $\mathcal{S}_K$ in a height-preserving fashion. Therefore, the terms of $\mathcal{S}_K$ are nice in $K_H(B)$. Since C is nice in H and B/C is finite, B is nice in H, hence by Lemma 6, all terms of $\mathcal{S}_K$ are nice in $S(H)$. □

Proposition 2. *Let $C \subseteq B \subseteq T \subseteq H' \subseteq H$, with C nice in H, and B/C of order p. Then there exists a nice composition series from $S(H')K_H(C)$ to $S(H')K_H(B)$.*

Proof. By Lemma 10, there exists a nice composition series from $K_{H'}(B)K_H(C)$ to $K_H(B)$. Take the product of each term of this series with $S(H')$ to obtain a composition series from $S(H')K_H(C)$ to $S(H')K_H(B)$ after deleting repeated terms. Lemma 8 guarantees that this will be a nice composition series once we observe that $S(H') \cap K_H(B) \subseteq S(H')K_H(C) \cap K_H(B) = (S(H') \cap K_H(B))K_H(C) = K_{H'}(B)K_H(C)$. □

5 Composition series for T

Assume in this section that T is totally projective of p-length $\lambda(T)$. Let $\lambda_1 < \lambda_2 < \ldots < \lambda_m$ be limit ordinals not exceeding $\lambda(T)$. Suppose that for every i with $1 \leq i \leq$

m, we have a sequence of ordinals $\mu_{i1} < \mu_{i2} < \ldots$ such that $\sup\{\mu_{ij}|1 \leq j\} = \lambda_i$. We may assume that $\lambda_{i-1} < \mu_{i1}$ for $2 \leq i \leq m$.

Lemma 11. *There exists a chain $N_1 \subseteq N_2 \subseteq \ldots$ of nice subgroups of T with $\bigcup_{1 \leq j} N_j = T$ such that:*

(i) $\{N_j\}$ *can be extended to a nice composition series for T; and*

(ii) *if $t \in N_j$ and $1 \leq i \leq m$, then $\mu_{ij} < |t| < \lambda_i$ does not hold.*

Proof. We induct on $\lambda = \lambda(T)$. For $\lambda = 0$, T is divisible and there are no λ_i, so take $N_j = T$ for all j. Now suppose that $\lambda = \sigma + 1$. Then $\lambda_m \leq \sigma$ and induction applies to T/T^σ. Therefore we may choose N_j with $T^\sigma \subseteq N_j \subseteq T$ such that $\{N_j/T^\sigma | 1 \leq j\}$ satisfies (i) and (ii) for T/T^σ. Clearly, $\bigcup_j N_j = T$ and (i) follows for T by pulling back a nice composition series from T/T^σ and noting that T^σ has a composition series nice in T. Further, (ii) is apparent for T since $|t| < \lambda_i$ implies that $|t|$ is the same as $|tT^\sigma|$ in T/T^σ.

Now assume that λ is a limit ordinal, hence $\lambda_m \leq \lambda$, and T/T^{λ_m} is a coproduct of totally projective groups of lengths $< \lambda_m$. Therefore we may write $T/T^{\lambda_m} = \coprod_{1 \leq k} A_k$, where $\lambda(A_k) \leq \mu_{mk}$ for all k. Moreover, since we allow $A_k = 1$, if $m > 1$ we may assume that for every $k \geq 1$, either $\lambda_{m-1} \leq \lambda(A_k)$ or else $A_k = 1$. By induction applied to each A_k, there exist $N_{k1} \subseteq N_{k2} \subseteq \ldots$, nice subgroups of A_k and therefore of T, such that: $\bigcup_j N_{kj} = A_k$; $\{N_{kj}|1 \leq j\}$ can be extended to a nice composition series for A_k; and if $t \in N_{kj}$ and $1 \leq i \leq m - 1$, then $\mu_{ij} < |t| < \lambda_i$ does not hold. (All this is clear if $A_k = 1$.)

Define $N_1 \subseteq N_2 \subseteq \ldots \subseteq T$ by $T^{\lambda_m} \subseteq N_j$ and $N_j/T^{\lambda_m} = \coprod_{1 \leq k \leq j} N_{kj}$. Then $\bigcup_j N_j = T$ is clear. Put $N_0 = T^{\lambda_m}$. Then N_0 has a composition series nice in T and we may interpolate a nice composition series at each step from N_j to N_{j+1} ($j \geq 0$), by using the series for the A_k.

Finally, if $1 \leq i \leq m - 1$ and $t \in N_j$, then $\mu_{ij} < |t| < \lambda_i$ does not hold since it does not hold in N_{kj} for $1 \leq k \leq j$. For $i = m$, $t \in N_j$ implies $\mu_{mj} < |t| < \lambda_m$ does not hold since $N_{kj} \subseteq A_k$ and $\lambda(A_k) \leq \mu_{mk} \leq \mu_{mj}$. □

We now modify the lemma.

Lemma 12. *Let $L_1 \subseteq L_2 \subseteq \ldots$ be a chain of finite subgroups of T. Then there exists a chain $T_0 \subseteq T_1 \subseteq T_2 \subseteq \ldots$ of nice subgroups of T with $\bigcup_{0 \leq j} T_j = T$ such that:*

(i) $L_1 = T_0$ *and $L_j \subseteq T_{j-1}$ for $1 \leq j$;*

(ii) *there is a nice composition series with union T containing $\{T_j\}$; and*

(iii) *for $t \in T_j$ ($1 \leq j$) and $1 \leq i \leq m$, $|t|$ can assume only finitely many values with $\mu_{ij} < |t| < \lambda_i$.*

Proof. Start with the $\{N_j\}$ from Lemma 11 and a nice composition series containing the N_j. Put $N_0 = 1$. For $0 \leq j$, put $T_j = N_j L_{j+1}$, thus (i) is clear and we have nice subgroups with union T. Next, for every N in the composition series such that $N_j \subseteq N \subset N_{j+1}$, replace N by NL_{j+1}. Adding terms where there is a jump of finite index more than p, we obtain a nice composition series satisfying (ii). Finally, for (iii), suppose that for some i and j, $|t|$ can assume infinitely many values with $\mu_{ij} < |t| < \lambda_i$ and $t \in T_j$. Since L_{j+1} is finite, there exist $t_1, t_2 \in N_j$ and $s \in L_{j+1}$ such that $\mu_{ij} < |t_1 s| < |t_2 s| < \lambda_i$. But then $\mu_{ij} < |t_1 t_2^{-1}| < \lambda_i$, contrary to (ii) of Lemma 11. □

For every torsion-free $g \in G$, define $\lambda(g)$ as follows. First assume that the height sequence of g has infinitely many gaps (we shall say there is a gap at $|g^{p^k}|$ if $|g^{p^k}| + 1 < |g^{p^{k+1}}|$). In this case, put $\lambda(g) = \sup\{|g^{p^k}| \mid 0 \leq k\}$. Now assume that k is minimal such that there is no gap at or above $|g^{p^k}|$. If $|g^{p^k}|$ is finite, put $\lambda(g) = 0$. If $|g^{p^k}| = \infty$, put $\lambda(g) = \infty$. If $|g^{p^k}| = \lambda + n$, where λ is a limit ordinal and $n < \omega$, put $\lambda(g) = \lambda$. In the case of infinitely many gaps, clearly $\lambda(g)$ has cofinality ω. In the case of finitely many gaps where $\lambda(g) \neq 0, \infty$, since we are assuming that T is totally projective, [5, Lemma 4] implies that $\lambda(g)$ has cofinality ω also in this case. Thus $\lambda(g)$ can assume values 0, ∞, or a limit ordinal of cofinality ω. Moreover, the value can be determined from the behavior of any terminal segment of the height sequence for g, thus $\lambda(g^n) = \lambda(g)$ for all $n \neq 0$.

Now to fix some notation, let $g_1, \ldots, g_k$ be independent torsion-free elements of G, and let P denote the free subgroup that they generate.

Lemma 13. $\lambda(g)$ assumes at most k distinct values for $g \in P$, $g \neq 1$.

Proof. Suppose that $h_1, h_2 \in P$ with $\lambda(h_1) < \lambda(h_2)$. We claim that $|h_1^{p^n}| < |h_2^{p^n}|$ for sufficiently large n. This is clear if $\lambda(h_2) = \infty$. If $\lambda(h_1) = 0$, then the only problem might be when $\lambda(h_2) = \omega$ and the height sequence of h_2 has all finite entries. But then there are infinitely many gaps in the height sequence of h_2, so these entries eventually exceed the corresponding ones for h_1. We are left with the case that $\lambda(h_1)$ and $\lambda(h_2)$ are both limit ordinals. Here we need to examine the situation where the height sequence of h_1 has finitely many gaps, that of h_2 has infinitely many gaps, and $\lambda(h_2) = \lambda(h_1) + \omega$. But this is just like the preceding case for 0 and ω, so the claim is justified.

By the claim, we have $|(h_1 h_2)^{p^n}| = |h_1^{p^n}|$ for sufficiently large n, thus $\lambda(h_1 h_2) = \lambda(h_1)$. Now suppose that $h_1, \ldots, h_m$ are torsion-free elements of P such that $\lambda(h_1) < \ldots < \lambda(h_m)$, and that $h_1^{e_1} = h_2^{e_2} \cdots h_m^{e_m}$ with all $e_i \neq 0$. But then $\lambda(h_2^{e_2} \cdots h_m^{e_m}) = \lambda(h_2^{e_2}) > \lambda(h_1^{e_1})$, a contradiction. Therefore, such elements must be independent, consequently $m \leq k$. □

Lemma 14. *Suppose that the values other than 0 or ∞ assumed by $\lambda(g)$ for $g \in P$, $g \neq 1$, are $\lambda_1 < \ldots < \lambda_m$. Let μ_{ij} be as at the beginning of this section. Then there exist finite subgroups $L_1 \subseteq L_2 \subseteq \ldots$ of T such that if $g \in P, j \geq 1$, and $\lambda(g) = \lambda_i$, then there exists $s_0 \in L_j$ such that $\mu_{ij} < |g s_0|$. If $\lambda(g) = \infty$, then there exists $s_0 \in L_1$ such that $|g s_0| = \infty$.*

Proof. Since P is finitely generated, if λ is an ordinal or ∞, then $\langle g \in P | \lambda(g) = \lambda \rangle$ can be generated by finitely many g with $\lambda(g) = \lambda$. Thus it is sufficient to show that for a single g and $\mu < \lambda$ (in case $\lambda \neq \infty$), there exists $s_0 \in T$ with either $\mu < |g s_0|$ or $|g s_0| = \infty$, depending on the case.

First suppose that the height sequence of g has infinitely many gaps. Then there exists n such that $\mu < |g^{p^n}|$. Choose q such that there exist n gaps strictly between $|g^{p^n}|$ and $|g^{p^{n+q}}|$. Then $|g^{p^{n+q}}| \geq |g^{p^n}| + n + q > \mu + n + q$. Therefore, there exists g_1 such that $g^{p^{n+q}} = g_1^{p^{n+q}}$ and $|g_1| > \mu$. Put $s_0 = g^{-1}g_1$.

Now suppose that g has finitely many gaps with n minimal such that there is no gap at or above $|g^{p^n}|$. If $|g^{p^n}| = \infty$, then there exists g_1 such that $g^{p^n} = g_1^{p^n}$ and $|g_1| = \infty$. Thus we can take $s_0 = g^{-1}g_1$. Therefore we may assume that $|g^{p^n}| = \lambda + q$ for λ a limit ordinal. If $n \leq q$, then $g^{p^n} = g_1^{p^n}$, $\mu < \lambda \leq |g_1|$, thus can take $s_0 = g^{-1}g_1$. If $n > q$, first choose g_1 such that $g^{p^n} = g_1^{p^q}$ and $|g_1| = \lambda$. Since λ is a limit ordinal and $\mu < \lambda$, we may choose g_2 such that $g_2^{p^{n-q}} = g_1$ and $\mu < |g_2|$. Put $s_0 = g^{-1}g_2$. □

The L_j will be employed via

Lemma 15. *Let $N \subseteq T$ be nice in T, $g \in G$ torsion-free, and $t \in T$. Put $\mu = \sup\{|gts| \mid s \in N\}$. Assume there exists $s_0 \in N$ such that $\mu < |gs_0|$. Then there exists $s \in N$ such that $\mu = |gts|$.*

Proof. Since $|gts| < |gs_0|$, we have $\mu = \sup\{|gts(gs_0)^{-1}| \mid s \in N\} = \sup\{|tss_0^{-1}|\}$. But ss_0^{-1} ranges over N, thus there exists $s \in N$ such that $\mu = |tss_0^{-1}| = |gts|$. □

Proposition 3. *Let $H = PT$. Then there exists a composition series beginning at 1, with union T, and such that all terms are nice in H.*

Proof. First suppose that $\lambda(g)$ is never a limit ordinal for $g \in P$, $g \neq 1$. Take L_1 by Lemma 14 to satisfy the condition for $\lambda(g) = \infty$ if that occurs, otherwise put $L_1 = 1$. Take any nice composition series for T containing L_1 as a term. If $\lambda(g)$ does assume limit ordinal values, then choose finite subgroups $L_1 \subseteq L_2 \subseteq \ldots$ according to Lemma 14, and take nice subgroups $T_0 \subseteq T_1 \subseteq \ldots$ of T given by Lemma 12. By Lemma 12, we may take a nice composition series with union T and containing the $\{T_j\}$. We shall show that the terms of this series are nice in H.

Let N be a term of the composition series and let $gt \in H$, where $g \in P$ and $t \in T$. We may assume that $g \neq 1$. Note that $\lambda(gt) = \lambda(g)$. If $\lambda(g) = 0$, then there is a last gap, say at $|g^{p^{n-1}}|$, and $|g^{p^n}|$ is finite. Thus $|gts|$ can assume only bounded finite values for $s \in N$, hence an element of maximum height exists in the coset gtN. Lemma 12 tells us that T_0 is a term of the composition series and is finite since $L_1 = T_0$. In particular, we may assume that $T_0 \subseteq N$. If $\lambda(g) = \infty$, then Lemmas 14 and 15 imply that the coset gtN has an element of maximum height ∞.

We may now assume that $\lambda(g) = \lambda_i$ is an infinite limit ordinal. Suppose that $\lambda_i \leq |gts|$ for some $s \in N$. Then g has a last gap, say at $|g^{p^{n-1}}|$, and $|g^{p^n}| = \lambda_i + m$, $m < \omega$. Then $|gts|$ can assume only finitely many values $\geq \lambda_i$, hence a maximum exists. Finally, we may suppose that $|gts| < \lambda_i$ for every $s \in N$. By Lemma 12 and $T_0 \subseteq N$, there exists $j \geq 1$ such that $T_{j-1} \subseteq N \subseteq T_j$. If $|gts|$ assumes any value $> \mu_{ij}$, then it can assume only finitely many since $N \subseteq T_j$, hence a maximum height exists. Consequently, we may assume that $|gts| \leq \mu_{ij}$ for every $s \in N$. Since $L_j \subseteq T_{j-1} \subseteq N$, by Lemmas 14 and 15, gtN has an element of maximum height. □

6 Proof of the Theorem

We may choose subgroups $T = H_0 \subseteq H_1 \subseteq \ldots$ with $\bigcup_{m<\omega} H_m = G$, such that every H_m/T is finitely generated. In particular, Proposition 3 and the corollary to Lemma 5 apply to every H_m. Thus we have a chain $S(T) = S(H_0) \subseteq S(H_1) \subseteq \ldots$ of nice subgroups of $S(G)$ with $\bigcup_{m<\omega} S(H_m) = S(G)$. Since $S(T) = V(T)$, by [4, Lemma 3] we know that $S(T)$ is isotype in $S(G)$, thus balanced in $S(G)$. If we can show that there is a nice composition series from $S(H_{m-1})$ to $S(H_m)$ for every $m \geq 1$, then $S(T)$ will be a direct factor of $S(G)$ with totally projective complement. Since $S(T)$ is known to be totally projective by [4], we will have that $S(G)$ is totally projective, as desired.

By Proposition 3, we may choose a composition series $\{T_i | i < \mu\}$ with union T such that all T_i are nice in H_m. Then $\{K_{H_m}(T_i) | i < \mu\}$ is a smooth chain of subgroups from $K_{H_m}(1) = 1$ to $\bigcup_{i<\mu} K_{H_m}(T_i) = S(H_m)$. By Proposition 2, for every $i < \mu$ we have a nice composition series from $S(H_{m-1})K_{H_m}(T_i)$ to $S(H_{m-1})K_{H_m}(T_{i+1})$. Linking these together, we obtain a nice composition series from $S(H_{m-1})$ to $S(H_m)$, proving the Theorem.

For Corollary 1, the first claim follows since T is a direct factor of $S(T)$ with totally projective complement (see [4]). Lemma 1 implies that $V(G) = GS(G)$, and clearly, $G \cap S(G) = T$, thus the second claim of the Corollary follows.

The proofs of Corollaries 2 and 3 go through just as those of [2, Corollaries 5.8, 5.9], where we avoid the restriction on the length of torsion by using Corollary 1. Note that the part of [2, Theorem 5.6] that is needed is valid for arbitrary length.

References

[1] L. Fuchs, *Infinite Abelian Groups, Vol. II*. Academic Press, New York, 1973.

[2] P. Hill and W. Ullery, *On commutative group algebras of mixed groups*, Comm. Algebra 25 (1997), pp. 4029–4038.

[3] W. May, *Commutative group algebras*, Trans. Amer. Math. Soc. 136 (1969), pp. 139–149.

[4] W. May, *Modular group algebras of simply presented abelian groups*, Proc. Amer. Math. Soc. 104 (1988), pp. 403–409.

[5] W. May, *Isomorphism of endomorphism algebras over complete discrete valuation rings*. Math. Z. 204 (1990), pp. 485–499.

[6] W. May, *Totally projective unit groups in modular abelian group algebras*, Forum Math. 18 (2006), pp. 603–609.

[7] W. Ullery, *On group algebras of p-mixed abelian groups*, Comm. Algebra 20 (1992), pp. 655–664.

Author information

Warren May, Department of Mathematics, University of Arizona, Tucson, Arizona 85721, U.S.A.
E-mail: may@math.arizona.edu

Pure covers in abelian p-groups

László Fuchs and Takashi Okuyama

Abstract. Generalizing the notion of pure hull, we discuss pure covers in abelian p-groups G: a pure cover of a subgroup A of G is a pure subgroup that contains A, but has no proper summands containing A. We describe the Ulm–Kaplansky invariants of the pure covers: they are the same for all pure covers. We also prove that every subgroup of G admits a pure cover if and only if G is the direct sum of a divisible and a quasi-complete p-group.

Pure covers are closely related to quasi-pure hulls discussed by the second author [10] for separable p-groups.

Key words. p-group, separable p-group, Ulm–Kaplansky invariants, overhang, pure hull, pure cover.

AMS classification. Primary: 20K10.

1 Introduction

In this note all groups are additively written abelian p-groups, where p is a fixed prime. For standard terminology and notation we refer to Fuchs [3].

A subgroup A of a group G is called *purifiable* if there is a pure subgroup H of G containing A such that H is minimal in the sense that there is no pure subgroup of G that contains A and is properly contained in H. Such an H is said to be a *pure hull* of A. Pure hulls have been studied by various authors; see e.g. Hill–Megibben [4], Benabdallah–Irwin [1], Okuyama [6–10].

Recently, the second author [10] has considered a more general concept. He called a pure subgroup H containing A a *quasi-pure hull* of A if A is *almost dense* in it, i.e. if $p^n H[p] \leq A + p^{n+1} H$ holds for all integers $n \geq 0$. He proved that in torsion-complete p-groups G, every subgroup admits a quasi-pure hull and its maximal quasi-pure hulls are isomorphic.

We are going to consider a concept that turns out to be equivalent to quasi-pure hull, using a different approach. We are looking for pure subgroups H containing the given subgroup A which are minimal in the sense that they do not contain any proper direct summand still containing A. Such groups H will be called *pure covers* of A. Our Lemma 3.3 below will show the equivalence of the two concepts.

We first study pure covers in separable abelian p-groups G, i.e. in p-groups without elements of infinite height. We prove a couple of characterizations for pure covers; see Lemmas 3.2–3.4. Theorem 6.2 tells us that, in general, in a p-group with unbounded basic subgroups, not all subgroups admit pure covers. We characterize the separable p-groups G in which pure covers exist for all subgroups: these are the quasi-complete p-groups (these-groups are defined by the property that p-adic closures of pure subgroups are again pure); see Theorem 6.2. Needless to say, pure covers need not be unique, not even up to isomorphism; however, we can claim that they all have something important in common: they share the same Ulm–Kaplansky invariants (Corollary 4.3). Hence if G happens to be a direct sum of cyclic groups, then all pure covers of a subgroup in G (if they exist) are isomorphic (Corollary 4.4).

Having settled the separable case, it is easy to describe all p-groups in which every subgroup admits a pure cover. In Theorem 7.2 we prove that these are exactly the direct sums of divisible p-groups and quasi-complete p-groups.

We wish to thank the referee for his/her useful comments.

2 Preliminaries

Recall a few important results concerning pure hulls.

Hill–Megibben [4] proved that a pure hull H of the subgroup A in a p-group G admits a direct decomposition
$$H = M \oplus N$$
where $M[p] = A[p]$ and N is a bounded subgroup. They also derived the conclusion that if the heights of the elements of A (computed in G) are bounded by an integer m, then A admits a pure hull (which is even a minimal summand of G containing A).

Benabdallah–Irwin [1] have shown that a subgroup A of a p-group G is purifiable if and only if there exists a pure subgroup H of G containing A such that $p^m H[p] \leq A$ for some integer m.

In what follows the Ulm–Kaplansky invariants (UK-invariants) will be considered in the $\mathbb{Z}/p\mathbb{Z}$-vector space form (rather than just their dimensions), i.e. for an ordinal α, the αth UK-invariant is $f_\alpha(G) = p^\alpha G[p]/p^{\alpha+1} G[p]$. More generally, for a subgroup A of G, we let
$$f_\alpha(A) = \frac{(A \cap p^\alpha G)[p]}{(A \cap p^{\alpha+1} G)[p]}.$$

Recall that the αth *relative Ulm–Kaplansky invariant* of A in G is defined as the $\mathbb{Z}/p\mathbb{Z}$-vector space
$$f_\alpha(G, A) = \frac{p^\alpha G[p]}{\{(A \cap p^\alpha G) + p^{\alpha+1} G\}[p]}.$$

We will find it useful to give a different form to the αth relative Ulm–Kaplansky invariant that is more suitable for some of our proofs. First, we write $(A \cap p^\alpha G) + p^{\alpha+1} G \cong p^\alpha G \cap (A + p^{\alpha+1} G)$ by the modular law, and then using the first Isomorphism Theorem, we obtain
$$f_\alpha(G, A) = \frac{p^\alpha G[p]}{p^\alpha G[p] \cap (A + p^{\alpha+1} G)[p]} \cong \frac{p^\alpha G[p] + (A + p^{\alpha+1} G)[p]}{(A + p^{\alpha+1} G)[p]}.$$

We shall need other invariants as well. The αth *overhang* of a p-group G over its subgroup A was defined by Benabdallah–Okuyama [2] as the $\mathbb{Z}/p\mathbb{Z}$-vector space
$$V_\alpha(G, A) = \frac{\{(A \cap p^\alpha G) + p^{\alpha+1} G\}[p]}{(A \cap p^\alpha G)[p] + p^{\alpha+1} G[p]}.$$

Okuyama [6] has shown that the nth UK-invariant $f_n(N)$ of the summand N of the pure hull H of A (see above) is given by the nth overhang $V_n(G, A)$ of G over A, i.e. $f_n(N) \cong V_n(G, A)$. By making use of this result and Hill–Megibben's mentioned above, we derive:

Lemma 2.1. *If the heights of elements in a subgroup A of a p-group G are all less than the integer n, then A admits a pure hull H. Every pure hull H of A in G satisfies*

$$p^{n-1}H = A \cap p^{n-1}G.$$

Proof. If $A \cap p^n G = 0$, then the overhangs $V_m(G, A) = 0$ for all $m \geq n$, so H contains no elements $\neq 0$ of heights $\geq n$. □

3 Pure covers

Let G be a p-group and A a proper subgroup of G. It is a very special situation when A has a pure hull in G. However, even if it admits no pure hull, it can very well happen that there exist proper pure subgroups H of G that contain A, but no proper direct summand of H contains A.

We start with the following definition.

Definition 3.1. Let A be a subgroup of the p-group G. A pure subgroup H of G containing A is called a *pure cover* of A if H has no proper direct summand containing A.

We now prove the following useful criterion which we are going to use most of the time.

Lemma 3.2. *In a p-group, a pure subgroup H containing the subgroup A is a pure cover of A if and only if H contains no quasi-cyclic summand disjoint from A and for any n, no cyclic summand of order p^n disjoint from $A + p^n H$.*

Proof. Suppose that H is not a pure cover of A. Then $H = B \oplus C$ with $A \leq B$ and $C \neq 0$. If C is not reduced, then it contains a quasicyclic summand which is then disjoint from A. If C is reduced, then it contains a cyclic summand C' of order p^n for some $n > 0$. We may assume that $C' = C$. Clearly, $p^n H = (B \cap p^n H) \oplus (C \cap p^n H) = B \cap p^n H$ implies that $p^n H$ is contained in B. Thus C is disjoint from $A + p^n H \leq B$.

Conversely, if C is a quasi-cyclic subgroup of H with $C \cap A = 0$, then C is a summand of H, so H can not be a pure cover of A. If $C \neq 0$ is a cyclic summand of order p^n in H such that $C \cap (A + p^n H) = 0$, then H has a decomposition $H = B \oplus C$ with $A + p^n H \leq B$; indeed, choose B to be any subgroup that is maximal with respect to the properties of containing $A + p^n H$ and not intersecting C. Again, H can not be a pure cover of A. □

It is clear that a cyclic summand $\langle c \rangle$ of order p^{n+1} in H is disjoint from $A + p^{n+1}H$ if and only if $p^n c \notin A + p^{n+1}H$. Since $p^n c \in p^n H[p]$, this means that $p^n H[p] \not\leq A + p^{n+1}H$. Since every element in $p^n H[p] \setminus (A + p^{n+1}H)$ supports a cyclic summand of H of order p^{n+1}, we conclude:

Lemma 3.3. *A pure subgroup H containing the subgroup A of a separable p-group G is a pure cover of A if and only if*

$$p^n H[p] \leq A + p^{n+1} H$$

for all integers n. In other words, if and only if H is a quasi-pure hull of A.

Suppose for a moment that A is a purifiable subgroup of the separable p-group G, and H is a pure hull of A in G. Then we have $\{(A \cap p^n H) + p^{n+1} H\}[p] = p^n H[p]$. Hence it follows that the nth relative UK-invariant $f_n(H, A)$ of A in H is 0 for every integer $n \geq 0$. The same holds for pure covers; moreover, it becomes a sufficient condition as well in separable p-groups:

Lemma 3.4. *In a separable p-group, a pure subgroup H containing the subgroup A is a pure cover of A, if and only if the relative UK-invariants satisfy*

$$f_n(H, A) = 0 \quad \text{for all } n \in \omega.$$

Proof. For the proof it suffices to observe that the vector space $f_n(H, A) \cong \{p^n H[p] + (A + p^{n+1} H)[p]\}/(A + p^{n+1} H[p])$ is 0 for every n if and only if the inclusion relation in Lemma 3.3 holds for every n. □

We point out an easy corollary to the definition; it has been proved in a different setting by Okuyama [10]. In this corollary, we say that a pure cover H of the subgroup A of the group G is a *maximal pure cover* if $K = H$ whenever K is a pure cover of A with $H \leq K$.

Corollary 3.5. *Every pure cover of A in a p-group G is contained in a maximal pure cover of A.*

Proof. As the union of a chain of pure covers may not contain any summand described by Lemma 3.2, the claim is evident. □

Next we give examples for pure covers.

Example 3.6. Every pure hull H of A is a pure cover of A. As Theorem 4.1 will show, so is every pure subgroup of G that contains H as a dense subgroup.

Example 3.7. The group G is a pure cover of its subgroup A if and only if A is almost dense in G. Since $f_n(G, A) = (p^n G[p])/\{(A + p^{n+1} G) \cap p^n G\}[p] = 0$ holds if and only if $(A + p^{n+1} G)[p] \cap p^n G[p] = p^n G[p]$ if and only if $p^n G[p] \leq A + p^{n+1} G$. The last inclusion is the definition of A being almost dense in G, cf. Okuyama [8].

Example 3.8. We exhibit an example where the subgroup A is not purifiable in a p-group G, but admits (several) pure covers in G. Let $\{G_i\}_{i \in I}$ be an infinite set of separable p-groups, and let A_i be a purifiable subgroup of G_i for each $i \in I$. Let $H_i = M_i \oplus N_i$ denote a pure hull of A_i in G_i where $M_i[p] = A_i[p]$ and N_i is bounded, $p^{m_i} N_i = 0$ with minimal m_i. If the set $\{m_i\}_{i \in I}$ is unbounded, then $A = \oplus_{i \in I} A_i$ is not

purifiable in $G = \oplus_{i \in I} G_i$ (Benabdallah–Irwin [1]), but $H = \oplus_{i \in I} H_i$ is a pure cover of A in G. In fact,
$$f_n(\oplus_{i \in I} H_i, \oplus_{i \in I} A_i) \cong \oplus_{i \in I} f_n(H_i, A_i) = 0$$
for every $n \in \omega$. Again, from Theorem 4.1 or Lemma 5.3 below we can conclude that pure dense subgroups of H containing A are likewise pure covers and so are all pure subgroups K of G that contain H as a dense subgroup.

4 The invariants of pure covers

The following result is crucial; it gives relevant information about pure covers containing other pure covers.

Theorem 4.1. *Let A be a subgroup of the separable p-group G, and $H \leq K$ pure subgroups of G containing A.*

(a) *For every $n \in \omega$, there is a natural monomorphism*
$$\phi_n : f_n(H, A) \to f_n(K, A)$$
induced by the embedding $H \to K$.

(b) *The map ϕ_n is surjective for every $n \in \omega$ if and only if H is a dense subgroup of K.*

Proof. (a) ϕ_n is defined in the obvious way:
$$\phi_n : \frac{p^n H[p] + (A + p^{n+1} H)[p]}{(A + p^{n+1} H)[p]} \to \frac{p^n K[p] + (A + p^{n+1} K)[p]}{(A + p^{n+1} K)[p]}.$$

Thus ϕ_n acts by sending the coset $x + (A + p^{n+1} H)[p]$ upon the coset $x + (A + p^{n+1} K)[p]$, where $x = p^n h$ with $h \in H$ and $px = 0$. Thus
$$\operatorname{Im} \phi_n = \frac{p^n H[p] + (A + p^{n+1} K)[p]}{(A + p^{n+1} K)[p]}.$$

Suppose that $x + (A + p^{n+1} H)[p] \in \operatorname{Ker} \phi_n$, which amounts to saying that $x = p^n u \in A + p^{n+1} K$ ($u \in H$), i.e. $p^n u = a + p^{n+1} y$ for some $a \in A, y \in K$. Then $p^n u - a = p^{n+1} y \in H \cap p^{n+1} K = p^{n+1} H$, and therefore $p^n u \in A + p^{n+1} H$ shows that $x \in (A + p^{n+1} H)[p]$, i.e. ϕ_n is monic.

(b) First assume that H is dense in K, i.e. $K = H + p^n K$ for each $n \geq 1$. To prove that $p^n K[p] \leq p^n H[p] + (A + p^{n+1} K)[p]$ for every $n \in \omega$, let $p^n y \in p^n K[p] \setminus p^{n+1} K$ ($y \in K$). In view of the purity and the density of H in K, we can write $y = u + pz$ with $u \in H, z \in K$, where u has the same order p^{n+1} as y. Thus $p^n y = p^n u + p^{n+1} z \in p^n H[p] + (A + p^{n+1} K)[p]$, establishing the surjectivity of ϕ_n.

Conversely, suppose that each ϕ_n is surjective, In other words, suppose that $p^n K[p] \leq p^n H[p] + (A + p^{n+1} K)[p]$ for every $n \in \omega$. Write $p^n y \in p^n K[p]$ in the form $p^n y = p^n x + (a + p^{n+1} z)$ with $x \in H, a \in A, z \in K$. It follows that the height of a is at least n, so a can be included in the term $p^n x$. We will induct on n to show that $K[p^n] \leq H + pK$, leading to the density relation $K = H + pK$. For $n = 1$, if $y \in K[p]$, then clearly $y = x + pz \in H + pK$. If $y \in K[p^{n+1}]$, then $p^n y \in K[p]$, and $p^n y = p^n x + p^{n+1} z$ implies that $y - x - pz \in K[p^n]$, so by induction hypothesis $y - x - pz \in H + pK$. Hence also $y \in H + pK$, completing the proof. □

We can calculate the UK-invariants of pure covers.

Theorem 4.2. *Let H be a pure cover of the subgroup A in a separable p-group G. Then the UK-invariants $f_n(H)$ of H satisfy:*

$$f_n(H) \cong f_n(A) \oplus V_n(H, A).$$

Proof. We have

$$f_n(A) = \frac{(p^n H \cap A)[p]}{(p^{n+1} H \cap A)[p]} = \frac{p^n H \cap A[p]}{p^{n+1} H \cap A[p]}$$
$$\cong \frac{(p^n H[p] \cap A) + p^{n+1} H[p]}{p^{n+1} H[p]} = \frac{p^n H[p] \cap (A[p] + p^{n+1} H[p])}{p^{n+1} H[p]}.$$

Furthermore, since the relative UK-invariant $f_n(H, A) = 0$, we clearly have $p^n H[p] = \{(A \cap p^n H) + p^{n+1} H\}[p]$, and therefore we can write

$$V_n(H, A) = \frac{\{(A \cap p^n H) + p^{n+1} H\}[p]}{(A \cap p^n H)[p] + p^{n+1} H[p]} = \frac{p^n H[p]}{p^n H[p] \cap (A[p] + p^{n+1} H[p])}.$$

From these it follows that $f_n(H) = p^n H[p]/p^{n+1} H[p]$ is an extension of $f_n(A)$ by $V_n(H, A)$, so it is indeed isomorphic to the direct sum of the vector spaces $f_n(A)$ and $V_n(H, A)$. □

Hence we conclude:

Corollary 4.3. *Every pure cover of the subgroup A in a separable p-group G has the same collection of UK-invariants. Pure covers of A have isomorphic basic subgroups.*

Proof. This is immediate from the preceding theorem. □

Since subgroups of direct sums of cyclic groups are again direct sums of cyclic groups, and since such groups are determined up to isomorphism by their UK-invariants, we can state without additional comments:

Corollary 4.4. *Let G be a p-group which is a direct sum of cyclic groups. All pure covers of a subgroup A of G (if they exist) are isomorphic.*

5 Pure covers and p-adic density

Our next goal is to show that dense subgroups behave similarly inasmuch as pure covers are concerned.

We start with the following lemma. This lemma is analogous to [9, Lemma 3.9] which shows that a p-subgroup is purifiable if and only if its p-adic closure is.

Lemma 5.1. *Suppose that B is a dense subgroup of the subgroup A and H is a pure subgroup of the p-group G that contains B. Then*

(i) $H^* = A + H$ *is a pure subgroup of G containing A;*

(ii) *if H is a pure cover of B, then H^* is a pure cover of A.*

Proof. (i) Let $a + x \in H^*$ be an element of height n where $a \in A, x \in H$. We can write $a = p^n c + b$ with $c \in A, b \in B$, so $a + x = p^n c + b + x$ where $b + x \in H$ must have a height $\geq n$. Thus by purity there exists a $y \in H$ such that $p^n y = b + x$. We obtain $a + x = p^n(c + y)$ with $c + y \in H^*$, establishing the purity of H^*.

(ii) Assume that H contains no summand of order p^n disjoint from $B + p^n H$. By way of contradiction, assume that there exists a summand $\langle a + x \rangle$ ($a \in A, x \in H$) of H^* of order p^n that is disjoint from $A + p^n H^* = A + p^n H$. Write $a = b + pc$ with $c \in A, b \in B$. We obtain $p^n(b+x) = p^n(a+x) - p^{n+1}c = -p^{n+1}c = p^{n+1}y$ for some $y \in H$ in view of the purity of H. Note that the height-sequence of $a + x$ must be $(0, 1, \ldots, n-1, \infty, \ldots)$, where the only gap occurs after $n - 1$. Hence it is clear that not only is $b + x - py$ annihilated by p^n, but it has the same height-sequence as $a + x$. This means that $\langle b + x - py \rangle$ is a summand of H of order p^n. To see that it is disjoint from $B + p^n H$, observe that $p^{n-1}b + p^{n-1}x - p^n y \in B + p^n H$ would imply $p^{n-1}a - p^n c + p^{n-1}x - p^n y \in A + p^n H$. Because of $p^n c + p^n y \in p^n H$ hence we would obtain that $0 \neq p^{n-1}a + p^{n-1}x \in A + p^n H$. This contradicts our hypothesis that $\langle a + x \rangle$ does not intersect $A + p^n H$. □

From these considerations we can derive:

Corollary 5.2. *In a separable p-group, a pure subgroup H containing A is a pure cover of A, if and only if it is a pure cover of any basic subgroup of A.* □

We shall need the following lemma.

Lemma 5.3. *Suppose that H is a pure subgroup of the separable p-group G which contains a subgroup A but it contains no cyclic direct summand disjoint from A. Then the p-adic closure $K = \bar{H}$ of H in G enjoys the same property.*

Proof. We verify that under the stated conditions K contains no cyclic summand $\langle x \rangle$ of order p^n with $\langle x \rangle \cap (A + p^n K) = 0$. Write $x = y + pz$ with $y \in H, z \in K$. We may assume that y is also of order p^n, for if not, then we may replace y by $y - pu$ and z by $z + u$, where $u \in H$ is chosen so as to satisfy $p^n x = -p^{n+1} z = p^{n+1} u$ (use the purity of H). Then $\langle y \rangle$ is a summand of order p^n in H, since y has the same height-sequence as x. By hypothesis, $\langle y \rangle$ can not be disjoint from $A + p^n H$, so there are $a \in A, v \in H$

such that $p^{n-1}y = a+p^n v$. But then $p^{n-1}x = a+p^n v+p^n z \in A+p^n H$ is not disjoint from $A + p^n H$. This contradiction completes the proof of our claim that K does not have any cyclic summand of order p^n disjoint from $A + p^n K$. □

6 Pure covers in quasi-complete p-groups

The main problem concerning pure covers consists in finding necessary and sufficient conditions for their existence. First we show that in general, subgroups of separable p-groups G do not admit pure covers.

Lemma 6.1. *If H is a pure subgroup of a separable p-group G such that its p-adic closure $\bar{H}$ is not pure in G, then $\bar{H}$ has no pure cover in G.*

Proof. Indeed, a pure cover K of $\bar{H}$ would have a larger basic subgroup than H, and any cyclic summand in this larger basic subgroup not in $\bar{H}$ would violate the condition of Lemma 3.2 for pure cover. □

We now prove the following theorem. The "if" part for torsion-complete p-groups has been proved by Okuyama [10].

Theorem 6.2. *Every subgroup of a separable p-group G admits pure covers if and only if G is a quasi-complete group.*

Proof. In order to prove necessity, assume that G is a separable p-group such that all subgroups of G admit pure covers. Working toward contradiction, let us suppose that G is not quasi-complete, i.e. it contains a pure subgroup H whose p-adic closure $\bar{H}$ is not pure in G. From Lemma 6.1 it follows that the subgroup $A = \bar{H}$ fails to admit a pure cover in G.

Turning to the proof of sufficiency, let A be a subgroup in the quasi-complete p-group G. We construct a countable sequence

$$0 = B_0 \leq B_1 \leq \cdots \leq B_n \leq \cdots$$

of subsocles of A along with a sequence

$$0 = H_0 \leq H_1 \leq \cdots \leq H_n \leq \cdots$$

of pure subgroups of G such that

$$B_n \leq H_n \quad \text{and} \quad p^n H_n = 0 \ (n \in \omega).$$

The starting point is $B_0 = 0 = H_0$. If, for some n, the sequence of the B_j and the sequence of the H_j have been constructed for all $j \leq n$ as required, then we define B_{n+1} to be a maximal subgroup with respect to the properties

$$B_n \subseteq B_{n+1} \subseteq A[p] \quad \text{and} \quad B_{n+1} \cap p^{n+1}G = 0.$$

Let $C_n = B_{n+1} \cap p^n G$, so that all the elements $\neq 0$ of C_n are of height n and $B_{n+1} = B_n \oplus C_n$. Height comparison leads to $C_n \cap H_n = 0$. Since the subgroup $C_n \oplus H_n$ is bounded by p^n, it admits a pure hull (Hill–Megibben [4]) which is our choice for H_{n+1}. We observe that H_{n+1} is a summand of G, bounded by p^{n+1}, and Lemma 2.1 implies that $p^n H_{n+1} = C_n$. Continuing this way, it is clear that $H' = \cup_{n \in \omega} H_n$ is a pure subgroup of G. Furthermore, $B' = \cup_{n \in \omega} B_n = \oplus_{n \in \omega} C_n$ is a dense subsocle of A. Though H' does not necessarily contain all of A, it will have the relevant property that it can not contain any cyclic summand of order p^n that is disjoint from $A + p^n H'$ as we shall see below.

By way of contradiction, suppose that H' contains a cyclic summand $\langle x \rangle$ of order p^n such that $\langle x \rangle \cap (A + p^n H') = 0$. There is an index m such that $x \in H_{m+1}$, and we may assume that $m+1$ is the minimal index for which such a summand exists. By construction, $\langle x \rangle$ can not be disjoint from $C_m + H_m + p^n H_{m+1}$, thus there exist elements $c \in C_m, y \in H_m, u \in H_{m+1}$ such that $p^{n-1} x = c + y + p^n u$. Here c is of height m, and $y \neq 0$ since by assumption $\langle x \rangle \cap (A + p^n H_{m+1}) = 0$, so $y \in H_m$ implies that the height of y is at most $m-1$. As $p^{n-1} x$ has height exactly $n-1$, from the equation we derive the equality $n-1 = m-1$. Hence we conclude that c can be written as $c = p^n v$ with $v \in H_{m+1}$. Consider the element $x' = x - pu - pv \in H_{m+1}$ which has the same height sequence as x and satisfies $p^{n-1} x' = y$. In view of the purity of H_m, there is a cyclic summand $\langle w \rangle \leq H_m$ with $p^{n-1} w = y$. The minimal choice of $m+1$ guarantees that $\langle w \rangle \cap (A + p^n H') \neq 0$. But then $p^{n-1} x = p^n v + p^{n-1} w + p^n u \in A + p^n H'$, so $\langle x \rangle$ is not disjoint from $A + p^n H'$, an obvious contradiction.

Consider the p-adic closure H of H' in G; by quasi-completeness, H is pure in G. Evidently, H must contain all of $A[p]$ (but not necessarily all of A). By virtue of Lemma 5.3, what we have proved in the preceding paragraph implies that H does not contain any forbidden cyclic summands.

We first settle the special case when G is torsion-complete. In view of Fuchs [3, Corollary 68.9] the subgroup $H = \bar{H}'$ is a summand of G. Let $G = H \oplus K$ be a decomposition of G. We are going to show that in this direct decomposition H can be replaced by a summand F that contains A. Observe that $F[p] = H[p]$ implies that F must enjoy the same property that we proved for H in the preceding paragraph. We note that the socle of $(A \oplus K)/A$ is nothing else than $(A \oplus K[p])/A$. If $x + A \in (A \oplus K[p])/A$ ($x \in K[p]$) is of height n in G/A, then x must have height n in K. We conclude that $(A \oplus K)/A$ is pure in G/A. It is isomorphic to K which is torsion-complete, so we obtain a decomposition $G/A = F/A \oplus (A \oplus K)/A$ for some subgroup F of G. This implies that $G = F \oplus K$. It follows that F is a pure cover of A in the torsion-complete p-group G.

Resuming the general case, we argue with the pure subgroup H constructed above. If H is bounded, there is nothing to prove. If it is unbounded, then G/H is a torsion-complete group (cf. Hill–Megibben [5] or Fuchs [3, Proposition 74.5]). By the torsion-complete case settled above it follows that the subgroup $A' = (A + H)/H$ of G/H admits a pure cover K/H. We claim that the subgroup K of G does not admit any cyclic summand $\langle x \rangle$ of order p^n such that $\langle x \rangle \cap (A + p^n K) = 0$. By way of contradiction, assume that $\langle x \rangle$ is such a summand. If $p^{n-1} x \notin H$, then $\langle (x + H)/H \rangle$ is a cyclic summand of K/H, and so $\langle (x + H)/H \rangle \cap (A' + p^n (K/H)) \neq 0$ implies that

$\langle x \rangle \cap (A + p^n K) \not\leq H$, a contradiction. This shows that $p^{n-1} x \in H$ must hold. But then H contains a summand $\langle y \rangle$ with $p^{n-1} y = p^{n-1} x$, and $\langle x \rangle \cap (A + p^n K) = 0$ implies $\langle y \rangle \cap (A + p^n H) = 0$ in contradiction to our argument above. Hence no such summand $\langle x \rangle$ may exist, and consequently, K is a pure cover of A. □

7 When every subgroup has a pure cover

We turn our attention to p-groups in general, and in the next theorem we give a characterization of all p-groups which enjoy the property that each of their subgroups has a pure cover. Our result should be compared to the well-known theorem stating that every subgroup of a p-group G has a pure hull exactly if G is a direct sum of a divisible p-group and a bounded p-group; see Hill–Megibben [4].

Lemma 7.1. *The first Ulm subgroup G^1 of a p–group G has a pure cover in G if and only if it is divisible.*

Proof. To prove necessity, if suffices to point out that if G^1 fails to be divisible, then the basic subgroup of any pure subgroup H of G that contains G^1 is non-trivial, so it has a non-zero cyclic summand. This summand is disjoint form G^1, so by Lemma 3.2, H cannot be a pure cover for G^1. □

Theorem 7.2. *A p-group G has the property that each of its subgroups admits a pure cover in G if and only if G is the direct sum of a divisible p-group and a quasi-complete p-group.*

Proof. ($\Rightarrow$) Suppose that G has the property that every subgroup in G admits a pure cover. Lemma 7.1 shows that then G^1 is divisible, so we can write $G = D \oplus C$ where D is the maximal divisible subgroup of G and C is a separable p-group. Let A be a subgroup in C, and H a pure cover of A in G. It is straightforward to verify that the projection of H into C is a pure cover of A in C. Thus also C has the property that all of its subgroups admit pure covers, and from Theorem 6.2 we conclude that C has to be a quasi-complete p-group.

($\Leftarrow$) Let $G = D \oplus C$ where D is a divisible p-group and C is a quasi-complete p-group. Given a subgroup A of G, write $D + A = D \oplus A_0$ where $A_0 \leq C$ may be chosen. By Theorem 6.2, A_0 has a pure cover H_0 in C. Evidently, there is a divisible subgroup $D_0 \leq D$ that contains $A \cap D$ as an essential subgroup. We now claim that

$$H = D_0 \oplus H_0$$

is a pure cover of A in G. Clearly, A is contained in H which is evidently pure in G. It is also obvious that H cannot have any cyclic or quasi-cyclic summand disjoint from A, because such a summand would then be disjoint from A_0 and $A \cap D$, respectively. Consequently, H is indeed a pure cover of A. □

References

[1] K. Benabdallah and J. Irwin, *On minimal pure subgroups*, Publ. Math. Debrecen 23 (1976), 111–114.

[2] K. Benabdallah and T. Okuyama, *On purifiable subgroups of primary abelian groups*, Comm. Algebra 19 (1991), 85–96.

[3] L. Fuchs, Infinite Abelian Groups, vol. 2 (Academic Press, 1973).

[4] P. Hill and C. Megibben, *Minimal pure subgroups in primary groups*, Bull. Math. Soc. France 92 (1964), 251–257.

[5] P. Hill and C. Megibben, *Quasi-closed primary groups*, Acta Math. Acad. Sci. Hungar. 16 (1965), 271–274.

[6] T. Okuyama, *On the existence of pure hulls in primary abelian groups*, Comm. Algebra 19 (1991), 3089–3098.

[7] T. Okuyama, *Note on purifiable subgroups of primary abelian groups*, Hokkaido Math. J. 24 (1995), 445–451.

[8] T. Okuyama, *On purifiable subgroups in arbitrary abelian groups*, Comm. Algebra 28 (2000), 121–139.

[9] T. Okuyama, *Purifiable subgroups. II*, Hokkaido Math. J. 34 (2005), 237–245.

[10] T. Okuyama, *Quasi-purifiable subgroups and minimal direct summands*, Comm. Algebra 35 (2007), 1155–1165.

Author information

László Fuchs, Department of Mathematics, Tulane University, New Orleans, Louisiana 70118, U.S.A.
E-mail: fuchs@tulane.edu

Takashi Okuyama, Department of Mathematical Sciences, Yamagata University, Yamagata 990-8560, Japan.
E-mail: okuyama@sci.kj.yamagata-u.ac.jp

Partially decomposable primary Abelian groups and the generalized core class property

Patrick W. Keef

Abstract. The primary Abelian group H satisfies the generalized core class property if for every non-negative integer n, H is not $p^{\omega+n}$-projective iff it has a subgroup which is a proper $p^{\omega+n+1}$-projective. A large class of primary Abelian groups, referred to as *partially decomposable*, is studied, and in particular, it is shown that every partially decomposable group has the generalized core class property.

Key words. Core class, partially decomposable, C-decomposable.

AMS classification. 20K10.

1 Introduction

It is an honor to have my work represented in this volume dedicated to the life and mathematical legacy of Tony Corner. I first encountered the beauty and power of his contributions to Abelian group theory by reading [5]. Later, after meeting Tony in person, I came to realize that the man was nearly as graceful and elegant as were his mathematical ideas.

In the following, by the term "group" we will mean an Abelian p-group, for some fixed prime p. We will use without comment the notation and terminology on Abelian groups found in [5], and for set-theoretic notions we will on occasion refer the reader to [4].

If $n < \omega$, then a group H is said to be $p^{\omega+n}$-*projective* if $p^{\omega+n}\mathrm{Ext}(H, X) = \{0\}$ for all groups X. This will be true iff there is a subgroup $P \subseteq H[p^n]$ such that H/P is Σ-cyclic (i.e., a direct sum of cyclic groups), so the p^{ω}-projectives are just the Σ-cyclic groups. A $p^{\omega+n}$-projective group is *proper* if it is not $p^{\omega+n-1}$- projective. It is an easy result that an arbitrary subgroup of a $p^{\omega+n}$-projective group is also $p^{\omega+n}$-projective. As a partial converse of this result, the "core class" theorem of [1], states that H fails to be a p^{ω}-projective ($= \Sigma$-cyclic) group iff it has a subgroup which is a proper $p^{\omega+1}$-projective. Various authors have worked on extending this for integers $n \in \omega$, i.e., if H fails to be $p^{\omega+n}$ projective, then it has a subgroup which is a proper $p^{\omega+n+1}$-projective. We will say that H satisfies the *generalized core-class property* if this implication holds for all $n < \omega$. The main result of this paper is to prove the generalized core class property for a large class of groups. As such, it can be viewed as an extension of the work contained in [2], [6] and [7].

A group H of infinite final rank γ will be said to be *partially decomposable* if there are groups G_1 and G_2 of final rank γ and a homomorphism $g : G_1 \oplus G_2 \to H$ such that the kernel and cokernel of g have cardinality strictly less than γ. For example, if H is *C-decomposable* (i.e., $H \cong H' \oplus C$, where C is a Σ-cyclic group of final rank γ), then H is trivially partially decomposable. In addition, any torsion-complete group is partially decomposable. We give several characterizations of the groups in this class (Proposition 2.7). It follows that any group with a countable unbounded basic subgroup is partially decomposable, so there are examples of partially decomposable groups of final rank $c = 2^{\aleph_0}$ which are not decomposable into two summands of final rank c (Proposition 3.1). It might be tempting to conjecture that all groups are partially decomposable. However, in the constructible universe (V = L), we provide an example of a separable group of final rank $\aleph_1$ which is not partially decomposable (Proposition 3.3).

The main result of this paper (Theorem 2.9) states that any group which is partially decomposable has the generalized core class property. This extends and unites many of the known results in the study of this property, such as Theorem 8 of [2], which states that any C-decomposable group has the generalized core class property. It is still unknown, however, whether every group has the generalized core class property.

2 Results

If γ is an infinite cardinal, and G and H are groups, then a homomorphism $g : G \to H$ will be said to be γ-*injective* if $|K| < \gamma$, where $K = \{x \in G : g(x) = 0\}$ is the kernel of g; g will said to be γ-*surjective* if $|C| < \gamma$, where $C = H/g(G)$ is the cokernel of g; and finally, g will be said to be γ-*bijective* if it is both γ-injective and γ-surjective. We begin with the following summary of some elementary properties of this terminology:

Proposition 2.1. *Suppose γ is an infinite cardinal, G, H and L are groups, and $g : G \to H$ and $h : H \to L$ are homomorphisms. Then,*

(a) *If g and h are γ-injective, then so is $h \circ g$;*

(b) *If g and h are γ-surjective, then so is $h \circ g$;*

(c) *If g and h are γ-bijective, then so is $h \circ g$.*

Proof. If K is the kernel of g, K' is the kernel of h and K'' is the kernel of $h \circ g$, then there is a natural left-exact sequence:

$$0 \to K \to K'' \to K',$$

so that (a) follows. Dually, if C is the cokernel of g, C' is the cokernel of h and C''' is the cokernel of $h \circ g$, then there is a natural right-exact sequence:

$$C \to C'' \to C' \to 0,$$

so that (b) follows. Finally, (c) is an immediate consequence of (a) and (b). □

One important property of bijectivity, however, is not preserved: If $g : G \to H$ is a bijective homomorphism, then its inverse $g^{-1} : H \to G$ is also a bijective homomorphism. Contrast that with the following:

Example. Let B be a countable unbounded Σ-cyclic group, and $G = \overline{B}$ be the torsion-completion of G. Then $|G| = 2^{\aleph_0} = c$ and G/B is a divisible group of cardinality c. If D is the divisible hull of B, then D/B is countable, and we can identify the divisible hull of G, which we denote by H, with $D + G$, where $D \cap G = B$. The inclusion map, $g : G \subseteq H$ is clearly c-injective, and since $H/G \cong (D+G)/G \cong D/(D \cap G) \cong D/B$ is countable, g is, in fact, c-bijective. However, since H is divisible and G is reduced, any homomorphism $h : H \to G$ is, in fact, 0. Therefore, there is no "c-inverse" for g.

Proposition 2.2. *Suppose γ is an infinite cardinal, G and H are groups, $g : G \to H$ is a homomorphism, and $n < \omega$ is a non-negative integer. Then,*

(a) *If g is γ-injective, then so is $g|_{p^n G} : p^n G \to p^n H$;*

(b) *If g is γ-surjective, then so is $g|_{p^n G} : p^n G \to p^n H$;*

(c) *If g is γ-bijective, then so is $g|_{p^n G} : p^n G \to p^n H$;*

(d) *If g is γ-injective, then so is $g|_{(p^n G)[p]} : (p^n G)[p] \to (p^n H)[p]$;*

(e) *If g is γ-bijective, then so is $g|_{(p^n G)[p]} : (p^n G)[p] \to (p^n H)[p]$.*

Proof. Let K and C be the kernel and cokernel of g, respectively. The kernel of $g|_{p^n G}$ is $K \cap p^n G$, so that (a) follows. Multiplication by p^n gives a surjection from C to the cokernel of $g|_{p^n G}$, so that (b) follows. Clearly, (c) follows from (a) and (b). In proving (d) and (e), by referring to (a) and (c), it suffices to assume that $n = 0$. Note that the kernel of $g|_{G[p]}$ is $K \cap G[p]$, so that (d) is clear. To establish (e), then, we need to show that if g is γ-bijective, then $g|_{G[p]} : G[p] \to H[p]$ is γ-surjective. Let I be the image of g. Then the natural surjection $G \to I$ and the natural inclusion $I \to H$, are γ-bijections. Observe that there is an exact sequence,

$$0 \to K[p] \to G[p] \to I[p] \to K/pK$$

(this can be viewed as part of the long-exact sequence for $\text{Tor}(-, \mathbf{Z}_p)$, $- \otimes \mathbf{Z}_p$; alternatively, if $y = g(x) \in I[p]$, then $px \in K$ and the assignment $I[p] \to K/pK$ given by $y \mapsto px + pK$ can be shown to be well-defined and to fit into the sequence as described). This implies that $G[p] \to I[p]$ is a γ-isomorphism. Finally, there is another left-exact sequence,

$$0 \to I[p] \to H[p] \to C[p],$$

which shows that $I[p] \to H[p]$ is also a γ-isomorphism. □

On the other hand, note that if G is a group of infinite final rank γ, then if $H = pG$ and $g : G \to H$ is multiplication by p, then g is clearly γ-surjective, but $g|_{G[p]}$ is, in fact, 0, so not γ-surjective.

Corollary 2.3. *If γ is an infinite cardinal, G and H are groups, $\text{fr}(G) \geq \gamma$ and $g : G \to H$ is a γ-bijective homomorphism, then $\text{fr}(G) = \text{fr}(H)$.*

Proof. Since for all $n < \omega$, the kernel and cokernel of the map $(p^n G)[p] \to (p^n H)[p]$ have rank strictly less than $\gamma \leq \mathrm{r}((p^n G)[p])$, it follows that $\mathrm{r}((p^n G)[p]) = \mathrm{r}((p^n H)[p])$, which clearly implies that $\mathrm{fr}(G) = \mathrm{fr}(H)$. □

Recall the following definition from [6]: If G is a group and γ is a cardinal, then a γ-homomorphism is a homomorphism $f : G \to B$, where B is Σ-cyclic and $f((p^n G)[p])$ has rank at least γ, for all $n < \omega$. If G has final rank γ, we say it is *far from thick* if it admits a γ-homomorphism.

Proposition 2.4. *If γ is an infinite cardinal, G and H are groups and $g : G \to H$ is a γ-bijection, then G admits a γ-homomorphism iff H admits a γ-homomorphism.*

Proof. Suppose B is Σ-cyclic and $f : H \to B$ is a γ-homomorphism. Given $n < \omega$, $g|_{(p^n G)[p]} : (p^n G)[p] \to (p^n H)[p]$ is γ-bijective, so there is a subgroup V of $(p^n H)[p]$ of rank strictly less than γ such that

$$(p^n H)[p] = V + g((p^n G)[p]).$$

It follows that,

$$f((p^n H)[p]) = f(V + g((p^n G)[p])) = f(V) + (f \circ g)((p^n G)[p]),$$

and since $f((p^n H)[p])$ has rank at least γ and $f(V)$ has rank strictly less than γ, we can conclude that $(f \circ g)((p^n G)[p])$ also has rank at least γ, so that $f \circ g$ is a γ-homomorphism.

Conversely, suppose $f : G \to B$ is a γ-homomorphism. Let K be the kernel of g and I be the image of g, so $|K| < \gamma$. Note that there is a decomposition $B = B_1 \oplus B_2$, where $|B_2| < \gamma$ and $f(K) \subseteq B_2$. If $f' : G \to B_1$ is the composition of f with the projection onto the summand B_1, then since $f'(K) = 0$, we can conclude that f' determines a homomorphism $f_1 : I \to B_1$. Since $f((p^n G)[p]) \subseteq f_1((p^n I)[p]) + B_2[p]$ for every $n < \omega$, it follows that f_1 is a γ-homomorphism. Let D_1 be a divisible hull for B_1. There is an extension of $f_1 : I \to B_1$ to a homomorphism $f_2 : H \to D_1$. Suppose $H = I + Y$, where Y has rank less than γ. There are decompositions $B_1 = B_3 \oplus B_4$, $D_1 = D_3 \oplus D_4$, where D_i is a divisible hull of B_i for $i = 3, 4$, $\mathrm{r}(D_4) = \mathrm{r}(B_4) < \gamma$ and $f_2(Y) \subseteq D_4$. Note that $f_2(H) = f_1(I) + f_2(Y) \subseteq B_3 \oplus D_4$. Let $f_3 : H \to B_3$ be the composition of f_2 with the projection $B_3 \oplus D_4 \to B_3$. Since $f_1((p^n I)[p]) \subseteq f_3((p^n H)[p]) + B_4[p]$, for every $n < \omega$, it follows that f_3 is a γ-homomorphism, proving the result. □

Corollary 2.5. *If γ is an infinite cardinal, G is a group of final rank γ and $f : G \to H$ is a γ-bijection, then G is far from thick iff H is far from thick.*

Recall that the group G is C-decomposable if there is a decomposition $G \cong G' \oplus B$, where B is Σ-cyclic and $\mathrm{fr}(G) = \mathrm{fr}(B)$.

Proposition 2.6. *Suppose γ is an infinite cardinal, G and H are groups, $\mathrm{fr}(G) \geq \gamma$ and $g : G \to H$ is a γ-bijection. Then G is C-decomposable iff H is C-decomposable.*

Proof. Note that if $\mathrm{fr}(G) = \aleph_0$, then $\gamma = \aleph_0$. In this case it is easy to check that G is C-decomposable iff the reduced part of G is infinite iff the reduced part of H is infinite iff H is C-decomposable. We can, therefore, assume that $\gamma = \mathrm{fr}(G) = \mathrm{fr}(H) > \aleph_0$.

If I is the image of g, then g is the composition of two γ-bijections: $G \to I$ and $I \to H$. So we break the argument into two components: First, we assume g is injective, and second, we assume that g is surjective.

So, assume g is injective and G is C-decomposable; in fact, assume $G \subseteq H$, $|H/G| < \gamma$ and $G = G' \oplus B$, where $\mathrm{fr}(B) = \mathrm{fr}(G) = \gamma$. Find a subgroup X of H such that $G + X = H$ and $|X| < \gamma$. Next, find a decomposition $B = B_1 \oplus B_2$ such that $(G' + X) \cap B \subseteq B_2$ and $|B_2| < \gamma$. Note that $\mathrm{fr}(B_1) = \mathrm{fr}(B) = \mathrm{fr}(G) = \gamma$. We will show that $H = (G' + X + B_2) \oplus B_1$, which will prove the result in this case: First, $(G' + X + B_2) + B_1 = G' + X + B = G + X = H$, and second, if $b_1 = g' + x + b_2 \in (G' + X + B_2) \cap B_1$, where each symbol is an element of the obvious group, then $b_1 - b_2 = g' + x \in (G' + X) \cap B \subseteq B_2$, which implies that $0 = b_1 = g' + x + b_2$, which is what we wanted to show.

Assume next that $G \subseteq H$, $|H/G| < \gamma$ and H is C-decomposable. Let $H = H' \oplus B$, where $\mathrm{fr}(B) = \mathrm{fr}(H) = \gamma$. If $B' = B \cap G$ and $G' = H' \cap G$, then $B' \cap G' \subseteq B \cap H' = \{0\}$, so that we may define $G_0 = G' \oplus B' \subseteq G$. Next, B/B' embeds in H/G, so $B' \subseteq B$ is a γ-bijection, so that $\mathrm{fr}(B') = \mathrm{fr}(B) = \mathrm{fr}(H) = \mathrm{fr}(G) = \mathrm{fr}(G_0)$, and so G_0 is C-decomposable. Since G/G_0 embeds in $H/(G' \oplus B') = (H' \oplus B)/(G' \oplus B') \cong (H'/G') \oplus (B/B')$, it clearly has cardinality less than γ so that $G_0 \subseteq G$ is a γ-bijection, and by the first part of the proof, G is C-decomposable.

Now, we assume γ is actually surjective, and in fact, we assume $H = G/K$, where $|K| < \gamma$. If G is C-decomposable, then $G = G' \oplus B$, where $\mathrm{fr}(B) = \mathrm{fr}(G) = \gamma$. Note there is a decomposition $B = B_1 \oplus B_2$, where $K \subseteq G' + B_2$ and $|B_2| < \gamma$. It follows that $\mathrm{fr}(B_1) = \mathrm{fr}(B) = \mathrm{fr}(G) = \mathrm{fr}(H)$, so if $H' = (G' + B_2)/K$, then the isomorphism $H \cong H' \oplus B_1$ shows that H is C-decomposable.

Finally, assume $H = G/K$, $|K| < \gamma$ and H is C-decomposable. Let $H = H' \oplus B$, where $\mathrm{fr}(B) = \mathrm{fr}(H) = \gamma$, and define subgroups G_0 and B_0 of G by the equations $K \subseteq G_0$, $G_0/K = H'$, $K \subseteq B_0$ and $B_0/K = B$. Using the uncountability of γ, by a standard "back-and-forth" argument, there is a subgroup K_0 of B_0 which satisfies (i) $|K_0| \leq \aleph_0|K| < \gamma$; (ii) K_0 is pure in B_0; (iii) K_0/K is a summand of $B_0/K = B$. Since this implies that $B_0/K_0 \cong (B_0/K)/(K_0/K)$ is Σ-cyclic, the purity of K_0 in B_0 implies that $B_0 = B_1 \oplus K_0$ for some $B_1 \cong B_0/K$, which is Σ-cyclic. Note also that $\mathrm{fr}(B_1) = \mathrm{fr}(B_0) = \mathrm{fr}(B) = \mathrm{fr}(H) = \gamma$. We will be done, then, if we can show that $G = (G_0 + K_0) \oplus B_1$. Note $(G_0 + K_0) + B_1 = G_0 + B_0 = G$, and if $b_1 = g_0 + k_0 \in (G_0 + K_0) \cap B_1$ (where each symbol is an element of the corresponding group), then $b_1 - k_0 = g_0 \in G_0 \cap (B_1 + K_0) = G_0 \cap B_0 = K \subseteq K_0$, so $g_0 + k_0 = b_1 = 0$, proving the result. □

Recall that a group H of infinite final rank γ is partially decomposable if there are groups G_1 and G_2 of final rank γ and a γ-bijection $g : G = G_1 \oplus G_2 \to H$. For example, suppose H has final rank γ and a basic subgroup B such that $|B| < \gamma$. It follows that H/B is divisible of rank γ, so there are divisible groups D_1, D_2 of rank

γ such that $H/B = D_1 \oplus D_2$. Let G_1 be the subgroup of H containing B defined by the equations $G_1/B = D_1$, and define G_2 similarly. If $G = G_1 \oplus G_2 \to H$ is the natural homomorphism (i.e., $g((g_1, g_2)) = g_1 + g_2$), then g is surjective and its kernel is isomorphic to $G_1 \cap G_2 = B$, so that g is a γ-bijection. The following shows that the above example is fairly generic:

Proposition 2.7. *Suppose H is a group of infinite final rank γ. Then the following are equivalent:*

(a) *H is partially decomposable.*

(b) *H has subgroups H_1 and H_2 of final rank γ such that $H_1 \cap H_2$ and $H/(H_1 + H_2)$ have cardinality less than γ.*

(c) *There is a pure-exact sequence,*
$$0 \to B \to H \to L_1 \oplus L_2 \to 0,$$
where B is Σ-cyclic, $|B| < \gamma$ and both L_1 and L_2 have final rank γ.

(d) *There is a pure-exact sequence,*
$$0 \to B \to G_1 \oplus G_2 \to H \to 0,$$
where B is Σ-cyclic, $|B| < \gamma$ and both G_1 and G_2 have final rank γ.

Proof. Note again that any group of countably infinite final rank is the direct sum of two groups of the same form, so that all of the various conditions stated above are true for any such group of final rank $\aleph_0$. Assume, therefore, that γ is uncountable.

(a) $\Rightarrow$ (b): Suppose $G = G_1 \oplus G_2$, and $g : G \to H$ is a γ-bijective homomorphism. If $H_i = g(G_i)$ for $i = 1, 2$, then $|H/(H_1 + H_2)| = |H/g(G)| < \gamma$. If K is the kernel of g, then let $M \subseteq H$ be the set of all h such that $h = g(x)$ for some $(x, -y) \in K$. Note that $h \in H_1 \cap H_2$ iff $h = g(x) = g(y)$ for some $x \in G_1$, $y \in G_2$ iff $h = g(x)$ for some $(x, -y) \in K$ iff $h \in M$. It follows that $|H_1 \cap H_2| = |M| \leq |K| < \gamma$.

(b) $\Rightarrow$ (c): Let X be a subgroup of H such that $H = H_1 + H_2 + X$ and $|X| < \gamma$. We claim that $|H_1 \cap (H_2 + X)| < \gamma$: If $x \in X$, $h_x = h_2 + x$ and $h'_x = h'_2 + x$, where $h_x, h'_x \in H_1$ and $h_2, h'_2 \in H_2$, then $y = h'_x - h_x = h'_2 - h_2 \in H_1 \cap H_2$, and $h'_x = h_x + y$. Since there are less than γ possible such x and y, we can conclude that there are less than γ such h_x, proving the claim. Therefore, replacing H_2 with $H_2 + X$, we may assume $H = H_1 + H_2$. Note that there is a decomposition
$$H/(H_1 \cap H_2) \cong H_1/(H_1 \cap H_2) \oplus H_2/(H_1 \cap H_2).$$

Using the uncountability of γ and a standard "back and forth" argument, construct a pure subgroup, Y, of H such that $H_1 \cap H_2 \subseteq Y$, $Y = (Y \cap H_1) + (Y \cap H_2)$ and $|Y| < \gamma$. Therefore, in the above decomposition, we have,
$$Y/(H_1 \cap H_2) \cong (Y \cap H_1)/(H_1 \cap H_2) \oplus (Y \cap H_2)/(H_1 \cap H_2).$$

It follows that,
$$H/Y \cong H_1/(Y \cap H_1) \oplus H_2/(Y \cap H_2).$$

Next, suppose B is a basic subgroup of Y, so that B is pure in H. Then Y/B is divisible, so it is a summand of H/B. Hence,
$$H/B \cong L_1 \oplus L_2,$$
where $L_1 \cong H_1/(Y \cap H_1)$ and $L_2 \cong Y/B \oplus H_2/(Y \cap H_2)$. Finally, the obvious homomorphisms $H_1 \to L_1$ and $H_2 \to H_2/(Y \cap H_2) \to L_2$ are γ-bijective, so fr$(L_i) = \gamma$ for $i = 1, 2$, as required.

(c) $\Rightarrow$ (d): Without loss of generality, assume $B \subseteq H$ and $H/B = L_1 \oplus L_2$. For $i = 1, 2$, let G_i be the (pure) subgroups of H containing B such that $G_i/B = L_i$. Note that $B \oplus B$ is a pure subgroup of $G_1 \oplus G_2$, and $B' = \{(b, -b) : b \in B\}$ is a pure subgroup of $B \oplus B$. Therefore,
$$0 \to B' \to G_1 \oplus G_2 \to (G_1 \oplus G_2)/B' \to 0$$
is pure-exact. Clearly $B' \cong B$ and since B' is the kernel of the sum map $G_1 \oplus G_2 \to H$, we conclude that $(G_1 \oplus G_2)/B' \cong H$. Finally, for $i = 1, 2$, the homomorphism $G_i \to G_i/B = L_i$ is a γ-bijection, so that fr$(G_i) = \gamma$, as required.

(d) $\Rightarrow$ (a): This is clear, since the map $G_1 \oplus G_2 \to H$ is a γ-bijection. □

In the above, the statement (c) shows that the γ-bijection used in the definition of partial-decomposability has, in fact, a "γ-inverse." We now pause for a simple observation:

Lemma 2.8. *If γ is an infinite cardinal and G is a group of final rank γ, then G has a subgroup M of final rank γ which is Σ-cyclic.*

Proof. Let B be a basic subgroup of G. If the final rank of B is γ, then let $M = B$. Otherwise, G/B is a divisible group of rank γ. Let M' be any Σ-cyclic group of rank and final rank γ. Then M' embeds in G/B, so if M is the subgroup of G containing B such that $M/B = M'$, then $M \cong M' \oplus B$ is a subgroup of the required form. □

Theorem 2.9. *If H is a partially decomposable group, then H has the generalized core class property.*

Proof. Suppose γ is the (infinite) final rank of H. If H admits a γ-homomorphism (i.e., if H is far from thick), then the result follows from Corollary 27 of [6]. So suppose H does not admit a γ-homomorphism.

By Proposition 2.7(b), there are subgroups H_1, H_2 of H of final rank γ such that the sum map $H_1 \oplus H_2 \to H$ is γ-bijective. It follows from Proposition 2.4 that $H_1 \oplus H_2$ does not admit a γ-homomorphism. Hence H_1 does not admit a γ-homomorphism. From Corollary 28 of [6], it follows that H_1 is not $p^{\omega+n}$-projective for any non-negative integer, n. By Lemma 2.8, there is a Σ-cyclic subgroup M of H_2 of final rank γ. Since $|H_1 \cap H_2| < \gamma$, there is a decomposition $M = N \oplus M'$, where $H_1 \cap M \subseteq M'$ and

$|M'| < \gamma$. Then the final rank of N is γ, and $H_1 \cap N = \{0\}$. Therefore, $H_3 = H_1 \oplus N$ is C-decomposable and not $p^{\omega+n}$-projective for any non-negative n. It follows from Theorem 8 of [2] that for every non-negative n, H_3, and hence H, has a subgroup which is a proper $p^{\omega+n}$-projective. This, however, clearly implies that H has the generalized core class property. □

3 Examples

Suppose $H = \overline{B}$ is torsion-complete of final rank γ, where B is an unbounded Σ-cyclic. Then there is a decomposition, $B = B_1 \oplus B_2$, where $\mathrm{fr}(B_1) = \mathrm{fr}(B_2) = \mathrm{fr}(B)$. This leads to a decomposition, $H = \overline{B_1} \oplus \overline{B_2}$, showing that H is not only partially decomposable, but that it actually decomposes into summands of final rank γ.

Proposition 3.1. *If H is any group with a countable unbounded basic subgroup B, then H is partially decomposable; however such a group may be essentially indecomposable (i.e., if $H \cong G_1 \oplus G_2$, then either G_1 or G_2 is bounded.)*

Proof. Either H is actually countable, so that it is the direct sum of two groups of countably infinite final rank, or H is uncountable, and $H/B \cong D_1 \oplus D_2$, where $\mathrm{r}(D_i) = \mathrm{r}(G)$ for $i = 1, 2$, so that H is partially decomposable by Proposition 2.7(c).

There are many examples of a separable group H with a countable unbounded basic subgroup which is essentially indecomposable (e.g., see Corollary 74.6 and the example of page 48 of Volume II of [5]). □

We now construct an example, at least in the context of the constructible universe (V = L), of a group of final rank $\aleph_1$ which fails to be partially decomposable. The group is defined using a variation on a now standard argument, and in the form presented here, is in large part borrowed from [3]. We begin with the following observation, which is clearly a reformulation of Lemma 5 of [3]. If A is a subgroup of G, we will use the notation $A \sqsubseteq G$ to signify that A is a direct summand of G.

Lemma 3.2. *Suppose $H = C \oplus C' \oplus D$ is Σ-cyclic, C' is unbounded and $C = \bigoplus_{k<\omega} C_k$ where each C_k is unbounded. Then H can be embedded as a pure subgroup in a Σ-cyclic group H' such that $(\bigoplus_{k \leq n} C_k) \oplus C' \oplus D \sqsubseteq H'$ for each $n < \omega$, but C is not contained in a complementary summand of C' in H'.*

Proposition 3.3 (V = L). *In the constructible universe, there is a separable group H of rank and final rank $\aleph_1$ which is not partially decomposable.*

Proof. Let $\{H_i\}_{i<\omega_1}$ be a smoothly ascending chain of countable sets such that $|H_{i+1} - H_i| = \aleph_0$ for each $i < \omega_1$ (e.g., let $H_i = \omega \cdot i \subseteq \omega_1$) and let $H = \bigcup_{i<\omega_1} H_i$. Let E be a stationary-costationary subset of ω_1 consisting of limit ordinals, and let $\{Y_i, Y'_i\}_{i \in E}$ be a $\diamond_E$-sequence for E (so $Y_i \subseteq H_i$ and $Y'_i \subseteq H'_i$ for all $i \in E$ and if $M, M' \subseteq H$ and $S \subseteq \omega_1$ is a CUB, then there is an $i \in S \cap E$ such that $M \cap H_i = Y_i$ and $M' \cap H'_i = Y'_i$). We inductively define group structures on H_j, $j < \omega_1$, so that the following hold for all $i < j$:

(a) H_j is Σ-cyclic;

(b) H_i is a pure subgroup of H_j and H_j/H_i is unbounded;

(c) $H_i \sqsubseteq H_j$ whenever $i \notin E$.

To define H_j, assuming H_i has been defined for all $i < j$, we proceed as follows: If j is a limit, then continuity forces the issue at $H_j = \bigcup i < jH_i$. If $j = i+1$ is isolated, then whenever the conditions (i)–(vi) below do not hold, we simply let $H_j = H_i \oplus X_i$, where X_i is some unbounded countable Σ-cyclic. On the other hand, if $j = i+1$ where,

(i) $i \in E$;

(ii) Y_i, Y_i' and $B = Y_i \cap Y_i'$ are pure subgroups of H_i;

(iii) i is the supremum of a strictly ascending sequence $\{\tau_k\}_{k<\omega}$ such that $\tau_k \notin E$, for all $k < \omega$, where:

(iv) $B = Y_i \cap Y_i' \subseteq H_{\tau_0}$;

and if we let $Z_k = H_{\tau_k} \cap Y_i$ and $Z_k' = H_{\tau_k} \cap Y_i'$, then for each $k < \omega$ we have:

(v) Z_{k+1}/Z_k and Z_{k+1}'/Z_k' are unbounded;

(vi) $H_{\tau_k} = Z_k + Z_k'$,

then we proceed to construct H_j as follows: Since $Y_i \cap Y_i' = Z_k \cap Z_k' = B \subseteq H_{\tau_0}$, for each $k < \omega$, condition (vi) implies

$$H_{\tau_k}/B = (Z_k/B) \oplus (Z_k'/B).$$

From this, since B is pure in H_i (and hence in H_{τ_k}), Z_k and Z_k' are pure as well. In the union, then, we have

$$H_i/B = (Y_i/B) \oplus (Y_i'/B).$$

It also follows that

$$H_{\tau_{k+1}}/H_{\tau_k} \cong (Z_{k+1}/Z_k) \oplus (Z_{k+1}'/Z_k').$$

Since $\tau_k \notin E$, by induction, these quotients are Σ-cyclic, so let $Z_{k+1} = C_k \oplus Z_k$ and $Z_{k+1}' = C_k' \oplus Z_k'$, where $C_k \subseteq Z_{k+1} = H_{\tau_{k+1}} \cap Y_i$, and similarly for C_k'. Now let $C = \bigoplus_{k<\omega} C_k$, $C' = \bigoplus_{k<\omega} C_k'$, $D = H_{\tau_0}$. Using this notation, we have

$$H_{\tau_n} = \left(\bigoplus_{k<n} C_k\right) \oplus \left(\bigoplus_{k<n} C_k'\right) \oplus D \text{ and } H_i = C \oplus C' \oplus D,$$

so we apply Lemma 3.2 to construct $H' = H_j = H_{i+1}$.

We next verify that in this case, conditions (a), (b) and (c) are still valid for this $j = i+1$. The first condition follows immediately from the lemma, and the second follows easily, as well: If $\ell < j$, then $\ell \le i$, so H_ℓ is pure in H_i, and by Lemma 3.2, H_i is a pure subgroup of H_j, so that (b) holds. For (c), suppose again that $\ell < j$

satisfies $\ell \notin E$. Since $i \in E$, $\ell < i$, so choose n such $\ell < \tau_n$. Then $H_\ell \sqsubseteq H_{\tau_n} = (\bigoplus_{k \leq n} C_k) \oplus (\bigoplus_{k \leq n} C'_n) \oplus D \sqsubseteq (\bigoplus_{k \leq n} C_k) \oplus C' \oplus D \sqsubseteq H_j$, establishing (c). Inducting through all j, we have defined $\bar{H} = \bigcup_{j<\omega_1} H_j$.

Suppose now that H is partially decomposable. By Proposition 2.7(c), there is a pure short exact sequence

$$0 \to B \to H \to L_1 \oplus L_2 \to 0,$$

where B is countable and L_1, L_2 have final rank $\aleph_1$. Let Y, Y' be subgroups of H containing B such that $Y/B = L_1$, $Y'/B = L_2$. Note that since B is pure in H and Y/B is pure in $H/B \cong (Y/B) \oplus (Y'/B)$, Y, and similarly Y', is pure in H. With the filtration $\{H_i\}_{i<\omega_1}$ of H constructed above, by choosing a CUB, $S_0 \subseteq \omega_1$, we may assume that for all $i \in S_0$,

(x) $B \subseteq H_i$;

(y) $(H_j \cap Y)/(H_i \cap Y)$ and $(H_j \cap Y')/(H_i \cap Y')$ are unbounded whenever $i, j \in S_0$ with $i < j$;

(z) $H_i = (H_i \cap Y) + (H_i \cap Y')$.

As above, (x) and (z) imply that $H_i/B \cong ((H_i \cap Y)/B) \oplus ((H_i \cap Y')/B)$, so $H_i \cap Y$ and $H_i \cap Y'$ are pure in H_i, and so in H. In addition, we note for future reference that if $\tau, \ell \in S_0$ with $\tau < \ell$, we have

$$(H_\ell \cap Y) \cap ((H_\ell \cap Y') + H_\tau)$$
$$= (H_\ell \cap Y) \cap ((H_\ell \cap Y') + ((H_\tau \cap Y) + (H_\tau \cap Y')))$$
$$= (H_\ell \cap Y) \cap ((H_\ell \cap Y') + (H_\tau \cap Y)) \qquad (*)$$
$$= ((H_\ell \cap Y) \cap (H_\ell \cap Y')) + (H_\tau \cap Y)$$
$$= B + (H_\tau \cap Y) = H_\tau \cap Y \subseteq H_\tau.$$

Let S be the limit points of $S_0 \cap (\omega - E)$, so S is also a CUB. Therefore, there is an $i \in S \cap E$ such that $Y_i = H_i \cap Y$ and $Y'_i = H_i \cap Y'$.

By the construction of S, i is the limit of an ascending sequence $\{\tau_k\}_{k<\omega}$ of ordinals $\tau_k \in S_0 - E$. Now, (x), (y) and (z) and the above remarks imply that (i)–(vi) are valid, so we have constructed $H_j = H_{i+1}$ as in that discussion. Let $\ell \in S_0$ be such that $\ell > j$. Note that $C' \oplus D \sqsubseteq H_j \sqsubseteq H_\ell$, so that $C' \oplus D \sqsubseteq (H_\ell \cap Y') + D$. Let $C' \oplus D \oplus W = (H_\ell \cap Y') + D$. We claim that

$$H_\ell = C' \oplus ((H_\ell \cap Y) + D + W).$$

First, note that

$C' + ((H_\ell \cap Y) + D + W) = (H_\ell \cap Y) + (C' + D + W) = (H_\ell \cap Y) + (H_\ell \cap Y') + D = H_\ell.$

Second, assume $c' = h_\ell + d + w$ (where each symbol is in the obvious group). By $(*)$, $h_\ell = c' - d - w \in (H_\ell \cap Y) \cap ((H_\ell \cap Y') + D) \subseteq D$, so that $c' = 0$, establishing the claim.

It follows, therefore, that

$$H_j = C' \oplus (((H_\ell \cap Y) + D + W) \cap H_j),$$

but since C is clearly a subgroup of the second factor, we have violated the construction of H_j. □

Note that the group H constructed above is actually strongly $\aleph_1$-Σ-cyclic, i.e., every countable subgroup is Σ-cyclic and if $X \subseteq H$ is countable, then there is a countable subgroup Y containing X such that Y is a summand of every countable subgroup Z of H that contains Y (just consider $Y = H_i$ for $i \notin E$).

Of course, there are many questions left unanswered by the above discussion. Three that immediately come to mind are the following:

(a) Is the use of the diamond principle really necessary in Proposition 3.3, i.e., is it provable in ZFC that there is a group which fails to be partially decomposable?

(b) Is every group which is far from thick also partially decomposable?

(c) Does every group satisfy the generalized core class property?

Acknowledgements. The author expresses his appreciation for the helpful suggestions of the referee, which significantly improved the exposition in this paper.

References

[1] K. Benabdallah, J. Irwin, and M. Rafiq, *A core class of Abelian p-groups*, Sympos. Math. **13** (1974), 195–206.

[2] D. Cutler, J. Irwin, and T. Snabb, *Abelian p-groups containing proper $p^{\omega+n}$-projective subgroups*, Comm. Math. Univ. St. Pauli, **33** No. 1 (1984), 95–97.

[3] D. Cutler, A. Mader, and C. Megibben, *Essentially indecomposable Abelian p-groups having a filtration of prescribed type*, Contemporary Mathematics, Vol. 87, American Mathematical Society, Providence, 1989, 43–50.

[4] P. Eklof and A. Mekler, *Almost Free Modules, Set Theoretic Methods*, Revised Edition, North-Holland, Amsterdam, 2002.

[5] L. Fuchs, *Infinite Abelian Groups*, Vol. 1 and 2, Academic Press, New York, 1970 and 1973.

[6] J. Irwin and P. Keef, *Primary Abelian groups and direct sums of cyclics*, J. Algebra, **159** No. 2 (1993), 387–399.

[7] J. Irwin, T. Snabb, and D. Cutler, *On $p^{\omega+n}$-projective p-groups*, Comm. Math. Univ. St. Pauli, **35** No. 1 (1986), 49–51.

Author information

Patrick W. Keef, Department of Mathematics, Whitman College, Walla Walla, WA, 99362, USA.
E-mail: keef@whitman.edu

The additive group of a finite local ring in which each ideal can be n-generated

K. Robin McLean

Abstract. We determine the finite abelian p-groups that can support a local ring in which each ideal can be n-generated.

Key words. Additive group, finite local ring, fully invariant subgroup, number of generators of an ideal.

AMS classification. 20K01, 16L30, 13M99.

1 Introduction

Tony Corner always appreciated elegant results, however simple. He once told me that every undergraduate who studies finite abelian groups should meet the result that each finite ring is the direct sum of rings on abelian p-groups. The more general form of this result, which holds for all rings on torsion groups, is given in the chapter on rings and their additive groups in L. Fuchs's classic [3], a book whose publication may have played a part in Tony's decision to switch his early research from topology to abelian groups.

The aim of this paper is to present a new result (Theorem 3.1) on the additive group of a finite ring. It is unusual both in having been overlooked in previous research and in drawing on the theories of abelian groups and associative rings in roughly equal proportions for its proof. A special case of the result (Theorem 2.2) is integral to the discussion.

It is natural to focus on local rings here, for any finite (or, indeed, artinian) ring, R, that is commutative has a decomposition

$$R = R_1 \oplus \cdots \oplus R_t$$

where each R_i is a local ring. These local summands are uniquely determined up to isomorphism. (See [1, Theorem 8.7].) It is easy to show that each ideal of R can be n-generated if and only if each ideal of R_i can be n-generated for $i = 1, \ldots, t$. The simplest case, $n = 1$, corresponds to R being a principal ideal ring. Also, since local rings are indecomposable, the additive group of a finite local ring is a p-group.

Theorem 2.2 gives necessary and sufficient conditions for a finite abelian p-group to support a local principal ideal ring. Commutative rings of this type were studied in [6], whilst [9] examined their non-commutative counterparts, although neither paper was restricted to finite rings. Their additive structure played a key role in [7], which solved one of Fuchs's problems in [3]. Local rings in which each ideal can be n-generated were the subject of [8], and a result from that paper, reproduced here as Lemma 3.2, is needed to prove Theorem 3.1.

The notation and terminology of abelian groups used here can be found in [3] and [4]. Every ring is taken to be associative and to have an identity element. Let G be an abelian group and R be a ring. If the additive group, R_+, of R is isomorphic to G, we say that G *supports* R and that R is a ring *on* G. We denote by $(x_1, \ldots, x_n)$ the right ideal generated by elements $x_1, \ldots, x_n \in R$, and by $\lambda(M)$ the length of a right R-module, M. We say that R is a *local* ring if $R \neq 0$ and R has a unique (possibly zero) maximal right ideal, $\mathbf{m}$. Several equivalent conditions to R being local are given in [5, Proposition 1, p. 75]. When R is local, $\mathbf{m}$ is a two-sided ideal and we often write R as $(R, \mathbf{m})$. Let I be a right ideal of a local ring $(R, \mathbf{m})$. It is easy to see that I is generated by $x_1, \ldots, x_n$ if and only if $x_1 + \mathbf{m}I, \ldots, x_n + \mathbf{m}I$ generate $I/\mathbf{m}I$ as a right vector space over the skew field $R/\mathbf{m}$. In the special case when $x_1 + \mathbf{m}I, \ldots, x_n + \mathbf{m}I$ is a basis for $I/\mathbf{m}I$, we say that $x_1, \ldots, x_n$ is a *minimal basis* for I. Clearly all minimal bases for I contain the same number of elements, $\nu(I)$, and

$$\nu(I) = \lambda(I/\mathbf{m}I) = \dim_{R/\mathbf{m}}(I/\mathbf{m}I).$$

2 Local principal ideal rings

Before stating Theorem 2.2, it may be helpful to note two simple results set out in the following lemma.

Lemma 2.1. *Let G ($\neq 0$) be a finite abelian p-group. Then*

(a) *G can support a local ring, and*

(b) *G can support a principal right ideal ring (PRI-ring).*

Proof. Let

$$G = a_1 C(p^{n_1}) \oplus \cdots \oplus a_k C(p^{n_k}) \tag{2.1}$$

where $aC(p^n)$ denotes the direct sum of a copies of the cyclic group of order p^n and $n_1 > n_2 > \cdots > n_k$. Let $r = a_1 + \cdots + a_k$. Then G has a basis $x_1, \ldots, x_r$ in which the additive orders of this sequence of elements are monotonic and non-increasing. To construct a local ring, R, on G, we define multiplication of our basis elements by $x_1 x_i = x_i x_1 = x_i$ for $i = 1, \ldots, r$ and $x_i x_j = 0$ for all $i, j > 1$, and then extend this to multiplication on the whole of G by the distributive laws. Our multiplication is associative, so R is a ring with the identity element x_1. Each element $x \in R$ has the form $x = \lambda x_1 + y$ for some integer λ and some element $y \in R$ such that $y^2 = 0$. If $\lambda \not\equiv 0 \pmod{p}$, then there is an integer μ such that $\lambda \mu \equiv 1 \pmod{p^{n_1}}$. In this case $x = \lambda(x_1 + \mu y)$, which has $\mu(x_1 - \mu y)$ as inverse. If $\lambda \equiv 0 \pmod{p}$, then $\lambda^{n_1} x_1 = 0$ and $x^{n_1+1} = \lambda^{n_1+1} x_1 + (n_1 + 1)\lambda^{n_1} x_1 y = 0$. In this case, $x_1 - x$ has the inverse $x_1 + x + x^2 + \cdots + x^{n_1}$. Hence R is a local ring by [5, Proposition 1, p. 75].

For assertion (b), let $aZ(p^n)$ denote the direct sum of a copies of the ring of integers modulo p^n. Then
$$a_1 Z(p^{n_1}) \oplus \cdots \oplus a_k Z(p^{n_k})$$
is a PRI-ring whose additive group is isomorphic to G. □

In striking contrast to Lemma 2.1, very few finite abelian p-groups can support a local PRI-ring. These groups are the subject of Theorem 2.2 and they constitute building blocks for the more general groups of Theorem 3.1. Some of Theorem 2.2 is implicit in [7] and the implication (2) ⇒ (4) can be found in the proof of [2, Theorem 6.2.1]. Although most of Theorem 2.2 can be viewed as a special case of Theorem 3.1, it deserves a separate brief proof. Recall Kaplansky's discussion of U-sequences and fully invariant subgroups of p-groups, summarized in [4, Theorem 25, p. 61]. Also note that if I is a fully invariant subgroup of an abelian group, G, and R is any ring on G, then I is an ideal of R. (For if $r \in R$, the maps $x \to xr$ and $x \to rx$ are endomorphisms of G, so I is closed under multiplication by elements of R.)

Theorem 2.2. *Let G ($\neq 0$) be a finite abelian p-group. Then the following conditions are equivalent:*

(1) *G can support a local PRI-ring.*

(2) *The lattice of fully invariant subgroups of G is totally ordered.*

(3) *The U-sequence of each fully invariant subgroup of G contains at most one gap.*

(4) *$G \cong aC(p^m) \oplus bC(p^{m-1})$ for some integers $a \geq 1$, $b \geq 0$ and $m \geq 1$.*

Proof. (1) ⇒ (2): Suppose that G is given by equation (2.1) above and that R is a local PRI-ring on G. Then the lattice of right ideals of R is a chain. (See [9, Theorem 1.2].) This gives (2).

(2) ⇒ (3): Let $\{\alpha_0, \alpha_1, \alpha_2, \dots\}$ be a U-sequence with gaps immediately after both α_i and α_j for some integers i, j such that $i < j$. Then $\alpha_i \leq \alpha_j - 2$. Moreover, both $\{\alpha_i, \infty, \infty, \dots\}$ and $\{\alpha_j - 1, \alpha_j, \infty, \infty, \dots\}$ are U-sequences. If (2) holds, then $\alpha_i \geq \alpha_j - 1$, contradicting our earlier inequality. Hence each U-sequence contains at most one gap.

(3) ⇒ (4): Let G be given by equation (2.1) above. If $n_k \leq n_1 - 2$, then $\{n_k - 1, n_1 - 1, \infty, \infty, \dots\}$ is a U-sequence with two gaps.

(4) ⇒ (1): Suppose that (4) holds. Let $r = a + b$ and $t = mr - b$, so that G contains exactly p^t elements. Let
$$R = Z_p[X]/(X^r - p, X^t)$$
where Z_p denotes the ring of p-adic integers. By [6, Theorem 3.2], R is a commutative local PRI-ring. Earlier parts of the current proof show that R_+ has the same form as G. Since R_+ has rank r and p^t elements, $R_+ \cong G$. □

Definition 2.3. A finite abelian p-group that satisfies the equivalent conditions of Theorem 2.2 will be called a *one-gap* group.

3 Local rings in which each ideal can be n-generated

Theorem 3.1. *Let $G(\neq 0)$ be a finite abelian p-group and n be a positive integer. Then the following conditions are equivalent:*

(1) *G can support a local ring in which each right ideal can be generated by at most n elements.*

(2) *The U-sequence of each fully invariant subgroup of G contains at most n gaps.*

(3) *G is the direct sum of at most n one-gap groups.*

To prove Theorem 3.1, we shall use a result (Lemma 3.2) that appeared as the Corollary on p. 330 in [8]. In order to make this account reasonably self-contained, a proof is given here.

Lemma 3.2. *Let R be a commutative noetherian local ring with maximal ideal $\mathbf{m}$ and let k and n be positive integers. If $\mathbf{m}^{k+1} = y\mathbf{m}^k$ for some element $y \in R$ and each ideal of $R/\mathbf{m}^{k+1}$ can be generated by n elements, then each ideal of R can be generated by n elements.*

Proof. Let I be an ideal of R. By the Artin–Rees lemma, $I \cap \mathbf{m}^i \subseteq \mathbf{m}I$ for all sufficiently large integers i. So, for all sufficiently large i,

$$I/\mathbf{m}I = I/\{\mathbf{m}I + (I \cap \mathbf{m}^{i+1})\} = I/\{I \cap \mathbf{m}(I + \mathbf{m}^i)\} \cong (I + \mathbf{m}^{i+1})/\mathbf{m}(I + \mathbf{m}^i)$$

and $\lambda(I/\mathbf{m}I) = \lambda\{(I + \mathbf{m}^{i+1})/\mathbf{m}(I + \mathbf{m}^i)\} \leq \lambda\{(I + \mathbf{m}^i)/\mathbf{m}(I + \mathbf{m}^i)\}$, whence

$$\nu(I) \leq \nu(I + \mathbf{m}^i) \qquad \text{for all sufficiently large } i. \tag{3.1}$$

Using bars to denote images under the natural map $R \to R/\mathbf{m}^{k+1} = \overline{R}$, let $\overline{a}_1, \ldots, \overline{a}_r$ be a minimal basis for $\overline{I}$ and $\overline{b}_1, \ldots, \overline{b}_s$ be a minimal basis for $\overline{I} \cap \overline{\mathbf{m}}^k$. It is easy to check that the $\overline{b}_j$'s can be extended to a minimal basis $\overline{b}_1, \ldots, \overline{b}_s, \overline{c}_1, \ldots, \overline{c}_t$ for $\overline{\mathbf{m}}^k$ and that $\overline{a}_1, \ldots, \overline{a}_r, \overline{c}_1, \ldots, \overline{c}_t$ is a minimal basis for $\overline{I} + \overline{\mathbf{m}}^k$. Now lift each $\overline{a}_j$ to $a_j \in I$, each $\overline{b}_j$ to $b_j \in (a_1, \ldots, a_r)$ and each $\overline{c}_j$ to $c_j \in \mathbf{m}^k$. Since $(I + \mathbf{m}^k) \cap \mathbf{m}^{k+1} \subseteq \mathbf{m}(I + \mathbf{m}^k)$, the elements $a_1, \ldots, a_r, c_1, \ldots, c_t$ constitute a minimal basis for $I + \mathbf{m}^k$. Also

$$\mathbf{m}^k = (b_1, \ldots, b_s, c_1, \ldots, c_t)$$

and $I + \mathbf{m}^{k+1} = (a_1, \ldots, a_r) + y\mathbf{m}^k = (a_1, \ldots, a_r, yc_1, \ldots, yc_t)$. Thus

$$\nu(I + \mathbf{m}^{k+1}) \leq r + t = \nu(I + \mathbf{m}^k).$$

By induction, $\nu(I + \mathbf{m}^{i+1}) \leq \nu(I + \mathbf{m}^i)$ for all $i \geq k$, so that

$$\nu(I + \mathbf{m}^i) \leq \nu(I + \mathbf{m}^k) \quad \text{for all } i \geq k. \tag{3.2}$$

From (3.1) and (3.2), $\nu(I) \leq \nu(I + \mathbf{m}^k) = r + t = \nu(\overline{I} + \overline{\mathbf{m}}^k)$. But each ideal of $\overline{R}$ can be generated by n elements, so $\nu(I) \leq n$ as desired. □

Proof of Theorem 3.1. (1) $\Rightarrow$ (2): Suppose that R is a local ring on G and that each right ideal of R can be generated by at most n elements. Let I be a fully invariant subgroup of G with U-sequence

$$U(I) = \{\alpha_0, \alpha_1, \alpha_2, \dots\}.$$

Suppose that $U(I)$ contains t gaps and that the ith gap occurs immediately after $\alpha_{g(i)}$. Then there is a fully invariant subgroup, I_i, whose U-sequence is

$$U(I_i) = \{\alpha_{g(i)} - g(i), \alpha_{g(i)} - g(i) + 1, \alpha_{g(i)} - g(i) + 2, \dots, \alpha_{g(i)}, \infty, \infty, \dots\}.$$

Then $U(I_i)$ contains a single gap and the term $\alpha_{g(i)}$ occurs in the same position in both $U(I)$ and $U(I_i)$. Now $I = I_1 + \cdots + I_t$ and, from [4, Lemma 24, p. 59], there is an element $a_i \in I_i$ with $U(I_i)$ as its Ulm sequence. This element lies outside the sum $I_1 + \cdots + I_{i-1} + I_{i+1} + \cdots + I_t$. Let $A = (a_1, \dots, a_t)$ be the right ideal generated by all the a_i's. As R is a local ring, the a_i's contain a minimal basis for A. Our choice of the a_i's ensures that A cannot be generated by fewer than t elements, so $t \le n$ as required.

(2) $\Rightarrow$ (3): Suppose that G is given by equation (2.1). Let G_1 be the subgroup of G defined by

$$G_1 = \begin{cases} a_1 C(p^{n_1}) \oplus a_2 C(p^{n_2}) & \text{if } n_2 = n_1 - 1, \\ a_1 C(p^{n_1}) & \text{otherwise.} \end{cases}$$

Then G_1 is a one-gap group and $G = G_1 \oplus G_1'$ for some subgroup G_1' of G such that $p^{n_1-2} G_1' = 0$. Let m_1, m_2 be the least integers such that $p^{m_1} G_1 = 0$ and $p^{m_2} G_1' = 0$. Then $n_1 = m_1 \ge m_2 + 2$. By applying this process to G_1' and repeating it sufficiently often we reach a decomposition

$$G = G_1 \oplus G_2 \oplus \cdots \oplus G_s \tag{3.3}$$

in which each G_i is a one-gap group and, if m_i is the least integer such that $p^{m_i} G_i = 0$, then $m_i \ge m_{i+1} + 2$ for $i = 1, \dots, s-1$. To show that this decomposition has the desired properties, we must prove that $s \le n$.

Since G possesses cyclic summands of each of the orders p^{m_i},

$$\{m_s - 1, m_{s-1} - 1, \dots, m_2 - 1, m_1 - 1, \infty, \infty, \dots\}$$

is a U-sequence. It contains s gaps. Then, [4, Theorem 25, p. 61] shows that there is a fully invariant subgroup with this U-sequence so, by condition (2), $s \le n$.

(3) $\Rightarrow$ (1): If (3) holds, then, by grouping cyclic direct summands appropriately, we may suppose that G has the decomposition of equation (3.3) for some integer $s \le n$, where each G_i has the form

$$G_i = a_i C(p^{m_i}) \oplus b_i C(p^{m_i - 1}) \quad (i = 1, \dots, s)$$

and $m_i \ge m_{i+1} + 2$ for $i = 1, \dots, s-1$. It is sufficient to exhibit a commutative local ring R on G in which each ideal can be generated by at most s elements. Let

$r_i = \text{rank}(G_i) = a_i + b_i$ and let $t_i = m_i r_i - b_i = (m_i - 1)r_i + a_i$, so that G_i contains exactly p^{t_i} elements. Let
$$R = Z_p[X_1, \ldots, X_s]/I,$$
where Z_p denotes the ring of p-adic integers and I is the ideal generated by the polynomials $p - X_1^{r_1} - X_2^{r_2} - \cdots - X_s^{r_s}$, $X_1^{t_1}, X_2^{t_2+1}, X_3^{t_3+1}, \ldots, X_s^{t_s+1}$ and all products $X_j X_k$ such that $1 \leq j, k \leq s$ and $j \neq k$.

We must show that $R_+ \cong G$. Let $e, x_1, \ldots, x_s$ be the images of $1, X_1, \ldots, X_s$ under the natural map $\theta : Z_p[X_1, \ldots, X_s] \to R$. Then
$$x_1^{r_1} = pe - x_2^{r_2} - x_3^{r_3} - \cdots - x_s^{r_s}. \tag{3.4}$$

Multiplying (3.4) by x_i and using the fact that $x_j x_k = 0$ (when $j \neq k$) gives
$$x_i^{r_i+1} = p x_i \quad \text{for } i = 1, 2, \ldots, s. \tag{3.5}$$

(3.4), (3.5) and the fact that $x_j x_k = 0$ (when $j \neq k$) show that each element of R is a linear combination of the elements
$$e, x_1, x_1^2, \ldots, x_1^{r_1-1}; x_2, x_2^2, \ldots, x_2^{r_2}; \ldots; x_s, x_s^2, \ldots, x_s^{r_s}. \tag{3.6}$$

I claim that these elements are linearly independent. Suppose that $\theta(f) = 0$, where
$$f = f(X_1, \ldots, X_s) = \lambda_0 + \sum_{j=1}^{r_1-1} \lambda_j X_1^j + \sum_{i=2}^{s} \sum_{j=1}^{r_i} \lambda_{ij} X_i^j$$
for some $\lambda_0, \lambda_j, \lambda_{ij} \in Z_p$. Then $f(X_1, \ldots, X_s) \in I$. Let J be the ideal of $Z_p[X_1, \ldots, X_s]$ generated by the powers $X_1^{t_1}, X_2^{t_2+1}, X_3^{t_3+1}, \ldots, X_s^{t_s+1}$ together with all products $X_j X_k$ such that $j \neq k$. Then $J \subseteq I$ and there are polynomials $g_1(X_1), \ldots, g_s(X_s)$ such that
$$f(X_1, \ldots, X_s) \equiv (p - X_1^{r_1} - \cdots - X_s^{r_s})\{g_1(X_1) + \cdots + g_s(X_s)\} \pmod{J}. \tag{3.7}$$

Let $g_1(X_1) = \mu_0 + \mu_1 X_1 + \mu_2 X_1^2 + \cdots$. Equating (as we may) constant terms and coefficients of $X_1, X_1^2, \ldots, X_1^{t_1-1}$ in (3.7), gives
$$\lambda_j = p\mu_j \quad \text{for } j = 0, 1, \ldots, r_1 - 1,$$
$$0 = p\mu_j - \mu_{j-r_1} \quad \text{for } j = r_1, r_1+1, \ldots, t_1-1.$$

Hence
$$\lambda_0 = p\mu_0 = p^2 \mu_{r_1} = p^3 \mu_{2r_1} = \cdots = p^{m_1} \mu_{(m_1-1)r_1}.$$

This and similar manipulations show that λ_j is divisible by p^{m_1} for $j = 0, 1, \ldots, a_1 - 1$ and by p^{m_1-1} for $j = a_1, a_1+1, \ldots, r_1-1$. Noting that $x_1^{t_1} = 0$ and making repeated use of (3.5) with $i = 1$, we see that
$$\lambda_1 x_1 = \lambda_2 x_1^2 = \cdots = \lambda_{r_1-1} x_1^{r_1-1} = 0. \tag{3.8}$$

Similarly, equating coefficients of $X_i, X_i^2, \ldots, X_i^{t_i}$ in (3.7) leads to

$$\lambda_{i1}x_i = \lambda_{i2}x_i^2 = \cdots = \lambda_{ir_i}x_i^{r_i} = 0 \quad \text{for } i = 2, \ldots, s. \tag{3.9}$$

Substituting (3.8) and (3.9) in $\theta(f) = 0$ gives $\lambda_0 e = 0$, so that (3.6) is a basis for R_+. We have also established that the subgroup spanned by $e, x_1, \ldots, x_1^{r_1-1}$ is isomorphic to G_1, whilst that spanned by $x_i, x_i^2, \ldots, x_i^{r_i}$ is isomorphic to G_i for $i = 2, \ldots, s$. This completes the proof that $R_+ \cong G$.

Clearly R is commutative. Let $\mathbf{m}$ be the ideal generated by the elements $x_1, \ldots, x_s$. Each of these elements is nilpotent, so $\mathbf{m}$ itself is nilpotent and lies in the Jacobson radical of R. But $R/\mathbf{m}$ is a field. Hence R is a local ring and $\mathbf{m}$ is its maximal ideal. (See [5, Proposition 1, p. 75].) Let $y = x_1 + x_2 + \cdots + x_s$. Since $x_j x_k = 0$ whenever $j \neq k$, we have $\mathbf{m}^2 = y\mathbf{m}$. Moreover, each ideal of $R/\mathbf{m}^2$ can be generated by at most s elements. By Lemma 3.2, each ideal of R can be s-generated, and we saw at the outset that $s \leq n$. □

References

[1] M. F. Atiyah and I. G. Macdonald, *Introduction to commutative algebra*, Addison-Wesley, Reading Mass., 1969.

[2] S. Feigelstock, *Additive groups of rings, Vol. II*, Pitman Research Notes in Mathematics No.169, Longman Scientific and Technical, Harlow New York, 1988.

[3] L. Fuchs, *Abelian groups*, Hungarian Academy of Sciences, Budapest, 1958.

[4] I. Kaplansky, *Infinite abelian groups*, University of Michigan Press, Ann Arbor, 1954 and 1969.

[5] J. Lambek, *Lectures on rings and modules*, Blaisdell, Waltham Mass., Toronto, London, 1966 and Chelsea Publishing Company, New York 1976.

[6] K. R. McLean, *Commutative artinian principal ideal rings*, Proceedings of the London Mathematical Society (3) 26 (1973), pp. 249–272.

[7] K. R. McLean, *p-Rings whose only right ideals are the fully invariant subgroups*, Proceedings of the London Mathematical Society (3) 30 (1975), pp. 445–458.

[8] K. R. McLean, *Local rings with bounded ideals*, Journal of Algebra 74 (1982), pp. 328–332.

[9] K. R. McLean, *Principal ideal rings and separability*, Proceedings of the London Mathematical Society (3) 45 (1982), pp. 300–318.

Author information

K. Robin McLean, Department of Mathematical Sciences, University of Liverpool, Liverpool L69 7ZL, United Kingdom.
E-mail: krmclean@liverpool.ac.uk

A Jacobson radical isomorphism theorem for torsion-free modules

Mary Flagg

Abstract. Let R be a complete discrete valuation domain. Wolfson showed that two torsion-free R-modules are isomorphic if their endomorphism rings are isomorphic R-algebras. If one of the modules is not divisible, an R-algebra isomorphism between the Jacobson radicals of the endomorphism rings of the respective modules is sufficient to imply that the modules are isomorphic.

Key words. Isomorphism theorem, Jacobson radical, complete discrete valuation domain, torsion-free modules.

AMS classification. 20K30.

1 Introduction

The Baer–Kaplansky theorem [2, 3] states that two torsion groups are isomorphic if and only if their endomorphism rings are isomorphic rings. Wolfson [6] proved that a similar theorem also holds for the class of torsion-free modules over a complete discrete valuation domain. The purpose of this paper is to show that, in the class of torsion-free modules over a complete discrete valuation domain which are not divisible, an algebra isomorphism between only the Jacobson radicals of the endomorphism rings of two modules is sufficient to imply that the modules are isomorphic.

Let R be a complete discrete valuation domain with prime p and field of quotients Q. All modules will be torsion-free left R-modules. For a module M, let $E(M)$ be the endomorphism algebra of M. Endomorphisms will be written as acting from the right. For an endomorphism ring E, define the set $pE = \{p\gamma : \gamma \in E\}$. Further, define the sets $E_0 = \{\gamma \in E : \text{rank}(M\gamma) < \infty\}$ and $E_{fg} = \{\gamma \in E : M\gamma \text{ is finitely generated}\}$. Denote by pE_0 and pE_{fg} the intersections $pE \cap E_0$ and $pE \cap E_{fg}$ respectively; note that pE, pE_0 and pE_{fg} are two-sided ideals of E.

For an endomorphism algebra E, let $J(E)$ denote the Jacobson radical of the ring E. Recall that the Jacobson radical of a ring with identity is the intersection of all maximal left (or right) ideals of the ring. An equivalent, and very useful, description of the Jacobson radical is that $J(E)$ is the sum of all right quasi-regular ideals of E, see [1, Theorem 15.3]. An element $\varepsilon \in E$ is right quasi-regular if $1 - \varepsilon$ has a right inverse in E. Then, $J(E) = \{\varepsilon \in E : \varepsilon\beta \text{ is right quasi-regular for all } \beta \in E\}$.

Primitive idempotents in the endomorphism ring are the key maps used to construct an isomorphism between two modules when their respective endomorphism rings are isomorphic. When only the Jacobson radical of the endomorphism ring is accessible, there are no nonzero idempotents available. This presents the challenge of identifying elements in the Jacobson radical of the endomorphism ring that provide the needed connection to cyclic direct summands of the modules. In order to establish that the desired endomorphisms exist in the Jacobson radical, the preliminary step in this proof is to establish upper and lower bounds for the Jacobson radical of the endomorphism ring of a torsion-free module.

2 Bounds for the Jacobson radical

For a module M with endomorphism ring E, the goal of this section is to show the relationship between the Jacobson radical of the ring E and the ideals pE, pE_0, and pE_{fg}. The following description of the ideal pE is a consequence of the fact that M is torsion-free.

Lemma 2.1. *Let M be a torsion-free R-module with endomorphism ring E. Then $pE = \{\gamma \in E : M\gamma \subseteq pM\}$.*

Proof. The fact that if $\gamma \in pE$, then $M\gamma \subseteq pM$ is clear. The other inclusion is not as trivial. Suppose $\gamma \in E$ and $M\gamma \subseteq pM$. Then, for every $x \in M$, there exists $y_x \in M$ such that $x\gamma = py_x$, and the y_x is unique by the fact that M is torsion-free. Define a map $\beta : M \to M$ by $x\beta = y_x$. Then, $\beta \in E$ and $\gamma = p\beta$, which completes the proof. □

First consider the case of a reduced module, and note that the module is not required to be free.

Theorem 2.2. *Let R be a complete discrete valuation domain and let M be a reduced torsion-free R-module. Let E be the endomorphism ring of M and let J be the Jacobson radical of E. Then $pE_0 \subseteq J \subseteq pE$.*

Proof. The proof of the inclusion $J \subseteq pE$ is accomplished by a contrapositive argument. Suppose $\gamma \in E\setminus pE$. Then, by Lemma 2.1, there exists $a \in M\gamma$ such that $a \notin pM$. Then Ra is a pure submodule of M which is complete in the p-adic topology, and hence Ra is a direct summand of M by [3, Theorem 23]. Let $b \in M$ be chosen such that $b\gamma = a$. There exists $C \leq M$ such that $M = Ra \oplus C$, and there exists a homomorphism $\phi \in E$ defined by $a\phi = b$ and $C\phi = 0$. Then, $b(1 - \gamma\phi) = b - a\phi = b - b = 0$. Thus, $1 + \gamma(-\phi)$ is not quasi-regular, which implies $\gamma \notin J$.

Conversely, if $\gamma \in E_0$, Lemma 3.2 of Liebert [4] implies there exists a decomposition $M = F \oplus K$ for some $F, K \leq M$ such that F is free of finite rank, $M\gamma \leq F$ and $K \leq \ker \gamma$. It is straightforward, using this lemma, to show that $pE_0 = \{p\gamma : \gamma \in E_0\}$. Then for any $p\gamma$, with $\gamma \in E_0$, $1 - p\gamma$ is invertible. The formal inverse $\phi = 1 + p\gamma + p^2\gamma^2 + p^3\gamma^3 + \cdots$ makes sense since $F\gamma \leq F$ and F is free of finite rank and hence complete in the p-adic topology. Therefore, $pE_0 \subseteq J$. □

When the module is no longer assumed to be reduced, the Jacobson radical of its endomorphism ring does not fit into the bounds described in the reduced case. However, the ability to manage the general case comes from the fact that a module over a

discrete valuation domain can always be decomposed as the direct sum of a reduced module and a divisible module. Let M be a torsion-free R-module. Then there exists a reduced submodule B and a divisible submodule D such that $M = B \oplus D$. Assuming the natural inclusions and projections, the endomorphism ring of M may be described as
$$E(M) = E(B) + \mathrm{Hom}_R(B,D) + E(D).$$
Then, the Jacobson radical can also be described in terms of this decomposition.

Proposition 2.3. *Let M be a torsion-free R-module. Then there exists a torsion-free reduced module B and a torsion-free divisible module D such that $M = B \oplus D$. Then,*
$$J(E(M)) = J(E(B)) + \mathrm{Hom}_R(B,D).$$

Proof. Let $I = J(E(B)) + \mathrm{Hom}_R(B,D) + J(E(D))$. Using the matrix representation of the ring $E(M)$, it is a simple exercise to show that I is a quasi-regular ideal of $E(M)$. Since $E(M)/I \cong E(B)/J(E(B)) \times E(D)/J(E(D))$, $J(E(M)/I) = 0$. This implies $I = J(E(M))$. As a torsion-free divisible R-module, D is also a vector space over Q. Then $E(D)$, the ring of R-endomorphisms of D, is the same as the ring of Q-endomorphisms of D and the latter is the set of all row-finite $\dim(D) \times \dim(D)$ matrices over Q. By the result of Patterson [5], $J(E(D))$ is contained in the ring of row-finite matrices over $J(Q)$; since this latter is 0, $J(E(D)) = 0$. □

Finally, it is important to examine the bounds for the Jacobson radical in the general case. Note that pE_0 is not contained in $J(E(M))$ when M is not reduced. However, the next result shows that pE_{fg} is a lower bound in the general case.

Theorem 2.4. *Let R be a complete discrete valuation domain and let M be a torsion-free R-module. Then $pE_{fg} \subseteq J(E(M)) \subseteq pE(M)$.*

Proof. Decompose M into the direct sum of a reduced module B and a divisible module D. By Theorem 2.2, $J(E(B)) \subseteq pE(B)$. Since D is divisible, $\mathrm{Hom}_R(B,D)$ is divisible. Therefore, if $\gamma \in J(E(M))$, then $\gamma \in pE(M)$. To prove the inclusion $pE_{fg} \subseteq J(E(M))$, let $\gamma \in pE_{fg}$. Using the decomposition $M = B \oplus D$, let $\gamma = p\alpha + p\beta + p\delta$ for some $\alpha \in E(B)$, $\beta \in \mathrm{Hom}_R(B,D)$, and $\delta \in E(D)$. Since $M\gamma$ is finitely generated, it is free of finite rank. So, $p\delta = 0$. Also, $B\alpha$ must be of finite rank, so $p\alpha \in pE(B)_0$. Hence, Theorem 2.2 implies $p\alpha \in J(E(B))$. Therefore, Proposition 2.3 shows that $\gamma \in J(E(M))$. □

3 The isomorphism theorem

The bounds for the Jacobson radical are needed to make the following observation about the Jacobson radical of the endomorphism ring of an arbitrary torsion-free module, M. The module M can be decomposed as $M = B \oplus D$ with B reduced and D divisible. Assume that M is not a divisible module, which is equivalent to assuming that $B \neq 0$. Since B is a reduced module, there exists $a \in B$ such that $a \neq 0$ and

$B = Ra \oplus B'$ for some $B' \leq B$. Let $C = B' \oplus D$, so that $M = Ra \oplus C$ and then let π be the projection onto Ra along C. Since $p\pi \in pE_{fg}$, Theorem 2.4 implies that $p\pi$ is in the Jacobson radical of the endomorphism ring of M.

The last piece of information needed to prove an isomorphism theorem with only the Jacobson radical is the following lemma that allows the map $p\pi$ to be recognized as p times a primitive idempotent from its properties in the Jacobson radical.

Proposition 3.1. *Let R be a complete discrete valuation domain and let M be a torsion-free R-module that is not divisible. Let $E = E(M)$, let $J = J(E(M))$, and let $0 \neq \omega \in J$. Then there exists $\beta \in E$ such that $\omega = p\beta$ and β is a primitive idempotent if and only if the following conditions hold:*

(1) $\omega^2 = p\omega$.

(2) *If there exists $\sigma_1, \sigma_2 \in J$ such that:* (i) $\omega = \sigma_1 + \sigma_2$, (ii) $\sigma_1\sigma_2 = 0 = \sigma_2\sigma_1$, *and* (iii) $\sigma_1^2 = p\sigma_1$ *and* $\sigma_2^2 = p\sigma_2$, *then* $\sigma_1 = 0$ *or* $\sigma_2 = 0$.

Proof. By Theorem 2.4, $\omega \in J$ implies that there exists $\beta \in E$ such that $\omega = p\beta$. First assume that β is a primitive idempotent. Then, by definition, $\beta^2 = \beta$. Hence, $\omega^2 = (p\beta)^2 = p^2\beta = p\omega$, which shows ω satisfies condition (1). To show that ω satisfies condition (2), suppose there exists $\sigma_1, \sigma_2 \in J$ that satisfy (i)–(iii). Since $J \subseteq pE$, there exists $\tau_1, \tau_2 \in E$ such that $\sigma_i = p\tau_i$ for $i = 1, 2$. Also, for $i = 1, 2$, $\sigma_i^2 = p\sigma_i$ implies that $p^2(\tau_i^2 - \tau_i) = 0$. Since E is torsion free, this implies that τ_1 and τ_2 are idempotents in E. Furthermore, by condition (ii), $p^2\tau_1\tau_2 = \sigma_1\sigma_2 = 0 = \sigma_2\sigma_1 = p^2\tau_2\tau_1$. This implies that τ_1 and τ_2 are orthogonal idempotents. By condition (i), $\omega = \sigma_1 + \sigma_2$, so $p(\beta - \tau_1 - \tau_2) = 0$. Thus $\beta = \tau_1 + \tau_2$. Since β is a primitive idempotent, $\tau_1 = 0$ or $\tau_2 = 0$ which implies $\sigma_1 = 0$ or $\sigma_2 = 0$.

Conversely, suppose that $\omega = p\beta \in J$ satisfies conditions (1) and (2). To complete the proof, it is necessary to show that β is an idempotent and then that it is primitive. The fact that β is an idempotent follows directly from the condition that $\omega^2 = p\omega$ since this implies $p^2(\beta^2 - \beta) = 0$, or $\beta^2 = \beta$.

Showing that β is a primitive idempotent is more complicated. Suppose there exist $\pi_1, \pi_2 \in E$ such that $\pi_1^2 = \pi_1$, $\pi_2^2 = \pi_2$, $\pi_1\pi_2 = 0 = \pi_2\pi_1$, and $\beta = \pi_1 + \pi_2$. Then to show β is a primitive idempotent, it is sufficient to show that $\pi_1 = 0$ or $\pi_2 = 0$. Observe that condition (2) appears to imply that β is primitive, but it does not because only maps in the Jacobson radical are tested. So, to use condition (2), it is necessary to show that $p\pi_1$ and $p\pi_2$ are in the Jacobson radical. Note that since it cannot be assumed that π_1 and π_2 are in E_0, this is not automatically true. For $i = 1, 2$, multiplying the equation $\beta = \pi_1 + \pi_2$ by π_i says that $\pi_i\beta = \pi_i\pi_1 + \pi_i\pi_2 = \pi_i^2 = \pi_i$, which implies $\pi_i p\beta = \pi_i\omega = p\pi_i$. Since $\omega \in J$ and J is a two-sided ideal of E, the last equation implies $p\pi_i \in J$ for $i = 1, 2$. Then, since $p\beta = p\pi_1 + p\pi_2$, condition (2) shows that $p\pi_1 = 0$ or $p\pi_2 = 0$. Since the Jacobson radical is a torsion-free ring, this completes the proof that β is a primitive idempotent. □

Proposition 3.1 gives criteria for recognizing p times primitive idempotents in the Jacobson radical, the following corollary shows that the only possible multiples of primitive idempotents in the Jacobson radical are those which are multiples of projections onto a cyclic direct summands.

Corollary 3.2. *Let M be a torsion-free R-module and let $J = J(E(M))$. Let $\omega \in J$ be such that $\omega \neq 0$ and ω satisfies conditions (1) and (2) of Proposition 3.1. Then, there exists $\beta \in E(M)$ such that $\omega = p\beta$ and β is an idempotent projection onto a summand isomorphic to R.*

Proof. By Proposition 3.1, there exists $\beta \in E(M)$ such that $\omega = p\beta$ and β is a primitive idempotent. There exists $B, D \leq M$ such that B is reduced and D is divisible and $M = B \oplus D$. Since $\gamma \in J$ implies that $D\gamma = 0$, $M\beta$ cannot be isomorphic to Q. Therefore, $M\beta$ is isomorphic to R. □

The technical tools are now available to state and prove the main theorem.

Theorem 3.3. *Let M and N be torsion-free R-modules. Suppose M is not divisible and that there exists an R-algebra isomorphism $\Phi : J(E(M)) \to J(E(N))$. Then, there exists an R-module isomorphism $\phi : M \to N$ that induces Φ.*

Proof. Denote $J(E(M))$ and $J(E(N))$ by J_M and J_N, respectively. Since M is not divisible, $J_M \neq 0$, and hence $J_N \neq 0$ and this implies that N is not divisible. For notational simplicity, for a map $\alpha \in J_M$, let $\alpha^* = \alpha\Phi \in J_N$. Further, the notation α' will indicate that $\alpha' \in E(N)$, but $\alpha' \notin J_N$.

Since M is not divisible, it has a nontrivial cyclic direct summand. So, there exists $a \in M$ such that Ra is isomorphic to R and $M = Ra \oplus C$ for some submodule $C \leq M$. Let $\beta \in E$ be the projection onto Ra associated with this decomposition, and note that β is a primitive idempotent. Theorem 2.4 shows $p\beta \in J_M$. Let $\omega = p\beta$, and note that ω satisfies conditions (1) and (2) of Proposition 3.1. Since the properties of conditions (1) and (2) of Proposition 3.1 are preserved under R-algebra isomorphisms, the map ω^* also satisfies these conditions. Theorem 2.4 implies that there exists $\beta' \in E_N$ such that $\beta' \neq 0$ and $\omega^* = p\beta'$. Then, Proposition 3.1 implies that β' is a primitive idempotent. Since $p\beta' \in J_N$, Corollary 3.2 implies that $N\beta'$ is isomorphic to R. Therefore, there exists $a^* \in N$ such that

$$N\beta' = Ra^* \text{ and } N = Ra^* \oplus C^*$$

for some $C^* \leq N$.

The mapping properties of the free summands $Ra \leq M$ and $Ra^* \leq N$ are the connectors between the endomorphism rings and the underlying modules. This connection needs to be made more precise to define an isomorphism. For every $x \in M$, there exists a unique $f_x \in E_M$ such that

$$af_x = x \quad \text{and} \quad Cf_x = 0.$$

Since $f_x \in E_{fg}$, Theorem 2.4 implies that $pf_x \in J_M$. Also, $p^2 f_x = p^2 \beta f_x = (p\beta)(pf_x) = \omega(pf_x)$, which shows that

$$a[\omega(pf_x)] = p^2 x.$$

Since $[\omega(pf_x)]^* = \omega^*(pf_x)^* = p\beta'(pf_x)^*$, the fact that $C^*\beta' = 0$ implies $C^*[\omega(pf_x)]^* = 0$. There exists $y^* \in N$ such that $a^*p\beta'(pf_x)^* = y^*$. It is necessary to show that $y^* \in p^2N$. By definition, $y^* = a^*(p\beta')(pf_x)^* = p(a^*(pf_x)^*)$. Since $(pf_x)^* \in J_N$, Theorem 2.4 implies $(pf_x)^* \in pE_N$. So, $a^*(pf_x)^* \in pN$ by Lemma 2.1. Hence, $y^* = p(a^*(pf_x)^*) \in p^2N$. Thus, there exists a unique $x^* \in N$ such that

$$a^*\omega^*(pf_x)^* = y^* = p^2 x^*.$$

Define a map $\phi : M \to N$ by letting $x\phi = x^*$ for every $x \in M$. The map ϕ is well-defined because $x^* \in N$ is uniquely determined for each $x \in M$. The fact that ϕ is an R-module homomorphism follows from its definition and the property that if $x, y \in M$ and $r, s \in R$, then $f_{rx+sy} = rf_x + sf_y$. The proof that ϕ is bijective is a straightforward exercise.

Finally, it remains to show that ϕ induces Φ. Let $x \in M$ and $\gamma \in J_M$. Then, notice that the definition of f_x implies $f_{x\gamma} = f_x\gamma$. So, $pf_{x\gamma} = (pf_x)\gamma$. Now, $(x\gamma)\phi = z^*$ if

$$\begin{aligned} p^2 z^* &= a^*(p\beta)^*(pf_{x\gamma})^* \\ &= a^*(p\beta)^*(pf_x\gamma)^* \\ &= a^*(p\beta)^*(pf_x)^*\gamma^* \\ &= [p^2 x\phi]\gamma^* = p^2[(x\phi)\gamma^*]. \end{aligned}$$

Thus, $p^2(x\gamma)\phi = p^2[(x\phi)\gamma^*]$, which implies ϕ induces Φ. □

References

[1] F. W. Anderson and K. R. Fuller, *Rings and categories of modules*, 2nd ed., Springer-Verlag, New York, 1992.

[2] R. Baer, *Automorphism rings of primary abelian operator groups*, Ann. Math. 44 (1943), pp. 192–227.

[3] I. Kaplansky, *Infinite abelian groups*, rev. ed., Univ. of Michigan Press, Ann Arbor, 1969.

[4] W. Liebert, *Endomorphism rings of reduced torsion-free modules over complete discrete valuation rings*, Trans. Amer. Math. Soc. 169 (1972), pp. 347–363.

[5] E. M. Patterson, *On the radicals of certain rings of infinite matrices*, Proc. Roy. Soc. Edinburgh 65 (1960), pp. 263–271.

[6] K. G. Wolfson, *Isomorphisms of the endomorphism rings of torsion-free modules*, Proc. Amer. Math. Soc. 13 (1962), pp. 712–714.

Author information

Mary Flagg, University of Houston, Houston, TX 77204-3008, USA.
E-mail: mflagg@math.uh.edu

Anti-isomorphisms and the failure of duality

A. L. S. Corner, B. Goldsmith and S. L. Wallutis

Abstract. Groups and modules with isomorphic endomorphism rings are known, in certain cases, to be necessarily isomorphic. When such a ring isomorphism is replaced by an anti-isomorphism, the modules are often determined only up to isomorphism of certain duals. This type of situation is examined in a number of cases with special emphasis on the situation for mixed Abelian groups, where it is shown that no reasonable duality may exist.

Key words. Anti-isomorphism, Baer–Kaplansky theorem, mixed Abelian groups.

AMS classification. 20K30, 20K21.

1 Introduction

A well-known theorem of Baer and Kaplansky ([1, 10]) states that Abelian p-groups are isomorphic if and only if their endomorphism rings are isomorphic. This theorem has been extended to other classes of Abelian groups and modules (see e.g. [9, 13]); in all cases the proofs are reasonably straightforward. The corresponding results for automorphism groups, i.e. that certain classes of modules are determined up to isomorphism by their automorphism groups, has also been the subject of a great deal of attention (see e.g. [11, 12, 2, 8]). It is a noticeable feature of these latter proofs that they are considerably more difficult than those relating to endomorphism algebras. Moreover the results contain an inbuilt 'duality' in that usually modules are determined not up to isomorphism but rather only up to isomorphism of the modules or their duals; the duals being a suitable group of homomorphisms. There is a 'halfway' case that has received some attention, *viz* the case of modules with anti-isomorphic endomorphism algebras. This situation is complex enough to admit the duality type outcome but is amenable, at least in some cases, to a more straightforward approach than is possible when dealing with automorphism groups. Note, of course, that if modules have anti-isomorphic endomorphism algebras, then composition of this anti-isomorphism with group inversion yields an isomorphism between the corresponding automorphism groups. Consequently, some of our results may be obtained from the corresponding results on automorphism groups; there are, however, situations, particularly involving the prime 2, where our approach yields results without amending the standard proof, whilst the corresponding result for automorphism groups is either quite complicated or unknown. We note that many of the results we display have been obtained previously

---

Tony Corner died on September 3rd 2006, prior to the completion of this work.

(see e.g. [4, 14]) but our approach is quite different and more reminiscent of Kaplansky's approach in [10, Theorem 28].

There is an immediate problem in extending Kaplansky's theorem from p-groups to mixed groups: the endomorphism rings of the quasi-cyclic group $\mathbb{Z}(p^\infty)$ and the group of p-adic integers $\widehat{\mathbb{Z}_p}$ are isomorphic commutative rings. This led Kaplansky to comment [10, Exercise 96, p. 73]: *This points up a critical difficulty in extending Theorem 28 to mixed modules. There are reasons for believing that a theory of duality is needed to clarify the situation. Perhaps it is true that when the rings of endomorphisms are isomorphic [or anti-isomorphic], the modules are isomorphic or "dual".* In the final section of this paper we shall show that such a belief is rather naive: there exists a family of $2^{\aleph_0}$ pairwise non-isomorphic groups $\{G_\alpha | \alpha \in \mathbb{R}\}$ with $G_\alpha \leq G_\beta$ if $\alpha \leq \beta$, such that any pair have isomorphic and anti-isomorphic endomorphism rings. It is hard to see how a duality of the kind envisaged could be reconciled with such a situation.

Finally, we note that our notation and terminology is standard and may be found in [6, 7]; an exception being that maps are written on the right.

2 Torsion-free homogeneous separable groups

There is, of course, a natural setting for anti-isomorphism in the context of Abelian group or module theory. Although our primary interest here is in Abelian groups, the module setting will be both natural and useful. Recall that if G is a left R-module and G^* denotes the R-module $\mathrm{Hom}_R(G, R)$, then every endomorphism $\alpha \in \mathrm{End}_R G$ induces a map $\alpha^* \in \mathrm{End}_R G^*$ by the rule

$$f\alpha^* = \alpha \circ f \quad (f \in G^*). \tag{2.1}$$

The mapping $()^* : \mathrm{End}_R G \to \mathrm{End}_R G^*$ is clearly an anti-homomorphism, so its composite with the corresponding $()^* : \mathrm{End}_R G^* \to \mathrm{End}_R G^{**}$ is a homomorphism $()^{**} : \mathrm{End}_R G \to \mathrm{End}_R G^{**}$. Moreover, there is a natural homomorphism $\iota : G \to G^{**}$ given by the evaluation map

$$f(g\iota) = gf \quad (g \in G, f \in G^*); \tag{2.2}$$

recall that G is said to be *reflexive* if this canonical map ι is an isomorphism. It is well known that these canonical homomorphisms constitute a natural transformation $\mathrm{id} \to ()^{**}$ in the sense that:

Lemma 2.1. *For any R-module G and $\alpha \in \mathrm{End}_R G$, the canonical homomorphism $\iota : G \to G^{**}$ satisfies*

$$(g\iota)\alpha^{**} = (g\alpha)\iota \quad (g \in G). \tag{2.3}$$

Proof. Since $\alpha^{**} \in \mathrm{End}_R G^{**}$ and $g\iota \in G^{**}$, we have $(g\iota)\alpha^{**} = \alpha^* \circ (g\iota) \in G^{**}$. Thus if $f \in G^*$, one has from (2.1) and (2.2) that

$$f((g\iota)\alpha^{**}) = (f\alpha^*)(g\iota) = g(\alpha \circ f) = (g\alpha)f = f((g\alpha)\iota).$$

Since f was arbitrary we have the desired result. □

It is now rather easy to establish the fundamental fact underlying all discussions of anti-isomorphisms in this context.

Proposition 2.2. *If G is a reflexive R-module, then $\operatorname{End}_R G$ and $\operatorname{End}_R G^*$ are anti-isomorphic.*

Proof. From equation (2.3) in Lemma 2.1 above, we have that $(g\iota)\alpha^{**} = (g\alpha)\iota$ ($g \in G$, $\alpha \in \operatorname{End}_R G$), and now, since G is assumed to be reflexive, we may identify G with G^{**} so that ι becomes the identity. Then (2.3) simply asserts that $\alpha^{**} = \alpha$, so that the composite $\operatorname{End}_R G \xrightarrow{()^*} \operatorname{End}_R G^* \xrightarrow{()^*} \operatorname{End}_R G$ is the identity. But G being reflexive implies that G^* is also reflexive and so, by symmetry, the two maps $()^*$ are inverse anti-isomorphisms. In particular, $\operatorname{End}_R G$ and $\operatorname{End}_R G^*$ are anti-isomorphic. □

Our principal objective in the remainder of this section is to establish a converse of the above Proposition when working with homogeneous separable groups; this restriction is quite natural since it is well known (see e.g. [3, IV Corollary 2.10]) that dual groups of the form $\operatorname{Hom}(G, \mathbb{Z})$ are always separable. It is also inevitable that in some situations one must invoke a further restriction requiring the type of the homogeneous group to be idempotent: if $S \leq \mathbb{Q}$ is a rational group which is not of idempotent type, then $S^{\aleph_0}$ is not separable homogeneous (see e.g. [7, Lemma 96.4]).

So suppose now that G, H are homogeneous separable groups of type R, S respectively with anti-isomorphic endomorphism rings. We need one final piece of notation: if X is a homogeneous group of type S then we set $X^* = \operatorname{Hom}(X, S)$ and $X_* = \operatorname{Hom}(S, X)$. The remainder of the section is devoted to showing:

Theorem 2.3. *Let G and H be homogeneous separable groups of types R, S respectively. If the endomorphism rings of G and H are anti-isomorphic then:*

(i) *the reduced types of R, S coincide: $R_0 = S_0$,*

(ii) $G_* \cong H^*$ *and* $H_* \cong G^*$,

(iii) *if the type R of G is idempotent then G is R-reflexive.*

Proof. Let G, H be as above and let $(r_1, \ldots, r_n, \ldots)$ and $(s_1, \ldots, s_n, \ldots)$ be characteristics representing R, S respectively. Then $r_n = \infty$ if and only if multiplication by $p_n 1_G$ is a unit in $\operatorname{End} G$. However since anti-isomorphisms also preserve units, we immediately deduce that $s_n = \infty$ if and only if $r_n = \infty$, which is equivalent to saying that $R_0 = S_0$.

Since G is homogeneous separable of type R, we may choose a direct decomposition $G = G_0 \oplus Rg_0$, where $g_0 \neq 0$. Let π be the projection of G onto Rg_0 along G_0. Then, in the standard way, we may identify $\operatorname{Hom}(Rg_0, G)$ with πE, where $E = \operatorname{End} G$. Similarly, $\operatorname{Hom}(G, Rg_0)$ may be identified with $E\pi$. Since there is an anti-isomorphism $()' : \operatorname{End} G \to \operatorname{End} H$, we have $\pi' \in E' = \operatorname{End} H$, and as π is an indecomposable idempotent in E, π' is an indecomposable idempotent in E'. Therefore the summand $H\pi'$ has the form Sh_0 for some $h_0 \neq 0$ and $H = H_0 \oplus Sh_0$, where $H_0 = \operatorname{Ker} \pi'$. Clearly, the anti-isomorphism maps πE isomorphically onto $E'\pi'$ and $E\pi$ isomorphically onto $\pi' E'$. Thus $G_* = \operatorname{Hom}(R, G) \cong \pi E \cong E'\pi' = \operatorname{Hom}(H, Sh_0) \cong H^*$.

Similarly, $H_* = \text{Hom}(S, H) \cong \pi'E' \cong E\pi \cong \text{Hom}(G, Rg_0) \cong G^*$. This establishes (ii).

Finally, suppose that R is an idempotent type so that $R = R_0$. Consider any $g \in G$, $f \in G^*$ and let $gf = r \in R$. Denote by $g\rho, f\lambda$ the unique elements of πE and $E\pi$ respectively, with

$$g_0(g\rho) = g \quad \text{and} \quad x(f\lambda) = (xf)g_0 \quad (x \in G, \; xf \in R). \tag{2.4}$$

Then $g_0(g\rho)(f\lambda) = g(f\lambda) = (gf)g_0 = rg_0$ i.e.

$$\text{if } g \in G, f \in G^* \text{ and } gf = r, \text{ then } (g\rho)(f\lambda) = r\pi. \tag{2.5}$$

Now apply the anti-isomorphism ()' to the identity $(g\rho)(f\lambda) = r\pi$ of (2.5) re-written in the form $n(g\rho)(f\lambda) = m\pi$, where $r = m/n$ and m, n are integers. Then one gets $n(f\lambda)'(g\rho)' = m\pi'$. However, since R is of idempotent type, the integer n consists only of prime factors p_n corresponding to the places in the type of R in which ∞ occurs. As the reduced types of R, S are equal by (i), the occurrence of an ∞ in R corresponds exactly to that in S and so one may divide across this last equation to obtain $(f\lambda)'(g\rho)' = r\pi'$.

Since R is of idempotent type, $G_* \cong G$ so we have an isomorphism $G \xrightarrow{\rho} \pi E$, which yields an isomorphism $\iota : G \to H^*$. Also $G^* \cong H_* \hookrightarrow H$, so we have a monomorphism $\kappa : G^* \hookrightarrow H$. But now it follows that $(f\kappa)(g\iota) = r$. Thus, if we identify G^* in H so that κ becomes the identity, we retrieve equation (2.2) above; in other words, the isomorphism ι is none other than the canonical homomorphism $G \to G^{**}$. Therefore G is reflexive. □

Remark 2.4. (i) It is not clear that one obtains reflexivity in the case where the type is not idempotent. Certainly the argument above fails at the key point where multiplication by the element $r \in R$ is preserved by the anti-isomorphism.

(ii) An examination of the proof of Theorem 2.3 shows that one does not require the full strength of the hypothesis that the groups be homogeneous and separable. In fact, the key property used is that, for each group, any rank-1 direct summand is of a fixed type (but not, of course, necessarily of the same type for the different groups). Thus, by a well-known theorem of Mishina (see e.g. [7, Proposition 96.2]), the result can be extended to include *inter alia* vector groups of the form $V = \prod R$, where R is a fixed rank-1 group, even when R is not of idempotent type; as noted above, in such circumstances V is neither homogeneous nor separable.

It is rather easy to show that one cannot replace the groups G_* and H_* in the theorem by G and H:

Example 2.5. Let R be a rank-1 group with type $(1, 1, 1, \ldots)$ and let $G = R^\omega$, $H = R^{(\omega)}$. Then there is an anti-isomorphism between End G and End H (via matrices) but $G^* = \text{Hom}(G, R) = \bigoplus_\omega \text{Hom}(R, R) = \mathbb{Z}^{(\omega)} \not\cong H$ and $H^* = \text{Hom}(\bigoplus_\omega R, R) \cong \prod_\omega \text{Hom}(R, R) = \mathbb{Z}^\omega \not\cong G$.

3 p-groups and mixed groups

The situation for separable p-groups is similar to, but more involved than, that for homogeneous separable torsion-free groups. Although the final outcome may be presented so that it seems not to involve a duality – two separable p-groups with anti-isomorphic endomorphism rings are necessarily isomorphic – there is, in fact, a strong duality at play here but the nature of p-groups allows one to actually deduce isomorphism from the duality. The result we present below is a special case of a complete characterization of the situation given by the first author in an unpublished manuscript dating back to the early 1960's. Since several other versions of the results for p-groups are now available (see in particular [4]), we present only an outline proof which is much influenced by, but somewhat simpler than, Liebert's approach to automorphism groups [12].

Theorem 3.1. *If G and H are unbounded reduced separable p-groups with* End G *anti-isomorphic to* End H, *then* $G \cong H^*$ *and* $H \cong G^*$, *where* $(-)^*$ *denotes the adjoint group* $t(\text{Hom}(-, \mathbb{Z}(p^\infty)))$.

As indicated above, one can actually deduce from this that G and H are isomorphic; in fact G, H must be torsion-complete p-groups with each Ulm invariant equal to 1 (see e.g. [5, Lemma 2.2]).

Proof. Since G is an unbounded reduced separable p-group, for each $i \geq 1$ we may choose, as in [10, Theorem 28], direct decompositions $G = S_1 \oplus \cdots \oplus S_i \oplus T_i$ where

(a) each S_i is cyclic of order $p^{n(i)}$, with $n(i)$ increasing monotonically;

(b) $T_1 > T_2 > \ldots$;

(c) $S_j < T_i$ for $j > i$.

Choose generators x_i for S_i and idempotents $e_i \in$ End G mapping S_i identically to S_i and annihilating the other summands. Since G is separable we may define endomorphisms e_{ij} such that e_{ij} maps x_i to x_j if $i > j$ and maps x_i to $p^{n(j)-n(i)}x_j$ if $i < j$; in either case the complementary summand is annihilated. Note that $e_i e_{ij} = e_{ij} e_j = e_{ij}$. For convenience, denote End G by A and End H by B and let Φ be the anti-isomorphism from A onto B. Set $e_i \Phi = f_i$ so that f_i is again an idempotent. Moreover, since $e_i A e_i \cong \text{End}(G e_i) = \text{End}(\langle x_i \rangle)$, it follows from $e_i A e_i \cong f_i B f_i \cong \text{End}(H f_i)$, that $H f_i$ must also be cyclic of the same order $p^{n(i)}$. Let $H f_i = \langle y_i \rangle$.

Now consider the directed system of groups $\{e_i A, \pi_{ij}\}$ where for $i < j$ π_{ij} is given by $e_i \alpha \mapsto e_{ji} e_i \alpha$ for each $\alpha \in A$. Notice that since $e_{ji} e_i = e_j e_{ji}$, the expression $e_{ji} e_i \alpha$ is actually in $e_j A$. As we have observed previously, $e_i A$ is isomorphic to the group $\text{Hom}(x_i, G)$ and hence the evaluation map $e_i \alpha \mapsto x_i e_i \alpha$ is a natural isomorphism of $e_i A$ onto $G[p^{n(i)}]$. The naturality of this map ensures that the directed systems $\{e_i A, \pi_{ij}\}$ and $\{G[p^{n(i)}], \iota_{ij}\}$, where ι_{ij} denotes inclusion, are isomorphic and hence the corresponding direct limits are isomorphic: $\varinjlim e_j A \cong \varinjlim G[p^{n(j)}] \cong G$.

Now define endomorphisms $f_{ij} \in B$ by setting $f_{ij} = e_{ji} \Phi$; note the change in order of the subscripts. By applying the anti-isomorphism Φ to the relations connecting the elements $e_{ij} \in A$, one obtains corresponding relations $f_i f_{ij} = f_{ij} f_j$. Consider the directed system $\{Bf_i, q_{ij}\}$ where the maps q_{ij} are given by $q_{ij} : \beta f_i \mapsto \beta f_i f_{ij} = \beta f_{ij} f_j$. We claim that the directed systems $\{e_i A, \pi_{ij}\}$ and $\{Bf_i, q_{ij}\}$ are isomorphic. This follows immediately from the commutivity, for each n, of the diagrams:

$$\begin{array}{ccc} e_n A & \xrightarrow{\pi_{n,n+1}} & e_{n+1} A \\ \downarrow \Phi & & \downarrow \Phi \\ Bf_n & \xrightarrow{q_{n,n+1}} & Bf_{n+1} \end{array}$$

Thus, if we can establish that the limit of the directed system $\{Bf_i, q_{ij}\}$ is isomorphic to $t(\mathrm{Hom}(H, \mathbb{Z}(p^\infty)))$, where this latter is regarded as the direct limit of the socles $H^*[p^{n(i)}]$ with inclusions as the connecting maps, we are finished.

Take a presentation of the quasi-cyclic group $\mathbb{Z}(p^\infty) = \langle z_1, z_2, \dots \rangle$ where $p^{n(1)} z_1 = 0$, $p^{n(2)-n(1)} z_2 = z_1, \dots$. Now define, for each n, a map $\theta_n : H^* \to \mathbb{Z}(p^\infty)$ by $y_n \theta_n = z_n$ with θ_n annihilating the complement of $\langle y_n \rangle$. Note that $\gamma_k : \beta f_k \mapsto \beta f_k \theta_k$ ($\beta \in B$) is a natural map taking Bf_k isomorphically onto $H^*[p^{n(k)}]$ since $Bf_k \cong \mathrm{Hom}(H, \langle y_k \rangle)$. The existence of the isomorphism we are seeking to establish is equivalent to showing that, for each k, the diagrams below are commutative:

$$\begin{array}{ccc} Bf_n & \xrightarrow{q_{k,k+1}} & Bf_{k+1} \\ \downarrow \gamma_k & & \downarrow \gamma_{k+1} \\ H^*[p^{n(k)}] & \xrightarrow{\iota} & H^*[p^{n(k+1)}] \end{array}$$

This follows from a simple diagram chase: if $h \in H$ and $\beta \in B$, then $h\beta$ can be expressed in the form $h\beta = r_k y_k + h'$, so that $h\beta f_k q_{k,k+1} = r_k p^{n(k+1)-n(k)} y_{n+1}$. Applying γ_{k+1} to this expression yields $r_k p^{n(k+1)-n(k)} z_{k+1} = r_k z_k$ and this is identical to the expression $h\beta f_k \gamma_k$. This completes the proof. $\square$

The situation for mixed groups with anti-isomorphic endomorphism rings is, however, vastly more complicated, even when the torsion part is a p-group. As noted in the introduction, Kaplansky felt that it was likely that the situation could be clarified by the use of a duality. Our next result shows that such a hope was essentially naive: the groups exhibited below have pairwise anti-isomorphic (and isomorphic) endomorphism rings, yet the containment relation between the groups is order-isomorphic to the real numbers.

Theorem 3.2. *There exists a mixed group G (which is even a p-adic module) together with a family of subgroups $G(\xi)$ indexed by a real parameter ξ, $0 < \xi \leq 1$ with the following properties:*

(a) $G(\xi) \not\cong G(\xi')$ *if* $0 < \xi < \xi' \leq 1$;

(b) $\mathrm{End}(G(\xi)) \cong \mathrm{End}\, G$ *for* $0 < \xi \leq 1$;

(c) $\mathrm{End}\, G$ *admits an anti-isomorphism.*

Proof. Let $P = \prod_{i=1}^{\infty} \langle a_i \rangle$ be the direct product of finite cyclic groups $\langle a_i \rangle$ of order p^i. As usual we identify a_i with the element $(0, \ldots, 0, a_i, 0, \ldots)$ of P, so that the elements a_i generate a subgroup S of P which may be identified with the direct sum of the cyclic groups $\langle a_i \rangle$. Let $\widehat{S}$ denote the p-adic completion of S; clearly $\widehat{S} \leq P$.

Define G and $G(\xi)$ $(0 < \xi \leq 1)$ as subsets of $\widehat{S}$ in the following way: set $G = \widehat{S}$ and define $G(\xi)$ by $g = (g_1, g_2, \ldots) \in G(\xi)$ if and only if there exists an integer k such that the height

$$h(g_{k+i}) \geq \xi i \quad \text{for } i = 1, 2, \ldots. \tag{3.1}$$

It is easily verified that the $G(\xi)$ are subgroups of $\widehat{S}$, that $T = G(1)$ is the torsion-completion of S, or equivalently the maximal torsion subgroup of $\widehat{S}$ and that we have the inclusion

$$S \subset T \subset G(\xi') \subset G(\xi) \subset G \subset P \quad \text{for } 0 < \xi < \xi' < 1. \tag{3.2}$$

To simplify matters we write also $G(0) = G$.

First we show that

(A) The quotient group $G(\xi)/S$ is divisible for $0 \leq \xi \leq 1$.

It will be enough to prove that if $g \in G(\xi)$, then there exists an element $g' \in G(\xi)$, with $g - pg' \in S$. We shall prove this only in the case $0 < \xi \leq 1$; the case $\xi = 0$ is immediate since $G(0) = \widehat{S}$. Suppose then that g satisfies condition (3.1) for some k. Let n be the least integer not less than ξ^{-1}. Then, for $i = 1, 2, \ldots$ we have

$$h(g_{k+n+i}) \geq \xi(n+i) \geq 1 + \xi i; \tag{3.3}$$

consequently $\langle a_{k+n+i} \rangle$ contains an element g'_{k+n+i} with $pg'_{k+n+i} = g_{k+n+i}$. It follows from (3.3) that $h(g'_{k+n+i}) \geq \xi i$ for $i = 1, 2, \ldots$ and therefore $G(\xi)$ contains the element $g' = (0, \ldots, g'_{k+n+1}, g'_{k+n+2}, \ldots)$. Clearly $g - pg' \in S$.

Let ϕ be an arbitrary homomorphism $S \to P$. Then

(B) for $0 \leq \xi \leq 1$, ϕ can be extended to a homomorphism $\bar{\phi} : G(\xi) \to P$ in precisely one way.

Since P is reduced and, by (A), $G(\xi)/S$ is divisible, there can be at most one such extension $\bar{\phi}$. Moreover, since $G(\xi) \leq G$ for $0 \leq \xi \leq 1$, we need only prove that ϕ can be extended to a homomorphism $\bar{\phi} : G \to P$. This, however, follows immediately by continuity since P is itself complete in the p-adic topology.

Now observe that

(C) if ϕ is an arbitrary homomorphism $G(\xi) \to P$ for some ξ, $0 \leq \xi \leq 1$, then $G(\xi)\phi \leq G(\xi)$.

Again we consider only the case $0 < \xi \leq 1$. Let $g \in G(\xi)$. Then g satisfies (3.1) for some integer k. It follows from (B) that for $j = 1, 2, \ldots$

$$(g\phi)_{k+j} = \sum_{i=1}^{\infty} (g_i \phi)_{k+j}$$

where all but a finite number of the terms of the sum are zero. We deduce that $h((g\phi)_{k+j}) \geq \min_i h((g_i \omega)_{k+j})$. For $1 \leq i \leq k$, we have $p^i g_i = 0$ and, therefore,

$p^i(g_i\phi)_{k+j} = 0$, whence $h((g_i\phi)_{k+j}) \geq k+j-i \geq j \geq \xi j$; for $k < i \leq k+j$, we have that $h(g_i) \geq x$, where x is the least integer not less than $\xi(i-k)$; it follows that $p^{i-x}g_i = 0$ and so $p^{i-x}(g_i\phi)_{k+j} = 0$; consequently $h((g_i\phi)_{k+j}) \geq k+j-(i-x) = k+j-i+x \geq (k+j-i)+\xi(i-k) \geq \xi(k+j-i)+\xi(i-k) = \xi j$. For $k+j < i$, we have $h((g_i\phi_{k+j}) \geq h(g_i) \geq \xi(i-k) > \xi j$. Hence $h((g\phi)_{k+j}) \geq \min_i h((g_i\phi)_{k+j}) \geq \xi j$; therefore $g\phi \in G(\xi)$.

We are now ready to prove (a), (b) and (c).

(a) Suppose, if possible that $\phi: G(\xi') \to G(\xi)$ is an isomorphism where $0 < \xi < \xi' \leq 1$. If we regard ϕ as a homomorphism of $G(\xi')$ into P, then it follows from (C) that $G(\xi) = G(\xi')\phi \leq G(\xi')$. Now $G(\xi)$ contains the element $g = (g_1, g_2, \dots)$ where for $i = 1, 2, \dots$, $g_i = p^{[\xi i+1]}a_i$; here, as usual, $[x]$ denotes the integer part of the real number x. It follows from the inclusion $G(\xi) \leq G(\xi')$ that $g \in G(\xi')$, i.e. for some integer k we have

$$[\xi(k+i)+1] = h(g_{k+i}) \geq \xi'i \quad \text{for } i = 1, 2, \dots.$$

Consequently, $\xi k+1 \geq (\xi'-\xi)i$ for $i = 1, 2, \dots$; since, by hypothesis $\xi'-\xi > 0$, this is impossible. Thus we have shown that $G(\xi) \not\cong G(\xi')$.

(b) Let $0 < \xi \leq 1$ and suppose that $\phi \in \text{End}(G(\xi))$. Denote the restriction of ϕ to T by ϕ' so that $\varepsilon: \phi \mapsto \phi'$ is a ring homomorphism $\text{End}(G(\xi)) \to \text{End } T$. By (A) the kernel of ε is 0: any element of $\text{Ker } \varepsilon$ induces a map from the divisible group $G((\xi))/T$ into the reduced group $G((\xi))$ and hence is identically zero. Moreover, it follows from (B) and (C) that any element of $\text{End } T$ lifts to an endomorphism of $G((\xi))$, so that ε is onto. Thus $\text{End}(G(\xi)) \cong \text{End } T$ as required.

(c) Since $\text{End } G \cong \text{End}(G(1))$, it suffices to show that $\text{End}(G(1)) = \text{End } T$ possesses an anti-automorphism. If $\phi \in \text{End } T$, then ϕ is uniquely determined by its restriction to S and thus we may represent ϕ as an $\omega \times \omega$ matrix obtained in the usual way from the equations $a_i\phi = \sum_{j=1}^{\infty} r_{ij}a_j$, where the right hand side is being interpreted as an element of P. Now let ϕ^* be the mapping $S \to P$ defined by $a_i\phi^* = \sum_{j=1}^{\infty} r_{ji}a_j$. Clearly ϕ^* is a well-defined homomorphism and corresponds to the classical transpose matrix of ϕ. Moreover, from (B), ϕ^* may be extended to a map which we continue to denote by $\phi^*: T = G(1) \to P$ and, as T is fully invariant, ϕ^* actually maps $T \to T$, i.e. $\phi^* \in \text{End } T$. Clearly the assignment $\phi \mapsto \phi^*$ is an anti-homomorphism of $\text{End } T$ and since $(\phi^*)^* = \phi$, it is an anti-automorphism. □

References

[1] R. Baer, *Automorphism rings of primary abelian operator groups*, Ann. Math. **44** (1943), 192–277.

[2] A. L. S. Corner and B. Goldsmith, *Isomorphic Automorphism Groups of p-adic Modules*, in Abelian Groups, Module Theory and Topology, (ed. D. Dikranjian and L. Salce), Lecture Notes in Pure and Applied Mathematics, **201**, 125–130, Marcel Dekker, New York (1998).

[3] P. Eklof and A. Mekler, *Almost Free Modules, Set-theoretic Methods*, North-Holland, 2002.

[4] G. D'Este, *Abelian Groups with Anti-Isomorphic Endomorphism Rings*, Rend. Sem. Mat. Univ. Padova **60** (1978), 55–75.

[5] K. Faltings, *On the Automorphism Group of a Reduced Primary Abelian Group*, Trans. Amer. Math. Soc. **165** (1972), 1–25.

[6] L. Fuchs, *Infinite Abelian groups* **I**, Academic Press, New York (1970).

[7] L. Fuchs, *Infinite Abelian groups* **II**, Academic Press, New York (1973).

[8] B. Goldsmith and P. Zanardo, *Endomorphism Rings and Automorphism Groups of Separable Torsion-Free Modules over Valuation Domains*, in Abelian Groups, Module Theory and Topology, (ed. D. Dikranjian and L. Salce), Lecture Notes in Pure and Applied Mathematics, **201**, 249–260, Marcel Dekker, New York (1998).

[9] G. J. Hauptfleisch, *Torsion-free abelian groups with isomorphic endomorphism rings*, Arch. Math. **24** (1973), 269–273.

[10] I. Kaplansky, *Infinite Abelian Groups*, University of Michigan Press, Ann Arbor, 1954 and 1969.

[11] H. Leptin, *Abelsche p-Gruppen und ihre Automorphismengruppen*, Math. Z. **73** (1960), 235–253.

[12] W. Liebert, *Isomorphic Automorphism Groups of Primary Abelian Groups*, in Abelian Group Theory, Proceedings of the 1985 Oberwolfach Conference, 9–31, Gordon and Breach, New York (1987).

[13] W. May, *Isomorphism of Endomorphism Algebras over Complete Discrete Valuation Rings*, Math. Z. **204** (1990), 485–499.

[14] K. J. Wolfson, *Anti-Isomorphisms of Endomorphism Rings of Locally Free Modules*, Math. Z. **202** (1989), 151–159.

Author information

A. L. S. Corner, formerly of Worcester College, Oxford, England.

B. Goldsmith, School of Mathematical Sciences, Dublin Institute of Technology, Aungier Street, Dublin 2, Ireland.
E-mail: `brendan.goldsmith@dit.ie`

S. L. Wallutis, Fachbereich 6, Mathematik, Universität Duisburg-Essen, 45117 Essen, Germany.
E-mail: `simone@wallutis.de`

Automorphism groups of models of first order theories

V. V. Bludov, M. Giraudet, A. M. W. Glass and G. Sabbagh

Abstract. Some years ago (see [5, Problem 12.14]), we asked if every first order theory having infinite models has a model whose automorphism group has undecidable theory. In this note, we prove that this is true in a strong form. It will be a consequence of our other theorem that there is a right orderable finitely presented group with insoluble word problem.

Key words. Finitely presented group, insoluble word problem, right ordered group, automorphism group, undecidable theory, existential theory.

AMS classification. 03B25, 06F15, 20B27.

1 Introduction

Recall that a group G is *right orderable* if there is a total order $\leq$ on G such that $f \leq g$ in G implies that $fh \leq gh$ for all $h \in G$.

We will prove

Theorem 1.1. *There is a right orderable finitely presented group with insoluble word problem.*

and deduce

Theorem 1.2. *Every first order theory having infinite models has a model whose automorphism group has undecidable existential theory. Moreover, such models exist in all infinite cardinals which are at least the cardinality of the language.*

Theorem 1.2 applies to theories of R-modules having infinite models. To achieve this, adjoin a family of unary operators $\{\mu_r : r \in R\}$ to the language of abelian groups with extra axioms for the interpretation $\mu_r(x) = rx$ $(r \in R)$. If desired, extra operators, relations and constants can be added to the language to include theories of modules with extra operations, relations and constants.

2 Proof that Theorem 1.1 implies 1.2

As noted in [6, Example 1.2.3], if G is any right ordered group, then G can be embedded (as a group) in $\mathrm{Aut}(G, \leq)$. But if $(\Omega, \leq)$ is any countable totally ordered set, then the group $\mathrm{Aut}(\Omega, \leq)$ can be embedded in $\mathrm{Aut}(\mathbb{Q}, \leq)$, where $(\mathbb{Q}, \leq)$ is the set of rational numbers with the usual standard ordering (see [3, Theorem 2.J]). Hence we obtain the following (folklore):

Lemma 2.1. *Every countable right orderable group can be embedded in* $\mathrm{Aut}(\mathbb{Q}, \leq)$.

We now use this to deduce Theorem 1.2 from Theorem 1.1.

Proof. Let G be any right orderable finitely presented group with insoluble word problem; say,
$$G := \langle x_1, \ldots, x_m : u_1(x_1, \ldots, x_m) = 1, \ldots, u_n(x_1, \ldots, x_m) = 1\rangle.$$
For any word $w(x_1, \ldots, x_m)$, let ψ_w be the sentence
$$\forall x_1, \ldots, x_m \Big[\Big(\bigwedge_{i=1}^n u_i(x_1, \ldots, x_m) = 1\Big) \to w(x_1, \ldots, x_m) = 1\Big].$$
Let H be any group containing G. If $H \models \psi_w$, then $G \models \psi_w$ since ψ_w is a universal sentence ([2, Theorem 5.2.4]). But each $u_j(x_1, \ldots, x_m) = 1$ in G ($j = 1, \ldots, n$), so $w(x_1, \ldots, x_m) = 1$ in G if $H \models \psi_w$. Conversely, suppose that $w(x_1, \ldots, x_m) = 1$ in G. Let $h_1, \ldots, h_m \in H$ with $u_j(h_1, \ldots, h_m) = 1$ in H ($j = 1, \ldots, n$). Then the subgroup of H generated by $h_1, \ldots, h_m$ is a homomorphic image of G. Hence $w(h_1, \ldots, h_m) = 1$ in H. Thus $H \models \psi_w$. So $w(x_1, \ldots, x_m) = 1$ in G if and only if $H \models \psi_w$. Since the word problem for G is insoluble, the universal theory of H is undecidable. Therefore the existential theory of H must also be undecidable. By Lemma 2.1, we may take H to be $\mathrm{Aut}(\mathbb{Q}, \leq)$; so the latter has undecidable existential theory. But if T is any first order theory having infinite models, then it has a model $\mathcal{M}$ built using Skolem functions with indiscernibles $(\mathbb{Q}, \leq)$; moreover, every element of $\mathrm{Aut}(\mathbb{Q}, \leq)$ extends uniquely to an element of $\mathrm{Aut}(\mathcal{M})$ (see [2, Section 3.3]). We therefore have embeddings of G into $\mathrm{Aut}(\mathbb{Q}, \leq)$ and $\mathrm{Aut}(\mathbb{Q}, \leq)$ into $\mathrm{Aut}(\mathcal{M})$. By letting H above be $\mathrm{Aut}(\mathcal{M})$, we see that $\mathrm{Aut}(\mathcal{M})$ has undecidable existential theory. Consequently, $\mathcal{M}$ is the desired model of T and the first part of Theorem 1.2 follows.

For the general case, let $\mathcal{L}$ be the language of T and $\kappa \geq ||\mathcal{L}||$ be an infinite cardinal thought of as an ordinal. Let $(X, \leq)$ be the disjoint union of $\mathbb{Q}$ and κ, totally ordered by letting every rational number be less than every element of κ. Build a model $\mathcal{M}$ of T using Skolem functions and indiscernibles $(X, \leq)$. It has cardinality κ. Now $\mathrm{Aut}(\mathbb{Q}, \leq) = \mathrm{Aut}(X, \leq)$ which can be embedded in $\mathrm{Aut}(\mathcal{M})$. By the same proof as in the previous paragraph, $\mathrm{Aut}(\mathcal{M})$ has undecidable existential theory. This completes the deduction of Theorem 1.2 from Theorem 1.1. □

3 Proof of Theorem 1.1

Let m be a positive integer with $m \geq 2$, and F_m be the free group on m free generators. Let $P := F_m \times F_m$. Then there is a finitely generated subgroup H of P such that the membership problem for H (in P) is insoluble ([7, Theorem 4.3]). That is, there is no algorithm which determines for an arbitrary element of P whether it belongs to H or not. Let H be generated by $h_1, \ldots, h_n$, say. Consider the group
$$G := \langle P, t : [t, h_i] = 1 \ (i = 1, \ldots, n)\rangle.$$
We establish Theorem 1.1 by proving

Proposition 3.1. *G is a right orderable finitely presented group with insoluble word problem.*

Proof. The proof relies on a result of George Bergman ([1, Theorem 35]):

Let $\{G_i : i \in I\}$ *be a family of right orderable groups having a common subgroup S. Suppose that there is a right orderable group U and embeddings $\varphi_i : G_i \to U$ with $s\varphi_i = s\varphi_j$ (for all $s \in S$ and $i, j \in I$). Then the free product of $\{G_i : i \in I\}$ with S amalgamated is right orderable.*

P can be given by the finite presentation

$$P = \langle a_1, \ldots, a_m, b_1, \ldots, b_m : [a_i, b_j] = 1 \ (i, j \in \{1, \ldots, m\}) \rangle.$$

Hence G is finitely presented. Next, G has insoluble word problem, since otherwise we could solve the membership problem for H in P (an element $c \in P$ belongs to H iff $[t, c] = 1$ in G; see [7, Section IV.2]). Finally, we need to show that G is right orderable. Now any free group is orderable ([4, Example 1.3.24]). Hence so is P (use the lexicographic order). Let N be the normal subgroup of G generated by P. Then N is a free product of orderable groups $\{t^{-k}Pt^k : k \in \mathbb{Z}\}$ with the same subgroup H of each amalgamated. For each $k \in \mathbb{Z}$, let $\varphi_k : t^{-k}Pt^k \to P$ be the natural isomorphism given by $t^{-k}pt^k \mapsto p$. By our hypotheses, all φ_k ($k \in \mathbb{Z}$) agree on H. By Bergman's result (with $I = \mathbb{Z}$, $G_k = t^{-k}Pt^k$, $S = H$ and $U = P$), we get that N is right orderable. Let $\leq_N$ be such a right order on N. Every element g of G can be written uniquely in the form $g = xt^\ell$ with $x \in N$ and $\ell \in \mathbb{Z}$. Let $h = yt^{\ell'}$ with $y \in N$ and $\ell' \in \mathbb{Z}$. Define $g \leq h$ in G if and only if $\ell < \ell'$ (in $\mathbb{Z}$) or both $\ell = \ell'$ and $x \leq_N y$. This gives a right ordering on G and establishes the theorem. □

For future reference we observe that we have proved

Corollary 3.2. *If G is a right orderable group with subgroup H, then the group $\langle G, t : [t, h] = 1 \ (h \in H) \rangle$ is right orderable.*

Also, as mentioned in the proof of the deduction of Theorem 1.2 from Theorem 1.1, we have that

Corollary 3.3. *Aut$(\mathbb{Q}, \leq)$ has undecidable existential theory.*

Acknowledgements. This solution to Problem 12.14 in [5] was obtained as a result of discussion at the "Colloque Giraudet" conference. We wish to thank Modnet MRTN-CT-2004-512234 for funds for this meeting and for covering the expenses of the third author.

References

[1] G. M. Bergman, *Ordering coproducts of groups and semigroups*, J. Algebra **133** (1990), 313–339.

[2] C. C. Chang and H. J. Keisler, *Model Theory*, Studies in Logic **73**, North-Holland Publishing Co., Amsterdam, 1973.

[3] A. M. W. Glass, *Ordered Permutation Groups*, London Math. Society Lecture Notes Series **55**, Cambridge University Press, Cambridge, 1981.

[4] A. M. W. Glass, *Partially Ordered Groups*, Series in Algebra **7**, World Scientific Publishing Co., Singapore, 1999.

[5] E. Khukhro (ed.), *Unsolved Problems in Group Theory: Kourovka Note Book*, Novosibirsk.

[6] V. M. Kopytov and N. Ya. Medvedev, *Right Ordered Groups*, Kluwer Publishing Co., Dordrecht, 1996.

[7] R. C. Lyndon and P. E. Schupp, *Combinatorial Group Theory*, Ergebnisse der Math. **89**, Springer, Heidelberg, 1977.

Author information

V. V. Bludov, Department of Mathematics, Physics, and Informatics, Irkutsk Teachers Training University, Irkutsk 664011, Russia.
E-mail: vasily-bludov@yandex.ru

M. Giraudet, Équipe de Logique, Université Paris 7, 2 Place Jussieu, 75251 Paris cedex 05, France.
E-mail: giraudet@logique.jussieu.fr

A. M. W. Glass, Department of Pure Mathematics and Mathematical Statistics, Centre for Mathematical Sciences, Wilberforce Rd., Cambridge CB3 0WB, England.
E-mail: amwg@dpmms.cam.ac.uk

G. Sabbagh, Équipe de Logique, Université Paris 7, 2 Place Jussieu, 75251 Paris cedex 05, France.
E-mail: sabbagh@logique.jussieu.fr

The classification problem for finite rank Butler groups

Simon Thomas

Abstract. In this paper, we shall consider the complexity of the isomorphism and quasi-isomorphism problems for the finite rank Butler groups, as well as the related question of the complexity of the classification problem for representations of finite posets over various fields.

Key words. Butler groups, Borel equivalence relations.

AMS classification. 03E05, 20K15.

1 Introduction

In 1937, Baer [3] introduced the notion of the type of an element in a torsion-free abelian group and showed that this notion provided a complete invariant for the classification problem for torsion-free abelian groups of rank 1. Since then, despite the efforts of such mathematicians as Kurosh [17] and Malcev [18], no satisfactory system of complete invariants has been found for the torsion-free abelian groups of finite rank $n \geq 2$. Consequently, it was natural to ask whether the classification problem for the higher rank groups was genuinely more difficult than that for the rank 1 groups. In 1998, Hjorth [13] proved that this was indeed the case; and soon afterwards, making essential use of the work of Adams–Kechris [1], Hjorth [13] and Zimmer [22], Thomas [21] proved that the complexity of the classification problem for the torsion-free abelian groups of rank n increases strictly with the rank n. Of course, abelian group theorists had long before reached the conclusion that the classification problem for the higher rank groups was hopelessly difficult and had shifted their attention from the study of arbitrary finite rank groups to various restricted subclasses, where further progress could be made. In this paper, we shall consider the complexity of the isomorphism and quasi-isomorphism problems for the class of finite rank Butler groups, which has been the main focus of recent research in torsion-free abelian groups of finite rank.

Definition 1.1. A finite rank torsion-free abelian group A is said to be a *Butler group* iff A can be expressed as the (not necessarily direct) sum

$$A = A_1 + \cdots + A_s$$

of finitely many rank 1 subgroups $A_1, \ldots, A_s$. (Equivalently, A can be expressed as a sum of finitely many pure rank 1 subgroups.)

Research partially supported by NSF Grants DMS 0100794 and 0600940.

Theorem 1.2. *The isomorphism and quasi-isomorphism problems for the finite rank Butler groups are both hyperfinite.*

This paper is organized as follows. In Section 2, we shall give a brief review of the basic theory of Borel equivalence relations. In particular, we shall define the notion of a hyperfinite equivalence relation and present some examples which arise naturally in the study of finite rank torsion-free abelian groups. The proof of Theorem 1.2 will be given in Section 3. Finally, in Section 4, we shall discuss the problem of classifying the representations rep(S, K) of a finite poset S over a field K.

2 Hyperfinite equivalence relations

In this section, we shall give a brief review of some of the basic notions of the theory of Borel equivalence relations, focussing on the class of hyperfinite equivalence relations. (For more detailed accounts, see Hjorth–Kechris [14] and Jackson–Kechris–Louveau [15].)

Let X be a standard Borel space; i.e. a Polish space equipped with its associated σ-algebra of Borel subsets. Then a *Borel equivalence relation* on X is an equivalence relation $E \subseteq X^2$ which is a Borel subset of X^2. The Borel equivalence relation E is said to be *countable* iff every E-equivalence class is countable. If E, F are Borel equivalence relations on the standard Borel spaces X, Y respectively, then we say that E is *Borel reducible* to F and write $E \leq_B F$ if there exists a Borel map $f : X \to Y$ such that $x \mathrel{E} y$ iff $f(x) \mathrel{F} f(y)$. Such a map f is called a *Borel reduction* from E to F. We say that E and F are *Borel bireducible* and write $E \sim_B F$ if both $E \leq_B F$ and $F \leq_B E$. Finally we write $E <_B F$ if both $E \leq_B F$ and $F \not\leq_B E$. All of the Borel equivalence relations that we shall consider in this paper arise from group actions as follows. Let G be a locally compact second countable group. Then a *standard Borel G-space* is a standard Borel space X equipped with a Borel action $(g, x) \mapsto g \cdot x$ of G on X. The corresponding G-orbit equivalence relation on X, which we shall denote by E_G^X, is a Borel equivalence relation. In fact, by Kechris [16], E_G^X is Borel bireducible with a countable Borel equivalence relation. Conversely, by Feldman–Moore [9], if E is an arbitrary countable Borel equivalence relation on the standard Borel space X, then there exists a countable group G and a Borel action of G on X such that $E = E_G^X$.

Example 2.1. Let $n \geq 1$ and let $\mathbb{Q}^n$ be the canonical n-dimensional vector space over $\mathbb{Q}$. Then

$$R(\mathbb{Q}^n) = \{A \leqslant \mathbb{Q}^n \mid A \text{ contains } n \text{ linearly independent elements}\}$$

is a standard Borel space. Note that the natural action of $GL_n(\mathbb{Q})$ on the vector space $\mathbb{Q}^n$ induces a Borel action on the space $R(\mathbb{Q}^n)$ of rank n groups; and that if A, $B \in R(\mathbb{Q}^n)$, then A, B are isomorphic iff there exists an element $\varphi \in GL_n(\mathbb{Q})$ such that $\varphi(A) = B$. It follows that the isomorphism relation $\cong_n$ on $R(\mathbb{Q}^n)$ is a countable Borel equivalence relation. Combining Hjorth [13] and Thomas [21], we have that

$$(\cong_1) <_B (\cong_2) <_B \cdots <_B (\cong_n) <_B \cdots.$$

In other words, with respect to Borel reducibility, the complexity of the classification problem for the torsion-free abelian groups of rank n increases strictly with the rank n.

The least complex Borel equivalence relations are those which are *smooth*; i.e. those Borel equivalence relations E on a standard Borel space X for which there exists a Borel map $f : X \to Y$ into a standard Borel space Y such that

$$x \mathrel{E} y \quad \text{iff} \quad f(x) = f(y).$$

By Burgess [5], if $E = E_G^X$ is the orbit equivalence relation of a Borel action of a Polish group G, then E is smooth iff there exists a Borel selector for E; i.e. a Borel map $s : X \to X$ which selects a fixed element from each E-class.

Next in complexity come those Borel equivalence relations E which are Borel bireducible with the *Vitali equivalence relation* E_0 defined on $2^{\mathbb{N}}$ by

$$x \mathrel{E_0} y \quad \text{iff} \quad x(n) = y(n) \text{ for all but finitely many } n.$$

More precisely, by Harrington–Kechris–Louveau [12], if E is a Borel equivalence relation, then E is nonsmooth iff $E_0 \leq_B E$. Furthermore, by Dougherty–Jackson–Kechris [8], if E is a countable Borel equivalence relation on a standard Borel space X, then the following are equivalent:

(1) $E \leq_B E_0$.

(2) E is *hyperfinite*; i.e. there exists an increasing sequence

$$F_0 \subseteq F_1 \subseteq \cdots \subseteq F_n \subseteq \cdots$$

of finite Borel equivalence relations on X such that $E = \bigcup_{n \in \mathbb{N}} F_n$. (Here an equivalence relation F is said to be *finite* iff every F-equivalence class is finite.)

Example 2.2. Let $\mathbb{P}$ be the set of primes. Recall that if A is a torsion-free abelian group and $0 \neq a \in A$, then the *characteristic* $\chi(a)$ of a is defined to be the sequence

$$\langle h_a(p) \mid p \in \mathbb{P} \rangle \in (\mathbb{N} \cup \{\infty\})^{\mathbb{P}},$$

where $h_a(p)$ is the p-height of a. Two characteristics $\chi_1, \chi_2 \in (\mathbb{N} \cup \{\infty\})^{\mathbb{P}}$ are said to *belong to the same type*, written $\chi_1 \equiv \chi_2$, iff the following conditions are satisfied:

- $\chi_1(p) = \chi_2(p)$ for all but finitely many primes p; and
- $\chi_1(p) = \infty$ iff $\chi_2(p) = \infty$ for all primes p.

It is easily checked that $\equiv$ is a countable Borel equivalence relation on the standard Borel space $(\mathbb{N} \cup \{\infty\})^{\mathbb{P}}$ and that $\equiv \sim_B E_0$. If $0 \neq a \in A$, then the *type* $\tau(a)$ of a is defined to be the corresponding $\equiv$-equivalence class $[\chi(a)]$. If $A \in R(\mathbb{Q})$ is a rank 1 group, then it is easily checked that $\tau(a) = \tau(b)$ for all $0 \neq a, b \in A$. Hence, in this case, we can define the *type* $\tau(A)$ of A to be $\tau(a)$, where a is any nonzero element of A. By Baer [3], if $A, B \in R(\mathbb{Q})$, then $A \cong B$ iff $\tau(A) = \tau(B)$. It follows that the isomorphism problem for the rank 1 groups is hyperfinite.

Example 2.3. Let $n \geq 1$. If $A, B \in R(\mathbb{Q}^n)$, then A and B are said to be *quasi-equal*, written $A \approx B$, iff $A \cap B$ has finite index in both A and B. It is easily checked that $\approx$ is a countable Borel equivalence relation on $R(\mathbb{Q}^n)$. (For example, see Thomas [21, Lemma 3.2].) By Thomas [21, Theorem 3.8], the quasi-equality relation $\approx$ on $R(\mathbb{Q}^n)$ is hyperfinite. Once again, this result is proved by finding a suitable complete invariant as follows. For each prime $p \in \mathbb{P}$, let $\mathbb{Z}_p$ be the ring of p-adic integers and let $\mathbb{Z}_{(p)} = \mathbb{Z}_p \cap \mathbb{Q}$. Regard $\mathbb{Q}^n$ as an additive subgroup of the n-dimensional vector space $\mathbb{Q}_p^n$ over the field $\mathbb{Q}_p$ of p-adic numbers. For each $A \in R(\mathbb{Q}^n)$, let

$$A_p = \mathbb{Z}_{(p)} \otimes A \leqslant \mathbb{Q}^n \quad \text{and} \quad \widehat{A}_p = \mathbb{Z}_p \otimes A \leqslant \mathbb{Q}_p^n.$$

By Fuchs [11, Lemma 93.3], there exists a decomposition

$$\widehat{A}_p = V_p(A) \oplus M_p(A),$$

where $V_p(A)$ is a $\mathbb{Q}_p$-subspace of $\mathbb{Q}_p^n$ and $M_p(A)$ is a free $\mathbb{Z}_p$-module. By Thomas [21, Section 4], $A, B \in R(\mathbb{Q}^n)$, then A and B are quasi-equal iff the following conditions are satisfied:

- $A_p = B_p$ for all but finitely many primes p; and
- $V_p(A) = V_p(B)$ for all primes p.

At this point, it should be fairly clear that this data can be encoded within the Vitali equivalence relation E_0 and hence that the quasi-equality relation $\approx$ is hyperfinite. For more details, see Thomas [21, Section 4].

The collection of hyperfinite countable Borel equivalence relations has (amongst others) the following basic closure properties, all of which will be needed in the proof of Theorem 1.2. (We have chosen $\mathbb{N}^+$ as the index set because we wish to reserve the symbol E_0 for the Vitali equivalence relation.)

Theorem 2.4 (Jackson–Kechris–Louveau [15]). *Let E, F and E_i for $i \in \mathbb{N}^+$ be countable Borel equivalence relations on the standard Borel spaces X, Y and X_i respectively.*

(a) *If $X = Y$, $E \subseteq F$ and F is hyperfinite, then E is also hyperfinite.*

(b) *If $E \leq_B F$ and F is hyperfinite, then E is also hyperfinite.*

(c) *If E_i is hyperfinite for each $i \in \mathbb{N}^+$, then the relation $\bigsqcup_i E_i$, defined on the standard Borel space $\bigsqcup_i X_i = \bigcup_i X_i \times \{i\}$ by*

$$(x, i) \bigsqcup_i E_i \, (y, j) \quad \text{iff} \quad i = j \text{ and } x \, E_i \, y,$$

is also hyperfinite.

(d) *If $E_1, \ldots, E_n$ are hyperfinite, then the relation $E_1 \times \cdots \times E_n$, defined on the standard Borel space $X_1 \times \cdots \times X_n$ by*

$$(x_1, \ldots, x_n) \, E_1 \times \cdots \times E_n \, (y_1, \ldots, y_n) \quad \text{iff} \quad x_i \, E_i \, y_i \text{ for all } 1 \leq i \leq n,$$

is also hyperfinite.

(e) If $X = Y$, $E \subseteq F$, E is hyperfinite and every F-equivalence class contains only finitely many E-equivalence classes, then F is also hyperfinite.

3 The proof of Theorem 1.2

In this section, we shall present the proof of Theorem 1.2. By Theorem 2.4(c), it is enough to show that for each $n \geq 1$, the isomorphism and quasi-isomorphism relations on
$$B(\mathbb{Q}^n) = \{A \in R(\mathbb{Q}^n) \mid A \text{ is a Butler group }\}$$
are both hyperfinite. Fix some $n \geq 1$ and let $\cong$ and $\sim$ denote the isomorphism and quasi-isomorphism relations on $B(\mathbb{Q}^n)$. (Recall that if A, $B \in R(\mathbb{Q}^n)$, then A, B are quasi-isomorphic iff there exists $\varphi \in GL_n(\mathbb{Q})$ such that $\varphi(A)$, $\varphi(A)$ are quasi-equal.) Then $\cong$ and $\sim$ are countable Borel equivalence relations on $B(\mathbb{Q}^n)$ and clearly $\cong \, \subseteq \, \sim$. Hence, by Theorem 2.4(a), it is enough to find a hyperfinite countable Borel equivalence relation E on $B(\mathbb{Q}^n)$ such that $\sim \, \subseteq E$.

For each $A \in R(\mathbb{Q}^n)$, let $T(A) = \{\tau(a) \mid 0 \neq a \in A\}$ be the typeset of A. Clearly if C, D are rank 1 groups, then
$$C \sim D \quad \text{iff} \quad C \cong D \quad \text{iff} \quad \tau(C) = \tau(D).$$
It follows that if A, $B \in R(\mathbb{Q}^n)$ and $A \sim B$, then $T(A) = T(B)$. Hence it is enough to show that the equivalence relation E defined on $B(\mathbb{Q}^n)$ by
$$A \, E \, B \quad \text{iff} \quad T(A) = T(B)$$
is a hyperfinite countable Borel equivalence relation.

Lemma 3.1 (Butler [6]). *For each $A \in B(\mathbb{Q}^n)$, the typeset $T(A)$ is finite.*

Proof. By Butler [6, Proposition 3], if A is a finite rank Butler group, then A is a pure subgroup of a completely decomposable torsion-free abelian group B of finite rank. It is easily checked that $T(B)$ is finite; and since A is a pure subgroup of B, it follows that $T(A) \subseteq T(B)$. □

The next result is implicitly contained in Butler [7]. The following proof is based upon the account in Arnold [2, Theorem 3.3.2]. (Recall that the set of types forms a distributive lattice under the partial ordering defined by: $\sigma \leq \tau$ iff there exist characteristics $\psi \in \sigma$, $\chi \in \tau$ such that $\psi(p) \leq \chi(p)$ for all primes $p \in \mathbb{P}$.)

Lemma 3.2. *If T is a finite set of types, then there exist at most countably many groups $A \in B(\mathbb{Q}^n)$ such that $T(A) = T$.*

Proof. For each $A \in B(\mathbb{Q}^n)$ and type $\sigma \in T(A)$, the σ-socle of A is the pure subgroup defined by
$$A(\sigma) = \{a \in A \mid \tau(a) \geq \sigma\} \cup \{0\}.$$

Let $\mathbb{Q}A(\sigma)$ be the $\mathbb{Q}$-vector subspace of $\mathbb{Q}^n$ generated by $A(\sigma)$. Clearly there are only countably many possibilities for $\mathbb{Q}A(\sigma)$. Hence the lemma is an immediate consequence of the following claim, together with the fact that the quasi-equality relation $\approx$ is a countable Borel equivalence relation on $B(\mathbb{Q}^n)$.

Claim 3.3. *Suppose that $A, B \in B(\mathbb{Q}^n)$ satisfy the following conditions:*

(i) $T(A) = T(B) = T$; *and*

(ii) $\mathbb{Q}A(\sigma) = \mathbb{Q}B(\sigma)$ *for all* $\sigma \in T$.

Then A and B are quasi-equal.

To see this, express $A = A_1 + \cdots + A_s$ as a sum of finitely many pure rank 1 subgroups. For each $1 \leq i \leq s$, let $\sigma_i = \tau(A_i) \in T$. Since $A_i \leqslant \mathbb{Q}A(\sigma_i) = \mathbb{Q}B(\sigma_i)$, there exists an integer $m_i \geq 1$ such that $m_i A_i \leqslant B(\sigma_i)$. Hence if $m = m_1 \cdots m_s$, then $mA \leqslant B$. Similarly, there exists an integer $m' \geq 1$ such that $m'B \leqslant A$ and so A, B are quasi-equal. □

Combining Lemmas 3.1 and 3.2, it follows that E is a countable Borel equivalence relation on $B(\mathbb{Q}^n)$.

Lemma 3.4. *E is hyperfinite.*

Proof. By Theorem 2.4(c), it is enough to show that $E \restriction B_k(\mathbb{Q}^n)$ is hyperfinite for each $k \geq 1$, where $B_k(\mathbb{Q}^n) = \{A \in B(\mathbb{Q}^n) \mid |T(A)| = k\}$. From now on, fix some $k \geq 1$. Let $X = (\mathbb{N} \cup \{\infty\})^{\mathbb{P}}$ be the space of characteristics and for each $\chi \in X$, let $[\chi]$ be the corresponding type. Then there exists a Borel map $f : B_k(\mathbb{Q}^n) \to X^k$,

$$A \mapsto (\chi_1(A), \ldots, \chi_k(A)),$$

such that $T(A) = \{[\chi_1(A)], \ldots, [\chi_k(A)]\}$ for each $A \in B_k(\mathbb{Q}^n)$. Clearly f is a Borel reduction from E to the countable Borel equivalence relation F on X^k defined by

$$(\chi_1, \ldots, \chi_k) \, F \, (\theta_1, \ldots, \theta_k) \quad \text{iff} \quad \{[\chi_1], \ldots, [\chi_k]\} = \{[\theta_1], \ldots, [\theta_k]\}.$$

Hence, by Theorem 2.4(b), it is enough to show that F is hyperfinite. To see this, note that Theorem 2.4(d) implies that the countable Borel equivalence relation F_0, defined on X^k by

$$(\chi_1, \ldots, \chi_k) \, F_0 \, (\theta_1, \ldots, \theta_k) \quad \text{iff} \quad ([\chi_1], \ldots, [\chi_k]) = ([\theta_1], \ldots, [\theta_k]),$$

is hyperfinite. Hence, by Theorem 2.4(e), since each F-equivalence class is the union of finitely many F_0-equivalence classes, it follows that F is also hyperfinite. □

This completes the proof of Theorem 1.2.

4 Representations of posets over fields

Unfortunately, the proof of Theorem 1.2 provides absolutely no new insights into the isomorphism and quasi-isomorphism problems for the finite rank Butler groups. To understand why this is the case, notice that the key step in the proof is Lemma 3.2, which allows us to shift our focus from the quasi-isomorphism relation to the much simpler relation of equality of typesets. As the experts in the field will immediately recognize, the gap between these two equivalence relations is precisely the classification problem for $\mathbb{Q}$-representations of finite posets. (For example, see Arnold [2, Section 3.3].) And since each finite poset has only countably many $\mathbb{Q}$-representations, the theory of Borel equivalence relations has nothing to say on this classification problem. This is particularly disappointing since, by Nazarova [19], most finite posets have wild representation type; and a number of mathematicians have asked whether the theory of Borel equivalence relations yields a nontrivial hierarchy within the class of wild representation types. In the remainder of this section, in order to partially address this question, we shall consider the complexity of the classification problems for representations of finite posets over local fields of characteristic 0; i.e. over the real numbers $\mathbb{R}$, the complex numbers $\mathbb{C}$ and the finite extensions of the fields $\mathbb{Q}_p$ of p-adic numbers.

We shall begin by recalling some basic definitions. Let F be a field and let S be a finite poset. Then $\mathrm{rep}(S, F)$ is the category with objects

$$U = (U_0, U_s \mid s \in S),$$

where U_0 is a finite dimensional F-vector space and $(U_s \mid s \in S)$ is a collection of subspaces of U_0 such that

$$\text{if } s \leq t, \text{ then } U_s \subseteq U_t,$$

together with the obvious morphisms. In particular, two representations

$$U = (U_0, U_s \mid s \in S) \quad \text{and} \quad U' = (U'_0, U'_s \mid s \in S)$$

are isomorphic iff there exists a vector space isomorphism $\varphi : U_0 \to U'_0$ such that $\varphi(U_s) = U'_0$ for all $s \in S$.

Suppose now that K is a local field of characteristic 0. Then, by restricting our attention to those representations $U = (U_0, U_s \mid s \in S)$ such that $U_0 = K^n$ for some $n \geq 1$, we can regard $\mathrm{rep}(S, K)$ as a standard Borel space.

Theorem 4.1. *Let K be a local field of characteristic 0 and let S be a finite poset. Then the classification problem for $\mathrm{rep}(S, K)$ is smooth.*

Proof. Let S be a finite poset and let K be a local field. In [4, Section 3], Beliskii–Sergeichuk present an explicit reduction of the classification problem for $\mathrm{rep}(S, K)$ to the problem of classifying pairs of square matrices up to simultaneous similarity over K (for arbitrary fields K, not just for local fields). It is easily checked that the Beliskii–Sergeichuk reduction defines a Borel map between the corresponding standard Borel spaces. Thus it is enough to show that for each $n \geq 1$, the problem of classifying pairs of $n \times n$ matrices up to simultaneous similarity over K is smooth.

To see this, fix some $n \geq 1$ and let P_n be the standard Borel space of ordered pairs (M, N) of $n \times n$ matrices. Regard P_n as a K-variety and consider the algebraic action of $GL_n(K)$ on P_n defined by

$$g \cdot (M, N) = (gMg^{-1}, gNg^{-1}).$$

Then clearly (M, N), $(M', N') \in P_n$ are simultaneously similar over K iff these pairs of matrices lie in the same $GL_n(K)$-orbit. By Zimmer [22, Theorem 3.1.3], every $GL_n(K)$-orbit on P_n is locally closed in the Hausdorff topology. Hence, by Zimmer [22, Proposition 2.1.12], the orbit equivalence relation of $GL_n(K)$ on P_n is smooth. □

By Theorem 4.1, if K is a local field of characteristic 0 and S is a finite poset, then the classification problem for rep(S, K) admits complete invariants in a suitably chosen standard Borel space. However, the proof gives no indication of how to actually compute such invariants. This problem has been solved in the case when $K = \mathbb{C}$ by Sergeichuk [20], who presented an algorithm for reducing the matrix associated with a $\mathbb{C}$-representation of a finite poset S to a canonical form. (A full system of invariants for pairs of complex matrices up to simultaneous similarity was obtained earlier by Friedland [11], making use of the basic notions of algebraic geometry.)

5 Concluding remarks

If E, F are Borel equivalence relations on the standard Borel spaces X, Y and $E \leq_B F$, then this relationship is often interpreted informally as meaning that the E-classification problem is less complicated than the F-classification problem (or that these problems are equally complicated if $E \sim_B F$). But while this interpretation is intuitively convincing in many cases, such as the classification problems for the torsion-free abelian groups of different finite ranks, it clearly breaks down for the classification problems considered in this paper. For example, let S be any finite poset such that the isomorphism relation $\cong_S$ on rep$(S, \mathbb{C})$ is wild and let $\cong_1$ be the isomorphism relation on the space $R(\mathbb{Q})$ of torsion-free abelian groups of rank 1. Then, combining Example 2.2 and Theorem 4.1, it follows that $\cong_S <_B \cong_1$. However, every abelian group theorist would agree that:

- Baer's classification of the rank 1 groups is the prototypical example of a satisfactory classification by invariants; while

- wild classification problems are much too complicated to admit a satisfactory classification by invariants.

Of course, the root of this apparent paradox lies in our use of vague terms such as "complicated" and "satisfactory". In rigorous mathematical terms, the theory of Borel equivalence relations is best understood as a study of the structure of the possible invariants for various classification problems (X, E); or, equivalently, a study of the structure of the corresponding quotient spaces X/E. And this interpretation of the relationship $\cong_S <_B \cong_1$ is entirely unproblematic:

- The quotient space $\text{rep}(S, \mathbb{C})/\cong_S$ is a standard Borel space; and there exists an algorithm for reducing the matrix associated with each $\mathbb{C}$-representation of S to a canonical form.

- The quotient space $R(\mathbb{Q})/\cong_1$ is not a standard Borel space; and it is impossible to define canonical forms within $R(\mathbb{Q})$.

Acknowledgements. I would like to thank David Arnold and Alexander Kechris for very helpful discussions concerning the material in this paper.

References

[1] S. R. Adams and A. S. Kechris, *Linear algebraic groups and countable Borel equivalence relations*, J. Amer. Math. Soc. 13 (2000), pp. 909–943.

[2] D. M. Arnold, *Abelian Groups and Representations of Finite Partially Ordered Sets*, CMS Books in Mathematics, Springer-Verlag, New York, 2000.

[3] R. Baer, *Abelian groups without elements of finite order*, Duke Math. Journal 3 (1937), pp. 68–122.

[4] G. R. Belitskii and V. V. Sergeichuk, *Complexity of matrix problems*, Linear Algebra Appl. 361 (2003), pp. 203–222.

[5] J. P. Burgess, *A selection theorem for group actions*, Pacific J. Math. 80 (1979), pp. 333–336.

[6] M. C. R. Butler, *A class of torsion-free abelian groups of finite rank*, Proc. London Math. Soc. (3) 15 (1965), pp. 680–698.

[7] M. C. R. Butler, *Torsion-free modules and diagrams of vector spaces*, Proc. London Math. Soc. (3) 18 (1968), pp. 635–652.

[8] R. Dougherty, S. Jackson and A. S. Kechris, *The structure of hyperfinite Borel equivalence relations*, Trans. Amer. Math. Soc. 341 (1994), pp. 193–225.

[9] J. Feldman and C. C. Moore, *Ergodic equivalence relations, cohomology and von Neumann algebras, I*, Trans. Amer. Math. Soc. 234 (1977), pp. 289–324.

[10] S. Friedland, *Simultaneous similarity of matrices*, Adv. Math. 50 (1983), pp. 189–265.

[11] L. Fuchs, *Infinite Abelian Groups*, Pure and Applied Mathematics Vol. 36, Academic Press, London New York, 1970.

[12] L. Harrington, A. S. Kechris and A. Louveau, *A Glimm-Effros dichotomy for Borel equivalence relations*, J. Amer. Math. Soc. 3 (1990), pp. 903–927.

[13] G. Hjorth, *Around nonclassifiability for countable torsion-free abelian groups*. Abelian groups and modules (Dublin, 1998), Trends Math., pp. 269–292. Birkhäuser, Basel, 1999.

[14] G. Hjorth and A. S. Kechris, *Borel equivalence relations and classification of countable models*, Annals of Pure and Applied Logic 82 (1996), pp. 221–272.

[15] S. Jackson, A. S. Kechris, and A. Louveau, *Countable Borel equivalence relations*, J. Math. Logic 2 (2002), pp. 1–80.

[16] A. S. Kechris, *Countable sections for locally compact group actions*, Ergodic Theory and Dynamical Systems 12 (1992), pp. 283–295.

[17] A. G. Kurosh, *Primitive torsionsfreie abelsche Gruppen vom endlichen Range*, Ann. Math. 38 (1937), pp. 175–203.

[18] A. I. Malcev, *Torsion-free abelian groups of finite rank* (Russian), Mat. Sbor. 4 (1938), pp. 45–68.

[19] L. A. Nazarova, *Partially ordered sets of infinite type*, Math. USSR Izvestija 9 (1975), pp. 911–938.

[20] V. V. Sergeichuk, *Canonical matrices for linear matrix problems*, Linear Algebra Appl. 317 (2000), pp. 53–102.

[21] S. Thomas, *The classification problem for torsion-free abelian groups of finite rank*, J. Amer. Math. Soc. 16 (2003), pp. 233–258.

[22] R. J. Zimmer, *Ergodic Theory and Semisimple Groups*, Birkhäuser, Basel, 1984.

Author information

Simon Thomas, Mathematics Department, Rutgers University, 110 Frelinghuysen Road, Piscataway, New Jersey 08854-8019, USA.
E-mail: sthomas@math.rutgers.edu

$\mathcal{L}$-groups

Theodore G. Faticoni

Abstract. The groups in this paper are abelian. As in [6] we use *rtffr* as a short way of denoting the chain of properties *reduced torsion-free finite rank*. The rtffr group G is an *Eichler* group if none of the simple ring factors of $\mathbb{Q}\mathrm{End}(G)$ is a totally definite quaternion algebra. The rtffr group G is called a $\mathcal{J}$-*group* if G is isomorphic to each subgroup of finite index in G, and G is called an $\mathcal{L}$-*group* if G is locally (= nearly) isomorphic to each subgroup of finite index in G. We show that Eichler $\mathcal{L}$-groups are $\mathcal{J}$-groups, and that $\mathrm{End}(G)$ is a domain if G is an indecomposable $\mathcal{L}$-group.

Key words. Torsion-free groups of finite rank, $\mathcal{J}$-group, $\mathcal{L}$-group, Eichler group, local isomorphism.

AMS classification. 20K15, 20K30.

1 $\mathcal{J}$-groups, $\mathcal{L}$-groups, and $\mathcal{S}$-groups

We follow [6] and write *locally isomorphic* instead of *near isomorphic* [3] or *the same genus class* [14]. Thus groups G and H are locally isomorphic iff for each integer $n \neq 0$ there are maps $f_n : G \longrightarrow H$ and $g_n : H \longrightarrow G$ and an integer $m \neq 0$ such that m and n are relatively prime, $f_n g_n = m 1_H$, and $g_n f_n = m 1_G$. The E-modules M and N are locally isomorphic iff for each integer $n \neq 0$ there are E-module maps $f_n : M \longrightarrow N$ and $g_n : N \longrightarrow M$ and an integer $m \neq 0$ such that m and n are relatively prime, $f_n g_n = m 1_N$, and $g_n f_n = m 1_M$.

The rtffr group G is an *Eichler* group if none of the simple ring factors of $\mathbb{Q}\mathrm{End}(G)$ is a totally definite quaternion algebra. The group G is called a $\mathcal{J}$-*group* if G is isomorphic to each subgroup of finite index in G, and G is called an $\mathcal{L}$-*group* if G is locally isomorphic to each subgroup of finite index in G.

Evidently $\mathcal{J}$-groups $\Rightarrow \mathcal{L}$-groups. C. Murley [11] proves that pure subgroups of the $\mathbb{Z}$-adic completion $\widehat{\mathbb{Z}}$ are $\mathcal{J}$-groups. The reader can show that $\mathbb{Z} \oplus K$ is a $\mathcal{J}$-group if K is a $\mathcal{J}$-group. $\mathcal{J}$-groups are introduced by D. M. Arnold in [4]. $\mathcal{L}$-groups are introduced in [6]. Further research on $\mathcal{J}$-groups can be found in [8] where we show that if $\mathbb{Q}\mathrm{End}(G)$ is commutative then G is a Murley group iff G is a $\mathcal{J}$-group. Also [8, Proposition III.6] states that if G is an Eichler group then G is a finitely faithful $\mathcal{S}$-group iff G is a $\mathcal{J}$-group.

Example 1.1. This is an example of a strongly indecomposable $\mathcal{J}$-group that is not a Murley group. The example is inspired by D. M. Arnold's example in [4].

Choose a prime $p \neq 2 \in \mathbb{Z}$ and let E be a classical maximal $\mathbb{Z}_p$-order in the Hamiltonian quaternions over $\mathbb{Q}$. Then each right ideal $I \subset E$ is principal, $I = xE$ for some $x \in E$.

Since $E_p/pE_p \cong \mathrm{Mat}_{2\times 2}(\mathbb{Z}/p\mathbb{Z})$, the completion $\widehat{E}_p$ satisfies $\widehat{E}_p \cong \mathrm{Mat}_{2\times 2}(\widehat{\mathbb{Z}}_p)$. Choose a pure and dense rtffr right E-submodule M such that

$$\begin{pmatrix} \mathbb{Z}_p & 0 \\ \mathbb{Z}_p & 0 \end{pmatrix} \subset M \subset \begin{pmatrix} \widehat{\mathbb{Z}}_p & 0 \\ \widehat{\mathbb{Z}}_p & 0 \end{pmatrix}$$

and using [6, Theorem 2.3.4] construct a short exact sequence

$$0 \longrightarrow M \longrightarrow G \longrightarrow \mathbb{Q}E \oplus \mathbb{Q}E \longrightarrow 0$$

of left E-modules such that $E = \mathrm{End}(G)$. Since $r_p(E) = 4 = r_p(G)^2$, G is not a Murley group.

However, M is pure and dense in G, and one proves that each subgroup of finite index in M/pM is a left E/pE-module. Then given $H \doteq G$ there is a right ideal $I \subset E$ such that $H = IG$. Since $I \cong E$, $H \cong G$, so that G is an indecomposable $\mathcal{J}$-group.

2 Eichler groups

2.1. Given a group G, the Wedderburn theorem states that $\mathbb{Q}\mathrm{End}(G) = B \oplus \mathcal{N}(\mathbb{Q}\mathrm{End}(G))$ for some semi-simple $\mathbb{Q}$-algebra B. Let

$$E(G) = \{b \,|\, b \oplus c \in \mathrm{End}(G) \text{ for some } c \in \mathcal{N}(\mathbb{Q}\mathrm{End}(G))\}.$$

Then $E(G)$ is a semi-prime subring of $\mathbb{Q}\mathrm{End}(G)$. By the Beaumont–Pierce theorem [3, Theorem 14.2], $\mathrm{End}(G) \doteq T \oplus \mathcal{N}(\mathrm{End}(G))$ for some semi-prime ring T such that $\mathbb{Q}T = B$. Thus $T \subset E(G)$ and $\mathrm{End}(G) \doteq E(G) \oplus \mathcal{N}(\mathrm{End}(G))$. One can show that $E(G) \cong \mathrm{End}(G)/\mathcal{N}(\mathrm{End}(G))$.

If G is an rtffr group then $E(G)$ is a semi-prime Noetherian ring, $E(G)$ is finitely generated by its center S, and $\mathbb{Q}E(G)$ is the semi-simple Artinian classical right ring of quotients of $E(G)$. The semi-prime ring E is *integrally closed* if given a ring $E \subset E' \subset \mathbb{Q}E$ such that E'/E is finite then $E = E'$. If E is a semi-prime rtffr ring then there is an integrally closed ring $\overline{E} \subset \mathbb{Q}E$ such that $E \subset \overline{E}$ and $\overline{E}/E$ is finite. The semi-prime rtffr ring E is integrally closed iff E is a finite product of classical maximal orders [3, 14].

Lemma 2.2. *Let G be an rtffr group. There is an rtffr group $\overline{G}$ such that $E(\overline{G})$ is an integrally closed ring, $G \subset \overline{G}$, $E(G) \subset E(\overline{G})$, $\overline{G}/G$ is finite, and $E(\overline{G})/E(G)$ is finite.*

Proof. By (2.1), $\mathrm{End}(G) \doteq E(G) \oplus \mathcal{N}(\mathrm{End}(G))$ and $E(G)$ is a homomorphic image of $\mathrm{End}(G)$. Thus there is an integer $n \neq 0$ and an integrally closed ring $\overline{E} \subset B$ such that $n\overline{E} \subset E(G) \subset \overline{E}$. We can make n large enough so that $n\overline{E} \subset \mathrm{End}(G)$. Let $\overline{G} = \overline{E}G \subset \mathbb{Q}G$. Then $G \subset \overline{G}$ and $n\overline{G} = (n\overline{E})G \subset \mathrm{End}(G)G = G$. Hence $\overline{G}/G$ is finite and $\overline{E} \subset \mathrm{End}(\overline{G})$, so as in (2.1), $E(\overline{G})/\overline{E}$ is finite. Since $\overline{E}$ is integrally closed, $\overline{E} = E(\overline{G})$. This completes the proof. □

2.3. Let $\mathbf{P}_o(G)$ denote the category of rtffr groups H such that $H \oplus H' \cong G^n$ for some group H' and some integer $n > 0$. Similarly, $\mathbf{P}_o(E(G))$ is the category of finitely generated projective right $E(G)$-modules. There is a functor, employed by D. M. Arnold [3, Corollary 9.6(a)], defined by

$$A(\cdot) : \mathbf{P}_o(G) \longrightarrow \mathbf{P}_o(E(G))$$

such that
$$A(\cdot) = \text{Hom}(G, \cdot) \otimes_{\text{End}(G)} E(G).$$

For each $W \in \mathbf{P}_o((E(G))$ there is an $H \in \mathbf{P}_o(G)$, unique up to isomorphism, such that $A(H) \cong W$. Thus $A(H) \cong A(K)$ iff $H \cong K$ for each $H, K \in \mathbf{P}_o(G)$. Observe that $E(G) \cong A(G)$ naturally.

Let i, j, k be such that $-1 = i^2 = j^2 = k^2 = ijk$ and let **k** be an algebraic number field. The **k**-algebra $D = \mathbf{k}1 \oplus \mathbf{k}i \oplus \mathbf{k}j \oplus \mathbf{k}k$ is called a *totally definite quaternion algebra* if D is a division algebra, and each embedding of **k** into the complex numbers is an embedding of **k** into the reals. If we let $\mathbf{k} = \mathbb{Q}$ then D is the $\mathbb{Q}$-algebra of Hamiltonian quaternions. Thus the Hamiltonian quaternions form a totally definite quaternion algebra.

Write
$$\mathbb{Q}E(G) = A_1 \times \cdots \times A_t$$

for some simple nonzero $\mathbb{Q}$-algebras $A_1, \ldots, A_t$. We say that G is an *Eichler group* if G is rtffr and if none of the simple factors $A_1, \ldots, A_t$ is a totally definite quaternion algebra.

Since $E(G)$ is an rtffr semi-prime ring there are classical maximal orders $\overline{E}_1, \ldots, \overline{E}_t$ in the simple Artinian $\mathbb{Q}$-algebras $A_1, \ldots, A_t$, respectively, such that

$$E(G) \subset \overline{E}_1 \times \cdots \times \overline{E}_t$$

and such that $(\overline{E}_1 \times \cdots \times \overline{E}_t)/E(G)$ is finite, [3, Theorem 9.10]. So $E(G)$ is integrally closed iff $E(G) = \overline{E}_1 \times \cdots \times \overline{E}_t$.

We leave the proof of the next lemma as an interesting exercise.

Lemma 2.4. *Let* $G \doteq G_1^{e_1} \oplus \cdots \oplus G_t^{e_t}$ *for some strongly indecomposable, pairwise non-quasi-isomorphic, rtffr groups* $G_1, \ldots, G_t$ *and some integers* $e_1, \ldots, e_t > 0$. *Then*

$$\mathbb{Q}E(G) = \mathbb{Q}E(G_1^{e_1}) \times \cdots \times \mathbb{Q}E(G_t^{e_t})$$

and
$$\mathbb{Q}E(G_i^{e_i}) = \text{Mat}_{e_i \times e_i} \mathbb{Q}E(G_i)$$

for each $i = 1, \ldots, t$. *In particular,* $\mathbb{Q}E(G_1^{e_1}), \ldots, \mathbb{Q}E(G_t^{e_t})$ *are the simple ring factors of* $\mathbb{Q}E(G)$.

Proof. See [3, Theorem 9.10]. □

Lemma 2.5. *Let* $G \doteq G_1^{e_1} \oplus \cdots \oplus G_t^{e_t}$ *for some strongly indecomposable, pairwise non-quasi-isomorphic, rtffr groups* $G_1, \ldots, G_t$ *and some integers* $e_1, \ldots, e_t > 0$. *The following are equivalent.*

(i) G is an Eichler group.

(ii) If $\mathbb{Q}E(G_i)$ is a totally definite quaternion algebra for some integer $i \in \{1, \ldots, t\}$ then $e_i > 1$.

Proof. ((ii) $\Rightarrow$ (i)) Suppose that $\mathbb{Q}E(G_j)$ is a totally definite quaternion algebra for some $j \in \mathcal{J} \subset \{1, \ldots, t\}$. Then $e_j > 1$ so that $\mathbb{Q}E(G_j^{e_j})$ is not a domain for each $j \in \mathcal{J}$. It follows that $\mathbb{Q}E(G_i^{e_i})$ is not a totally definite quaternion algebra for each $i \in \{1, \ldots, t\}$. Lemma 2.4 shows us that the rings $\mathbb{Q}E(G_i^{e_i})$ are the simple factors of the ring $\mathbb{Q}E(G)$, so that G is an Eichler group.

((i) $\Rightarrow$ (ii)) We prove the contrapositive. Suppose that $\mathbb{Q}E(G_1)$ is a totally definite quaternion algebra and that $e_1 = 1$. Then $\mathbb{Q}E(G_1)$ is a simple factor of the ring $\mathbb{Q}E(G)$, so that G is not an Eichler group. This concludes the proof. □

Compare the next result to Example 2.7. The group G constructed there is not an Eichler group since $\mathbb{Q}\mathrm{End}(G) =$ the Hamiltonian quaternions.

Theorem 2.6. *Let G be an Eichler group, suppose that $E(G)$ is an integrally closed ring, and let H be an rtffr group. If $G \oplus G \cong G \oplus H$ then $G \cong H$.*

Proof. The referee suggests that this result, without the integrally closed hypothesis, can be found as [13, Proposition 6]. We offer a different proof.

Suppose that $E(G)$ is integrally closed, and assume that $G \oplus G \cong G \oplus H$. Write $\mathbb{Q}E(G) = A_1 \times \cdots \times A_t$ where for each $i = 1, \ldots, t$, $A_i = \mathbb{Q}E(G_i^{e_i})$ is a simple ring factor of $\mathbb{Q}E(G)$. Since $E(G)$ is integrally closed, $E(G) = \overline{E}_1 \times \cdots \times \overline{E}_t$ where $\overline{E}_i$ is a classical maximal order in A_i.

Because G and H are rtffr groups, Jonsson's theorem and $G \oplus G \cong G \oplus H$ implies that $G \doteq H$, so that $E(G) = A(G) \doteq A(H)$. That is, $A(H)$ is a fractional right ideal of $E(G)$ such that

$$E(G) \oplus E(G) \cong A(G) \oplus A(G) \cong A(G) \oplus A(H) \cong E(G) \oplus A(H).$$

Let $A(H)\overline{E}_i = I_i$. We then have $\overline{E}_i \oplus \overline{E}_i \cong \overline{E}_i \oplus I_i$ for each $i = 1, \ldots, t$.

Let S_i be the center of $\overline{E}_i$ for each $i = 1, \ldots, t$. Given a right ideal J of finite index in $\overline{E}_i$ let $\mathrm{nr}(J)$ be the *reduced norm of J* in S_i. (See [14, page 214].) Then $\mathrm{nr}(\overline{E}_i) = S_i$. From $\overline{E}_i \oplus \overline{E}_i \cong \overline{E}_i \oplus I_i$ and [14, page 311, Corollary 35.11(ii)] we see that

$$\mathrm{nr}(\overline{E}_i) = \mathrm{nr}(\overline{E}_i) \cdot \mathrm{nr}(\overline{E}_i) \cong \mathrm{nr}(\overline{E}_i) \cdot \mathrm{nr}(I_i) \cong \mathrm{nr}(I_i).$$

Inasmuch as G is an Eichler group, the simple ring factor $\mathbb{Q}E(G_i^{e_i})$ of $\mathbb{Q}E(G)$ is not a totally definite quaternion algebra for any $i = 1, \ldots, t$. Then by [14, page 311, Corollary 35.11(iii)], $\overline{E}_i \cong I_i$. Consequently, $A(G) = E(G) \cong A(H)$, and so $G \cong H$ by (2.3). □

In the last line of the above theorem, we conclude from $\mathrm{nr}(E_i) = \mathrm{nr}(I_i)$ that $E_i \cong I_i$. This matter is peculiar to Eichler groups, as the next example shows. The example is a consequence of [13, Theorem 8]. We offer a proof that refers to [14].

Example 2.7. Let D denote a totally definite quaternion $\mathbb{Q}$-algebra. By [14, midpage 305] there is a maximal $\mathbb{Z}$-order $\mathcal{O} \subset D$ and a non-principal right ideal $I \subset \mathcal{O}$ such that $\mathrm{nr}(\mathcal{O}) = \mathrm{nr}(I)$. The reduced norm is taken in $S = \mathrm{center}(\mathcal{O})$. Obviously, $\mathcal{O} \not\cong I$ but $\mathrm{nr}(\mathcal{O}) \cdot \mathrm{nr}(\mathcal{O}) \cong \mathrm{nr}(\mathcal{O}) \cdot \mathrm{nr}(I)$. Then by [14, page 311, Corollary 35.11(iv)], $\mathcal{O} \oplus \mathcal{O} \cong \mathcal{O} \oplus I$.

Use Corner's theorem to construct an rtffr group G such that $\mathcal{O} = \mathrm{End}(G)$ and let $H = IG$. Repeated uses of the Arnold–Lady theorem [5] prove the following implications. Since I is projective, $\mathrm{Hom}(G, H) = I$. Since $\mathcal{O}$ is locally isomorphic to I, G is locally isomorphic to H. Since $\mathcal{O} \oplus \mathcal{O} \cong \mathcal{O} \oplus I$, $G \oplus G \cong G \oplus H$. Since $\mathcal{O}$ and I are not isomorphic, G and H are not isomorphic. Thus G is locally isomorphic to H but not isomorphic to H.

3 Direct sums of $\mathcal{L}$-groups

We introduce a bit of power.

Theorem 3.1 ([3, R. B. Warfield, Jr., Theorem 13.9]). *Let G and H be rtffr groups. Then G and H are locally isomorphic iff $G^{(e)} \cong H^{(e)}$ for some integer $e > 0$.*

Theorem 3.2 ([3, E. L. Lady, Corollary 7.17]). *Let H, K, and L be rtffr groups. If $H \oplus L$ is locally isomorphic to $K \oplus L$ then H is locally isomorphic to K.*

Theorem 3.3 ([3, D. M. Arnold, Corollary 12.9(b)]). *Let G, H, and K be rtffr groups. If G is locally isomorphic to $H \oplus K$ then $G \cong H' \oplus K'$ for some groups H' and K' which are locally isomorphic to H and K, respectively.*

Lemma 3.4. *Let G be an $\mathcal{L}$-group. Then $E(G)$ is an integrally closed ring.*

Proof. Let $E = E(G)$. By Lemma 2.2 there is a group $\overline{G} \doteq G$ such that $E(\overline{G})$ is an integrally closed ring. Because G is an $\mathcal{L}$-group G is locally isomorphic to $\overline{G}$ and then Theorem 3.1 states that $G^{(e)} \cong \overline{G}^{(e)}$ for some integer $e > 0$. Then

$$\mathrm{Mat}_{e \times e}(\mathrm{End}(G)) \cong \mathrm{End}(G^{(e)}) \cong \mathrm{End}(\overline{G}^{(e)}) \cong \mathrm{Mat}_{e \times e}(\mathrm{End}(\overline{G}))$$

so that

$$\mathrm{Mat}_{e \times e}(E(G)) \cong \mathrm{Mat}_{e \times e}(E(\overline{G}))$$

is an integrally closed ring. Since *maximal order* is a Morita invariant property, [14], $E(G)$ is integrally closed. This completes the proof. □

Lemma 3.5. *Let G be an rtffr $\mathcal{L}$-group. If G is locally isomorphic to $H \oplus K$ then H is an $\mathcal{L}$-group.*

Proof. Let G be locally isomorphic to $H \oplus K$ and suppose that $H' \doteq H$. Then G, $H \oplus K$, and $H' \oplus K$ are quasi-isomorphic. Since G is assumed to be an $\mathcal{L}$-group $H \oplus K$ and $H' \oplus K$ are locally isomorphic groups. Then Theorem 3.2 states that H is locally isomorphic to H', whence H is an $\mathcal{L}$-group. This completes the proof. □

Lemma 3.6. *Let G be an rtffr $\mathcal{L}$-group. Then $G = G_1 \oplus \cdots \oplus G_t$ for some strongly indecomposable $\mathcal{L}$-groups $G_1, \ldots, G_t$.*

Proof. By Jonsson's theorem there are strongly indecomposable rtffr groups $G'_1, \ldots, G'_t$ such that $G \doteq G'_1 \oplus \cdots \oplus G'_t$. Since G is an $\mathcal{L}$-group G is locally isomorphic to $G'_1 \oplus \cdots \oplus G'_t$. Then by an induction on the number of summands in Theorem 3.3 there are rtffr strongly indecomposable $G_1, \ldots, G_t$ such that G_i is locally isomorphic to G'_i for each $i = 1, \ldots, t$, and $G = G_1 \oplus \cdots \oplus G_t$. By Lemma 3.5 each G_i is an $\mathcal{L}$-group. This completes the proof. □

Theorem 3.7. *Let $e > 0$ be an integer, let G be an indecomposable rtffr group, and assume that G^e is an $\mathcal{L}$-group. Then $\mathrm{End}(G^e)$ is a prime ring.*

Proof. Let G be an indecomposable rtffr group, and assume that G^e is an $\mathcal{L}$-group. By Lemma 3.5, G is an $\mathcal{L}$-group, so by Lemma 3.6 we have a strongly indecomposable rtffr $\mathcal{L}$-group G. Hence $E(G)$ is a domain, so that

$$\mathrm{End}(G^e)/\mathcal{N}(\mathrm{End}(G^e)) = \mathrm{Mat}_{e \times e}(\mathrm{End}(G)/\mathcal{N}(\mathrm{End}(G))) = \mathrm{Mat}_{e \times e}(E(G))$$

is a prime ring. It then suffices to show that $\mathcal{N}(\mathrm{End}(G)) = 0$. We assume for the sake of contradiction that $\mathcal{N}(\mathrm{End}(G)) \neq 0$. The proof is a series of numbered implications and their proofs.

Let $E = \mathrm{End}(G)$, let $\mathcal{N} = \mathcal{N}(E)$, let $E(G) = E/\mathcal{N}$, and let $\mathcal{M} = \{r \in E \,|\, \mathcal{N}r = 0\}$. Then $E(G)$ is a torsion-free ring, and $\mathcal{M}$ is an ideal in E that is pure as a subgroup of E.

3.8. Since G is reduced and of finite rank the reader will prove that there is an integer $m > 0$ such that $\bigcap_{k>0} m^k G_m = 0$. Let n be some power of m.

3.9. By hypothesis G is strongly indecomposable. In particular, $\mathbb{Q}E(G)$ is a division algebra and $\mathbb{Q}\mathcal{N} \neq 0$ is the unique largest right ideal in $\mathbb{Q}E$. Specifically, because $\mathbb{Q}\mathcal{N}$ is nilpotent, $0 \neq \mathbb{Q}\mathcal{M} \subset \mathbb{Q}\mathcal{N}$. Thus $0 \neq E \cap \mathbb{Q}\mathcal{M} = \mathcal{M} \subset \mathcal{N}$.

3.10. Let $I = nE + \mathcal{N}$ and let $J = \mathrm{Hom}(G, IG)$. We will show that $J_n = xE_n$ for some regular (= nonzero divisor) element $x \in E_n$. Since I is an ideal of E, J is an ideal of E. Since G is an $\mathcal{L}$-group and since $nG \subset IG \subset G$, IG is locally isomorphic to G. By Theorem 3.1, there is an integer $f \neq 0$ such that $(IG)^f \cong G^f$, so $J^f \cong E^f$ as right E-modules. By the Arnold–Lady theorem and Theorem 3.1, J is a projective right E-module that is locally isomorphic to E as a right module. Thus, there is an integer k relatively prime to n, and there are E-module maps $f_n : E \longrightarrow J$ and $g_n : J \longrightarrow E$ such that $f_n g_n = k1_{IG}$ and $g_n f_n = k1_G$. Localizing f_n and g_n at n shows that $E_n \cong J_n$ as right E_n-modules, so that $J_n = xE_n$ for some regular $x \in E_n$.

3.11. Because G is strongly indecomposable and because $\mathcal{N} \subset I$,

$$\mathcal{N} = \mathrm{Hom}(G, \mathcal{N}G) \subset \mathrm{Hom}(G, IG) = J.$$

The element x was found in (3.10). The reader will show that because each $u \in \mathbb{Q}E \setminus \mathbb{Q}\mathcal{N}$ is a unit in $\mathbb{Q}E$, $y \in \mathcal{N}_n$ iff $xy \in \mathcal{N}_n$. By (3.10), $J_n = xE_n$ so that

$\mathcal{N}_n = x K_n$ for some right ideal $K_n \subset E_n$. Hence $\mathcal{N}_n = K_n$ so that $\mathcal{N}_n = x\mathcal{N}_n$. By (3.9), $\mathcal{M} \subset \mathcal{N}$, so $\mathcal{M}_n = x\mathcal{M}_n$ in an analogous manner.

3.12. We claim that $xE(G)_n = E(G)_n x$. By (3.10) and our choice of I there are quasi-equal ideals $nE_n \subset I_n \subset J_n = xE_n \subset E_n$, so

$$E_n \subset \mathrm{End}_{E_n}(J_n) = xE_n x^{-1}$$

are quasi-equal rings. Viewing these rings as subrings of $\mathbb{Q}E$ we have quasi-equal rings

$$E(G)_n \cong \frac{E_n + \mathbb{Q}\mathcal{N}}{\mathbb{Q}\mathcal{N}} \subset \frac{xE_n x^{-1} + \mathbb{Q}\mathcal{N}}{\mathbb{Q}\mathcal{N}} \cong xE(G)_n x^{-1}.$$

By Lemma 3.4, $E(G)$, and hence $E(G)_n$, are integrally closed rings. Thus $E(G)_n = xE(G)_n x^{-1}$ and hence $xE(G)_n = E(G)_n x$, as claimed. In other words, $xE_n + \mathcal{N}_n = E_n x + \mathcal{N}_n$.

3.13. We claim that $\mathrm{Hom}_{E_n}(E_n, \mathcal{M}_n) = \mathrm{Hom}_{E_n}(J_n, \mathcal{M}_n)$. The inclusion

$$\mathrm{Hom}_{E_n}(E_n, \mathcal{M}_n) \subset \mathrm{Hom}_{E_n}(J_n, \mathcal{M}_n)$$

is the restriction map. Conversely, let $f \in \mathrm{Hom}_{E_n}(J_n, \mathcal{M}_n)$. By (3.10), $J_n = xE_n$, and by (3.11), $\mathcal{M}_n = x\mathcal{M}_n$, so $f \in \mathrm{Hom}_E(xE, x\mathcal{M}_n) = x\mathcal{M}_n x^{-1} \subset \mathbb{Q}\mathcal{M}$. By (3.9) and because $E_n \doteq J_n$ we have

$$f(\mathbb{Q}\mathcal{N}_n) \subset \mathbb{Q}\mathcal{N}_n \cdot f(\mathbb{Q}E_n) \subset \mathbb{Q}\mathcal{N}_n \cdot \mathbb{Q}\mathcal{M}_n = 0.$$

Hence $f(\mathcal{N}_n) = 0$. Then by (3.10) and (3.12)

$$\begin{aligned}
\mathrm{Hom}_{E_n}(J_n, \mathcal{M}_n) &= \mathrm{Hom}_{E_n}(J_n/\mathcal{N}_n, \mathcal{M}_n) \\
&\stackrel{(3.10)}{=} \mathrm{Hom}_{E_n}((xE_n + \mathcal{N}_n)/\mathcal{N}_n, \mathcal{M}_n) \\
&\stackrel{(3.12)}{=} \mathrm{Hom}_{E_n}(xE(G)_n, \mathcal{M}_n) \\
&= \mathrm{Hom}_{E_n}(E(G)_n x, \mathcal{M}_n).
\end{aligned}$$

Thus there is a $u \in \mathcal{M}_n$ such that $f(rx) = ru$ for each $r \in E(G)_n$. By (3.11), $\mathcal{M}_n = x\mathcal{M}_n$, so the product $x^{-1}u$ exists in $\mathcal{M}_n$. The map f then lifts to a map $\bar{f} \in \mathrm{Hom}_{E_n}(E_n, \mathcal{M}_n)$ such that $\bar{f}(r) = r(x^{-1}u)$ for each $r \in E_n$. As claimed, $\mathrm{Hom}_{E_n}(E_n, \mathcal{M}_n) = \mathrm{Hom}_{E_n}(J_n, \mathcal{M}_n)$.

3.14. By (3.10) and (3.11), $xE_n = J_n \subset E_n$ and $\mathcal{M} = x\mathcal{M}$. Thus

$$\begin{aligned}
\mathcal{M}_n = \mathrm{Hom}_{E_n}(E_n, \mathcal{M}_n) &\stackrel{(3.13)}{=} \mathrm{Hom}_{E_n}(J_n, \mathcal{M}_n) \\
&= \mathrm{Hom}_{E_n}(xE_n, x\mathcal{M}_n) = x\mathcal{M}_n x^{-1}.
\end{aligned}$$

Then $\mathcal{M}_n x = x\mathcal{M}_n = \mathcal{M}_n$.

3.15. Finally,
$$\mathcal{M}_n J_n G \stackrel{(3.10)}{=} \mathcal{M}_n(xG) \stackrel{(3.14)}{=} \mathcal{M}_n G$$
and $\mathcal{M}_n$ is nilpotent, (3.9), so $\mathcal{M}_n G$ is superfluous in $J_n G$. Then
$$J_n G = I_n G = nE_n G + \mathcal{M}_n G = nE_n G = nG_n.$$
Hence $\mathcal{M}_n G \subset nG_n$.

By (3.8), $n = m^k$ for some integer $k > 0$, so that $X_n = X_m$ for any torsion-free group X. Then by (3.15), $\mathcal{M}_m G_m = \mathcal{M}_n G_n \subset nG_n = m^k G_m$. Furthermore, $\bigcap_{k>0} m^k G_m = 0$. Thus $\mathcal{M} G \subset \mathcal{M}_m G_m = 0$, whence $\mathcal{M} = 0$. This contradiction to (3.9) shows us that $\mathcal{N} = 0$, i.e. E is a semi-prime ring, which completes the proof. □

Theorem 3.16. *Let G be an indecomposable rtffr $\mathcal{L}$-group. Then $\text{End}(G)$ is a classical maximal order in the division ring $\mathbb{Q}\text{End}(G)$.*

Proof. Since G is indecomposable, Lemma 3.6 implies that G is strongly indecomposable. Then $\mathbb{Q}\text{End}(G)$ is a local ring, and so $\mathbb{Q}E(G)$ is a division algebra. Since G is an indecomposable rtffr $\mathcal{L}$-group, Theorem 3.7 states that $\mathcal{N}(\text{End}(G)) = 0$. Then $E(G) = \text{End}(G)/\mathcal{N}(\text{End}(G)) = \text{End}(G)$, and by Lemma 3.4, $E(G)$ is integrally closed. Hence $\text{End}(G)$ is a classical maximal order in the division algebra $\mathbb{Q}\text{End}(G)$. This completes the proof. □

4 Eichler $\mathcal{L}$-groups are $\mathcal{J}$-groups

Evidently, a $\mathcal{J}$-group, is an $\mathcal{L}$-group, is an $\mathcal{S}$-group. If $\text{End}(G)$ is semi-prime then Lemma 4.1 shows us that an $\mathcal{L}$-group is a finitely faithful $\mathcal{S}$-group.

Lemma 4.1. *Let G be an rtffr $\mathcal{L}$-group. If $\text{End}(G)$ is semi-prime then G is a finitely faithful E-flat group.*

Proof. Suppose that $\text{End}(G)$ is semi-prime and let I be a right ideal of $\text{End}(G)$. By Lemma 3.4, $E(G) = \text{End}(G)$ is integrally closed, hence an hereditary ring, whence I is a projective (= flat) right $\text{End}(G)$-module. An application of $I \otimes_{\text{End}(G)} \cdot$ to the exact sequence
$$0 \to G \longrightarrow \mathbb{Q}G \longrightarrow \mathbb{Q}G/G \to 0$$
produces the exact sequence
$$0 = \text{Tor}^1_{\text{End}(G)}(I, \mathbb{Q}G/G) \longrightarrow I \otimes_{\text{End}(G)} G \longrightarrow I \otimes_{\text{End}(G)} \mathbb{Q}G$$
of groups. Consequently $I \otimes_{\text{End}(G)} G \longrightarrow I \otimes_{\text{End}(G)} \mathbb{Q}G : r \otimes x \longmapsto r \otimes x$ is an injection. Moreover, $\mathbb{Q}\text{End}(G)$ is a semi-simple Artinian algebra, so that $\mathbb{Q}G$ is a projective left $\mathbb{Q}\text{End}(G)$-module. Since $\mathbb{Q}\text{End}(G)$ is a flat left $\text{End}(G)$-module $\mathbb{Q}G$ is a flat left $\text{End}(G)$-module. Thus the canonical map
$$I \otimes_{\text{End}(G)} \mathbb{Q}G \longrightarrow \mathbb{Q}G : r \otimes x \longmapsto rx$$

is an injection, so that the composite map $I \otimes_{\text{End}(G)} G \to G : r \otimes x \mapsto rx$ is an injection. Hence G is a flat left $\text{End}(G)$-module.

Next suppose that I is a maximal right ideal of finite index in $\text{End}(G)$ such that $IG = G$. Since I is a finitely generated projective right $\text{End}_R(G)$-module the Arnold–Lady theorem [5] implies that $I \cong \text{End}(G)$, say $I = x\text{End}(G)$ for some $x \in \text{End}(G)$. Then
$$G = IG = x\text{End}(G)G = xG.$$
Since G has finite rank x is an automorphism of G. That is $I = \text{End}(G)$ so that G is finitely faithful. This completes the proof. □

The next three results examine the implications $\mathcal{S}$-group $\Rightarrow \mathcal{L}$-group $\Rightarrow \mathcal{J}$-group.

Theorem 4.2. *Let $e > 0$ be an integer, let G be an indecomposable group, and let G^e be an Eichler group. The following are equivalent.*

(i) G^e *is a $\mathcal{J}$-group.*

(ii) G^e *is an $\mathcal{L}$-group.*

(iii) G^e *is a finitely faithful $\mathcal{S}$-group.*

Proof. $\mathcal{J}$-group $\Rightarrow \mathcal{L}$-group $\Rightarrow \mathcal{S}$-group. Furthermore, the $\mathcal{L}$-group G^e has semi-prime endomorphism ring by Theorem 3.7. Thus ((i) $\Rightarrow$ (ii) $\Rightarrow$ (iii)) follows from Lemma 4.1.

((iii) $\Rightarrow$ (i)) Since G^e is an Eichler group, this implication follows from [8, Proposition III.6]. This completes the proof. □

Theorem 4.3. *Eichler $\mathcal{L}$-groups are $\mathcal{J}$-groups.*

Proof. Let G be an rtffr $\mathcal{L}$-group and let $H \doteq G$. By Jónsson's theorem $G \doteq G' = G_1^{e_1} \oplus \cdots \oplus G_t^{e_t}$ for some nonzero, pairwise non-quasi-isomorphic, strongly indecomposable groups $G_1, \ldots, G_t$ and some integers $e_1, \ldots, e_t > 0$. Since G is an $\mathcal{L}$-group, G' is an $\mathcal{L}$-group, so that H is locally isomorphic to G'. By Arnold's theorem 3.3, $H = H_1 \oplus \cdots \oplus H_t$ where $H_1, \ldots, H_t$ are rtffr groups such that $G_i^{e_i}$ and H_i are locally isomorphic for $i = 1, \ldots, t$. By Lemma 3.5, $G_i^{e_i}$ is an $\mathcal{L}$-group, and since $e_i > 1$ if $\mathbb{Q}E(G_i)$ is a totally definite quaternion algebra (Lemma 2.5), we see that $\mathbb{Q}E(G_i^{e_i})$ is not a totally definite quaternion algebra. Then $G_i^{e_i}$ is an Eichler $\mathcal{L}$-group and G_i is indecomposable. An appeal to Theorem 4.2 shows us that $G_i^{e_i}$ is a $\mathcal{J}$-group. Hence $G_i^{e_i} \cong H_i$ for each $i = 1, \ldots, t$, so that $G' \cong H$. Therefore G' is a $\mathcal{J}$-group. But then $G \cong G'$ is a $\mathcal{J}$-group. This completes the proof. □

Corollary 4.4. *An Eichler group G is a $\mathcal{J}$-group iff it is an $\mathcal{L}$-group.*

Theorem 4.5. *The following are equivalent for the rtffr group G.*

(i) $\text{End}(G)$ *is semi-prime and G is an $\mathcal{L}$-group.*

(ii) G *is a finitely faithful $\mathcal{S}$-group.*

Proof. Assume part (i): G is an $\mathcal{L}$-group with semi-prime endomorphism ring. By Lemma 4.1, G is finitely faithful. $\mathcal{L}$-groups are $\mathcal{S}$-groups, so we have proved part (ii).

Assume part (ii). Let G be a finitely faithful $\mathcal{S}$-group. By [4, Theorem III], $\text{Ext}^1_{\mathbb{Z}}(G,G)$ is a torsion-free group, hence $\text{Ext}^1_{\mathbb{Z}}(G^2,G^2)$ is a torsion-free group, whence G^2 is a finitely faithful $\mathcal{S}$-group. By Lemma 2.5, G^2 is an Eichler group, so by [8, Proposition III.6], G^2 is a $\mathcal{J}$-group. Then by Lemma 3.5, G is an $\mathcal{L}$-group.

Because G is a finitely faithful $\mathcal{S}$-group, [4, Theorem III] states that $\text{End}(G)/p\text{End}(G)$ is a simple Artinian algebra for each prime $p \in \mathbb{Z}$. Thus $\mathcal{N}(\text{End}(G)) \subset p\text{End}(G)$ for each prime $p \in \mathbb{Z}$. Consequently, $\mathcal{N}(\text{End}(G)) = p\mathcal{N}(\text{End}(G))$ for each prime $p \in \mathbb{Z}$. Since G is reduced, $\mathcal{N}(\text{End}(G)) = 0$, which completes the proof. □

Example 4.6. Here is an example of an $\mathcal{L}$-group with non-zero nilradical. Let H be an indecomposable $\mathcal{J}$-group as in Example 1.1. Then $\text{Hom}(H, \mathbb{Z}) = 0$ so that $G = \mathbb{Z} \oplus H$ is a $\mathcal{J}$-group with non-zero nilradical $\text{Hom}(\mathbb{Z}, H)$.

References

[1] U. Albrecht, *Baer's lemma and Fuchs' problem 84a*, Trans. Am. Math. Soc. **293**, (1986), 565–582.

[2] D. M. Arnold, Abelian Groups and Representations of Finite Partially Ordered Sets, Canadian Mathematical Society: Books in Mathematics, Springer, New York, (2000).

[3] D. M. Arnold, Finite Rank Abelian Groups and Rings, Lecture Notes in Mathematics **931**, Springer-Verlag, New York, (1982).

[4] D. M. Arnold, *Endomorphism rings and subgroups of finite rank torsion-free abelian groups*, Rocky Mt. J. Math. **12**, No. 2, (1982), 241–256.

[5] D. M. Arnold and E. L. Lady, *Endomorphism rings and direct sums of torsion-free abelian groups*, Trans. Am. Math. Soc. **211**, (1975), 225–237.

[6] T. G. Faticoni, Direct Sum Decompositions of Torsion-Free Finite Rank Groups, Taylor and Francis Group, Chapman and Hall/CRC, Boca Raton, New York, August, (2006).

[7] T. G. Faticoni and H. P. Goeters, *Examples of torsion-free groups flat as modules over their endomorphism rings*, Comm. Algebra **19**, (1), (1991), 1–28.

[8] T. G. Faticoni and H. P. Goeters, *On torsion-free Ext*, Comm. Algebra **16**, (9), (1988), 1853–1876.

[9] L. Fuchs, Infinite Abelian Groups I, II, Academic Press, New York-London, (1969, 1970).

[10] E. L. Lady, *Nearly isomorphic torsion-free abelian groups*, J. Algebra **35**, (1975), 235–238.

[11] C. Murley, *The classification of certain classes of torsion-free abelian groups*, Pac. J. Math. **40**, (1972), 647–665.

[12] K. C. O'Meara and C. Vinsonhaler, *Separative cancellation and multiple isomorphism in torsion-free abelian groups*, J. Algebra **221**, (1999), 536–550.

[13] K. C. O'Meara and C. Vinsonhaler, *Generalizations of Isomorphism in Torsion-free Abelian Groups*, Contemporary Mathematics, 273, (2001).

[14] I. Reiner, Maximal Orders, Academic Press Inc, New York, (1975).

[15] J. Seltzer, *A cancellation criterion for finite rank torsion-free abelian groups*, Proc. Amer. Math. Soc. **94**, (1985), 363–368.

Author information

Theodore G. Faticoni, Department of Mathematics Fordham University, Bronx, New York 10458, U.S.A.
E-mail: `faticoni@fordham.edu`

Co-local subgroups of nilpotent groups of class 2

Joshua Buckner and Manfred Dugas

Abstract. The notion of a cellular cover of a group M was introduced in [6] to be a pair (G, c) where G is a group, $c : G \to M$ is a homomorphism such that for any homomorphism $\psi : G \to M$ there is a unique homomorphism $\varphi : G \to G$ such that $\psi = c \circ \varphi$. Of particular interest are the subgroups K of the form $K = \ker(c)$. In the case of abelian groups G these are the co-local subgroups as investigated in [2] and [4]. The purpose of this paper is to answer in the negative a question posed in [6]. There are cellular covers of nilpotent groups (of class 2) whose kernels are not torsion-free, in contrast to the case of abelian groups, where such kernels are (co)torsion-free.

Key words. Kernels of cellular covers, co-local subgroups, endomorphisms of nilpotent groups.

AMS classification. 20F18, 2K21, 20K20.

1 Introduction

In [6], the notion of a cellular cover of a group M was introduced. The pair (G, c) is a cellular cover of the group M, if G is a group and $c : G \to M$ is a homomorphism such that for each homomorphism $\psi : G \to M$ there is a *unique* homomorphism $\varphi : G \to G$ such that $\psi = c \circ \varphi$. It was shown in [6] that one may restrict attention to the case where c is a surjective homomorphism. In that case, (G, c) is a cellular cover of $M \approx G/K$, $K = \ker(c)$, if and only if K is a co-local subgroup of G. Here we call a normal subgroup K of G a co-local subgroup of the group G, if for all homomorphisms $\psi : G \to G/K$ there is a unique endomorphism $\varphi : G \to G$ such that $\psi(x) = \varphi(x)K$ for all $x \in G$. We say that ψ is induced by φ. Co-local subgroups of abelian groups have been introduced in [2] and further studied in [4]. This notion was extended to the quasi-category of abelian groups in [3]. It was shown in [4] that co-local subgroups K of abelian groups are cotorsion-free, i.e., if $\widehat{\mathbb{Z}}$ denotes the $\mathbb{Z}$-adic completion of the additive group $\mathbb{Z}$ of integers, then $\mathrm{Hom}(\widehat{\mathbb{Z}}, K) = 0$. Note that cotorsion-free abelian groups are torsion-free. It was shown in [6] that any co-local subgroup of a group G is contained in the center of G. Moreover, co-local subgroups K of nilpotent groups G are investigated in [6] and it was shown that if K is not torsion-free, then G has to satisfy certain restrictions. This made it natural to pose the following question in [6, 1.C. Questions (3)]: Is the kernel of any cellular map $c : G \to M$, where M is nilpotent, a torsion-free group?

We rephrase this question: Are all co-local subgroups of nilpotent groups torsion-free?

We will modify a construction in [5] to prove

Theorem 1.1. *Let p be an odd prime integer and let K be an abelian group such that K is 2-torsion-free, p-reduced and p-adically complete. Then there exists a nilpotent group G of class 2 such that K is a co-local subgroup of G. Moreover, K is contained in the center of G.*

This shows that co-local subgroups of nilpotent groups need not be cotorsion-free (or torsion-free) at all. Indeed, all bounded p-groups, p-adic completions of separable p-groups or reduced p-local abelian groups, are all co-local subgroups of nilpotent groups of class 2. This is a dramatic difference between co-local subgroups of abelian and nilpotent groups!

2 The key result

Let p be a prime integer and $Q^{(p)} = \{\frac{z}{p^n} : z \in \mathbb{Z}, n \in \mathbb{N}\}$, a subring of $\mathbb{Q}$. Note that as an abelian group, $Q^{(p)}$ is generated by $a_n = \frac{1}{p^n}$ and relations $pa_{n+1} = a_n$ for all $n \in \mathbb{N}$. The following technical result is the key to the entire paper.

Lemma 2.1. *Let D be an abelian group and K a subgroup of D such that K is p-reduced and p-adically complete. Then the short exact sequence of abelian groups $0 \to K \to D \xrightarrow{\pi} D/K \to 0$ induces*

$$0 = \mathrm{Hom}(Q^{(p)}, K) \to \mathrm{Hom}(Q^{(p)}, D) \xrightarrow{\pi^*} \mathrm{Hom}(Q^{(p)}, D/K) \to 0,$$

i.e., each $\varphi \in \mathrm{Hom}(Q^{(p)}, D/K)$ is induced by a unique $\psi \in \mathrm{Hom}(Q^{(p)}, D)$.

Proof. Let $\psi \in \mathrm{Hom}(Q^{(p)}, D/K)$. There exist $y_n \in D$ such that $\psi(a_n) = y_n + K$ and $p\psi(a_{n+1}) - \psi(a_n) = 0$. It follows that $py_{n+1} = y_n + k_n$ for some $k_n \in K$. Consider $\ell_n = \sum_{\alpha=n}^{\infty} p^{\alpha-n} k_\alpha$, and note that $\ell_n \in K$ since K is p-adically complete. Set $b_n = y_n + \ell_n$ and compute

$$pb_{n+1} - b_n = py_{n+1} + p \sum_{\alpha=n+1}^{\infty} p^{\alpha-n-1} k_\alpha - y_n - \sum_{\alpha=n}^{\infty} p^{\alpha-n} k_\alpha$$

$$= py_{n+1} - y_n + \sum_{\alpha=n+1}^{\infty} p^{\alpha-n} k_\alpha - \sum_{\alpha=n}^{\infty} p^{\alpha-n} k_\alpha$$

$$= k_n - k_n = 0.$$

This shows that setting $\varphi(a_n) = b_n$ defines $\varphi \in \mathrm{Hom}(Q^{(p)}, D)$ such that $\psi(a_n) = y_n + K = b_n + K = \varphi(a_n) + K$. Thus π^* is surjective, and we are done. □

Now we can prove an essential

Theorem 2.2. *Let A be a $Q^{(p)}$-module, K a p-reduced, p-adically complete subgroup of the abelian group D. Then $\mathrm{Hom}(A, D)$ is naturally isomorphic to $\mathrm{Hom}(A, D/K)$.*

Proof. Let $0 \to L \to F \xrightarrow{\pi} A \to 0$ be a free resolution of the $Q^{(p)}$-module A, i.e., F is a free $Q^{(p)}$-module. Let $\psi \in \text{Hom}(A, D/K)$. By Lemma 2.1, there is some $\varphi' \in \text{Hom}(F, D)$ such that $(\psi \circ \pi)(x) = \varphi'(x) + K$ for all $x \in F$. Let $x \in L$. Then $0 = (\psi \circ \pi)(x) = \varphi'(x) + K$, and it follows that $\varphi' \restriction_L : L \to K$. Now K is p-reduced and L is a $Q^{(p)}$-submodule of F, i.e., L is p-divisible, and we infer $\varphi' \restriction_L = 0$.

Define $\varphi(a) = \varphi'(x)$ for $a = \pi(x)$. It follows that

$$\psi(a) = (\psi \circ \pi)(x) = \varphi'(x) + K = \varphi(a) + K$$

and $\varphi : A \to D$. □

3 Nilpotent groups of class two

In this section, we recall a construction of nilpotent groups from abelian groups that goes back to [1].

Notation. When talking about a "group X", we will always mean an additive abelian group X. If X is a (possibly non-abelian) multiplicative group, we write $X^\bullet$, to indicate that the binary operation on X is a multiplication.

Let G be a 2-torsion-free, non-trivial abelian group. A map $f : G \times G \to G$ is called an alternating bilinear map if $f(g, \_) : G \to G$ is a homomorphism for each $g \in G$ and $f(x, y) = -f(y, x)$ for all $x, y \in G$. It follows that $f(x, x) = 0$ for all $x \in G$.

Let $\text{Ker}(f) = \{g \in G : f(g, \_) = 0\}$. We will assume that $\text{Ker}(f) = f(G, G) = \{f(x, y) : x, y \in G\}$. If $\text{Ker}(f) = f(G, G) = G$ then $f = 0$ and $G = \{0\}$, a contradiction. This shows that f is not surjective, i.e., $G \neq f(G, G)$.

On the set G, define a multiplication by $xy = x + y + f(x, y)$ for all $x, y \in G$. It is easy to see that this multiplication is associative, $g^n = ng$, and $g^{-1} = -g$ for all $g \in G$. This shows that G, endowed with this multiplication, is a group which we denote by $G^\bullet$.

Let $Z(G^\bullet)$ denote that center of $G^\bullet$. Then $a \in Z(G^\bullet)$ if and only if $xa = ax$ for all $x \in G^\bullet$. This holds if and only if $x + a + f(x, a) = xa = ax = a + x + f(a, x) = a + x - f(x, a)$. Thus $a \in Z(G^\bullet)$ if and only if $2f(x, a) = 0$ for all $x \in G^\bullet$, which is true if and only if $f(x, a) = 0$, since G is 2-torsion-free. This shows that $Z(G^\bullet) = \text{Ker}(f) = f(G, G) \neq G$, i.e., $G^\bullet$ is not abelian.

Let $[x, y] = x^{-1} y^{-1} xy$ be the commutator of $x, y \in G^\bullet$. Then

$$[x, y] = x^{-1} y^{-1} xy = x^{-1} y^{-1} + xy + f(x^{-1} y^{-1}, xy)$$

$$= x^{-1} + y^{-1} + f(x^{-1}, y^{-1}) + x + y + f(x, y) + f(x^{-1} y^{-1}, xy)$$

$$= -x - y + f(-x, -y) + x + y + f(x, y) + f(x^{-1} y^{-1}, xy)$$

$$= f(x, y) + f(x, y) + f(-x - y + f(x, y), x + y + f(x, y))$$

$$= 2f(x, y) - f(x + y - f(x, y), x + y + f(x, y))$$

$$= 2f(x, y).$$

Now let $K^\bullet$ be a subgroup of $G^\bullet$ such that $K^\bullet \subseteq Z(G^\bullet)$, and let K be the set of elements of $K^\bullet$. Let $x, y \in K$. Then $x+y = xy - f(x,y) = xy$ since $f(Z(G^\bullet), Z(G^\bullet)) = 0$. We infer that K is a subgroup of G such that $K \subseteq f(G,G)$. Moreover, for $g \in G$, we have $gK^\bullet = \{gk : k \in K\} = \{g + k + f(g,k) : k \in K\} = \{g + k : k \in K\} = g + K$, i.e., as sets, the cosets of $K^\bullet$ in $G^\bullet$ are the same as the cosets of the subgroup K of G. Since $K^\bullet \subseteq Z(G^\bullet)$, $K^\bullet$ is a normal subgroup of $G^\bullet$ and we have, for $g_1, g_2 \in G$,

$$(g_1 K^\bullet)(g_2 K^\bullet) = \{g_1 g_2 k : k \in K\}$$
$$= \{g_1 + g_2 k + f(g_1, g_2 k) : k \in K\}$$
$$= \{g_1 + g_2 + k + f(g_1, g_2 + k + f(g_2, k)) : k \in K\}$$
$$= \{g_1 + g_2 + f(g_1, g_2) + k : k \in K\}.$$

Note that $f(g_1 + K, g_2 + K) = \{f(g_1 + k_1, g_2 + k_2) : k_1, k_2 \in K\} = \{f(g_1, g_2)\}$. This shows that the map $f_K : G/K \times G/K \to G/K$ with $f_K(x + K, y + K) = f(x,y) + K$ is well-defined, and it is easy to see that f_K is a bilinear map. Moreover, from

$$(g_1 + K)(g_2 + K) = (g_1 K^\bullet)(g_2 K^\bullet)$$
$$= g_1 + g_2 + f(g_1, g_2) + K$$
$$= (g_1 + K) + (g_2 + K) + f(g_1 + K, g_2 + K),$$

we infer that $G^\bullet/K^\bullet$, similar to $G^\bullet$, is defined on the abelian group G/K by the bilinear map f_K.

For future reference, we state

Lemma 3.1. *Let G be a nontrivial 2-torsion-free abelian group and $f : G \times G \to G$ an alternating bilinear map such that $\mathrm{Ker}(f) = f(G,G)$. Define the group $G^\bullet$ as above. Let $K^\bullet \subseteq Z(G^\bullet)$ be a subgroup of $G^\bullet$ with set of elements K. Thus $K \subseteq f(G,G)$ is a subgroup of G and $G^\bullet/K^\bullet$ is a group with G/K as the set of elements with $(x + K)(y + K) = (x + K) + (y + K) + f_K(x + K, y + K)$ where f_K is defined as above. Assume that G/K is 2-torsion-free as well. Then, as sets,*

$$\mathrm{Hom}(G^\bullet, G^\bullet/K^\bullet) = \{\varphi \in \mathrm{Hom}(G, G/K) : f(x,y)\varphi = f_K(x\varphi, y\varphi) \text{ for all } x, y \in G\}$$

Proof. Let $\varphi : G^\bullet \to G^\bullet/K^\bullet$ be a homomorphism and $x, y \in G$. As we have seen $[x,y] = 2f(x,y)$ and it follows in the same fashion that $[x\varphi, y\varphi] = 2f_K(x\varphi, y\varphi)$. Then $2(f(x,y)\varphi) = (2f(x,y))\varphi = [x,y]\varphi = [x\varphi, y\varphi] = 2f_K(x\varphi, y\varphi)$ and $f(x,y)\varphi = f_K(x\varphi, y\varphi)$ follows.

Now

$$(x+y)\varphi = (xy - f(x,y))\varphi = (xyf(x,y)^{-1})\varphi$$
$$= (x\varphi)(y\varphi)(f(x,y)^{-1}\varphi)$$
$$= [(x\varphi) + (y\varphi) + f_K(x\varphi, y\varphi)][f(x,y)^{-1}\varphi]$$
$$= x\varphi + y\varphi + f_K(x\varphi, y\varphi) - f(x,y)\varphi$$
$$+ f_K(x\varphi + y\varphi + f_K(x\varphi, y\varphi), -f(x,y)\varphi)$$

$$= x\varphi + y\varphi + f_K(x\varphi + y\varphi + f_K(x\varphi, y\varphi), -f(x,y)\varphi)$$
$$= x\varphi + y\varphi,$$

since $f(x,y)\varphi = f_K(x\varphi, y\varphi) \in \text{Ker}(f_K)$.

Therefore $\varphi \in \text{Hom}(G, G/K)$. □

4 Abelian groups with bilinear maps

Here we recall the construction in [5] and make a minor modification.

Let κ, λ be cardinals such that $\kappa < \lambda$. We identify a cardinal λ with the set of ordinals $\upsilon < \lambda$. Define:
- F_1 is the free $Q^{(p)}$-module with basis $\{e_\upsilon : \upsilon < \lambda\}$.
- L is the free $Q^{(p)}$-module with basis $\{e_\upsilon : \upsilon < \kappa\}$.
- F_1^- is the free $Q^{(p)}$-module with basis $\{e_\upsilon : \kappa \leq \upsilon < \lambda\}$.
- F_0 is the free $Q^{(p)}$-module with basis $\{d_{\upsilon\mu} : \upsilon < \mu < \lambda, \upsilon \geq \kappa \text{ or } \mu \geq \kappa\}$.

Now we define a bilinear map $f : F \times F \to F$ where $F = F_1 \oplus F_0$ as follows:

$$f(e_\alpha, e_\beta) = \begin{cases} 0 & \text{if } \alpha, \beta < \kappa, \\ d_{\alpha\beta} & \text{if } \alpha < \beta \text{ and } \alpha \geq \kappa \text{ or } \beta \geq \kappa, \\ -d_{\beta\alpha} & \text{if } \alpha > \beta \text{ and } \alpha \geq \kappa \text{ or } \beta \geq \kappa, \\ 0 & \text{if } \alpha = \beta, \end{cases}$$

and $\text{Ker}(f) = F_0 = f(F, F)$.

Virtually the same construction as in [5] yields a cotorsion-free abelian group H such that $F \subseteq H \subseteq \widehat{F}$ where $\widehat{F}$ is the q-adic completion of F for some fixed prime $q \neq p, 2$ and

$$\boxed{\text{End}(H) = Q^{(p)} \oplus f_H},$$

where $f_H = \{f(g, \_) : g \in H\}$ and f is (extended to) an alternating bilinear map on H. Actually, f extends to a bilinear map $\widehat{f} : \widehat{F} \times \widehat{F} \to \widehat{F}$ such that $\widehat{F_0} \subseteq \text{Ker}(\widehat{f})$ and $\widehat{f}(\widehat{L}, \widehat{L}) = \{0\}$. We will identify $\widehat{f}$ and f. Note that, by construction, $H \cap \widehat{F_0} = f(H, H)$ and is thus pure in H. The reader might want to redo this construction using the "Strong Black Box" as presented in [8].

5 The main construction

Let K be a p-reduced and p-adically complete abelian group and D a 2-torsion-free divisible group containing K. Since $p > 2$, and K is p-adically complete, we have that K is 2-divisible and thus 2-pure in D. We infer that D/K is 2-torsion-free.

Let H be the group constructed in the previous section and set $G = H \oplus D$. Let L_0 be a free $Q^{(p)}$-module with basis $\{d_{\upsilon\mu} : \upsilon < \mu < \kappa\}$ and define an alternating bilinear map $g' : L \times L \to L_0$ by $g'(e_\nu, e_\mu) = d_{\nu\mu}$ if $\nu < \mu < \kappa$ and $g'(e_\nu, e_\mu) = -d_{\mu\nu}$ if $\mu < \nu < \kappa$. Choose the cardinal κ big enough such that there is an epimorphism $\sigma : L_0 \to D \to 0$. This map extends to some epimorphism from $\widehat{L}$ onto D, which we denote by σ again.

Now define $g : \widehat{L} \times \widehat{L} \to D$ by $g(x, y) = \sigma(g'(x, y))$ for all $x, y \in \widehat{L}$. This is a bilinear map, which we extend to $G \times G$ by $\widehat{F_1^- \oplus F_0} \oplus D \subseteq \text{Ker}(g)$. Note that $f(D \times G) = \{0\} = f(D \times G)$. Recall that $f(\widehat{L}, \widehat{L}) = \{0\}$ and set $\gamma = g + f : G \times G \to G$, another bilinear map. Note that $\gamma(x + x', y + y') = \gamma(x, y)$ whenever $x', y' \in \text{Ker}(\gamma)$. We will use this fact frequently without reference. Note that $\gamma(G \times G) = f(H, H) \oplus g(H, H) = f(H, H) \oplus D = \text{Ker}(\gamma) = \text{Ker}(f) \oplus D$.

Let $G^\bullet$ be the nilpotent group defined by G and γ, i.e., $xy = x + y + \gamma(x, y)$ for all $x, y \in G^\bullet$. We will show that $G^\bullet$ is our desired group.

To this end, observe that $G/K = H \oplus (D/K)$ and

$\text{Hom}(G, G/K) =$

$$\left\{ \begin{bmatrix} s + f_h & \beta \\ 0 & \delta \end{bmatrix} : s \in Q^{(p)}, h \in H, \beta \in \text{Hom}(H, D/K), \delta \in \text{Hom}(D, D/K) \right\},$$

where the matrix operates from the right and $f_h = f(h, \cdot)$.

By Lemma 2, $\text{Hom}(G^\bullet, G^\bullet/K^\bullet)$ is a subset of that set. Let

$$\varphi = \begin{bmatrix} s + f_h & \beta \\ 0 & \delta \end{bmatrix} \in \text{Hom}(G, G/K)$$

and assume that $\varphi : G^\bullet \to G^\bullet/K^\bullet$ is also a homomorphism of the multiplicative groups.

Let $\kappa \leq \nu < \mu < \lambda$. Then $\gamma(e_\nu, e_\mu)\varphi = d_{\nu\mu}\varphi = [d_{\nu\mu}s + f(h, d_{\nu\mu})] + d_{\nu\mu}\beta = d_{\nu\mu}s + d_{\nu\mu}\beta \in H \oplus D/K = G/K$. On the other hand, this equals

$$\begin{aligned} \gamma_K(e_\nu\varphi, e_\mu\varphi) &= \gamma(e_\nu\varphi, e_\mu\varphi) + K \\ &= \gamma(e_\nu s + f_h(h, e_\nu) + e_\nu\beta, e_\mu s + f_h(h, e_\mu) + e_\mu\beta) + K \\ &= f(e_\nu s, e_\mu s) + K = f(e_\nu, e_\mu)s^2 + K = d_{\nu\mu}s^2 + K \end{aligned}$$

and we infer that $s = s^2$ and $\beta \upharpoonright_{F_0} = 0$. (Note that $\gamma(d + K, d' + K) = \gamma(d, d') + K$ is well defined.) Since $s \in Q^{(p)}$ we have $s \in \{0, 1\}$.

Now repeat this computation for any $(x, y) \in H \times H$:

$$\gamma(x, y)\varphi = [f(x, y) + g(x, y)]\varphi = f(x, y)s + [f(x, y)\beta + g(x, y)\delta]$$

and

$$\gamma(x\varphi, y\varphi) + K = \gamma(xs + x', ys + y') + K = \gamma(xs, ys) + K = f(x, y)s^2 + g(x, y)s^2 + K,$$

where x', y' are some elements from $\text{Ker}(\gamma)$.

Then $f(x,y)\beta + g(x,y)\delta = g(x,y)s^2 + K$. Let $d \in D$. Then there are elements $a, b \in L$ such that $d = g(a,b)$ and $f(a,b) = 0$. This shows $d\delta = ds + K$ and $s \in \{0,1\}$ as seen above. We infer that $\delta = s\pi$ where $\pi : D \to D/K$ is the natural map and s is 0 or 1. Moreover, $f(x,y)\beta \in K$. Thus $\beta \restriction_{f(H,H)} = 0$, i.e. β induces a map $\beta' : H/f(H,H) \to D/K$.

This shows that either $\delta = 0$ or $\delta = \pi$ and, as a set,

$$\mathrm{Hom}(G^\bullet, G^\bullet/K^\bullet) =$$

$$\left\{ \begin{bmatrix} \varepsilon + f_h & \beta \\ 0 & \varepsilon\pi \end{bmatrix} : h \in H, \varepsilon \in \{0,1\}, \beta \in \mathrm{Hom}(H, D/K), \beta \restriction_{f(H,H)} = 0 \right\}.$$

Note that such a map β induces $\beta' : H/f(H,H) \to D/K$ and $f(H,H) = H \cap \widehat{F_0}$, which shows that $H/f(H,H)$ is torsion-free and p-divisible, i.e., a $Q^{(p)}$-module. By Theorem 2, β' is induced by some $\beta'' : H/f(H,H) \to D$, which defines a map $\beta''' : H \to D$ such that $x\beta = x\beta''' + K$ for all $x \in H$. This shows that $K^\bullet$ is a co-local subgroup of $G^\bullet$. For the special case of $K^\bullet$ the trivial subgroup, we have

$$\mathrm{End}(G^\bullet) = \left\{ \begin{bmatrix} \varepsilon + f_h & \beta \\ 0 & \varepsilon \end{bmatrix} : h \in H, \varepsilon \in \{0,1\}, \beta \in \mathrm{Hom}(H,D), \beta \restriction_{f(H,H)} = 0 \right\}.$$

In conclusion, we summarize what we have shown in the following:

Theorem 5.1. *Let $G^\bullet = (H \oplus D)^\bullet$ be the nilpotent class 2 group just constructed and p the odd prime used in the construction. If K is any subgroup of the 2-torsion-free, divisible group D such that K is complete (and Hausdorff) in the p-adic topology on K, then $K^\bullet$ is a co-local subgroup of $G^\bullet$ and thus the canonical map $\pi : G^\bullet \to G^\bullet/K^\bullet$ is a cellular cover.*

References

[1] R. Baer, *Groups with Abelian central quotient groups*, Trans. Amer. Math. Soc. 44 (1938), pp. 357–386.

[2] J. Buckner and M. Dugas, *Co-local subgroups of abelian groups*, in: *Abelian groups, rings, modules, and homological algebra*, pp. 29–37, Lect. Notes Pure and Appl. Math., 249, Chapman&Hall/CRC, Boca Raton, FL, 2006.

[3] J. Buckner and M. Dugas, *Quasi-co-local subgroups of abelian groups*, J. Pure Appl. Algebra 221 (2007), pp. 392–399.

[4] M. Dugas, *Co-local subgroups of abelian groups II*, J. Pure Appl. Algebra 208 (2007), pp. 117–126.

[5] M. Dugas and R. Göbel, *Automorphisms of torsion-free nilpotent groups of class two*, Trans. Amer. Math. Soc. 332(2) (1992), pp. 633–646.

[6] E. D. Farjoun, R. Göbel and Y. Segev, *Cellular covers of groups*, J. Pure Appl. Algebra 208 (2007), pp. 61–76.

[7] L. Fuchs, *Infinite Abelian Groups*, vol. I, 1970; vol. II, 1973, Academic Press, New York and London.

[8] R. Göbel and S. Wallutis, *An algebraic version of the black box*, Algebra Discrete Math. 3 (2003), pp. 7–45.

Author information

Joshua Buckner, Department of Mathematics, Clayton State University, Atlanta, GA 30260, USA.
E-mail: jbuckner@clayton.edu

Manfred Dugas, Department of Mathematics, Baylor University, Waco, TX 76798, USA.
E-mail: Manfred_Dugas@baylor.edu

Divisible envelopes and cotorsion pairs over integral domains

Luigi Salce

Abstract. We give a proof, based on the analysis of the divisible envelopes of flat modules, of the fact that an integral domain R is almost perfect if and only if all divisible R-modules are weak-injective, if and only if the class of modules of projective dimension ≤ 1 is closed under direct limits. We derive that the cotorsion pairs generated by the finitely presented modules and by the modules of weak dimension ≤ 1 over the integral domain R coincide if and only if R is a Dedekind domain. We also prove that, if R is an almost maximal Prüfer domain, then every R-module has a divisible envelope if and only if R is a Dedekind domain.

Key words. Divisible modules, divisible envelopes, weak-injective modules, FP-injective modules, almost perfect domains, almost maximal Prüfer domains.

AMS classification. Primary: 13A05. Secondary: 13C11, 13F05.

1 Introduction

It is well known that the class $D(R)$ of the divisible modules over an integral domain R cogenerates the cotorsion pair $(CS(R), D(R))$, where $CS(R)$ is the class of all direct summands of *cp*-filtered modules, that is, of modules admitting a filtration of cyclically presented modules (see [9, p. 136]); these modules are called semi-Baer modules in [10]. The cotorsion pair $(CS(R), D(R))$ is complete but, in general, not perfect. This means, in particular, that every R-module M has a special divisible preenvelope, i.e., there exists an exact sequence

$$0 \to M \to D \to C \to 0$$

with D divisible and $C \in CS(R)$, but no divisible envelope of M exists in general for an arbitrary M.

Thus the following problem naturally arises:

Problem 1. Characterize the integral domains R such that the class $D(R)$ of the divisible R-modules is an enveloping class, that is, such that every R-module has a divisible envelope.

Notice that Dedekind domains trivially satisfy this condition, as over these domains divisible modules are injective, and injective envelopes always exist, by a well-known result due to Eckmann and Schopf [6].

This research was supported by MIUR, PRIN 2005.

A necessary condition in order that $D(R)$ is an enveloping class was found by Trlifaj [14] (see also [9, Corollary 6.3.16]), who proved that R, or any free R-module F, has a divisible envelope if and only if R is a Matlis domain, that is, the field of quotients Q of R has projective dimension at most 1; actually, in this case the embedding of F in its injective envelope is a divisible envelope of F. Recall that, if R is a Matlis domain, then divisible modules are h-divisible, namely, quotients of injective modules; thus one can easily deduce (see [8, VII.2.5]) that the cotorsion pair $(CS(R), D(R))$ coincides with the cotorsion pair $(P_1(R), P_1(R)^\perp)$, where $P_1(R)$ is the class of modules of projective dimension ≤ 1 and $P_1(R)^\perp$ its right Ext-orthogonal. Very recently, as an application of their investigation in [2] of when the cotorsion pairs generated by modules of bounded projective dimension are of finite type, Bazzoni and Herbera proved the remarkable result that these two cotorsion pairs are the same for any integral domain R, that is, semi-Baer modules are exactly the modules of projective dimension ≤ 1.

A cotorsion pair, which is in general bigger than $(P_1(R), P_1(R)^\perp)$ (in the ordering relation of the cotorsion pairs adopted in [9], which is the opposite of the original ordering defined in [13]), is $(F_1(R), F_1(R)^\perp)$, where $F_1(R)$ is the class of modules of weak (i.e., flat) dimension ≤ 1; the modules in its right Ext-orthogonal class $F_1(R)^\perp$ are called *weak-injective* by S. B. Lee, who investigated them recently in [12] and [11]; thus, from now on, we will denote the class $F_1(R)^\perp$ by $WI(R)$. It was proved by Trlifaj in [14] (see also [9, Theorem 4.1.3]) that the cotorsion pair $(F_1(R), WI(R))$ is perfect. Hence weak injective envelopes always exist (see also [12, Theorem 4.1]).

Therefore, if one characterizes the class of integral domains R such that the two classes of divisible module, $D(R)$, and of weak-injective modules, $WI(R)$, coincide, one gets another class of integral domains satisfying the condition of Problem 1. Recall that Angeleri Hügel and Trlifaj proved in [1] that, given an integral domain R, the closure under direct limits of the class $P_1(R)$ is the class $F_1(R)$. This fact generalizes to dimension 1 (but to integral domains only) the well-known theorem, due to Lazard and Govorov (see [8, VI.9.7]), that a module is flat if and only if it is a direct limit of projective modules.

Thus we are led to the following second problem.

Problem 2. Characterize the integral domains R such that $P_1(R)$ coincides with $F_1(R)$, or, equivalently, such that the class $P_1(R)$ is closed under direct limits.

Note that Dedekind domains trivially satisfy also this condition, as all modules over these domains have projective dimension ≤ 1. Furthermore, some conditions equivalent to $P_1(R) = F_1(R)$ are given in [12, Lemma 3.6], but all of module-theoretical type (see Lemma 2.3 in the next Section 2).

The solution to Problem 2 was given very recently by Fuchs and Lee in [7, Theorem 6.4], where they proved that an integral domain that is not a field has global weak-injective dimension 1 if and only if it is almost perfect. These domains have been introduced by Bazzoni and the author [3] in their investigation of the existence of strongly flat covers. They are those integral domains R such that every proper quotient R/I ($0 \neq I \leq R$) is a perfect ring. In [3] and [4] many characterizations of these domains have been obtained (see Theorem 2.4 in the next Section 2).

The first goal of this note is to provide in Section 2 an alternative proof of Theorem 6.4 in [7], which is based on the analysis of the divisible envelopes of flat modules. As an application of this result and of Theorem 3.2 in [12], which states that all weak-injective modules over an integral domain R are FP-injective (i.e., absolutely pure, see [8, IX.3]) if and only if R is a Prüfer domain, we deduce that the two cotorsion pairs $(F_1(R), WI(R))$ and $(FS(R), FI(R))$ coincide if and only if R is a Dedekind domain (recall that $(FS(R), FI(R))$ is the complete cotorsion pair generated by the finitely presented modules, see [9, 4.1]).

In connection to Problem 1, in Section 3 we improve Trlifaj's result [14, Proposition 4.8], by showing that an integral domain R is Matlis if and only if, for every strongly flat R-module M, the embedding into its injective envelope $E(M)$ is a divisible envelope. We also prove that every countably generated ideal of an almost maximal Matlis valuation domain, which is obviously flat, has no divisible envelope. From this result we derive that, for an almost maximal Prüfer domain R, $D(R)$ is an enveloping class if and only if R is a Dedekind domain.

Our general references are the monographs [8] and [9].

2 Domains whose divisible modules are weak-injective

We start by recalling the notion of envelope (see [9, 2.1.1]). Given a class $C(R)$ of modules over a ring R, closed under isomorphisms and direct summands, and given a module M, a map $f\colon M \to C \in C(R)$ is a $C(R)$-preenvelope of M if, for each $C' \in C(R)$ the induced map $\operatorname{Hom}_R(C, C') \to \operatorname{Hom}_R(M, C')$ is surjective. The $C(R)$-preenvelope f is a $C(R)$-envelope provided that every endomorphism g of C satisfying $f = gf$ is an automorphism. The rings considered in this note are always (commutative) integral domains.

We are interested in $D(R)$-preenvelopes and $D(R)$-envelopes (called also divisible preenvelopes and envelopes, respectively), where $D(R)$ is the class of divisible modules over an integral domain R, that is, of the modules M satisfying the equality $M = rM$ for every $0 \neq r \in R$.

Lemma 2.1. *Let M be a module and $i\colon M \to E$ the embedding in its injective envelope. If E is a divisible preenvelope of M, then it is a divisible envelope of it.*

Proof. Every map $g\colon E \to E$ which acts as the identity on iM must be obviously an automorphism. □

Recall that a module M is cotorsion (in Enochs' sense) if $\operatorname{Ext}^1_R(F, M) = 0$ for all flat modules F, while it is weakly cotorsion (in Matlis' sense) if $\operatorname{Ext}^1_R(Q, M) = 0$, where Q denotes the field of quotients of R. As Q is flat, cotorsion modules are weakly cotorsion. Recall also that the Matlis category equivalence, holding over any integral domain R, sends the h-divisible torsion R-module T to the torsionfree complete R-module $\operatorname{Hom}_R(K, T)$, where $K = Q/R$, and the torsionfree complete R-module M to the h-divisible torsion R-module $M \otimes_R K$ (see [8, VIII.2]).

We will need the following results proved by S. B. Lee and by Bazzoni–Salce.

Lemma 2.2 ([12, Lemma 3.6]). *For an integral domain R the following are equivalent:*

(a) $F_1(R) = P_1(R)$;

(b) *all h-divisible modules are weak-injective;*

(c) *all divisible modules are weak-injective;*

(d) *epic images of weak-injective modules are weak-injective.* □

Note that, in view of the result in [1] quoted in the Introduction, each one of the conditions in Lemma 2.2 is equivalent to saying that the class $P_1(R)$ is closed under direct limits.

Theorem 2.3 ([3, Theorem 4.5]). *For an integral domain R the following are equivalent:*

(i) R *is almost perfect;*

(ii) *every flat module is strongly flat;*

(iii) *every module has a strongly flat cover;*

(iv) R *is h-local and torsion modules are semiartinian;*

(v) *the class of strongly flat modules is closed under direct limits;*

(vi) *every weakly cotorsion module is cotorsion.* □

We can prove now the main result of this Section, which shows that the conditions in Lemma 2.2 are equivalent to the conditions in Theorem 2.3 and to some further conditions. The equivalence of (1), (2) and (3) in the next theorem has been already proved in [7]; it is worthwhile to remark that our proof of (1) implies (2), due to Silvana Bazzoni (many thanks!), is more direct than the proof in [7] which needs a result in [11] on the correspondence in the Matlis equivalence of cotorsion torsionfree modules and weak-injective torsion modules.

Theorem 2.4. *For an integral domain R the following are equivalent:*

(1) R *is almost perfect;*

(2) *all divisible modules are weak-injective;*

(3) *the two cotorsion pairs $(P_1(R), D(R))$ and $(F_1(R), WI(R))$ coincide;*

(4) *the embedding $i\colon F \to E$ of every flat module F in its injective envelope E is a divisible envelope of F;*

(5) *for every flat module F, p.d. $E/iF \leq 1$, where $i\colon F \to E$ is the embedding of F in its injective envelope.*

Proof. (1) ⇒ (2). We will prove that $F_1(R) = P_1(R)$. Let M be a module of weak dimension ≤ 1. By [8, VII.2.5], it is enough to prove that $\mathrm{Ext}^1_R(M, D) = 0$ for every h-divisible module D. Consider the exact sequence:

$$0 \to \mathrm{Hom}_R(Q/R, D) \to \mathrm{Hom}_R(Q, D) \to D \to \mathrm{Ext}^1_R(Q/R, D) = 0$$

where $\mathrm{Ext}^1_R(Q/R, D) = 0$ because $Q/R \in P_1(R)$ and D is h-divisible, $W = \mathrm{Hom}_R(Q/R, D)$ is weakly cotorsion by [8, VIII.6.5], and $E = \mathrm{Hom}_R(Q, D)$ is torsionfree divisible, hence injective. We deduce the long exact sequence

$$0 = \mathrm{Ext}^1_R(M, E) \to \mathrm{Ext}^1_R(M, D) \to \mathrm{Ext}^2_R(M, W) \to 0,$$

hence $\mathrm{Ext}^1_R(M, D) \cong \mathrm{Ext}^2_R(M, W)$. Let us consider now a free presentation of M:

$$0 \to F \to \bigoplus R \to M \to 0$$

where F is flat, as w.d. $M \leq 1$. The hypothesis that R is almost perfect ensures that F is strongly flat, hence $\mathrm{Ext}^1_R(F, C) = 0$ for every weakly cotorsion module C. In particular, $\mathrm{Ext}^1_R(F, W) = 0$. Thus the above free presentation induces the exact sequence:

$$0 = \mathrm{Ext}^1_R(F, W) \to \mathrm{Ext}^2_R(M, W) \to \mathrm{Ext}^2_R(\bigoplus R, W) = 0$$

hence $0 = \mathrm{Ext}^2_R(M, W) \cong \mathrm{Ext}^1_R(M, D)$, as desired.

(2) ⇒ (3) is obvious.

(3) ⇒ (4). First we show the $i \colon F \to E$ is a divisible preenvelope of F. Let D be an arbitrary divisible module. Then we have the induced exact sequence

$$\mathrm{Hom}_R(E, D) \to \mathrm{Hom}_R(F, D) \to \mathrm{Ext}^1_R(E/iF, D)$$

where the Ext^1 vanishes, since $E \cong \bigoplus Q$ is flat, so clearly w.d. $E/iF \leq 1$, i.e., $E/F \in F_1(R)$, which is the left Ext-orthogonal of $D(R)$ by assumption. Thus the map between the two Hom's is epic, as desired. The proof follows now from Lemma 2.1.

(4) ⇒ (5). Divisible envelopes are special divisible preenvelopes (see [9, p. 98]), therefore $E/iF \in P_1(R)$, hence p.d. $E/iF \leq 1$.

(5) ⇒ (1). We will prove that every flat module F is strongly flat, thus Theorem 2.3 will give the conclusion. Let C be a weakly cotorsion module and consider the exact sequence induced by the embedding $i \colon F \to E$:

$$\mathrm{Ext}^1_R(E, C) \to \mathrm{Ext}^1_R(F, C) \to \mathrm{Ext}^2_R(E/iF, C).$$

The first Ext^1 vanishes, since $E \cong \bigoplus Q$; Ext^2 also vanishes, as p.d. $E/iF \leq 1$ by the hypothesis. Therefore $\mathrm{Ext}^1_R(F, C) = 0$; this proves that F is strongly flat. □

From Theorem 2.4 we derive the following consequences.

Corollary 2.5. *If all torsionfree weakly cotorsion modules over an integral domain R are cotorsion, then all weakly cotorsion modules are such.*

Proof. The result in [11] quoted before Theorem 2.5 states that a complete torsionfree module is cotorsion if and only if its Matlis equivalent is weak-injective. Under our assumption, all torsionfree weakly cotorsion modules are cotorsion, hence all h-divisible modules are weak-injective. Then Lemma 2.2 shows that all divisible modules are weak-injective. Consequently R is almost perfect by Theorem 2.4. Therefore weakly cotorsion modules are cotorsion by Theorem 2.3. □

Note that Corollary 2.5 appears as Lemma 6.3 in [7], where it is needed to prove the main Theorem 6.4.

Corollary 2.6. *Let R be an almost perfect domain. Then a subprojective flat module is projective.*

Proof. If $0 \to F \to P \to P/F \to 0$ is an exact sequence with F flat and P projective, then w.d. $P/F \leq 1$, hence p.d. $P/F \leq 1$ by Theorem 2.4. Therefore F is projective. □

Corollary 2.6 should be compared with Proposition 2.5 in [5], where R is assumed to be only a Matlis domain, but F is strongly flat and P/F is torsion.

Recall that a module M is said to be *FP*-injective if $\text{Ext}^1_R(F, M) = 0$ for all finitely presented modules F. Megibben characterized these modules by the property of being pure in their injective envelope (see [8, IX.3.1]). The cotorsion pair generated by the finitely presented modules is denoted by $(FS(R), FI(R))$, and the modules in $FS(R)$ are called *fp*-filtered modules (see [9, 4.1.4]). S. B. Lee proved [12, Theorem 3.2] that all weak-injective modules over an integral domain R are *FP*-injective if and only if R is a Prüfer domain. This result, together with Theorem 2.4, easily gives the following

Corollary 2.7. *Given an integral domain R, the two cotorsion pairs $(F_1(R), WI(R))$ and $(FS(R), FI(R))$ coincide if and only if R is a Dedekind domain.*

Proof. The sufficiency is clear, since for Dedekind domains both weak-injective and *FP*-injective modules, being divisible, are injective.

Conversely, assume that $(F_1(R), WI(R)) = (FS(R), FI(R))$. Since $WI(R) \leq FI(R)$, Lee's result quoted above gives that R is Prüfer. But then $F_1(R) = \text{Mod}(R)$, so $FI(R)$ is the class of the injective modules, which is therefore closed under direct sums; consequently R is Noetherian, hence it is a Dedekind domain. □

3 Divisible envelopes over Prüfer domains

In the next result, which improves Proposition 4.8 in [14], we consider the embedding of a module M in its injective envelope E as an inclusion.

Proposition 3.1. *For an integral domain R the following are equivalent:*

(1) R is a Matlis domain;

(2) the injective envelope E of a strongly flat module M is the divisible envelope of M.

Proof. Assume R Matlis. Trlifaj has shown [14, Proposition 4.8] that the injective envelope of a free module is its divisible envelope. If M is a strongly flat module, then there exists an exact sequence

$$0 \to F \to M \oplus N \to X \to 0$$

where N is a suitable module, F is free and X is torsionfree divisible. The injective envelope E of M is divisible torsionfree, hence injective of projective dimension 1. If D is a divisible module, in view of Lemma 2.1 it is enough to prove that in the long exact sequence

$$\mathrm{Hom}_R(E, D) \to \mathrm{Hom}_R(M, D) \to \mathrm{Ext}^1_R(E/M, D) \to \mathrm{Ext}^1_R(E, D)$$

the restriction map between the two Hom's is epic. Since $\mathrm{Ext}^1_R(E, D) = 0$, by [8, VII.2.5], we must prove that $\mathrm{Ext}^1_R(E/M, D) = 0$. Consider the exact sequence

$$0 \to E' \to Y \to X \to 0$$

obtained through the pushout derived from the exact sequence $0 \to F \to M \oplus N \to X \to 0$ with respect to the embedding of $i \colon F \to E'$ of F into its injective envelope E'. If $\varepsilon \colon M \oplus N \to Y$ is the induced embedding, we have that $Y/\mathrm{Im}(\varepsilon) \cong E'/F$. From Trlifaj's result we have that p.d. $E'/F \leq 1$, hence $\mathrm{Ext}^1_R(E'/F, D) = 0$, so $\mathrm{Ext}^1_R(Y/\mathrm{Im}(\varepsilon), D) = 0$. But it is easy to check that $\varepsilon \colon M \oplus N \to Y$ is the embedding of $M \oplus N$ into its injective envelope, hence E/M is a summand of $Y/\mathrm{Im}(\varepsilon)$, therefore $\mathrm{Ext}^1_R(E/M, D) = 0$. The converse follows by [14, Proposition 4.8]. □

Proposition 3.1 cannot be extended from strongly flat modules to flat modules, as Theorem 2.4 shows. The following example considers the case of non-principal ideals of valuation domains.

Example 3.2. Let R be a non-Noetherian valuation domain. Let J be a non-principal ideal of R; J is flat but not projective. The field of quotients Q of R is the injective envelope of J, but Q is not a special divisible preenvelope of J, hence it is not a divisible envelope. In fact, p.d. $Q/J \geq 2$ by [8, VI. Exercise 3.3], hence Q/J does not belong to $P_1(R) = {}^\perp D(R)$. Recall that a divisible preenvelope D of J can be obtained as follows: If $0 \to H \to \bigoplus_\alpha x_\alpha R \to J \to 0$ is a free presentation of J, where H is a submodule of $\bigoplus_\alpha x_\alpha R$, then $D = \bigoplus_\alpha x_\alpha Q/H$. □

The preceding example gives a hint in the search for modules over valuation domains which do not admit divisible envelope. We need the following technical result.

Lemma 3.3. *Let R be a Matlis valuation domain and J a countably infinitely generated ideal of R. If $f \colon J \to X = \bigoplus_{1 \leq i \leq n} U_i$ is a non-zero map, where $U_i \cong Q/R$ for each i, then p.d. $X/f(J) = 2$.*

Proof. By induction on n. For $n = 1$, $X/f(J) \cong Q/J$, so the claim follows from [8, VI. Exercise 3.3]. Let us assume that $n > 1$. Since $f(J)$ is a uniserial submodule of $\bigoplus_{1\leq i\leq n} U_i$, by [8, XII.2.1] we can assume that $f(J) \cap U_n$ is pure in $f(J)$. By [8, XII.2.3] this implies that either $f(J) \cap U_n = 0$, or $f(J) \cap U_n = f(J)$. In the first case we have an induced non-zero map $g: J \to Y = \bigoplus_{1\leq i\leq n-1} U_i$, so by the inductive hypothesis p.d. $Y/g(J) = 2$; but $Y/g(J) \cong X/(f(J) \oplus U_n)$, and we can consider the exact sequence

$$0 \to U_n \cong (f(J) \oplus U_n)/f(J) \to X/f(J) \to X/(f(J) \oplus U_n) \to 0.$$

Assume, by way of contradiction, that p.d. $X/f(J) = 1$. By [8, VII.3.5] $X/f(J)$ is isomorphic to a finite direct sum of copies of Q/R, so, by [8, XII.2.3], the above exact sequence splits; but this is absurd, since p.d. $X/(f(J) \oplus U_n) = 2$. In the latter case that $f(J) \cap U_n = f(J)$, i.e., if $f(J) \leq U_n$, we deduce that

$$X/f(J) = (\bigoplus_{1\leq i\leq n-1} U_i) \oplus (U_n/f(J)),$$

therefore the conclusion follows from the case $n = 1$. □

We can now prove that, for a countably infinitely generated ideal J of an almost maximal Matlis valuation domain, not only Q is not its divisible envelope, as Example 3.2 shows, but that a divisible envelope of J does not exist at all.

Proposition 3.4. *A countably infinitely generated ideal J of an almost maximal Matlis valuation domain R has no divisible envelope.*

Proof. First note that, if $\varepsilon: J \to D$ is a special divisible preenvelope of J, then p.d. $D = 1$, since p.d. $J = 1$ and p.d. $D/\varepsilon J = 1$. Thus, if $\varepsilon: J \to D$ is a divisible envelope of J, $D = (\bigoplus_\alpha Q) \oplus (\bigoplus_\beta Q/R)$ (see [8, VII.3.5]), where α, β are cardinal numbers. First we prove that $\alpha = 1$. As $\varepsilon J \cap \bigoplus_\beta Q/R = 0$, εJ embeds into $\bigoplus_\alpha Q$ and its image is contained in a direct summand $Q' \cong Q$ of it. There follows that $\varepsilon J \leq Q' \oplus (\bigoplus_\beta Q/R)$, which is a summand of D. Thus the hypothesis that D is an envelope ensures that $D = Q' \oplus (\bigoplus_\beta Q/R)$. We claim now that $\beta \geq \aleph_0$. Let us assume, by way of contradiction, that $\beta = n$ is finite; note that Example 3.2 shows that $n \geq 1$, so that the projection of εJ into $\bigoplus_\beta Q/R$, call it V, is not zero. By Lemma 3.3, p.d.$(\bigoplus_\beta Q/R)/V = 2$. Note that

$$(\bigoplus_\beta Q/R)/V \cong (Q' \oplus (\bigoplus_\beta Q/R))/(\varepsilon J + Q').$$

On the other hand, consider the exact sequence

$$0 \to (\varepsilon J + Q')/\varepsilon J \to (Q' \oplus (\bigoplus_\beta Q/R))/\varepsilon J \to (Q' \oplus (\bigoplus_\beta Q/R))/(\varepsilon J + Q') \to 0.$$

The middle term has projective dimension 1, since D is an envelope of εJ, hence from [8, VII.3.5] we deduce that it is isomorphic to a finite direct sum of copies of Q/R. There follows that the exact sequence splits, by [8, XII.2.3], since it is pure, being $(\varepsilon J + Q')/\varepsilon J$ divisible. Looking at the projective dimensions, this gives the desired contradiction. Thus, being J countably infinitely generated, we can assume that $D = Q \oplus (\bigoplus_{n<\omega} U_n)$ (with $U_n \cong Q/R$ for all n) is a divisible envelope of J containing it, thus all the projections of εJ on the U_n's are non-zero.

We claim that there exists a map $g \colon D \to D$ which is not monic whose restriction to εJ is the identity map of εJ. Let εJ be generated by the sequence of elements $x_1, x_2, \ldots, x_k, \ldots$ subjected to the relations $r_{k+1} x_{k+1} = x_k$ for all k ($r_k \in R$). Let $x_1 = q_1 + v_1$, where $q_1 \in Q$, $v_1 \in W = \bigoplus_{i \in \Lambda} U_i$ and Λ is a finite non-empty subset of ω. For every $k > 1$ let $x_k = q_k + v_k + w_k$, where $q_k \in Q$, $v_k \in W$ and $w_k \in \bigoplus_{i \notin \Lambda} U_i$.

Let I be the submodule of Q generated by the elements q_k ($k \geq 1$); the map $\varphi \colon I \to Q \oplus W$ which sends each q_k to $q_k + v_k$ is well defined; since R is almost maximal, $Q \oplus W$ is injective, hence the map φ extends to a map $\psi \colon Q \to Q \oplus W$. Then define the map $g \colon D \to D$ as follows: $g(W) = 0$, $g_{|Q} = \psi$ and $g_{|U_i} = id_{U_i}$ for every $i \notin \Lambda$. Then obviously g is not monic and a straightforward computation shows that $g_{|\varepsilon J} = id_{\varepsilon J}$. Thus we have proved that D cannot be a divisible envelope, as claimed. □

We can now derive the main result of this section.

Theorem 3.5. *Let R be an almost maximal Prüfer domain. Then the class of divisible R-modules is an enveloping class if and only if R is a Dedekind domain.*

Proof. The sufficiency is obvious. Conversely, R must be a Matlis domain by [14, Prop. 4.8]. Let P be a maximal ideal of R. We claim that R_P is a DVR. If not, it contains a countably infinitely generated ideal J, which is of the form $J = I_P$ for a countably infinitely generated ideal I of R. Let $i \colon I \to D$ be a divisible envelope of I in $\mathrm{Mod}(R)$. Tensoring by R_P the exact sequence $0 \to I \xrightarrow{i} D \to D/iI \to 0$ we get the exact sequence
$$0 \to J \xrightarrow{\varepsilon} D_P \to (D/iI) \otimes R_P \to 0.$$
It is easily seen that $\varepsilon \colon J \to D_P$ is a divisible preenvelope in $\mathrm{Mod}(R_P)$. R_P is a Matlis almost maximal valuation domain, hence Proposition 3.4 ensures that there exists a map $g \colon D_P \to D_P$ such that $g \varepsilon = \varepsilon$, and which fails to be an automorphism of D_P. The fact that R is a Matlis domain ensures that $D = T \oplus D'$, where T is torsion and D' torsionfree, and the fact that R is almost maximal Prüfer ensures that $T = \bigoplus_{P'} T_{P'}$, where P' ranges over the maximal spectrum $\mathrm{Max}(R)$. Then obviously $D_P = T_P \oplus D'$, thus D_P is a direct summand of D. Therefore the map
$$h = g \oplus id_{\bigoplus_{P' \neq P} T_{P'}} \colon D \to D$$
is an endomorphism of D which fails to be an automorphism and satisfies the equality $hi = i$. This gives a contradiction to the fact that $i \colon I \to D$ is a divisible envelope of I, hence R_P is a DVR. As R is h-local and locally Noetherian, we deduce that R is Noetherian, hence a Dedekind domain. □

The question naturally arises whether Theorem 3.5 extends to arbitrary Prüfer domains, in particular, to non almost maximal valuation domains and to non h-local Prüfer domains. If the answer is in the positive, then one could formulate the following:

Conjecture. The integral domains R such that the class $\mathcal{D}(R)$ of the divisible R-modules is an enveloping class are exactly the almost perfect domains.

References

[1] Angeleri Hügel, L. and Trlifaj, J., *Direct limits of modules of finite projective dimension*, in Rings, modules, algebras, and abelian groups, Lecture Notes in Pure and Appl. Math. 236, Dekker, New York, 2004, 27–44.

[2] Bazzoni, S. and Herbera, D., *Cotorsion pairs generated by modules of bounded projective dimension*, preprint.

[3] Bazzoni, S. and Salce, L., *Strongly flat covers*, J. London Math. Soc. (2) 66 (2002) 276–294.

[4] Bazzoni, S. and Salce, L., *Almost perfect domains*, Colloq. Math. 95 (2003) 285–301.

[5] Bazzoni, S. and Salce, L., *On strongly flat modules over integral domains*, Rocky Mountain J. Math. 34 (2004) 417–439.

[6] Eckmann, B. and Schopf, A., *Über injektive Moduln*, Arch. Math. 4 (1953) 75–78.

[7] Fuchs, L. and Lee, S. B., *Weak-injectivity and almost perfect domains*, preprint.

[8] Fuchs, L. and Salce, L., *Modules over Non-Noetherian domains*, Mathematical Surveys and Monographs, 84, American Mathematical Society, Providence, RI, 2001.

[9] Göbel, R. and Trlifaj, J., *Approximations and endomorphism algebras of modules*, de Gruyter Expositions in Mathematics 41, de Gruyter, Berlin, 2006.

[10] Lee, S. B., *Semi-Baer modules over domains*, Bull. Austral. Math. Soc. 64 (2001) 21–26.

[11] Lee, S. B., *A note on the Matlis category equivalence*, J. Algebra 299 (2006) 854–862.

[12] Lee, S. B., *Weak-injective modules*, Comm. Algebra 34 (2006) 361–370.

[13] Salce, L., *Cotorsion theories for abelian groups*, in Symposia Mathematica, Vol. XXIII (Conf. Abelian Groups and their Relationship to the Theory of Modules, INDAM, Rome, 1977), Academic Press, 1979, 11–32.

[14] Trlifaj, J., *Covers, Envelopes and Cotorsion Theories*, Cortona Notes, 2000.

Author information

Luigi Salce, Dipartimento di Matematica Pura e Applicata, Università di Padova, Via Trieste 63, 35121 Padova, Italy.
E-mail: salce@math.unipd.it

The first Brauer–Thrall conjecture

Claus Michael Ringel

Abstract. Let Λ be an artin algebra and M a Λ-module. We show: If M is not the direct sum of copies of a finite number of indecomposable modules of finite length, then M has indecomposable submodules as well as indecomposable factor modules of arbitrarily large finite length. This improves the assertion of the first Brauer–Thrall conjecture as established by Roiter in 1968: Any artin algebra with infinitely many isomorphism classes of indecomposable modules of finite length has indecomposable modules of arbitrarily large finite length.

Key words. Artin algebra, modules of finite length, indecomposability, Krull–Remak–Schmidt–Azumaya theorem, Gabriel–Roiter measure, Gabriel–Roiter comeasure, Brauer–Thrall conjectures.

AMS classification. 16G60.

Let Λ be an artin algebra. The modules to be considered are left Λ-modules, and not necessarily of finite length. A module M is said to be *of finite type,* provided M is the direct sum of (arbitrarily many) copies of a finite number of indecomposable modules of finite length.

Theorem 1. *A module M which is not of finite type contains indecomposable submodules of arbitrarily large finite length.*

Remarks. (1) Recall that Λ is said to be *representation-finite* provided there are only finitely many isomorphism classes of indecomposable Λ-modules of finite length, otherwise Λ is called *representation-infinite*. The first Brauer–Thrall conjecture asserts that *a representation-infinite artin algebra has indecomposable submodules of arbitrarily large finite length.* The conjecture was solved by Roiter [7] in 1968. The theorem can be seen as a strengthening: Assume that there are infinitely many isomorphism classes of indecomposable modules M_i; take the direct sum $M = \bigoplus M_i$. The (Krull–Remak–Schmidt–)Azumaya theorem shows that M is not of finite type, thus we obtain indecomposable modules of arbitrarily large finite length as submodules of this particular module M.

(2) In order to provide a proof of the first Brauer–Thrall conjecture, it is sufficient to deal with the case where there are infinitely many isomorphism classes of indecomposable modules M_i of a fixed length and consider the direct sum $M = \bigoplus M_i$. This case has been discussed in [4, Appendix A]. There, we have shown that there are large indecomposable modules of finite length which are cogenerated by M (but they are not necessarily submodules of M, but only of the direct sum of countably many copies of M).

The essential part of the present proof will consist in dealing precisely with the special case of M being the direct sum of infinitely many pairwise non-isomorphic indecomposable modules of equal length.

(3) A weaker assertion is the "direct sum theorem" (4.3) of [5]: *Assume that M is a module with only finitely many isomorphism classes of indecomposable submodules of finite length. Then M is a direct sum of finite length modules* (and thus obviously of finite type). The first part of the proof presented below will incorporate the corresponding arguments used in [5].

(4) Observe that the "direct sum theorem" (and thus our theorem) implies the following result (see [8, Corollary 9.5] or [6], and also [1, Corollary 4.8]): *If an artin-algebra Λ is of finite type, then any Λ-module is of finite type.* We should stress that the converse implication is an obvious consequence of the (Krull–Remak–Schmidt–) Azumaya theorem.

Theorem 2. *A module M which is not of finite type has indecomposable factor modules of arbitrarily large finite length.*

Remarks. (5) Let us assume that M is a Λ-module which is not of finite type. This implies that Λ is representation-infinite, thus according to Auslander ([2], see also [4]) there are indecomposable Λ-modules of infinite length. Theorems 1 and 2 assert that M contains indecomposable submodules and factor modules of arbitrarily large finite length, but M may not contain an indecomposable submodule or factor module of infinite length, as the following examples show:

Let Λ be the Kronecker algebra. Let P_i, be the preprojective indecomposable modules, let Q_i be the preinjective indecomposable modules, such that $\operatorname{Hom}(P_i, P_{i+1})$ and $\operatorname{Hom}(Q_{i+1}, Q_i)$ are non-zero, for all $i \in \mathbb{N}$.

First, consider $M = \bigoplus_{i \in \mathbb{N}} P_i$. We show that any indecomposable submodule U of M is of finite length. Let $M_j = \bigoplus_{j \le i} P_i$. Assume U is any submodule of M. If U is contained in all M_j, then $U = 0$. Thus assume that U is contained in M_j, but not in M_{j+1}. We get a non-zero map $U \to M_j/M_{j+1} = P_j$. According to [3], U splits off a direct summand of the form P_i with $i \le j$.

Second, let $M' = \bigoplus_{i \in \mathbb{N}} Q_i$. We claim that any indecomposable factor module X of M' is of finite length. Let $p : M' \to X$ be the projection map. Let $M'_j = \bigoplus_{i \le j} Q_i$. Since M' is the union of these submodules, there has to be some j such that the restriction of p to M'_j is non-zero. But $\operatorname{Hom}(M'_j, X) \ne 0$ implies that X has an indecomposable preinjective direct summand, see [3]. Since we assume that X is indecomposable, it follows that X is of finite length.

We do not know whether there may exist a module which is not of finite type such that all its indecomposable submodules as well as all its indecomposable factor modules are of finite length.

(6) For the sake of completeness, let us note that for any representation-infinite artin algebra, there always do exist also modules M of **finite** type which have indecomposable submodules of arbitrarily large finite length and indecomposable factor modules of arbitrarily large finite length. For example, let P be any generator (for example $P = {}_\Lambda\Lambda$), and M the direct sum of countably many copies of P. Then M is of finite type and any Λ-module of finite length occurs as a factor module of M. Similarly, if

I is a cogenerator (for example $I = D(\Lambda_\Lambda)$) and M the direct sum of countably many copies of I, then M is of finite type and any Λ-module of finite length occurs as a submodule of M.

Proof of Theorem 1. We assume that the module M is not of finite type and show the existence of indecomposable submodules of arbitrarily large length. Thus, assume that the indecomposable submodules of M of finite length are of bounded length, thus there are only finitely many possible Gabriel–Roiter measures, see [4] and [5]. Assume that the indecomposable submodules of M of finite length have Gabriel–Roiter measure $\gamma_1 < \gamma_2 < \cdots < \gamma_s$. We show by induction on s that M is of finite type. The case $s = 1$ is well known and easy to see: If any indecomposable submodule of M of finite length is simple, then M has to be semi-simple, thus of finite type.

Assume now that $s \geq 2$. Consider a submodule M' of M which is a direct sum of modules of Gabriel–Roiter measure γ_s, and maximal with this property. If M' is of finite type, then [5, Theorem 4.2] asserts that M' is Σ-pure injective in $\mathcal{D}(\gamma_s)$, and of course M' is a pure submodule of M, thus M' is a direct summand of M, say $M = M' \oplus M''$ for some module M''. However, the indecomposable submodules of M'' of finite length have Gabriel–Roiter measure $\gamma_1, \ldots, \gamma_{s-1}$ (note that γ_s cannot occur by the maximality of M'), thus by induction M'' is of finite type. Then also $M = M' \oplus M''$ is of finite type.

Thus we can assume that there is a submodule $M^1 = \bigoplus_{i \geq 1} M_i$ of M which is an infinite direct sum of pairwise non-isomorphic indecomposable modules M_i with Gabriel–Roiter measure γ_s, indexed over $\mathbb{N}$. For any $r \in \mathbb{N}$, let $M^r = \bigoplus_{i \geq r} M_i$.

The modules M_i have all the same length, say length t. Let $\mathcal{U}_r$ be the set of isomorphism classes of indecomposable submodules of M^r of length at most $t - 1$.

(a) *The set $\mathcal{U}_r$ is finite for almost all r.* Otherwise, choose inductively pairwise non-isomorphic submodules U_j of M^1 of length at most $t - 1$ such that $U = \sum_{j \in \mathbb{N}} U_j$ is the direct sum of the modules U_j. (Namely, assume we have found $U_1, \ldots, U_s$ with $U' = \bigoplus_{j=1}^{s} U_j \subseteq M^1$, then $U' \subseteq \bigoplus_{i=1}^{r-1} M_i$ for some r. If $\mathcal{U}_r$ is infinite, we find inside M^r an indecomposable submodule U_{s+1} of length at most $t - 1$ which is not isomorphic to any of the $U_1, \ldots, U_s$. Since $\bigoplus_{i=1}^{r-1} M_i$ and M^r intersect in zero, we see that $\sum_{j=1}^{s+1} U_j$ is a direct sum.) As a submodule of M, all the indecomposable submodules of U of finite length have Gabriel–Roiter measure γ_i with $1 \leq i \leq s$ and actually γ_s does not occur as a Gabriel–Roiter measure (since such a submodule would be a direct summand of U, impossible). By induction, U has to be of finite type – but by construction, $U = \bigoplus_{j \in \mathbb{N}} U_j$ is not of finite type.

Let $\mathcal{U} = \bigcap_r \mathcal{U}_r$. As we have seen, this is a finite set of isomorphism classes, and of course non-empty. There is some r' with $\mathcal{U} = \mathcal{U}_{r'}$ and without loss of generality, we can assume that $r' = 1$ (replacing M^1 by $M^{r'}$). Thus we deal with the following situation: $M^1 = \bigoplus_{i \geq 1} M_i$ is an infinite direct sum of pairwise non-isomorphic indecomposable modules M_i with Gabriel–Roiter measure γ_s, and any indecomposable submodule of M^1 of length at most $t - 1$ is also a submodule of $M^r = \bigoplus_{i \geq r} M_i$ for any r.

(b) *Any indecomposable module of length at most $t - 1$ and cogenerated by M^1 is isomorphic to a submodule of M^1.* Assume that N is of length at most $t - 1$ and cogenerated by M^1, thus there is a finite number of maps $\pi : N \to M_i$ such that the kernels of these maps intersect in zero. These maps π cannot be surjective, since N is of length at most $t-1$, whereas M_i is of length t. If we decompose the images $\pi(N)$ of these maps, we obtain indecomposable submodules N_j of M_i of length at most $t-1$, and such submodules N_j occur frequently inside M^1, namely inside M^r, for any r. This shows that N is a submodule of M^1.

(c) In particular, we see that there are only finitely many isomorphism classes of modules which are cogenerated by M^1 and of length at most $t-1$. Let S be the direct sum of all the simple modules. As in [4], we consider the class $\mathcal{N}$ of all indecomposable modules cogenerated by $M^1 \oplus S$ and not isomorphic to any M_i. Clearly, this class is again closed under cogeneration and still finite. For any module M_i, let $f^\mathcal{N} M_i$ be the maximal factor module of M_i which belongs to add $\mathcal{N}$. Since M_i does not belong to $\mathcal{N}$, we see that $f^\mathcal{N} M_i$ is a module of length at most $t-1$ and cogenerated by $M^1 \oplus S$, thus there are only finitely many possibilities. It follows that there is a module Q in add N such that $f^\mathcal{N} M_i = Q$ for infinitely many i. Without loss of generality, we even may assume that $f^\mathcal{N} M_i = Q$ for all i (by deleting the remaining factors). For any module M_i, fix a projection $q_i : M_i \to Q$ and let K be the kernel of the map $(f_i)_i : M^1 \to Q$. Roiter's coamalgamation lemma (see [4]) asserts that K has no direct summand isomorphic to M_i, thus no submodule of Gabriel–Roiter measure γ_s (since M^1 belongs to $\mathcal{D}(\gamma_s)$ and M_i is relative injective in $\mathcal{D}(\gamma_s)$). By induction we see that K has to be of finite type. But this contradicts the Ext-lemma [4]: for any extension of the form

$$0 \to K \to X \to Q \to 0$$

with Q of finite length and K of infinite length and of finite type, the modules K and X will have common indecomposable direct summands. For $X = M^1$, the indecomposable direct summands have Gabriel–Roiter measure γ_s, but K has not even a submodule of measure γ_s. □

Proof of Theorem 2. Here we will need the Gabriel–Roiter comeasure. Now the Gabriel–Roiter measure concerns the existence of chains of indecomposable submodules of a given finite length module X, similarly, the comeasure measures the existence of chains of indecomposable factor modules of X, see [4]. In order to define the comeasure of the (left) Λ-module X of finite length, we just look at the dual DX of X, this a right Λ-module, thus a left Λ^{op}-module. By definition, the comeasure of X is $\gamma^*(X) = -\gamma(DX)$, where $\gamma(DX)$ is the Gabriel–Roiter measure of DX (as a Λ^{op}-module). Note that the minus sign is used so that the natural order of the (here now negative) rational numbers corresponds to the categorical structure of mod Λ; in particular, if $f : X \to Y$ is an epimorphism between finite length modules, then $\gamma^*(X) \leq \gamma^*(Y)$.

We consider now a Λ-module M and we assume that the indecomposable factor modules of M of finite length are of bounded length, thus there are only finitely many possible Gabriel–Roiter comeasures. Assume that the indecomposable factor modules of M of finite length have Gabriel–Roiter comeasures $\delta_s < \delta_{s-1} < \cdots < \delta_1$. We show by induction on s that M is of finite type. The case $s = 1$ is well known and easy to see: if any indecomposable factor module of M of finite length is simple, then M has to be semi-simple, thus of finite type (here we use that Λ is perfect).

Assume now that $s \geq 2$ and let $\delta = \delta_s$. Thus M has a factor module of finite length with comeasure δ, and any factor module of M of finite length has comeasure $\delta' \geq \delta$.

(a) *Let X be indecomposable of finite length with comeasure δ. Then any epimorphism $M \to X$ splits.* Namely, let $f : M \to X$ be an epimorphism with kernel U. We show that U is a pure submodule. Let $U' \subseteq U$ be a submodule with U/U' of finite length. Then the canonical map $M/U' \to M/U$ splits, according to the main property of the comeasure. According to [3] this means that U is a pure submodule. A pure submodule of finite colength is always a direct summand.

(b) *Let X be indecomposable of finite length with comeasure δ. Then $M = M' \oplus U$ where, on the one hand, U is a direct sum of copies of X, whereas, on the other hand, there is no surjective map $M' \to X$.* To see this let U be a maximal pure submodule of M which is a direct sum of copies of X (one obtains such a U by transfinite induction, splitting off copies of X; the existence of a maximal submodule of this kind comes from Zorn's lemma). Since X is Σ-algebraically compact, we see that U is even a direct summand, say $M = U \oplus M'$. Assume that there exists an epimorphism $M' \to X$. Then by (a), this epimorphism splits, thus $M' = X \oplus M''$, and $M = U \oplus M' = U \oplus X \oplus M''$. But, this contradicts the maximality of U.

We consider now pairwise non-isomorphic indecomposable factor modules of M of comeasure δ_s. If possible, we construct inductively submodules

$$M = U_0 \supset U_1 \supset U_2 \supset \cdots$$

such that the factors U_{i-1}/U_i are indecomposable with comeasure δ_s and pairwise non-isomorphic. If this process stops, then (b) asserts that we can write M as a direct sum of copies of finitely many indecomposables with comeasure δ_s and a module M' which has no factor module with comeasure δ_s. Since for M', the number of comeasures of factor modules has decreased by 1, we can use induction: By induction, M' is of finite type, thus also M is of finite type.

Thus consider the case where the sequence does not stop. Let $X_i = U_{i-1}/U_i$, then (a) shows that we get a pure submodule $V = \bigoplus X_i$ in M. Consider the Λ^{op}-module $V' = \bigoplus X_i^*$. This is a Λ^{op}-module which is not of finite type. Fix a natural number b. According to Theorem 1, there is an indecomposable submodule N of V' of length greater than b. Now N is a finite length submodule of V', thus a submodule of $\bigoplus_{i=1}^t X_i^*$ for some t. Dualizing, we see that $\bigoplus_{i=1}^t X_i$ has N^* as factor module. But $\bigoplus_{i=1}^t X_i$ is a direct summand of M, thus a factor module of M, therefore N^* is a factor module of M, and of length $|N^*| = |N| > b$. This contradicts the assumption that the factor modules of M are of bounded length! Thus the case where the sequence does not stop, cannot occur. This completes the proof. □

References

[1] M. Auslander. Representation theory of artin algebras II. Communications Algebra 1 (1974), 269–310.

[2] M. Auslander. Large modules over artin algebras. In: Algebra, Topology and Category Theory. Academic Press (1976), 1–17.

[3] C. M. Ringel. Infinite dimensional representations of finitedimensional hereditary algebras. Symposia Math. XXIII. Istituto Naz. Alta Mat. (1979), 321–412.

[4] C. M. Ringel. The Gabriel–Roiter measure. Bull. Sci. Math. 129 (2005), 726–748.

[5] C. M. Ringel. Foundation of the Representation Theory of Artin Algebras, Using the Gabriel–Roiter Measure. In: Trends in Representation Theory of Algebras and Related Topics. Edited by de la Peña and Bautista. Contemporary Math. 406. Amer. Math. Soc. (2006), 105–135.

[6] C. M. Ringel and H. Tachikawa. QF-3 rings. J. Reine Angew. Math. 272 (1975), 49–72.

[7] A. V. Roiter. Unboundedness of the dimension of the indecomposable representations of an algebra which has infinitely many indecomposable representations. Izv. Akad. Nauk SSSR. Ser. Mat. 32 (1968), 1275–1282.

[8] H. Tachikawa. Quasi-Frobenius Rings and Generalizations. Springer Lecture Notes in Mathematics 351 (1973).

Author information

Claus Michael Ringel, Fakultät für Mathematik, Universität Bielefeld, Germany.
E-mail: ringel@math.uni-bielefeld.de

A version of the Baer splitting problem for noetherian rings

Cornelius Greither, Dolors Herbera and Jan Trlifaj

Abstract. We call a module M over a commutative noetherian ring R *quasi-Baer* provided that $\operatorname{Ext}^1_R(M, T) = 0$ for each locally artinian (= semiartinian) module T. We prove that all quasi-Baer modules are projective in case R has finite Krull dimension, or R is of cardinality $< \aleph_\omega$.

Key words. Commutative noetherian rings, quasi-Baer modules, locally artinian modules, Mittag-Leffler modules.

AMS classification. Primary: 13E05, 13D07. Secondary: 16E30, 16D40.

Introduction

The recent solution in [1] of Kaplansky's integral domain version of the Baer splitting problem says that all Baer modules over any integral domain R are projective. Here, a module M is called *Baer* provided that $\operatorname{Ext}^1_R(M, T) = 0$ for each torsion module T. In other words, the class of all torsion modules over any integral domain is a test class for projectivity.

Motivated by a stimulating talk of Lidia Angeleri Hügel at the Kasch Festkolloquium at LMU in 2006, the first-named author raised the question of whether there are smaller test classes available in the particular case of noetherian integral domains. One always needs a test class for projectivity rather than a test set: by [4], it is consistent with ZFC + GCH that for any non-right perfect ring R and any uncountable cardinal λ of cofinality ω there is a λ^+-generated module M such that $\operatorname{Ext}^1_R(M, N) = 0$ for *any* module N of cardinality $< \lambda$, but the projective dimension of M equals 1.

The test class proposed was the class of all locally artinian modules. We recall that a right module T over a ring R is *locally artinian* provided that every finitely generated submodule of T is artinian.

Each locally artinian module is semiartinian, and the converse holds when R is right noetherian. If R is a commutative noetherian ring then the locally artinian modules T

First author: acknowledges support from DFG.
Second author: a large part of this research was done during the visits of the second- and third-named authors to CRM Barcelona supported by the Research Programme on Discrete and Continuous Methods of Ring Theory. The second author is partially supported by the DGI and the European Regional Development Fund, jointly, through Project MTM2005–00934, and by the Comissionat per Universitats i Recerca of the Generalitat de Catalunya, Project 2005SGR00206.
Third author: supported by GAČR 201/06/0510 and MSM 0021620839.

are also characterized by the localization T_p being zero for each non-maximal prime ideal p of R.

The class of locally artinian modules is in general much smaller than the class of torsion modules: over a commutative noetherian ring a nonzero module is locally artinian if and only if all its associated primes are maximal ideals. So over a commutative noetherian domain R that is not a field the class of torsion modules coincides with the class of locally artinian modules if and only if R has Krull dimension 1.

Let R be a commutative noetherian ring. We will call a module M *quasi-Baer* provided that $\operatorname{Ext}^1_R(M,T) = 0$ for each locally artinian module T. A ring R is said to be *quasi-Baer* if all quasi-Baer R-modules are projective.

The question of whether each commutative noetherian ring R is quasi-Baer appears to be open. In this paper, we provide a positive answer in the case when R has finite Krull dimension (so in particular, when R is local), and when R has cardinality $< \aleph_\omega$. We also present several results on quasi-Baer and projective modules that may be helpful for answering the question in full generality in the future.

1 Lemmas and main results

Let $\mathcal{C}$ be a class of modules. A module M is called $\mathcal{C}$-*filtered* provided there is a chain $(M_\alpha \mid \alpha \leq \sigma)$ of submodules of M such that $M_0 = 0$, $M_\alpha \subseteq M_{\alpha+1}$ and $M_{\alpha+1}/M_\alpha$ is isomorphic to some element of $\mathcal{C}$ for each $\alpha < \sigma$, $M_\alpha = \bigcup_{\beta < \alpha} M_\beta$ for each limit ordinal $\alpha \leq \sigma$, and $M_\sigma = M$.

In general, $\mathcal{C}$-filtered modules have a complex structure. However, if the class $\mathcal{C}$ consists of projective modules then each $\mathcal{C}$-filtered module is easily seen to be projective.

Let R be a commutative noetherian ring. We denote by $E = \bigoplus_{m \in \operatorname{mSpec}(R)} E(R/m)$ the minimal injective cogenerator for $\operatorname{Mod} R$. For a module N, we denote by N^* the *dual* module $N^* = \operatorname{Hom}_R(N, E)$.

For a module T, $i < \omega$, and a prime ideal p, let $\mu_i(p, T)$ denote the corresponding Bass invariant (= the multiplicity of $E(R/p)$ in the i-th term of the minimal injective coresolution of M).

Our basic reference on commutative ring theory will be [6] and [8], and we refer to [7] for results on set-theoretic homological algebra.

We start with a characterization of locally artinian modules:

Lemma 1.1. *Let R be a commutative noetherian ring. Then the following conditions are equivalent for a module T:*

(i) *T is locally artinian;*

(ii) *$\mu_i(p, T) = 0$ for all $i < \omega$ and p non-maximal;*

(iii) *T is semiartinian;*

(iv) *There is a set I such that T is isomorphic to a submodule of $E^{(I)}$.*

In particular, the dual of any finitely generated module is locally artinian.

Proof. If T is locally artinian and p is not maximal then $T_p = 0$, so

$$\mu_i(p, T) = \dim \operatorname{Ext}^i_{R_p}(R_p/p_p, T_p) = 0$$

by a classical formula of Bass [8, Theorem 18.7]. The latter implies that all injectives occurring in the minimal injective coresolution of T are direct sums of copies of $E(R/m)$ where m runs over all maximal ideals. Since $E(R/m)$ is artinian and $\{R/m\}$-filtered (see [8, Theorems 18.4(v), 18.4(vi), and 18.6(v)]), we infer that T is semiartinian. If T is semiartinian, then $E(T)$ is a direct sum of copies of $E(R/m)$ where m runs over all maximal ideals, so $E(T) \subseteq E^{(I)}$ for some set I. Finally, as E is artinian, a finitely generated submodule of $E^{(I)}$ is also artinian so any submodule of $E^{(I)}$ is locally artinian.

In particular, if F is an n-generated module (where n is finite), then the dual module $F^* = \operatorname{Hom}_R(F, E)$ embeds into $\operatorname{Hom}_R(R^n, E) \cong E^n$, hence F^* is locally artinian. □

We will also need a lemma on vanishing of $\operatorname{Ext}^1_R(M, -)$ for countably presented modules M. It is based on properties of Mittag-Leffler inverse systems discovered in [1] and [2] (notably on [2, Theorem 2.5] and [1, Proposition 1.2]). Its formulation here is taken from [10, Proposition 2.7]:

Lemma 1.2. *Let R be a ring, M a countably presented module, and $\mathcal{D}$ a class of modules closed under countable direct sums. Assume that $\operatorname{Ext}^1_R(M, D) = 0$ for each $D \in \mathcal{D}$. Let $(D_i \mid i \in I)$ be a family of modules from $\mathcal{D}$, and N be a pure submodule of $\prod_{i \in I} D_i$. Then $\operatorname{Ext}^1_R(M, N) = 0$.*

Now we are ready to settle the countable case:

Lemma 1.3. *Let R be a commutative noetherian ring. Then each quasi-Baer module is flat, and each countably generated quasi-Baer module is projective.*

Proof. First, we claim that any module N is a pure submodule in a direct product of duals of some finitely generated modules. Indeed, N purely embeds into N^{**}; moreover, N^* is a direct limit of finitely generated R-modules, hence a pure-epimorphic image of a direct sum of finitely generated modules. So N^{**} is a direct summand in a direct product of duals of finitely generated modules, and the claim follows.

By Lemma 1.1, our claim implies that any module N purely embeds into a direct product of locally artinian modules.

In particular, if N is pure-injective then this pure embedding splits. It follows that for each quasi-Baer module M, we have $\operatorname{Ext}^1_R(M, N) = 0$ for any pure-injective module N, and the latter just says that M is flat (see e.g. [7, 2.2.3]).

Let C be a countably generated quasi-Baer module. By Lemma 1.2, we have $\operatorname{Ext}^1_R(C, N) = 0$ for any module N which is a pure submodule of a direct product of locally artinian modules. However, the latter holds for any module N, so C is projective. □

The argument that countably generated quasi-Baer modules are projective can alternatively be given as follows: If C is a countably generated quasi-Baer module, then

C is flat by the first part of Lemma 1.3 and hence of projective dimension ≤ 1 by [6, VI.9.8]. A straightforward generalization of the Mittag-Leffler argument used in [1] then shows that C is projective.

In order to deal with arbitrary modules of finite projective dimension, we need the following result (see [11, Theorem 18] or [7, 4.3.10]):

Lemma 1.4. *Let R be a right noetherian ring, $n < \omega$, $\mathcal{T}$ be a class of modules closed under direct sums, and*

$$\mathcal{A} = \{M \mid \mathrm{Ext}_R^i(M,T) = 0 \text{ for all } 0 < i < \omega \text{ and all } T \in \mathcal{T}\}.$$

Denote by $\mathcal{C}_n$ the class of all countably generated modules from $\mathcal{A}$ that have projective dimension $\leq n$. Let $M \in \mathcal{A}$ be a module of projective dimension $\leq n$. Then M is $\mathcal{C}_n$-filtered.

Combining Lemmas 1.3 and 1.4, we obtain

Lemma 1.5. *Let R be a commutative noetherian ring and M be a quasi-Baer module of finite projective dimension. Then M is projective.*

Proof. Let $\mathcal{T}$ denote the class of all locally artinian modules. By Lemma 1.1, if $T \in \mathcal{T}$ then all the cosyzygy modules of T in its minimal injective coresolution are also locally artinian. So the class $\mathcal{A} = \{M \mid \mathrm{Ext}_R^i(M,T) = 0 \text{ for all } 0 < i < \omega \text{ and all } T \in \mathcal{T}\}$ coincides with the class of all quasi-Baer modules.

Let M be a quasi-Baer module of projective dimension $n < \omega$. By Lemma 1.4, M is $\mathcal{C}_n$-filtered where $\mathcal{C}_n$ is the class of all countably generated quasi-Baer modules of projective dimension $\leq n$. However, all modules in $\mathcal{C}_n$ are projective by Lemma 1.3, hence M is projective. □

Now, we can prove our main results:

Theorem 1.6. *Let R be a commutative noetherian ring of finite Krull dimension. Then each quasi-Baer module is projective.*

Proof. This follows from Lemmas 1.3 and 1.5, since the projective dimension of all flat modules is bounded by the Krull dimension of R by [9, Corollaire 3.2.7]. □

Theorem 1.7. *Let R be a commutative noetherian ring.*

(i) *Let M be a $< \aleph_\omega$-generated quasi-Baer module. Then M is projective.*

(ii) *Assume that R has cardinality $< \aleph_\omega$. Then each quasi-Baer module is projective.*

Proof. (i) By [6, VI.9.8], each flat $\leq \aleph_n$-generated module has projective dimension $\leq n+1$. So M is projective by Lemmas 1.3 and 1.5.

(ii) Let $\kappa = \max(\mathrm{card}(R), \aleph_0)$, so $\kappa = \aleph_n$ for some $n < \omega$ by assumption. Let F be a flat module. Then F is $\mathcal{C}$-filtered, where $\mathcal{C}$ denotes the class of all flat modules of cardinality $\leq \kappa$ (see e.g. [7, 3.2.7]). As above, each module $C \in \mathcal{C}$ has projective dimension $\leq n+1$, so F has projective dimension $\leq n+1$ by the Auslander Lemma. Thus the conclusion again follows by Lemmas 1.3 and 1.5. □

Remark 1.8. It is interesting to see to what extent Theorems 1.6 and 1.7 hold in the non-commutative setting. We do not have the coincidence of the conditions (i)–(iv) in Lemma 1.1 in general, but conditions (i) and (iii) are easily seen to be equivalent when R is right noetherian. So it appears natural to extend our definition to right R-modules over a right noetherian ring R as follows:

A right R-module M is *quasi-Baer* provided that $\operatorname{Ext}_R^1(M, T) = 0$ for each semiartinian or, equivalently, locally artinian right R-module T. Moreover, R is *right quasi-Baer* if all quasi-Baer right R-modules are projective. Then we have the following:

(1) If R is right artinian then R is right quasi-Baer, i.e., the claim of Theorem 1.6 holds for all right noetherian rings of right Krull dimension 0. This follows from the fact that in this case a right R-module M is projective whenever $\operatorname{Ext}_R^1(M, N) = 0$ for each simple right R-module N (see [7, Lemma 4.1.8]).

(2) There exist left and right noetherian rings R of left and right Krull dimension 1 such that *all* (left, right) R-modules are quasi-Baer, hence the claim of Theorem 1.6 fails completely.

For an example, consider the ring $R = k[y, D]$ of all differential polynomials of one variable y over a universal differential field k of characteristic 0 with the differentiation D. This is a left and right PID, and all proper cyclic (left, right) R-modules are homogenous completely reducible and injective. In particular, all semiartinian (left, right) R-modules are injective, so *all* (left, right) R-modules are quasi-Baer. Note that R has left and right Krull dimension 1 because R/I is artinian for each non-zero one sided ideal I. For more details on this example we refer to [5, 7.37–7.46].

(3) The example in (2) can be seen in the more general context of noetherian V-rings. We recall that a ring is said to be a *right V-ring* if all simple right R-modules are injective. In particular, over a right V-ring all artinian right modules are injective and, hence, all locally artinian right modules are direct limits of injective modules. Therefore if R is a right noetherian right V-ring, the Baer criterion yields that all locally artinian right R-modules are injective, so again we conclude that all right modules are quasi-Baer. For more information and further examples of noetherian V-rings we refer to [3].

(4) The example in (2) also yields failure of Theorem 1.7 in the non-commutative noetherian setting. Indeed, by the Löwenheim–Skolem theorem, the universal differential field k has a countable universal differential subfield, so w.l.o.g., $R = k[y, D]$ can be taken countable in (2). But then both parts (i) and (ii) of Theorem 1.7 fail for R.

2 More results on quasi-Baer and projective modules

Lemma 2.1. *Let $\varphi \colon R \to S$ be a ring epimorphism. Let M_R be a right R-module, and let T_S be a right S-module. If $\operatorname{Ext}_R^1(M_R, T_R) = 0$ then $\operatorname{Ext}_S^1(M \otimes_R S, T_S) = 0$.*

Proof. Consider an exact sequence of S-modules
$$0 \to T \to K \xrightarrow{\pi} M \otimes_R S \to 0. \tag{$*$}$$

We want to show that $(*)$ splits. Let K' be the R-module that is the pull-back of π and $\operatorname{Id} \otimes_R \varphi \colon M \to M \otimes_R S$. By assumption, the exact sequence of R-modules

$0 \to T \to K' \to M \to 0$ splits, and as φ is an epimorphism the exact sequence we obtain when we apply the functor $- \otimes_R S$ is isomorphic to $(*)$. Thus $(*)$ is also a split exact sequence. □

Lemma 2.2. *Let R be a commutative noetherian ring, and let M be a quasi-Baer R-module.*

(i) *If $\{m_i\}_{i \in \Lambda}$ is a finite nonempty set of maximal ideals of R and $\Sigma = R \setminus \bigcup_{i \in \Lambda} m_i$ then M_Σ is a quasi-Baer R_Σ-module.*

(ii) *If I is an ideal of R then M/MI is a quasi-Baer R/I-module.*

Proof. Both results follow applying Lemma 2.1 to the ring epimorphisms $R \to R_\Sigma = S_1$ and $R \to R/I = S_2$, respectively. One needs to make sure that for $j = 1, 2$, if an S_j-module T is semiartinian, then it is also semiartinian when considered as an R-module. This is trivial for $j = 2$.

For $j = 1$, consider a $\mathcal{C}$-filtration $\mathcal{T} = (T_\alpha \mid \alpha \leq \sigma)$ of T where $\mathcal{C} = \text{simp-}R_\Sigma$ is a representative set of all simple R_Σ-modules. Each $N \in \mathcal{C}$ is isomorphic as R_Σ-module to R_Σ/m_Σ for a maximal ideal m of R such that $m \cap \Sigma = \emptyset$, that is, for some $m \in \{m_i \mid i \in \Lambda\}$. But $N \cong (R/m)_\Sigma \cong R/m$ as R-modules, so we infer that $\mathcal{T}$ is also a simp-R-filtration of T proving that T is semiartinian as R-module. □

Corollary 2.3. *Let R be a commutative noetherian ring. Let $n \geq 1$ and $\{p_1, \ldots, p_n\}$ be a set of prime ideals of R and set $S = R \setminus \bigcup_{i=1}^n p_i$. If M is a quasi-Baer R-module then M_S is projective.*

In particular, if M is a quasi-Baer module and m a maximal ideal of R then $mM = M$ implies $M_m = 0$. If, in addition, R is a domain and M is nonzero then, for any ideal I, $IM = M$ implies $I = R$.

Proof. For $i = 1, \ldots, n$, let m_i be a maximal ideal of R such that $p_i \subseteq m_i$. Since a localization of a projective module is projective over the localization, it is enough to prove the claim for $S = R \setminus \bigcup_{i=1}^n m_i$.

By Lemma 2.2, M_S is a quasi-Baer module over R_S. As R_S has finite Krull dimension, the claim follows from Theorem 1.6.

Let now m be a maximal ideal of R. Then $mM = M$ implies that $M_m = mM_m = J(M_m)$. As M_m is projective, this implies that $M_m = 0$.

Assume finally that $M \neq 0$ and R is a domain. As M is flat, $M \hookrightarrow M_m$ for any maximal ideal M. If I is a proper ideal of R then it is contained in a maximal ideal m, hence $IM \subseteq mM \subsetneq M$. □

The following lemma is a consequence of the results in [9]; we include its proof for the sake of completeness, and to make clear that the commutativity hypothesis in [9] is not necessary.

We recall that a right R-module M is said to be Mittag-Leffler if the canonical map $M \otimes_R \prod_{i \in \Lambda} Q_i \to \prod_{i \in \Lambda}(M \otimes_R Q_i)$ is injective for any family of left R-modules $\{Q_i\}_{i \in \Lambda}$.

Lemma 2.4. *Let R be a ring, and let M be a flat right R-module. Then:*

(i) *If I is a nilpotent (two-sided) ideal of R, M/MI is a Mittag-Leffler right R/I-module if and only if M is a Mittag-Leffler R-module.*

(ii) *If I and J are two (two-sided) ideals of R, $M/M(I \cap J)$ is a Mittag-Leffler right $R/I \cap J$-module if and only if M/MI is a Mittag-Leffler R/I-module and M/MJ is a Mittag-Leffler R/J-module.*

Proof. Throughout the proof we shall use that for a two-sided ideal I, $M/MI \cong M \otimes_R R/I$ is Mittag-Leffler as R/I-module if and only if the canonical map $\rho \colon M \otimes_R \prod_{i \in \Lambda} Q_i \to \prod_{i \in I} M \otimes_R Q_i$ is injective for any family of left R/I-modules $\{Q_i\}_{i \in \Lambda}$. From this observation, it follows immediately that if M is a Mittag-Leffler R-module then M/MI is a Mittag-Leffler R/I-module.

To prove the converse of (i), assume that I is a nilpotent ideal of R, and let $\{Q_i\}_{i \in \Lambda}$ be a family of left R-modules. The exact sequences

$$0 \to IQ_i \to Q_i \to Q_i/IQ_i \to 0$$

and the flatness of M yield a commutative diagram

$$\begin{array}{ccccccccc} 0 & \to & M \otimes_R \prod_{i \in \Lambda} IQ_i & \to & M \otimes_R \prod_{i \in \Lambda} Q_i & \to & M \otimes_R \prod_{i \in \Lambda} Q_i/IQ_i & \to & 0 \\ & & \rho_1 \downarrow & & \rho \downarrow & & \rho_2 \downarrow & & \\ 0 & \to & \prod_{i \in \Lambda}(M \otimes_R IQ_i) & \to & \prod_{i \in \Lambda}(M \otimes_R Q_i) & \to & \prod_{i \in \Lambda}(M \otimes_R Q_i/IQ_i) & \to & 0 \end{array}$$

that allows to prove the injectivity of ρ, using that ρ_2 is injective by hypothesis and that ρ_1 is injective by an inductive argument on the nilpotency index of I.

To prove the converse of (ii), let $\{Q_i\}_{i \in \Lambda}$ be a family of $R/I \cap J$-left modules. The family of exact sequences

$$0 \to IQ_i \to Q_i \to Q_i/IQ_i \to 0$$

yields a commutative diagram like the one above. Now ρ_1 is injective because, for any $i \in \Lambda$, IQ_i is a left R/J-module; and ρ_2 is injective because, for any $i \in \Lambda$, Q_i/IQ_i is a left R/I-module. Hence ρ is also injective. □

Proposition 2.5. *Let R be a ring, and let M be a flat right R-module. Then:*

(i) *If I is a nilpotent (two-sided) ideal of R, M/MI is a projective right R/I-module if and only if M is a projective R-module.*

(ii) *If $I_1, \ldots, I_n$ are (two-sided) ideals of R such that $\bigcap_{i=1}^n I_i = 0$, then M is a projective right R-module if and only if, for each $i = 1, \ldots, n$, M/MI_i is a projective R/I_i-module.*

Proof. Notice that by Lemma 2.4, in both cases, the module M is a flat Mittag-Leffler module. So (i) and (ii) are a consequence of [9, Théorème 3.1.3] applied to the morphisms $R \to R/I$ and $R \hookrightarrow \prod_{i=1}^n R/I_i$, respectively; this is because in [9, Théorème 3.1.3] it is shown that, in both cases, the module M can be filtered by countably generated projective modules, so it must be projective. □

Remark 2.6. Assume that R is a commutative noetherian ring which has a nonprojective quasi-Baer module M (by our definition at the beginning of the paper we may equivalently say: which is not a quasi-Baer ring); and recall that by Lemma 1.3 any such module M must be flat. Let P be a proper ideal of R which is maximal with respect to the property that R/P is not quasi-Baer. We claim that P is a prime ideal.

First we observe that $S = R/P$ is reduced, because otherwise if M is a nonprojective quasi-Baer S-module and N denotes the nilradical of S, then by Lemma 2.2 M/NM is a quasi-Baer module over S/N. The maximality of P implies that M/NM is projective, but then M_S is a flat module that is projective modulo a nilideal. By Proposition 2.5(i) M is projective, which is a contradiction.

Let $P_1, \ldots, P_n$ be the minimal prime ideals of S. As S is reduced, $\bigcap_{i=1}^{n} P_i = 0$. If $n > 1$ then, for any $i = 1, \ldots, n$, M/P_iM is projective as S/P_i-module. By applying Proposition 2.5(ii) we conclude that M is projective also as an S-module, and this is again a contradiction. Hence $n = 1$, so S is a domain. Therefore P is a prime ideal as claimed.

Concluding comments: (1) The claim established in the preceding remark shows: If there is a commutative noetherian ring that is not quasi-Baer, then there is also such a ring which is a domain (pass from R to R/P in the notation of the first paragraph of the Remark), and with the extra property that every proper factor ring is quasi-Baer (this is the maximality property of P). Hence the general problem of showing that all commutative noetherian rings are quasi-Baer reduces to the following: show that the quasi-Baer module M is R-projective, under the assumption that R is a domain and M/IM is R/IR-projective for every nonzero ideal I of R.

(2) By 1.6 we know that every local noetherian ring is quasi-Baer, and Proposition 2.5(ii) looks a little like a local-global argument. But we seem to be far away from an argument that would reduce our main problem to the local case; the usual methods from commutative algebra and algebraic geometry strongly rely on restrictive finiteness conditions.

Acknowledgements. The authors would like to thank Lidia Angeleri Hügel, who established the contact between the three authors; without her initiative, this collaboration would never have arisen.

References

[1] Angeleri Hügel, L., Bazzoni, S. and Herbera, D.: *A solution to the Baer splitting problem*. Trans. Amer. Math. Soc. 360 (2008), 2409–2421.

[2] Bazzoni, S and Herbera, D.: *One dimensional tilting modules are of finite type*. Algebras and Repres. Theory 11 (2008), 43–61

[3] Cozzens, J. and Faith, C.: *Simple Noetherian Rings*. CTM 69, Cambridge Univ. Press, Cambridge 1975.

[4] Eklof, P. C. and Shelah, S.: *On Whitehead modules*. J. Algebra 142 (1991), 492–510.

[5] Faith, C.: Algebra: *Rings, Modules, and Categories I*. GMW 190, Springer, New York 1973.

[6] Fuchs, L. and Salce, L.: *Modules over Non-Noetherian Domains*. MSM 84, AMS, Providence 2001.

[7] Goebel, R. and Trlifaj, J.: *Approximations and Endomorphism Algebras of Modules*. GEM 41, Walter de Gruyter, Berlin 2006.

[8] Matsumura, H.: *Commutative Ring Theory*. CSAM 8, Cambridge Univ. Press, Cambridge 1990.

[9] Raynaud, M. and Gruson, L.: *Critères de platitude et de projectivité*, Invent. Math. 13 (1971), 1–89.

[10] Šaroch, J. and Šťovíček, J.: *The countable telescope conjecture for module categories*. Advances in Math. 219 (2008), 1002–1036.

[11] Šťovíček, J. and Trlifaj, J.: *All tilting modules are of countable type*. Bull. London Math. Soc. 39 (2007), 121–132.

Author information

Cornelius Greither, Fakultät Informatik, Universität der Bundeswehr München, 85577 Neubiberg, Germany.
E-mail: cornelius.greither@unibw.de

Dolors Herbera, Departament de Matemàtiques, Universitat Autònoma de Barcelona, 08193 Bellaterra (Barcelona), Spain.
E-mail: dolors@mat.uab.cat

Jan Trlifaj, Department of Algebra, Faculty of Mathematics and Physics, Charles University, Sokolovská 83, 186 75 Prague 8, Czech Republic.
E-mail: trlifaj@karlin.mff.cuni.cz

Characterizations and constructions of h-local domains

Bruce Olberding

Abstract. H-local domains arise in a variety of module- and ideal-theoretic applications. In the first half of the paper, we survey some of these applications, and collect a number of characterizations and examples of h-local domains. In the second half, we show how diverse examples of h-local Prüfer domains arise as overrings of Noetherian domains and polynomial rings in finitely many variables.

Key words. h-local domain, Prüfer domain, group of divisibility.

AMS classification. 13F05, 13G05.

1 Introduction

All rings in this article are commutative and have identity. Mostly we restrict to commutative integral domains, and when this is the case, we denote the quotient field of the domain by Q. A domain R has *finite character* if each nonzero element of R is contained in at most finitely many maximal ideals of R. The domain R is *h-local* if R has finite character and each nonzero prime ideal of R is contained in a unique maximal ideal. Thus every quasilocal domain (i.e., domain having a unique maximal ideal) is an h-local domain, and hence h-locality is uninteresting as a local property. As a global property however, it is a very desirable hypothesis under which to work because, although not immediately apparent from the definition, it allows one to pass with relative ease between a domain and its localizations. It manifests in a number of different guises, which we discuss in Section 2. In fact, one of the interesting aspects of this class of rings is how understanding of the h-local property evolved from close study of both non-Noetherian ideal theory and module theory, especially with an eye toward capturing certain essential aspects of abelian group theory.

One early such application was Matlis' classification of D-rings. Matlis remarks in a discussion of indecomposable modules that "an initial hypothesis, arising from a naive attempt to generalize the theory of finite-dimensional vector spaces, might be that rank 1 modules are the only indecomposable ones. Unfortunately, it can be shown that this hypothesis fails, even for abelian groups" [36, p. 1]. Here Matlis surely has in mind the sort of pathological decompositions of torsion-free finite rank modules exemplified by Corner's constructions in [14]. Matlis continues, "A more profitable approach is to turn the problem around and try to characterize the integral domains that do have this property." Matlis accomplishes such a characterization for both integrally closed domains and Noetherian domains, and he notes, "It is an unexpected phenomenon that the theory of h-local rings provides a key link in the chain of solving the problem" [36, p. 3].

The class of D-rings, those domains for which every indecomposable torsion-free finite rank module has rank one, is of course very narrow. However, the class of h-local domains is, at least relative to the class of D-rings, or even to the ring of integers, quite large. We give some examples later to justify this assertion. H-local domains capture a fundamental aspect of the ring of integers: Every torsion R-module T is isomorphic to the direct sum of primary components, that is, $T \cong \bigoplus_M T_M$, where M ranges over the maximal ideals of R (see Section 2). In classifying the D-rings, Matlis used this characterization, along with several others, to prove that such domains are necessarily h-local. A larger class of domains, the class of *reflexive domains*, those domains R for which $\text{Hom}_R(-, R)$ induces a duality on submodules of finite rank free R-modules, were also shown by Matlis in his 1968 article [34] to be h-local. (Independently, Heinzer showed in this same year that the domains for which every nonzero ideal is reflexive are h-local [25]). More recently, in this same vein Warfield's duality for torsion-free finite rank abelian groups has been extended to modules over domains and studied by several authors; see Fuchs–Salce [20, Chapter 15]. These domains are reflexive, and hence h-local.

One other important and well-known application of h-locality having its origins in abelian group theory is the classification of *FGC rings*, those rings for which every finitely generated module is a direct sum of cyclic modules. One of the major obstacles in obtaining this classification was showing that FGC domains are h-local; see [10] and [56].

For the moment we confine ourselves to the above applications, and postpone till later some brief mentions of applications to module theory and multiplicative ideal theory. In any case, given the range of applications of h-locality to module theory and ideal theory, and the desirability of working under such a hypothesis, it is of interest to have a good stock of examples to illustrate the theory. To this end, we give several constructions that yield finite character and h-local domains. Sections 2 and 3 are mostly expository, while Sections 4 and 5 contain new examples, with an emphasis on examples that occur as overrings of Noetherian domains. Not surprisingly, given the strength of the hypotheses involved, non-obvious examples are not so easy to obtain. One of our main class of examples occurs as overrings of polynomial rings over fields, where the challenge in this context is that affine algebras of dimension > 1 are about as far from being h-local as a domain can be. Despite this, we are able to give a realization theorem for h-local Prüfer overrings of polynomial rings in terms of the group of invertible fractional ideals of the domain.

I thank the referee for helpful comments.

2 Characterizations

In this section we compile some characterizations of h-local domains from the literature. We postpone till after the statement of each theorem a discussion of the background and references for the characterizations. We require first some terminology. An ideal I of the domain R is *unidirectional* if R/I is a quasilocal ring. If X and Y are

R-submodules of the quotient field Q, then $[Y : X] = \{q \in Q : qX \subseteq Y\}$. If P is a prime ideal of R, we view X_P as a submodule of Q.

Theorem 2.1. *The following conditions are equivalent for a domain R.*

(1) *R is an h-local domain.*

(2) *R/I is a finite product of quasilocal rings for every nonzero proper ideal I of R.*

(3) *Every nonzero proper ideal of R is a finite intersection (product) of unidirectional ideals.*

(4) *Every nonzero prime ideal of R contains an invertible unidirectional ideal.*

(5) *Every nonzero prime ideal of R contains a nonzero principal ideal that is a product of unidirectional ideals.*

(6) *For every maximal ideal M and $0 \neq x \in M$, $xR_M \cap R$ is an invertible unidirectional ideal.*

(7) *Every nonzero prime ideal of R is contained in a unique maximal ideal, and for every ideal I, if I_M is finitely generated for each maximal ideal M, then I is a finitely generated ideal.*

(8) *If $\{I_\alpha\}$ is a collection of ideals of R having nontrivial intersection and M is a maximal ideal of R such that $\bigcap_\alpha I_\alpha \subseteq M$, then $I_\alpha \subseteq M$ for some α.*

(9) *$(\bigcap_\alpha X_\alpha)_M = \bigcap_\alpha (X_\alpha)_M$ for every maximal ideal M of R and collection $\{X_\alpha\}$ of R-submodules of Q having nontrivial intersection.*

(10) *$(\bigcap_{N \neq M} R_N)R_M = Q$ for each maximal ideal M of R, where N ranges over the maximal ideals of R distinct from M.*

(11) *$Q/R \cong \bigoplus_M Q/R_M$, where M ranges over the maximal ideals of R.*

(12) *$\widetilde{R} \cong \prod_M \widetilde{R}_M$, where $\widetilde{R}$ denotes the completion of R in the R-topology and M ranges over the maximal ideals of R.*

(13) *$T \cong \bigoplus_M T_M$ for every torsion R-module T, where M ranges over the maximal ideals of R.*

(14) *$T \cong \bigoplus_M T_M$ for every cyclic torsion R-module T, where M ranges over the maximal ideals of R.*

(15) *$C \cong \prod_M \mathrm{Hom}_R(R_M, C)$ for each cotorsion R-module C, where M ranges over the maximal ideals of R.*

(16) *For all torsion R-modules T and R-submodules T_1 of T such that T/T_1 is a finitely generated R-module, T_1 is a supplemented R-submodule of T.*

(17) *For all nonzero ideals I of R, I is a supplemented R-submodule of R.*

(18) *For all R-modules A, A/t(A) does not contain a maximal R-submodule if and only if for all $B \subseteq A$ such that A/B is a finitely generated R-module, B is a supplemented R-submodule of A.*

(19) *For all cyclic R-modules A, $t(A)$ is the sum of all local submodules B of A.*

(20) *Every finitely generated torsion R-module has a strongly flat cover.*

Notes on Theorem 2.1. The characterization in (2) has been proved independently by several authors; see for example [2, Lemma 4.5] and [5, Theorem 4.9]. Characterization (3) was obtained by Jaffard in Théorème 6 of [29] and Matlis in Theorem 2.3 of [35][1]. Characterizations (4)–(7) appear in Corollary 3.4 of [4]; for (4) see also Remark 5.4 of [44]. In [4], Anderson and Zafrullah place these characterizations in a more general setting involving star operations.

Characterizations (8) and (9) are taken from Proposition 3.1 of [51]. Characterizations (10)–(13) can be found in Theorem 22 of [36]. A very general view on the fundamental characterization in (13) was obtained by McAdam in [39]: *Let T be an R-module. Then $T = \bigoplus_M T_M$ if and only if $R/Ann_R(t)$ is a finite direct product of quasilocal rings for each nonzero $t \in T$.* For some issues closely related to this characterization, see also the articles of Fuchs and Lee [18, 19]. Characterization (14) is proved by Brandal in Proposition 1 of [9]. Characterization (15) appears in Theorem 2.3 of [35]; recall that an R-module C is *cotorsion* if $\text{Hom}_R(Q, C) = \text{Ext}^1_R(Q, C) = 0$.

Characterizations (16)–(19) are proved in Theorem 4.6 of [2]. The notation $t(A)$ denotes the torsion submodule of A. If $A \subseteq B$ are R-submodules, then A has a *supplement* C in B if C is minimal in the collection of R-submodules D of B such that $A + D = B$. An R-module B is *local* if it is not a sum of its proper submodules.

Bazzoni and Salce classify in [5] the domains for which the strongly flat modules form a covering class, and they prove in Theorem 4.9 of this article the characterization in (20). Regarding the terminology here, an R-module is *strongly flat* if it is a direct summand of an extension of a free module by a torsion-free divisible module. Let A be an R-module and X be a strongly flat R-module. An R-homomorphism $\phi : X \to A$ is a *strongly flat precover* of A if for every R-homomorphism $\phi' : X' \to A$, with X' a strongly flat module, there exists a homomorphism $f : X' \to X$ such that $\phi' = \phi f$. The precover $\phi : X \to A$ is a *strongly flat cover* of A if every endomorphism f of X such that $\phi = \phi f$ is an automorphism.

Although we do not include it here, the interesting case of h-local domains such that $\text{Pic}(R) = 0$ is treated in Theorem 2.1 of [3] and Corollary 3.6 of [4].

Recall that an integral domain R is a Prüfer domain if every nonzero finitely generated ideal of R is invertible; equivalently, R_M is a valuation domain for each maximal ideal M of R. The classes of finite character and h-local Prüfer domains arise in many

---

[1] As discussed in [20, p. 148], Jaffard, in his 1952 article [29] and its sequel [30], appears to have been the first author to recognize the usefulness of pairing the two defining properties of h-local domains. He called the rings satisfying Theorem 2.1(3) "rings of Dedekind type," and showed that (3) was equivalent to the domain being, in our terminology, h-local. In the sixties, Matlis, evidently unaware of Jaffard's work, came at the concept from the module-theoretic point of view of Theorem 2.1(13), and according to his student Brandal, he introduced in 1964 the terminology "h-local," where "h" referred to "homological" [10, p. 17].

ideal- and module-theoretic contexts; see for example the monograph of Fuchs and Salce [20], where these classes of rings figure prominently in a number of applications.

Theorem 2.2. *The following conditions are equivalent for a domain R.*

(1) *R is an h-local Prüfer domain.*

(2) *For every maximal ideal M of R and nonzero $x, y \in M$, $(x,y)R_M \cap R$ is an invertible unidirectional ideal.*

(3) *For all ideals $I, J_1, J_2, \ldots, J_n$ of R,*

$$\left[I : \left[I : \bigcap_{i=1}^{n} J_i\right]\right] = \bigcap_{i=1}^{n} [I : [I : J_i]] \quad \text{and} \quad \left[I : \left[I : \sum_{i=1}^{n} J_i\right]\right] = \sum_{i=1}^{n} [I : [I : J_i]].$$

(4) *$[I : J \cap K] = [I : J] + [I : K]$ for all ideals I, J and K of R.*

(5) *$[M : J \cap K] = [M : J] + [M : K]$ for all ideals J and K of R and maximal ideals M.*

(6) *$[X : Y \cap W] = [X : Y] + [X : W]$ for all R-submodules X, Y and W of Q such that $Y + W \subseteq X$.*

(7) *$[I + J : K] = [I : K] + [J : K]$ for all ideals I, J and K of R.*

(8) *$[Y + W : X] = [Y : X] + [W : X]$ for all R-submodules X, Y and W of Q such that $X \subseteq Y \cap W$.*

(9) *$[R : I \cap J] = [R : I] + [R : J]$ for all ideals I and J of R; R is integrally closed; and each maximal ideal M contains a finitely generated unidirectional ideal.*

(10) *$[R : I \cap J] = [R : I] + [R : J]$ for all ideals I and J of R; R is integrally closed; and no subspace of $\mathrm{Max}\, R$ is homeomorphic to the Stone–Čech compactification $\beta\mathbb{N}$ of $\mathbb{N}$.*

(11) *Every short exact sequence of torsion-free modules,*

$$0 \to A \to I_1 \oplus I_2 \oplus \cdots \oplus I_n \to B \to 0,$$

where $I_1, \ldots, I_n$ are ideals of R, splits.

(12) *For all (maximal) ideals I of R, I is injective with respect to every short exact sequence of torsion-free modules,*

$$0 \to A \to J_1 \oplus J_2 \oplus \cdots \oplus J_n \to B \to 0,$$

where $J_1, \ldots, J_n$ are ideals of R.

Notes on Theorem 2.2. Characterization (2) is quoted from Proposition 4.5 in [4]. Characterizations (3)–(5) and (9)–(12) are taken from Theorems 3.2 and 6.10 of [44]; (6)–(8) can be found in Theorem 2.6 and Corollary 3.5 of [49]. In the article [44], an ideal I of R is defined to be *colon-splitting* if $[I : J \cap K] = [I : J] + [I : K]$ for all ideals J and K of R. This notion is motivated by the fact that every ideal of injective dimension 1 is a colon-splitting ideal [44, Lemma 7.1]. Thus from characterization (5) we deduce that if every maximal ideal of a domain R has injective dimension 1, then R is an h-local Prüfer domain. It follows that a domain R is an almost maximal Prüfer domain if and only if every maximal ideal has injective dimension 1; see [37] and [44]. A number of characterizations and decomposition results for colon-splitting ideals are given in [44]. With regards to (9) and (10), there does exist an integrally closed domain R such that R is a colon-splitting R-ideal but R is not an h-local domain. Such an example is discussed in Example 1.1 of [44]; it has its origins in an article of Brandal [9, Example 2.1], where he proves that a certain overring of the ring of entire functions has injective dimension 1 as a module over itself.

To state (5) of Theorem 2.3, we introduce some ideas developed by Brewer and Klinger in [11] that will also be important in Sections 4 and 5. For a Prüfer domain R, let $\mathfrak{J}(R)$ denote the group of invertible fractional nonzero ideals of R. Then $\mathfrak{J}(R)$ can be partially ordered by $A \leq B$ if and only if $B \subseteq A$ for all $A, B \in \mathfrak{J}(R)$. In fact, since R is a Prüfer domain, then $\mathfrak{J}(R)$ is a lattice-ordered group [11, Theorem 2]. Write the set of maximal ideals of R as $\{M_i\}$. For each i, let v_i denote the valuation corresponding to the valuation ring R_{M_i}, and let G_i denote the value group of R_{M_i}. The group $\prod_{i \in I} G_i$ is lattice-ordered with respect to the pointwise ordering, and, when equipped with this ordering, is called the *cardinal product* of the G_i (cf. [20, p. 108]). Then the mapping

$$\Phi : \mathfrak{J}(R) \to \prod_{i \in I} G_i : A \mapsto (v_i(A))_{i \in I}$$

is an embedding of lattice-ordered groups [11, Theorem 2]. (Here $v_i(A) = \min\{v_i(a) : a \in A\}$. Since A is finitely generated such a minimum exists.) Moreover, the Prüfer domain R has finite character if and only if the image of Φ is a subgroup of the *cardinal sum* $\bigoplus_{i \in I} G_i$ [11, Theorem 2].

Theorem 2.3. *The following statements are equivalent for a Prüfer domain R.*

(1) *R is an h-local domain.*

(2) *For each maximal ideal M of R, $\bigcap_{N \neq M} R_N$ is not a fractional ideal of R, where N ranges over the maximal ideals of R distinct from M.*

(3) *For each maximal ideal M and nonzero prime ideal P contained in M, $\bigcap_{N \neq M} R_N \not\subseteq R_P$, where N ranges over the maximal ideals of R distinct from M.*

(4) *For every finite collection $\{M_1, \ldots, M_n\}$ of maximal ideals of R and every choice of nonnegative elements $g_i \in G_i$, there is a finitely generated ideal A of R such that $v_i(A) = g_i$ for $1 \leq i \leq n$, and $v_j(A) = 0$ for all other maximal ideals M_j of R.*

(5) *Φ maps onto $\bigoplus_{i \in I} G_i$.*

(6) *$[J : I]_M = [J_M : I_M]$ for all ideals I and J of R and maximal ideals M.*

(7) *$[R : I]_M = [R_M : I_M]$ for all ideals I of R and maximal ideals M.*

(8) *For all torsion-free R-modules G and H such that the canonical homomorphism $\mathrm{Hom}_R(G,H) \otimes_R Q \to \mathrm{Hom}_R(G, Q \otimes_R H)$ is an isomorphism, the canonical homomorphism $\mathrm{Hom}_R(G,H) \otimes_R R_M \to \mathrm{Hom}_{R_M}(G_M, H_M)$ is an isomorphism for each maximal ideal M of R.*

(9) *There exists an R-submodule X of Q such that $[X : [X : Y]] = Y$ for all R-submodules Y of X (equivalently, for all ideals Y of R).*

(10) *For each nonzero ideal I of R, I is divisorial if and only if I_M is a divisorial ideal of R_M for each maximal ideal M of R.*

(11) *For each nonzero ideal I of R, $I = (I^{-1})^{-1} M_1 \cdots M_n$, where $M_1, \ldots, M_n$ are precisely the nondivisorial maximal ideals of R which contain I for which I_M is nondivisorial and this factorization is unique in the sense that no M_i can be omitted.*

(12) *Max R is a Noetherian space, and every nonzero prime ideal of R is contained in a unique maximal ideal.*

(13) *Every nonzero ideal of R has finitely many minimal prime ideals, each of which is contained in a unique maximal ideal of R.*

Notes on Theorem 2.3. Characterizations (2) and (3) appear in Corollary 3.2 of [51]; (6) and (7) are taken from Theorem 3.10 of this same article. The characterization in (6) has been placed in a more general setting by Anderson and Zafrullah in [4]. Characterizations (4) and (5) are proved by Brewer and Klingler in Proposition 4 and Theorem 5 of [11]. For another point of view on (5), see [41, Proposition 4.10], where McGovern characterizes among Bézout domains those that are h-local using the group of divisibility of the domain. Characterization (8) is taken from Lemma 4.4 in [48]; it is deduced there from an argument similar to the one Goeters gives in Theorem 1.4 of [22].

For (9), see Lemmas 4.4 and 4.6 of [50]. Characterizations (10) and (11) are proved by Fontana, Huckaba and Lucas in Theorem 1.12 of [17], where it is shown in fact that (10) is also a valid characterization of h-local Prüfer domains if it is only assumed that locally divisorial ideals are divisorial. (Recall that an ideal I is *divisorial* if $I = (I^{-1})^{-1}$.)

The characterization in (13) appears in Proposition 3.4 of [51]. The characterization in (12) follows from this same proposition and Corollaries 1.3 and 1.5 of [53]; the topology here is the subspace topology on Max R induced by the Zariski topology on Spec R.

Regardless of whether R is a Prüfer domain, statements (2), (3), (6), (7), (8), (10) and (12) always hold for an h-local domain; see the above-mentioned references.

3 Examples

We collect next some examples of finite character and h-local domains.

Example 3.1. *If R is a Noetherian domain of Krull dimension* 1, *then R is an h-local domain.* This is because R/I is an Artinian ring for each nonzero ideal I of R, so that every nonzero ideal of R is contained in at most finitely many maximal ideals of R. More generally, a domain R is *almost perfect* if for every nonzero ideal I of R, R/I has the descending chain condition for principal ideals. Bazzoni and Salce prove in Theorem 4.5 of [5] that a domain is almost perfect if and only if it is h-local and locally almost perfect. Interesting examples of almost perfect, hence h-local, non-Noetherian domains can be found in [7], [54] and [58]. □

Example 3.2. *There exist non-local Noetherian domains D such that D is h-local and every maximal ideal of D has height* 2. In Theorem 2 of [57], S. Wiegand proves the following realization theorem for posets: *Let X be a countable partially ordered set with unique minimal element* 0 *and finitely many maximal elements $m_1, m_2, \ldots, m_n$, all of height 2, and suppose that each element of X is contained in a maximal element. For each $i \leq n$, let $G(m_i) = \{p \in X : p \leq m_i\}$. Then X is order isomorphic to* Spec R *for some countable semilocal Noetherian domain R if and only if for each $i \leq n$, the set $G(m_i) \setminus \bigcup_{j \neq i} G(m_j)$ is infinite.* By choosing X so that $n > 1$ and the sets $G(m_i) \setminus \{0\}$ are pairwise disjoint, we obtain from Wiegand's theorem an h-local Noetherian domain of Krull dimension 2 that is not local. □

If R is a domain and $\mathcal{F}$ is a collection of prime ideals of R, then $\mathcal{F}$ has *finite character* if every nonzero element of R is contained in at most finitely many prime ideals in $\mathcal{F}$.

Example 3.3. *For each $n > 0$, there exists a Noetherian h-local domain having infinitely many maximal ideals and Krull dimension n.* In Theorem 5.3 and Corollary 5.4 of [24], Gulliksen gives an example of a finite character Noetherian domain R whose regular locus and Cohen-Macaulay locus coincide but are not constructible in Spec R. He does so by providing a general method of constructing Noetherian domains as localizations of polynomial rings in infinitely many variables. Let K be a field, and let $D = K[X_i : i \in I]$ be a polynomial ring over a set $\{X_i : i \in I\}$ of indeterminates for K. Suppose that $\mathcal{F}$ is a partition of $\{X_i : i \in I\}$ into finite sets. For each $A \in \mathcal{F}$, define $\mathfrak{p}_A = AD$. Observe that each $0 \neq f \in D$ is in at most finitely many of the $\mathfrak{p}_A$, $A \in \mathcal{F}$, so that $\{\mathfrak{p}_A\}$ is a finite character collection of prime ideals of D. Define $S = D \setminus (\bigcup_{A \in \mathcal{F}} \mathfrak{p}_A)$ and $R = D_S$. Then by Theorems 3.6 and 4.1 of [24], R is a finite character Noetherian domain whose set of maximal ideals is $\{AR : A \in \mathcal{F}\}$. Thus if $\mathcal{F}$ has infinitely many members, and all but one of these members, say A_1, is a singleton, then R is an h-local Noetherian domain having infinitely many maximal ideals, and the Krull dimension of R is equal to the number of elements in A_1. This is because each maximal ideal of R other than $A_1 R$ is a height 1 ideal of R. □

If R is a domain with quotient field Q, then the *group of divisibility* of R is the group of nonzero principal fractional ideals of R. This is a partially ordered group with respect to the ordering defined by $xR \leq yR$ if and only if $yR \subseteq xR$ for all nonzero $x, y \in R$. The group of divisibility of the domain R encodes a good deal of the ideal-theoretic structure of R; see for example Chapter III of [20]. Recall that an integral domain R is a *Bézout domain* if every finitely generated ideal of R is a principal ideal, and note that if R is a Bezout domain, then the group of divisibility is lattice-ordered and coincides with $\mathfrak{I}(R)$. The Jaffard–Kaplansky–Ohm Theorem asserts that every lattice-ordered group can be realized as the group of divisibility of a Bézout domain [20, Theorem 5.3]. This is the key idea behind the next example.

Example 3.4. *Every cardinal sum of totally ordered abelian groups occurs as the group of divisibility of an h-local Bézout domain.* Let $\{G_i\}$ be a collection of totally ordered abelian groups, and define G to be the cardinal sum $\bigoplus_i G_i$. Then G is a lattice-ordered abelian group, so there exists an h-local Bézout domain whose group of divisibility is isomorphic to G [20, Example III.5.4]. □

A Prüfer domain R is *strongly discrete* if $P \neq P^2$ for every nonzero prime ideal P of R. It is implicit in the article [6] of Bazzoni and Salce that a Prüfer domain R is h-local and strongly discrete if and only if every nonzero ideal of every overring S of R is a divisorial ideal of S. In another direction, it is shown in [46] (but see also [47]) that a domain is an h-local strongly discrete Prüfer domain if and only if every proper ideal is a product of prime and finitely generated ideals.

Example 3.5. *There exist h-local strongly discrete Bézout domains of arbitrarily large Krull dimension, and having arbitrarily large numbers of maximal ideals.* Let $(X, \leq)$ be a Noetherian tree having a least element 0, and suppose that for each $0 \neq p \in X$ there exists a unique maximal element m of X such that $p \leq m$. Then there exists an h-local Bézout domain R such that Spec R is isomorphic as a partially ordered set to X. This is proved in [51, Proposition 5.5], using a result of Facchini [16, Theorem 5.3] that depends ultimately on the Jaffard–Kaplansky–Ohm Theorem. □

Let D be a domain, and let Σ be a collection of valuation overrings of D. Then Σ has *finite character* if each nonzero element of D is a unit in all but at most finitely many members of Σ. The next examples show how certain finite character collections of valuation rings give rise to finite character and h-local Prüfer domains. If R is the intersection of finitely many *independent* valuation rings all sharing the same quotient field, then R is an h-local Prüfer domain; this is a consequence of a theorem of Nagata [42, Theorem 11.11]. The next example discusses a situation where "independent" is redundant. Recall the notion of a reflexive domain discussed in the introduction. A valuation domain V is reflexive if and only if V has a principal maximal ideal and Q/V is an injective V-module; see for example [20, Theorem XV.9.5].

Example 3.6. *If $V_1, \ldots, V_n$ are reflexive valuation domains, all of which have the same quotient field, then $R = V_1 \cap \cdots \cap V_n$ is an h-local domain.* In fact, R is reflexive domain [50, Corollary 4.3]. The content of this assertion is that such valuation rings must be independent, a fact that can be traced through a result of Facchini [16, Theorem 4.3] back to Vámos [55]. □

In general, a finite character intersection of reflexive valuation domains need not be h-local; consider for example the ring $K[X,Y]$, where K is a field and X and Y are indeterminates. Here $K[X, Y]$ is a finite character intersection of DVRs (i.e., quasilocal PIDs), but $K[X, Y]$ is neither h-local nor a Prüfer domain. However, working with power series in infinitely many variables, it is possible to construct h-local Prüfer overrings as finite character intersections of infinitely many reflexive valuation domains. This is done in [45].

Example 3.7. *A "small" finite character intersection of valuation rings is a finite character Bézout domain.* Let D be a domain containing a field of cardinality α, and let Σ be a collection of valuation overrings of D. If Σ has cardinality $< \alpha$, then $R := \bigcap_{V \in \Sigma} V$ is a Bézout domain [52, Theorem 6.6]. If in addition Σ has finite character, then R is a finite character domain. This follows from the fact that a Prüfer domain that is a finite character intersection of valuation overrings is necessarily of finite character [12, Corollary 2]. Note that R is h-local if the valuation rings in Σ are pairwise independent. □

Example 3.8. *A finite character intersection of valuation overrings of a domain containing a non-algebraically closed field K is a finite character Prüfer domain if each valuation ring has residue field K. If also these valuation rings are independent, then R is an h-local domain.* Let D be a domain containing a non-algebraically closed field K. Suppose that Σ is a finite character collection of valuation overrings of D, each having residue field K. Then $R = \bigcap_{V \in \Sigma} V$ is a Prüfer domain, a fact attributable to several authors; see [33] and [43] for discussion and references for this result. For simplicity we have stated only a special case, and by consulting [33] or [43] the reader may easily generalize our example. As in Example 3.7, we see that if Σ has finite character, then R is a finite character domain. If in addition the valuation rings in Σ are independent, then R is h-local. □

For the constructions in the previous two examples to be useful, it is necessary to be able to locate interesting collections of finite character collections of valuation overrings. One straightforward way to do this is the following. Let D be a domain, let Σ be a collection of valuation overrings of D and let $\mathcal{F}$ be the collection of prime ideals of D on which the valuation rings in Σ are centered. Then Σ has finite character if and only if $\mathcal{F}$ has finite character and for each prime ideal P in $\mathcal{F}$, there are at most finitely many members of Σ centered on P. Thus, given a finite character collection $\mathcal{F}$ of prime ideals, a finite character collection of valuation overrings can be constructed by selecting for each prime ideal P in $\mathcal{F}$, finitely many valuations overrings centered on P. (This is in fact what we do in Sections 4 and 5.) We give in the following proposition a simple criterion for when a countable domain has an infinite finite character collection of prime ideals. More interesting examples are given in Section 5.

Proposition 3.9. *Let D be a countable domain. Suppose that $\mathcal{F}$ is a collection of prime ideals with $\bigcap_{\mathfrak{p} \in \mathcal{F}} \mathfrak{p} = 0$. Then there exists an infinite finite character collection of prime ideals in $\mathcal{F}$.*

Proof. For each $d \in D$, let $V(d)$ denote the set of prime ideals in $\mathcal{F}$ that contain d. Enumerate the nonzero nonunits of D as $d_1, d_2, d_3, \ldots$ We claim: $(\star)$ *for all $N > 0$, there exists $t > N$ such that $V(d_t) \nsubseteq V(d_1 d_2 \cdots d_{t-1})$.* For otherwise there is $N > 0$

such that for all $i > 0$, $V(d_{N+i}) \subseteq V(d_1 d_2 \cdots d_{N+i-1})$. But then for each $i > 0$, $V(d_{N+i}) \subseteq V(d_1 d_2 \cdots d_N) \cup V(d_{N+1}) \cup \cdots \cup V(d_{N+i-1}) = V(d_1 d_2 \cdots d_N)$, and it follows that $\mathcal{F} = V(d_1) \cup V(d_2) \cup \cdots \cup V(d_N)$, so that $d_1 d_2 \cdots d_N \in \bigcap_{\mathfrak{p} \in \mathcal{F}} \mathfrak{p} = 0$, contrary to the assumption that the d_i are nonzero elements of D. This proves $(\star)$.

Using the claim $(\star)$, we now define a sequence of positive numbers $t_1 < t_2 < \cdots < t_n < \cdots$ in the following way. Let t_1 be the smallest positive number such that $V(d_{t_1}) \not\subseteq V(d_1 \cdots d_{t_1-1})$. For each $k > 0$, define (via $(\star)$) t_k to be the smallest number such that $t_k > t_{k-1}$ and $V(d_{t_k}) \not\subseteq V(d_1 d_2 \cdots d_{t_k-1})$, and let $\mathfrak{p}_k \in V(d_{t_k}) \setminus V(d_1 d_2 \cdots d_{t_k-1})$. Then for each $i > 0$, $d_i \in \mathfrak{p}_k$ only if $t_k \leq i$. In particular, d_i is contained in at most finitely many of the $\mathfrak{p}_k$. □

There is an interesting connection here to a couple of topics, one from module theory and the other from ideal theory.

Slender rings. In Theorem 2 of [32], E. L. Lady proves that if a commutative ring R has an infinite finite character collection of maximal ideals, then R is a *slender ring*, meaning that for every countable family $\{A_i\}$ of R-modules and every homomorphism $\phi : \prod_{i=1}^{\infty} A_i \to R$, there exists $n > 0$ such that $\phi(\prod_{i \geq n} A_i) = 0$. Lady shows in Examples 5 and 6 of [32] that every domain that is finitely generated as an algebra over a field has a countably infinite finite character collection of maximal ideals. We derive this same fact in Corollary 5.3, but from a somewhat more general point of view (namely, Proposition 5.1).

Conforming spectra. Motivated by applications in [28] and [38, Chapters 3, 7 and 14], McAdam considers in the unpublished manuscript [40] a stronger version of what we consider here[2]. Let R be a commutative ring, let $P \in \operatorname{Spec} R$ and let $X \subseteq \operatorname{Spec} R$ such that every prime ideal in X properly contains P. Then (P, X) is a *conforming pair* if X is infinite and $\{L/P : L \in X\}$ is a finite character collection of prime ideals of R/P. Thus a domain R has finite character and infinitely many maximal ideals if and only if $(0, \operatorname{Max} R)$ is a conforming pair. The ring R has a *conforming spectrum* if for all prime ideals P of R and $X \subseteq \operatorname{Spec} R$ such that every prime ideal in X properly contains P and such that $\bigcap_{L \in X} L = P$, there exists $Y \subseteq X$ such that (P, Y) is a conforming pair. Thus if a domain R has Jacobson radical 0 and a conforming spectrum, then R has an infinite finite character collection of maximal ideals. Among a number of interesting results in [40], McAdam proves: *If a ring R is finitely or countably generated over a field or R is an integral extension of a ring with a conforming spectrum, then R has a conforming spectrum.* He asks whether every Noetherian ring has a conforming spectrum, and he indicates that he suspects the answer is no. However, in the other direction, he shows that to prove that every Noetherian ring of finite Krull dimension has a conforming spectrum, it suffices to show that every Noetherian domain R of Krull dimension 2 has the property that whenever Y is a set of nonzero prime ideals of R with $\bigcap_{P \in Y} P = 0$, then there exists an infinite finite character collection $X \subseteq Y$.

---

[2] I thank Steve McAdam for showing me this manuscript, and for some very helpful conversations on these topics.

4 H-local Prüfer overrings of Noetherian rings

Our goal in this section and the next is to construct h-local Prüfer overrings of Noetherian domains. Recall from the discussion preceding Theorem 2.3 the lattice-ordered group $\mathfrak{I}(R)$ of invertible fractional ideals of a Prüfer domain R. In this section we identify precisely which lattice-ordered groups occur as the group of invertible fractional ideals of an h-local Prüfer overring of a Noetherian domain. As indicated by Theorem 2.3 and Example 3.4, a lattice-ordered abelian group is the group of invertible fractional ideals of an h-local Prüfer domain if and only if it is a cardinal sum $\bigoplus_{i \in I} G_i$, where $\{G_i : i \in I\}$ is a collection of totally ordered abelian groups. As explained in Example 3.4, one way that a given cardinal sum $\bigoplus_{i \in I} G_i$ can be realized as the group of invertible fractional ideals of a domain R is via the Jaffard–Kaplansky–Ohm Theorem. However, the ring R arising from this construction is very large – it occurs as an overring of a polynomial ring in infinitely many variables. In this section we show that a similar realization theorem can be proved for overrings of Noetherian domains, and thus one obtains a diverse set of examples of h-local Prüfer overrings of rings of classical interest.

Our method of construction is via intersections of valuation rings "hidden" over finite character collections of maximal ideals. We establish in Lemma 4.2 that such intersections are indeed h-local and Prüfer. This is a consequence of a technical lemma:

Lemma 4.1. *Let D be a domain, let Σ be a finite character collection of valuation overrings of D and let $X = \{P \cap D : P \text{ is a nonzero prime ideal of some member } V \text{ of } \Sigma\}$. If X satisfies the descending chain condition and any two incomparable prime ideals in X are comaximal in D, then $R := \bigcap_{V \in \Sigma} V$ is a finite character Prüfer domain.*

Proof. We show first that R is a Prüfer domain by proving that R_M is a valuation domain for each maximal ideal M of R. Let M be a maximal ideal of R, and set $\mathfrak{m} = M \cap D$. Since Σ has finite character and localization commutes with finite character intersections, we have $R_\mathfrak{m} = \bigcap_{V \in \Sigma} V_\mathfrak{m}$ [26, Lemma 1]. Let $\Gamma = \{V \in \Sigma : V_\mathfrak{m} \neq Q\}$. We claim that Γ is a finite set. For once this is established, it follows that $R_\mathfrak{m}$ is a finite intersection of valuation overrings, and hence $R_\mathfrak{m}$ is a Prüfer domain [42, Theorem 11.11]. Since then R_M is a quasilocal overring of the Prüfer domain $R_\mathfrak{m}$, it must be that R_M is a valuation domain, proving the claim. Define $Y = \{P \cap D : P \text{ is a nonzero prime ideal of some member of } \Gamma \text{ such that } P \cap D \subseteq \mathfrak{m}\}$. Since by assumption incomparable elements of X are comaximal, Y is a chain, and hence, by assumption, has a smallest element, say $\mathfrak{p}$. However, since Σ has finite character, $\mathfrak{p}$ survives in only finitely many members of Σ, and this then implies that Γ is finite. For suppose that $V \in \Gamma$. Then there exists a nonzero prime P of V such that $P \cap D \subseteq \mathfrak{m}$, and hence $P \cap D \in Y$. But then $\mathfrak{p} \subseteq P$, so that $\mathfrak{p}$ survives in V, and since $\mathfrak{p}$ survives in only finitely many members of Σ, we conclude that Γ is a finite set. This proves that R is a Prüfer domain. Moreover, as a Prüfer domain that is a finite character intersection of valuation domains, R is a finite character domain [12, Corollary 2]. □

Let V be a valuation overring of the domain D. Then V is a *hidden* valuation overring of D, if for every nonzero prime ideal P of V, $P \cap D$ is a maximal ideal of D. Examples of hidden valuation overrings of Noetherian domains abound. For example, suppose that $(x_1, \ldots, x_n)D$ is a maximal ideal of the Noetherian domain D with $x_1 \neq 0$, and set $R = D[x_1, x_2 x_1^{-1}, \ldots, x_n x_1^{-1}]$. If V is a valuation overring of R such that x_1 is in every nonzero prime ideal of V, then V is a hidden valuation overring of D; indeed, it is hidden over $(x_1, \ldots, x_n)D$. In this way, one can construct diverse examples of hidden valuation overrings; see Lemma 4.3 and the discussion that precedes it.

Lemma 4.2. *Let D be a domain, and let Σ be a finite character collection of hidden valuation overrings of D. Then $R := \bigcap_{V \in \Sigma} V$ is a finite character Prüfer domain. If also distinct valuation rings in Σ are centered on distinct maximal ideals of D, then R is an h-local domain and $\Sigma = \{R_M : M \text{ is a maximal ideal of } R\}$.*

Proof. The set X (where X is as in Lemma 4.1) consists of maximal ideals of D, and hence meets the requirements of the lemma. Thus in view of the lemma, R is a finite character Prüfer domain. Suppose also that distinct valuation rings in Σ are centered on distinct maximal ideals of R, and that M and N are distinct maximal ideals of R containing a nonzero prime ideal P of R. Then since R is a Prüfer domain and Σ has finite character, we have $R_M, R_N \in \Sigma$ [12, Theorem 1]. Thus, since R_M and R_N are valuation domains, we have $PR_M = PR_P = PR_N$. But then $PR_M \cap D = PR_P \cap D = PR_N \cap D$, contrary to the assumption that distinct members of Σ are centered on distinct maximal ideals of D. We conclude that R is an h-local Prüfer domain.

We have already observed that $\{R_M : M \text{ is a maximal ideal of } R\} \subseteq \Sigma$. To see that the reverse inclusion holds, let $V \in \Sigma$. Since R is a Prüfer domain, there exists a prime ideal P of R such that $V = R_P$. Let M be a maximal ideal of R containing P. Then, as we have established, $R_M \in \Sigma$. Since each member of Σ is hidden above D and centered on a distinct maximal ideal of D, it follows that the members of Σ are incomparable with respect to inclusion. Therefore, since $R_M \subseteq R_P = V$, this forces $R_M = V$, which proves the lemma. □

Recall that the *rational rank* of a totally ordered abelian group G is the $\mathbb{Q}$-dimension of $\mathbb{Q} \otimes_{\mathbb{Z}} G$. The *rank* of G is the order type of the set of proper convex subgroups of G. The group G is *discrete* if it is order-isomorphic to the Hahn product of copies of $\mathbb{Z}$. In case G has finite rank, then G is discrete if and only if G is order-isomorphic to a direct sum of copies of $\mathbb{Z}$ with the lexicographic ordering. Recently, F. V. Kuhlmann has proved very general existence theorems for valuation overrings of polynomial rings. We quote only a very special case that is needed here: *Let K be a field, and let w be a valuation on K with value group G_w and residue field k. Let G be a countably generated ordered abelian group extension of G_w such that G/G_w is an infinite torsion group. Then there exists an extension v of w from K to $K(X)$ such that the value group of v is G and the residue field of v is k* [31, Proposition 3.17].

Lemma 4.3. *Let $D = K[X_1, \ldots, X_n]$, where K is a field and $X_1, \ldots, X_n$ are indeterminates for K. If $\mathfrak{m}$ is a maximal ideal of D having residue field K and G is a* *totally ordered abelian group of rational rank $< n$ or a discrete group of rank n, then there exists a hidden valuation overring V of D centered on $\mathfrak{m}$ such that V has value group G.*

Proof. Since $\mathfrak{m}$ has residue field K, there exist $k_1, \ldots, k_n \in K$ such that $\mathfrak{m} = (X_1 - k_1, \ldots, X_n - k_n)D$. Thus after a change of variables we may assume without loss of generality that $\mathfrak{m} = (X_1, \ldots, X_n)D$. Let $g_1, \ldots, g_k$ be a maximal linearly independent set of positive elements of G (so that $k =$ rational rank of G), and such that g_1 is not in any proper convex subgroup of G. Define $H = \mathbb{Z}g_1 + \cdots + \mathbb{Z}g_k$. We may in addition assume that if G is a finitely generated abelian group, then $g_1, \ldots, g_k$ are chosen so that $G = H$. Moreover, by replacing $\{g_1, g_2, \ldots, g_k\}$ with the set $\{g_1, g_2 + g_1, \ldots, g_k + g_1\}$, which also is a maximal linearly independent subset of G that generates H, we may assume that $g_1 < g_i$ for all $i > 1$.

We will use Kuhlmann's theorem to show there exists a valuation v on $K(X_1, \ldots, X_n)$ with value group G such that v is trivial on K, $v(X_i) = g_i$ for all $1 \leq i \leq k$ and $v(X_i) = g_1$ for all $k < i \leq n$. For given such a valuation v with corresponding valuation ring V, $X_i V \subseteq X_1 V$ for all $1 \leq i \leq n$, and since g_1 is not in any proper convex subgroup of G and $v(X_1) = g_1$, X_1, hence each X_i, is an element of every nonzero prime ideal of V [15, Lemma 2.3.1]. Thus V is a valuation ring hidden over the maximal ideal $\mathfrak{m}$ of D and having value group G.

To construct such a valuation v, we first observe that since the elements $g_1, \ldots, g_k$ are linearly independent in G, there exists a valuation $w : K(X_1, \ldots, X_k) \to G \cup \{\infty\}$ such that $w(X_i) = g_i$ for all $1 \leq i \leq k$; cf. [8, VI.10.3, Theorem 1] or [31, Lemma 2.6]. Thus if G is a finitely generated abelian group, then by assumption $H = G$, and we may extend w to a valuation $v : K(X_1, \ldots, X_n) \to G \cup \{\infty\}$ such that $v(X_i) = g_i$ for all $1 \leq i \leq k$ and $v(X_j) = g_1$ for all $k < j \leq n$ [15, Theorem 2.2.1]. Therefore, the claim holds when G is a finitely generated abelian group.

On the other hand, suppose that G is not a finitely generated abelian group. Then by assumption the rational rank k of G is less than n. Let w' be a valuation extension of w to $K(X_1, \ldots, X_{n-1})$ such that $w'(X_j) = g_1$ for all $k < j \leq n - 1$ [15, Theorem 2.2.1]. Then w' has the same value group H as w. Now, since G is a torsion-free group and H is a finitely generated torsion-free group but G is not, it must be that G/H is an infinite group. Moreover, since H is generated by a maximal linearly independent subset of G, G/H is a torsion group. Thus by the theorem of Kuhlmann discussed above, there exists an extension v of w' to $K(X_1, \ldots, X_n)$ such that v has value group G. Thus we have verified the claim in all cases, and the proof is complete. □

We use the Kronecker function ring construction in the proof of Theorem 4.4. Let R be an integrally closed domain with quotient field Q, and let T be an indeterminate for Q. If V is a valuation overring of R, then V^b denotes the trivial extension of V to $Q[T]$ (i.e., V^b is the valuation domain of $Q(T)$ determined by the valuation that assigns to a polynomial in $Q[T]$ the minimum of the values of its coefficients). We define $R^b = \bigcap_V V^b$, where V ranges over the valuation overrings of R. Then R^b is a Bézout domain [21, Theorem 32.7]. If R is an h-local Prüfer domain, then R^b is an h-local Bézout domain; for example, apply Corollary 3.4 of [4]. Moreover, if R is a Prüfer domain, then the mapping $\gamma : \mathfrak{I}(R) \to \mathfrak{I}(R^b)$ defined by $\gamma(A) = AR^b$ is an order-isomorphism since every finitely generated ideal of R^b can be generated by elements of R, and $AR^b \cap R = A$ for all finitely generated ideals A of R [21, Theorem 32.7].

Theorem 4.4. *If R is an h-local Prüfer overring of a Noetherian domain D, then $\mathfrak{I}(R)$ is a cardinal sum $\bigoplus_{i \in I} G_i$, where $|I| \leq |D|$ and $\{G_i : i \in I\}$ is a collection of totally ordered abelian groups such that each G_i has finite rational rank. If in addition D has finite Krull dimension n, then each G_i has rational rank $< n$ or is a discrete group of rank n. Conversely, let $\{G_i : i \in I\}$ be a collection of totally ordered abelian groups, each of finite rational rank. Then there exists a Noetherian domain D with $|D| = \max\{\aleph_0, |I|\}$ and an h-local Bézout overring R of D such that $\mathfrak{I}(R)$ is order-isomorphic to the cardinal sum $\bigoplus_{i \in I} G_i$.*

Proof. Suppose that R is an h-local Prüfer overring of a Noetherian domain D. Since each ideal of D is finitely generated, it follows there are at most $|D|$ many ideals of D. Let $\Sigma = \{R_M : M \text{ is a maximal ideal of } R\}$. Then since R is a finite character domain, Σ is a finite character collection of valuation overrings of R. Hence there are at most finitely many members of Σ centered on a given prime ideal of R, and so there are at most $|D|$ many members of Σ. Let $V \in \Sigma$, and let M be the maximal ideal of V. Then V is a valuation overring of the Noetherian domain $D_{M \cap D}$. Let n denote the Krull dimension of $D_{M \cap D}$. Then since $D_{M \cap D}$ is a Noetherian domain, the value group G of V is either a discrete group of rank n or has rational rank $< n$ [1, Theorem 1]. Thus the claim is a consequence of Theorem 2.3(5).

To prove the converse, let K be a countable field, and let $\{X_{ij} : i \in I, j \in \mathbb{N}\}$ be a collection of indeterminates for K. Define $D = K[X_{ij} : i \in I, j \in \mathbb{N}]$. For each $i \in I$, let $n_i = 1 + \text{rank } G_i$, and define $\mathfrak{p}_i = (X_{i1}, X_{i2}, \ldots, X_{in_i})D$ and $F_i = K(X_{i'j} : i' \neq i$ or $j \in \mathbb{N} \setminus \{1, 2, \ldots, n_i\})$. Then $D_{\mathfrak{p}_i} = F_i[X_{i1}, \ldots, X_{in_i}]_{\mathfrak{p}_i}$, so by Lemma 4.3, there exists a hidden valuation overring V_i of $D_{\mathfrak{p}_i}$ with value group G_i. Define $S = D \setminus (\bigcup_{i \in I} \mathfrak{p}_i)$. As discussed in Example 3.3, D_S is a finite character Noetherian domain whose maximal ideals are of the form $\mathfrak{p}_i D_S$, $i \in I$. Since $\{V_i : i \in I\}$ is a collection of hidden valuation overrings of D_S, each centered on a distinct maximal ideal, we have by Lemma 4.2 that $R := \bigcap_{i \in I} V_i$ is an h-local Prüfer domain with $\{V_i : i \in I\} = \{R_M : M \in \text{Max} R\}$. Thus by Theorem 2.3(5), $\mathfrak{I}(R)$ is order-isomorphic to the cardinal sum $\bigoplus_{i \in I} G_i$. Also, since K is countable, $|D_S| = |K(X_{ij} : i \in I, j \in \mathbb{N})| = |I|$. Finally, as discussed above, the ring R^b is an h-local Bézout domain, and $\mathfrak{I}(R)$ is order isomorphic to $\mathfrak{I}(R^b)$. Thus the claim is proved, since $|D_S[T]| = |D_S| = |I|$ and R^b is an overring of the Noetherian domain $D_S[T]$. □

From the theorem, we deduce precisely which lattice-ordered groups are realizable as the group divisibility of an h-local Bézout overring of a Noetherian domain; compare this to Example 3.4.

Corollary 4.5. *Let G be a lattice-ordered group. Then G is order-isomorphic to the group of divisibility of an h-local Bézout overring of a Noetherian domain if and only if G is a cardinal sum of totally ordered abelian groups, each having finite rational rank.* □

5 H-local Prüfer overrings of polynomial rings

We next consider the problem of realizing lattice-ordered groups as the group of invertible fractional ideals of an h-local Prüfer overring of a polynomial ring in *finitely many* variables. Thus, in comparison to the last section, we seek a realization theorem for a very specific sort of Noetherian ring. As we see in this section, the obstacle here is finding infinite finite character collections of maximal ideals. Proposition 3.9 provides some simple examples of domains possessing an infinite finite character set of maximal ideals, but the approach in the proposition does not allow one to control the residue fields of the maximal ideals, something that we need to be able to do in Theorem 5.8. Thus we give in Proposition 5.1 an argument based on a technique of R. Heitmann in [27], where it is shown that if D is a countable PID of characteristic 0, then $D[X]$ has, in our terminology, a finite character set of maximal ideals with specified residue fields. For a related application, see the construction of Dedekind domains given by Goldman in [23].

Proposition 5.1. *If D is a countable domain with a finite character set $\{\mathfrak{m}_i : i \in \mathbb{N}\}$ of distinct maximal ideals, and $X_1, \ldots, X_n$ are indeterminates for D, then there exists a finite character set $\{\mathfrak{n}_i\}$ of distinct maximal ideals of $R := D[X_1, \ldots, X_n]$ such that for each i, $\mathfrak{m}_i = \mathfrak{n}_i \cap D$ and $R/\mathfrak{n}_i \cong D/\mathfrak{m}_i$.*

Proof. It is enough to prove the proposition for the case $n = 1$, since an inductive argument then finishes the proof. We first show there exist only finitely many positive integers i such that $|D/\mathfrak{m}_i| < n$. Suppose there exists $N > 0$ such that $|D/\mathfrak{m}_i| < N$ for infinitely many i. Let d be a nonunit in D. By assumption, $|D/\mathfrak{m}_i| < N$ for infinitely many i, so $d^{N!} + \mathfrak{m}_i = 1 + \mathfrak{m}_i$ for infinitely many i. Thus $d^{N!} - 1 \in \mathfrak{m}_i$ for infinitely many i, and since $\{\mathfrak{m}_i\}$ has finite character, $d^{N!} = 1$, implying d is a unit. This contradiction means that for each $n > 0$, there exist only finitely many positive integers i such that $|D/\mathfrak{m}_i| < n$. In particular, for each $n > 0$, there exists $i > 0$ such that $|D/\mathfrak{m}_j| > n$ for all $j > i$.

Now we show the proposition holds for the ring $D[X]$, where X is an indeterminate, by constructing a finite character set of maximal ideals of $D[X]$ of the form $(\mathfrak{m}_i, X - d_i)$, $d_i \in D$. (Here we use Heitmann's technique from Lemma 2.5 of [27].) Observe that for $f(X) \in D[X]$, then $f(X) \in (\mathfrak{m}_i, X - d_i)$ if and only if $f(d_i) \in \mathfrak{m}_i$. Thus we seek elements $d_1, d_2, d_3, \ldots$ of D such that for all $f(X) \in D[X]$, $f(d_i) \in \mathfrak{m}_i$ for only finitely many i.

Since $D[X]$ is countable, we may enumerate its elements: $f_1(X), f_2(X), f_3(X), \ldots$. We define an ascending sequence of positive integers $t_1, t_2, t_3, \ldots$ in the following manner. As we have shown, we may select t_1 such that $|D/\mathfrak{m}_t| > (\deg f_1(X)) + 1$ for all $t > t_1$. More generally, we may select $t_n > t_{n-1}$ such that

$$|D/\mathfrak{m}_t| > \Big(\sum_{j=1}^{n} \deg f_j(X)\Big) + 1$$

for all $t > t_n$. In this way we obtain a sequence $\{t_n\}$ of positive integers having the property that if $t > t_n$, then the image of $g_n(X) := f_1(X) \cdots f_n(X)$ in $(D/\mathfrak{m}_t)[X]$ is not identically zero for all elements of the field $D/\mathfrak{m}_t$. Indeed, the degree of $g_n(X)$ is less than $|D/\mathfrak{m}_{t_1}| - 1$ for all $t > t_n$, yet $g(X)$ has at most $\deg g(X)$ roots modulo $\mathfrak{m}_t$. In particular, for each n and $t > t_n$, there exists $d \in D$ such that $g_n(d) \notin \mathfrak{m}_t$.

Arbitrarily choose $d_1, \ldots, d_{t_1}$. If $m > 0$, select $d_{t_m+1}, \ldots, d_{t_{m+1}}$ such that $g_m(d_t) \notin \mathfrak{m}_t$ for all t such that $t_m + 1 \le t \le t_{m+1}$. Then for all $n > 0$, $f_n(d_t) \notin \mathfrak{m}_t$ for all $t > t_n$. Hence $f_n \notin (\mathfrak{m}_t, X - d_t)$ for all $t > t_n$. For each t, set $\mathfrak{n}_t = (\mathfrak{m}_t, X - d_t)$. Then it follows that $\{\mathfrak{n}_t\}_{t>0}$ is a finite character set of maximal ideals of $D[X]$. Moreover, for each t, $\mathfrak{m}_t = \mathfrak{n}_t \cap D$ and $D[X]/\mathfrak{n}_t \cong D/\mathfrak{m}_t$. This proves the proposition. □

In Corollary 5.3, we extend the proposition to affine domains over countable domains. This is done via:

Lemma 5.2. *Let $D \subseteq R$ be an integral extension of domains, and let $\{\mathfrak{p}_i\}_{i \in I}$ be a finite character collection of distinct prime ideals of D. If $\{\mathfrak{q}_i\}_{i \in I}$ is a collection of prime ideals of R such that for each $i \in I$, $\mathfrak{q}_i$ lies over $\mathfrak{p}_i$, then $\{\mathfrak{q}_i\}_{i \in I}$ has finite character.*

Proof. Suppose r is a nonzero element of R. Since R is integral over D, there exists a monic polynomial $g(X) \in D[X]$ such that $g(r) = 0$. Write $g(X) = X^n + a_{n-1}X^{n-1} + \cdots + a_0$. Since R is integral domain, we may assume without loss of generality that $a_0 \ne 0$. Let $f(X) = X^{n-1} + a_{n-1}X^{n-2} + \cdots + a_1$. Now $0 = g(r) = rf(r) + a_0$, and we have $rf(r) = -a_0$. If r is an element of infinitely many of the $\mathfrak{q}_i$, then so is $-a_0 = rf(r)$. But this implies that $0 \ne a_0 \in \mathfrak{q}_i \cap D = \mathfrak{p}_i$ for infinitely many i, contrary to assumption. □

Corollary 5.3. *Let D be a countable domain having a collection $\mathcal{F}$ of maximal ideals such that $\bigcap_{\mathfrak{m} \in \mathcal{F}} \mathfrak{m} = 0$. If R is a domain that is finitely generated as a D-algebra, then there exists an infinite finite character collection of maximal ideals of R, each of which lies over a distinct member of $\mathcal{F}$.*

Proof. By Proposition 3.9, there exists an infinite finite character collection $\{\mathfrak{m}_i\}$ of maximal ideals of D contained in $\mathcal{F}$. Let Q denote the quotient field of D. By applying Noether Normalization to $Q \otimes_D R$, we obtain Q-algebraically independent elements $x_1, x_2, \ldots, x_n$ of R such that $Q \otimes_D R$ is module-finite over $Q[x_1, x_2, \ldots, x_n]$. Thus there is an element $d \in D$ such that $R[d^{-1}]$ is integral over $D[d^{-1}, x_1, x_2, \ldots, x_n]$. Since $\{\mathfrak{m}_i\}$ has finite character, $D[d^{-1}]$ has an infinite finite character set of maximal ideals, namely those $\mathfrak{m}_i D[d^{-1}]$ such that $\mathfrak{m}_i$ survives in $D[d^{-1}]$. Hence by Proposition 5.1, $D[d^{-1}, x_1, \ldots, x_n]$ has an infinite finite character set of maximal ideals, each lying over some member of $\mathcal{F}$. By Lemma 5.2, $R[d^{-1}]$ also has an infinite finite character set of maximal ideals, say $\{\mathfrak{n}_k\}_{k>0}$, each lying over a maximal ideal in $\{\mathfrak{m}_i D[d^{-1}]\}$. Thus $\{\mathfrak{n}_k \cap R\}$ is an infinite character set of maximal ideals of R. □

Our specific interest in Proposition 5.1 is when $R = K[X_1, \ldots, X_n]$, with K a field. It is this case, which we treat in the next corollary, that is needed in the proof of Theorem 5.8.

Lemma 5.4. *Let $k \subseteq K$ be fields, and let $X_1, X_2, \ldots, X_n$ be a set of indeterminates for K. Define $D = k[X_1, X_2, \ldots, X_n]$ and $R = K[X_1, \ldots, X_n]$. If D has a finite* character set $\mathcal{F}$ of maximal ideals, each with residue field k, then $\{\mathfrak{m}R : \mathfrak{m} \in \mathcal{F}\}$ is a finite character set of maximal ideals of R, each with residue field K.

Proof. Write $\mathcal{F} = \{\mathfrak{m}_\alpha\}$. For each α, we may write $\mathfrak{m}_\alpha = (X_1 - p_{\alpha 1}, \ldots, X_n - p_{\alpha n})D$, with $p_{\alpha i} \in k$ for all $i \leq n$. For each α, let $p_\alpha = (p_{\alpha 1}, \ldots, p_{\alpha n})$. Since $\mathcal{F}$ is a finite character collection, we have that for each $0 \neq g(X_1, \ldots, X_n) \in D$, $g(p_\alpha) = 0$ for only finitely many α. For each α, let $\mathfrak{n}_\alpha := \mathfrak{m}_\alpha R$, and observe that $\mathfrak{n}_\alpha$ is a maximal ideal of R with residue field K. Let $0 \neq f \in R$, and define $V \subseteq K$ to be the k-vector space generated by the coefficients of f. Let $\{e_1, e_2, \ldots, e_m\}$ denote a k-basis of V. Then $f = \sum_{i=1}^m e_i g_i$, for some $g_1, g_2, \ldots, g_m \in D$. Then $f \in \mathfrak{n}_\alpha$ if and only if $f(p_\alpha) = 0$; if and only if $g_i(p_\alpha) = 0$ for all $i = 1, 2, \ldots, m$. However, for each i such that $g_i \neq 0$, there exist only finitely many α such that $g_i(p_\alpha) = 0$. Thus, if $f \neq 0$, then $f(p_\alpha) = 0$ for only finitely many α, and it follows that f is contained in only finitely many of the $\mathfrak{n}_\alpha$. □

Corollary 5.5. *Let K be an infinite field, and let $X_1, \ldots, X_n$ be indeterminates for K. Define $D = K[X_1, \ldots, X_n]$. Then D has an infinite finite character set of maximal ideals, each having residue field K.*

Proof. If $n = 1$, then D is a PID and the claim is clear, so suppose $n > 1$. Let k be a countably infinite subfield of K. Then $k[X_1]$ is countable and has an infinite finite character set of maximal ideals having residue field k. Thus by Proposition 5.1, $k[X_1][X_2, \ldots, X_n]$ has an infinite finite character set of maximal ideals, all having residue field k. Lemma 5.4 now completes the proof. □

The method of finding finite character collections in Propositions 3.9 and 5.1 is inductive; that is, we select maximal ideals one at a time until we obtain a countable finite character set of maximal ideals. By contrast, in Proposition 5.6 we show that for affine K-domains, where $K = \mathbb{C}$ or $K = \mathbb{R}$, one encounters infinite finite character sets of maximal ideals in a more natural way, as points along transcendental curves.

Proposition 5.6. *Let $K = \mathbb{C}$ or $\mathbb{R}$, let $X_1, \ldots, X_n$ be indeterminates for K, and let $D = K[X_1, \ldots, X_n]$. Suppose that $\phi_1, \ldots, \phi_n$ are K-algebraically independent elements of the ring of analytic functions $\phi : \mathbb{C} \to \mathbb{C}$ such that $\phi(K) \subseteq K$. Let I be an infinite compact subset of K, and define for each $t \in I$, $\mathfrak{m}_t := (X_1 - \phi_1(t), X_2 - \phi_2(t), \ldots, X_n - \phi_n(t))D$. Then $\{\mathfrak{m}_t : t \in I\}$ is a finite character set of maximal ideals of D.*

Proof. In order to show that $f \in D$ is contained in at most finitely many of the maximal ideals $\mathfrak{m}_t$, it suffices to check that $f(\phi_1(t), \phi_2(t), \ldots, \phi_n(t)) = 0$ for only finitely many $t \in I$. Let $f \in D$, and suppose f is contained in infinitely many of the $\mathfrak{m}_t$. Then there exists an infinite set $\{t_1, t_2, t_3, \ldots\}$ such that for $h(z) := f(\phi_1(z), \phi_2(z), \ldots, \phi_n(z))$, we have $h(t_k) = 0$ for all k. Since I is bounded, there is a limit point t_0 for the sequence $\{t_k\}$. The function $h(z)$, as a sum of products of analytic functions, is an analytic, hence continuous, function, so $h(t_0) = 0$. Thus t_0 is a cluster point of the zero set of h, which forces $h \equiv 0$ [13, Theorem 3.7]. Thus $f(\phi_1(z), \phi_2(z), \ldots, \phi_n(z)) = 0$ for all complex numbers z. Since $\{\phi_1, \phi_2, \ldots, \phi_n\}$ are K-algebraically independent, this forces $f = 0$. □

Corollary 5.7. *Let K be a field containing $\mathbb{R}$, and let $n > 0$. If D is an n-dimensional affine K-domain, then D has a finite character set of continuum many maximal ideals.*

Proof. Apply Noether Normalization to obtain a collection of algebraically independent elements $x_1, \ldots, x_n \in D$ such that D is an integral extension of $K[x_1, \ldots, x_n]$. If $K[x_1, \ldots, x_n]$ has a finite character set of continuum many maximal ideals, then by Lemma 5.2, so does D. Thus we may apply Lemma 5.4 and Proposition 5.6 to complete the proof. □

We prove now our main theorem of this section:

Theorem 5.8. *Let K be an infinite field, and let $X_1, \ldots, X_n$ be indeterminates for K. Let $D = K[X_1, \ldots, X_n]$, and let $\{G_i : i \in I\}$ be a collection of totally ordered abelian groups, where each G_i is a discrete group of rank n or has rational rank $< n$. If (a) I is a countable set or (b) $\mathbb{R} \subseteq K$ and I has cardinality no more than the continuum, then there exists an h-local Prüfer overring R of D such that $\mathfrak{I}(R)$ is order-isomorphic to the cardinal sum $\bigoplus_{i \in I} G_i$.*

Proof. In either case (a) or case (b), there exists by Corollary 5.5 or Proposition 5.6 a finite character collection $\{\mathfrak{m}_i : i \in I\}$ of $|I|$-many maximal ideals of D having residue field K. By Lemma 4.3, there exists for each $i \in I$ a valuation overring V_i of D having value group G_i, and such that V_i is hidden over $\mathfrak{m}_i$. By Lemma 4.2, $R := \bigcap_{i \in I} V_i$ is an h-local Prüfer domain with $\{V_i : i \in I\} = \{R_M : M \text{ is a maximal ideal of } R\}$. Thus by Theorem 2.3(5), since each V_i has value group G_i, $\mathfrak{I}(R)$ is order-isomorphic to the cardinal sum $\bigoplus_{i \in I} G_i$. □

It would be interesting to know whether for each field K the index set I can be chosen to be of the same cardinality as K. (By Theorem 4.4, it is necessary that $|I| \leq |K|$.) The proof of Theorem 5.8 shows that the answer is affirmative if also the following question has an affirmative answer.

Question. *Let K be a field, and let $X_1, \ldots, X_n$ be indeterminates for K. Does there exist a finite character collection $\mathcal{F}$ of maximal ideals of $K[X_1, \ldots, X_n]$, each having residue field K, such that $|\mathcal{F}| = |K|$?*

This question is equivalent to the curious geometric question: *Given a field K, does there exist a collection $\mathcal{P}$ of points in K^n such that $|\mathcal{P}| = |K|$ and for all nonzero $f \in K[X_1, \ldots, X_n]$, $f(p) = 0$ for at most finitely many points $p \in \mathcal{P}$?*

References

[1] S. Abhyankar, *On the valuations centered in a local domain*, Amer. J. Math. 78 (1956), 321–348.

[2] R. Alizade, G. Bilhan and P. F. Smith, *Modules whose maximal submodules have supplements*. Comm. Algebra 29 (2001), no. 6, 2389–2405.

[3] D. D. Anderson and T. Dumitrescu, *Condensed domains*. Canad. Math. Bull. 46 (2003), no. 1, 3–13.

[4] D. D. Anderson and M. Zafrullah, *Independent locally-finite intersections of localizations.* Houston J. Math. 25 (1999), no. 3, 433–452.

[5] S. Bazzoni and L. Salce, *Strongly flat covers.* J. London Math. Soc. (2) 66 (2002), no. 2, 276–294.

[6] S. Bazzoni and L. Salce, *Warfield domains.* J. Algebra 185 (1996), no. 3, 836–868.

[7] S. Bazzoni and L. Salce, *Almost perfect domains.* Colloq. Math. 95 (2003), no. 2, 285–301.

[8] N. Bourbaki, *Commutative algebra, Chapters 1–7,* 1985.

[9] W. Brandal, *On h-local integral domains.* Trans. Amer. Math. Soc. 206 (1975), 201–212.

[10] W. Brandal, *Commutative rings whose finitely generated modules decompose.* Lecture Notes in Mathematics, 723. Springer, Berlin, 1979.

[11] J. Brewer and L. Klingler, *The ordered group of invertible ideals of a Prüfer domain of finite character.* Comm. Algebra 33 (2005), no. 11, 4197–4203.

[12] J. Brewer and J. Mott, *Integral domains of finite character.* J. Reine Angew. Math. 241 1970 34–41.

[13] J. Conway, *Functions of one complex variable.* Second edition. Graduate Texts in Mathematics, 11. Springer, New York-Berlin, 1978.

[14] A. L. S. Corner, *A note on rank and direct decompositions of torsion-free Abelian groups.* Proc. Cambridge Philos. Soc. 57 (1961) 230–233.

[15] A. J. Engler and A. Prestel, *Valued fields.* Springer Monographs in Mathematics. Springer, Berlin, 2005.

[16] A. Facchini, *Generalized Dedekind domains and their injective modules.* J. Pure Appl. Algebra 94 (1994), no. 2, 159–173.

[17] M. Fontana, E. Houston and T. Lucas, *Factoring ideals in Prüfer domains,* J. Pure Appl. Algebra, to appear.

[18] L. Fuchs and S. B. Lee, *Primary decompositions over domains.* Glasgow Math. J. 38 (1996), no. 3, 321–326.

[19] L. Fuchs and S. B. Lee, *Primary decompositions of torsion modules over domains.* Mathematika 44 (1997), no. 1, 88–99.

[20] L. Fuchs and L. Salce, *Modules over non-Noetherian domains.* Mathematical Surveys and Monographs, 84. American Mathematical Society, Providence, RI, 2001.

[21] R. Gilmer, *Multiplicative ideal theory.* Corrected reprint of the 1972 edition. Queen's Papers in Pure and Applied Mathematics, 90. Queen's University, Kingston, ON, 1992.

[22] H. P. Goeters, *Warfield duality and extensions of modules over an integral domain,* in: Abelian Groups and Modules, Proceedings of the Padova Conference of 1995, Kluwer Academic Press, Dordrecht, 1995.

[23] O. Goldman, *On a special class of Dedekind domains.* Topology 3 (1964), suppl. 1, 113–118.

[24] T. Gulliksen, *The Krull ordinal, coprof, and Noetherian localizations of large polynomial rings.* Amer. J. Math. 96 (1974), 324–339.

[25] W. Heinzer, *Integral domains in which each non-zero ideal is divisorial.* Mathematika 15 (1968), 164–170.

[26] W. Heinzer and J. Ohm, *Noetherian intersections of integral domains.* Trans. Amer. Math Soc. 167 (1972), 291–308.

[27] R. Heitmann, *PID's with specified residue fields,* Duke Math. J. **41** (1974), 565–582.

[28] E. Houston and S. McAdam, *Rank in Noetherian rings.* J. Algebra 37 (1975), no. 1, 64–73.

[29] P. Jaffard, *Théorie arithmétique des anneaux du type de Dedekind.* Bull. Soc. Math. France 80, (1952). 61–100.

[30] P. Jaffard, *Théorie arithmétique des anneaux du type de Dedekind. II.* Bull. Soc. Math. France 81, (1953). 41–61.

[31] F. V. Kuhlmann, *Value groups, residue fields, and bad places of rational function fields.* Trans. Amer. Math. Soc. 356 (2004), no. 11, 4559–4600.

[32] E. L. Lady, *Slender rings and modules.* Pacific J. Math. 49 (1973), 397–406.

[33] K. A. Loper, *Constructing examples of integral domains by intersecting valuation domains.* Non-Noetherian commutative ring theory, 325–340, Math. Appl., 520, Kluwer Acad. Publ., Dordrecht, 2000.

[34] E. Matlis, *Reflexive domains.* J. Algebra 8 (1968) 1–33.

[35] E. Matlis, *Decomposable modules.* Trans. Amer. Math. Soc. 125 (1966) 147–179.

[36] E. Matlis, *Torsion-free modules.* Chicago Lectures in Mathematics. The University of Chicago Press, Chicago-London, 1972.

[37] E. Matlis, *Ideals of injective dimension* 1. Michigan Math. J. 29 (1982), no. 3, 335–356.

[38] S. McAdam, *Primes associated to an ideal.* Contemporary Mathematics, 102. American Mathematical Society, Providence, RI, 1989.

[39] S. McAdam, *Deep decompositions of modules.* Comm. Algebra 26 (1998), no. 12, 3953–3967.

[40] S. McAdam, *Conforming spectra,* unpublished manuscript.

[41] W. McGovern, *Neat rings.* J. Pure Appl. Algebra 205 (2006), no. 2, 243–265.

[42] M. Nagata, *Local rings.* Interscience Tracts in Pure and Applied Mathematics, No. 13, John Wiley & Sons, New York–London, 1962.

[43] B. Olberding, *Holomorphy rings of function fields.* Multiplicative ideal theory in commutative algebra, 331–347, Springer, New York, 2006.

[44] B. Olberding, *Injective and colon properties of ideals of integral domains,* Forum Math. 19 (2007), 1047–1074.

[45] B. Olberding, *A geometric setting for some properties of torsion-free modules.* Proceedings of the Second Honolulu Conference on Abelian Groups and Modules (Honolulu, HI, 2001). Rocky Mountain J. Math. 32 (2002), no. 4, 1281–1297.

[46] B. Olberding, *Factorization into prime and invertible ideals.* J. London Math. Soc. (2) 62 (2000), no. 2, 336–344.

[47] B. Olberding, *Factorization into prime and invertible ideals II,* J. London Math. Soc., to appear.

[48] B. Olberding, *Modules of injective dimension one over Prüfer domains.* J. Pure Appl. Algebra 153 (2000), no. 3, 263–287.

[49] B. Olberding, *Prüfer domains and pure submodules of direct sums of ideals.* Mathematika 46 (1999), no. 2, 425–432.

[50] B. Olberding, *Almost maximal Prüfer domains.* Comm. Algebra 27 (1999), no. 9, 4433–4458.

[51] B. Olberding, *Globalizing local properties of Prüfer domains.* J. Algebra 205 (1998), no. 2, 480–504.

[52] B. Olberding and M. Roitman, *The minimal number of generators of an invertible ideal.* Multiplicative ideal theory in commutative algebra, 349–367, Springer, New York, 2006.

[53] D. Rush and L. Wallace, *Noetherian maximal spectrum and coprimely packed localizations of polynomial rings,* Houston J. Math. 28 (2002) 437–448.

[54] L. Salce and P. Zanardo, *Loewy length of modules over almost perfect domains.* J. Algebra 280 (2004), no. 1, 207–218.

[55] P. Vámos, *Multiply maximally complete fields.* J. London Math. Soc. (2) 12 (1975/76), no. 1, 103–111.

[56] R. Wiegand and S. Wiegand, *Commutative rings whose finitely generated modules are direct sums of cyclics.* Abelian group theory (Proc. Second New Mexico State Univ. Conf., Las Cruces, N.M., 1976), pp. 406–423. Lecture Notes in Math., Vol. 616, Springer, Berlin, 1977.

[57] S. Wiegand, *Intersections of prime ideals in Noetherian rings.* Comm. Algebra 11 (1983), no. 16, 1853–1876.

[58] P. Zanardo, *Almost perfect local domains and their dominating Archimedean valuation domains.* J. Algebra Appl. 1 (2002), no. 4, 451–467.

Author information

Bruce Olberding, Department of Mathematical Sciences, New Mexico State University, Las Cruces, NM 88003-8001, USA.
E-mail: olberdin@nmsu.edu

Variations on Whitehead's problem and the structure of Ext

C. U. Jensen

Abstract. For a countable flat module A over a countable ring R it is proved that either A is projective or $\operatorname{Ext}_R^1(A, R)$ has cardinality $2^{\aleph_0}$. A module over a commutative ring R is called i-realizable if it is isomorphic to $\operatorname{Ext}_R^i(B, C)$ for suitable R-modules B and C. If R is an integral domain and not a field any torsion-free divisible R-module is 1-realizable. For $i > 1$ some necessary conditions for i-realizability are given.

Key words. Projective modules, flat modules, injective modules, projective limits.

AMS classification. 13D07, 16E30, 13C11, 18A30.

1 Introduction and statement of the results

If A is a free abelian group, clearly $\operatorname{Ext}_\mathbb{Z}^1(A, \mathbb{Z}) = 0$. The classical Whitehead's Problem for abelian groups asks whether the converse holds, i.e. is the abelian group A free if $\operatorname{Ext}_\mathbb{Z}^1(A, \mathbb{Z}) = 0$? It is easy to see that A is necessarily torsion-free. If A is countable, K. Stein [13] in 1951 proved that A is free. He actually proved something more: If A is a countable torsion-free abelian group, then either A is free or $\operatorname{Ext}_\mathbb{Z}^1(A, \mathbb{Z})$ has the cardinality of the continuum.

If A is uncountable, S. Shelah [12] in 1973 proved that the statement "$\operatorname{Ext}_\mathbb{Z}^1(A, \mathbb{Z}) = 0 \Rightarrow A$ is free" can be neither proved nor disproved, i.e. both the above statement and its negation are consistent with the standard ZFC axiom system for sets.

In this note we shall consider the analogue of Whitehead's problem for modules over an arbitrary ring R, i.e. we ask whether a module A over a ring R is projective, if $\operatorname{Ext}_R^1(A, R) = 0$. Here we must put some restrictions on the module, since it is easy to give examples of countable commutative Noetherian rings R such that $\operatorname{Ext}_R^1(R/J, R) = 0$ for some non-trivial ideal J. Primarily, we consider the case where A is a flat R-module.

Theorem 1. *Let R be a countable (not necessarily commutative) ring and A a flat countably generated (left) R-module. Then either A is projective or $\operatorname{Ext}_R^1(A, R)$ has the cardinality $2^{\aleph_0} (=$ the cardinality of the continuum$)$. In particular, A is projective if and only if $\operatorname{Ext}_R^1(A, R) = 0$.*

Corollary 2. *Let R be a (commutative) countable Prüfer domain and A a countably generated torsion-free R-module. Then A is projective if and only if the cardinality of $\operatorname{Ext}_R^1(A, R)$ is $< 2^{\aleph_0}$ if and only if $\operatorname{Ext}_R^1(A, R) = 0$.*

Remark 3. The countability assumption on the ring R in the above results is important. Indeed, if R is a commutative complete local Noetherian ring (not a field), there exist countably generated non-projective flat R-modules A such that $\operatorname{Ext}^1_R(A,R) = 0$.

The Whitehead problem has an interesting counterpart in the "Baer" problem, where $\mathbb{Z}$, resp. R, is placed by arbitrary torsion groups, resp. arbitrary torsion modules. In particular, one should mention the striking result in [3] that a module A over a commutative domain R is projective if (and, of course, only if) $\operatorname{Ext}^1_R(A,T) = 0$ for any torsion R-module T. The proof in [3] makes use of inverse limits and its derived functors just as the proof of Theorem 1 does. For a detailed account of this whole complex of problems we refer to the monograph [2].

The following results concern the structure of the modules $\operatorname{Ext}^i_R(B,C)$ where B and C are modules over a commutative ring R. A module of such a form is called i-realizable.

In Nunke [7] it was proved that any divisible torsion-free R-module is 1-realizable if R is a Dedekind domain. This is generalized in

Theorem 4. *Let R be a commutative integral domain which is not a field. Then any divisible torsion-free R-module is 1-realizable.*

Concerning i-realizability for $i > 1$ the following holds.

Theorem 5. *Let R be either an arbitrary commutative domain or a Noetherian commutative ring with no non-zero nilpotent elements and let K be the quotient field, resp. the full ring of quotients of R, (i.e. the ring of quotients with respect to all non-zero divisors). Assume that the projective dimension $\operatorname{dh}_R(K)$ of K as an R-module is ≤ 2.*

If $\operatorname{dh}_R(K) = 1$, then an i-realizable R-module D, $i \geq 2$, is never injective unless it is zero. More precisely, $\operatorname{Hom}_R(K,D) = \operatorname{Ext}^1_R(K,D) = 0$.

If $\operatorname{dh}_R(K) = 2$, then an i-realizable R-module D, $i \geq 2$, satisfies the condition $\operatorname{Ext}^1_R(K,D) = 0$.

In the case where R is a domain and $\operatorname{dh}_R(K) = 1$, Theorem 5 is a theorem of Matlis [6]. Note that the condition $\operatorname{dh}_R(K) = 1$ holds if R is countable and that the condition $\operatorname{dh}_R(K) = 2$ holds if the cardinality of R is $\leq \aleph_1$.

Theorem 5 is, in some sense, best possible as is shown in

Remark 6. For any $i \geq 2$, there exists a commutative Noetherian domain R and a non-zero i-realizable R-module which is torsion-free divisible (in particular injective).

2 Proofs of Theorem 1 and Corollary 2

For the proofs we need some preparations about direct limits, projective limits and the derived functors of projective limits over directed partially ordered sets. The directed posets needed in our applications will be countable. Since a countable poset contains $\mathbb{N}$ (with its natural ordering) as a cofinal subset in the following we shall only consider direct and projective limits with index set $\mathbb{N}$. The notion of a stable projective system (or a Mittag-Leffler system) plays an important role in the following. Let $(A_n, f_{mn};$ $f_{mn} : A_n \mapsto A_m)$, $m \leq n$ be a projective system over $\mathbb{N}$ of abelian groups. Such a projective system is called *stable* if for each $m \in \mathbb{N}$ there exists $n \in \mathbb{N}$, such that $m \leq n$ and $\text{Im}(f_{mn}) = \text{Im}(f_{mr})$ for every $r \in \mathbb{N}$, $r \geq n$. For more literature on inverse limits and their derived functors cf. [1, 4, 10, 11, 14].

The first lemma is just a standard result from set theory.

Lemma 7. *Let (A_n, f_{mn}) be a projective system over $\mathbb{N}$ of countable abelian groups. If all the maps $f_{m(m+1)}$ are surjective and infinitely many of them are not bijections, then the cardinality of the projective limit $\varprojlim A_n$ is $2^{\aleph_0}$.*

By $\varprojlim^{(i)}$ we denote the ith right derived functor of $\varprojlim$. For countable index sets, in particular for $\mathbb{N}$, it is well known ([4, 10]) that $\varprojlim^{(i)}$ vanishes for $i > 1$. It is also known that $\varprojlim^{(1)}$ vanishes for a stable projective system of abelian groups over $\mathbb{N}$, (see for instance [4, p. 92] or [14, Proposition 3.5.7]). If the abelian groups are countable the converse holds. This is shown – in a slightly stronger form – in the next lemma which relates the size of $\varprojlim^{(1)}$ for a projective system to the stability of the system.

Lemma 8. *Let (A_n, f_{mn}) be a projective system over $\mathbb{N}$ of countable abelian groups. If $\varprojlim^{(1)}(A_n)$ has cardinality $< 2^{\aleph_0}$, then (A_n, f_{mn}) is stable. In particular, if $\varprojlim^{(1)}(A_n) = 0$, then (A_n, f_{mn}) is stable.*

Proof. For any fixed natural number m we have to show that the descending chain $f_{mn}A_n$, $n \geq m$, of subgroups of A_m is stationary. To this purpose we consider the following projective systems.

Let $\mathcal{P}$ be the tail of the given projective system starting at A_m, consisting of the groups A_n, $n \geq m$, and the given maps $f_{nr}, r \geq n$; let $\mathcal{P}_0$ be the projective system where all the groups are A_m, and the maps the identity; let $\mathcal{P}_1$ be the projective system where the groups are $f_{mn}(A_n)$, $n \geq m$, and the maps the natural injections; and let $\mathcal{P}_2$ be the projective system where the groups are the quotients $A_m/f_{mn}(A_n)$ and the maps the canonical surjections.

Now assume the descending chain $f_{mn}(A_n)$, $n \geq m$, were not stationary. By Lemma 7 this would imply that the cardinality of $\varprojlim \mathcal{P}_2$ would be $2^{\aleph_0}$. From the short exact sequence
$$0 \to \mathcal{P}_1 \to \mathcal{P}_0 \to \mathcal{P}_2 \to 0$$
(with the obvious maps) we get a long exact sequence
$$0 \to \varprojlim \mathcal{P}_1 \to \varprojlim \mathcal{P}_0 \to \varprojlim \mathcal{P}_2 \to \varprojlim{}^{(1)} \mathcal{P}_1 \to \varprojlim{}^{(1)} \mathcal{P}_0 = 0.$$
Since $\varprojlim \mathcal{P}_0 = A_m$ is countable, the cardinality of $\varprojlim^{(1)} \mathcal{P}_1$ is $2^{\aleph_0}$.

Now $\varprojlim^{(2)}$ vanishes on projective systems over $\mathbb{N}$, hence $\varprojlim^{(1)}$ is right exact on projective systems over $\mathbb{N}$. With the obvious maps we get a surjection $\mathcal{P} \to \mathcal{P}_1 \to 0$ which induces a surjection
$$\varprojlim{}^{(1)} \mathcal{P} \to \varprojlim{}^{(1)} \mathcal{P}_1 \to 0.$$

Since cardinality of $\varprojlim{}^{(1)} \mathcal{P}_1$ is $2^{\aleph_0}$, the cardinality of $\varprojlim{}^{(1)} \mathcal{P}$ is $\geq 2^{\aleph_0}$ (actually $= 2^{\aleph_0}$). But $\mathcal{P}$ is a tail of the given projective system, hence $\varprojlim{}^{(1)} \mathcal{P} = \varprojlim{}^{(1)}(A_n)$. This gives the desired contradiction. □

The following is an easy consequence of a result (Corollaire 2.2.2) in [9].

Lemma 9. *Let A be a countably generated flat (left) R-module written as a direct limit of finitely generated free (left) R-modules F_n, $n \in \mathbb{N}$, over $\mathbb{N}$. Then A is projective if and only if the projective system of dual right R-modules $\mathrm{Hom}_R(F_n, R)$ is stable.*

Proof. The dual modules $F_n^* = \mathrm{Hom}(F_n, R)$ with maps $f_{mn}^* = \mathrm{Hom}(f_{mn}, 1_R)$ form a projective system over $\mathbb{N}$. In view of proposition (2.1.4) in [9] A is projective if and only if the projective system $\mathrm{Hom}_R(F_n, R^\mathbb{N})$ with the obvious maps is stable. But $\mathrm{Hom}_R(F_n, R^\mathbb{N}) = (F^*)^\mathbb{N}$ and clearly the system (F_n^*, f_{mn}^*) is stable if and only if $((F^*)^\mathbb{N}, (f_{mn}^*)^\mathbb{N})$ is stable. □

We are now in a position to prove Theorem 1. We write the flat R-module A as the direct limit $\varinjlim F_n$ of finitely generated free left R-modules F_n the index set being $\mathbb{N}$. Assume A is not projective, then by Lemma 9 the projective system of dual right R-modules $\mathrm{Hom}_R(F_n, R)$ is not stable. Hence by Lemma 8 the cardinality of $\varprojlim{}^{(1)} \mathrm{Hom}_R(F_n, R)$ is $\geq 2^{\aleph_0}$. There is a spectral sequence (cf. [4] or [10])

$$E_2^{p,q} = \varprojlim{}^{(p)} \mathrm{Ext}_R^q(F_n, R)) \Rightarrow \mathrm{Ext}_R^n(\varinjlim F_n, R).$$

Since $\varprojlim{}^{(2)}$ vanishes on projective systems over $\mathbb{N}$ and the F_n' s are projective the spectral sequence degenerates into an isomorphism

$$\mathrm{Ext}^1(\varinjlim F_n, R) \simeq \varprojlim{}^{(1)}(\mathrm{Hom}_R(F_n, R)).$$

But $A = \varinjlim F_n$ implies that $\mathrm{Ext}_R^1(A, R)$ has cardinality $\geq 2^{\aleph_0}$ (actually exactly $2^{\aleph_0}$). This completes the proof of Theorem 1. □

Corollary 2 is an immediate consequence of Theorem 1 since any torsion-free module over a Prüfer domain is flat.

3 Proofs of Theorem 4, Theorem 5 and Remark 6

In this section we only consider commutative rings. As for Theorem 4 we only consider commutative domains. For a domain R we need some basic facts about cotorsion modules. For an R-module A the submodules $\{IA\}$ of A, I running through the non-zero ideals of R form a directed system under containment relation. Hence we can form the projective limit $\tilde{A} = \varprojlim A/IA$, I running through the non-zero ideals of R. If R is not a field and A a free R-module the canonical mapping from A to $\tilde{A}$ is injective. In [6] it is proved that $\tilde{A}$ is a cotorsion module meaning that $\mathrm{Hom}_R(K, \tilde{A}) = \mathrm{Ext}_R^1(K, \tilde{A}) = 0$, where K is the quotient field of R.

We are now in a position to prove Theorem 4. Let D be a torsion-free divisible R-module. D is the homomorphic image of a free R-module A. The above remarks imply that there is an injective mapping $A \to \tilde{A}$, where $\tilde{A}$ is cotorsion. Since the torsion-free divisible R-module D is injective, D is also a homomorphic image of $\tilde{A}$. We thus get a short exact sequence of R-modules

$$0 \to B \to \tilde{A} \to D \to 0$$

which in turn gives rise to the exact sequence

$$\operatorname{Hom}_R(K, \tilde{A}) \to \operatorname{Hom}_R(K, D) \to \operatorname{Ext}^1_R(K, B) \to \operatorname{Ext}^1_R(K, \tilde{A}).$$

Since D is torsion-free and divisible $\operatorname{Hom}_R(K, D) \simeq D$. Moreover since $\tilde{A}$ is cotorsion $\operatorname{Hom}_R(K, \tilde{A}) = \operatorname{Ext}^1_R(K, \tilde{A}) = 0$, Consequently $\operatorname{Ext}^1_R(K, B) \simeq D$, so D is 1-realizable. □

Finally we give a proof of Theorem 5. If R is a commutative Noetherian ring with no non-zero nilpotent elements and K is its full ring of quotients, K is a semi-simple Artinian ring (cf. e.g. [5, Satz 4.23]). This implies that for any R-module B the tensor product $K \otimes_R B$ is a direct sum of direct summands of K. Hence $\operatorname{dh}_R(K \otimes_R B) \leq \operatorname{dh}_R(K)$ and thus $\operatorname{Ext}^j_R(K \otimes_R B, -) = 0$ for all $j \geq \operatorname{dh}_R(K) + 1$. Furthermore, $\operatorname{Tor}^R_q(K, -) = 0$ for all $q > 0$ since K is a flat R-module.

Evidently these statements also hold true in the case where R is an arbitrary commutative domain.

Now, let D be an i-realizable R-module, i.e. $D = \operatorname{Ext}^i_R(B, C)$ for some R-modules B and C.

We have to show
a) $\operatorname{Hom}_R(K, D) = \operatorname{Ext}^1_R(K, D) = 0$ if $\operatorname{dh}_R(K) = 1$ and $i \geq 2$.
b) $\operatorname{Ext}^1_R(K, D) = 0$ if $\operatorname{dh}_R(K) = 2$ and $i \geq 2$.

For the proofs we use that the canonical isomorphism between $\operatorname{Hom}_R(X \otimes Y, Z)$ and $\operatorname{Hom}_R(X, \operatorname{Hom}_R(Y, Z))$ gives rise to two spectral sequences having the same limit H^*:

$$E_2^{p,q} = \operatorname{Ext}^p_R(K, \operatorname{Ext}^q_R(B, C)) \underset{p}{\Rightarrow} H^*$$

and

$$E_2^{'p,q} = \operatorname{Ext}^p_R(\operatorname{Tor}^R_q(K, B), C) \underset{p}{\Rightarrow} H^*.$$

Assertion a): On the diagonal $p + q = i, i \geq 2$, the terms $E_2^{'p,q}$ vanish since $\operatorname{Ext}^j_R(K \otimes_R B, -) = 0$ for all $j \geq 2$ and $\operatorname{Tor}^R_q(K, -) = 0$ for all $q > 0$. Therefore the ith cohomology of the total complex vanishes for $i \geq 2$.

On the diagonal $p + q = i, i \geq 2$, of the first spectral sequence we have $E_2^{p,q} = 0$ whenever $p \neq 0$ and $p \neq 1$. Consequently $E_2^{0,i} = E_3^{0,i} = \cdots = E_\infty^{0,i} =$ a subquotient of H^i. Therefore $\operatorname{Hom}_R(K, \operatorname{Ext}^i_R(B, C)) = \operatorname{Hom}_R(K, D) = 0$ for $i \geq 2$. Similarly, on the diagonal $p + q = i, i \geq 3$, we have $E_2^{p,q} = 0$ whenever $p \neq 0$ and $p \neq 1$. Just as above we conclude that $E_2^{1,i-1} =$ a subquotient of H_i for $i \geq 3$. Therefore

$\operatorname{Ext}_R^1(K, \operatorname{Ext}_R^{i-1}(B,C)) = 0$ when $i - 1 \geq 2$, so $\operatorname{Ext}_R^1(K,D) = 0$ if D is i-realizable for $i \geq 2$.

Assertion b): Proceeding as above we see that on the diagonal $p + q = i, i \geq 3$, the terms $E_2'^{p,q}$ vanish since $\operatorname{dh}_R(K) = 2$. Hence the ith cohomology H^i of the total complex vanishes for $i \geq 3$.

In the first spectral sequence the terms $E_2^{p,q}$ vanish if $p \geq 3$. Arguing as in case a) we conclude that $E_2^{1,i}$ is a subquotient of H^{i+1} for $i \geq 2$. Hence $E_2^{1,i} = \operatorname{Ext}_R^1(K, \operatorname{Ext}_R^i(B,C)) = \operatorname{Ext}_R^1(K,D) = 0$. □

As for Remark 6 we need the following result in [8]. Let F be a field of cardinality $\aleph_k$ and let R be the polynomial ring $F[x_1, \ldots, x_n]$. Then the projective dimension of the quotient field $K = F(x_1, \ldots, x_n)$ viewed as an R-module is $\min(n, k + 1)$. Therefore if $k = i - 1$ and $n \geq i$, there is an R-module M such that $\operatorname{Ext}_R^i(K,M) \neq 0$. But $\operatorname{Ext}_R^i(K,M)$ is a non-zero vector space over K and thus a divisible and torsion-free R-module, in particular, an injective R-module. □

References

[1] I. Emmanouil, *Mittag-Leffler condition and the vanishing of* $\varprojlim^1$, Topology **35** (1995), 267–271.

[2] R. Göbel and J. Trlifaj, *Approximations and endomorphism algebras of modules*, de Gruyter Expositions in Mathematics, Vol. 41, Walter de Gruyter, Berlin, 2006.

[3] L. A. Hügel, S. Bazzoni and D. Herbera, *A solution to the Baer splitting problem*, arXiv:math/0602272, 13 pp.

[4] C. U. Jensen, *Les foncteurs dérivés de* $\varprojlim$ *et leurs applications en théorie des modules*, Lecture Notes in Mathematics, Vol. 254. Springer-Verlag, Berlin, New York, 1972.

[5] E. Kunz, *Einführung in die kommutative Algebra und algebraische Geometrie*, Vieweg, Braunschweig, 1980.

[6] E. Matlis, *Cotorsion modules*, Mem. Amer. Math. Soc. No. **49** (1964), 66 pp.

[7] R. J. Nunke, *Modules of extensions over Dedekind rings*, Illinois J. Math. **3** (1959), 222–241.

[8] B. L. Osofsky, *Homological dimension and the continuum hypothesis*, Trans.Amer. Math.Soc. **132** (1968), 217–230.

[9] M. Raynaud and L. Gruson, *Critères de platitude et projectivité. Techniques de "platification" d'un module*, Invent. Math. **13** (1971), 1–89.

[10] J.-E. Roos, *Sur les foncteurs dérivés de* $\varprojlim$. *Applications*, C. R. Acad. Sci. Paris **252** (1961), 3702–3704.

[11] J.-E. Roos, *Derived functors of inverse limits revisited*, J. London Math. Soc. (2) **73** (2006), 65–83.

[12] S. Shelah, *Whitehead problem and some constructions*, Israel J. Math. **18** (1974), 243–256.

[13] K. Stein, *Analytische Funktionen mehrerer komplexer Veränderlichen zu vorgegebenen Periodizitätsmoduln und das zweite Cousinsche Problem*, Math. Ann. **123** (1951), 201–222.

[14] C. A. Weibel, *An introduction to homological algebra*, Cambridge Studies in Advanced Mathematics, 38, Cambridge University Press, 1994.

Author information

C. U. Jensen, Department of Mathematics, University of Copenhagen, Universitetsparken 5, 2100 Copenhagen, Denmark.
E-mail: `cujensen@math.ku.dk`

The lattice of torsionfree precover classes

Ladislav Bican

Abstract. The purpose of this brief note is to show that the set of all hereditary torsion theories, the torsionfree classes of which are precover classes, is a sublattice of the lattice of all hereditary torsion theories for the category R-mod.

Key words. Precover class, hereditary torsion theory, relative purity.

AMS classification. 16D80, 16D90, 16S90.

Throughout this note R stands for an associative ring with the identity element and R-mod denotes the category of all unital left R-modules. It follows immediately from the proof of the existence of the injective hull $E(M)$ of a module M that an upper bound for the cardinality of $E(M)$ is $|M|^{|R|}$. For the basic properties of rings and modules we refer to [1].

Recall that a *hereditary torsion theory* $\tau = (\mathbf{T}_\tau, \mathbf{F}_\tau)$, or simply $\tau = (\mathbf{T}, \mathbf{F})$, for the category R-mod consists of two abstract classes $\mathbf{T}$ and $\mathbf{F}$, the τ-*torsion class* and the τ-*torsionfree class*, respectively, such that $\mathrm{Hom}\,(T, F) = 0$ whenever $T \in \mathbf{T}$ and $F \in \mathbf{F}$, the class $\mathbf{T}$ is closed under submodules, factor-modules, extensions and arbitrary direct sums, the class $\mathbf{F}$ is closed under submodules, extensions and arbitrary direct products and for each module M there exists an exact sequence $0 \to T \to M \to F \to 0$ such that $T \in \mathbf{T}$ and $F \in \mathbf{F}$. It is easy to see that every module M contains the unique largest τ-torsion submodule (isomorphic to T), which is called the τ-*torsion part of the module* M and it is usually denoted by $\tau(M)$. Recall, that a submodule K of the module M is said to be τ-*pure* in M, if the factor-module M/K is τ-torsionfree. Associated to each hereditary torsion theory τ is the *Gabriel filter* $\mathbf{L}_\tau$ of all the left ideals $I \leq R$ such that $R/I \in \mathbf{T}$.

Let $\varrho = (\mathbf{T}_\varrho, \mathbf{F}_\varrho)$ and $\sigma = (\mathbf{T}_\sigma, \mathbf{F}_\sigma)$ be hereditary torsion theories for the category R-mod. Recall, that the *meet* $\tau = \varrho \cap \sigma$ is defined by $\mathbf{T}_\tau = \mathbf{T}_\varrho \cap \mathbf{T}_\sigma$, while the *join* $\tau = \varrho \vee \sigma$ is defined by $\mathbf{F}_\tau = \mathbf{F}_\varrho \cap \mathbf{F}_\sigma$. For further details on hereditary torsion theories see e.g. [9] or [6].

If $\mathbf{G}$ is an abstract class (closed under isomorphic copies) of modules, then a homomorphism $\varphi : G \to M$, $G \in \mathbf{G}$, is called a $\mathbf{G}$-*precover* of the module M, if for each homomorphism $f : F \to M$ with $F \in \mathbf{G}$ there is a homomorphism $g : F \to G$ such that $\varphi g = f$. A $\mathbf{G}$-precover $\varphi : G \to M$ is a $\mathbf{G}$-*cover* of M if each endomorphism f of G with $\varphi g = \varphi$ is an automorphism of G. An abstract class $\mathbf{G}$ of modules is called a *precover class* (*cover class*), if every module has a $\mathbf{G}$-precover ($\mathbf{G}$-cover). For further properties of precovers and covers we refer to [11].

We shall start with the definition of the condition (*) for an abstract class **G** of modules. It should be noted that this condition has been introduced in [3] as the **G**-property of the class **G**. If **G** is an abstract class of modules satisfying the condition (*), then the class Coprod (**G**) of all direct sums of members of **G** is a precover class ([3, Theorem 3]). Conversely, if **G** is a hereditary (closed under submodules) precover class of modules, then it satisfies the condition (*) ([3, Theorem 5]). In this note we shall use this property in order to show that the subset of the lattice of all hereditary torsion theories for the category R-mod consisting of all hereditary torsion theories, the torsionfree classes of which are precover classes, forms the sublattice.

Definition 1. Let **G** be an abstract class of modules. We say that **G** satisfies the *condition* (*) if to any infinite cardinal λ there exists a cardinal $\kappa > \lambda$ such that for each module $F \in \mathbf{G}$ with $|F| > \kappa$ and each its submodule K with $|F/K| \leq \lambda$ there exists a submodule $L \leq K$ such that $F/L \in \mathbf{G}$ and $|F/L| \leq \kappa$.

Remark 2. It should be noted that if κ is a cardinal corresponding to an infinite cardinal λ under the condition (*) then κ can be obviously replaced by any cardinal $\kappa' > \kappa$ (see the proofs in [3] for more details). Further, if $\tau = (\mathbf{T}, \mathbf{F})$ is a hereditary torsion theory for the category R-mod such that the class **F** satisfies the condition (*) then the cardinal κ (a suitable one, possibly) corresponding to the infinite cardinal λ will be often denoted by $\kappa(\lambda, \tau)$. Finally, if μ is an infinite cardinal, then there are cardinals κ such that $\kappa^\mu = \kappa$ owing to the fact that $(\kappa^\mu)^\mu = \kappa^\mu$.

We start with some preliminary results, some of which are well known, but we shall include it here for the convenience of the reader.

Lemma 3. *Let T be a submodule of a module F and let $H \leq F$ be maximal with respect to $H \cap T = 0$. Then*

(i) $H \oplus T$ *is essential in* F;

(ii) H *has no essential extension in* F;

(iii) *for each $u \in F \setminus (H \oplus T)$ the annihilator left ideal $I = ((H \oplus T) : u)$ is essential in R;*

(iv) *for each $u \in F \setminus H$ the inclusion $(H : u) \subsetneq ((H \oplus T) : u)$ is proper;*

(v) $(H \oplus T)/H$ *is essential in* F/H.

Proof. (i) Assuming the existence of $0 \neq V \leq F$ with $(H \oplus T) \cap V = 0$ we get $(H \oplus V) \cap T \neq 0$ by the choice of H. So there is $0 \neq t = h + v \in T \cap (H \oplus V)$, where $v \neq 0$, for otherwise we shall come to a contradiction $0 \neq t = h \in T \cap H = 0$. Thus, $(H \oplus T) \cap V \neq 0$ and the essentiality of $H \oplus T$ in F is proved.

(ii) Let $\tilde{H}$ be a proper essential extension of H in F. Then $\tilde{H} \cap T \neq 0$ by the choice of H and so each element $0 \neq u \in \tilde{H} \cap T$ has a non-zero multiple $0 \neq ru \in H \cap T$, which is impossible.

(iii) If $r \in R \setminus I$ is an arbitrary element, then $ru \notin H \oplus T$ and consequently $0 \neq sru \in H \oplus T$ for some $s \in R$ by (i). Hence $0 \neq sr \in I$, as desired.

(iv) From $u \notin H$ we have $(H + Ru) \cap T \neq 0$ by the choice of H and so there is an element $0 \neq t = h + ru \in (H + Ru) \cap T$. Assuming on the contrary, that $(H : u) = ((H \oplus T) : u)$ we see that the equality $ru = t - h$ means that $r \in (H : u)$. However, then $t = h + ru \in T \cap H = 0$, which is a contradiction finishing the proof.

(v) If $u \in F \setminus (H \oplus T)$ is arbitrary, then $(H : u) \subsetneq ((H \oplus T) : u)$ by (iv) and so for each $r \in ((H \oplus T) : u) \setminus (H : u)$ we have $0 \neq r(u + H) \in (H \oplus T)/H$, as we wished to show. □

Lemma 4. *Let $\mu = \max(|R|, \aleph_0)$ and let κ be an infinite cardinal such that $\kappa^\mu = \kappa$. If F is a module such that $|F| > \kappa$ and if $T \leq F$ is such that $|T| \leq \kappa$, then for each submodule H of F maximal with respect to $H \cap T = 0$ it is $|F/H| \leq \kappa$.*

Proof. It is well known (see e.g. [1]) that the size of the injective hull $E(T)$ of T is at most $\kappa^\mu = \kappa$. Now the monomorphism $(H \oplus T)/H \cong T \to E(T)$ extends to a monomorphism $f : F/H \to E(T)$ in view of the fact that $(H \oplus T)/H$ is essential in F/H by Lemma 3(v) and the rest is obvious. □

Lemma 5. *Let ϱ, σ and $\tau = \varrho \cap \sigma$ be hereditary torsion theories for the category R-mod and let $F \in \mathbf{F}_\tau$ be arbitrary. If $T = \varrho(F)$ and $H \leq F$ is maximal with respect to $H \cap T = 0$, then H is σ-pure in F.*

Proof. Proving indirectly, let $0 \neq u + H \in \sigma(F/H)$ be arbitrary. Then $I = (H : u) \in \mathbf{L}_\sigma$ and $I \subsetneq J = ((H \oplus T) : u)$ by Lemma 3(iv). Now let $r \in J$ be an arbitrary element. Then $(I : r) = (H : ru) \in \mathbf{L}_\sigma$ and $ru = h + t$, $h \in H$, $t \in T$. Taking $s \in (I : r)$ arbitrarily we have $sru = sh + st \in H$ and consequently $st = sru - sh \in T \cap H = 0$. Thus $(I : r)t = 0$, which means that $t \in \sigma(F)$. However, $t \in \varrho(F)$ gives that $t \in \varrho(F) \cap \sigma(F) = \tau(F) = 0$ and so $t = 0$. Thus $ru = h$ means that $r \in I$ and the equality $J = I$ yields a contradiction finishing the proof. □

Lemma 6. *Let ϱ, σ and $\tau = \varrho \cap \sigma$ be hereditary torsion theories for the category R-mod such that $\mathbf{F}_\varrho$ and $\mathbf{F}_\sigma$ are precover classes. Let λ be an infinite cardinal and let κ be a cardinal corresponding to λ under the condition (*) for the class $\mathbf{F}_\varrho$ such that $\kappa^\mu = \kappa$, where $\mu = \max(|R|, \aleph_0)$. If K is a submodule of a module $F \in \mathbf{F}_\tau$ such that $|F| > \kappa$, $|F/K| \leq \lambda$ and if $|T| \leq \kappa$, where $T = \varrho(F)$, then there is a submodule $L \leq K$ such that $F/L \in \mathbf{F}_\tau$ and $|F/L| \leq \kappa$.*

Proof. If $H \leq F$ is a submodule maximal with respect to $H \cap T = 0$, then $\varrho(H) \subseteq \varrho(F) \cap H = T \cap H = 0$ and $H/(H \cap K) \cong (H + K)/K \leq F/K$ shows that $|H/(H \cap K)| \leq \lambda$. By the condition (*) there is $L \leq H \cap K$ with $H/L \in \mathbf{F}_\varrho$ and $|H/L| \leq \kappa$. Furthermore, the short exact sequence $0 \to H/L \to F/L \to F/H \to 0$ gives that $|F/L| \leq \kappa$ in view of the fact that $|F/H| \leq \kappa$ by Lemma 4. Finally, L is ϱ-pure in H and H is σ-pure in F by Lemma 5 and consequently L is τ-pure in F and we are through. □

Lemma 7. *Let ϱ, σ and $\tau = \varrho \cap \sigma$ be hereditary torsion theories for the category R-mod such that the classes $\mathbf{F}_\varrho$ and $\mathbf{F}_\sigma$ satisfy the condition (*). If K is a submodule of a module F such that $F \in \mathbf{F}_\tau$, $|F| > \kappa$, $|F/K| \leq \lambda$, where λ is an infinite cardinal* and κ is a regular cardinal such that $\kappa > \kappa(\lambda, \varrho)$, $\kappa > \kappa(\lambda, \sigma)$, $\kappa^\mu = \kappa$, where $\mu = \max(|R|, \aleph_0)$ and if $T = \varrho(F)$ is such that $|F/T| \leq \kappa$, then there is a submodule $L \leq K$ such that $F/L \in \mathbf{F}_\tau$ and $|F/L| \leq \kappa$.

Proof. From $T/(T \cap K) \cong (T + K)/K \leq F/K$ we get $|T/(T \cap K)| \leq \lambda$ and since $|T| > \kappa$ and $T \in \mathbf{F}_\sigma$, we obtain the existence of a submodule $L \leq T \cap K \leq T$ such that $T/L \in \mathbf{F}_\sigma$ and $|T/L| \leq \kappa$ and it suffices to note that from the short exact sequence $0 \to T/L \to F/L \to F/T \to 0$ and $T/L \in \mathbf{F}_\sigma$, $F/T \in \mathbf{F}_\varrho$ we obtain $F/L \in \mathbf{F}_\tau$ and $|F/L| \leq \max(|T/L|, |F/T|) \leq \kappa$. □

Now we are ready to prove one of the main results, namely that the set of all hereditary torsion theories, the torsionfree class of which is a precover class, is closed under intersections.

Theorem 8. *Let ϱ, σ and $\tau = \varrho \cap \sigma$ be hereditary torsion theories for the category R-mod such that $\mathbf{F}_\varrho$ and $\mathbf{F}_\sigma$ are precover classes. Then $\mathbf{F}_\tau$ is a precover class, too.*

Proof. Let λ be an infinite cardinal and let κ be a regular cardinal such that $\kappa > \kappa(\lambda, \varrho)$, $\kappa > \kappa(\lambda, \sigma)$ and $\kappa^\mu = \kappa$, where $\mu = \max(|R|, \aleph_0)$. Further, let K be a submodule of a module F such that $F \in \mathbf{F}_\tau$, $|F| > \kappa$ and $|F/K| \leq \lambda$. If either $|\varrho(F)| \leq \kappa$ or $|\sigma(F)| \leq \kappa$, then it suffices to use Lemma 6. Similarly, if either $|F/\varrho(F)| \leq \kappa$ or $|F/\sigma(F)| \leq \kappa$, then Lemma 7 applies. So, let $T = \varrho(F)$ be such that $|T| > \kappa$ and $|F/T| > \kappa$. In this case $T/(T \cap K) \cong (T+K)/K \leq F/K$ yields $|T/(T \cap K)| \leq \lambda$ and so there is $\tilde{L} \leq T \cap K$ such that $|T/\tilde{L}| \leq \kappa$ and $T/\tilde{L} \in \mathbf{F}_\sigma$. Further, $\varrho(F/\tilde{L}) = T/\tilde{L}$ is of the size at most κ and Lemma 6 yields the existence of a submodule $L/\tilde{L} \leq K/\tilde{L}$ which is τ-pure in $F/\tilde{L}$ and such that $|F/L| \leq \kappa$ and we are through. □

Now we proceed to show that the set of all hereditary torsion theories, the torsion-free class of which is a precover class, is closed under joins.

In the following remark we shall give the construction of a cardinal κ which will correspond to $\kappa(\lambda, \tau)$ for an infinite cardinal λ and the join $\varrho \vee \sigma$.

Remark 9. Let $\varrho = (\mathbf{T}_\varrho, \mathbf{F}_\varrho)$ and $\sigma = (\mathbf{T}_\sigma, \mathbf{F}_\sigma)$ be hereditary torsion theories for the category R-mod such that $\mathbf{F}_\varrho$ and $\mathbf{F}_\sigma$ satisfy the condition (*) and let λ be an infinite cardinal. We put $\kappa_0 = \lambda$ and we define by induction: $\kappa_{2i+1} = \kappa(\kappa_{2i}, \varrho)$, $\kappa_{2i+2} = \kappa(\kappa_{2i+1}, \sigma)$, $i = 0, 1, \ldots$. Finally we take κ to be the first regular cardinal such that $\kappa_i < \kappa$ for all $i = 0, 1, \ldots$ and $\kappa^\mu = \kappa$, where $\mu = \max(|R|, \aleph_0)$.

Theorem 10. *Let $\varrho = (\mathbf{T}_\varrho, \mathbf{F}_\varrho)$ and $\sigma = (\mathbf{T}_\sigma, \mathbf{F}_\sigma)$ be hereditary torsion theories for the category R-mod such that $\mathbf{F}_\varrho$ and $\mathbf{F}_\sigma$ are precover classes. Then $\mathbf{F}_\tau$, where $\tau = \varrho \vee \sigma$, is a precover class, too.*

Proof. With respect to [3, Theorem 9] it suffices to verify that $\mathbf{F}_\tau$ satisfies the condition (*). So, let λ be an arbitrary infinite cardinal and let $\kappa > \lambda$ be the cardinal corresponding to λ under the above construction. Further, let $F \in \mathbf{F}_\tau$ be such that $|F| > \kappa$ and let $K \leq F$ be a submodule of F with $|F/K| \leq \lambda = \kappa_0$. Then there is $L_1 \leq L_0 = K$ such that $F/L_1 \in \mathbf{F}_\varrho$ and $|F/L_1| \leq \kappa_1$. Similarly, there is $L_2 \leq L_1$

with $F/L_2 \in \mathbf{F}_\sigma$ and $|F/L_2| \leq \kappa_2$. Continuing by the induction we obtain an infinite non-increasing sequence $L_0 \supseteq L_1 \supseteq \cdots$ of submodules of F such that $F/L_{2i} \in \mathbf{F}_\sigma$, $F/L_{2i+1} \in \mathbf{F}_\varrho$ and $|F/L_i| \leq \kappa_i$ for each $i < \omega$. Setting $L = \bigcap_{i<\omega} L_i$ we see that $L = \bigcap_{i<\omega} L_{2i} = \bigcap_{i<\omega} L_{2i+1}$, hence L is ϱ-pure and σ-pure in F and consequently it is τ-pure in F and it remains to verify the cardinality property. However, there is the natural monomorphism $\alpha : F/L \to \prod_{i<\omega} F/L_i$ and the inequalities $|F/L_i| \leq \kappa_i < \kappa$ yield $|\prod_{i<\omega} F/L_i| \leq \kappa^{\aleph_0} \leq \kappa^\mu = \kappa$ by the hypothesis. Thus $|F/L| \leq \kappa$ and we are through. □

Summarizing, we can present the main result of this note.

Theorem 11. *The set of all hereditary torsion theories, the torsionfree classes of which are precover classes, is the sublattice of the lattice of all hereditary torsion theories for the category R-mod.*

Proof. It follows immediately from Theorems 8 and 10. □

Acknowledgements. This research has been partially supported by the Grant Agency of the Czech Republic, grant #GAČR 201/06/510 and also by the institutional grant MSM 0021620839.

References

[1] F. W. Anderson and K. R. Fuller, *Rings and Categories of Modules*, Graduate Texts in Mathematics **13**, Springer-Verlag (1974).

[2] B. Balcar and P. Štěpánek, *Teorie množin*, Academia Praha (1986).

[3] L. Bican, *On precover classes*, Ann. Univ. Ferrara, Sez. VII, Sc. Mat **LI** (2005), 61–67.

[4] L. Bican, *On injective hulls*, Contributions to General Algebra, Proceedings of the Vienna Conference 2005 (AAA70), Verlag Johannes Heyn, Klagenfurt 17 (2006), 25–29.

[5] L. Bican, R. El Bashir and E. Enochs, *All modules have flat covers*, Proc. London Math. Society 33 (2001), 649–652. Proc

[6] L. Bican, T. Kepka and P. Němec, *Rings, Modules, and Preradicals*, Marcel Dekker, New York (1982).

[7] L. Bican and B. Torrecillas, *Precovers*, Czech. Math. J. 53 (128) (2003), 191–203.

[8] L. Bican and B. Torrecillas, *On covers*, J. Algebra 236 (2001), 645–650.

[9] J. Golan, *Torsion Theories*, Pitman Monographs and Surveys in Pure and Applied Matematics, **29**, Longman Scientific and Technical (1986).

[10] T. Jech, *Set theory*, Academic Press New York (1978).

[11] J. Xu, *Flat Covers of Modules*, Lecture Notes in Mathematics 1634, Springer-Verlag Berlin Heidelberg New York (1996).

Author information

Ladislav Bican, KA MFF UK, Sokolovská 83, 186 76 Praha 8, Karlín, Czech Republic.
E-mail: `bican@karlin.mff.cuni.cz`

Non-singular rings of injective dimension 1

Ulrich Albrecht

Abstract. A singular module S has the essential extension property if there are a projective module P and an essential extension M of P such that $M/P \cong S$. All singular right and left modules over a semi-prime right and left Goldie ring R have the essential extension property if and only if R is right and left Noetherian and has injective dimension at most 1 as a right and left R-module. Several related results and examples are presented.

Key words. Essential extension, injective dimension, Goldie-ring, Utumi-ring.

AMS classification. 16D10.

1 Introduction

The *singular submodule* of a right R-module M is $Z(M) = \{x \in M \mid xI = 0 \text{ for some essential right ideal } I \text{ of } R\}$. A module M is *non-singular (singular)* if $Z(M) = 0$ ($Z(M) = M$), while R is *right non-singular* if R_R is a non-singular module (see [6] and [9] for further details). Right non-singular rings are precisely the rings with a regular, right self-injective maximal right ring of quotients Q^r. The embedding $R \subseteq Q^r$ gives rise to the R-R-bimodule $K^r = Q^r/R$. The right and left non-singular rings for which $Q^\ell = Q^r$ are the right and left Utumi-rings ([6, Theorem 2.38]) for which we write Q and K instead of Q^r and K^r. We are particularly interested in the case that R is a semi-prime right and left Goldie-ring. In this case, R has a classical right and left ring of quotients which also is its maximal right and left ring of quotients.

A singular right R-module S has the *(finite) essential extension property* if there is an essential extension M of a (finitely generated) projective submodule P with $S \cong M/P$. Although every torsion module has the essential extension property if R is a Dedekind domain, this is not the case for Prüfer domains or unique factorization domains which are not Dedekind (Corollary 3.3 and Proposition 3.6). Theorem 3.1 shows that all singular right and left R-modules over a semi-prime right and left Goldie-ring have the essential extension property if and only if R is right and left Noetherian and has injective dimension at most 1 as a right and left R-module. Therefore, the Noetherian reflexive domains studied in [5] and [8] are the integral domains for which every torsion module has the essential extension property. In particular, any such ring which is not Dedekind also is an example of a domain which is not hereditary, but all whose torsion modules have the essential extension property. The last results of this section investigate the finite essential extension property for right and left Utumi-rings of injective dimension at most 1.

2 The essential extension property

Let R be a right non-singular ring. If M is a submodule of Q_R^r, then $(R:M)_\ell = \{q \in Q^r \mid qM \subseteq R\}$, while $(R:N)_r = \{q \in Q^r \mid Nq \subseteq R\}$ whenever N is a submodule of $_RQ^r$. Furthermore, the symbol $E(M)$ indicates the injective hull of M.

The class of modules with the (finite) essential extension property is closed with respect to submodules and (finite) direct sums. Furthermore, it is easy to see that a (finitely generated) singular submodule of a module M with $p.d.(M) \leq 1$ has the (finite) essential extension property if R has finite right Goldie-dimension.

Lemma 2.1. *Let R be a right non-singular ring of finite right Goldie dimension.*

(a) *M has the (finite) essential extension property if and only if M is isomorphic to a submodule of $(K^r)^{(I)}$ for some (finite) index-set I.*

(b) *If I is an essential right ideal of R, then R/I has the essential extension property if and only if $I = (R:U)_r$ for some finitely generated submodule U of $_RQ^r$ containing R.*

(c) *The class of modules with the finite essential extension property is closed with respect to essential extensions if and only if $E(K^r)$ has this property.*

Proof. (a) If M has the finite essential extension property, then there are a non-singular module N and a finitely generated projective submodule P of N with $M \cong N/P$. Without loss of generality, we may assume that P is free, say $P = R^n$ for some $n < \omega$. Since $(Q^r)^n = E(P)$, N is isomorphic to a submodule of $E(P)$, and M is isomorphic to a submodule of $(K^r)^n$. The case of modules with the essential extension property is handled in the same way observing only that the finite Goldie-dimension of R_R is needed to ensure that $(Q^r)^{(I)}$ is the injective hull of $R^{(I)}$ if I is infinite [9, Proposition XIII.3.3]. The converse is obvious.

(b) If I is an essential right ideal I of R such that R/I has the finite essential extension property, then there is a monomorphism $\alpha : R/M \to (K^r)^n$ for some $n < \omega$ by (a). Write $\alpha(1+I) = (u_1 + R, \ldots, u_n + R)$ for some $u_1, \ldots, u_n \in Q^r$. Let $U = R + Ru_1 + \ldots + Ru_n$, and observe that $u_iI \subseteq R$ yields $I \subseteq (R:U)_r$. Conversely, if $Uq \subseteq R$ for some $q \in Q^r$, then $q \in R$ since $R \subseteq U$. Moreover, $0 = (u_1q + R, \ldots, u_nq + R) = \alpha(q+I)$ yields $q \in I$. Hence, $I = (R:U)_r$.

Conversely, let $I = (R:U)_r$ for a finitely generated submodule U of $_RQ^r$ containing R. If $U = Ru_1 + \ldots + Ru_n$, then define $\alpha : R \to (K^r)^n$ by $\alpha(1) = (u_1 + R, \ldots, u_n + R)$. Clearly $r \in \ker \alpha$ if and only if $u_1r, \ldots, u_nr \in R$. Since this happens exactly if $Ur \subseteq R$, we have $\ker \alpha = (R:U)_r = I$.

To see (c), assume that $E = E(K^r)$ has the finite essential extension property. If M is an essential extension of a submodule U which has the finite essential extension property, then there is a monomorphism $U \to (K^r)^n$ for some $n < \omega$. This map extends to a monomorphism $M \to E^n$. Since the class of modules with the finite essential extension property is closed with respect to submodules and finite direct sums, M has this property too. The converse is obvious. □

Let R be a regular and right self-injective ring, which is not semi-simple Artinian. Since R is its own maximal right ring of quotients, there exist no singular modules with the finite essential extension property since all finitely generated projective modules are injective.

A ring R satisfies the *restricted right minimum condition* if R/I is Artinian for every essential right ideal I of R. Every right and left Noetherian, hereditary ring satisfies the restricted right minimum condition by [4, Theorem 8.21]. If M is an R-module, then the *socle of M* which is denoted by $\text{soc}(M)$ is the sum of all simple submodules of M.

Theorem 2.2. *Let R be a right Noetherian right non-singular ring:*

(a) *All singular right R-modules M such that $\text{soc}(M)$ is essential in M have the essential extension property if and only if $Q^r \oplus K^r$ contains an injective cogenerator E of $\mathcal{M}_R$.*

(b) *All singular right R-modules have an essential socle if and only if R satisfies the restricted right minimum condition.*

Proof. (a) Assume that all singular modules with an essential socle have the essential extension property. For a simple module S, choose a maximal right ideal I of R with $S = R/I$. If I is not essential in R, then $R = I \oplus I'$, and $I' \cong S$ is simple. Because Q_R^r is injective, it contains $E(S)$. On the other hand, if I is essential, then S is singular. Then, $E(S)$ has the essential extension property, and there are an index-set J and a monomorphism $\phi : E(S) \to (K^r)^{(J)}$ by Lemma 2.1. For $j \in J$, let $\pi_j : (K^r)^{(J)} \to K^r$ be the projection onto the jth coordinate. There is $j_0 \in J$ such that $\pi_{j_0}\phi(S) \neq 0$. Since S is simple, $\pi_{j_0}\phi : E(S) \to K^r$ is a monomorphism. Consequently, $Q^r \oplus K^r$ contains a copy of $E(S)$ for every simple singular module S. Since R is right Noetherian, $\oplus_S E(S) \subseteq Q^r \oplus K^r$ is injective.

Conversely, let E be an injective cogenerator contained in $Q^r \oplus K^r$. Consider a singular right R-module M such that soc(M) is an essential submodule of M. Write $\text{soc}(M) = \oplus_I S_i$ where each S_i is simple. There is a monomorphism $\alpha : \text{soc}(M) \to E^{(I)}$. Because R is right Noetherian, $E^{(I)}$ is injective, and α extends to a monomorphism $\beta : M \to (Q^r)^{(I)} \oplus (K^r)^{(I)}$. If $\pi : (Q^r)^{(I)} \oplus (K^r)^{(I)} \to (K^r)^{(I)}$ is a projection with kernel $(Q^r)^{(I)}$, then $\beta(M) \cap \ker \pi = 0$ since M is singular. Thus, $\pi\beta : M \to (K^r)^{(I)}$ is one-to-one.

(b) Suppose that R satisfies the restricted right minimum condition. Let M be a singular R-module. If $0 \neq x \in M$, then there is an essential right ideal I of R with $R/I \cong xR$. Because R satisfies the restricted right minimum condition, xR is an Artinian R-module. Therefore, xR contains a simple submodule V, and $\text{soc}(M)$ is essential in M.

Conversely, assume that every singular right R-module has an essential socle. Let I be a proper essential right ideal of R, and assume that one has constructed an ascending chain $I = I_0 \subseteq I_1 \subseteq \ldots \subseteq I_n$ of right ideals of R such that I_k/I_{k-1} is a non-zero Artinian module. If $I_n \neq R$, then R/I_n has an essential socle. Choose a right ideal I_{n+1} of R containing I_n with $I_{n+1}/I_n = \text{soc}(R/I_n)$. Since R is right Noetherian, I_{n+1} is finitely generated, and I_{n+1}/I_n is Artinian as a finite sum of simple modules. Using the fact that R is right Noetherian once more, yields that the process has to stop. □

3 Right and left Utumi-rings

If M is a right (left) R-module, then the symbol M^* denotes the left (right) R-module $\text{Hom}_R(M, R)$. If R is right non-singular, then $M^* \cong (R : M)_\ell/\text{ann}_\ell(M)$ for all submodules M of Q_R^r where $\text{ann}_\ell(M) = \{q \in Q^r \mid qM = 0\}$ since Q_R^r is right self-injective. Because $\text{ann}_\ell(M) = 0$ if M is essential in Q_R^r, we identify M^* with $(R : M)_\ell$ in this case.

We want to remind the reader that a right non-singular ring R is a *right Utumi-ring* if every right ideal I of R, for which R/I is non-singular, is a right annihilator, i.e. there is a subset S of R with $I = \{r \in R \mid Sr = 0\}$. For a right and left Utumi-ring R, we identify N^* and $(R : N)_r$ for all essential submodules N of $_RQ$ in the same way. Since $_RR$ is essential in $_RQ$, we obtain that $U^* = (R : U)_r$ if U is a finitely generated submodule U of $_RQ^r$ which contains R. Thus, U^* is an essential right ideal of R, and $U^{**} = (R : U^*)_\ell$ is a submodule of $_RQ$ which contains U. In the same way, I^* is a submodule of $_RQ$ which contains R whenever I is an essential right ideal of R. Hence, $I^{**} = (R : I^*)_r \supseteq I$.

An R-module M is *torsion-less* if the natural map $\psi_M : M \to M^{**}$ is a monomorphism. It is *reflexive* if ψ_M is an isomorphism. For a right and left Utumi-ring R, an essential right ideal I of R is reflexive if and only if $I = I^{**}$. In the same way, a finitely generated submodule U of $_RQ$ which contains R is reflexive if and only if $U^{**} = U$.

Theorem 3.1. *The following are equivalent for a semi-prime right and left Goldie-ring R:*

(a) *R is a right and left Noetherian ring such that K is injective as a right and as a left R-module.*

(b) *All cyclic singular right and left R-modules have the essential extension property.*

(c) *All singular right and left R-modules have the essential extension property.*

Furthermore, any such ring satisfies the restricted right and left minimum conditions.

Proof. Let Q denote the classical right and left ring of quotients of R. We first show that a semi-prime, right and left Noetherian ring with the property that all right and left ideals are reflexive satisfies the restricted right and left minimum conditions. For this, consider an essential right ideal I of R, and assume that there exist right ideals $\{I_n\}_{n<\omega}$ such that $I \subseteq \ldots \subseteq I_{n+1} \subseteq I_n \subseteq \ldots \subseteq I_0 = R$. We obtain an ascending chain $R^* = I_0^* \subseteq \ldots \subseteq I_n^* \subseteq I_{n+1}^* \subseteq \ldots \subseteq I^*$ of submodules of $_RQ$. Since R is right Noetherian, there is a finitely generated projective module P and an epimorphism $\pi : P \to I$. Hence, $\pi^* : I^* \to P^*$ is a monomorphism. Since P^* is a finitely generated left R-module and R is left Noetherian, I^* is a Noetherian R-module. Hence, there is $m < \omega$ such that $I_{m+k}^* = I_m^*$ for all $k < \omega$. Since the I_j's are reflexive, $I_{m+k} = I_{m+k}^{**} = I_m^{**} = I_m$, and R/I is Artinian. By symmetry, R also satisfies the restricted left minimum condition.

(a) $\Rightarrow$ (c): By the first corollary on page 72 of [7], every right and left ideal of R is reflexive. Therefore, R satisfies the restricted right and left minimum condition because of what was shown in the first paragraph. Let I be an essential right ideal of R. Then,

$I = I^{**} = (R : I^*)_r$. Moreover, there is an epimorphism $\pi : R^n \to I$ for some $n < \omega$ since R is right Noetherian. But then $\pi^* : I^* \to (R^n)^*$ is a monomorphism. Since R is left Noetherian, I^* is finitely generated. By Lemma 2.1, R/I has the essential extension property. In particular, R/M can be embedded into $Q \oplus K$ for all maximal right ideals M of R. Since $Q \oplus K$ is injective, it is an injective cogenerator of $\mathcal{M}_R$. By Theorem 2.2(a), all singular R-modules have the essential extension property since each of them has an essential socle by Theorem 2.2(b). By symmetry, the same is true for all singular left modules.

Since (c) $\Rightarrow$ (b) is obvious, we turn to (b) $\Rightarrow$ (a): We consider an essential right ideal I of R. Since R/I has the essential extension property, there is a finitely generated submodule U of $_RQ$ containing R with $(R : U)_r = I$. Because U is an essential submodule of $_RQ$, $(R : U)_r = U^*$. By the third dual theorem in [7], the canonical map $\psi_{U^*} : U^* \to U^{***}$ splits. Since $I = U^*$, we have $U^{***} = I^{**}$. Hence, $I^{**} = I \oplus W$ for some submodule W of Q_R. However, I is an essential submodule of I^{**} since I is essential in R_R and R_R is essential in Q_R. This is only possible if $I = I^{**}$. Since every right ideal is a direct summand of an essential right ideal, all right ideals are reflexive. By symmetry, the same holds for all left ideals.

Observe that $R \subseteq U \subseteq Rc^{-1}$ for some regular element c of R since Q is a classical right ring of quotients of R. Thus, $Rc \subseteq Uc \subseteq R$, and U is isomorphic to an essential left ideal J of R since R has finite left Goldie-dimension. By what has been shown in the last paragraph, J, and therefore U, are reflexive. Then, $I^* = U^{**} \cong U$ is a finitely generated submodule of $_RQ$ which contains R and is isomorphic to J. Since R/J has the essential extension property by (b), there is a finitely generated submodule V of Q_R which contains R with $J = (R : V)_\ell = V^*$. Since V is isomorphic to a right ideal of R because Q is a classical left ring of quotients of R, V is reflexive. Thus, $I \cong I^{**} \cong J^* = V^{**} \cong V$ is finitely generated. Therefore, R is a right Noetherian ring. By symmetry, R is left Noetherian. Hence, R satisfies the restricted right and left minimum condition by what was shown in the first paragraph of this proof.

It remains to show that R_R has injective dimension at most 1. For this, again consider an essential right ideal I of R. The standard homological arguments yield $\text{Ext}^1_R(R/I, K) \cong \text{Ext}^1_R(I, R)$. Therefore, it suffices to show that the latter vanishes. Since R is a semi-prime right Goldie-ring, I contains cR for some regular element c of R. Since R/cR is Artinian and Noetherian, there exist right ideals $R \cong cR = I_0 \subseteq I_1 \subseteq \ldots \subseteq I_n = R$ of R such that I_{i+1}/I_i is non-zero and simple. Following [8], we inductively show that $\text{Ext}^1_R(I_k, R) = 0$ for all k by considering the exact sequence $I^*_{i+1} \to I^*_i \xrightarrow{\delta} \text{Ext}^1_R(I_{i+1}/I_i, R) \to \text{Ext}^1_R(I_{i+1}, R) \to \text{Ext}^1_R(I_i, R) = 0$. All these maps are left R-module homomorphisms since they are induced by the functor $\text{Hom}_R(-, R)$ [7]. The arguments, which are used in the proof of [8, Theorem 3.8] to show that δ is onto, carry over to the non-commutative setting except that $\text{Ext}^1_R(I_{i+1}/I_i, R)$ need not be isomorphic to I_{i+1}/I_i. However, it is simple as a left R-module. To see this, choose an essential maximal right ideal M of R such that $R/M \cong I_{i+1}/I_i$. Then, $\text{Ext}^1_R(R/M, R) \cong M^*/R^*$ as a left R-module. Arguing as before, M^* is a finitely generated submodule of Q containing R because R is right and left Noetherian. Hence, M^* is isomorphic to a left ideal of R, and consequently, every submodule of M^* is reflexive. If $R^* \subseteq U \subseteq M^*$, then $M = M^{**} \subseteq U^* \subseteq R^{**} = R$, and one obtains $M = U^*$ or $R = U^*$. Applying the duality once more yields $U = U^{**} = M^*$ or $U = U^{**} = R^*$. Therefore, $\text{Ext}_R^1(I_{i+1}/I_i, R)$ is a simple left R-module. As in [8, Theorem 3.8], δ is onto, and $\text{Ext}_R^1(I_{i+1}, R) = 0$. □

Corollary 3.2. *Let R be an integral domain. All torsion modules have the essential extension property if and only if R is Noetherian and Q/R is injective.*

[5, Exercise IX.4.5] states a result of Kaplansky that the subring R of the formal power series ring over a field consisting of those power series, whose first degree terms are missing, is a local Noetherian domain which is not Dedekind, but has the property that Q/R is injective. On the other hand, we have

Corollary 3.3. *Let R be a unique factorization domain. Every cyclic singular R-module has the essential extension property if and only if R is a principal ideal domain.*

Proof. Let R be a unique factorization domain with field of quotients Q which is not a principal ideal domain. If all torsion R-modules have the essential extension property, then R is Noetherian by Theorem 3.1. There exist $c_1, c_2 \in R$ such that $c_1 R + c_2 R$ is not principal. Observe that c_1 and c_2 are not units. Let d be the greatest common divisor of c_1 and c_2, and write $c_i = db_i$ for $i = 1, 2$. Multiplication by d induces an isomorphism $b_1 R + b_2 R \to c_1 R + c_2 R$. In particular, $b_1 R + b_2 R$ is a proper ideal of R, and there is a maximal ideal M which contains $b_1 R + b_2 R$. Suppose that there is a finitely generated submodule U of Q containing R such that $(R : U) = M$. If U is generated by $u_1, \ldots, u_n$, then, for $i = 1, \ldots, n$, select $r_i, s_i \in R$ such that $u_i = r_i s_i^{-1}$ and r_i and s_i have no common prime divisors. Since $b_1, b_2 \in (R : U)$, there are $t_{i1}, t_{i2} \in R$ with $r_i b_j = t_{ij} s_i$ for $j = 1, 2$. If p is a prime dividing s_i, then p does not divide r_i. Hence, p has to divide b_1 and b_2 which is not possible. Thus, s_i is a unit, and $U = R$, a contradiction. By Lemma 2.1, R/M does not have the essential extension property. □

Corollary 3.4. *If R is an integral domain such that every ideal is generated by at most two elements, then all torsion modules have the essential extension property.*

Proof. Since Q is an injective R-module, $\text{Ext}_R^1(R/I, Q/R) \cong \text{Ext}_R^1(I, R)$ for all ideals I of R. By [3, Proposition 1.5], every exact sequence $0 \to R \to X \to I \to 0$ splits. Hence, K is injective. Now apply Theorem 3.1. □

For instance, the ring $R = \mathbb{Z} + 2i\mathbb{Z}$ has the property that every ideal is generated by at most 2 elements. Non-commutative examples of rings, to which Theorem 3.1 applies, are a direct consequence of

Proposition 3.5. *The class of semi-prime right and left Noetherian rings, such that K is injective as a right and left R-module, is closed with respect to Morita equivalence.*

Proof. Suppose that the ring S is equivalent to R. By [2] and [6, page 12], S is a semi-prime, right and left Noetherian ring. Moreover, every finitely generated projective R-module P fits into an exact sequence $0 \to P \to E_1 \to E_2 \to 0$ since R has right and left injective dimension at most 1. Since Morita-equivalence preserves injectives, S has injective dimension at most 1. □

While the last results were concerned with Noetherian rings of injective dimension of most one, there also are non-Noetherian rings with this property, for instance the almost maximal Prüfer domains which are not Dedekind [5].

Proposition 3.6. *Let R be a right and left Utumi-ring which has finite right Goldie-dimension.*

(a) *If M is an essential maximal right ideal of R, then R/M has the essential extension property if and only if M is reflexive.*

(b) *If R is right semi-hereditary, then a finitely generated singular module M has the essential extension property if and only if $p.d.(M) = 1$. In particular, all cyclic singular modules have the essential extension property if and only if R is right hereditary.*

Proof. (a) Suppose that M has the essential extension property. Then, $M = U^*$ for some finitely generated submodule of $_RQ$ which contains R. As in the proof of (b) $\Rightarrow$ (a) of Theorem 3.1, one concludes $M = M^{**}$.

Conversely, if R/M cannot be embedded into K, then $\operatorname{Hom}_R(R/M, K) = 0$. We obtain the exact sequence

$$0 = \operatorname{Hom}_R(R/M, K) \to \operatorname{Ext}^1_R(R/M, R) \to \operatorname{Ext}^1_R(R/M, Q) = 0$$

from which $\operatorname{Ext}^1_R(R/M, R) = 0$ follows. Because $0 = (R/M)^* \to R^* \to M^* \to \operatorname{Ext}(R/M, R) = 0$ is exact, the map $\iota^{**} : M^{**} \to R^{**}$ is an isomorphism. Since M is reflexive, $R = M$, a contradiction.

(b) Suppose that M is a finitely generated module M with the essential extension property. By Lemma 2.1, M is isomorphic to a submodule U of K^n for some $n < \omega$. Select a finitely generated submodule V of Q^n containing R^n with $U = V/R^n$. By [9, Proposition XI.5.4], R is right strongly non-singular. Therefore, one can embed V into a free module W. Since R is right semi-hereditary, W is projective.

If all cyclic singular modules have the essential extension property, then R/I has projective dimension 1 whenever I is an essential right ideal, i.e. R is right hereditary. □

In particular, a Prüfer domains is Dedekind if all cyclic torsion modules have the essential extension property.

Corollary 3.7. *Let R be a right and left Utumi-ring of finite right and left Goldie-dimension such that $i.d.(R_R), i.d.(_RR) \leq 1$. Then:*

(a) *Finitely generated non-singular R-modules are reflexive.*

(b) *If I is an essential right or left ideal of R, then R/I has the essential extension property if and only if I is reflexive and I^* is finitely generated.*

(c) *If M is a maximal right or left ideal, which is finitely generated and essential, then R/M has the essential extension property.*

Proof. (a) If M is a finitely generated non-singular module, then there is a finitely generated free module F with $M \subseteq F$ by [9, Proposition XI.5.4]. Consequently, ψ_M is one-to-one.

The exact sequence $0 = \text{Ext}^1_R(F, R) \to \text{Ext}^1_R(M, R) \to \text{Ext}^2_R(F/M, R) = 0$ yields $\text{Ext}^1_R(M, R) = 0$. If $0 \to U \to P \xrightarrow{\pi} M \to 0$ is an exact sequence with P finitely generated and projective, then the induced sequence $0 \to M^* \xrightarrow{\pi^*} P^* \to U^* \to \text{Ext}^1_R(M, R) = 0$ yields that U^* is a finitely generated torsion-less left R-module. By [9, Proposition XI.5.4], U^* can be embedded into a free module. Hence, $\text{Ext}^1_R(U^*, R) = 0$ by symmetry. Thus, the sequence $0 \to U^{**} \to P^{**} \xrightarrow{\pi^{**}} M^{**} \to 0$ is exact. Since $\psi_M \pi = \pi^{**} \psi_P$, we obtain that ψ_M is also onto.

(b) Suppose that I is an essential right ideal of R such that R/I has the essential extension property. There exists a finitely generated submodule U of $_RQ$ with $U^* = I$ by Lemma 2.1(b). By (a), U is reflexive. Then, $I^* \cong U^{**} \cong U$ is finitely generated and reflexive.

Conversely, let I be an essential right ideal which is reflexive such that I^* is finitely generated. Then, $I^* = \{q \in Q \mid qI \subseteq R\}$ contains R and $I^{**} = (R : I^*)_r$. Thus, $I = (R : I^*)_r$ since I is reflexive, and R/I has the desired property by Lemma 2.1.

(c) is a direct consequence of Theorem 3.6(b). □

Similar to Proposition 3.5, the class of right and left Utumi-rings of finite right Goldie dimension, such that R has right and left injective dimension at most 1, is closed under Morita equivalence [2].

Example 3.8. There exists a right and left Utumi-ring R of finite right and left Goldie dimension such that $i.d.(_RR) \leq 1$, but $i.d.(R_R) \geq 2$. Moreover, all singular left R-modules have the essential extension property, while there exists a cyclic singular right R-module which does not have it.

Proof. Consider the ring $R = \left\{ \begin{pmatrix} n & 0 \\ x & y \end{pmatrix} \mid n \in \mathbb{Z}, x, y \in \mathbb{Q} \right\}$ whose maximal right and left ring of quotients is $Q = \text{Mat}_2(\mathbb{Q})$ [6]. As in [6, Chapter 4A], we obtain that R is a left Noetherian, left hereditary ring which is not right Noetherian, and hence not right hereditary [4]. Since R is left hereditary, all singular left R-modules have the essential extension property and $i.d.(_RR) \leq 1$.

Since Q is an injective R-module, $i.d.(R_R) \geq 2$ if and only if $\text{Ext}^1_R(I, R) \neq 0$ for some right ideal I of R. The idempotent $e = \begin{pmatrix} 1 & 0 \\ 0 & 0 \end{pmatrix}$ satisfies $eR = eRe$. Hence, the map $R \to \mathbb{Z}$ which sends $r \in R$ to the left upper corner of the matrix ere is a ring homomorphism. Thus, the $\mathbb{Z}$-module structure of an abelian group G induces a right R-module structure on G. In particular, homomorphisms between abelian groups are also R-linear. Moreover, $\mathbb{Q}$ is a right R-module which is isomorphic to $N(R)$, the nilradical of R, because of $\begin{pmatrix} 0 & 1 \\ 0 & 0 \end{pmatrix} \begin{pmatrix} 0 & 0 \\ \mathbb{Q} & 0 \end{pmatrix} = \begin{pmatrix} \mathbb{Q} & 0 \\ 0 & 0 \end{pmatrix}$. Since $\mathbb{Z} \cong eR$, we obtain $\text{Ext}^1_R(N(R), eR) \cong \text{Ext}^1_\mathbb{Z}(\mathbb{Q}, \mathbb{Z}) \neq 0$.

By [4, Chapter 8], R is right semi-hereditary. Therefore, if I is an essential right ideal of R such that R/I has the essential extension property, then I is projective by

[9, Proposition XI.5.4]. Consequently, $R/N(R)$ cannot have the essential extension property since $\text{Ext}_R^1(N(R), R) \neq 0$. □

References

[1] Albrecht, U., Dauns, J., and Fuchs, L.; *Torsion-freeness and non-singularity over right p.p.-rings*; J. of Alg. 285 (2005); 98–119.

[2] Ancerson, F., and Fuller, K.; *Rings and Categories of Modules*; Graduate Texts in Mathematics 13; Springer-Verlag (1992).

[3] Bass, H.; *Torsion-free and projective modules*; Trans. Amer. Math. Soc. 102(2) (1962); 319–327.

[4] Chatters, A. W., and Hajarnavis, C. R.; *Rings with Chain Conditions*; Pitman Advanced Publishing 44; Boston, London, Melbourne (1980).

[5] Fuchs, L., and Salce, L.; *Modules over Non-Noetherian Domains*; AMS 84 (2001).

[6] Goodearl, K.; *Ring Theory*; Marcel Dekker; New York, Basel (1976).

[7] Jans, J.; *Rings and Homology*; Holt, Rinehart, and Winston (1963).

[8] Matlis, E.; *Reflexive Domains*; J. Alg. 8 (1968); 1–33.

[9] Stenström, B.; *Rings of Quotients*; Lecture Notes in Math. 217; Springer-Verlag, Berlin, Heidelberg, New York (1975).

Author information

Ulrich Albrecht, Department of Mathematics, Auburn University, Auburn, AL 36849, U.S.A.
E-mail: `albreuf@auburn.edu`

On classes defining a homological dimension

Francesca Mantese and Alberto Tonolo

Abstract. A class $\mathcal{F}$ of objects of an abelian category $\mathcal{A}$ is said to define a *homological dimension* if for any object in $\mathcal{A}$ the length of any $\mathcal{F}$-resolution is uniquely determined. In the present paper we investigate classes satisfying this property.

Key words. Homological dimension, abelian categories, cotorsion pairs.

AMS classification. 18G20, 16E10.

Introduction

In general the class of the objects of a given abelian category $\mathcal{A}$ is too complex to admit any satisfactory classification. Starting from a known subclass $\mathcal{F}$ of $\mathcal{A}$, one may try to approximate arbitrary objects by the objects in $\mathcal{F}$. This approach has successfully been followed over the past few decades for categories of modules through the theory of precovers and preenvelopes, or left and right approximations (see [6] or [8] for a detailed list of references).

Another point of view could be to measure the "distance" of any object in $\mathcal{A}$ from the class $\mathcal{F}$, introducing a notion of *dimension* with respect to the class $\mathcal{F}$, computed by means of $\mathcal{F}$-resolutions. In this framework, the notions of projective dimension, weak dimension, Gorenstein dimension of modules have been deeply studied.

Our aim is to define a good concept of dimension with respect to a wide family of classes of objects. We say that a class $\mathcal{F}$ of objects of an abelian category $\mathcal{A}$ defines a *homological dimension* if for any object in $\mathcal{A}$, the length of any $\mathcal{F}$-resolution is uniquely determined (see Definition 1.6). In such a way to each object in $\mathcal{A}$ one can associate an $\mathcal{F}$-invariant number which represents locally the relevance of $\mathcal{F}$.

In the first section we study several properties of classes defining a homological dimension; in particular we discuss their closure properties and the connection with precover classes and cotorsion pairs. In the second section, using tools from derived categories, we generalize the Auslander notion of Gorenstein dimension to arbitrary abelian categories. We consider a homological dimension associated to an adjoint pair $(\varPhi, \varPsi)$ of contravariant functors, obtaining again the classical Gorenstein dimension on R-modules in case $\varPhi = \varPsi = \text{Hom}(-, R)$ for a commutative noetherian ring R.

Second author: Research supported by grant CDPA048343 of Padova University.

1 Homological dimension

Definition 1.1 (cf. [2]). Let $\mathcal{F}$ be a class of objects in an abelian category $\mathcal{A}$. We say that an object M in $\mathcal{A}$ has *left $\mathcal{F}$-dimension* $\leq \alpha$, $\alpha \in \mathbb{N} \cup \{\infty\}$, if there exists a long exact sequence

$$\ldots \to F_i \to F_{i-1} \to \ldots \to F_1 \to F_0 \to M \to 0$$

with $F_i \in \mathcal{F} \cup \{0\}$, and $F_i = 0$ for $i > \alpha$. We denote by $\mathcal{F}_\alpha$ the class of objects M of left $\mathcal{F}$-dimension $\leq \alpha$ (shortly $\mathcal{F}\dim M \leq \alpha$), and by $\mathcal{F}_{<\infty}$ the class of objects of finite left $\mathcal{F}$-dimension.

In general there exist objects which have not a left $\mathcal{F}$-dimension: in particular all objects which are not quotients of objects in $\mathcal{F}$. We denote by $\overline{\mathcal{F}}$ the class of all objects in $\mathcal{A}$ which are homomorphic image of objects in $\mathcal{F}$.

Remark 1.2. If $\mathcal{A}$ has enough projectives and $\mathcal{F}$ is closed under direct summands, then $\overline{\mathcal{F}} = \mathcal{A}$ if and only if $\mathcal{F}$ contains all projective objects.

In particular, if $\mathcal{A} = R$-Mod, denoted by $\mathcal{P}$ and $\mathcal{F}l$ the classes of projective and flat modules respectively, then $\overline{\mathcal{P}} = R$-Mod and $\overline{\mathcal{F}l} = R$-Mod, and left $\mathcal{P}$- and left $\mathcal{F}l$-dimensions are the usual projective and flat (or weak) dimensions of a module.

Definition 1.3. We say that $\mathcal{A}$ has *global left $\mathcal{F}$-dimension* $\leq \alpha$ (resp. $< \infty$), $\alpha \in \mathbb{N} \cup \{\infty\}$, if for each object M in $\mathcal{A}$ we have $\mathcal{F}\dim M \leq \alpha$ (resp. $< \infty$).

Clearly $\mathcal{A}$ has global left $\mathcal{F}$-dimension $\leq \infty$ if and only if $\mathcal{A} = \overline{\mathcal{F}}$.

In any abelian category $\mathcal{A}$ it is possible (see [13, Ch. VII]) to define, for any pair of object $A, B \in \mathcal{A}$, the family $\text{Ext}^i_\mathcal{A}(A, B)$ of equivalence classes of exact sequences of length i with left end B and right end A, with respect to the Yoneda equivalence relation. The family $\text{Ext}^i_\mathcal{A}(A, B)$ in general is not a set (see [7, Ch. VI]); nevertheless it can be equipped with an additive structure and become a *big abelian group*. The big abelian groups are defined in the same way as ordinary abelian groups, except than the underlying class need not be a set. Quoting [13], "[...] we are prevented from talking about the category of big abelian groups because the class of morphisms between a given pair of big groups need not be a set. Nevertheless this will not keep us from talking about kernels, cokernels, images, exact sequences, etc., for big abelian groups." If $\mathcal{A}$ has enough injectives or projectives, then $\text{Ext}^i_\mathcal{A}(A, B)$ is an abelian group for each $A, B \in \mathcal{A}$.

Given a class of objects $\mathcal{G}$, we denote by

$$\mathcal{G}^{\perp_m} = \{M \in \mathcal{A} : \text{Ext}^i_\mathcal{A}(G, M) = 0, \forall 1 \leq i \leq m, G \in \mathcal{G}\};$$

the intersection $\bigcap_{m \geq 1} \mathcal{G}^{\perp_m}$ will be denoted by $\mathcal{G}^{\perp_\infty}$. Dually, we denote by

$$^{\perp_m}\mathcal{G} = \{M \in \mathcal{A} : \text{Ext}^i_\mathcal{A}(M, G) = 0, \forall 1 \leq i \leq m, G \in \mathcal{G}\};$$

the intersection $\bigcap_{m \geq 1} {}^{\perp_m}\mathcal{G}$ will be denoted by $^{\perp_\infty}\mathcal{G}$.

Definition 1.4 ([13, Ch. VI.6]). Let A be an object of an abelian category $\mathcal{A}$. The *cohomological dimension* ch.dim A of A is the least integer n such that the one variable functor $\operatorname{Ext}_{\mathcal{A}}^n(-, A)$ is not zero.

If $\mathcal{A}$ has enough injective objects (e.g., if $\mathcal{A}$ is a Grothendieck category) the cohomological dimension of an object coincides with its injective dimension.

Proposition 1.5. *Assume that $\mathcal{A}$ has enough projectives.*

(i) *If* $\operatorname{gl}\mathcal{F}\dim\mathcal{A} \leq n$, $n \in \mathbb{N}$, *then* ch.dim $Y \leq n$ *for each* $Y \in \mathcal{F}^{\perp n+1}$.

(ii) *If* $\mathcal{F} = {}^{\perp_m}\mathcal{G}$ *for a class $\mathcal{G}$ of modules of cohomological dimension less or equal than $n \in \mathbb{N}$, then* $\operatorname{gl}\mathcal{F}\dim\mathcal{A} \leq n$.

Proof. Let M be an arbitrary object in $\mathcal{A}$.
 (i) Since $\operatorname{gl}\mathcal{F}\dim\mathcal{A} \leq n$, there exists an exact sequence
$$0 \to F_n \to F_{n-1} \to \cdots \to F_1 \to F_0 \to M \to 0.$$

Applying the contravariant functor $\operatorname{Hom}(-, Y)$, since $Y \in \mathcal{F}^{\perp n+1}$, by dimension shift we get $\operatorname{Ext}_{\mathcal{A}}^{n+1}(M, Y) \cong \operatorname{Ext}_{\mathcal{A}}^1(F_n, Y) = 0$. Since $\operatorname{Ext}_{\mathcal{A}}^{n+1}(M, Y) = 0$ for each object M in $\mathcal{A}$, and the latter has enough projectives, then $\operatorname{Ext}_{\mathcal{A}}^{n+i}(M, Y) = 0$ for each $i \geq 1$, i.e. ch.dim $Y \leq n$.
 (ii) Consider an exact sequence
$$0 \to K_n \to P_{n-1} \to \cdots \to P_1 \to P_0 \to M \to 0$$

with P_i projective for $i = 0, \ldots, n-1$. Since $P_i \in \mathcal{F}$ it is enough to prove that K_n belongs to $\mathcal{F}$. So let $G \in \mathcal{G}$; then $\operatorname{Ext}_{\mathcal{A}}^i(K_n, G) \cong \operatorname{Ext}_{\mathcal{A}}^{n+i}(M, G) = 0$ for $1 \leq i$. Therefore $K_n \in {}^{\perp_\infty}\mathcal{G} \subseteq {}^{\perp_m}\mathcal{G} = \mathcal{F}$. □

In order to introduce a good measure of the distance between an object of $\mathcal{A}$ and a given class $\mathcal{F}$, the length of a $\mathcal{F}$-resolution has to be uniquely determined.

Definition 1.6. We say that the left $\mathcal{F}$-dimension associated to a class $\mathcal{F}$ is *homological* (or that the class $\mathcal{F}$ defines a *homological dimension*) if

(i) for any short exact sequence $0 \to K \to F \to M \to 0$ with $F \in \mathcal{F}$ and $M \in \mathcal{F}_\infty$, the object K belongs to $\mathcal{F}_\infty$;

(ii) for any exact sequence
$$0 \to K_n \to F_{n-1} \to \cdots \to F_1 \to F_0 \to X \to 0$$

with $F_i \in \mathcal{F}$, $i = 0, 1, \ldots, n-1$, and $X \in \mathcal{F}_n$, the object K_n belongs to $\mathcal{F}$.

Clearly if $\mathcal{A} = \overline{\mathcal{F}}$ we have $\mathcal{A} = \mathcal{F}_\infty$, and the first condition is empty.

Example 1.7. If $\mathcal{A} = R$-Mod, the classes $\mathcal{P}$ and $\mathcal{F}l$ define a homological dimension. The class of free modules defines a homological dimension if and only if it coincides with the class of projective modules (see Proposition 1.9), e.g. if R is local.

If $\mathcal{A}$ is the category of coherent sheaves on a noetherian scheme X, the classes of the locally free sheaves $\mathcal{LF}$ and of the invertible sheaves both define a homological dimension (see [9, Chp. 2 §5, Chp. 3 §6]). If X is quasi-projective over $\operatorname{Spec} R$, where R is a noetherian commutative ring, then $\overline{\mathcal{LF}} = \mathcal{A}$.

Note that the notion of homological dimension can be easily dualized obtaining a notion of *homological codimension*; for instance, if $\mathcal{A} = R$-Mod, the class of injective modules defines a homological codimension. Most of the results we obtain in this paper could be reformulated for this dual concept.

In the sequel we study closure properties of classes defining a homological dimension.

Let $\mathcal{F}$ be a class of modules and $0 \to A \to F \to C \to 0$ be an exact sequence with $F \in \mathcal{F}$. Thus, for any $i \geq 1$ in $\mathbb{N}$, if $A \in \mathcal{F}_{i-1}$ then $C \in \mathcal{F}_i$.

Lemma 1.8. *Let $\mathcal{F}$ be a class of objects in $\mathcal{A}$ and $0 \to A \to F \to C \to 0$ be an exact sequence with $F \in \mathcal{F}$. If $\mathcal{F}$ defines a homological dimension and $C \in \mathcal{F}_i$, then $A \in \mathcal{F}_{i-1}$. In particular $\mathcal{F}$ is closed under kernels of epimorphisms.*

Proof. By the definition of homological dimension, A belongs to $\mathcal{F}_\infty$. Therefore consider an exact sequence

$$0 \to K_{i-1} \to F_{i-2} \to \cdots \to F_1 \to F_0 \to A \to 0$$

with $F_j \in \mathcal{F}$. Since

$$0 \to K_{i-1} \to F_{i-2} \to \cdots \to F_1 \to F_0 \to F \to C \to 0$$

is an $\mathcal{F}$-resolution for C and $\mathcal{F}\dim C \leq i$, we get that $K_{i-1} \in \mathcal{F}$. □

Proposition 1.9. *Let $\mathcal{F}$ be a class of objects defining a homological dimension. If $\mathcal{F}$ is closed under countable direct sums, then $\mathcal{F}$ is closed under direct summands.*

Proof. Let $L \oplus M = F \in \mathcal{F}$; consider the short exact sequence

$$0 \to L \to L \oplus (M \oplus L)^{(\omega)} \to (M \oplus L)^{(\omega)} \to 0;$$

since both $(M \oplus L)^{(\omega)}$ and $L \oplus (M \oplus L)^{(\omega)} \cong (L \oplus M)^{(\omega)}$ belong to $\mathcal{F}$, also L belongs to $\mathcal{F}$. □

In the next theorem we compare the $\mathcal{F}$-dimension of objects in a short exact sequence.

Theorem 1.10. *Assume $\mathcal{F}$ defines a homological dimension and it is closed under finite direct sums. Let $0 \to A \to B \to C \to 0$ be a short exact sequence. Then for each $i \in \mathbb{N}$ we have that*

(1_i) *if B and C belong to $\mathcal{F}_i$ then A belongs to $\mathcal{F}_i$;*

(2_i) *if A and B belong to $\mathcal{F}_i$ then C belongs to $\mathcal{F}_{i+1}$.*

If $\overline{\mathcal{F}}$ is closed under extensions, then

(3_i) *if A and C belong to $\mathcal{F}_{i+1}$, then B belongs to $\mathcal{F}_{i+1}$;*

(4_i) *if $B \in \mathcal{F}_i$ and $C \in \mathcal{F}_{i+1}$, then A belongs to $\mathcal{F}_i$.*

Proof. (1)–(2): If $i = 0$, (2_0) is clearly true by definition and (1_0) follows by $\mathcal{F}\dim C = 0 \leq 1$ and the fact that $\mathcal{F}$ defines a homological dimension. Assume (1_{i-1}) and (2_{i-1}) true for $i - 1 \geq 0$. Let us consider the pullback diagram

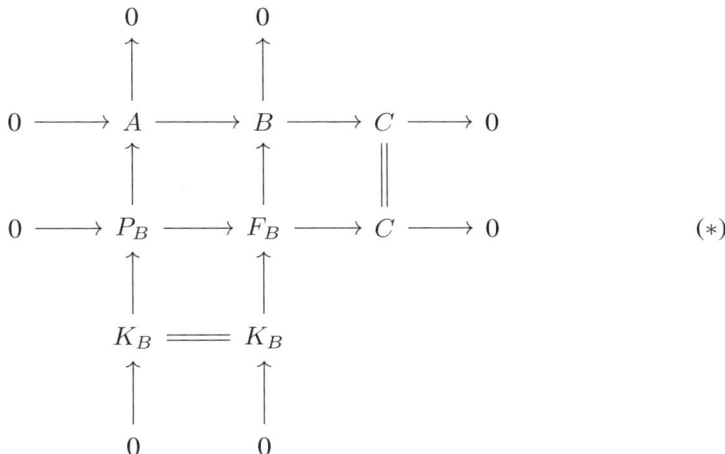

$(*)$

with F_B in $\mathcal{F}$.

(1_i): By Lemma 1.8 both K_B and P_B in diagram $(*)$ belong to $\mathcal{F}_{i-1}$, and so by induction $A \in \mathcal{F}_i$.

(2_i): Let now A and B be in $\mathcal{F}_i$; there exist $F_B \in \mathcal{F}$ and an epimorphism $\pi : F_B \to B$. Consider the following pullback diagram:

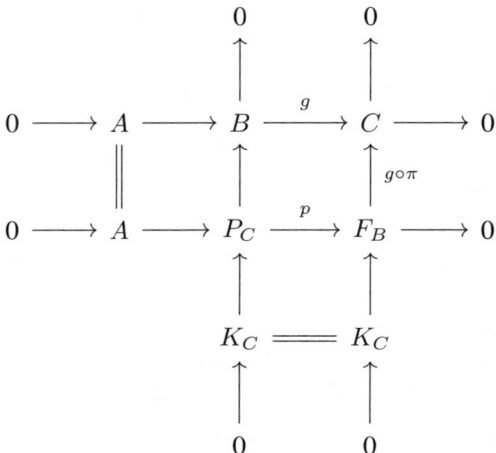

Since P_C is a pullback, there exists $j : F_B \to P_C$ such that $p \circ j = 1_{F_B}$. Then the middle exact sequence splits, and therefore $P_C = A \oplus F_B$; since $\mathcal{F}$ is closed under finite direct sums, P_C belongs to $\mathcal{F}_i$. Therefore by (1_i) we have $K_C \in \mathcal{F}_i$ and hence C belongs to $\mathcal{F}_{i+1}$.

(3)–(4): If $i = 0$, (4_0) follows by the definition of homological dimension. Since $\overline{\mathcal{F}}$ is closed under extensions, if A and C are in $\overline{\mathcal{F}}$, also B belongs to $\overline{\mathcal{F}}$. Then, if A and C belong to $\mathcal{F}_1$, we can consider the pullback diagram $(*)$ with F_B in $\mathcal{F}$. Since C belongs to $\mathcal{F}_1$, then P_B belongs to $\mathcal{F}$; since A belongs to $\mathcal{F}_1$, then also K_B belongs to $\mathcal{F}$, and therefore B belongs to $\mathcal{F}_1$. Assume (3_{i-1}) and (4_{i-1}) true for $i - 1 \geq 0$.

(4_i): Let us consider the pullback diagram

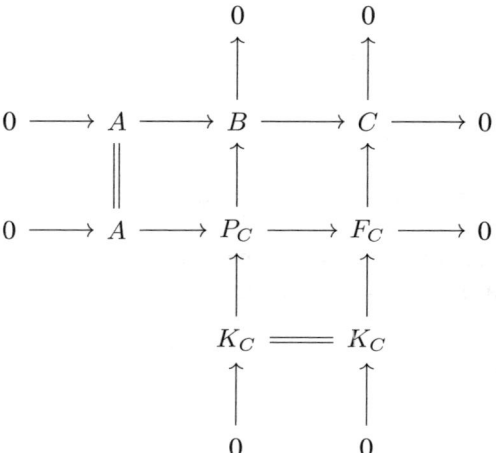

with $F_C \in \mathcal{F}$; then K_C belongs to $\mathcal{F}_i$. Since B belongs to $\mathcal{F}_i$, by (3_{i-1}) we have that $P_C \in \mathcal{F}_i$, and hence, by (1_i), A belongs to $\mathcal{F}_i$.

(3_i): Since $\overline{\mathcal{F}}$ is closed under extensions, we can consider the pullback diagram $(*)$ with F_B in $\mathcal{F}$. By Lemma 1.8, P_B belongs to $\mathcal{F}_i$; then $K_B \in \mathcal{F}_i$ by (4_i), and hence B belongs to $\mathcal{F}_{i+1}$. □

Remark 1.11. It follows that if $\mathcal{F}$ is closed under finite direct sums and $\overline{\mathcal{F}}$ is closed under extensions, then

- the class $\mathcal{F}_{<\infty}$ is closed under extensions, kernels of epimorphisms and cokernels of monomorphisms;
- the classes $\mathcal{F}_i$, $i \geq 0$, are closed under kernels of epimorphisms; if $i \geq 1$, they are closed also under extensions.

Proposition 1.12. *Assume $\mathcal{F}$ defines a homological dimension, it is closed under finite direct sums, and $\overline{\mathcal{F}} = \mathcal{A}$. Then also $\mathcal{F}_i$ and $\mathcal{F}_{<\infty}$ define a homological dimension for any $i \geq 1$.*

Proof. Since $\overline{\mathcal{F}} = \mathcal{A}$, also $\overline{\mathcal{F}_i} = \mathcal{A} = \overline{\mathcal{F}_{<\infty}}$. Therefore condition 1 in Definition 1.6 is empty in both the cases. Let M be an object admitting an $\mathcal{F}_i$-resolution

$$0 \to F_{i,n} \to F_{i,n-1} \to \cdots \to F_{i,0} \to M \to 0.$$

Consider an exact sequence $0 \to K \to F'_{i,n-1} \to \cdots \to F'_{i,0} \to M \to 0$ with $F'_{i,j} \in \mathcal{F}_i$. From the first sequence, applying recursively Theorem 1.10(2_i), we get that $M \in \mathcal{F}_{n+i}$. Applying recursively Theorem 1.10(4_i) to the second exact sequence we obtain that $K \in \mathcal{F}_i$. Since each finite $\mathcal{F}_{<\infty}$ resolution is actually an $\mathcal{F}_m$ resolution for a suitable $m \in \mathbb{N}$, we conclude that also $\mathcal{F}_{<\infty}$ defines a homological dimension. □

In case the abelian category $\mathcal{A}$ has enough projectives, a relevant family of classes defining a homological dimension is given by the left orthogonal of any class.

Proposition 1.13. *Assume $\mathcal{A}$ has enough projectives, and let $\mathcal{G}$ be a class of objects in $\mathcal{A}$. Then $\mathcal{F} = {}^{\perp_m}\mathcal{G}$, $1 \leq m \in \mathbb{N}$, defines a homological dimension if and only if*

$$\mathcal{F} = {}^{\perp_\infty}\mathcal{G}.$$

In such a case $\mathcal{A} = \overline{\mathcal{F}}$.

Proof. Assume $\mathcal{F} = {}^{\perp_m}\mathcal{G}$ defines a homological dimension. Let us prove that $\mathcal{F} = {}^{\perp_{m+1}}\mathcal{G}$; then we conclude inductively. Consider an arbitrary object $F \in \mathcal{F}$. Consider a short exact sequence

$$0 \to K \to P \to F \to 0$$

with P projective; since P belongs to $\mathcal{F}$, by Lemma 1.8 we have that also $K \in \mathcal{F}$. Therefore for each $G \in \mathcal{G}$ we have

$$\operatorname{Ext}_{\mathcal{A}}^{m+1}(F, G) \cong \operatorname{Ext}_{\mathcal{A}}^{m}(K, G) = 0,$$

because $K \in \mathcal{F}$.

Conversely, let us prove that $\mathcal{F} = {}^{\perp_\infty}\mathcal{G}$ defines a homological dimension. Clearly, containing $\mathcal{F}$ the projectives, each object has left $\mathcal{F}$-dimension $\leq \infty$. Let M be an object with $\mathcal{F}\dim M \leq n$, $n \in \mathbb{N}$. Then there exists an exact sequence

$$0 \to F'_n \to F'_{n-1} \to \cdots \to F'_1 \to F'_0 \to M \to 0$$

with $F'_i \in \mathcal{F}$ for $i = 0, \ldots, n$. Let us consider an exact sequence

$$0 \to K_n \to F_{n-1} \to \cdots \to F_1 \to F_0 \to M \to 0$$

with $F_i \in \mathcal{F}$ for $i = 0, \ldots, n-1$. Let us show that $K_n \in \mathcal{F}$. In fact, let $X \in \mathcal{G}$. Then $\mathrm{Ext}^i_\mathcal{A}(K_n, X) \cong \mathrm{Ext}^{n+i}_\mathcal{A}(M, X) \cong \mathrm{Ext}^i_\mathcal{A}(F'_n, X) = 0$ for each $i \geq 1$. □

Example 1.14. (i) Since $\mathbb{Z}$ has global dimension 1, the class $\mathcal{W} = {}^{\perp_1}\mathbb{Z} = {}^{\perp_\infty}\mathbb{Z}$ of Whitehead abelian groups defines a homological dimension. By Proposition 1.5(ii) we have $\mathrm{gl}\mathcal{W}\dim \mathbb{Z} \leq 1$.

(ii) Any torsion free class in a category of modules defines a homological dimension, since it is closed under submodules. In general it is not the left orthogonal of any class. Consider for example the class $\mathcal{R}$ of reduced abelian groups; since $\mathcal{R}^{\perp_\infty}$ is the class of divisible groups, ${}^{\perp_\infty}(\mathcal{R}^{\perp_\infty})$ is the whole class of abelian groups. Therefore $\mathcal{R}$ cannot be the left orthogonal of a class, otherwise ${}^{\perp_\infty}(\mathcal{R}^{\perp_\infty})$ would be equal to $\mathcal{R}$.

In the following results we are interested in giving necessary or sufficient conditions for a class defining a homological dimension to be a left orthogonal.

Lemma 1.15. *Assume $\mathcal{A}$ has enough projectives. If $\mathcal{F}$ defines a homological dimension and it contains the projectives, then $\mathcal{F}^{\perp_1} = \mathcal{F}^{\perp_\infty}$.*

Proof. Let M be an object in $\mathcal{F}^{\perp_1}$ and $F \in \mathcal{F}$. Consider a short exact sequence $0 \to F' \to P \to F \to 0$ with P projective; since $\mathcal{F}$ defines a homological dimension also F' belongs to $\mathcal{F}$. Applying $\mathrm{Hom}_\mathcal{A}(-, M)$ we get $\mathrm{Ext}^{i+1}_\mathcal{A}(F, M) \cong \mathrm{Ext}^i_\mathcal{A}(F', M)$; then $\mathrm{Ext}^2_\mathcal{A}(F, M) = 0$ and we conclude by induction. □

Theorem 1.16. *Assume $\mathcal{A}$ has enough projectives, and let $\mathcal{F}$ be a special precover class. Then $\mathcal{F}$ defines a homological dimension if and only if $\mathcal{F} = {}^{\perp_\infty}(\mathcal{F}^{\perp_\infty})$.*

Proof. If $\mathcal{F} = {}^{\perp_\infty}(\mathcal{F}^{\perp_\infty})$, by Proposition 1.13 we get that $\mathcal{F}$ defines a homological dimension.

Conversely, suppose that $\mathcal{F}$ defines a homological dimension. Let us prove that $\mathcal{F} = {}^{\perp_1}(\mathcal{F}^{\perp_\infty})$. Of course $\mathcal{F} \subseteq {}^{\perp_1}(\mathcal{F}^{\perp_\infty})$. Let now $M \in {}^{\perp_1}(\mathcal{F}^{\perp_\infty})$; consider a special $\mathcal{F}$-precover $0 \to K \to F \to M \to 0$. Since by the previous lemma $K \in \mathcal{F}^{\perp_1} = \mathcal{F}^{\perp_\infty}$, we get $\mathrm{Ext}^1_R(M, K) = 0$. Since the special precover classes are closed under direct summands [8, Section 2.1], then $M \leq^\oplus F$ belongs to $\mathcal{F}$. Again by Proposition 1.13 we conclude that $\mathcal{F} = {}^{\perp_\infty}(\mathcal{F}^{\perp_\infty})$. □

Most of the examples of classes defining a homological dimension give special precovers. Nevertheless observe that this is not always the case: Eklof and Shelah in [5] proved that, consistently with ZFC, the class of Whitehead abelian groups, which defines a homological dimension (see Example 1.14), does not provide precovers. In particular they proved that $\mathbb{Q}$, which has $\mathcal{W}$-dimension 1, does not admit $\mathcal{W}$-precover.

Remark 1.17. If $\mathcal{F}$ is a special precover class and it defines a homological dimension, then for each module M it is possible to get an $\mathcal{F}$-resolution

$$\cdots \to F_i \to \cdots \to F_1 \to F_0 \to M \to 0$$

such that, denoted by $\Omega^i_{\mathcal{F}}(M)$ the i-th $\mathcal{F}$ syzygy of M, the induced map $F_j \to \Omega^{j-1}_{\mathcal{F}}(M)$ is a special $\mathcal{F}$-precover of $\Omega^{j-1}_{\mathcal{F}}(M)$. Therefore, in such a case our definition of $\mathcal{F}$-dimension coincides with the definition given by Enochs and Jenda (see [6, Definition 8.4.1]).

Other significative classes defining a homological dimension are those studied by Auslander–Buchweitz in [2]. In that paper they introduced the notion of Ext-*injective cogenerator* for an additively closed exact subcategory $\mathcal{F}$ of $\mathcal{A}$: an additively closed subcategory $\omega \subseteq \mathcal{F}$ is an Ext-injective cogenerator for $\mathcal{F}$ if $\omega \subseteq \mathcal{F}^{\perp_\infty}$ and for any $F \in \mathcal{F}$ there exists an exact sequence $0 \to F \to X \to F' \to 0$ where $F' \in \mathcal{F}$ and $X \in \omega$.

Proposition 1.18 ([2, Propositions 2.1, 3.3]). *Let $\mathcal{F}$ be an additively closed exact subcategory of $\mathcal{A}$ closed under kernels of epimorphisms. If $\mathcal{F}$ admits an Ext-injective cogenerator ω, then $\mathcal{F}$ defines a homological dimension. Moreover, if any object has finite $\mathcal{F}$-dimension, then $\mathcal{F} = {}^{\perp_\infty}\mathcal{G}$, where $\mathcal{G}$ is the class of objects in $\mathcal{A}$ of finite ω-dimension.*

We conclude this section remarking the connection between classes defining a homological dimension and cotorsion pairs in categories of modules. So we assume $\mathcal{A} = R\text{-Mod}$, the category of left R-modules over a ring R.

Definition 1.19. Let $\mathfrak{A}$ and $\mathfrak{B}$ be two classes of modules. The pair $(\mathfrak{A}, \mathfrak{B})$ is called a *cotorsion pair* if $\mathfrak{A} = {}^{\perp_1}\mathfrak{B}$ and $\mathfrak{A}^{\perp_1} = \mathfrak{B}$. The pair $(\mathfrak{A}, \mathfrak{B})$ is called an *hereditary cotorsion pair* if $\mathfrak{A} = {}^{\perp_\infty}\mathfrak{B}$ or equivalently $\mathfrak{A}^{\perp_\infty} = \mathfrak{B}$.

We stress that, by Proposition 1.13, the hereditary cotorsion pairs are exactly the cotorsion pairs $(\mathfrak{A}, \mathfrak{B})$ such that $\mathfrak{A}$ defines a homological dimension.

Example 1.20. Let R be a commutative domain. A module M is *Matlis cotorsion* provided that $\text{Ext}^1_R(Q, M) = 0$, where Q is the quotient field of R. Since Q is flat, the class $\mathcal{MC}$ of Matlis cotorsion modules contains the class $\mathcal{EC} := \mathcal{F}l^{\perp_1}$ of Enochs cotorsion modules. Denoted by $\mathcal{TF}$ the class of torsion-free modules, the latter class $\mathcal{EC}$ contains the class $\mathcal{WC} := \mathcal{TF}^\perp$ of Warfield cotorsion modules. Thus we have the following chain of cotorsion pairs, ordered with respect to the inclusion on the first class:

$$({}^{\perp_1}\mathcal{MC}, \mathcal{MC}) \leq (\mathcal{F}l = {}^{\perp_1}\mathcal{EC}, \mathcal{EC}) \leq (\mathcal{TF} = {}^{\perp_1}\mathcal{WC}, \mathcal{WC}).$$

The modules in $^{\perp_1}\mathcal{MC}$ are called *strongly flat*. The Enochs and Warfield cotorsion pairs $(\mathcal{Fl}, \mathcal{EC})$ and $(\mathcal{TF}, \mathcal{WC})$ are hereditary and the classes of flat and torsion free modules, as well known, define a homological dimension. In general the Matlis cotorsion pair $(^{\perp_1}\mathcal{MC}, \mathcal{MC})$ is not hereditary and therefore strongly flat modules do not define a homological dimension; precisely, the Matlis cotorsion pair is hereditary, and so strongly flat modules define a homological dimension, if and only if the quotient field Q of R has projective dimension ≤ 1, i.e. R is a Matlis domain [12, Section 10].

2 Generalizing the Gorenstein dimension

Auslander in [1] introduced the notion of Gorenstein dimension for finite modules over a commutative noetherian ring. More precisely, let R be a commutative noetherian ring; following [4, Definition 1.1.2] we say that a finite R-module M belongs to the *G-class* $G(R)$ if

(i) $\operatorname{Ext}_R^m(M, R) = 0$ for $m > 0$;

(ii) $\operatorname{Ext}_R^m(\operatorname{Hom}_R(M, R), R) = 0$ for $m > 0$;

(iii) the canonical morphism $\delta_M : M \to \operatorname{Hom}_R(\operatorname{Hom}_R(M, R), R)$, $\delta_M(x)(\psi) = \psi(x)$, is an isomorphism.

Any finite module admitting a $G(R)$-resolution of length n is said to have *Gorenstein dimension* at most n. In [4, Theorem 1.2.7] it is shown that $G(R)$ defines a homological dimension on the category of finite R-modules.

Given an abelian category $\mathcal{A}$, we denote by $\mathcal{K}(\mathcal{A})$ (resp. $\mathcal{K}^+(\mathcal{A})$, $\mathcal{K}^-(\mathcal{A})$, $\mathcal{K}^b(\mathcal{A})$) the homotopy category of unbounded (resp. bounded below, bounded above, bounded) complexes of objects of $\mathcal{A}$ and by $\mathcal{D}(\mathcal{A})$ (resp. $\mathcal{D}^+(\mathcal{A})$, $\mathcal{D}^-(\mathcal{A})$, $\mathcal{D}^b(\mathcal{A})$) the associated derived category. In the sequel with $\mathcal{D}^*(\mathcal{A})$ or $\mathcal{D}^\dagger(\mathcal{A})$ we will denote any of these derived categories.

Consider a right adjoint pair of contravariant functors (Φ, Ψ) between the abelian categories $\mathcal{A}$ and $\mathcal{B}$, with the natural morphisms η and ξ as unities. Following [9, Theorem 5.1], to guarantee the existence of the derived functors $\mathbf{R}^*\Phi : \mathcal{D}^*(\mathcal{A}) \to \mathcal{D}(\mathcal{B})$ and $\mathbf{R}^\dagger\Psi : \mathcal{D}^\dagger(\mathcal{B}) \to \mathcal{D}(\mathcal{A})$, we assume the existence of triangulated subcategories $\mathcal{P}$ of $\mathcal{K}^*(\mathcal{A})$ and $\mathcal{Q}$ of $\mathcal{K}^\dagger(\mathcal{B})$ such that:

- every object of $\mathcal{K}^*(\mathcal{A})$ and every object of $\mathcal{K}^\dagger(\mathcal{B})$ admits a quasi-isomorphism into objects of $\mathcal{P}$ and $\mathcal{Q}$, respectively;

- if P and Q are exact complexes in $\mathcal{P}$ and $\mathcal{Q}$, then also $\Phi(P)$ and $\Psi(Q)$ are exact.

Given complexes $X \in \mathcal{D}^*(\mathcal{A})$ and $Y \in \mathcal{D}^\dagger(\mathcal{B})$, we have $\mathbf{R}^*\Phi X = \Phi P$ and $\mathbf{R}^\dagger\Psi Y = \Psi Q$, where P is a complex in $\mathcal{P}$ quasi-isomorphic to X, and Q is a complex in $\mathcal{Q}$ quasi-isomorphic to Y.

The functor Φ has *cohomological dimension* $\leq n$ if, for each A in $\mathcal{A}$, we have $H^i(\mathbf{R}^*\Phi A) = 0$ for $|i| > n$.

An object A in $\mathcal{A}$ is called Φ-*acyclic* if $H^i(\mathbf{R}^*\Phi A) = 0$ for any $i \neq 0$. Similarly, Ψ-acyclic objects in $\mathcal{B}$ are defined.

Definition 2.1. We say that an object $A \in \mathcal{A}$ belongs to the class $\mathcal{G}_{\Phi\Psi}$ if

(i) A is Φ-acyclic;

(ii) $\Phi(A)$ is Ψ-acyclic;

(iii) the morphism $\eta_A \colon A \to \Psi\Phi(A)$ is an isomorphism.

Note that, since the category of modules over a ring R has enough projectives, the total derived functor $\mathbf{R}\operatorname{Hom}(-, R)$ always exists (see [14]). Thus the class $\mathcal{G}_{\Phi\Psi}$ for the adjoint pair $(\Phi, \Psi) = (\operatorname{Hom}(-, R), \operatorname{Hom}(-, R))$ in the category of finite R-modules, coincides with the $G(R)$-class introduced above if R is a commutative noetherian ring.

We want to prove that the class $\mathcal{G}_{\Phi\Psi}$ associated to the right adjoint pair (Φ, Ψ) always defines a homological dimension.

First we prove that the $\mathcal{G}_{\Phi\Psi}$-dimension can be computed using the cohomology groups $H^i(\mathbf{R}^*\Phi)$. As a consequence it follows that, when the category $\mathcal{A}$ has enough projectives, the $\mathcal{G}_{\Phi\Psi}$-dimension can be compared with the projective dimension (cf. [4, Proposition 1.2.10]).

Proposition 2.2. *Let A be an object in $\mathcal{A}$ of finite $\mathcal{G}_{\Phi\Psi}$-dimension. Then:*

(a) $\mathcal{G}_{\Phi\Psi}$*-dim* $A = \sup\{i : H^i(\mathbf{R}^*\Phi A)\} \neq 0$.

(b) *If $\mathcal{A}$ has enough projectives, then $\mathcal{G}_{\Phi\Psi}$-dim $A \leq \operatorname{pd} A$.*

Proof. (a) Let $\mathcal{G}_{\Phi\Psi}$-dim $A = n$. Therefore there exists an exact sequence $0 \to G_n \to G_{n-1} \to \ldots \to G_0 \to A \to 0$ with $G_i \in \mathcal{G}_{\Phi\Psi}$, $i = 0, 1, \ldots, n$. By shift dimension we get $H^i(\mathbf{R}^*\Phi A) = 0$ for each $i > n$. If $\sup\{i : H^i(\mathbf{R}^*\Phi A) \neq 0\} < n$, let K be the cokernel of $G_n \to G_{n-1}$. We will prove that K belongs to $\mathcal{G}_{\Phi\Psi}$ contradicting the assumption $\mathcal{G}_{\Phi\Psi}$-dim $A = n$. Indeed, K is Φ-acyclic since $H^i(\mathbf{R}^*\Phi K) \cong H^{(i+n-1)}(\mathbf{R}^*\Phi A) = 0$ for each $i > 0$; applying Ψ to the short exact sequence $0 \to \Phi K \to \Phi G_{n-1} \to \Phi G_n \to 0$ and comparing it with the short exact sequence $0 \to G_n \to G_{n-1} \to K \to 0$, we get that ΦK is Ψ-acyclic and the unity η_K is an isomorphism.

(b) If $\mathcal{A}$ has enough projectives, then any object A in $\mathcal{A}$ admits a projective resolution P. Since the projectives are Φ-acyclic, we have $\mathbf{R}^*\Phi A = \Phi P$ and then

$$\sup\{i : H^i(\mathbf{R}^*\Phi A) \neq 0\} = \sup\{i : H^i(\Phi P) \neq 0\} \leq \operatorname{pd} A. \qquad \square$$

Observe that, differently from the $G(R)$-dimension, the inequality between the $\mathcal{G}_{\Phi\Psi}$-dimension and the projective dimension can be strict also for objects of finite projective dimension (cf. [4, Proposition 1.2.10]).

Example 2.3. Let Λ be the path algebra of the quiver

$$1 \longrightarrow 2 \longrightarrow 3.$$

Let us consider the module $_\Lambda U = \begin{smallmatrix}1\\2\\3\end{smallmatrix} \oplus \begin{smallmatrix}2\\3\end{smallmatrix} \oplus 2$ and let $S = \operatorname{End}_\Lambda(U)$. Consider the adjoint pair $(\operatorname{Hom}_\Lambda(-, U), \operatorname{Hom}_S(-, U))$: since $\operatorname{Ext}^1_\Lambda(U, U) = 0$, $\operatorname{Ext}^1_S(S, U) = 0$ and $U \cong \operatorname{Hom}_S(\operatorname{Hom}_\Lambda(U, U), U)$, the Λ-module U belongs to $\mathcal{G}_{\Phi\Psi}$, where $(\Phi, \Psi) = (\operatorname{Hom}_\Lambda(-, U), \operatorname{Hom}_S(-, U))$. Thus U has projective dimension one, but obviously $\mathcal{G}_{\Phi\Psi}$-dimension 0.

In order to prove that the class $\mathcal{G}_{\Phi\Psi}$ defines a homological dimension, we also need to recall some notions and results on derived categories. By [10, Lemma 13.6] we know that, in our assumptions, $(\mathbf{R}^*\Phi, \mathbf{R}^\dagger\Psi)$ is a right adjoint pair in the derived categories $\mathcal{D}^*(\mathcal{A})$ and $\mathcal{D}^\dagger(\mathcal{B})$, with unities $\hat{\eta}$ and $\hat{\xi}$ naturally inherited from the unities η and ξ. In [11] a complex $X \in \mathcal{D}^*(\mathcal{A})$ is called $\mathcal{D}$-*reflexive* if the morphism $\hat{\eta}_X$ is an isomorphism in $\mathcal{D}^*(\mathcal{A})$. An object $A \in \mathcal{A}$ is called $\mathcal{D}$-*reflexive* if it is $\mathcal{D}$-reflexive as a stalk complex.

Lemma 2.4. *Let $X \in \mathcal{A}$ such that X is Φ-acyclic and $\Phi(X)$ is Ψ-acyclic. Then $\hat{\eta}_X$ is a quasi-isomorphism if and only if η_X is an isomorphism. In particular any object in $\mathcal{G}_{\Phi\Psi}$ is $\mathcal{D}$-reflexive.*

Proof. In general, if $C \in \mathcal{D}^*(\mathcal{A})$ and L is a complex quasi-isomorphic to C such that any term L_i of L is Φ-acyclic and $\Phi(L_i)$ is Ψ-acyclic, then $\hat{\eta}_C$ coincides with η_L, where η_L is the term-to-term extension of the unity η to the triangulated category $\mathcal{K}^*(\mathcal{A})$ (cf. [11]). Then we easily get the statement. □

Corollary 2.5. *Any object A in $\mathcal{A}$ of finite $\mathcal{G}_{\Phi\Psi}$-dimension is $\mathcal{D}$-reflexive.*

Proof. Let $\mathcal{G}_{\Phi\Psi}$-dim $A = n$. Therefore there exists an exact sequence $0 \to G_n \to G_{n-1} \to \cdots \to G_0 \to A \to 0$ with $G_i \in \mathcal{G}_{\Phi\Psi}$, $i = 0, 1, \ldots, n$. Therefore in the bounded derived category $\mathcal{D}^b(\mathcal{A})$, A is quasi-isomorphic to the complex $G := 0 \to G_n \to G_{n-1} \to \cdots \to G_0 \to 0$. Since G is a complex with $\mathcal{D}$-reflexive terms by Lemma 2.4, we conclude by [11, Theorem 3.1 (1)] that A is $\mathcal{D}$-reflexive. □

Proposition 2.6. *If $X \in \mathcal{A}$ is Φ-acyclic and $\mathcal{D}$-reflexive, then X belongs to $\mathcal{G}_{\Phi\Psi}$.*

Proof. Since X is Φ-acyclic, $\mathbf{R}^*\Phi X$ is quasi isomorphic to the stalk complex $\Phi(X)$. Moreover, for X is $\mathcal{D}$-reflexive, we get that $\mathbf{R}^\dagger\Psi(\Phi X) \cong \mathbf{R}^\dagger\Psi(\mathbf{R}^*\Phi X)$ is quasi-isomorphic to X. Thus $H^i(\mathbf{R}^\dagger\Psi(\Phi X)) = 0$ for any $i \neq 0$ and so ΦX is Ψ-acyclic. Finally we conclude since, by the previous lemma, η_X is an isomorphism. □

Theorem 2.7. *The class $\mathcal{G}_{\Phi\Psi}$ defines a homological dimension.*

Proof. Let us consider a long exact sequence $0 \to G_n \to G_{n-1} \to \cdots \to G_0 \to X \to 0$ with $G_i \in \mathcal{G}_{\Phi\Psi}$. By Corollary 2.5, X is $\mathcal{D}$-reflexive. Consider now a long exact sequence $0 \to X_n \to F_{n-1} \to \cdots \to F_0 \to X \to 0$ with $F_i \in \mathcal{G}_{\Phi\Psi}$. The $\mathcal{D}$-reflexive objects are a thick subcategory of $\mathcal{A}$ (see [11]), i.e., if two terms of a short exact sequence in $\mathcal{A}$ are $\mathcal{D}$-reflexive, then also the third is $\mathcal{D}$-reflexive. Therefore, by induction, it follows that X_n is $\mathcal{D}$-reflexive. Since by Proposition 2.2 $H^i(\mathbf{R}^*\Phi X) = 0$ for each $i > n$, by shift dimension X_n is Φ-acyclic, and so we conclude that X_n belongs to $\mathcal{G}_{\Phi\Psi}$. □

In [11], the authors were interested in characterizing the $\mathcal{D}$-reflexive objects associated to a given adjoint pair (Φ, Ψ). Assume $\mathcal{A}$ is a module category and denote by FP_n the class of modules A which have an exact resolution

$$P_{n-1} \to \cdots \to P_1 \to P_0 \to A \to 0,$$

where the P_i's are finitely generated projectives. In particular FP_1 is the class of finitely generated modules. Then the $\mathcal{D}$-reflexive modules in FP_n can be characterized through their $\mathcal{G}_{\Phi\Psi}$-dimension.

Theorem 2.8. *Let $\mathcal{A} = R$-Mod for an arbitrary ring R. Assume $_RR$ to be $\mathcal{D}$-reflexive and Φ of cohomological dimension $\leq n$. Then a module $M \in FP_n$ is $\mathcal{D}$-reflexive if and only if it has $\mathcal{G}_{\Phi\Psi}$-dimension $\leq n$.*

Proof. The sufficiency of the finiteness of the $\mathcal{G}_{\Phi\Psi}$-dimension is proved in Corollary 2.5. Conversely, suppose M to be a $\mathcal{D}$-reflexive module in FP_n. Let $0 \to K \to P_{n-1} \to \cdots \to P_0 \to M \to 0$ be an exact sequence with the P_i's finitely generated projectives. Since $_RR$ is assumed to be $\mathcal{D}$-reflexive, any P_i is $\mathcal{D}$-reflexive, and so we get that K is $\mathcal{D}$-reflexive. Since Φ has cohomological dimension $\leq n$, by shift dimension we get that K is Φ-acyclic. Then, by Proposition 2.6, we conclude that K belongs to $\mathcal{G}_{\Phi\Psi}$. □

Acknowledgements. We are pleased to thank the referee for her/his valuable suggestions.

References

[1] M. Auslander, *Anneaux de Gorenstein, et torsion en algèbre commutative.* Séminaire d'Algèbre Commutative dirigé par Pierre Samuel, 1966/67. Texte rédigé, d'après des exposés de Maurice Auslander, Marguerite Mangeney, Christian Peskine et Lucien Szpiro. École Normale Supérieure de Jeunes Filles Secrétariat mathématique, Paris 1967.

[2] M. Auslander and R. O. Buchweitz, *The homological theory of maximal Cohen-Macaulay approximations*, Mém. Soc. Math. France (N.S.) No. 38 (1989), 5–37.

[3] S. Bazzoni and L. Salce, *On strongly flat modules over integral domains*, RockyMountain J. Math. 34 (2004), no. 2, 417–439.

[4] L. W. Christensen, *Gorenstein Dimension.* Lecture Notes in Mathematics, 1747. Springer-Verlag, Berlin.

[5] P. C. Eklof and S. Shelah, *On the existence of precovers*, Illinois J. Math. 47 (2003), no. 1–2, 173–188.

[6] E. E. Enochs and O. M. G. Jenda, *Relative homological algebra.* Expositions in Math. 30, Walter de Gruyter 2000.

[7] P. Freyd, *Abelian categories.* Harper's Series in Modern Mathematics, Harper and Row, New York, Evanston, London, 1964.

[8] R. Göbel and J. Trlifaj, *Approximations and endomorphism algebras of modules.* Expositions in Math. 41, Walter de Gruyter 2006.

[9] R. Hartshorne, *Algebraic geometry.* Graduate Texts in Mathematics 52, Springer-Verlag, New York, Heidelberg, 1977.

[10] B. Keller, *Derived Categories and their use*, Handbook of Algebra. Vol. 1. Edited by M. Hazewinkel, North-Holland Publishing Co., Amsterdam, 1996.

[11] F. Mantese and A. Tonolo, *Reflexivity in derived categories*, arXiv:0705.2537.

[12] E. Matlis, *Torsion-free modules.* Chicago Lectures in Mathematics. The University of Chicago Press, Chicago, London, 1972.

[13] B. Mitchell, *Theory of categories.* Pure and Applied Mathematics, Academic Press, New York, London, 1965.

[14] N. Spaltenstein, *Resolutions of unbounded complexes*, Compositio Math. 65 (1988), no. 2, 121–154.

Author information

Francesca Mantese, Dipartimento di Informatica, Università degli Studi di Verona, strada Le Grazie 15, 37134 Verona, Italy.
E-mail: `francesca.mantese@univr.it`

Alberto Tonolo, Dipartimento di Matematica Pura ed Applicata, Università degli Studi di Padova, via Trieste 63, 35121 Padova, Italy.
E-mail: `alberto.tonolo@unipd.it`

Truncated path algebras are homologically transparent

A. Dugas, B. Huisgen-Zimmermann and J. Learned

Abstract. It is shown that path algebras modulo relations of the form $\Lambda = KQ/I$, where Q is a quiver, K a coefficient field, and $I \subseteq KQ$ the ideal generated by all paths of a given length, can be readily analyzed homologically, while displaying a wealth of phenomena. In particular, the syzygies of their modules, and hence their finitistic dimensions, allow for smooth descriptions in terms of Q and the Loewy length of Λ. The same is true for the distributions of projective dimensions attained on the irreducible components of the standard parametrizing varieties for the modules of fixed K-dimension.

Key words. Representation theory of associative algebras and quivers, homological dimensions, algebraic varieties parametrizing representations.

AMS classification. 16G10, 16G20, 16E10.

1 Introduction and notation

The problem of opening up general access roads to the finitistic dimensions of a finite dimensional algebra Λ, given through quiver and relations, is quite challenging. This is witnessed, for instance, by the fact that the longstanding question "Is the (left) little finitistic dimension of Λ,

$$\operatorname{fin dim} \Lambda = \sup\{\operatorname{p dim} M \mid M \in \mathcal{P}^{<\infty}(\Lambda\text{-mod})\},$$

always finite?" (Bass 1960) has still not been settled. Here $\operatorname{p dim} M$ is the projective dimension of a module M, and $\mathcal{P}^{<\infty}(\Lambda\text{-mod})$ denotes the category of finitely generated (left) Λ-modules of finite projective dimension.

In [1], Babson, the second author, and Thomas showed that truncated path algebras of quivers are particularly amenable to geometric exploration, while nonetheless displaying a wide range of interesting phenomena. This led the authors of the present paper to the serendipitous discovery that the same is true for the homology of such algebras. By a *truncated path algebra* we mean an algebra of the form KQ/I, where KQ is the path algebra of a quiver Q with coefficients in a field K and $I \subseteq KQ$ is the ideal generated by all paths of a fixed length $L + 1$. In particular, truncated path algebras are monomial algebras. In this case, the finitistic dimensions are known to be finite (see [6]). Our goal here is to show how much more is true in the truncated scenario.

The second author was partly supported by a grant from the National Science Foundation.

Roughly, our three main results (Theorems 2.6, 3.2, 3.6) show the following for a truncated path algebra Λ:

- The little and big finitistic dimensions of Λ coincide and can be determined through a straightforward computation from Q and L. Moreover, from a minimal amount of structural data for a Λ-module M, namely the radical layering

$$\mathbb{S}(M) = \left(J^l M / J^{l+1} M\right)_{0 \le l \le L}$$

(or, alternatively, any "skeleton" of M), one can determine the syzygies and projective dimension of M in a purely combinatorial fashion. (See Theorems 2.2, 2.6, and the first part of Theorem 3.2 for finer information.)

- The "generic projective dimension" of any irreducible component $\mathcal{C}$ of one of the classical module varieties (see beginning of Section 3) is readily obtainable from graph-theoretic data as well. So is the full spectrum of values of the function p dim attained on the class of modules parametrized by $\mathcal{C}$. In particular, it turns out that the supremum of the finite values among the generic finitistic dimensions of the various irreducible components equals fin dim Λ. (See Theorems 3.2 and 3.6 for detail.)

The picture emerging from the main theorems will be supplemented in a sequel, where it will be shown that the category $\mathcal{P}^{<\infty}(\Lambda\text{-mod})$ is contravariantly finite in the full category of finitely generated Λ-modules, whenever Λ is a truncated path algebra.

Conventions. We fix a positive integer L. Throughout, Λ denotes a truncated path algebra of Loewy length $L + 1$, that is, $\Lambda = KQ/I$, where K is a field, Q a quiver, and I the ideal generated by all paths of length $L + 1$. The Jacobson radical J of Λ satisfies $J^{L+1} = 0$ by construction. A (*nonzero*) *path in* Λ is the I-residue of a path in $KQ \setminus I$, that is, the I-residue of a path p in KQ of length at most L; so, in particular, any path in Λ is a *nonzero* element of Λ under this convention. Clearly, the paths in Λ form a K-basis for Λ. Due to the fact that I is homogeneous with respect to the path-length grading of KQ, defining the *length* of such a path $p + I$ to be that of p, yields an unambiguous concept of length for the elements of this basis. A distinguished role is played by the paths $e_1, \ldots, e_n$ of length zero in Λ: They constitute a full set of orthogonal primitive idempotents, which is in obvious one-to-one correspondence with the vertices of Q. We will identify each e_i with the corresponding vertex, and whenever we refer to a primitive idempotent in Λ, we will mean one of the e_i. Then the left ideals Λe_i and their radical factors $S_i = \Lambda e_i / J e_i$, for $1 \le i \le n$, constitute full sets of isomorphism representatives for the indecomposable projective and simple left Λ-modules, respectively.

Finally, we say that a path p in Λ or in KQ is an *initial subpath* of a path q if there is a path p' with $q = p'p$; here the product $p'p$ stands for "p' after p."

2 The standard homological dimensions of Λ-Mod

The (left) *big finitistic dimension* of Λ is the supremum, Fin dim Λ, of the projective dimensions of all left Λ-modules of finite projective dimension; for the *little finitistic dimension* consult the introduction. We start by recording some prerequisites established in [1]. As was shown in [1], the well-known fact that all *second* syzygies of modules over a monomial algebra are direct sums of cyclic modules generated by paths of positive length (see [7] and [2]), can be improved for truncated path algebras so as to cover *first* syzygies as well. In particular, this makes the big and little finitistic dimensions of Λ computable from a finite set of cyclic test modules.

More sharply: Given any left Λ-module M, we can explicitly pin down a decomposition of the syzygy $\Omega^1(M)$ into cyclics. This description of $\Omega^1(M)$ relies on a *skeleton of* M. Roughly speaking, this is a path basis for M with the property that the path lengths respect the radical layering, $(J^l M/J^{l+1} M)_{0 \le l \le L}$. The concept of a skeleton, defined in [1] in full generality, can be significantly simplified for a truncated path algebra Λ.

Definition 2.1 (Skeleton of a Λ-module M). Fix a projective cover P of M, say $P = \bigoplus_{r \in R} \Lambda z_r$, where each z_r is one of the primitive idempotents in $\{e_1, \ldots, e_n\}$, tagged with a place number r (the index set R may be infinite). A *path of length l* in P is any element $p z_r \in P$, where p is a path of length l in Λ which starts in z_r (in particular, the paths in P are again nonzero). Identify M with an isomorphic factor module of P, say $M = P/C$.

(a) A *skeleton* of $M = P/C$ is a set σ of paths in P such that for each $l \le L$, the residue classes $q + J^l M$ of the paths q of length l in σ form a K-basis for $J^l M/J^{l+1} M$. Moreover, we require that σ be closed under initial subpaths, that is, if $q = p' p z_r \in \sigma$, then $p z_r \in \sigma$.
(b) A path q in $P \setminus \sigma$ is called σ-*critical* if it is of the form $q = \alpha p z_r$, where α is an arrow and $p z_r$ a path in σ.

In particular, the definition entails that, for any skeleton σ of $M = P/C$, the full set of residue classes $\{q + C \mid q \in \sigma\}$ forms a basis for M. Furthermore, it is easily checked that every Λ-module M has at least one skeleton, and only finitely many when M is finitely generated (as long as we keep the projective cover P fixed).

Theorem 2.2 (Known facts [1, Lemma 5.10]). *If M is any nonzero left Λ-module with skeleton σ, then*

$$\Omega^1(M) \cong \bigoplus_{q \ \sigma\text{-critical}} \Lambda q.$$

In particular, $\Omega^1(M)$ is isomorphic to a direct sum of cyclic left ideals generated by nonzero paths of positive length in Λ.

Consequently, Fin dim $\Lambda =$ fin dim $\Lambda = s + 1$, *where*

$$s = \max\{\text{p dim}\, \Lambda q \mid q \text{ a path of positive length in } \Lambda \text{ with } \text{p dim}\, \Lambda q < \infty\},$$

provided that the displayed set is nonempty, and $s = -1$ otherwise. □

We briefly point out another nice consequence concerning Auslander's notion of the representation dimension of an algebra. In [9] Ringel shows that any algebra with only finitely many indecomposable torsionless modules up to isomorphism has representation dimension at most 3. Since the above theorem classifies the indecomposable torsionless Λ-modules as those modules isomorphic to Λq for some nonzero path q, of which there are only finitely many, we obtain the following.

Corollary 2.3. *The representation dimension of any truncated path algebra is at most* 3.

We illustrate the results on finitistic dimension with an example which will accompany us throughout.

Example 2.4. Let $\Lambda = KQ/I$ be the truncated path algebra of Loewy length $L+1 = 4$ based on the following quiver Q.

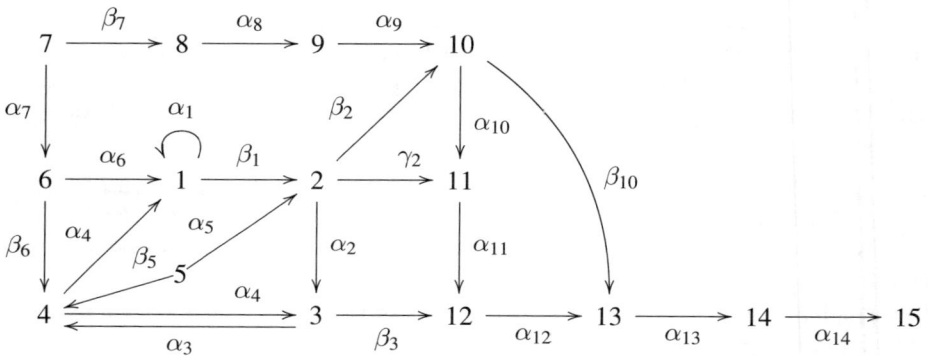

Then the indecomposable projective left Λ-modules Λe_1 and Λe_3 have the following layered and labeled graphs (in the sense of [7] and [8]):

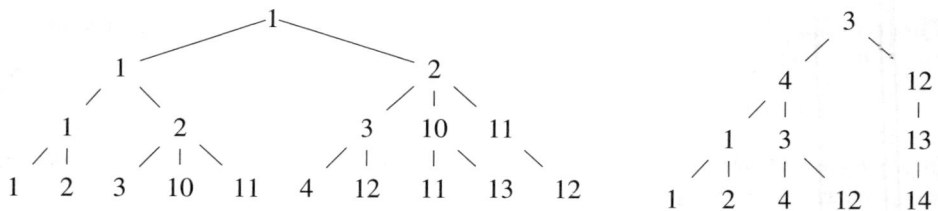

If $P = \Lambda z_1$ with $z_1 = e_i$ in the notation of Definition 2.1, each of the modules Λe_i has a unique skeleton, which can be read off the graph: It is the set of all initial subpaths of the edge paths in the graph, read from top to bottom. The skeleton of Λe_1, for instance, consists of the paths $z_1 = e_1$ of length zero in P, the paths $\alpha_1 z_1, \beta_1 z_1$ of length 1, the paths $\alpha_1^2 z_1, \beta_1 \alpha_1 z_1, \alpha_2 \beta_1 z_1, \gamma_2 \beta_1 z_1, \beta_2 \beta_1 z_1$ of length 2, together with all edge paths of length 3.

For a sample application of Theorem 2.2, we consider the module M determined by the following graph:

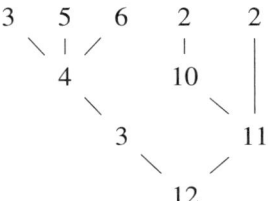

A projective cover of M is $P = \Lambda e_3 \oplus \Lambda e_5 \oplus \Lambda e_6 \oplus (\Lambda e_2)^2$, where $z_1 = e_3$, $z_2 = e_5$ and so on. A skeleton σ of M (in this case there are several), together with the σ-critical paths is communicated by the following graph, in which the solid and dashed edges play different roles, as explained below:

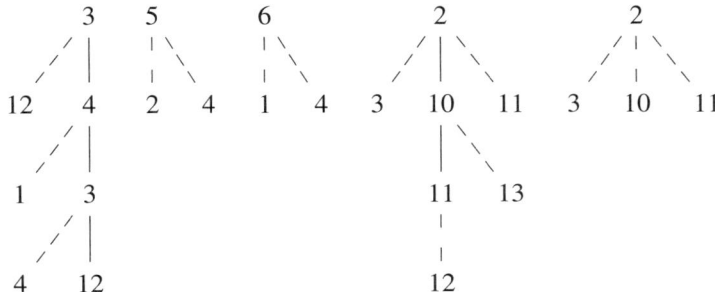

As above, the paths in σ correspond to the initial subpaths of the solidly drawn edge paths, including all paths of length zero – e.g., $\beta_4\alpha_3 z_1$, z_3 and $\alpha_{10}\beta_2 z_4$. The σ-critical paths are all the paths in the graph (again read from top to bottom) which terminate in a dashed edge; for instance, $\alpha_3\beta_4\alpha_3 z_1$ and $\alpha_5 z_2$ are σ-critical. Since $\Omega^1(M) \cong \bigoplus_{q\ \sigma\text{-critical}} \Lambda q$, we find this syzygy to be the direct sum

$$\Lambda\beta_3 \oplus \Lambda\alpha_4\alpha_3 \oplus \Lambda\alpha_3\beta_4\alpha_3 \oplus \Lambda\alpha_5 \oplus \Lambda\beta_5 \oplus \Lambda\alpha_6 \oplus \Lambda\beta_6 \oplus \Lambda\alpha_2 \oplus \cdots.$$

The graphs of $\Lambda\alpha_4\alpha_3$, $\Lambda\beta_3$, and $\Lambda\alpha_2$ are respectively

```
        1           12           3
       / |          |           / \
      1  2          13         4   12
                    |          |\   |
                    14         1 3  13
```

The main result of this section provides the projective dimensions of the building blocks for the syzygies of arbitrary Λ-modules; compare with Theorem 2.2.

Definition 2.5. Let l be a nonnegative integer $\leq L$, and c any nonnegative integer. We define
$$l\text{-deg}(c) = \left[\frac{c}{L+1}\right] + \left[\frac{c+l}{L+1}\right].$$

Here $[x]$ stands for the largest integer smaller than or equal to x. Moreover, we set $l\text{-deg}(\infty) = \infty$.

The l-degree defines a nondecreasing function $\mathbb{N} \cup \{0, \infty\} \to \mathbb{N} \cup \{0, \infty\}$ for any $l \leq L$. Moreover, for $0 \leq l \leq l' \leq L$ and arbitrary $c \in \mathbb{N} \cup \{0\}$, the difference $l'\text{-deg}(c) - l\text{-deg}(c)$ belongs to the set $\{0, 1\}$. This observation will entail the final claim of the upcoming theorem, once the first – displayed – equality is established.

Theorem 2.6. *Suppose $q \in \Lambda$ is a path of length $l > 0$ in Λ (i.e., the I-residue of a path of length at most L in KQ) with terminal vertex e. Let $c = c(e)$ be the supremum of the lengths of the paths in KQ starting in e. Then*

$$\text{p dim} \, \Lambda q = l\text{-deg}(c).$$

In particular, $\text{p dim} \, \Lambda q < \infty$ if and only if $c(e) < \infty$ (meaning that there is no path starting in e and terminating on an oriented cycle).

Moreover, if q' is another path in Λ that ends in e such that $L \geq \text{length}(q') \geq \text{length}(q) \geq 1$, then

$$\text{p dim} \, \Lambda q \leq \text{p dim} \, \Lambda q' \leq 1 + \text{p dim} \, \Lambda q.$$

In Example 2.4, $c(e_7)$ is infinite, for instance, while $c(e_{10}) = 5$; the latter shows that $\text{p dim}(\Lambda \alpha_9 \alpha_8 \beta_7) = 3\text{-deg}(5) = 3$. The argument backing Theorem 2.6 is purely combinatorial, the intuitive underpinnings being of a graphical nature. We start with two definitions setting the stage. The first is clearly motivated by the statement of Theorem 2.6.

Definition 2.7. We call a vertex e of the quiver Q (alias a primitive idempotent of Λ) *cyclebound* in case there is a path from e to a vertex lying on an oriented cycle. In case e is cyclebound, we also call the simple module $\Lambda e/Je$ cyclebound.

Next, we consider the following partial order on the set of paths in KQ. Namely, given paths p and p' in KQ, we define

$$p' \leq p \iff p' \text{ is an initial subpath of } p;$$

recall that the latter amounts to the existence of a path p'' with the property that $p = p''p'$. Hence, any two paths which are comparable have the same starting point, and $e \leq p$ for any path p starting in the vertex e. Clearly, this partial order induces a partial order on the set of paths in Λ.

Finally, we introduce a class of modules, which will turn out to tell the full homological story of Λ. The left ideals of the form Λq – the basic building blocks of all syzygies of Λ-modules – are among them.

Definition 2.8 (Tree modules and branches. Comments). Any module $\mathcal{T}$ of the form $\mathcal{T} \cong \Lambda e/V$, where e is a vertex of Q and $V = \sum_{v \in \mathfrak{V}} \Lambda v$ is generated by some set $\mathfrak{V}$ of paths of positive length in Λe (possibly empty), will be called a *tree module with root e*. In particular, Λe is a tree module with root e, the unique candidate of maximal dimension among the tree modules with root e, in fact; the simple module $\Lambda e/Je$ is the tree module with root e that has minimal positive dimension.

The terminology is motivated by the fact that the *graphs* of tree modules are trees "growing downwards" from their roots. Note that tree modules are determined up to isomorphism by their graphs.

Given a tree module T as above, let $b_1, \ldots, b_r \in \Lambda$ be the maximal paths in Λe – in the above partial order – which are not contained in V. The b_i are uniquely determined by the isomorphism class of T and are called the *branches* of T. Conversely, if we know T to be a tree module, then the branches of T pin T down up to isomorphism.

If $T \cong \Lambda e/Je$ is the simple tree module with root e, then e is the only branch of T. By contrast, if $T = \Lambda e/V$ is a nonsimple tree module, then all branches of T have positive length. Moreover, it is straightforward to see that T has a basis of the following form:

$$\{e + V\} \cup \{q + V \mid q \text{ is an initial subpath of positive length of one of } b_1, \ldots, b_r\},$$

where $b_1, \ldots, b_r$ are the branches of T. If we pull back this basis to a set of paths in the projective cover Λe of T, then σ is a skeleton of T in the sense of Definition 2.1 (the only one).

Apart from M, all the modules displayed in Example 2.4 are tree modules. Their branches are precisely the maximal edge paths in their graphs, read from top to bottom. The proof of the next lemma is straightforward and we leave it to the reader.

Lemma 2.9. *Whenever q is a path in Λ ending in e, not necessarily of positive length, the cyclic left ideal Λq is a tree module with root e. More precisely: If $l = \text{length}(q)$, let $b_1, \ldots, b_r$ be the maximal candidates among the paths of length $\leq L - l$ starting in e. Then $\Lambda q = \Lambda e/V$, where*

$$V = \Omega^1(\Lambda q) = \bigoplus_{\beta \text{ an arrow}, i \leq r} \Lambda \beta b_i,$$

and the b_i are the branches of Λq.

In particular, if $l > 0$, then $\text{p dim} \Lambda q < \infty$ if and only if e is non-cyclebound. □

Combined with Theorem 2.2, Lemma 2.9 shows that all syzygies of Λ-modules are direct sums of tree modules. Contrasting the final statement for $l > 0$, we see that, for the path $q = e$ of length zero, $\Lambda q = \Lambda e$ is projective, irrespective of the positioning of e in Q. As for the other extreme: By Lemma 2.9, the simple module $S = \Lambda e/Je$ has infinite projective dimension precisely when it is cyclebound. In Example 2.4, the vertices $e_1, \ldots, e_7$ are cyclebound, while $e_8, \ldots, e_{15}$ are not. Hence $S_1, \ldots, S_7$ are precisely the simple modules of infinite projective dimension.

Note that the only potential branches b_i of length $< L - l$ of a tree module Λq as in Lemma 2.9 end in a sink of the quiver Q.

Proof of Theorem 2.6. As in the statement of the theorem, let q be a path of positive length $l \leq L$ in Λ, which ends in the vertex e. In light of the remark preceding Theorem 2.6, we only need to show the equality $\text{p dim} \Lambda q = l\text{-deg}(c)$, where $c = c(e)$

is the supremum of the lengths of the paths in KQ starting in e. If e is cyclebound, this equality follows from Lemma 2.9. So let us assume that e is non-cyclebound – meaning $c < \infty$ – and induct on c. If $c \le L - l$, all of the branches of the tree module Λq end in sinks of the quiver Q. We infer that $\Lambda q \cong \Lambda e$ in that case, whence $\operatorname{p dim} \Lambda q = 0 = l\text{-deg}(c)$.

Now suppose $c > L - l$, and assume that $\operatorname{p dim} \Lambda p' = l'\text{-deg}(c(e'))$ for all paths p' of length $l' \le L$ in Λ that end in a non-cyclebound vertex e' of Q with $c(e') < c$. Using the notation of Lemma 2.9, we obtain $\Omega^1(\Lambda q) = \bigoplus_{\beta,\, i \le r} \Lambda \beta b_i$, where the b_i are the branches of the tree module Λq and the β are arrows. Since the lengths of the b_i are bounded from above by $L - l \le L - 1$, the paths in KQ of the form βb_i where β is an arrow, have length at most L; therefore each of them gives rise to a path in Λ. By the definition of c, there exists a path u of length c in KQ which starts in the vertex e, and by the definition of the branches of Λq, there exists an index j such that b_j is an initial subpath of u. Necessarily, $\operatorname{length}(b_j) = L - l$, because $\operatorname{length}(u) > L - l$. In fact, $c > L - l$ guarantees that $u = u'\beta_j b_j$ in KQ for some arrow β_j and a suitable path u' of length $c' = \operatorname{length}(u) - (L - l) - 1 = c - (L - l) - 1 \le c - 1$. Since u starts in the non-cyclebound vertex e, the terminal vertex of $\beta_j b_j$ – call it e' – is again non-cyclebound. Moreover, the maximality property of u entails that $c' = c(e')$ is the maximal length of a path in KQ starting in e'. Therefore, our induction hypothesis guarantees that $\operatorname{p dim} \Lambda \beta_j b_j = (L - l + 1)\text{-deg}(c')$. This degree in turn equals

$$\left[\frac{c'}{L+1}\right] + \left[\frac{c' + L - l + 1}{L+1}\right] = \left[\frac{c + l - (L+1)}{L+1}\right] + \left[\frac{c}{L+1}\right] = l\text{-deg}(c) - 1;$$

the final equality follows from $\frac{c+l-(L+1)}{L+1} = \frac{c+l}{L+1} - 1$. Analogous applications of the induction hypothesis, combined with the basic properties of the degree function, yield $\operatorname{p dim} \Lambda \beta b_i \le \operatorname{p dim} \Lambda \beta_j b_j$ for any path βb_i appearing in the decomposition of $\Omega^1(\Lambda q)$. We conclude that $\operatorname{p dim} \Lambda q = 1 + \operatorname{p dim} \Lambda \beta_j b_j = l\text{-deg}(c)$ as required. □

The following dichotomy for the finitistic dimension of Λ results from a combination of Theorems 2.2 and 2.6 with Lemma 2.9.

Corollary 2.10. *Suppose that $S_1, \ldots, S_t$ are precisely the non-cyclebound simple left Λ-modules. Then either*

$$\operatorname{fin dim} \Lambda = \max_{1 \le i \le t} \operatorname{p dim} S_i \ \ or \ \ \operatorname{fin dim} \Lambda = 1 + \max_{1 \le i \le t} \operatorname{p dim} S_i,$$

and

$$\max_{1 \le i \le t} \operatorname{p dim} S_i = 1 + 1\text{-deg}(m - 1),$$

where m is the maximum of the lengths of the paths in Q starting in one of the vertices $e_1, \ldots, e_t$. □

Both options for $\operatorname{fin dim} \Lambda$ occur in concrete instances (see below); of course, the smaller value equals the global dimension whenever the quiver Q is acyclic. For the decision process in specific instances, combine Theorems 2.2 and 2.6. To contrast

Corollary 2.10 with the homology of more general algebras: Recall that arbitrary natural numbers occur as finitistic dimensions of monomial algebras all of whose simple modules have infinite projective dimension. So the corollary again attests to the degree of simplification that occurs when the paths factored out of KQ have uniform length.

Example 2.4 revisited. With the aid of Corollary 2.10, the finitistic dimension of Λ can, in a first step, be computed up to an error of 1, through a simple count. Here $m = 7$, and $L = 3$, whence the maximum of the projective dimensions of the non-cyclebound simple modules (here $S_8, \ldots, S_{15}$) is $1 + 1\text{-deg}(6) = 3$.

To obtain the precise value of the finitistic dimension, we further observe: The arrow β_7 ends in the vertex e_8 with maximal finite length $c(e_8) = 7$ of departing paths, and hence $\operatorname{p\,dim}(\Lambda e_7/\Lambda \beta_7) = 1 + 1\text{-deg}(7) = 4$. Consequently, Fin dim Λ = fin dim $\Lambda = 4$.

3 Generic behavior of the homological dimensions

Recall that, for any finite dimensional algebra Δ and $d \in \mathbb{N}$, the following affine variety $\operatorname{Mod}_d(\Delta)$ parametrizes the d-dimensional Δ-modules: Let $a_1, \ldots, a_r$ be a set of algebra generators for Δ over K. For instance, if Δ is a path algebra modulo relations, then the primitive idempotents (alias vertices of the quiver), together with the (residue classes in Δ of the) arrows constitute such a set of generators. For $d \in \mathbb{N}$,

$$\operatorname{Mod}_d(\Delta) = \left\{ (x_i) \in \prod_{1 \le i \le r} \operatorname{End}_K(K^d) \mid \text{the } x_i \text{ satisfy all relations satisfied by the } a_i \right\}.$$

As is well known, the isomorphism classes of d-dimensional (left) Λ-modules are in one-to-one correspondence with the orbits of $\operatorname{Mod}_d(\Lambda)$ under the GL_d-conjugation action. Indeed, the orbits coincide with the fibres of the map from $\operatorname{Mod}_d(\Delta)$ to the set of isomorphism classes of d-dimensional left Δ-modules, which maps a point x to the class of K^d, endowed with the Δ-multiplication $a_i v = x_i(v)$. If $\mathcal{C}$ is a subvariety of $\operatorname{Mod}_d(\Delta)$, we refer to the modules represented by the points in $\mathcal{C}$ as *the modules in $\mathcal{C}$*.

It is, moreover, a standard fact that the homological dimensions of the d-dimensional modules, such as p dim, are generically constant on any irreducible component of $\operatorname{Mod}_d(\Delta)$ (for a proof, see [4, Lemma 4.3] or [10, Theorem 5.3], where the result is attributed to Bongartz). In fact, it is known that, given any irreducible subvariety $\mathcal{C}$ of $\operatorname{Mod}_d(\Delta)$, there exists a dense open subset $U \subseteq \mathcal{C}$ such that the function p dim is constant on U. Moreover, this *generic projective dimension on $\mathcal{C}$* is the minimum of the projective dimensions attained on the modules in $\mathcal{C}$. In most interesting cases, the projective dimension fails to be constant on all of $\mathcal{C}$, however. (Think, e.g., of the path algebra Δ of the quiver $1 \to 2$, and let $\mathcal{C}$ be the irreducible component of $\operatorname{Mod}_2(\Delta)$, whose points correspond to the modules with composition factors S_1, S_2; here the generic projective dimension is 0, while $\operatorname{p\,dim}(S_1 \oplus S_2) = 1$.) This raises the question of how the following generic variant of the finitistic dimension relates to the classical little finitistic dimension of Δ.

Definition 3.1. The *generic left finitistic dimension* of a finite dimensional algebra Δ is the supremum of the finite numbers gen-p dim$(\mathcal{C})$, where $\mathcal{C}$ traces the irreducible components of the varieties $\mathbf{Mod}_d(\Delta)$; here gen-p dim$(\mathcal{C})$ is the generic value of the function p dim, restricted to the modules in $\mathcal{C}$.

Clearly, the (left) generic finitistic dimension of an algebra Δ is always bounded above by fin dim Δ. When are the two dimensions equal? Given an irreducible component $\mathcal{C} \subseteq \mathbf{Mod}_d(\Delta)$, what is the spectrum of values attained by the projective dimension on $\mathcal{C}$?

The completeness with which these questions can be answered in the case of a truncated path algebra Λ came as a surprise to us. The resulting picture underscores the pivotal role played by tree modules and supplements the fact that, in the truncated scenario, the irreducible components are fairly well understood. They are in one-to-one correspondence with certain sequences of semisimple modules, as follows:

Recall that, given a finitely generated left Λ-module M, its *radical layering* is $\mathbb{S}(M) = (J^l M / J^{l+1} M)_{0 \leq l \leq L}$. We will identify isomorphic semisimple modules so that the radical layerings of isomorphic Λ-modules become identical. That the K-dimension of M be d evidently translates into the equality $\sum_{0 \leq l \leq L} \dim_K J^l M / J^{l+1} M = d$. For each sequence $\mathbb{S} = (\mathbb{S}_0, \ldots, \mathbb{S}_L)$ of semisimple modules $\mathbb{S}_l$ with total dimension d, let $\mathbf{Mod}(\mathbb{S})$ be the subset of $\mathbf{Mod}_d(\Lambda)$ consisting of those points which correspond to the modules with radical layering $\mathbb{S}$. Then the locally closed subvariety $\mathbf{Mod}(\mathbb{S})$ of $\mathbf{Mod}_d(\Lambda)$ is irreducible by [1, Theorem 5.3], whence so is its closure in $\mathbf{Mod}_d(\Lambda)$. The maximal candidates among the closures $\overline{\mathbf{Mod}(\mathbb{S})}$, where $\mathbb{S}$ traces the sequences $\mathbb{S}$ of total dimension d, are therefore the irreducible components of $\mathbf{Mod}_d(\Lambda)$; indeed, there are only finitely many such sequences. It is, moreover, easy to recognize whether a given sequence $\mathbb{S}$ of semisimple modules as above arises as the radical layering of a Λ-module, that is, whether $\mathbf{Mod}(\mathbb{S}) \neq \varnothing$ (see [1]). Namely, suppose that $\mathbb{S}_l = \bigoplus_{0 \leq l \leq L} S_i^{s(i,l)}$ and let P be the projective cover of $\mathbb{S}_0$. Then $\mathbf{Mod}(\mathbb{S}) \neq \varnothing$ if and only if there exists a set σ of paths in P, which is closed under initial subpaths, such that σ is *compatible with* $\mathbb{S}$ in the following sense: For each $i \in \{1, \ldots, n\}$ and each $l \in \{0, 1, \ldots, L\}$, the set σ contains precisely $s(i, l)$ paths of length l which end in the vertex e_i. Observe that, whenever M is a module with radical layering $\mathbb{S}(M) = \mathbb{S}$, any skeleton of M is compatible with $\mathbb{S}$. Consequently, the requirement that $\mathbf{Mod}(\mathbb{S}) \neq \varnothing$ implies that the l-th layer $\mathbb{S}_l$ of $\mathbb{S}$ be a direct summand of the l-th layer $J^l P / J^{l+1} P$ in the radical layering of P.

Theorem 3.2. *Let* $\mathbb{S} = (\mathbb{S}_0, \mathbb{S}_1, \ldots, \mathbb{S}_L)$ *be a sequence of semisimple Λ-modules such that* $\mathbf{Mod}(\mathbb{S}) \neq \varnothing$, *and let P a projective cover of $\mathbb{S}_0$. Moreover, suppose*

$$J^l P / J^{l+1} P = \left(\bigoplus_{1 \leq i \leq n} S_i^{s(i,l)} \right) \oplus \left(\bigoplus_{1 \leq i \leq n} S_i^{r(i,l)} \right)$$

for suitable nonnegative integers $r(i, l)$; here $s(i, l)$ is the multiplicity of S_i in $\mathbb{S}_l$ as above.

(1) *The projective dimension of a module M depends only on its radical layering $\mathbb{S}(M)$. In other words, the projective dimension is constant on each of the varieties $\mathbf{Mod}(\mathbb{S})$. This constant value, denoted $\operatorname{p\,dim}\mathbb{S}$, is the generic projective dimension of the irreducible subvariety $\overline{\mathbf{Mod}(\mathbb{S})}$ of $\mathbf{Mod}_d(\Lambda)$.*

(2) *If $\operatorname{p\,dim}\mathbb{S} > 0$, then*

$$\operatorname{p\,dim}\mathbb{S} = 1 + \sup\{l\text{-}\deg(c(e_i)) \mid i \leq n,\, l \leq L \text{ with } r(i,l) \neq 0\}.$$

(We adopt the standard convention "$1 + \infty = \infty$".) In particular, $\operatorname{p\,dim}\mathbb{S}$ is finite if and only if $r(i,l) = 0$ for all cyclebound vertices e_i, that is, if and only if every simple module of infinite projective dimension has the same composition multiplicity in P as in $\bigoplus_{0 \leq l \leq L} \mathbb{S}_l$.

(3) *The generic finitistic dimension of Λ coincides with $\operatorname{fin\,dim}\Lambda$. It is the projective dimension of a tree module $\mathcal{T}$ – of dimension d say – whose orbit closure is an irreducible component of $\mathbf{Mod}_d(\Lambda)$.*

Computing $\operatorname{p\,dim}\mathbb{S}$ in concrete examples amounts to performing at most n counts: Indeed, if $r(i,l) \neq 0$ for some l, then $l\text{-}\deg(c(e_i)) \leq l_i\text{-}\deg(c(e_i))$, where l_i is maximal with $r(i,l_i) \neq 0$. Observe moreover that the event $\operatorname{p\,dim}\mathbb{S} = 0$ is readily recognized: It occurs if and only if $\mathbb{S} = \mathbb{S}(P)$; in this case, $\mathbf{Mod}(\mathbb{S})$ consists of the GL_d-orbit of P only.

We smooth the road towards a proof of Theorem 3.2 with two preliminary observations.

Observation 3.3. *Given any finitely generated Λ-module with skeleton σ, there exists a direct sum of tree modules with the same skeleton.*

In particular, the syzygy of any finitely generated Λ-module is isomorphic to the syzygy of a direct sum of tree modules, and all projective dimensions in the set $\{0, 1, \ldots, \operatorname{fin\,dim}\Lambda\}$ are attained on tree modules.

Proof. Let M be any finitely generated left Λ-module, $P = \bigoplus_{1 \leq r \leq t} \Lambda z_r$ a projective cover of M with $z_r = e(r) \in \{e_1, \ldots, e_n\}$, and $\sigma \subseteq P$ a skeleton of M. For fixed $r \leq t$, let $\sigma^{(r)}$ be the subset of σ consisting of all paths in σ of the form pz_r. Then $\mathcal{T}^{(r)} := \Lambda z_r / (\sum_{q\ \sigma^{(r)}\text{-critical}} \Lambda q)$ is a tree module whose branches are precisely the maximal paths in $\sigma^{(r)}$ relative to the "initial subpath order". Hence, $\bigoplus_{1 \leq r \leq t} \mathcal{T}^{(r)}$ is a direct sum of tree modules, again having skeleton σ. Since, by Theorem 2.2, any skeleton of a module determines its syzygy up to isomorphism, the remaining claims follow. □

The next observation singles out candidates for the tree module postulated in Theorem 3.2(3). Let ϵ be the sum of all non-cyclebound primitive idempotents in the full set $e_1, \ldots, e_n$. (In Example 2.4, we have $\epsilon = e_8 + \cdots + e_{15}$.) Clearly, the left ideal $\Lambda\epsilon \subseteq \Lambda$ of finite projective dimension equals $\epsilon\Lambda\epsilon$. In particular, given any left Λ-module M, the subspace ϵM is a sub*module* of M.

Observation 3.4. *Let e_i be any vertex of Q. Then* $\operatorname{p dim} \epsilon J e_i < \infty$, *and*

$$\operatorname{p dim} \epsilon J e_i \geq \operatorname{p dim} \Lambda q$$

for every nonzero path q of positive length in Λ with starting vertex e_i such that $\operatorname{p dim} \Lambda q < \infty$.

Moreover: The factor module $T_i = \Lambda e_i / \epsilon J e_i$ is a tree module. If $\dim_K T_i = d_i$, and $\mathbb{S}(T_i) = \mathbb{S}^{(i)}$ is the radical layering of T_i, then the subvariety $\mathbf{Mod}(\mathbb{S}^{(i)})$ of $\mathbf{Mod}_{d_i}(\Lambda)$ coincides with the GL_{d_i}-orbit of T_i and is open in $\mathbf{Mod}_{d_i}(\Lambda)$.

Proof. We first address the second set of claims. Let p_{ij}, $1 \leq j \leq t_i$, be the different paths of positive length in Λ which start in e_i, end in a non-cyclebound vertex, and are minimal with these properties in the "initial subpath order"; that is, every proper initial subpath of positive length of one of the p_{ij} ends in a cyclebound vertex. Clearly, $\epsilon J e_i = \bigoplus_{1 \leq j \leq t_i} \Lambda p_{ij}$, which shows in particular that T_i is a tree module. Moreover, any module $\widetilde{M}$ sharing the radical layering of T_i also has projective cover Λe_i, and a comparison of composition factors shows that every epimorphism $\Lambda e_i \to M$ has kernel $\epsilon J e_i$. Thus $M \cong T_i$, which shows $\mathbf{Mod}(\mathbb{S}^{(i)})$ to equal the GL_{d_i}-orbit of T_i. Moreover, it is readily checked that $\operatorname{Ext}^1_\Lambda(T_i, T_i) = 0$, whence the orbit $\mathbf{Mod}(\mathbb{S}^{(i)})$ of T_i is open in $\mathbf{Mod}_{d_i}$ (see [5, Corollary 3]), and the proof of the final assertions is complete.

For the first claim, let $q = q e_i$ be a nonzero path of positive length in Λ with $\operatorname{p dim} \Lambda q < \infty$. Then q ends in a non-cyclebound vertex by Lemma 2.9 – call it e – and hence q has an initial subpath q' among the paths p_{ij}; let e' be the (non-cyclebound) terminal vertex of q'. If l and l' are the lengths of q and q', respectively, $c(e') - c(e) \geq l - l' \geq 0$, and hence $c(e') + l' \geq c(e) + l$. This shows

$$\operatorname{p dim} \Lambda q' = l'\text{-deg}(c(e')) \geq l\text{-deg}(c(e)) = \operatorname{p dim} \Lambda q,$$

which yields the desired inequality. □

Proof of Theorem 3.2. (1) Let M be a module with radical layering $\mathbb{S}$ and σ any skeleton of M. By [1, Theorem 5.3], the points in $\mathbf{Mod}_d(\Lambda)$ parametrizing the modules that share this skeleton constitute a dense open subset of $\mathbf{Mod}(\mathbb{S})$. All modules represented by this open subvariety have the same projective dimension as M, because any skeleton of a module pins down its syzygy up to isomorphism. Therefore, $\operatorname{p dim} M$ is the generic value of the function $\operatorname{p dim}$ on the irreducible subvariety $\overline{\mathbf{Mod}(\mathbb{S})}$ of $\mathbf{Mod}_d(\Lambda)$.

(2) Suppose that $\operatorname{p dim} \mathbb{S} > 0$, which means $r(i, l) > 0$ for some pair (i, l). Let M be any module with $\mathbb{S}(M) = \mathbb{S}$. By part (1), $\operatorname{p dim} \mathbb{S} = \operatorname{p dim} M$. To scrutinize the projective dimension of M, let $\widehat{\sigma}$ be a skeleton of P and $\sigma \subset \widehat{\sigma}$ a skeleton of M. We have $\Omega^1(M) \cong \bigoplus_{q\ \sigma\text{-critical}} \Lambda q$ by Theorem 2.2. Since $r(i, l) > 0$ whenever q is a σ-critical path of length l ending in e_i, we glean that $\operatorname{p dim} M$ is bounded above by the supremum displayed in part (2) of Theorem 3.2. For the reverse inequality, choose any pair (i, l) with $r(i, l) > 0$. This inequality amounts to the existence of a path $p z_r$ of length l in $\widehat{\sigma} \setminus \sigma$ which ends in e_i. Denote by $p' z_r$ the maximal initial subpath of $p z_r$ which belongs to σ. Since $p z_r \notin \sigma$, there is a unique arrow α such that $\alpha p' z_r$ is

in turn an initial subpath of pz_r. In particular, if $q = \alpha p'$, then qz_r is a σ-critical path ending in some vertex e_j. Invoking once again the above decomposition of $\Omega^1(M)$, we deduce that the cyclic left ideal Λq is isomorphic to a direct summand of $\Omega^1(M)$. By Theorem 2.6, it therefore suffices to show that the length(q)-degree of $c(e_j)$ is larger than or equal to l-deg($c(e_i)$). For that purpose, we write $pz_r = q'qz_r$ for a suitable path q' in Λ. Since $c(e_j) \geq c(e_i) + \text{length}(q')$, we obtain $c(e_j) \geq c(e_i)$, and consequently $c(e_j) + \text{length}(q) \geq c(e_i) + l$. We conclude

$$\left[\frac{c(e_j)}{L+1}\right] + \left[\frac{c(e_j) + \text{length}(q)}{L+1}\right] \geq \left[\frac{c(e_i)}{L+1}\right] + \left[\frac{c(e_i) + l}{L+1}\right] = l\text{-deg}(c(e_i)).$$

Thus p dim $M - 1 \geq l$-deg($c(e_i)$) as required. The final equivalence under (2) is an immediate consequence.

(3) By construction, the tree modules $\mathcal{T}_i$ of Observation 3.4 all have finite projective dimension. Combining the first part of this observation with the final statement of Theorem 2.2, we moreover see that fin dim Λ equals the maximum of these dimensions. The final statement of Observation 3.4 now completes the proof of (3). □

Let $\mathbb{S} = (\mathbb{S}_0, \ldots, \mathbb{S}_L)$ again be a sequence of semisimple modules of total dimension d such that $\mathbf{Mod}(\mathbb{S}) \neq \varnothing$. As we saw, the projective dimension p dim $\mathbb{S}$ holds some information about path lengths in KQ; namely on the lengths of paths starting in vertices that belong to the support of $\Omega^1(M)$, where M is any module in $\mathbf{Mod}(\mathbb{S})$. To obtain a tighter correlation between Q and the homology of Λ, we will next explore the full spectrum of values of the function p dim attained on the closure $\overline{\mathbf{Mod}(\mathbb{S})}$. While those ranges of values are better gauges of how the vertices corresponding to the simples in the various layers $\mathbb{S}_l$ of $\mathbb{S}$ are placed in the quiver Q, the refined homological data still do not account for the intricacy of the embedding of $\overline{\mathbf{Mod}(\mathbb{S})}$ into $\mathbf{Mod}_d(\Lambda)$ in general. (See the comments following the next theorem.) On the other hand, for p dim $\mathbb{S} < \infty$ and small L, far more of this picture is preserved in the homology than in the hereditary case.

We first recall from [1, Section 2.B] that, for any M in $\overline{\mathbf{Mod}(\mathbb{S})}$, the sequence $\mathbb{S}(M)$ is larger than or equal to $\mathbb{S}$ in the following partial order: Suppose that $\mathbb{S}$ and $\mathbb{S}'$ are semisimple modules with $\bigoplus_{0 \leq l \leq L} \mathbb{S}_l = \bigoplus_{0 \leq l \leq L} \mathbb{S}'_l$. Then "$\mathbb{S}' \geq \mathbb{S}$" means that $\bigoplus_{l \leq r} \mathbb{S}_l$ is a direct summand of $\bigoplus_{l \leq r} \mathbb{S}'_l$, for all $r \geq 0$. In intuitive terms this says that, in the passage from $\mathbb{S}$ to $\mathbb{S}'$, the simple summands of the $\mathbb{S}_l$ are only upwardly mobile relative to the layering.

Lemma 3.5. *If $\mathbb{S}' \geq \mathbb{S}$ and $\mathbf{Mod}(\mathbb{S}') \neq \varnothing$, then p dim $\mathbb{S}' \geq$ p dim $\mathbb{S}$.*

Proof. Let P be a projective cover of $\mathbb{S}_0$ as before and P' a projective cover of $\mathbb{S}'_0$. Decompose the radical layers of P' in analogy with the decomposition given for P above:

$$J^l P' / J^{l+1} P' = \bigoplus_{1 \leq i \leq n} S_i^{s'(i,l)} \oplus \bigoplus_{1 \leq i \leq n} S_i^{r'(i,l)},$$

where $\mathbb{S}'_l = \bigoplus_{1 \leq i \leq n} S_i^{s'(i,l)}$. It follows immediately from the definition of the partial order on sequences of semisimples that, whenever $r(i,l) > 0$, there exists $l' \geq l$ with $r'(i,l') > 0$. In light of Theorem 3.2, this proves the lemma. □

We give two descriptions of the range of values of p dim on the closure $\overline{\mathbf{Mod}(\mathbb{S})}$. For a combinatorial version, we keep the notation of Theorem 3.2 and the proof of Lemma 3.5: Namely, $\mathbb{S}_l = \bigoplus_{1 \le i \le n} S_i^{s(i,l)}$, and P is a projective cover of $\mathbb{S}_0$. From $\mathbf{Mod}(\mathbb{S}) \ne \varnothing$, one then obtains $J^l P/J^{l+1} P = \mathbb{S}_l \oplus \bigoplus_{1 \le i \le n} S_i^{r(i,l)}$. In our graph-based description of the values $\operatorname{p\,dim} M > \operatorname{p\,dim} \mathbb{S}$, where M traces $\overline{\mathbf{Mod}(\mathbb{S})}$, the exponents $s(i,l)$ take over the role played by the $r(i,l)$ relative to the generic projective dimension, $\operatorname{p\,dim} \mathbb{S}$: Recall from Theorem 3.2 that, whenever $\operatorname{p\,dim} \mathbb{S}$ is nonzero, it is the maximum of the values $1 + l\text{-deg}(c(e_i)) \in \mathbb{N} \cup \{0, \infty\}$ which accompany the pairs (i,l) with $r(i,l) > 0$. (Note: In view of $\mathbb{S}_0 = P/JP$, the inequality $r(i,l) > 0$ entails $l \ge 1$.)

Now, we consider the different candidates $n_1, \ldots, n_v$ among those elements in $\mathbb{N} \cup \{0, \infty\}$ which have the form

$$1 + l\text{-deg}(c(e_j)), \quad l \ge 1, \; S_j \subseteq \mathbb{S}_l$$

and are *strictly larger* than $\operatorname{p\,dim} \mathbb{S}$. In other words,

$$\{n_1, \ldots, n_v\} = [\operatorname{p\,dim} \mathbb{S} + 1, \, \infty] \cap \{1 + l\text{-deg}(c(e_j)) \mid l \ge 1, \, s(j,l) > 0\}.$$

Theorem 3.6. *Let $\mathbb{S}$ be a semisimple sequence of total dimension d with $\mathbf{Mod}(\mathbb{S}) \ne \varnothing$. The range of values,*

$$\{\operatorname{p\,dim} M \mid M \text{ in } \overline{\mathbf{Mod}(\mathbb{S})}\},$$

of the function $\operatorname{p\,dim}$ on the closure of $\mathbf{Mod}(\mathbb{S})$ in $\mathbf{Mod}_d$, is equal to the following coinciding sets:

$$\{\operatorname{p\,dim} \mathbb{S}' \mid \mathbb{S}' \ge \mathbb{S}, \, \mathbf{Mod}(\mathbb{S}') \ne \varnothing\} = \{\operatorname{p\,dim} \mathbb{S}\} \cup \{n_1, \ldots, n_v\}.$$

In general, describing the closure of $\mathbf{Mod}(\mathbb{S})$ in $\mathbf{Mod}_d(\Lambda)$ is an intricate representation-theoretic task, a fact not reflected by the homology. For instance: • When $\mathbb{S}'$ is a sequence of semisimple modules such that $\mathbb{S}' \ge \mathbb{S}$ and $\mathbf{Mod}(\mathbb{S}') \ne \varnothing$, the intersection $\overline{\mathbf{Mod}(\mathbb{S})} \cap \overline{\mathbf{Mod}(\mathbb{S}')}$ may still be empty. • The condition $\overline{\mathbf{Mod}(\mathbb{S})} \cap \overline{\mathbf{Mod}(\mathbb{S}')} \ne \varnothing$ does not imply $\mathbf{Mod}(\mathbb{S}') \subseteq \overline{\mathbf{Mod}(\mathbb{S})}$. See the final discussion of our example for illustration.

Proof. Set $\mathcal{P} = \{\operatorname{p\,dim} M \mid M \text{ in } \overline{\mathbf{Mod}(\mathbb{S})}\}$. We already know that

$$\mathcal{P} \subseteq \{\operatorname{p\,dim} \mathbb{S}' \mid \mathbb{S}' \ge \mathbb{S}\};$$

indeed, this is immediate from Lemma 3.5 and the remarks preceding it.

Suppose that $\mathbb{S}'$ is a sequence of semisimple modules with $\mathbb{S}' \ge \mathbb{S}$ and $\mathbf{Mod}(\mathbb{S}') \ne \varnothing$. Assume $\operatorname{p\,dim} \mathbb{S}' > \operatorname{p\,dim} \mathbb{S}$, which, in particular, implies $\operatorname{p\,dim} \mathbb{S}' > 0$. To show that $\operatorname{p\,dim} \mathbb{S}'$ equals one of the n_k, we adopt the notation used in the proof of Lemma 3.5. By Theorem 3.2, $\operatorname{p\,dim} \mathbb{S}' = 1 + a\text{-deg}(c(e_i))$ for some pair (i, a) with $r'(i,a) > 0$. Again invoking Theorem 3.2, we moreover infer that $r(i,a) = 0$ from $\operatorname{p\,dim} \mathbb{S} < \operatorname{p\,dim} \mathbb{S}'$. If $s(i,a) > 0$, we are done, since necessarily $a \ge 1$. So let us suppose that also

$s(i, a) = 0$, meaning that S_i fails to be a summand of the a-th layer $J^a P / J^{a+1} P$ of P. In light of $S_i \subseteq J^a P' / J^{a+1} P'$, this entails the existence of a simple $S_j \subseteq \mathbb{S}_0' / \mathbb{S}_0$ with the property that $S_i \subseteq J^a e_j / J^{a+1} e_j$. Consequently, $c(e_j) \geq c(e_i) + a$. On the other hand, $S_j \subseteq \bigoplus_{l \geq 1} \mathbb{S}_l$, because the total multiplicities of the simple summands of $\mathbb{S}$ and $\mathbb{S}'$ coincide. This means $s(j, k) > 0$ for some $k \geq 1$. In light of $\mathbb{S}_0 \oplus S_j \subseteq \mathbb{S}_0'$ and $\bigoplus_{0 \leq l \leq L} \mathbb{S}_l = \bigoplus_{0 \leq l \leq L} \mathbb{S}_l'$, we deduce that $r'(j, b) > 0$ for some pair (j, b) with $b \geq 1$ and $s(j, b) > 0$. Another application of Theorem 3.2 thus yields

$$\operatorname{p dim} \mathbb{S}' - 1 \geq b\text{-deg}(c(e_j)) = \left\lceil \frac{c(e_j)}{L+1} \right\rceil + \left\lceil \frac{c(e_j) + b}{L+1} \right\rceil$$

$$\geq \left\lceil \frac{c(e_i) + a}{L+1} \right\rceil + \left\lceil \frac{c(e_i) + a + b}{L+1} \right\rceil \geq a\text{-deg}(c(e_i)) = \operatorname{p dim} \mathbb{S}' - 1.$$

We conclude that all inequalities along this string are actually equalities, that is,

$$b\text{-deg}(c(e_j)) = a\text{-deg}(c(e_i)).$$

This shows that $\operatorname{p dim} \mathbb{S}' = 1 + b\text{-deg}(c(e_j))$ for a pair (j, b) with $s(j, b) > 0$ as required.

Finally, we verify that each of the numbers n_k belongs to $\mathcal{P}$. By definition, n_k is of the form $1 + l\text{-deg}(c(e_i))$ for some pair (i, l) with $l \geq 1$ and $s(i, l) > 0$. Let D be any direct sum of tree modules with $\mathbb{S}(D) = \mathbb{S}$; in light of $\mathbf{Mod}(\mathbb{S}) \neq \varnothing$, such a module D exists by Observation 3.3. Then there is a tree direct summand $\mathcal{T}$ of D with a branch that contains an initial subpath q of length l ending in the vertex e_i. The direct sum of tree modules $D' = (\mathcal{T}/\Lambda q) \oplus \Lambda q \oplus D/\mathcal{T}$ belongs to $\overline{\mathbf{Mod}(\mathbb{S})}$. In fact, D' is well known to belong to the closure of the orbit of D in $\mathbf{Mod}_d(\Lambda)$; see, e.g., [3, Section 3, Lemma 2]. Therefore $\operatorname{p dim} D' \in \mathcal{P}$. As for the value of this projective dimension: Up to isomorphism, Λq is a direct summand of the syzygy of the tree module $\mathcal{T}/\Lambda q$: indeed, q is σ-critical relative to the obvious skeleton σ of $\mathcal{T}/\Lambda q$ consisting of all initial subpaths of the branches (see Theorem 2.2 and the comments accompanying Definition 2.8). Theorems 2.6 and 3.2 moreover yield

$$\operatorname{p dim} D' - 1 \geq \operatorname{p dim} \Lambda q = l\text{-deg}(c(e_i)) = n_k - 1 > \operatorname{p dim} \mathbb{S} - 1 = \operatorname{p dim} D - 1,$$

whence $\operatorname{p dim}(\mathcal{T}/\Lambda q) = n_k = \operatorname{p dim} D'$. This shows n_k to belong to $\mathcal{P}$ and completes the argument. □

A final visit to Example 2.4. (a) First, let $\mathbb{S} = \mathbb{S}(\Lambda e_1)$ be the radical layering of the projective tree module $\mathcal{T} = \Lambda e_1$. By Theorems 3.2 and 3.6, the values of $\operatorname{p dim}$ on $\overline{\mathbf{Mod}(\mathbb{S})}$ are $\operatorname{p dim} \mathbb{S} = 0, 2, 3, 4, \infty$. For instance, $\operatorname{p dim}((\mathcal{T}/\Lambda\beta_3\alpha_2\beta_1) \oplus (\Lambda\beta_3\alpha_2\beta_1))$ equals 2. Note that the value 1, on the other hand, is not attained.

(b) Next we justify the comments following the statement of Theorem 3.6. Let $\mathbb{S}$ and $\mathbb{S}'$ be the radical layerings of the modules M and M' with the following graphs, respectively:

Then $\mathbb{S}' \geq \mathbb{S}$, while $\overline{\mathbf{Mod}(\mathbb{S})} \cap \mathbf{Mod}(\mathbb{S}') = \varnothing$.

On the other hand, if $\mathbb{S} := \mathbb{S}(N)$, $\mathbb{S}' := \mathbb{S}(N')$, and $\mathbb{S}'' := \mathbb{S}(N'')$ where N, N', and N'' are given by the graphs

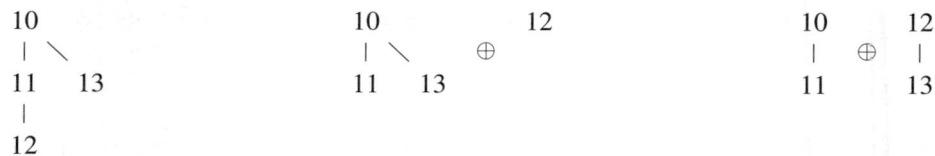

then $\mathbb{S}' = \mathbb{S}'' \geq \mathbb{S}$, and the intersection $\overline{\mathbf{Mod}(\mathbb{S})} \cap \mathbf{Mod}(\mathbb{S}')$ contains N', but not N''.

References

[1] E. Babson, B. Huisgen-Zimmermann, and R. Thomas, *Generic representation theory of quivers with relations*. Preprint (2007). Available at http://www.math.ucsb.edu/~birge/papers.html.

[2] M. C. R. Butler, *The syzygy theorem for monomial algebras*. Trends in the Representation Theory of Finite Dimensional Algebras (Seattle, Washington 1997), E.L. Green and B. Huisgen-Zimmermann, Eds. Contemp. Math. Vol. 229, pp. 111–116. Amer. Math. Soc., Providence, 1998.

[3] W. Crawley-Boevey, *Lectures on representations of quivers*. Course Notes (1992). Available at http://www.maths.leeds.ac.uk/~pmtwc/.

[4] W. Crawley-Boevey and J. Schröer, *Irreducible components of varieties of modules*, J. reine angew. Math. 553 (2002), pp. 201–220.

[5] J. A. de la Peña, *Tame algebras: some fundamental notions*, Ergänzungsreihe E95-010 Sonderforschungsbereich 343. Universität Bielefeld, 1995.

[6] E. L. Green, E. E. Kirkman, and J. J. Kuzmanovich, *Finitistic dimensions of finite dimensional monomial algebras*, J. Algebra 136 (1991), pp. 37–50.

[7] B. Huisgen-Zimmermann, *Predicting syzygies over monomial relation algebras*, Manuscripta Math. 70 (1991), pp. 157–182.

[8] B. Huisgen-Zimmermann, *Homological domino effects and the first finitistic dimension conjecture*, Invent. Math. 108 (1992), pp. 369–383.

[9] C. M. Ringel, *The torsionless modules of an artin algebra*, Lectue Notes (2008). Available at http://www.mathematik.uni-bielefeld.de/~ringel/opus/torsionless.pdf.

[10] S. O. Smalø, *Lectures on Algebras* (Mar del Plata, Argentina, March 2006). Revista de la Unión Matemática Argentina, 2007.

Author information

A. Dugas, Department of Mathematics, University of California, Santa Barbara, CA 93106-3080, USA.
E-mail: asdugas@math.ucsb.edu

B. Huisgen-Zimmermann, Department of Mathematics, University of California, Santa Barbara, CA 93106-3080, USA.
E-mail: birge@math.ucsb.edu

J. Learned, Department of Mathematics, College of the Desert, Palm Desert, CA 92260, USA.
E-mail: jlearned@collegeofthedesert.edu

Representations of the category of serial modules of finite Goldie dimension

Alberto Facchini and Pavel Příhoda

Abstract. We study the category SUsr of all serial right modules of finite Goldie dimension over a fixed ring R. This category has natural valuations m-dim$_U$ and e-dim$_U$ for every uniserial right R-module U, because the number m-dim$_U(A)$ (e-dim$_U(A)$) of modules in the same monogeny (epigeny) class as U in any indecomposable direct-sum decomposition of an object A of SUsr is uniquely determined by A. These valuations m-dim$_U$, e-dim$_U \colon V(\mathrm{SUsr}) \to \mathbb{N}_0$ are not essential valuations in general, so that the quotient categories of SUsr corresponding to the valuations m-dim$_U$, e-dim$_U$ are Krull–Schmidt categories but not IBN categories. With these valuations, we obtain a natural representation SUsr $\to \coprod_{i \in I} \mathcal{A}_i$ that is an isomorphism reflecting and direct-summand reflecting additive functor. Here the categories $\mathcal{A}_i$ are quotient Krull–Schmidt categories of SUsr. If, instead of considering all the valuations m-dim$_U$, e-dim$_U$ of SUsr that come from the monogeny classes and the epigeny classes of all uniserial right R-modules, we consider only those that are essential valuations, then we get a divisor theory of the monoid $V(\mathrm{SUsr})$ and, correspondingly, a representation SUsr $\to \coprod_{i \in J} \mathcal{A}_i$ in which the categories $\mathcal{A}_i$, $i \in J$, are IBN categories.

Key words. Representations of additive categories, uniserial modules, serial modules.

AMS classification. 16D90, 18B15, 18E05.

1 Introduction

An additive skeletally small category $\mathcal{A}$ in which idempotents split is a Krull–Schmidt category (that is, every object of $\mathcal{A}$ is a direct sum of finitely many indecomposable objects in an essentially unique way, i.e., in a unique way up to isomorphism and up to a permutation of the direct summands) if and only if there is a weak equivalence $\mathcal{A} \to \coprod_{i \in I} \mathcal{A}_i$ for which the categories $\mathcal{A}_i$ are IBN categories, that is, categories equivalent to the category of all finitely generated free modules over an IBN ring [4]. The proof of this fact is based on considering suitable valuations $v_i \colon V(\mathcal{A}) \to \mathbb{N}_0$ of the monoid $V(\mathcal{A})$ of all isomorphism classes of objects of $\mathcal{A}$ into the additive monoid $\mathbb{N}_0$ of nonnegative integers. Valuations $V(\mathcal{A}) \to \mathbb{N}_0$ of the monoid $V(\mathcal{A})$ are also called valuations of the category $\mathcal{A}$ and can be viewed as mappings $\mathrm{Ob}(\mathcal{A}) \to \mathbb{N}_0$. The categories $\mathcal{A}_i$ are the quotient categories $\mathcal{A}/\mathcal{I}_i$ of $\mathcal{A}$ modulo the ideal $\mathcal{I}_i$ of all morphisms

---

First author: Partially supported by Ministero dell'Università e della Ricerca (Prin 2005 "Perspectives in the theory of rings, Hopf algebras and categories of modules"), by Gruppo Nazionale Strutture Algebriche e Geometriche e loro Applicazioni of Istituto Nazionale di Alta Matematica, and by Università di Padova (Progetto di Ateneo CDPA048343 "Decomposition and tilting theory in modules, derived and cluster categories").

Second author: Partially supported by Research Project MSM 0021620839.

of $\mathcal{A}$ that factor through some object C of $\mathcal{A}$ with $v_i(C) = 0$. A similar idea applies not only to Krull–Schmidt categories $\mathcal{A}$, case in which the commutative monoid $V(\mathcal{A})$ is free, but also, more generally, to the case of the additive skeletally small categories $\mathcal{A}$ in which idempotents split and for which $V(\mathcal{A})$ is a Krull monoid (because for these monoids there are sufficiently many valuations $v_i \colon V(\mathcal{A}) \to \mathbb{N}_0$; cf. [5, Theorem 6.1]).

In this paper, we develop the same idea and apply the same technique to the special case in which $\mathcal{A}$ is the category SUsr of all serial right modules of finite Goldie dimension over a fixed ring R. The category SUsr is additive, skeletally small and with split idempotents. In this case, there are natural valuations m-dim$_U$, e-dim$_U \colon V(\text{SUsr}) \to \mathbb{N}_0$ [8]. They are, for any uniserial right R-module U, the number m-dim$_U(A)$ (e-dim$_U(A)$) of modules in the same monogeny (epigeny) class as U in any indecomposable direct-sum decomposition of the object A of SUsr. By the Weak Krull–Schmidt Theorem for uniserial modules [2], this number is uniquely determined by A. The problem is that these valuations m-dim$_U$, e-dim$_U \colon V(\text{SUsr}) \to \mathbb{N}_0$ are not essential valuations in general, so that we cannot apply the results of [5] directly to obtain IBN quotient categories. In fact, the quotient categories of SUsr corresponding to the valuations m-dim$_U$, e-dim$_U$, which we obtain in this case, are not IBN categories in general, they are only Krull–Schmidt categories. Thus we obtain a natural representation SUsr $\to \coprod_{i \in I} \mathcal{A}_i$, that is, an isomorphism reflecting and direct-summand reflecting additive functor, in which the categories $\mathcal{A}_i$ are quotient Krull–Schmidt categories of SUsr. If, instead of considering all the valuations m-dim$_U$, e-dim$_U$ of SUsr that come from the monogeny classes and the epigeny classes of all uniserial right R-modules, we consider only those that are essential valuations, then we get a divisor theory $V(\text{SUsr}) \to \mathbb{N}_0^{(J)}$ [7, Section 4] and, correspondingly, a representation SUsr $\to \coprod_{i \in J} \mathcal{A}_i$ with the categories $\mathcal{A}_i$, $i \in J$, that are IBN categories.

Let us explain in more detail the terminology used above. Recall that an R-module U is *uniserial* if, for any submodules V and W of U, we have $V \subseteq W$ or $W \subseteq V$, and a module is *serial* if it is a direct sum of uniserial modules. Thus a module is serial of finite Goldie dimension if and only if it is a direct sum of finitely many uniserial modules. For a fixed ring R, we shall denote by Mod-R the category of all right R-modules, and by SUsr the full subcategory of Mod-R whose objects are all serial modules of finite Goldie dimension. The second author proved in [13] that direct summands of serial modules of finite Goldie dimension are serial, so that idempotents split in the category SUsr.

Given any valuation $v \colon \text{Ob}(\mathcal{A}) \to \mathbb{N}_0$ of an additive category $\mathcal{A}$, consider the ideal $\mathcal{I}_v$ of $\mathcal{A}$ defined, for every $A, B \in \text{Ob}(\mathcal{A})$, by $\mathcal{I}_v(A, B) := \{\, f \in \mathcal{A}(A, B) \mid \text{there exist } C \in \text{Ob}(\mathcal{A}), g \in \mathcal{A}(A, C) \text{ and } h \in \mathcal{A}(C, B) \text{ with } v(C) = 0 \text{ and } f = hg \,\}$. It is possible to construct the quotient category $\mathcal{A}/\mathcal{I}_v$, and we are interested in the structure of this category, because, when v ranges in the set of all valuations of $\mathcal{A}$, we get different representations of $\mathcal{A}$ via the canonical functors $\mathcal{A} \to \mathcal{A}/\mathcal{I}_v$. In the particular case in which idempotents split in $\mathcal{A}$ and v is an *essential* valuation of $\mathcal{A}$ (that is, for every $A, B \in \text{Ob}(\mathcal{A})$ with $v(A) \leq v(B)$, there exist $C, D \in \text{Ob}(\mathcal{A})$ with $v(D) = 0$ and $A \oplus C \cong B \oplus D$), we get that the quotient category $\mathcal{A}/\mathcal{I}_v$ turns out to be an IBN category, that is, a Krull–Schmidt category with only one indecomposable object up to isomorphism [5, Theorem 4.5].

If A is a module, we shall denote by $\langle A \rangle$, $[A]_m$ and $[A]_e$ the *isomorphism class*, the *monogeny class* and the *epigeny class* of A, respectively, that is, $\langle A \rangle = \{\, B \in \text{Mod-}R \mid A \cong B \,\}$, $[A]_m = \{\, B \in \text{Mod-}R \mid \text{there exist a monomorphism } A \to B \text{ and a monomorphism } B \to A \,\}$, and $[A]_e = \{\, B \in \text{Mod-}R \mid \text{there exist an epimorphism } A \to B \text{ and an epimorphism } B \to A \,\}$. The first author proved in [2, Theorem 1.9] that if $U_1, \ldots, U_n, V_1, \ldots, V_t$ are non-zero uniserial modules over an arbitrary ring, the direct sums $U_1 \oplus \cdots \oplus U_n$ and $V_1 \oplus \cdots \oplus V_t$ are isomorphic modules if and only if $n = t$ and there are two permutations σ, τ of $\{1, 2, \ldots, n\}$ such that $[U_i]_m = [V_{\sigma(i)}]_m$ and $[U_i]_e = [V_{\tau(i)}]_e$ for every $i = 1, 2, \ldots, n$ (Weak Krull–Schmidt Theorem for uniserial modules). This theorem allowed to solve a problem posed by Warfield in 1975, showing that Krull–Schmidt fails for finitely presented modules over a serial ring, because finitely presented modules over serial rings are serial modules of finite Goldie dimension.

Thus the uniserial direct summands of a serial module A of finite Goldie dimension, when we write A as a direct sum of non-zero uniserial modules, are not determined by A up to isomorphism, but their monogeny classes and epigeny classes are uniquely determined. Hence there are natural valuations of the category SUsr of all serial modules of finite Goldie dimension over a fixed ring R defined as follows. Fix a uniserial module U. Given any module $A \in \text{Ob(SUsr)}$, decompose A as a direct sum $A = U_1 \oplus \cdots \oplus U_n$ of uniserial modules U_i, and set m-$\dim_U(A) = |\{\, i \mid i = 1, 2, \ldots, n,\ [U_i]_m = [U]_m \,\}|$ and e-$\dim_U(A) = |\{\, i \mid i = 1, 2, \ldots, n,\ [U_i]_e = [U]_e \,\}|$. Thus, for each uniserial module U, we get two valuations m-$\dim_U$, e-$\dim_U \colon \text{Ob(SUsr)} \to \mathbb{N}_0$ of the category SUsr, which, as we have said above, are not essential in general. When we restrict our attention to the valuations m-$\dim_U$, e-$\dim_U$ that are essential, we get a divisor theory for the Krull monoid $V(\text{SUsr})$. Canceling repetitions (m-$\dim_U$ can be equal to e-$\dim_U$ for some U), we get that the divisor theory is defined, in a canonical way, by suitable valuations v_W ($W \in K$), valuations m-$\dim_U$ ($U \in M'$) and valuations e-$\dim_V$ ($V \in E'$), where K is a set of representatives up to isomorphism of the Krull–Schmidt uniserial modules, M' is a set of representatives up to monogeny of the non-zero uniserial modules U that are not Krull–Schmidt modules, and E' is a set of representatives up to epigeny of the non-zero uniserial right R-modules V that are not Krull–Schmidt modules. Factoring out the corresponding ideals of the category SUsr, we get IBN quotient categories $\text{SUsr}/\mathcal{I}_W$, $\text{SUsr}/\mathcal{M}_U$ and $\text{SUsr}/\mathcal{E}_V$, whose coproduct category $\mathcal{D}$ is a Krull–Schmidt category. Thus we get a representation $\text{SUsr} \to \mathcal{D}$ that induces a divisor theory of the Krull monoid $V(\text{SUsr})$ into the free commutative monoid $V(\mathcal{D})$.

2 Notation and terminology

In this paper, all modules will be unital right modules over a fixed associative ring R with identity $1 \neq 0$. We shall denote by $\mathbb{N}_0$ the set of all nonnegative integers.

For a category $\mathcal{A}$, we shall denote by $\text{Ob}\,\mathcal{A}$ the class of objects of $\mathcal{A}$. An *additive category* is a preadditive category with a zero object and finite products. If $\mathcal{A}$ is an additive category and $A, B \in \operatorname{Ob} \mathcal{A}$, then the product $A \prod B$ is canonically isomorphic to the coproduct $A \coprod B$. We will denote it as $A \oplus B$, and call it the *direct sum* (or *biproduct*) of A and B.

Let $\mathcal{A}$ be an additive category. Recall that an *ideal* of $\mathcal{A}$ is a subfunctor of the two variable functor $\mathcal{A}(-,-)$ [12, p. 18]. That is, $\mathcal{I}$ is an ideal of $\mathcal{A}$ if, for every pair of objects $A, B \in \operatorname{Ob} \mathcal{A}$, $\mathcal{I}(A, B)$ is a subgroup of $\mathcal{A}(A, B)$ such that, for every morphism $\varphi \colon C \to A$, $\psi \colon A \to B$ and $\omega \colon B \to D$ with $\psi \in \mathcal{I}(A, B)$, one has that $\omega \psi \varphi \in \mathcal{I}(C, D)$. If $\mathcal{A}$ is a skeletally small preadditive category, the ideals of $\mathcal{A}$ form a set, otherwise, in general, they form a class $\mathcal{L}(\mathcal{A})$. This class $\mathcal{L}(\mathcal{A})$ can be partially ordered by setting $\mathcal{I} \subseteq \mathcal{J}$ if $\mathcal{I}(A, B) \subseteq \mathcal{J}(A, B)$ for every pair (A, B) of objects of $\mathcal{A}$.

For every object A of an additive category $\mathcal{A}$, let $\langle A \rangle$ denote the *isomorphism class* of A, that is, the subclass of $\operatorname{Ob} \mathcal{A}$ whose elements are all $B \in \operatorname{Ob} \mathcal{A}$ that are isomorphic to A. Let $V(\mathcal{A}) := \{ \langle A \rangle \mid A \in \operatorname{Ob} \mathcal{A} \}$ be the class of all isomorphism classes of the objects of $\mathcal{A}$, so that $V(\mathcal{A})$ is a set if and only if the additive category $\mathcal{A}$ is skeletally small. Define an addition on the class $V(\mathcal{A})$ by $\langle A \rangle + \langle B \rangle := \langle A \oplus B \rangle$ for every $A, B \in \operatorname{Ob} \mathcal{A}$. With this operation, the class $V(\mathcal{A})$ becomes a "large" additive commutative monoid, where "large" is in the sense that it is a class, and not a set in general.

All the monoids considered in this paper will be commutative additive monoids, that is, commutative additive semigroups with a zero element. Let M be a (commutative additive) monoid. We denote by $U(M)$ the subgroup of M consisting of all elements $a \in M$ with an additive inverse $-a$ in M, and we call M *reduced* if $U(M) = \{0\}$, that is, if $a + b = 0$ implies $a = b = 0$ for every $a, b \in M$. For a monoid M, we denote by M_{red} the quotient monoid $M/U(M)$, whose elements are all cosets $x + U(M)$ with $x \in M$. The monoid M_{red} is always a reduced monoid.

A *translation-invariant pre-order* on a monoid M is a reflexive and transitive relation $\leq$ on the set M such that $x, y, z \in M$ and $x \leq y$ imply $x + z \leq y + z$. There is a natural translation-invariant pre-order on any commutative additive monoid M, defined by $x \leq y$ if there exists $z \in M$ such that $x + z = y$. It is called the *algebraic pre-order* on M. A submonoid M' of a commutative additive monoid M is said to be *divisor-closed* if $x \in M$, $y \in M'$ and $x \leq y$ in M implies $x \in M'$. A *prime ideal* of a commutative monoid M is a proper subset P of M such that $M \setminus P$ is a divisor-closed submonoid; that is, for any $x, y \in M$ one has $x + y \in P$ if and only if either $x \in P$ or $y \in P$. The set $\operatorname{Spec}(M)$ of all prime ideals of M, partially ordered by set inclusion, is a complete lattice whose greatest element is the prime ideal $M \setminus U(M)$ and whose least element is the empty ideal $\emptyset$. For a prime ideal P of M, the *localization* M_P of M at P is the monoid whose elements are all formal differences $x - s$ with $x \in M$ and $s \in M \setminus P$, and in which, for all $x, x' \in M$ and $s, s' \in M \setminus P$, $x - s = x' - s'$ if and only if there exists $t \in M \setminus P$ such that $x + s' + t = x' + s + t$ [10, §4]. The monoid $(M_P)_{\mathrm{red}} = M_P/U(M_P)$ is called the *reduced localization* of M at P. If $x, x' \in M$ and $s, s' \in M \setminus P$, then $x - s + U(M_P) = x' - s' + U(M_P)$ in $(M_P)_{\mathrm{red}}$ if and only if there exist elements $t, t' \in M \setminus P$ such that $x + t = x' + t'$. Notice that the canonical homomorphism $\varphi \colon M \to (M_P)_{\mathrm{red}}$, defined by $x \mapsto x - 0 + U(M_P)$, is surjective.

The localization $M_\emptyset$ of M at the empty prime ideal $\emptyset$ is an abelian group, usually called the *Grothendieck group* of M, or the *group of differences* of M, and denoted by $G(M)$. There is a canonical monoid homomorphism $\psi_M\colon M \to G(M)$, and $G(-)$ turns out to be a functor of the category of commutative monoids into the category of abelian groups. The Grothendieck group $G(M)$ of M is a pre-ordered group with positive cone $G(M)^+ := \operatorname{Im} \psi_M$.

Lemma 2.1. *Let $\mathcal{A}$ be an additive category in which idempotents split. There is an order-reversing injective mapping*

$$\varphi\colon \operatorname{Spec}(V(\mathcal{A})) \to \mathcal{L}(\mathcal{A})$$

of the lattice $\operatorname{Spec}(V(\mathcal{A}))$ of all prime ideals of the monoid $V(\mathcal{A})$ into the class $\mathcal{L}(\mathcal{A})$ of all ideals of the category $\mathcal{A}$. It associates to every prime ideal $P \in \operatorname{Spec}(V(\mathcal{A}))$ the ideal $\mathcal{I}_P \in \mathcal{L}(\mathcal{A})$ of all the morphisms of $\mathcal{A}$ that factor through an object C of $\mathcal{A}$ with $\langle C \rangle \notin P$. That is, $\mathcal{I}_P(A,B) = \{\, f \in \mathcal{A}(A,B) \mid \text{there exist } C \in \operatorname{Ob} \mathcal{A},\ g \in \mathcal{A}(A,C)$ and $h \in \mathcal{A}(C,B)$ with $\langle C \rangle \notin P$ and $f = hg \,\}$ for every $A, B \in \operatorname{Ob}(\mathcal{A})$.

Proof. It is easily checked that $\mathcal{I}_P$ is an ideal of $\mathcal{A}$ for every $P \in \operatorname{Spec}(V(\mathcal{A}))$ and that φ is order-reversing. In order to prove that φ is injective, assume that $P, P' \in \operatorname{Spec}(V(\mathcal{A}))$ and $P \neq P'$. Without loss of generality we can suppose that there exists $A \in \operatorname{Ob}(\mathcal{A})$ with $\langle A \rangle \in P$ and $\langle A \rangle \notin P'$. Then $1_A \in \mathcal{I}_{P'}(A,A)$. To show that $\mathcal{I}_P \neq \mathcal{I}_{P'}$ it suffices to show that $1_A \notin \mathcal{I}_P(A,A)$. Assume the contrary, so that there exist $C \in \operatorname{Ob} \mathcal{A}$, $g \in \mathcal{A}(A,C)$ and $h \in \mathcal{A}(C,A)$ with $\langle C \rangle \notin P$ and $1_A = hg$. Then A is isomorphic to a direct summand of C [5, Lemma 2.1]. Thus $\langle A \rangle \leq \langle C \rangle$ in the monoid $V(\mathcal{A})$. But P is prime, $\langle A \rangle \leq \langle C \rangle$ and $\langle C \rangle \notin P$ imply $\langle A \rangle \notin P$, contradiction. □

Recall that if $\mathcal{I}$ is an ideal of $\mathcal{A}$, the factor category $\mathcal{A}/\mathcal{I}$ has the same objects of $\mathcal{A}$ and, for $A, B \in \operatorname{Ob}(\mathcal{A}) = \operatorname{Ob}(\mathcal{A}/\mathcal{I})$, the Hom-sets are $(\mathcal{A}/\mathcal{I})(A,B) := \mathcal{A}(A,B)/\mathcal{I}(A,B)$. We shall denote an object A of $\mathcal{A}$ as $\overline{A}$ when we view it as an object of $\mathcal{A}/\mathcal{I}$. Thus $(\mathcal{A}/\mathcal{I})(\overline{A},\overline{B}) = \mathcal{A}(A,B)/\mathcal{I}(A,B)$ and $1_{\overline{A}} = 1_A + \mathcal{I}(A,A)$. Observe that $\overline{A}$ and $\overline{0}$ are different objects of $\mathcal{A}/\mathcal{I}$ provided A and 0 are not the same object of $\mathcal{A}$. The expression $\overline{A} = \overline{0}$ means that $\overline{A}$ is a zero object in $\mathcal{A}/\mathcal{I}$.

Lemma 2.2. *Let $\mathcal{A}$ be an additive category in which idempotents split, let P be a prime ideal in $V(\mathcal{A})$ and let A be an object of $\mathcal{A}$. Then $\overline{A} = \overline{0}$ in $\mathcal{A}/\mathcal{I}_P$ if and only if $\langle A \rangle \notin P$.*

Proof. If $\overline{A} = \overline{0}$ in $\mathcal{A}/\mathcal{I}_P$, then $1_A + \mathcal{I}_P(A,A) = 0_A + \mathcal{I}_P(A,A)$, so that $1_A \in \mathcal{I}_P(A,A)$. Now the proof continues like for the previous lemma. There exist $C \in \operatorname{Ob} \mathcal{A}$, $g \in \mathcal{A}(A,C)$ and $h \in \mathcal{A}(C,A)$ with $\langle C \rangle \notin P$ and $1_A = hg$. Then A is isomorphic to a direct summand of C [5, Lemma 2.1]. Thus $\langle A \rangle \leq \langle C \rangle$ in the monoid $V(\mathcal{A})$. But P prime, $\langle A \rangle \leq \langle C \rangle$ and $\langle C \rangle \notin P$ imply $\langle A \rangle \notin P$.

Conversely, if $\langle A \rangle \notin P$, the identity of A factors trivially through an object not in P, so that $1_A \in \mathcal{I}_P(A,A)$. It follows that the unique morphism $A \to 0$ and the unique morphism $0 \to A$ become one the inverse of the other in the quotient category $\mathcal{A}/\mathcal{I}_P$. □

Remark 2.3. The notion of prime ideal P in the monoid $V(\mathcal{A})$ and the corresponding ideal $\mathcal{I}_P$ of the category $\mathcal{A}$ turn out to be very natural. Given any additive functor

$F\colon \mathcal{A} \to \mathcal{B}$ between two skeletally small additive categories $\mathcal{A}$ and $\mathcal{B}$, the set of all $\langle A \rangle \in V(\mathcal{A})$ with $F(A) \neq 0$ is always a prime ideal of $V(\mathcal{A})$, and the ideal $\mathcal{I}_P$ is always contained in the ideal $\mathcal{K}erF$ of $\mathcal{A}$, defined by $\mathcal{K}erF(A,B) = \{\, f \in \mathcal{A}(A,B) \mid F(f) = 0 \,\}$ for every $A, B \in \mathrm{Ob}\,\mathcal{A}$.

There is a one-to-one correspondence between the set of all submonoids of an abelian group G and the set of all translation-invariant pre-orders $\leq$ on G [9]. This one-to-one correspondence associates to every translation-invariant pre-order $\leq$ on G the *positive cone* $G^+ = \{\, a \in G \mid 0 \leq a \,\}$. A *pre-ordered (abelian) group* $(G, +, \leq)$ is an (abelian) group $(G, +)$ with a translation-invariant pre-order $\leq$ on G. A subgroup H of a pre-ordered group G is a pre-ordered group defining $H^+ := H \cap G^+$. A *convex subset* of a pre-ordered group G is any subset H with the property that whenever $x, z \in H$ and $y \in G$ with $x \leq y \leq z$, then $y \in H$. A *convex subgroup* of a pre-ordered group G is any subgroup H of G which is also a convex subset of G. Clearly a subgroup H of G is convex if and only if whenever $0 \leq a \leq b$ with $b \in H$ and $a \in G$, then $a \in H$; see [9, p. 8]. Recall that a subgroup H of a pre-ordered abelian group G is *directed* in case $H = H^+ - H^+$. Denote by $\mathcal{L}(G)$ the set of all directed convex subgroups of the pre-ordered group G. We say that a commutative monoid M is *directly finite* if for every $x, y \in M$, $x + y = y$ implies $x = 0$. If M is a commutative monoid, $G(M)$ denotes its Grothendieck group, $\mathrm{Spec}(M)$ is the set of prime ideals of M and $\mathrm{Spec}'(M)$ denote the set of all $P \in \mathrm{Spec}(M)$ with $(M_P)_\mathrm{red}$ directly finite, then there is an order reversing bijection $f\colon \mathrm{Spec}'(M) \to \mathcal{L}(G(M))$ defined by $f(P) := \psi_M(M \setminus P) - \psi_M(M \setminus P)$ for every $P \in \mathrm{Spec}'(M)$ [1, Proposition 6.2].

Let M, N be monoids. We say that a monoid homomorphism $\varphi\colon M \to N$ is *essential* if, for every $x, y \in M$ with $\varphi(x) \leq \varphi(y)$, there exist $z, t \in M$ with $\varphi(z) = 0$ and $x + t = y + z$ [5]. In particular, this definition will be applied to valuations of M, that is, monoid homomorphisms $M \to \mathbb{N}_0$.

Proposition 2.4. *Let $\mathcal{A}$ be an additive category in which idempotents split, and let $P \in \mathrm{Spec}'(V(\mathcal{A}))$.*

(a) *If A and B are objects of $\mathcal{A}$, then $\overline{A} \cong \overline{B}$ in $\mathcal{A}/\mathcal{I}_P$ if and only if there exist $C, D \in \mathrm{Ob}(\mathcal{A})$ with $\langle C \rangle, \langle D \rangle \notin P$ and $A \oplus C \cong B \oplus D$ in $\mathcal{A}$.*

(b) $V(\mathcal{A}/\mathcal{I}_P) \cong (V(\mathcal{A})_P)_\mathrm{red}$.

Proof. (a) From $\overline{A} \cong \overline{B}$, it follows that there exist morphisms $f\colon A \to B$ and $g\colon B \to A$ with $1_A - gf \in \mathcal{I}_P(A, A)$. Thus there exist $D \in \mathrm{Ob}(\mathcal{A})$, $h\colon A \to D$ and $\ell\colon D \to A$ with $1_A - gf = \ell h$ and $\langle D \rangle \notin P$. The composite mapping of

$$\begin{pmatrix} f \\ h \end{pmatrix} \colon A \to B \oplus D \quad \text{and} \quad \begin{pmatrix} g & \ell \end{pmatrix} \colon B \oplus D \to A$$

is the identity 1_A of A. By [5, Lemma 2.1], A is isomorphic to a direct summand of $B \oplus D$. Thus $A \oplus C \cong B \oplus D$ for some object $C \in \mathrm{Ob}(\mathcal{A})$. Hence $\langle A \rangle + \langle C \rangle = \langle B \rangle + \langle D \rangle$ in the monoid $V(\mathcal{A})$, so that $\langle A \rangle + \langle C \rangle = \langle B \rangle$ in the monoid $(V(\mathcal{A})_P)_\mathrm{red}$. Interchanging the roles of A and B one finds that $\langle B \rangle + \langle C' \rangle = \langle A \rangle$ in the monoid $(V(\mathcal{A})_P)_\mathrm{red}$ for a suitable $C' \in \mathrm{Ob}(\mathcal{A})$. As $(V(\mathcal{A})_P)_{\mathrm{red}}$ is directly finite, we get that $\langle C \rangle + \langle C' \rangle = 0$ in $(V(\mathcal{A})_P)_{\mathrm{red}}$. But $(V(\mathcal{A})_P)_{\mathrm{red}}$ is reduced, so $\langle C \rangle = 0$ in $(V(\mathcal{A})_P)_{\mathrm{red}}$. Therefore there exist $\langle E \rangle, \langle E' \rangle \in V(\mathcal{A}) \setminus P$ with $\langle C \rangle + \langle E \rangle = \langle E' \rangle$. As P is a prime ideal, we get that $\langle C \rangle \notin P$, as desired.

For the converse it suffices to observe that if $C, D \in \mathrm{Ob}(\mathcal{A})$, $A \oplus C \cong B \oplus D$ and $\langle C \rangle, \langle D \rangle \notin P$, then $\overline{C} = \overline{D} = \overline{0}$ in $\mathcal{A}/\mathcal{I}_P$ by Lemma 2.2.

(b) Since $P \in \mathrm{Spec}'(V(\mathcal{A}))$, the canonical projection $V(\mathcal{A}) \to (V(\mathcal{A})_P)_{\mathrm{red}}$ is an essential monoid homomorphism and $(V(\mathcal{A})_P)_{\mathrm{red}}$ is directly finite and reduced [5, Proposition 4.7]. Statement (b) now follows immediately from [5, Theorem 4.5]. □

3 Krull–Schmidt objects and valuations

Let $\mathcal{A}$ be an additive category. We say that an object A of $\mathcal{A}$ is *indecomposable* if it is non-zero and $A \cong B \oplus C$ in $\mathcal{A}$ implies that either $B = 0$ or $C = 0$. We say that A *cancels from direct sums* if $A \oplus B \cong A \oplus C$ in $\mathcal{A}$ implies $B \cong C$. The additive category $\mathcal{A}$ is a *Krull–Schmidt category* if every object of $\mathcal{A}$ is a direct sum of finitely many indecomposable objects in an essentially unique way, that is, in a unique way up to isomorphism and up to a permutation of the direct summands. By a *representation* of an additive category $\mathcal{A}$ into a Krull–Schmidt category $\mathcal{C}$ we mean an additive functor $F \colon \mathcal{A} \to \mathcal{C}$.

We say that an object A of an additive category $\mathcal{A}$ is a *Krull–Schmidt object* if it is an indecomposable object, cancels from direct sums, and, for every $B, C \in \mathrm{Ob}(\mathcal{A})$ with A isomorphic to a direct summand of $B \oplus C$, one has that either A is isomorphic to a direct summand of B or A isomorphic to a direct summand of C. (This third conditions could be stated as "$\langle A \rangle + V(\mathcal{A})$ is a prime ideal of the monoid $V(\mathcal{A})$", or "$\langle A \rangle$ is a prime element in the monoid $V(\mathcal{A})$". Thus we could say that A is a Krull–Schmidt object of $\mathcal{A}$ if and only if the element $\langle A \rangle$ of the monoid $V(\mathcal{A})$ is an atom, is a cancellative element and is a prime element.) If $\mathcal{A}$ is an additive category in which idempotents split, every object with a local endomorphism ring is a Krull–Schmidt object.

If $\mathcal{A}$ is an additive category in which every object of $\mathcal{A}$ is a direct sum of finitely many indecomposables, then $\mathcal{A}$ is a Krull–Schmidt category if and only if every indecomposable object in $\mathcal{A}$ is a Krull–Schmidt object.

Lemma 3.1. *Let $\mathcal{A}$ be an additive category in which every object is a direct sum of finitely many indecomposables and let A be an object of $\mathcal{A}$. Then A is a Krull–Schmidt object if and only if the cyclic submonoid of $V(\mathcal{A})$ generated by $\langle A \rangle$ is a direct summand of $V(\mathcal{A})$ isomorphic to $\mathbb{N}_0$.*

The proof is trivial. If A is a Krull–Schmidt object, a direct complement of the cyclic submonoid $\mathbb{N}_0 \langle A \rangle$ of $V(\mathcal{A})$ generated by $\langle A \rangle$ is the divisor-closed submonoid $V(\mathcal{A}) \setminus (\langle A \rangle + V(\mathcal{A}))$, which is the set of all elements of $V(\mathcal{A})$ that are sums of finitely many indecomposable elements different from $\langle A \rangle$. For elementary facts concerning direct-sum decompositions of monoids, see [6, Proposition 6.1].

Recall that a *valuation* of an additive category $\mathcal{A}$ is an onto mapping $v\colon \mathrm{Ob}(\mathcal{A}) \to \mathbb{N}_0$ of the class $\mathrm{Ob}(\mathcal{A})$ of objects of $\mathcal{A}$ onto the additive monoid $\mathbb{N}_0$ such that: (1) if A and B are isomorphic objects of $\mathcal{A}$, then $v(A) = v(B)$; and (2) $v(A \oplus B) = v(A) + v(B)$ for every pair A, B of objects of $\mathcal{A}$. For instance, let $\mathcal{A}$ be an additive category in which every object is a direct sum of finitely many indecomposables and let A be a Krull–Schmidt object of $\mathcal{A}$. Then $V(\mathcal{A}) = \mathbb{N}_0 \langle A \rangle \oplus V(\mathcal{A}) \setminus ((\langle A \rangle + V(\mathcal{A}))$, so that the canonical projection $V(\mathcal{A}) \to \mathbb{N}_0 \langle A \rangle$ composed with the unique isomorphism $\mathbb{N}_0 \langle A \rangle \to \mathbb{N}_0$ is a valuation $v_A \colon \mathrm{Ob}(\mathcal{A}) \to \mathbb{N}_0$.

If $\mathcal{A}$ is an additive category and $v\colon \mathrm{Ob}(\mathcal{A}) \to \mathbb{N}_0$ is a valuation of $\mathcal{A}$, then $P_v := \{\, \langle A \rangle \in V(\mathcal{A}) \mid v(A) > 0 \,\}$ is a prime ideal of $V(\mathcal{A})$, so that, according to Lemma 2.1, there is a corresponding ideal $\mathcal{I}_v$ in $\mathcal{A}$. It consists of all morphisms of $\mathcal{A}$ that factor through some object B of $\mathcal{A}$ with $v(B) = 0$. It is therefore possible to construct the category $\mathcal{A}/\mathcal{I}_v$.

Lemma 3.2. *Let $\mathcal{A}$ be an additive category in which idempotents split and $v\colon \mathrm{Ob}(\mathcal{A}) \to \mathbb{N}_0$ a valuation of $\mathcal{A}$. If A is an object of $\mathcal{A}$ and $v(A) = 1$, then $\overline{A}$ is indecomposable in $\mathcal{A}/\mathcal{I}_v$.*

Proof. Let A be an object of $\mathcal{A}$ with $v(A) = 1$ and $\overline{A} \cong \overline{B} \oplus \overline{C}$ for objects B, C of $\mathcal{A}$. By Proposition 2.4(a), there exist $D, E \in \mathcal{A}$ with $v(D) = v(E) = 0$ such that $A \oplus D \cong B \oplus C \oplus E$ in $\mathcal{A}$. As $v(A \oplus D) = 1$, it follows that either $v(B) = 0$ or $v(C) = 0$. By Lemma 2.2, either $\overline{B} = \overline{0}$ or $\overline{C} = \overline{0}$. □

Now let $\mathcal{A}$ be an additive category in which idempotents split and in which every object is a direct sum of finitely many indecomposables. It is easily checked that any valuation $v\colon \mathrm{Ob}(\mathcal{A}) \to \mathbb{N}_0$ of $\mathcal{A}$ induces a monoid isomorphism $V(\mathcal{A})_{P_v} \to \mathbb{N}_0$, so that, a fortiori, $(V(\mathcal{A})_{P_v})_{\mathrm{red}} \cong \mathbb{N}_0$. Thus $P_v \in \mathrm{Spec}'(V(\mathcal{A}))$, from which $V(\mathcal{A}/\mathcal{I}_v) \cong \mathbb{N}_0$ (Proposition 2.4(b)).

We say that a category $\mathcal{B}$ is IBN if it is additive and the monoid $V(\mathcal{B})$ is isomorphic to the monoid $\mathbb{N}_0$. A category $\mathcal{B}$ is IBN if and only if there exists an IBN ring R with $\mathcal{B}$ equivalent to the full subcategory $\mathcal{F}_R$ of Mod-R whose objects are all finitely generated free right R-modules [4].

Combining the above remarks, we obtain:

Proposition 3.3. *Let $\mathcal{A}$ be an additive category in which idempotents split and in which every object is a direct sum of finitely many indecomposables. Let A be a Krull–Schmidt object of $\mathcal{A}$ and let $\mathcal{I}_A$ be the ideal of $\mathcal{A}$ whose morphisms are the morphisms of $\mathcal{A}$ that factor through objects of $\mathcal{A}$ that are direct sums of indecomposable objects not isomorphic to A. Then $\mathcal{A}/\mathcal{I}_A$ is an IBN category.*

Proof. Since A is a Krull–Schmidt object, there is an associated valuation $v_A \colon V(\mathcal{A}) \to \mathbb{N}_0$, for which $v_A(B)$ is the number of direct summands isomorphic to A in an indecomposable direct-sum decomposition of B for every object B of $\mathcal{A}$. It follows that $P_A := \{\, \langle B \rangle \in V(\mathcal{A}) \mid v(B) > 0 \,\}$ is a prime ideal of $V(\mathcal{A})$ (notice that an object isomorphic to A appears in an indecomposable direct-sum decomposition of B if and only if an object isomorphic to A appears in *every* indecomposable direct-sum decomposition of B, because A is a Krull–Schmidt object). The ideal $\mathcal{I}_A$ of $\mathcal{A}$ is the ideal corresponding to the prime ideal P_A of $V(\mathcal{A})$ according to Lemma 2.1. □

4 Three full subcategories of Mod-R

We shall denote by Mod-R both the category of all right modules over a fixed ring R and the class of objects of this category. Recall that a module is *uniform* if it has Goldie dimension 1, and is *uniserial* if its lattice of submodules is linearly ordered. Consider the following subclasses of Mod-R:

fGd = { right R-modules of finite Goldie dimension },
SUfm = { direct sums of finitely many uniform right R-modules },
SUsr = { direct sums of finitely many uniserial right R-modules }.

Notice that fGd $\supseteq$ SUfm $\supseteq$ SUsr. We shall view these three subclasses of Mod-R as full subcategories of Mod-R, and we shall apply the results of the previous sections to these three full subcategories of Mod-R. Notice that in these three categories all objects are direct sums of finitely many indecomposables and that these three categories are skeletally small, because if $\mathcal{M}$ is a set of representatives of the cyclic modules up to isomorphism, then every object of fGd is isomorphic to a submodule of an injective envelope of the module $\oplus_{M \in \mathcal{M}} M^{(\aleph_0)}$.

Notice that the class SUfm is closed under finite direct sums, but not under direct summands. For instance, let R be a commutative integral domain with an indecomposable projective module P of torsion-free rank 2, e.g., the ring $R = \mathbb{R}[x, y, z]/(x^2 + y^2 + z^2 - 1)$ [11, Example 2.10], for which $R^3 \cong R \oplus P$. Then P is a direct summand of a free module, but $P \notin$ SUfm. It follows that SUfm, viewed as a full subcategory of Mod-R, is an additive category in which not all idempotents split. Therefore we shall sometimes consider the category add(SUfm) of all R-modules that are isomorphic to direct summands of modules in SUfm.

The class SUsr is closed for direct summands [13], so that idempotents split in the category SUsr. Trivially, idempotents split in the category fGd as well.

In the next proposition we show that the definition of Krull–Schmidt object we have introduced in Section 3 is compatible with the terminology of [2, Definition 1.10]. For every right R-module A, we denote by $[A]_m$ the *monogeny class* of A, that is, the class of all right R-modules B for which there exist a monomorphism $A \to B$ and a monomorphism $B \to A$. Dually, the *epigeny class* $[A]_e$ of A is the class of all right R-modules B for which there exist an epimorphism $A \to B$ and an epimorphism $B \to A$.

Proposition 4.1. *The Krull–Schmidt objects in the category* SUsr *are exactly the non-zero uniserial modules A such that either*

(a) *for every uniserial module B, $[A]_m = [B]_m$ implies $A \cong B$, or*

(b) *for every uniserial module B, $[A]_e = [B]_e$ implies $A \cong B$.*

Proof. Let A be a Krull–Schmidt object in the category SUsr. Since A is indecomposable, A must be a non-zero uniserial module. If (a) and (b) do not hold, then there exist uniserial modules B and C with $[A]_m = [B]_m$, $A \not\cong B$, $[A]_e = [C]_e$ and $A \not\cong C$.

Then the module A is isomorphic to a direct summand of $B \oplus C$. Cf. [2, proof of Corollary 1.11].

For the converse, see [2, Corollaries 1.3 and 1.11]. □

Now let U be a uniform right R-module and let A be an arbitrary right R-module. Set m-dim$_U(A) = \sup\{\, i \in \mathbb{N}_0 \mid$ there exist morphisms $f \colon U^i \to A$ and $g \colon A \to U^i$ with gf a monomorphism $\}$. Notice that if gf is a monomorphism, then f is necessarily a monomorphism, so that m-dim$_U(A)$ is less or equal to the Goldie dimension $\dim(A)$ of A. If U, V are uniform modules with $[U]_m = [V]_m$, then m-dim$_U(A) =$ m-dim$_V(A)$ for every module A.

Let $\mathcal{A}$ be any of the three categories fGd, add(SUfm) or SUsr. Fix a uniform module U in $\mathrm{Ob}(\mathcal{A})$. Then m-dim$_U \colon \mathrm{Ob}(\mathcal{A}) \to \mathbb{N}_0$ is a valuation of the category $\mathcal{A}$ [8, Theorem 2.5]. If U, V are uniform modules, then m-dim$_U(V) = 1$ if $[U]_m = [V]_m$ and m-dim$_U(V) = 0$ otherwise. Therefore the definition of m-dim$_U$ we have given in the previous paragraph is compatible with the definition we gave in the Introduction for SUsr.

We shall see in Lemma 5.3 that the valuation m-dim$_U$ is not, in general, an essential homomorphism in the sense defined in Section 2, so that we cannot apply [5, Theorem 4.5] directly.

Consider the ideal $\mathcal{M}_U$ in the category $\mathcal{A}$ corresponding to the valuation m-dim$_U$. The ideal $\mathcal{M}_U$ consists of all morphisms in $\mathcal{A}$ that can be factored through objects of $\mathcal{A}$ with m-dim$_U$ value equal to zero. More precisely, let $\mathcal{Z}_U$ be the class of all modules $C \in \mathrm{Ob}(\mathcal{A})$ with m-dim$_U(C) = 0$. Then, for all A, B in $\mathcal{A}$, set $\mathcal{M}_U(A, B) = \{\, f \mid$ there exist a module $C \in \mathcal{Z}_U$ and R-homomorphisms $g \colon A \to C$ and $h \colon C \to B$ with $f = hg\,\}$. Construct the quotient category $\mathcal{A}/\mathcal{M}_U$. Notice that the set $P_U = \{\, \langle A \rangle \mid A \in \mathcal{Z}_U\,\}$ is a prime ideal of the additive monoid $V(\mathcal{A})$. By Lemma 2.4, if A, B are objects of $\mathcal{A}$, then $\overline{A} \cong \overline{B}$ in $\mathcal{A}/\mathcal{M}_U$ if and only if there exist $C, D \in \mathcal{Z}_U$ with $A \oplus C \cong B \oplus D$ in $\mathcal{A}$. Notice that if A, B are objects of $\mathcal{A}$ and $\overline{A} \cong \overline{B}$ in $\mathcal{A}/\mathcal{M}_U$, then m-dim$_U(A) =$ m-dim$_U(B)$. By Lemma 2.2, if A is an object of $\mathcal{A}$, $\overline{A} = \overline{0}$ in $\mathcal{A}/\mathcal{M}_U$ if and only if m-dim$_U(A) = 0$.

Proposition 4.2. *Let $\mathcal{A}$ be any of the three categories* fGd, add(SUfm) *or* SUsr, *and let U and V be uniform right R-modules in* $\mathrm{Ob}(\mathcal{A})$. *Then:*

(a) *$\overline{V}$ is indecomposable in $\mathcal{A}/\mathcal{M}_U$ if $[V]_m = [U]_m$;*

(b) *$\overline{V} = \overline{0}$ in $\mathcal{A}/\mathcal{M}_U$ if $[V]_m \neq [U]_m$.*

Proof. If $[V]_m = [U]_m$, then m-dim$_U(V) = 1$, so that $\overline{V}$ is indecomposable by Lemma 3.2. If $[V]_m \neq [U]_m$, then m-dim$_U(V) = 0$, so that $\overline{V} = \overline{0}$ by Lemma 2.2. □

The dual results also hold. Namely, we can consider the full subcategories fdGd $\supseteq$ add(SCufm) $\supseteq$ SCufm $\supseteq$ SUsr of Mod-R, where the objects of fdGd are the modules of finite dual Goldie dimension and the objects of SCufm are the finite direct sums of couniform modules. For any fixed couniform module U, the restriction of the mapping e-dim$_U$ is a valuation of the categories fdGd, add(SCufm), SCufm and SUsr.

Let $\mathcal{A}$ be any of the three categories fdGd, add(SCufm) or SUsr, and let U be a couniform module in $\mathcal{A}$. Define e-dim$_U(A) = \sup\{\, i \in \mathbb{N}_0 \mid$ there exist morphisms

$f\colon U^i \to A$ and $g\colon A \to U^i$ with gf an epimorphism $\}$. Consider the ideal $\mathcal{E}_U$ in the category $\mathcal{A}$ corresponding to the valuation e-dim$_U$, that is, let the ideal $\mathcal{E}_U$ consist of all morphisms in $\mathcal{A}$ that can be factored through objects of $\mathcal{A}$ with e-dim$_U$ value equal to zero. Construct the quotient category $\mathcal{A}/\mathcal{E}_U$. By Lemma 2.4, if A and B are objects of $\mathcal{A}$, then $\overline{A} \cong \overline{B}$ in $\mathcal{A}/\mathcal{E}_U$ if and only if there exist $C, D \in \mathcal{A}$ with $A \oplus C \cong B \oplus D$ in $\mathcal{A}$ and e-dim$_U(C)$ = e-dim$_U(D) = 0$.

Let U and V be couniform right R-modules in Ob($\mathcal{A}$). Then $\overline{V}$ is indecomposable in $\mathcal{A}/\mathcal{E}_U$ if $[V]_e = [U]_e$, and $\overline{V} = \overline{0}$ if $[V]_e \neq [U]_e$.

Remark 4.3. The categories fdGd, add(SCufm) (and SUsr) are skeletally small. To see it, set $\xi = \max\{\aleph_0, |E|\}$, where $|E|$ denotes the cardinality of the minimal injective cogenerator in Mod-R. It suffices to prove that every right R-module M of finite dual Goldie dimension n is a homomorphic image of the free module $R_R^{(\xi)}$ of rank ξ. If a module M has finite dual Goldie dimension n, then there exist submodules $N_1, \ldots, N_n$ of M such that $N = N_1 \cap \cdots \cap N_n$ is superfluous in M and $M/N \cong \oplus_{i=1}^n M/N_i$ is a direct sum of n couniform modules [3, p. 57]. For each $i = 1, \ldots, n$, fix an element $x_i \in M \setminus N_i$ and a submodule $N'_i \supseteq N_i$ of M maximal with respect to $x_i \notin N'_i$, so that M/N'_i is a subdirectly irreducible module. Set $N' = N'_1 \cap \cdots \cap N'_n$. By [3, Prop. 2.42], the modules M/N'_i are couniform, the module $M/N' \cong \oplus_{i=1}^n M/N'_i$ has dual Goldie dimension n, and N' is superfluous in M. As far as cardinalities are concerned, the cardinality of each M/N'_i is $\leq \xi$, so that the same is true for M/N'. Thus there is an epimorphism $R_R^{(\xi)} \to M/N'$, which can be lifted to a homomorphism $f\colon R_R^{(\xi)} \to M$ with $f(R_R^{(\xi)}) + N' = M$. But N' is superfluous in M, so that f is an epimorphism.

5 The graph of uniserial modules

We now recall the construction of a graph G that describes the behaviour of uniserial right R-modules for every fixed ring R. For further details of this construction, see [8, Section 7]. Let M, E and L, respectively, be the set of all monogeny classes, epigeny classes and isomorphism classes, respectively, of all non-zero uniserial right R-modules. The graph $G = (M \cup E, L)$ is bipartite because there are no edges between two vertices in M or between two vertices in E. For every edge $\langle U \rangle \in L$, $\langle U \rangle$ connects the vertices $[U]_m \in M$ and $[U]_e \in E$. By [2, Proposition 1.6], the graph G has no multiple edges.

Recall that a *full* subgraph of a graph G without multiple edges is a subgraph G' of G such that any two vertices of G' adjacent in G are adjacent in G' as well. We shall view the connected components of G as full subgraphs of G. We say that two non-zero uniserial right R-modules U and V *are in the same connected component*, and write $U \sim V$, if the two edges $\langle U \rangle$ and $\langle V \rangle$ belong to the same connected component of G.

Recall that a graph is called a *complete bipartite graph* if there is a partition $X \cup Y$ of its set of vertices for which $X \neq \emptyset$, $Y \neq \emptyset$, and there are no edges between any two vertices in X, no edges between any two vertices in Y, and exactly one edge between any vertex in X and any vertex in Y. For any ring R, the connected components of the graph G are complete bipartite graphs [8, Proposition 7.1].

The *degree* $d(v)$ of a vertex v of a graph is the cardinality of the set of edges incident to v. In particular, every vertex $[U]_m$ (or $[U]_e$) of our graph G has a non-zero degree, possibly an infinite cardinal. By Proposition 4.1, a non-zero uniserial module U is a Krull–Schmidt object in the category SUsr if and only if either $d([U]_m) = 1$ or $d([U]_e) = 1$, that is, if $\langle U \rangle$ is a leaf edge of the graph G.

The degree of a vertex v does not change restricting our attention to the connected component of v. Moreover, in a complete bipartite graph in which the partition of the set of vertices is $X \cup Y$, all vertices in X have the same degree, and all vertices in Y have the same degree. Since the connected components of the graph G are complete bipartite graphs, we get that:

Proposition 5.1. *Let U and V be two non-zero uniserial modules in the same connected component. Then $d([U]_m) = d([V]_m)$ and $d([U]_e) = d([V]_e)$. In particular, U is a Krull–Schmidt module if and only if V is a Krull–Schmidt module.*

Thus a non-zero uniserial module is a Krull–Schmidt module if and only if its connected component in the graph G is a star, that is, a connected graph in which all vertices except at most one have degree one. By [3, Theorem 9.1], if U is a non-zero uniserial module, the set of all non-injective endomorphisms of U and the set of all non-surjective endomorphisms of U are two ideals in the ring $S = \text{End}_R(U)$. If these two ideals are comparable by inclusion, then S is a local ring with Jacobson radical equal to biggest of these two ideals. Otherwise, these are the only two (left, right, two-sided) maximal ideals of S. Obviously, any uniserial module with a local endomorphism ring is a Krull–Schmidt module.

Lemma 5.2. *Let U, V be nonzero uniserial modules in the same connected component. Then U has a local endomorphism ring if and only if V has a local endomorphism ring.*

Proof. It suffices to show that if the endomorphism ring of U is not local, then the endomorphism ring of V is not local. Moreover, it is sufficient to consider the two cases $[U]_m = [V]_m$ and $[U]_e = [V]_e$.

Assume $[U]_m = [V]_m$, so that there are monomorphisms $f \colon U \to V$ and $g \colon V \to U$. As $\text{End}_R(U)$ contains a monomorphism φ that is not an epimorphism, it follows that $f\varphi g \in \text{End}_R(V)$ is a monomorphism that is not an epimorphism [3, Lemma 6.26]. We need check that $\text{End}_R(V)$ also contains an epimorphism that is not a monomorphism. We can suppose that $V \subseteq U$ and that g is the inclusion. Let $\psi \colon U \to U$ be an epimorphism that is not a monomorphism. Let $V' = \psi^{-1}(V)$ and let ψ' be the restriction of ψ to V'. Since ψ is an epimorphism, ψ' is an epimorphism of V' onto V. On the other hand, $V' \subseteq U$ and $f \colon U \to V$ is a monomorphism. Therefore there exists a monomorphism $\varepsilon \colon V' \to V$. Since ψ' is not a monomorphism, either ε or $\psi' + \varepsilon$ is an isomorphism by [3, Lemma 9.2]. In any case, there exists an isomorphism $\alpha \colon V' \to V$. Since ψ' is not a monomorphism, $\psi' \alpha^{-1}$ is an epimorphism that is not a monomorphism.

Now suppose $[U]_e = [V]_e$, so that there are epimorphisms $f \colon U \to V$ and $g \colon V \to U$. As $\text{End}_R(U)$ contains an epimorphism ψ that is not a monomorphism, it follows that $f\psi g \in \text{End}_R(V)$ is an epimorphism that is not a monomorphism. We need find a monomorphism in $\text{End}_R(V)$ that is not an epimorphism. We can assume that $f\colon U \to V = U/X$ is the canonical projection for some $X \subseteq U$. Let $\varphi\colon U \to U$ be a monomorphism that is not an epimorphism. Then $U/X \cong \varphi(U)/\varphi(X)$, so that there is a monomorphism $\nu\colon U/X \to U/\varphi(X)$ that is not an epimorphism. Clearly, $[U]_e = [V]_e$ implies that there is an epimorphism $U/X = V \to U$. The composite mapping of this epimorphism and the canonical projection $U \to U/\varphi(X)$ is an epimorphism $\pi'\colon U/X \to U/\varphi(X)$. Now ν is a monomorphism not an epimorphism and π' epimorphism, so that either π' or $\nu + \pi'$ is an isomorphism, α say, of U/X onto $U/\varphi(X)$. Then $\alpha^{-1}\nu$ is an injective endomorphism of $V = U/X$ which is not an epimorphism. □

As in Section 3, if U is a Krull–Schmidt uniserial module, we will denote by $v_U\colon V(\text{SUsr}) \to \mathbb{N}_0$ the corresponding valuation of SUsr.

Lemma 5.3. *Let U, V be non-zero uniserial R-modules.*

(a) *If U is a Krull–Schmidt uniserial module, the corresponding valuation $v_U\colon V(\text{SUsr}) \to \mathbb{N}_0$ is essential.*

(b) *m-dim$_U$ = e-dim$_V$ if and only if $U \cong V$ and $d([U]_m) = d([V]_e) = 1$; equivalently, if and only if $[U]_m$ and $[V]_e$ are in the same connected component, which consists just of one edge and two vertices.*

(c) *The valuation m-dim$_U\colon V(\text{SUsr}) \to \mathbb{N}_0$ is essential if and only if either $d([U]_m) = 1$ or $d([U]_e) \neq 1$.*

(d) *The valuation e-dim$_U\colon V(\text{SUsr}) \to \mathbb{N}_0$ is essential if and only if either $d([U]_e) = 1$ or $d([U]_m) \neq 1$.*

(e) *If $d([U]_m) = 1$, then m-dim$_U = v_U$. If $d([U]_e) = 1$, then e-dim$_U = v_U$.*

Proof. (a) If U is a Krull–Schmidt uniserial module, the cyclic submonoid of $V(\text{SUsr})$ generated by $\langle U \rangle$ is a direct summand of $V(\text{SUsr})$ isomorphic to $\mathbb{N}_0$ (Lemma 3.1), and a direct complement is the divisor-closed submonoid $V(\text{SUsr}) \setminus (\langle U \rangle + V(\text{SUsr}))$. The valuation $v_U\colon V(\text{SUsr}) \to \mathbb{N}_0$ is then the canonical projection, which is obviously an essential valuation.

(b) m-dim$_U$ = e-dim$_V$ if and only if m-dim$_U(A)$ = e-dim$_V(A)$ for every object A of SUsr, that is, if and only if m-dim$_U(W)$ = e-dim$_V(W)$ for every non-zero uniserial R-module W. Now m-dim$_U(W) = 1$ if and only if $[W]_m = [U]_m$, and similarly for e-dim$_V(W)$. It follows that m-dim$_U$ = e-dim$_V$ if and only if, for every non-zero uniserial R-module W, $[W]_m = [U]_m \Leftrightarrow [W]_e = [V]_e$, that is, if and only if an edge $\langle W \rangle$ is adjacent to $[U]_m$ if and only if it is adjacent to $[V]_e$.

(e) Assume $d([U]_m) = 1$, so that U is a Krull–Schmidt module. Then, for every non-zero uniserial module V, $[U]_m = [V]_m$ if and only if $U \cong V$. Equivalently, m-dim$_U(V) = 1$ if and only if $v_U(V) = 1$. It follows that m-dim$_U = v_U$. Similarly for $d([U]_e) = 1$.

(c) Assume $d([U]_m) \neq 1$ and $d([U]_e) = 1$. Then there exists a non-zero uniserial module W with $[W]_m = [U]_m$ and $[W]_e \neq [U]_e$. Thus m-dim$_U(U) \leq$ m-dim$_U(W)$, but for all $C, D \in \text{Ob}(\text{SUsr})$, $U \oplus C \cong W \oplus D$ implies m-dim$_U(D) \neq 0$, because from $U \oplus C \cong W \oplus D$ and that fact that U is a Krull–Schmidt module we get that either U is isomorphic to a direct summand of V or U is isomorphic to a direct summand of D. Now the first case can not take place because U is not isomorphic to W, and the second case implies m-dim$_U(D) \neq 0$. Thus the valuation m-dim$_U : V(\text{SUsr}) \to \mathbb{N}_0$ is not essential.

Assume $d([U]_m) = 1$. Then U is a Krull–Schmidt module and m-dim$_U = v_U$ by (e). We conclude that m-dim$_U$ is essential by (a).

Assume $d([U]_e) \neq 1$. Let A, B be objects of SUsr with m-dim$_U(A) \leq$ m-dim$_U(B)$. Then A and B have direct-sum decompositions into non-zero uniserials of the form $A = U_1 \oplus \cdots \oplus U_s \oplus V_1 \oplus \cdots \oplus V_t$ and $B = U_{s+1} \oplus \cdots \oplus U_{s+n} \oplus V_{t+1} \oplus \cdots \oplus V_{t-r}$ with $[U_i]_m = [U]_m$ for every $i = 1, \ldots, s+n$, $[V_j]_m \neq [U]_m$ for every $j = 1, \ldots, t+r$, and $s \leq n$. From $d([U]_e) \neq 1$, it follows that there is a non-zero uniserial module W in the connected component of U with $[U]_m \neq [W]_m$. Since the connected components are complete bipartite graphs, for every $i = 1, \ldots, s+n$ there exists a uniserial module W_i with $[W]_m = [W_i]_m$ and $[U_i]_e = [W_i]_e$. It follows that $U_i \oplus W_{s+i} \cong U_{s+i} \oplus W_i$ for every $i = 1, \ldots, s$. Then the direct-sum decompositions

$$A \oplus U_{2s+1} \oplus \cdots \oplus U_{s+n} \oplus W_{s+1} \oplus \cdots \oplus W_{2s} \oplus V_{t+1} \oplus \cdots \oplus V_{t+r}$$
$$\cong B \oplus W_1 \oplus \cdots \oplus W_s \oplus V_1 \oplus \cdots \oplus V_t$$

show that m-dim$_U$ is an essential valuation.

(d) is similar to (c). □

Let $G = (V \cup W, L)$ be a bipartite graph and let $F := \mathbb{N}_0^{(V)} \oplus \mathbb{N}_0^{(W)}$ be the free commutative monoid with free set of generators the disjoint union $V \cup W$. The elements of F are tuples of nonnegative integers, almost all zero, indexed in $V \cup W$. We will write the elements of F in the form $(a_v)_{v \in V} \cup (b_w)_{w \in W}$. Here the a_v's and the b_w's belong to $\mathbb{N}_0$ and are almost all zero. Let S_G be the submonoid of F whose elements are all $(a_v)_{v \in V} \cup (b_w)_{w \in W} \in F$ with $\sum_{v \in V_C} a_v = \sum_{w \in W_C} b_w$ for every connected component $C = (V_C \cup W_C, L_C)$ of G. The monoid S_G is generated by its elements $f_{\overline{v}, \overline{w}} := (\delta_{\overline{v},v})_{v \in V} \cup (\delta_{\overline{w},w})_{w \in W} \in F$, where δ is the Kronecker delta and $\{\overline{v}, \overline{w}\}$ ranges in the set L of all edges of G.

Theorem 5.4 ([8, Theorem 7.3]). *For the graph $G = (M \cup E, L)$, the monoids $V(\text{SUsr})$ and S_G are isomorphic via the isomorphism $V(\text{SUsr}) \to S_G$, defined by*

$$\langle A \rangle \mapsto (\text{m-dim}_U(A))_{[U]_m \in M} \cup (\text{e-dim}_U(A))_{[U]_e \in E}$$

for every module A in SUsr.

In the next proposition, as a corollary of Theorem 5.4, we determine all prime ideals of the monoid $V(\text{SUsr})$. Recall that if $v: V(\text{SUsr}) \to \mathbb{N}_0$ is an essential valuation, then $P = \{ \langle A \rangle \in V(\text{SUsr}) \mid v(A) > 0 \}$ is a minimal non-empty prime ideal of $V(\text{SUsr})$. Since it is easier to deal with divisor-closed submonoids, we prefer to determine all divisor-closed submonoids of $V(\text{SUsr})$, which are the complements of the prime ideals.

In the statement of the proposition, we consider the set $\mathcal{B}$ of all subgraphs of G whose connected components are complete bipartite graphs. We consider that in $\mathcal{B}$ there are both the improper subgraph G of G and the empty subgraph of G.

Proposition 5.5. *Let $G = (V \cup W, L)$ be a bipartite graph whose connected components are complete bipartite graphs. Let $\mathcal{B}$ be the set of all subgraphs of G whose connected components are complete bipartite graphs. Let $\mathcal{D}(S_G)$ be the set of all divisor-closed submonoids of the monoid S_G. Then there is a one-to-one order-preserving correspondence between $\mathcal{B}$ and $\mathcal{D}(S_G)$. It associates to every subgraph $G' = (V' \cup W', L')$ in $\mathcal{B}$ the divisor-closed submonoid $D_{G'}$ of S_G whose elements are all $(a_v)_{v \in V} \cup (b_w)_{w \in W} \in S_G$ with $a_v = 0$ for every $v \in V \setminus V'$, $b_w = 0$ for every $w \in W \setminus W'$, and $\sum_{v \in V_{C'}} a_v = \sum_{w \in W_{C'}} b_w$ for every connected component $C' = (V_{C'} \cup W_{C'}, L_{C'})$ of G'.*

Proof. It is easily checked that $D_{G'}$ is a divisor-closed submonoid of S_G isomorphic to $S_{G'}$ for every $G' \in \mathcal{B}$. In order to show that the correspondence $\mathcal{B} \to \mathcal{D}(S_G)$ is injective, assume $G', G'' \in \mathcal{B}$ and $G' \neq G''$. In a complete bipartite graph, all vertices have degree ≥ 1. Hence from $G' = (V' \cup W', L'), G'' = (V'' \cup W'', L'') \in \mathcal{B}$ and $G' \neq G''$, it follows that $L' \neq L''$. By symmetry, we can suppose that there is an edge $l' \in L', l' \notin L''$. The edge l' is adjacent to $\overline{v} \in V'$ and $\overline{w} \in W'$, say. Thus $\overline{v}$ and $\overline{w}$ are in the same connected component in G', but they are in different connected components in G''. Then $f_{\overline{v},\overline{w}}$ is in $S_{G'}$, but not in $S_{G''}$. This shows that the correspondence $\mathcal{B} \to \mathcal{D}(S_G)$ is injective.

Let us prove that the correspondence is surjective. Let D be a divisor-closed submonoid of S_G. As D is divisor-closed, it is generated, as a monoid, by the set of all $f_{v,w}$ with $v \in V$, $w \in W$ and $f_{v,w} \in D$. Let L' be the set of all edges $\{v, w\} \in L$ with $v \in V$, $w \in W$ and $f_{v,w} \in D$. Let V' be the set of all vertices $v \in V$ with v incident to some edge in L' and W' be the set of all vertices $w \in W$ with w incident to some edge in L'. Set $G' = (V' \cup W', L')$, so that G' is a subgraph of G. In order to show that the connected components of G' are complete bipartite graphs, we will prove that if two vertices $v \in V'$ and $w \in W'$ are in the same connected component of G', then $\{v, w\} \in L'$. Arguing by induction on the length of a path in G' from v to W, it suffices to show that if in G' there is a path of length three from v to w, then $\{v, w\} \in L'$. A path of length three from v to w is of the form $\{v, w_1\}, \{w_1, v_1\}, \{v_1, w\}$, with $\{v, w_1\}, \{w_1, v_1\}, \{v_1, w\}$ in L'. Thus $f_{v,w_1}, f_{w_1,v_1}, f_{v_1,w}$ belong to D. Then $f_{v,w}$ belongs to D, because $f_{v,w} \leq f_{v,w_1} + f_{w_1,v_1} + f_{v_1,w}$ and D is divisor-closed. Thus $\{v, w\} \in L'$. This shows that $G' \in \mathcal{B}$ and, clearly, $D_{G'} = D$. □

We can now determine a divisor theory for the Krull monoid $V(\text{SUsr})$ [7, Section 4]. Notice that, by Proposition 5.1, if U and V are two non-zero uniserial modules in the same monogeny class, then $d([U]_m) = d([V]_m)$ and $d([U]_e) = d([V]_e)$, so that U is Krull–Schmidt if and only if V is Krull–Schmidt. Therefore we can fix a set M' of representatives up to monogeny of the non-zero uniserial right R-modules U that are not Krull–Schmidt modules. Similarly, we can fix a set E' of representatives up to epigeny of the non-zero uniserial right R-modules V that are not Krull–Schmidt modules. Also, recall that for a uniserial Krull–Schmidt module W, we denote by v_W the valuation of the category SUsr associated to W.

Theorem 5.6. *Fix a set K of representatives up to isomorphism of the Krull–Schmidt uniserial right R-modules, a set M' of representatives up to monogeny of the non-zero uniserial right R-modules U that are not Krull–Schmidt modules, and a set E' of representatives up to epigeny of the non-zero uniserial right R-modules Z that are not Krull–Schmidt modules. Then the map $V(\text{SUsr}) \to \mathbb{N}_0^{(K)} \oplus \mathbb{N}_0^{(M')} \oplus \mathbb{N}_0^{(E')}$ defined by the v_W's ($W \in K$), the m-dim$_U$'s ($U \in M'$) and the e-dim$_V$'s ($V \in E'$) is a divisor theory.*

Proof. Apply [7, Proposition 4.3] to the divisor homomorphism $V(\text{SUsr}) \to \mathbb{N}_0^{(M)} \oplus \mathbb{N}_0^{(E)}$ [8, Corollary 7.5], that is, to the valuations m-dim$_U$, $U \in M$ and e-dim$_V$, $V \in E$. Take, among these valuations, one copy of the essential ones (Lemma 5.3). □

Let U be a non-zero uniserial module and let $F: \text{SUsr} \to \text{SUsr}/\mathcal{M}_U$ be the canonical functor.

Proposition 5.7. *Let V, W be uniserial modules with $[V]_m = [W]_m = [U]_m$, and let $f: V \to W$ be a homomorphism. Then $F(f)$ is an isomorphism if and only if either f is an isomorphism, or f is a monomorphism and $d([U]_e) \neq 1$.*

Proof. ($\Rightarrow$) Assume that $F(f)$ is an isomorphism and f is not an isomorphism. Let $g: W \to V$ be such that $F(g)$ is an inverse of $F(f)$. Then $1_V - gf$ factors through a module in SUsr with m-dim$_U$ equal to zero. That is, there exist uniserial modules $U_1, \ldots, U_n$ and morphisms $f_i: V \to U_i$ and $g_i: U_i \to V$ with $[U_i]_m \neq [U]_m$ for every i such that $1_V - gf = g_1 f_1 + \cdots + g_n f_n$. Now, for every $i = 1, \ldots, n$, either g_i or f_i is not injective, so that $g_i f_i$ is never injective [2, Lemma 1.1(a)]. Hence $g_1 f_1 + \cdots + g_n f_n$ is not injective. Thus $1_V - gf$ is not injective, from which gf must be injective, hence f must be injective. Therefore f is not onto, so that gf is not onto [2, Lemma 1.1(b)]. It follows that $1_V - gf$ is onto. From $1_V - gf = g_1 f_1 + \cdots + g_n f_n$ onto it follows that there exists an index $\bar{i}$ with $g_{\bar{i}} f_{\bar{i}}$ onto. Then $[V]_e = [U_{\bar{i}}]_e$. Thus $d([V]_e) \neq 1$. But $d([U]_e) = d([V]_e)$ by Proposition 5.1.

($\Leftarrow$) If f is an isomorphism, the implication is trivial. Therefore we can suppose that f is a monomorphism, that $d([U]_e) \neq 1$ and that f is not an isomorphism. Thus $d([V]_e) \neq 1$ (Proposition 5.1), i.e., there exists a uniserial module X with $[X]_e = [V]_e$ and $[X]_m \neq [V]_m$. Then there are two epimorphisms $h: X \to V$ and $h': V \to X$, which cannot be both monomorphisms. As $[V]_m = [W]_m$, there exists a monomorphism $g: W \to V$. Thus hh' and $h'h$ are epimorphisms but not monomorphisms, and fg, gf are monomorphisms that are not epimorphisms. Thus $gf + h'h$ and $fg + hh'$ are both automorphisms, so $F(f)$ is an isomorphism. □

Corollary 5.8. *Let U, V, W be non-zero uniserial right R-modules with $[V]_m = [W]_m = [U]_m$. Then $\overline{V} \cong \overline{W}$ in the category $\text{SUsr}/\mathcal{M}_U$ if and only if either $V \cong W$ or $d([U]_e) \neq 1$.*

Proposition 5.9. *The category* $\operatorname{SUsr}/\mathcal{M}_U$ *is a Krull–Schmidt category for any non-zero uniserial module* U.

Proof. Every object of $\operatorname{SUsr}/\mathcal{M}_U$ is a direct sum of finitely many objects of the form $\overline{V}$, where V is a non-zero uniserial module. By Proposition 4.2, every object of $\operatorname{SUsr}/\mathcal{M}_U$ is a direct sum of finitely many indecomposable objects of the form $\overline{V}$, where V is a uniserial module with $[V]_m = [U]_m$. Let $V_1, \ldots, V_n, V_1', \ldots, V_m'$ be uniserial modules, all of monogeny class $[U]_m$, and suppose $\overline{V_1} \oplus \cdots \oplus \overline{V_n} \cong \overline{V_1'} \oplus \cdots \oplus \overline{V_m'}$. By Lemma 2.4, there are $C, D \in \operatorname{SUsr}$ with $\operatorname{m-dim}_U(C) = \operatorname{m-dim}_U(D) = 0$ such that $V_1 \oplus \cdots \oplus V_n \oplus C \cong V_1' \oplus \cdots \oplus V_m' \oplus D$. Since $\operatorname{m-dim}_U(V_i) = \operatorname{m-dim}_U(V_j') = 1$ for every i and j, we get that $n = m$.

Now we have two cases. If $d([U]_e) \neq 1$, then $\overline{V_i} \cong \overline{V_j'} \cong \overline{U}$ for every i and j by Corollary 5.8, and therefore we have the desired direct-sum uniqueness. If $d([U]_e) = 1$, then $d([V_i]_e) = d([V_j']_e) = 1$ for every $i = 1, \ldots, n$ and every $j = 1, \ldots, n$. Let $C = V_{n+1} \oplus \cdots \oplus V_{n+t}$ and $D = V_{n+1}' \oplus \cdots \oplus V_{n+t'}'$ be direct-sum decompositions into uniserial modules. By the Weak Krull–Schmidt Theorem [2, Theorem 1.9], from $V_1 \oplus \cdots \oplus V_{n+t} \cong V_1' \oplus \cdots \oplus V_{n+t'}'$ we get that $n+t = n+t'$ and that there is a permutation τ of $\{1, 2, \ldots, n+t\}$ such that $[V_i]_e = [V_{\tau(i)}']_e$ for every $i = 1, 2, \ldots, n+t$. Now, $d([V_i]_e) = 1$ for $i = 1, \ldots, t$, so that for every uniserial module V, $[V_i]_e = [V]_e$ implies $V_i \cong V$. Therefore $V_i \cong V_{\tau(i)}'$ for every $i = 1, 2, \ldots, n$. From $\operatorname{m-dim}_U(V_i) = 1$ for $i = 1, \ldots, n$ and $\operatorname{m-dim}_U(V_j') = 0$ for $j = n+1, \ldots, n+t$, we get that $\tau(i) \geq n+1$ for every $i = n+1, \ldots, n+t$. Thus τ induces by restriction a permutation τ' of $\{1, 2, \ldots, n\}$ and $V_i \cong V_{\tau(i)}' \cong V_{\tau'(i)}'$ for every $i = 1, 2, \ldots, n$. □

Notice that, by Corollary 5.8, we have two cases. If $d([U]_e) \neq 1$, then $V(\operatorname{SUsr}/\mathcal{M}_U) \cong \mathbb{N}_0$, that is, $V(\operatorname{SUsr}/\mathcal{M}_U)$ is an IBN category. If $d([U]_e) = 1$, then $V(\operatorname{SUsr}/\mathcal{M}_U) \cong \mathbb{N}_0^{(d([U]_m))}$.

Dually, the category $\operatorname{SUsr}/\mathcal{E}_U$ is a Krull–Schmidt category for any non-zero uniserial module U.

6 The two representations

In this final section, we construct two representations of the category SUsr into two Krull–Schmidt categories. These constructions are the main aim of this paper.

Fix a set M of representatives of the non-zero uniserial right R-modules up to monogeny, and a set E of representatives of the non-zero uniserial right R-modules up to epigeny. For every $U \in M$ consider the valuation $\operatorname{m-dim}_U$ of SUsr, and for every $W \in E$ consider the valuation $\operatorname{e-dim}_W$ of SUsr. Thus, for every $U \in M$ we have a quotient category $\operatorname{SUsr}/\mathcal{M}_U$, and for every $W \in E$ we have a quotient category $\operatorname{SUsr}/\mathcal{E}_W$. All these quotient categories are Krull–Schmidt categories (Proposition 5.9).

Now we can form the coproduct category

$$\mathcal{C} = \Big(\coprod_{U \in M} \operatorname{SUsr}/\mathcal{M}_U \Big) \coprod \Big(\coprod_{W \in E} \operatorname{SUsr}/\mathcal{E}_W \Big)$$

(for the notion of coproduct category, see [6]). This coproduct category $\mathcal{C}$ is clearly a Krull–Schmidt category.

Let $F\colon \operatorname{SUsr} \to \mathcal{C} = \left(\coprod_{U\in M} \operatorname{SUsr}/\mathcal{M}_U\right) \coprod \left(\coprod_{W\in E} \operatorname{SUsr}/\mathcal{E}_W\right)$ be the canonical functor. Notice that, for every object V of SUsr, $F(V)$ is really an object of the Krull–Schmidt category $\mathcal{C}$. In order to prove it for an arbitrary V, it suffices to prove it for V non-zero uniserial. Now, when V is non-zero uniserial, m-$\dim_U(V) = 0$ for all $U \in M$ except for the $U \in M$ with $[U]_m = [V]_m$, so that $\overline{V} = \overline{0}$ in $\operatorname{SUsr}/\mathcal{M}_U$ for all $U \in M$ except one. Similarly, e-$\dim_W(V) = 0$ for all $W \in E$ except for the $W \in E$ with $[W]_e = [V]_e$. Thus, for every non-zero uniserial module V, the object $F(V)$ of $\mathcal{C}$ is the coproduct of two indecomposable objects. In the first part of this Section we study the additive functor F. Recall that a functor $F\colon \mathcal{A} \to \mathcal{B}$ between preadditive categories is said to be *isomorphism reflecting* (resp., *direct-summand reflecting*) if $F(A) \cong F(B)$ (resp., $F(A)$ is a direct summand of $F(B)$) implies $A \cong B$ (resp., A is a direct summand of B) for any objects A, B of $\mathcal{A}$.

Proposition 6.1. *The functor $F\colon \operatorname{SUsr} \to \mathcal{C}$ is direct-summand reflecting and isomorphism reflecting.*

Proof. The canonical functor $\operatorname{SUsr} \to \operatorname{SUsr}/\mathcal{M}_U$ induces a monoid homomorphism $V(\operatorname{SUsr}) \to V(\operatorname{SUsr}/\mathcal{M}_U)$, and $V(\operatorname{SUsr}/\mathcal{M}_U)$ is isomorphic either to $\mathbb{N}_0$ (if $d([U]_e) \neq 1$) or to $\mathbb{N}_0^{(d([U]_m))}$ (if $d([U]_e) = 1$). In both cases, taking the sum of the coordinates, we get a mapping $V(\operatorname{SUsr}/\mathcal{M}_U) \to \mathbb{N}_0$ that composed with the monoid homomorphism $V(\operatorname{SUsr}) \to V(\operatorname{SUsr}/\mathcal{M}_U)$ yields the valuation m-$\dim_U\colon V(\operatorname{SUsr}) \to \mathbb{N}_0$. Now the functor $F\colon \operatorname{SUsr} \to \mathcal{C}$ induces a monoid homomorphism

$$V(F)\colon V(\operatorname{SUsr}) \to V\left(\left(\coprod_{U\in M} \operatorname{SUsr}/\mathcal{M}_U\right) \coprod \left(\coprod_{W\in E} \operatorname{SUsr}/\mathcal{E}_W\right)\right),$$

and this last monoid is isomorphic to

$$\left(\bigoplus_{U\in M} V(\operatorname{SUsr}/\mathcal{M}_U)\right) \oplus \left(\bigoplus_{W\in E} V(\operatorname{SUsr}/\mathcal{E}_W)\right).$$

Thus the mappings $V(\operatorname{SUsr}/\mathcal{M}_U) \to \mathbb{N}_0$ and $V(\operatorname{SUsr}/\mathcal{E}_V) \to \mathbb{N}_0$ determine a monoid homomorphism $\left(\bigoplus_{U\in M} V(\operatorname{SUsr}/\mathcal{M}_U)\right) \oplus \left(\bigoplus_{W\in E} V(\operatorname{SUsr}/\mathcal{E}_W)\right) \to \mathbb{N}_0^{(M)} \oplus \mathbb{N}_0^{(E)}$, which composed with $V(F)$ gives the injective homomorphism $V(\operatorname{SUsr}) \to \mathbb{N}_0^{(M)} \oplus \mathbb{N}_0^{(E)}$ considered in [8, Corollary 7.5]. Hence $V(\operatorname{SUsr}) \to \mathbb{N}_0^{(M)} \oplus \mathbb{N}_0^{(E)}$ injective implies $V(F)$ injective, i.e., F is isomorphism reflecting. Also, since $V(\operatorname{SUsr}) \to \mathbb{N}_0^{(M)} \oplus \mathbb{N}_0^{(E)}$ is a divisor homomorphism, we get that $V(F)$ is a divisor homomorphism, i.e., F is direct-summand reflecting. □

The following lemma is implicit in the proof of [13, Theorem 7].

Lemma 6.2. *Let $V, U_1, \ldots, U_n$ be non-zero uniserial right R-modules, $f_i\colon V \to U_i$ a homomorphism for every $i = 1, \ldots, n$, and $f = \begin{pmatrix} f_1 \\ \vdots \\ f_n \end{pmatrix}\colon V \to U_1 \oplus \cdots \oplus U_n$. Then:*

(a) *f is a monomorphism if and only if f_i is a monomorphism for some $i = 1, \ldots, n$.*

(b) *f is a split monomorphism if and only if there exist $i, j \in \{1, \ldots, n\}$ such that $[U_i]_m = [V]_m$, $[U_j]_e = [V]_e$, $f_i \colon V \to U_i$ is a monomorphism and $f_j \colon V \to U_j$ is an epimorphism.*

Proof. Statement (a) is an easy consequence of the fact that V is uniform. For (b), let f be a split monomorphism. Then there exist homomorphisms $g_i \colon U_i \to V$ with $1_V = \sum_{i=1}^n g_i f_i$. Thus there are $i, j = 1, \ldots, n$ (not necessarily distinct) such that $g_i f_i$ is a monomorphism and $g_j f_j$ is an epimorphism. Then g_i and f_i are monomorphisms, and g_j and f_j are epimorphisms. In particular, $[U_i]_m = [V]_m$ and $[U_j]_e = [V]_e$.

For the converse, assume that $[U_i]_m = [V]_m$, $[U_j]_e = [V]_e$, $f_i \colon V \to U_i$ is a monomorphism and $f_j \colon V \to U_j$ is an epimorphism, so that f is a monomorphism by (a). Let $p_k \colon U_1 \oplus \cdots \oplus U_n \to U_k$ denote the canonical projection for every k. If f_i is an isomorphism, then $f_i^{-1} p_i \colon U_1 \oplus \cdots \oplus U_n \to V$ is a left inverse for f, so that f is a split monomorphism. Similarly if f_j is an isomorphism. Thus we can suppose that f_i is not an epimorphisms and f_j is not a monomorphism, so that in particular $i \neq j$. Let $g_i \colon U_i \to V$ be a monomorphism and $g_j \colon U_j \to V$ be an epimorphism. Then $g_i f_i + g_j f_j$ is an automorphism of V, so that $(g_i f_i + g_j f_j)^{-1}(g_i p_i + g_j p_j) \colon U_1 \oplus \cdots \oplus U_n \to V$ is a left inverse for f. □

Recall that a functor $F \colon \mathcal{A} \to \mathcal{B}$ is *local*, if for every morphism $f \colon A \to A'$, $F(f)$ isomorphism in $\mathcal{B}$ implies f isomorphism in $\mathcal{A}$.

Proposition 6.3. *The functor $F \colon \mathrm{SUsr} \to \mathcal{C}$ is local.*

Proof. We know that $X \cong Y$ because F is isomorphism reflecting. Thus we can assume $X = Y$. The proof of the proposition is by induction on the Goldie dimension of X.

Suppose X uniserial. If $f \colon X \to X$ is a morphism in the category SUsr and $F(f)$ is an automorphism, then the image of f in the category $\mathrm{SUsr}/\mathcal{M}_X$ is an automorphism and the same is true for the image of f in $\mathrm{SUsr}/\mathcal{E}_X$. Therefore f is an isomorphism by Proposition 5.7 and its dual.

Now suppose $X = U_1 \oplus \cdots \oplus U_n$ and $n \geq 2$. Respect to this direct-sum decomposition, f can be written in matrix notation as $f = (f_{ij})_{i,j}$, where $f_{ij} \colon U_j \to U_i$. Let $g = (g_{ij})_{i,j} \colon F(X) \to F(X)$ be the inverse of $F(f)$. If we project the equality $1_{F(U_1)} = \sum_{i=1}^n g_{1i} F(f_{i1})$ into $\mathrm{SUsr}/\mathcal{M}_{U_1}$, we see that there exist $g'_i \colon U_i \to U_1$ such that 1_{U_1} and $\sum_{i=1}^n g'_i f_{i1}$ are equal in $\mathrm{SUsr}/\mathcal{M}_{U_1}$. By Proposition 5.7, $\sum_{i=1}^n g'_i f_{i1} \colon U_1 \to U_1$ is a monomorphism. Thus $g'_i f_{i1}$ is a monomorphism for some i, so that $f_{i1} \colon U_1 \to U_i$ is a monomorphism and $[U_1]_m = [U_i]_m$. Similarly, there exists an index j with $f_{j1} \colon U_1 \to U_j$ an epimorphism and $[U_1]_e = [U_j]_e$. By Lemma 6.2(b), $f_1 := \begin{pmatrix} f_{11} \\ \vdots \\ f_{n1} \end{pmatrix} \colon U_1 \to U_1 \oplus \cdots \oplus U_n$ is a split monomorphism, so that $X = f_1(U_1) \oplus Y$.

Put $X' = U_2 \oplus \cdots \oplus U_n$. With respect to the direct-sum decompositions $X = U_1 \oplus X'$ and $X = f(U_1) \oplus Y$, f can be written in matrix notation as $f = \begin{pmatrix} f'_{11} & f'' \\ 0 & f' \end{pmatrix}$, where $f'_{11} = \pi_{f_1(U_1)} f_1$. Thus f is an isomorphism if and only if f'_{11} and f' are isomorphisms. Now $f_{11}\colon U_1 \to f_1(U_1)$ is an isomorphism because f_1 is a splitting monomorphism. By the inductive hypothesis, $f'\colon X' \to Y$ is an isomorphism. This allows us to conclude that f is an isomorphism. □

Now we shall consider a different functor G, much less natural than the functor F, but which has the advantages of representing SUsr into a coproduct of IBN categories, of inducing a divisor theory for the Krull monoid $V(\text{SUsr})$, and of having a "partial" universal property (Proposition 6.6 and Remak 6.7). As in Theorem 5.4, let K be a set of representatives up to isomorphism of the Krull–Schmidt uniserial right R-modules W, M' a set of representatives up to monogeny of the non-zero uniserial right R-modules U that are not Krull–Schmidt modules, and E' a set of representatives up to epigeny of the non-zero uniserial right R-modules V that are not Krull–Schmidt modules. For every $W \in K$ consider the valuation v_W of SUsr that "counts the number of direct summands isomorphic to W" (Section 3), for every $U \in M'$ consider the valuation m-dim$_U$ of SUsr, and for every $V \in E'$ consider the valuation e-dim$_V$. Thus, for every $W \in K$ we have a quotient category SUsr$/\mathcal{I}_W$, for every $U \in M'$ we have a quotient category SUsr$/\mathcal{M}_U$, and for every $V \in E'$ we have a quotient category SUsr$/\mathcal{E}_V$. All these quotient categories are IBN categories, and we can take their coproduct category

$$\mathcal{D} = \Big(\coprod_{W \in K} \text{SUsr}/\mathcal{I}_W \Big) \coprod \Big(\coprod_{U \in M'} \text{SUsr}/\mathcal{M}_U \Big) \coprod \Big(\coprod_{V \in E'} \text{SUsr}/\mathcal{E}_V \Big).$$

The coproduct category $\mathcal{D}$ is a Krull–Schmidt category. Let $G\colon \text{SUsr} \to \mathcal{D}$ be the canonical functor.

Proposition 6.4. *The functor $G\colon \text{SUsr} \to \mathcal{D}$ is direct-summand reflecting, isomorphism reflecting, local and full (that is, if $A, B \in \text{SUsr}$ and $f'\colon G(A) \to G(B)$, then there exists $f\colon A \to B$ with $f' = G(f)$).*

Proof. For $W \in K$, the canonical functor $\text{SUsr} \to \text{SUsr}/\mathcal{I}_W$ induces a monoid homomorphism $V(\text{SUsr}) \to V(\text{SUsr}/\mathcal{I}_W) \cong \mathbb{N}_0$. Similarly for $U \in M'$ and $V \in E'$. Thus the functor $G\colon \text{SUsr} \to \mathcal{D}$ induces the monoid homomorphism $V(\text{SUsr}) \to \mathbb{N}_0^{(K)} \oplus \mathbb{N}_0^{(M')} \oplus \mathbb{N}_0^{(E')}$, which is a divisor theory (Theorem 5.6). Therefore G is direct-summand reflecting and isomorphism reflecting.

The proof that the functor G is local is similar to the proof of Proposition 6.3.

We shall now show that G is full. As any object of SUsr is a finite direct sum of indecomposable objects and G preserves finite direct sums, we can suppose that A and B are uniserial modules. Observe that there exists a nonzero morphism between $G(A)$ and $G(B)$ only if $A \sim B$. If A, B are Krull–Schmidt or $A \not\cong B$, then it is easy to see that $G(A)$ and $G(B)$ are both nonzero in at most one component. Therefore any morphism between $G(A)$ and $G(B)$ is an image of some morphism in SUsr.

Finally, suppose that $A = B$ is a uniserial module that is not Krull–Schmidt. Then $\mathcal{M}_A(A, A)$ contains a monomorphism that is not an epimorphism and, similarly, $\mathcal{E}_A(A, A)$ contains an epimorphism that is not a monomorphism, in particular $\mathcal{M}_A(A, A) + \mathcal{E}_A(A, A) = \mathrm{End}_R(A)$. Therefore the canonical morphism

$$\mathrm{End}_R(A) \to \mathrm{End}_R(A)/\mathcal{M}_A(A, A) \times \mathrm{End}_R(A)/\mathcal{E}_A(A, A)$$

is onto, i.e., any endomorphism of $G(A)$ is the image of an endomorphism of A. □

In the proof we have seen that the induced monoid homomorphism $V(G) \colon V(\mathrm{SUsr}) \to V(\mathcal{D})$ is a divisor theory of the Krull monoid $V(\mathrm{SUsr})$ into the free commutative monoid $V(\mathcal{D})$.

Remark 6.5. If $v \colon \mathrm{Ob}(\mathrm{SUsr}) \to \mathbb{N}_0$ is a valuation of the category SUsr, then $D_v = \{\,\langle A \rangle \in V(\mathrm{SUsr}) \mid v(A) = 0\,\}$ is a divisor-closed submonoid of $V(\mathrm{SUsr})$. If v is an essential valuation, then D_v is a maximal proper divisor-closed submonoid of $V(\mathrm{SUsr})$ [7, Lemma 4.1]. Therefore, in the notation of Proposition 5.5, D_v corresponds to a suitable element of $\mathcal{B}$, that is, to a suitable subgraph of the bipartite graph G.

When v is the valuation m-dim$_U$ for some non-zero uniserial module U, then D_v corresponds to the element of $\mathcal{B}$ obtained from G canceling all edges adjacent to $[U]_m$, that is, the star centered in $[U]_m$. Similarly when v is the valuation e-dim$_U$.

When v is the valuation v_W for some uniserial module $W \in K$, then D_v corresponds to the element of $\mathcal{B}$ obtained from G canceling the edge $\langle W \rangle$.

In the final part of the paper, we consider the validity of a "partial" universal property of the functor G among full additive functors of SUsr into IBN categories. More precisely, the universal property we consider is the following: given any full additive functor $H \colon \mathrm{SUsr} \to \mathcal{K}$, where $\mathcal{K}$ is an IBN category, is it possible to factor $H \colon \mathrm{SUsr} \to \mathcal{K}$ through one of the canonical functors $\mathrm{SUsr} \to \mathrm{SUsr}/\mathcal{I}_W$, $\mathrm{SUsr} \to \mathrm{SUsr}/\mathcal{M}_U$ or $\mathrm{SUsr} \to \mathrm{SUsr}/\mathcal{E}_V$ for some $W \in K, U \in M'$ or $V \in E'$? Notice that if $\mathcal{K}er H$ denotes the ideal of SUsr defined by $f \in \mathcal{K}er H$ if and only if $H(f) = 0$, then H factors through one of the canonical functors $\mathrm{SUsr} \to \mathrm{SUsr}/\mathcal{I}_W$, $\mathrm{SUsr} \to \mathrm{SUsr}/\mathcal{M}_U$ or $\mathrm{SUsr} \to \mathrm{SUsr}/\mathcal{E}_V$ for suitable $W \in K$, $U \in M'$ or $V \in E'$ if and only if either $\mathcal{I}_W \subseteq \mathcal{K}er H$, or $\mathcal{M}_U \subseteq \mathcal{K}er H$ or $\mathcal{E}_V \subseteq \mathcal{K}er H$. In the next proposition we examine this situation. If H is the zero functor, the statement is trivial, and we shall omit that case. Thus we can assume $H \neq 0$, so that there exists a uniserial module U with $H(U) \neq 0$.

Proposition 6.6. *Let $\mathcal{K}$ be an IBN category and let $H \colon \mathrm{SUsr} \to \mathcal{K}$ be a full additive functor. Let $\mathcal{K}er H$ be the ideal of SUsr defined by $f \in \mathcal{K}er H$ if $H(f) = 0$. Let U be a uniserial module such that $H(U) \neq 0$. Then:*

(a) *The object $H(U)$ of $\mathcal{K}$ is the indecomposable object in $\mathcal{K}$ (unique up to isomorphism). Therefore, every object of $\mathcal{K}$ is isomorphic to $H(U^n)$ for some $n \in \mathbb{N}$ and has a semilocal endomorphism ring.*

(b) *If V is a uniserial module such that $[V]_m \neq [U]_m$ and $[V]_e \neq [U]_e$, then $H(V) = 0$.*

(c) If U is not a Krull–Schmidt module, then either $\mathcal{M}_U \subseteq \mathcal{K}er H$ or $\mathcal{E}_U \subseteq \mathcal{K}er H$.

(d) If $d([U]_m) = 1$ and $d([U]_e) = 1$, then $\mathcal{I}_U \subseteq \mathcal{K}er H$.

(e) If $\operatorname{End}_R(U)$ is a local ring, then $\mathcal{I}_U \subseteq \mathcal{K}er H$.

Proof. (a) Let X be an indecomposable object of $\mathcal{K}$, so that $H(U) \cong X^n$ for some $n \in \mathbb{N}$. Since H is full, there exists a surjective ring morphism $\varphi \colon \operatorname{End}_R(U) \to \operatorname{Mat}_n(\operatorname{End}_{\mathcal{K}}(X))$. By [3, Theorem 9.1], $\operatorname{End}_R(U)/J(\operatorname{End}_R(U))$ is a product of one or two division rings. As φ induces an epimorphism of $\operatorname{End}_R(U)/J(\operatorname{End}_R(U))$ onto $\operatorname{Mat}_n(\operatorname{End}_{\mathcal{K}}(X)/J(\operatorname{End}_{\mathcal{K}}(X)))$, we see that we must have $n = 1$. The rest is easy.

(b) Let V be a uniserial module with $[V]_m \neq [U]_m$, $[V]_e \neq [U]_e$ and $H(V) \neq 0$. By (a) applied to V, it follows that $H(U) \cong H(V)$, so that there exist morphisms $f \colon U \to V$ and $g \colon V \to U$ with $H(f), H(g)$ mutually inverse isomorphisms in $\mathcal{K}$. Thus $H(1 - gf) = 0$. As $[U]_m \neq [V]_m$ and $[U]_e \neq [V]_e$, the morphism gf is neither a monomorphism nor an epimorphism [3, Lemma 6.26], therefore $1 - gf$ is an automorphism. Thus $H(1 - gf) = 0$ is an automorphism of $H(U) \neq 0$, which is not possible.

(c) Suppose that U is not a Krull–Schmidt module, so that there are uniserial modules U', V, V' not isomorphic to V with $U \oplus U' \cong V \oplus V'$. From $H(U) \neq 0$ it follows that either $H(V) \neq 0$ or $H(V') \neq 0$. Possibly interchanging the roles of V and V', we can assume that $V \not\cong U$ is a uniserial module with $H(V) \neq 0$. By (b), either $[V]_m = [U]_m$ or $[V]_e = [U]_e$ (but not both). Suppose $[V]_m = [U]_m$. Let W be any uniserial module such that $[W]_e = [U]_e$ and $W \not\cong U$ (so $[W]_m \neq [U]_m$). Then there exists a uniserial module W' such that $U \oplus W' \cong V \oplus W$. Necessarily $[W']_m = [W]_m$ and $[W']_e = [V]_e$. By (b), $H(W') = 0$, so that $H(W) = 0$. This shows that, for every uniserial module W, $H(W) \neq 0$ implies $[W]_e \neq [U]_e$ or $W \cong U$, so that $[W]_m = [U]_m$ by (b). Let us prove that $\mathcal{M}_U \subseteq \mathcal{K}er H$. If $f \in \mathcal{M}_U$, then f factors through an object C of SUsr with m-$\dim_U(C) = 0$. Thus C is the direct sum of finitely many uniserial modules W with $[W]_m \neq [U]_m$. Hence $H(C) = 0$, from which $f \in \mathcal{K}er H$. Similarly, if we suppose $[V]_e = [U]_e$, then, for every uniserial module W, $H(W) \neq 0$ implies $[W]_e = [U]_e$, and from this $\mathcal{E}_U \subseteq \mathcal{K}er H$ follows.

(d) Assume $d([U]_m) = 1$, $d([U]_e) = 1$ and $f \in \mathcal{I}_U$. Then f factors through a direct sum of objects $W_i \not\cong U$. Thus $[W_i]_m \neq [U]_m$ and $[W_i]_e \neq [U]_e$ for every i, so that $H(W_i) = 0$ by (b). Hence $H(f) = 0$.

(e) The ring $\operatorname{End}_R(U)$ is local if and only if every monomorphism in $\operatorname{End}_R(U)$ is an isomorphism or every epimorphism in $\operatorname{End}_R(U)$ is an isomorphism [3, Theorem 9.1]. Assume that every epimorphism in $\operatorname{End}_R(U)$ is an isomorphism. In view of (d), we can assume that the connected component of U has at least two edges. This implies $d([U]_e) = 1$ and $d([U]_m) > 1$. Let V be a uniserial module not isomorphic to U such that $H(V) \neq 0$. Then $[U]_e \neq [V]_e$ and $H(U) \cong H(V)$, so that there exist $f \colon U \to V$ and $g \colon V \to U$ with $H(1_U - gf) = 0$. From $[V]_e \neq [U]_e$ it follows that the homomorphism gf is not onto, so that $1_U - gf$ is an epimorphism. By our assumption, $1_U - gf$ is an automorphism, so that $H(1_U - gf) = 0$ is an automorphism of $H(U)$, hence $H(U) = 0$, a contradiction. Therefore $H(V) = 0$ for every uniserial module $V \not\cong U$, from which $\mathcal{I}_U \subseteq \mathcal{K}er H$. The case in which every monomorphism of $\operatorname{End}_R(U)$ is an automorphism is handled similarly. □

Remark 6.7. The previous proposition does not exhaust all possible cases for the module U, because the case of U a Krull–Schmidt module with non-local endomorphism ring and whose connected component has at least two edges is not considered in the statement. Let us explain why.

Let U be a uniserial module with the above properties, for instance a module with $d([U]_m) > 1$, $d([U]_e) = 1$ and $\operatorname{End}_R(U)$ not local. Consider the ideal $\mathcal{K}$ of SUsr defined by $\mathcal{K}(A, B) = \{\, f \colon A \to B \mid \ker f \text{ is an essential submodule of } A \,\}$ for every $A, B \in \mathrm{Ob}(\mathrm{SUsr})$. Set $\mathcal{I} = \mathcal{M}_U + \mathcal{K}$. Let $H \colon \mathrm{SUsr} \to \mathrm{SUsr}/\mathcal{I}$ be the canonical functor. Let us prove that the category $\mathrm{SUsr}/\mathcal{I}$ is an IBN category. If V is uniserial and $[V]_m \neq [U]_m$, then $\overline{V} = \overline{0}$ in $\mathrm{SUsr}/\mathcal{M}_U$, hence a fortiori $H(V) = 0$ in $\mathrm{SUsr}/\mathcal{I}$. If V is uniserial, $[V]_m = [U]_m$ and $f \in \mathcal{M}_U(V, V)$, then $\operatorname{End}_R(V)$ is not local (Lemma 5.2), and f factors through a finite direct sum of uniserial modules V_i with $[V_i]_m \neq [U]_m$. The composite morphism of a morphism $V \to V_i$ and a morphism $V_i \to V$ cannot be a monomorphism, otherwise $[V_i]_m = [V]_m = [U]_m$, a contradiction. Hence such a composite morphism is not a monomorphism, from which f is not a monomorphism. This proves that $\mathcal{M}_U(V, V) \subseteq \mathcal{K}(V, V)$, so that the endomorphism ring of the object $\overline{V}$ of $\mathrm{SUsr}/\mathcal{I}$ is the division ring $\operatorname{End}_R(V)/\mathcal{K}(V, V)$. Let us prove that $\overline{V} \cong \overline{U}$ in $\mathrm{SUsr}/\mathcal{I}$ for every such uniserial module V with $[V]_m = [U]_m$. There exist monomorphisms $f \colon V \to U$ and $g \colon U \to V$. Hence $\mathcal{I}(U, U) = \mathcal{K}(U, U)$ implies $H(gf) \neq 0$. As the endomorphism ring of $H(U)$ in $\mathrm{SUsr}/\mathcal{I}$ is a division ring, its non-zero element $H(gf)$ is invertible. Thus $H(f)$ is left invertible. Similarly, $\mathcal{I}(V, V) = \mathcal{K}(V, V)$ implies $H(fg) \neq 0$, and $H(f)$ is right invertible. Thus $H(V) \cong H(U)$ whenever $[V]_m = [U]_m$. In particular, every object of $\mathrm{SUsr}/\mathcal{I}$ is isomorphic to $H(U)^n$ for some $n \in \mathbb{N}_0$. To conclude the proof that $\mathrm{SUsr}/\mathcal{I}$ is an IBN category, assume that $m \geq n > 0$ are such that $H(U^n) \cong H(U^m)$. Then there are $f \colon U^m \to U^n$ and $g \colon U^n \to U^m$ with $1_{U^m} - gf \in \mathcal{I}(U^m, U^m)$. Hence the entries of the matrix representing $1_{U^m} - gf$ are in $\mathcal{I}(U, U) = \mathcal{K}(U, U)$. It follows that $1_{U^m} - gf \in \mathcal{K}(U^m, U^m)$, so that gf is a monomorphism. Therefore f is a monomorphism and $m \leq n$, i.e., $m = n$. Thus H is a full additive functor of SUsr into an IBN category. But $\mathcal{I}_U \not\subseteq \mathcal{K}er H$, because there exists V uniserial with $[V]_m = [U]_m$ and $V \not\cong U$, and for such a V one has $\mathcal{I}_U(V, V) = \operatorname{Hom}_R(V, V) \not\subseteq \mathcal{K}er H(V, V) = \mathcal{I}(V, V) = \mathcal{K}(V, V)$. Finally, observe that $\mathcal{I}$ contains $\mathcal{M}_U$ properly, because $\mathrm{SUsr}/\mathcal{M}_U$ is not an IBN category.

References

[1] P. Ara and A. Facchini, *Direct sum decompositions of modules, almost trace ideals, and pullbacks of monoids*, Forum Math. **18** (2006), 365–389.

[2] A. Facchini, *Krull–Schmidt fails for serial modules*, Trans. Amer. Math. Soc. **348** (1996), 4561–4575.

[3] A. Facchini, *Module Theory. Endomorphism rings and direct sum decompositions in some classes of modules*, Progress in Math. **167**, Birkhäuser Verlag, Basel, 1998.

[4] A. Facchini, *A characterization of additive categories with the Krull–Schmidt property*, in "Algebra and Its Applications", D. V. Huynh, S. K. Jain and S. R. López-Permouth Eds., Contemporary Math. Series **419**, Amer. Math. Soc., 2006, pp. 125–129.

[5] A. Facchini, *Representations of additive categories and direct-sum decompositions of objects,* Indiana Univ. Math. J. **56** (2007), 659–680.

[6] A. Facchini and R. Fernández-Alonso, *Subdirect products of preadditive categories and weak equivalences,* Appl. Categ. Structures **15** (2007).

[7] A. Facchini and F. Halter-Koch, *Projective modules and divisor homomorphisms,* J. Algebra Appl. **2** (2003), 435–449.

[8] A. Facchini and P. Příhoda, *Monogeny dimension relative to a fixed uniform module,* J. Pure Appl. Algebra **212** (2008), 2092–2104.

[9] K. R. Goodearl, *Partially ordered abelian groups with interpolation,* Math. Surveys and Monographs **20**, Amer. Math. Soc., Providence, RI, 1986.

[10] F. Halter-Koch, *Ideal Systems. An Introduction to Multiplicative Ideal Theory,* Marcel Dekker, New York, 1998.

[11] Tsit-Yuen Lam, *Modules with Isomorphic Multiples and Rings with Isomorphic Matrix Rings,* Monographie **35**, L'Enseignement Mathématique, Genève, 1999.

[12] B. Mitchell, *Rings with several objects,* Advances in Math. **8** (1972), 1–161.

[13] P. Příhoda, *Weak Krull–Schmidt theorem and direct sum decompositions of serial modules of finite Goldie dimension,* J. Algebra **281**(1) (2004), 332–341.

Author information

Alberto Facchini, Dipartimento di Matematica Pura e Applicata, Università di Padova, 35121 Padova, Italy.
E-mail: facchini@math.unipd.it

Pavel Příhoda, Department of Algebra, Faculty of Mathematics and Physics, Charles University in Prague, Sokolovská 83, 186 75 Prague, Czech Republic.
E-mail: prihoda@karlin.mff.cuni.cz

Commutativity modulo small endomorphisms and endomorphisms of zero algebraic entropy

L. Salce and P. Zanardo

Abstract. Let G be an Abelian p-group. We investigate the relationships between the commutativity of $\text{End}(G)$ modulo the ideal of small endomorphisms, the fact that the endomorphisms of G of zero entropy form a subring of $\text{End}(G)$, and the property of G of being quasi finitely indecomposable.

Key words. Entropy, endomorphism rings, Abelian p-groups, small endomorphisms.

AMS classification. 20K30, 20K10.

1 Introduction

In his pioneering 1963 paper [P] on homomorphism groups and endomorphism rings of Abelian p-groups, Dick Pierce proved that, given an unbounded reduced p-group G, its endomorphism ring $\text{End}(G)$ decomposes as a J_p-module into $\text{End}(G) = A \oplus E_s(G)$, where A is a reduced and complete torsionfree J_p-module, and $E_s(G)$ is the two-sided ideal of the small endomorphisms.

Recall that a p-group G is determined by its endomorphism ring. In fact, the celebrated Baer–Kaplansky theorem (see [F, 108.1]) states that, given two p-groups G and H, every ring isomorphism $\Phi : \text{End}(G) \to \text{End}(H)$ is induced by a group isomorphism $\theta : G \to H$, that is, $\Phi(\phi) = \theta \cdot \phi \cdot \theta^{-1}$ for every $\phi \in \text{End}(G)$. A close inspection to the proof of the Baer–Kaplansky theorem shows that the ideal $E_s(G)$ of small endomorphisms is enough to determine G, so the J_p-module A is a component of $\text{End}(G)$ which is "disengaged" from G.

With these results at hand, and starting from a construction of Crawley [Cr], Tony Corner proved in 1969 [C1] one of his milestone realization theorems. He showed that, given a torsionfree complete J_p-algebra A whose additive group A^+ satisfies suitable conditions, there exist 2^c non-isomorphic separable p-groups G such that $\text{End}(G)$ is a split extension of A by $E_s(G)$. This means that there exists a ring embedding $\varepsilon : A \to \text{End}(G)$ and a surjective ring homomorphism $\pi : \text{End}(G) \to A$ such that $\pi \cdot \varepsilon = 1_A$ and $\text{Ker}(\pi) = E_s(G)$. It follows, identifying A with εA, that $\text{End}(G) = A \oplus E_s(G)$ as J_p-modules.

Corner himself [C2], and then Dugas–Göbel [DG], Corner–Göbel [CG], Goldsmith [G] and Behler–Göbel–Mines [BGM] extended and generalized the results in [C1] to wider classes of algebras and p-groups. It is worth remarking that in all these papers

---

Research supported by MIUR, PRIN 2005.

only the additive structure of the J_p-algebra $A \cong \mathrm{End}(G)/E_s(G)$ is considered and plays a role in the realization theorems which are obtained.

Recently the two authors, jointly with Dikran Dikranjan and Brendan Goldsmith, thoroughly extended in [DGSZ] the study of the algebraic entropy of endomorphisms of Abelian groups, which was introduced in a 1965 paper by Adler–Konheim–McAndrew [AKM], and was reconsidered in 1975 by Weiss [W]. Algebraic entropy is mostly significant when the Abelian group G is a p-group. In this case, the study of algebraic entropy is strongly related to the multiplicative structure of the J_p-algebra $\mathrm{End}(G)/E_s(G)$, which then plays a fundamental role, differently from the case of realization theorems. For instance, in [DGSZ] it is proved that, given a semi-standard unbounded p-group G, if $\mathrm{End}(G)/E_s(G)$ is integral over J_p, then every endomorphism of G has zero algebraic entropy.

The main goal of this note is to investigate when the set of the endomorphisms of zero entropy, denoted by $\mathrm{Ent}_0(G)$, is a ring. It turns out that the most natural multiplicative property of $\mathrm{End}(G)/E_s(G)$, namely, its commutativity, is enough to ensure that $\mathrm{Ent}_0(G)$ is a subring of $\mathrm{End}(G)$, whenever the p-group G is semi-standard. This assumption on G is unavoidable, since we prove that, when $\mathrm{Ent}_0(G)$ is a ring (or just an Abelian group), then, necessarily, G is quasi finitely indecomposable (for the definitions of semi-standard and quasi finitely indecomposable p-groups see the next section).

In the final section we consider the ideal of $\mathrm{End}(G)$, denoted by $E_0(G)$, which consists of the endomorphisms whose restrictions to the socle $G[p]$ of G have finite image. We have $p\,\mathrm{End}(G) \subseteq E_0(G) \subseteq \mathrm{Ent}_0(G)$, and the inclusion $E_s(G) \subseteq E_0(G)$ holds exactly when G is semi-standard. We show that $\mathrm{End}(G)/E_0(G)$ commutative implies that $\mathrm{Ent}_0(G)$ is a subring of $\mathrm{End}(G)$. This result is strictly more general than the analogous one for $\mathrm{End}(G)/E_s(G)$. In fact we provide an example of a p-group G with $\mathrm{End}(G)/E_0(G)$ commutative and $\mathrm{End}(G)/E_s(G)$ non-commutative.

2 Preliminaries

In what follows, all the Abelian groups considered will be reduced p-groups. Our terminology and notations are standard and any undefined term may be found in the text [F].

We recall the notion of *Ulm–Kaplansky invariants* (of finite index) of a p-group G. For each integer $n \geq 0$, the n-th Ulm–Kaplansky invariant $\alpha_n(G)$ of G is the dimension of the F_p-vector space $p^n G[p]/p^{n+1} G[p]$ (F_p is the field with p elements). We recall a crucial property of the Ulm–Kaplansky invariants: For each $n \geq 0$, G contains a direct summand isomorphic to a direct sum of $\alpha_n(G)$ copies of $\mathbb{Z}/p^{n+1}\mathbb{Z}$. A p-group G is said to be *semi-standard* if $\alpha_n(G)$ is finite for all $n \geq 0$.

A reduced p-group G is *essentially indecomposable* if, whenever $G = H \oplus K$, either H or K is bounded (see [F, II p. 55] and [M]); G is said to be *essentially finitely indecomposable* if, whenever $G = C \oplus K$, with C a direct sum of cyclic groups, C is necessarily bounded (see [CI]). Obviously, essentially indecomposable groups are essentially finitely indecomposable.

We call a p-group G *quasi finitely indecomposable* if $G = C \oplus K$, with C a direct sum of cyclic groups, implies that C is finite. This notion will be crucial in the present paper. It is readily seen that G is quasi finitely indecomposable if and only if G is essentially finitely indecomposable and semi-standard.

A major role in the structure of the endomorphism rings of Abelian p-groups is played by small endomorphisms, introduced by Pierce in his seminal paper [P]. Recall that an endomorphism ϕ of the p-group G is *small* if, given an arbitrary positive integer k, there exists an integer $n \geq 0$ such that $\phi(p^n G[p^k]) = 0$. Obvious example of small endomorphisms are furnished by the bounded endomorphisms.

The small endomorphisms of a reduced unbounded group G form a two-sided ideal $E_s(G)$ of the ring $\text{End}(G)$; recall that we have the direct decomposition of J_p-modules

$$\text{End}(G) = A \oplus E_s(G),$$

where A is the completion of a free J_p-module (see [P, Theorem 7.5]). However, it is important to remark that for the p-groups G, constructed in Corner's realization theorems [C1], the above decomposition is a split extension, i.e., A is a J_p-algebra which is a subring of $\text{End}(G)$.

Let us now recall the notion of algebraic entropy, as described in [DGSZ]. In fact, all our discussion on entropies will be based on that paper.

Let G be an Abelian group and denote by $\mathcal{F}(G)$ the family of its finite subgroups. If $\phi : G \to G$ is an endomorphism of G, for every positive integer n and every $F \in \mathcal{F}(G)$ we set

$$T_n(\phi, F) = F + \phi F + \phi^2 F + \cdots + \phi^{n-1} F.$$

The subgroup of F, $T(\phi, F) = \sum_{n>0} T_n(\phi, F) = \sum_{n \geq 0} \phi^n F$ will be called the ϕ-*trajectory* of F. The ϕ-trajectory of an element x is just the ϕ-trajectory of the cyclic subgroup $\mathbb{Z}x$, i.e., the smallest ϕ-invariant subgroup of G containing x, simply denoted by $T(\phi, x)$. It is clear that the ϕ-trajectory of the finite group F is finite if and only if the ϕ-trajectory of each $x \in F$ is such.

Given the finite subgroup F and the endomorphism ϕ of G, for each $n \geq 1$ we define the real number:

$$H_n(\phi, F) = \log |T_n(\phi, F)|.$$

Clearly we have the increasing sequence of real numbers

$$0 < H_1(\phi, F) \leq H_2(\phi, F) \leq H_3(\phi, F) \leq \cdots.$$

Now define

$$H(\phi, F) = \lim_{n \to \infty} \frac{H_n(\phi, F)}{n}.$$

The next proposition (see [DGSZ, Prop. 1.3] for the proof) shows that this is a good definition, namely, the limit exists, it is finite and we can compute its exact value.

Proposition 2.1. *Let* $\phi \in \text{End}(G)$ *and let* F *be a finite subgroup of* G. *Then* $|T_{n+1}(\phi, F)/T_n(\phi, F)| = p^k$ *is constant for all* n *large enough, and, for such* n, *we have*

$$H(\phi, F) = \log |T_{n+1}(\phi, F)| - \log |T_n(\phi, F)| = k \log p.$$

Moreover, $H(\phi, F) = 0$ if and only if the ϕ-trajectory $T(\phi, F)$ of F is finite.

Following Weiss [W], we define the *algebraic entropy* of an endomorphism ϕ of G as
$$\text{ent}(\phi) = \sup_{F \in \mathcal{F}(G)} H(\phi, F);$$
the *algebraic entropy* of G is defined as in [DGSZ], by
$$\text{ent}(G) = \sup_{\phi \in \text{End}(G)} \text{ent}(\phi).$$

Henceforth the term "entropy" will always mean "algebraic entropy". Of course, Proposition 2.1 implies that $\text{ent}(G) = 0$ whenever G is finite. Moreover, by Proposition 2.1, both the algebraic entropy of ϕ and the algebraic entropy of G are either an integral multiple of $\log p$ or the symbol ∞.

The following two results are the main tools for verifying when an endomorphism has zero entropy (see Propositions 1.18 and 2.4 in [DGSZ]).

Proposition 2.2. *Let G be a p-group and let $G[p]$ be its socle. If $\phi \in \text{End}(G)$ is such that $\text{ent}(\phi) > 0$, then $\text{ent}(\phi|_{G[p]}) > 0$.*

An endomorphism ϕ of the p-group G is said to be *point-wise integral* if, for every $x \in G$, there exists a monic polynomial $g(X) \in J_p[X]$ (depending on x), such that $g(\phi)(x) = 0$. Obviously every integral endomorphism of G is point-wise integral.

Proposition 2.3. *Let ϕ be an endomorphism of the p-group G. The following conditions are equivalent:*

(1) *ϕ is point-wise integral;*

(2) *the ϕ-trajectory of every $x \in G$ is finite;*

(3) *$\text{ent}(\phi) = 0$.*

3 Endomorphisms of zero entropy

Throughout the paper, we will denote by $\text{Ent}_0(G)$ the set of the endomorphisms of G which have zero entropy. We want to investigate when $\text{Ent}_0(G)$ is a subring of $\text{End}(G)$, or, more generally, when $\text{Ent}_0(G)$ is closed under multiplication or sum.

We start showing that if $\text{Ent}_0(G)$ is a subring of $\text{End}(G)$, then the p-group G is necessarily quasi finitely indecomposable. This result is based on the following technical lemma. Note that there is no requirement that the sequence $\{r_i\}$ below, be strictly increasing.

Proposition 3.1. *Let $B = \bigoplus_{n \geq 0} \langle b_n \rangle$, with $\langle b_n \rangle \cong \mathbb{Z}/p^{r_n}\mathbb{Z}$, where $r_0 \leq r_1 \leq \cdots \leq r_n \leq \cdots$ is an ascending sequence of positive integers. Then there exist $\phi, \psi \in \text{End}(B)$ such that $\text{ent}(\phi) = 0 = \text{ent}(\psi)$ but $\text{ent}(\phi\psi) \neq 0 \neq \text{ent}(\phi + \psi)$. The same holds for the torsion completion $\bar{B}$ of B.*

Proof. For $n \geq 0$, we set $c_n = p^{r_n-1}b_n \in B[p]$; then $B[p] = \bigoplus_{n\geq 0}\langle c_n \rangle$. We define $\phi \in \text{End}(B)$ by the following assignments, for $k \geq 0$,

$$\phi : b_{2k+1} \mapsto p^{s_k}b_{2k+2} \, , \, b_{2k} \mapsto 0$$

where $s_k = r_{2k+2} - r_{2k+1}$. We easily see that $\phi^2 = 0$, which implies $\text{ent}(\phi) = 0$, by Proposition 2.3. Note also that $\phi(c_{2k+1}) = c_{2k+2}$ and $\phi(c_{2k}) = 0$.

We now define $\psi \in \text{End}(B)$ by the following assignments, for $k \geq 0$,

$$\psi : b_{2k} \mapsto p^{t_k}b_{2k+1} \, , \, b_{2k+1} \mapsto 0$$

where $t_k = r_{2k+1} - r_{2k}$. We have $\psi^2 = 0$, whence $\text{ent}(\psi) = 0$. Symmetrically with respect to ϕ, we see that $\psi(c_{2k}) = c_{2k+1}$, and $\psi(c_{2k+1}) = 0$, for all $k \geq 0$.

We now examine $\phi\psi$. Using the definitions we get, for $k \geq 0$, $\phi\psi(c_{2k}) = c_{2k+2}$. Therefore every c_{2k} has infinite $\phi\psi$-trajectory, which shows that $\text{ent}(\phi\psi) \neq 0$, again by Proposition 1.3. (We note that $\phi\psi(c_{2k+1}) = 0$, hence c_{2k+1} has finite $\phi\psi$-trajectory.)

Let us now examine $\phi + \psi$. For $k \geq 0$, we have $(\phi + \psi)(c_{2k}) = \psi(c_{2k}) = c_{2k+1}$ and $(\phi + \psi)(c_{2k+1}) = \phi(c_{2k+1}) = c_{2k+2}$. It follows that any c_{2k} has infinite $(\phi + \psi)$-trajectory, and therefore $\text{ent}(\phi + \psi) \neq 0$.

Let now ϕ_1 and ψ_1 be the extensions to $\bar{B}$ of ϕ, ψ, respectively. Let z be an arbitrary element of $\bar{B}[p] = \prod_{n\geq 0}\langle c_n \rangle$. It is then easy to check that $\phi_1^2(z) = 0 = \psi_1^2(z)$. Since $z \in \bar{B}[p]$ was arbitrary, it follows that $\text{ent}(\phi_1) = 0 = \text{ent}(\psi_1)$, while $\text{ent}(\phi\psi) \neq 0 \neq \text{ent}(\phi + \psi)$ implies $\text{ent}(\phi_1\psi_1) \neq 0 \neq \text{ent}(\phi_1 + \psi_1)$. □

Proposition 3.2. *Let G be a p-group such that $\text{Ent}_0(G)$ is closed either under product or under sum. Then G is quasi finitely indecomposable (hence, in particular, semi-standard).*

Proof. Let us assume that G is not quasi finitely indecomposable. Then G must have a direct summand B like in the statement of the preceding lemma, say $G = B \oplus C$. In the notation of that result, we extend the endomorphisms ϕ, ψ of B to endomorphisms ϕ_1, ψ_1 of G in the obvious way, setting $\phi_1(C) = 0 = \psi_1(C)$. Then $\text{ent}(\phi_1) = 0 = \text{ent}(\psi_1)$ and $\text{ent}(\phi_1\psi_1) \neq 0 \neq \text{ent}(\phi_1 + \psi_1)$. This shows that $\text{Ent}_0(G)$ is neither closed under product nor under sum. □

The following immediate corollary settles the easier case when G is a bounded p-group.

Corollary 3.3. *Let G be a bounded p-group. Then $\text{Ent}_0(G)$ is a ring if and only if G is finite.*

Small endomorphisms of a semi-standard p-group have zero entropy. This was proved in Theorem 5.2 of [DGSZ]. We give a slightly more general result, useful for our purposes.

Proposition 3.4. *Let θ be a small endomorphism of the semi-standard p-group G. Then $\theta(G[p])$ is finite. In particular, $E_s(G) \subseteq \text{Ent}_0(G)$.*

Proof. Since θ is small, there exists $k > 0$ such that $\theta(p^k G[p]) = 0$. Since G is semi-standard, we have $G[p] = V \oplus p^k G[p]$, where V is finite. It follows that $\theta(G[p]) = \theta(V)$ is finite. In particular, from Proposition 2.2 it follows that $\text{ent}(\theta) = 0$. Since θ was arbitrary, we get $E_s(G) \subseteq \text{Ent}_0(G)$. □

The next technical lemma is a main ingredient for proving our main result.

Lemma 3.5. *Let R be a ring, and let $\phi, \psi, \theta \in R$ be such that $\psi\phi = \phi\psi + \theta$. Let $\mu = a_1 a_2 \cdots a_n \in R$, where each factor a_i equals either ϕ or ψ. Then μ can be written in the following form*
$$\mu = \phi^h \psi^k + \sum_{i,j} \phi^i \psi^j \theta b_{ij},$$
for suitable nonnegative integers h, k, i, j and elements $b_{ij} \in R$.

Proof. For convenience, we denote by Z the set of the sums
$$Z = \Big\{ \sum_{i,j} \phi^i \psi^j \theta b_{ij} : i, j \geq 0, b_{ij} \in R \Big\}.$$

Of course, Z is closed with respect to the sum in R. We make induction on n, the case $n = 1$ being obvious. Assume the result true for $n \geq 1$. Then, by induction, we have
$$\mu' = a_1 a_2 \cdots a_n a_{n+1} = (\phi^h \psi^k + b) a_{n+1},$$
where $b \in Z$. If now $a_{n+1} = \psi$, then $\mu' = \phi^h \psi^{k+1} + b\psi$ has the required form, since $b\psi \in Z$.

Let us then assume that $a_{n+1} = \phi$. If $k = 0$, then $\mu' = \phi^{h+1} + b\phi$ has the required form, since $c = b\phi \in Z$. Let $k > 0$. Then
$$\mu' = \phi^h \psi^{k-1}(\psi\phi) + c = \phi^h \psi^{k-1}(\phi\psi + \theta) + c = \phi^h \psi^{k-1}\phi\psi + \phi^h \psi^{k-1}\theta + c.$$

Note that $d = \phi^h \psi^{k-1}\theta + c \in Z$. If now $k - 1 = 0$, we see that $\mu' = \phi^{h+1}\psi + d$ has the required form. Otherwise we repeat the procedure, and, since Z is closed for the sums and $\phi^h \psi^{k-r-1}\theta \psi^r \in Z$ for all $r \geq 0$, after k steps we get $\mu' = \phi^{h+1}\psi^k + e$, with $e \in Z$, as desired. □

Of course, in the notation of the preceding lemma, we easily see that $h + k = n$ and $i \leq h, j \leq k$. However, these facts are not needed in our next theorem.

We can now prove the main result of this section.

Theorem 3.6. *Let G be a reduced semi-standard p-group, such that $\text{End}(G)/E_s(G)$ is commutative. Then $\text{Ent}_0(G)$ is a subring of $\text{End}(G)$.*

Proof. Let ϕ, ψ be two arbitrary endomorphisms with zero entropy. We have to show that $\phi\psi$ and $\phi + \psi$ have zero entropy, as well.

We first consider $\phi + \psi$. Since the endomorphisms commute modulo $E_s(G)$, we can write $\psi\phi = \phi\psi + \theta$, for a suitable small endomorphism θ. We will show that the $(\phi + \psi)$-trajectory of every finite subgroup F of $G[p]$ is finite, which implies $\text{ent}(\phi + \psi) = 0$, by Proposition 2.2. It is enough to show that there is a finite subgroup F_0 of $G[p]$ such that $(\phi + \psi)^n(F) \leq F_0$, for every $n \geq 0$. To that purpose, it is enough to show that for any product $\mu = a_1 a_2 \cdots a_n$ of n factors, where each a_i equals either ϕ or ψ, we have $\mu(F) \leq F_0$.

Since θ is small, by the preceding proposition $\theta(G[p]) = W$ is finite. Since ψ and ϕ have zero entropy, the ψ-trajectory W_1 of W is finite, and the ϕ-trajectory W_2 of W_1 is finite, as well. Moreover the ψ-trajectory F_1 of F is finite, and the ϕ-trajectory F_2 of F_1 is finite.

Now, using the preceding lemma, we get

$$\mu = \phi^h \psi^k + \sum_{i,j} \phi^i \psi^j \theta b_{ij},$$

for suitable nonnegative integers h, k, i, j and endomorphisms b_{ij} of G. In view of the above discussion on the trajectories, we see that $\phi^h \psi^k(F)$ is contained in F_2, while, for all i, j, $\phi^i \psi^j \theta b_{ij}(F)$ is contained in W_2, since $\theta b_{ij}(F) \leq W$ and $\psi^j(W) \leq W_1$. We conclude that $\mu(F) \leq F_2 + W_2 = F_0$, as required.

In order to show that $\mathrm{ent}(\phi\psi) = 0$, it suffices to prove that for every finite subgroup $F \leq G[p]$ there exists a finite subgroup F_0 such that $(\phi\psi)^n(F) \leq F_0$, for every $n \geq 0$. Since $(\phi\psi)^n = a_1 \cdots a_{2n}$, where all the a_i are either ϕ or ψ, we are exactly in the same situation examined above, and the desired conclusion follows. □

Remark 3.1. (1) Recall that Corner noted in [C1] that for semi-standard groups the first Kaplansky Test Problem has positive solution (G summand of H and H summand of G imply $G \cong H$), but not the second Kaplansky Test Problem.

(2) In the hypothesis of Theorem 3.6, $\mathrm{Ent}_0(G)$ is a J_p-subalgebra of $\mathrm{End}(G)$ containing $p\,\mathrm{End}(G)$ (every $p\phi$ obviously annihilates $G[p]$, hence $\mathrm{ent}(p\phi) = 0$), hence we can compute the dimension over F_p of $\mathrm{End}(G)/\mathrm{Ent}_0(G)$, which "measures" how far is G from being of zero entropy.

The remainder of this section is devoted to showing that Theorem 3.6 cannot be improved.

Let us first note that our main result is not reversible.

Proposition 3.7. *There exist p-groups G such that* $\mathrm{Ent}_0(G) = \mathrm{End}(G)$, *and* $\mathrm{End}(G)/E_s(G)$ *is not commutative.*

Proof. Let A be any non-commutative reduced J_p-algebra of finite rank. Corner's Theorem 4.1 in [C1] (see also Theorem 5.10 of [DGSZ]) shows that there exists a semi-standard separable p-group G such that $\mathrm{End}(G) = A \oplus E_s(G)$ is a split extension (Corner's theorem is always applicable when A has finite rank). Since A is a free J_p-module of finite rank, then A is integral over J_p, and therefore, by Corollary 5.7 of [DGSZ], we get $\mathrm{Ent}_0(G) = \mathrm{End}(G)$, while $\mathrm{End}(G)/E_s(G) = A$ is not commutative. □

Now let us see that, in Theorem 3.6, the hypothesis that G is semi-standard cannot be deleted (actually, G has to be quasi finitely indecomposable).

Proposition 3.8. *There exist p-groups G which are not semi-standard and such that $\operatorname{End}(G)/E_s(G)$ is commutative. In this case, $\operatorname{Ent}_0(G)$ cannot be a ring.*

Proof. We start with any commutative algebra A which is the completion, in the p-adic topology, of a free J_p-module of rank $\leq \aleph_0$ (e.g., take $A = J_p$). We invoke Theorem 1.1 of Corner's paper [C1], which allows us to construct a p-group G with basic group of final rank $2^{\aleph_0}$ such that $\operatorname{End}(G) = A \oplus E_s(G)$ is a split extension. Then $A = \operatorname{End}(G)/E_s(G)$ is commutative, and G cannot be semi-standard, since its final rank is uncountable. □

Finally, it is convenient to provide an example of a p-group G such that $\operatorname{Ent}_0(G)$ is a proper subring of $\operatorname{End}(G)$ and $\operatorname{End}(G)/E_s(G)$ is commutative.

Example 3.2. Our example is based on the proof of Theorem 4.4 of [DGSZ], to which we refer for the check of unexplained facts. Let $B = \bigoplus_{n>0} \langle b_n \rangle$ be the standard group, where $\langle b_n \rangle$ is cyclic of order p^n, for all $n > 0$. We denote by σ the endomorphism of B determined by the assignments $b_n \mapsto p b_{n+1}$; it has a unique extension to the torsion completion $\bar{B}$ of B, which we continue to denote by σ. Now consider the sub-J_p-algebra $J_p[\sigma] = R_\sigma$ of $\operatorname{End}(\bar{B})$ generated by $1, \sigma$. Then R_σ is isomorphic to the ring of polynomials with coefficients in J_p; denote by Φ its p-adic completion, still contained in $\operatorname{End}(\bar{B})$. As in [DGSZ, Theorem 4.4], we see that we can apply Corner's Theorem 2.1 of [C1]. Then there exists a group G, pure in $\bar{B}$ and containing B, such that $\operatorname{End}(G) = \Phi \oplus E_s(G)$ is a split extension. Now $\Phi = \operatorname{End}(G)/E_s(G)$ is a commutative integral domain, and therefore $\operatorname{Ent}_0(G)$ is a ring. Moreover, $\operatorname{Ent}_0(G)$ is a proper subring of $\operatorname{End}(G)$, since σ is an endomorphism of G of infinite entropy, having infinite entropy when restricted to B.

When G is semi-standard and $\theta \in E_s(G)$, then θ is point-wise integral over J_p, since $\operatorname{ent}(\theta) = 0$ (Propositions 2.3 and 3.4). In the following example, we see that θ is not necessarily integral over J_p, not even when G is quasi finitely indecomposable.

Example 3.3. We consider the standard module $B = \bigoplus_{n>0} \langle b_n \rangle$, where $\langle b_n \rangle \cong \mathbb{Z}/p^n\mathbb{Z}$. Let $\theta \in \operatorname{End}(B)$ be defined by the assignments $\theta : b_{m^2} \mapsto p^m b_{m^2}$ ($m > 0$) and $\theta : b_n \mapsto 0$, when n is not the square of an integer.

Note that $\theta^m(b_{m^2}) = 0$. We want to show that θ is a small endomorphism of B, which is not integral over J_p.

For any fixed $e > 0$, we will prove that $\theta(p^{e^2} B[p^e]) = 0$. Then $\theta \in E_s(B)$ will follow. In fact, $B[p^e] = \bigoplus_{n>0} \langle p^{n-e} b_n \rangle$, whence $p^{e^2} B[p^e] = \bigoplus_{n-e \geq e^2} \langle p^{n-e} b_n \rangle$. To verify that $\theta(p^{e^2} B[p^e]) = 0$, it is enough to show that $\theta(p^{m^2-e} b_{m^2}) = p^{m+m^2-e} b_{m^2} = 0$ for all m such that $m^2 - e \geq e^2$. Since $m^2 \geq e^2 + e$ implies $m + m^2 - e \geq m^2$, we get at once the desired conclusion.

Let us now prove that θ is not integral over J_p. Assume, for a contradiction, that

$$f(\theta) = \theta^k + a_{k-1}\theta^{k-1} + \cdots + a_1 \theta + a_0 = 0$$

where the $a_i \in J_p$. Let $j \leq k$ be the minimum index such that $a_j \neq 0$ (we set $a_k = 1$). Denote by v_p the p-adic valuation on J_p and fix an integer $m > k$ such that $v_p(a_j) < m$.

When $j < k$ we get

$$v_p(a_j) + jm < im \leq v_p(a_i) + im \quad (j < i \leq k).$$

Moreover, when $j = k$ we get $v_p(a_k) + km = km < m^2$.

We have

$$f(\theta) : b_{m^2} \mapsto \Big(\sum_{i=j}^{k} a_i p^{im}\Big) b_{m^2}.$$

The choice of m yields $v_p(\sum_{i=j}^{k} a_i p^{im}) = v_p(a_j) + jm < m^2$, implying $f(\theta)(b_{m^2}) \neq 0$, since the annihilator of b_{m^2} is $p^{m^2}\mathbb{Z}$. It follows that $f(\theta) \neq 0$, a contradiction.

Now consider the group $G = \bar{B}$, the torsion completion of B. Note that G is quasi finitely indecomposable. Let θ_1 be the unique extension of θ to G. Since G is the torsion subgroup of $\prod_{n>0} \langle b_n \rangle$, it is straightforward to verify that $\theta_1 \in E_s(G)$. Of course, θ_1 cannot be integral over J_p, since θ is not integral.

4 The ideal of socle-finite endomorphisms

An endomorphism ϕ of G is said to be *socle-finite* if the image of the socle $\phi(G[p])$ is finite. We denote by $E_0(G)$ the set of socle-finite endomorphisms.

It is readily seen that $E_0(G)$ is a two-sided ideal of $\mathrm{End}(G)$; it contains $p\,\mathrm{End}(G)$, since $p\psi(G[p]) = 0$ for all $\psi \in \mathrm{End}(G)$. It follows that $\mathrm{End}(G)/E_0(G)$ is an algebra over the field F_p. Note also that $E_0(G) \subseteq \mathrm{Ent}_0(G)$, in view of Proposition 2.2. But for the trivial case when $G[p]$ is finite, the last inclusion is always proper, since the identity $1 \in \mathrm{Ent}_0(G)$, while 1 is not a socle-finite endomorphism.

The next result improves Proposition 3.4.

Proposition 4.1. *The p-group G is semi-standard if and only if $E_0(G) \supseteq E_s(G)$.*

Proof. Let G be semi-standard. Then Proposition 3.4 shows that $E_s(G) \subseteq E_0(G)$.

Conversely, let us assume that G is not semi-standard. Then G has an infinite direct summand of the form $C = \bigoplus_{i \geq 0} \langle x_i \rangle$, where $\langle x_i \rangle \cong \mathbb{Z}/p^m\mathbb{Z}$ for any $i \geq 0$. Let us consider the shift σ determined by the assignments $x_i \mapsto x_{i+1}$. We know that $\mathrm{ent}(\sigma) = m \log p$ (see [DGSZ, Sec. 1.2]). If we extend, in the obvious way, σ to an endomorphism σ_1 of G, we see that $\mathrm{ent}(\sigma_1) = m \log p$, as well, hence $\sigma_1 \notin \mathrm{Ent}_0(G)$ implies $\sigma_1 \notin E_0(G)$. However, σ_1 is a small endomorphism, since $p^m \sigma_1 = 0$. □

We are interested in the commutativity of the F_p-algebra $\mathrm{End}(G)/E_0(G)$.

Proposition 4.2. *Let G be a p-group such that $\mathrm{End}(G)/E_0(G)$ is commutative. Then G is quasi finitely indecomposable.*

Proof. Let us assume, for a contradiction, that G is not quasi finitely indecomposable. Then G has a direct summand B like in the statement of Proposition 3.1. In the notation of that proposition, take the two maps ϕ and ψ. Then $\phi\psi - \psi\phi$ is not a socle-finite endomorphism of B, since $(\phi\psi - \psi\phi)(c_{2k}) = \phi\psi(c_{2k}) = c_{2k+2}$, for $k \geq 0$. Extending

ϕ, ψ to G in the obvious way, we easily deduce that $\text{End}(G)/E_0(G)$ is not commutative, a contradiction. □

Theorem 4.3. *Let G be a reduced p-group such that $\text{End}(G)/E_0(G)$ is commutative. Then $\text{Ent}_0(G)$ is a subring of $\text{End}(G)$.*

Proof. In the notation of the proof of Theorem 3.6, we just ask that the endomorphism θ is socle-finite, and not small, in general. Note that the property of θ, required for the argument to work, is exactly that $\theta(G[p])$ is finite. Therefore the proof of the present theorem is equal to that of Theorem 3.6, modulo the obvious changes. □

A point of interest of the above theorem is that we can avoid the hypothesis that G is semi-standard (unavoidable in Theorem 3.6), since we get this condition for free by Proposition 4.2.

The preceding result is, in fact, strictly more general than Theorem 3.6, as the following example shows, where $\text{End}(G)/E_0(G)$ is commutative, while $\text{End}(G)/E_s(G)$ is not commutative.

Example 4.1. Let A be the algebra of the 2×2 matrices over J_p having the following form

$$T = \begin{pmatrix} a & b \\ 0 & a' \end{pmatrix}$$

where $a, a', b \in J_p$ and $a \equiv a'$ modulo pJ_p. A direct check shows that, if $T_1, T_2 \in A$, then for a suitable $c \in J_p$ (which is non-zero, in general) we have

$$T_1 T_2 - T_2 T_1 = \begin{pmatrix} 0 & pc \\ 0 & 0 \end{pmatrix} = p \begin{pmatrix} 0 & c \\ 0 & 0 \end{pmatrix} \in pA.$$

The above formula shows that A is not commutative, while A/pA is commutative.

Since A is a free J_p-module of finite rank (the rank is 3, in fact), then A is complete in the p-adic topology. We can apply Corner's Theorem 4.1 in [C1]: there exists a semi-standard separable p-group G such that $\text{End}(G) = A \oplus E_s(G)$ is a split extension. Here $\text{End}(G)/E_s(G) = A$ is not commutative; however $\text{End}(G)/E_0(G)$ is a quotient of the commutative algebra A/pA, since $E_0(G) \supset p\text{End}(G)$, hence it is also commutative. In fact, we actually have $\text{End}(G)/E_0(G) = A/pA$, see the remark that follows.

Remark 4.2. As noted above, for every semi-standard p-group G we have the inclusion $E_0(G) \supseteq pA \oplus E_s(G)$. We do not have examples where this inclusion is proper. In fact, all the semi-standard p-groups G constructed by Corner in [C1], which realize a torsionfree complete J_p-algebra A as the component "disengaged from G" of $\text{End}(G) = A \oplus E_s(G)$, satisfy the equality $E_0(G) = p\text{End}(G) + E_s(G)$.

To prove this equality, first note that $p\text{End}(G) + E_s(G) = pA \oplus E_s(G)$, and an easy computation shows that $E_0(G) = pA \oplus E_s(G)$ is equivalent to $E_0(G) \cap A = pA$. Moreover, in Corner's constructions, the J_p-algebra A is a subalgebra of $\text{End}(\bar{B})$, where $\bar{B}$ is the torsion-completion of a basic subgroup B of G. The crucial assumption of the theorems in [C1] is that A satisfies the so-called Crawley's condition:

(C) If $\phi \in A$ and $\phi(p^n \bar{B}[p]) = 0$ for some n, then $\phi \in pA$.

It is readily seen that the Crawley's condition (C) is equivalent to the equality $E_0(\bar{B}) \cap A = pA$, which implies $E_0(G) \cap A = pA$, since $E_0(G) \subseteq E_0(\bar{B})$.

References

[AKM] R. L. Adler, A. G. Konheim and M. H. McAndrew, *Topological entropy*, Trans. Amer. Math. Soc. 114 (1965), 309–319.

[BGM] R. Behler, R. Göbel and R. Mines, *Endomorphism rings of p-groups having length cofinal with ω*, Abelian groups and noncommutative rings, 33–48, Contemp. Math., 130, Amer. Math. Soc., Providence, RI, 1992.

[C1]   A. L. S. Corner, *On endomorphism rings of primary Abelian groups*, Quart. J. Math. Oxford (2) 20 (1969), 277–296.

[C2]   A. L. S. Corner, *On endomorphism rings of primary Abelian groups II*, Quart. J. Math. Oxford (2) 27 (1976), 5–13.

[CG]   A. L. S. Corner and R. Göbel, *Prescribing endomorphism algebras, a unified treatment*, Proc. London Math. Soc. 50 (1985), 447–479.

[Cr]   P. Crawley, *Solution of Kaplansky's test problem for primary Abelian groups*, J. Algebra 2 (1965), 413–431.

[CI]   D. Cutler and J. Irwin *Essentially finitely indecomposable abelian p-groups*, Quaest. Math. 9 (1986), 135–148.

[DGSZ] D. Dikranjan, B. Goldsmith, L. Salce and P. Zanardo, *Algebraic entropy for Abelian groups*, to appear.

[DG]   M. Dugas and R. Göbel, *On endomorphism rings of primary Abelian groups*, Math. Annalen 261 (1982), 359–385.

[F]   L. Fuchs, *Infinite Abelian Groups*, Vol. I and II, Academic Press, 1970 and 1973.

[G]   B. Goldsmith, *On endomorphism rings of non-separable Abelian p-groups*, J. Algebra 127 (1989), 73–79.

[M]   G. S. Monk, *Essentially indecomposable Abelian p-groups*, J. London Math. Soc. (2) 3 (1971), 341–345.

[P]   R. S. Pierce, *Homomorphisms of primary Abelian groups*, in Topics in Abelian Groups, Scott Foresman (1963), 215–310.

[W]   M. D. Weiss, *Algebraic and other entropies of group endomorphisms*, Math. System Theory, 8 (1974/75), no. 3, 243–248.

Author information

L. Salce, Dipartimento di Matematica Pura e Applicata, Via Trieste 63, 35121 Padova, Italy.
E-mail: salce@math.unipd.it

P. Zanardo, Dipartimento di Matematica Pura e Applicata, Via Trieste 63, 35121 Padova, Italy.
E-mail: pzanardo@math.unipd.it